Abstract Algebraic L

An Introductory Textbook

Studies in Logic

Mathematical Logic and Foundations

Volume 60

Volume 49
Introduction to Propositional Satisfiability
Victor Marek

Volume 50
Intuitionistic Set Theory
John L. Bell

Volume 51
Metalogical Contributions to the Nonmonotonic Theory of Abstract Argumentation
Ringo Baumann

Volume 52
Inconsistency Robustness
Carl Hewitt and John Woods, eds.

Volume 53
Aristotle's Earlier Logic
John Woods

Volume 54
Proof Theory of N4-related Paraconsistent Logics
Norihiro Kamide and Heinrich Wansing

Volume 55
All about Proofs, Proofs for All
Bruno Woltzenlogel Paleo and David Delahaye, eds.

Volume 56
Dualities for Structures of Applied Logics
Ewa Orlowska, Anna Maria Radzikowsa and Ingrid Rewitzky

Volume 57
Proof-theoretic Semantics
Nissim Francez

Volume 58
Handbook of Mathematical Fuzzy Logic. Volume 3
Petr Cintula, Christian G. Fermüller and Carles Noguera, eds.

Abstract Algebraic Logic

An Introductory Textbook

Josep Maria Font

College Publications
London

Josep Maria Font
Departament de Matemàtiques i Informàtica
Universitat de Barcelona (UB)

ISBN 978-1-84890-207-7

College Publications
Scientific Director: Dov Gabbay
Managing Director: Jane Spurr
Department of Informatics
King's College London, Strand, London WC2R 2LS, UK

www.collegepublications.co.uk

Original cover design by Orchid Creative www.orchidcreative.co.uk
This cover produced by Laraine Welch
Printed by Lightning Source, Milton Keynes, UK

To my wife *Anna*
and my sons *Francesc* and *Bernat*,
in the hope that, finally, I will be able
to give them back the time spent on this book

The devil is in the details

(POPULAR)

Le bon Dieu est dans le détail

(attr. to GUSTAVE FLAUBERT)

Short contents

Detailed contents

A letter to the reader

Verba volant, scripta manent
(Caius Titus)

Dear reader,

Let me be clear from the start: this book is an *introductory textbook on abstract algebraic logic*. This straightforward phrase condenses the distinctive features of the book: its subject matter, its character and style, and its intended readership.

The subject is *abstract algebraic logic* or, if you prefer, simply *algebraic logic*; the **Introduction** begins (page xix) with a short discussion of the topic itself. I hesitated over whether to delete the word "abstract" from the book's title, but finally decided to keep it, thus emphasizing that the book is about the modern approach to algebraic logic. I believe, however, that *abstract algebraic logic is simply algebraic logic for the twenty-first century*, and I modestly hope that the book will help to spread this view and that one day in the future the term will become synonymous with "algebraic logic".

In character, the book is *introductory* to the subject. This means I do not intend to cover every aspect or to arrive at the deepest results on the notions I treat. I privilege *breadth* over *depth*, intending to present a broad view of a largely unknown territory without intending to reach all its borders—a task for explorers rather than for the public at large. If you find important results missing (even some of a basic character), this indicates that it is probably not the book for you, and that you will know where to find them. If you are new to the subject and after reading or studying the book you become interested in learning more, the final part (pages xxviiff.) of the **Reading Guide** is intended to provide clues to help you find your way in the literature.

The book was also conceived as a *textbook*, that is, mainly with students (or beginners in the field) in mind, and it takes a bottom-up approach, that is, going from the particular to the general. I do not intend to start by presenting the most general possible concepts, proving theorems at their highest level of generality and then providing specificity through down-to-earth cases. I prefer to guide readers by means of successive steps of generalization and abstraction. Moreover, since there are very few formal courses on algebraic logic in the university curricula round the world, I anticipate that the book will be used mostly for self-study. Advanced readers will have enough good sense to go more quickly through any deliberately detailed or verbose explanations and to choose only what may be of interest to them.

However, there are a few more advanced sections or subsections of a *survey* character, in which I want to offer a taste of some of the farther edges of recent research. To do this I have preferred a rather cursory presentation in order to be able to discuss some beautiful results which would be impossible to fully prove within this book, in the available space. I hope I have found a reasonable compromise that will motivate some readers to study more advanced material from other sources.

The book has been written as a *workbook*: very often you will be asked to provide proofs of minor, auxiliary results and to complete *exercises*, which I hope are not particularly difficult. They have been carefully designed to help you gradually master the material; only occasionally do they supplement the exposition of the main text with additional results. You should work them out as soon as you reach them on your way through the book; in some cases the answer appears, more or less hidden, in a later comment, or the properties they contain are superseded by later results, but this does not diminish the usefulness of doing them at the right time. Besides, you should convince yourself that all unproven statements not explicitly formulated as exercises are indeed true, providing a real proof if in doubt.

There is a certain amount of *repetition*: some topics are introduced early but are studied thoroughly only in later chapters. My guess is that a significant group of readers will study only part of the book, and yet there are some nice results which they should be aware of, even if they do not reach the chapters in which these results are treated in their proper framework (and where the result in question appears again, sometimes proved in a different way, sometimes proved fully for the first time). In my early days as a teacher of statistics I learned a popular *dictum* in information theory, which says that *without redundancy, there is no information*.

You will find a few things anticipated: at some points you will be asked to use or consult a theorem that is proved much later in the book. This is mainly done for illustration purposes, often in examples, rather than to prove other results, and I have ensured that this does not compromise the logical flow of the mathematical argumentation.

Historical notes and references are included here and there rather unsystematically. I have tried to highlight the origin of the most central ideas and theorems and acknowledge their authors, but only to put the subject in context, rather than aim at an exhaustive historical survey.

A warning is in order: *this book is not for beginners in logic*. Mastery of a standard first course in mathematical logic is essential, and an elementary background in universal algebra, model theory and the theory of ordered sets, lattices and Boolean algebras is highly recommended; among the best textbooks on these topics are [24, 51, 78, 89], where all the undefined notions of these fields may be found. A good, succint exposition of the logical and universal algebraic background needed to start the study of algebraic logic appears in Chapter 1 of [123]. There are a few sections requiring some proficiency with these subjects, and I have inserted appropriate warnings and given further references.

In addition, to understand the framework, and even to grasp its very first general notion (that of a sentential logic), it is undoubtedly necessary for you to build upon a certain body of *knowledge of particular logics*. This is instrumental in order to appreciate the theory: it is a general theory of logics, so an insight into real logics, how different they can be and what they have in common, is essential. Thus I assume (not as prerequisites in the mathematical sense) that you have some previous knowledge of some sentential logics, typically classical logic and a few non-classical ones. Many of these are described explicitly as *examples*, summarizing their properties and giving references; the level of detail varies, and I have made no attempt at uniformity in this regard. In any case, this book is no substitute for a textbook on non-classical logics; for that, you can read [26, 55, 128, 147, 187, 216, 238] and other relevant chapters of handbooks such as [27, 60, 119].

❧ ❧ ❧

I have felt the need for a book such as this one mainly when teaching. The present book originates (and hopefully benefits) from previous experience in teaching introductory courses or tutorials to graduate students and scholars of different background and nationality. My main teaching task in this area has been in the Master and Ph. D. programmes *Logic and Foundations of Mathematics* and *Pure and Applied Logic*, which have been running now for more than twenty years, and in the more recent Master in *Advanced Mathematics*, either at the University of Barcelona or as joint programmes of the Catalan universities. I have also taught shorter courses and tutorials round the world, and have written a few survey-style or historical papers [94, 96, 97, 102, 110]. All this activity has contributed to shape a certain view on the exposition of the subject. Often, when asked to recommend reading, I have realized that the material that constitutes the new approach is scattered over many research and survey papers and a few monographs, which are usually either at a rather hard level or presented sketchily, and that no systematic treatment exists at an introductory level.

Moreover, I wanted to convey *my personal view* of the subject. This explains the choice of some topics, and the (sometimes non-standard) paths followed to reach some results. At any rate, I do not claim to choose the "best" way: it is just my taste, and I offer it for you to try, and to agree or disagree with.

❧ ❧ ❧

This book is undoubtedly a consequence of my intellectual adventure over a whole career, and many people has had a direct, personal influence on my scholarly development. In the first place, I owe a great deal to Ramon Jansana and Don Pigozzi, with whom I shared an enthusiasm for abstract algebraic logic and the initiatives to spread it, and from whom I have learned a lot about logic, mathematics and scholarship. I also want to thank my mentors, colleagues, coauthors, collaborators and former students from Catalonia (Francesc d'A. Sales Vallès, Josep Pla, Ventura Verdú, Antoni Torrens, Gonzalo Rodríguez, Miquel Rius, Àngel Gil, Jordi Rebagliato, Román Adillón, Raimon Elgueta, Pilar Dellunde, Joan Roselló, Francesc Esteva, Lluís Godo, Joan Gispert, Félix Bou, Carles Noguera, Àngel Garcia-Cerdanya), from Spain (Antonio-Jesús Rodríguez, José Luis García Lapresta, José Gil-Férez, David Gracia, María Esteban) and from the rest of the world (José Coch, Robert Bull, Jerzy Perzanowski, Hiroakira Ono, Kosta Došen, Isabel Loureiro, Wim Blok, Andrzej Wroński, Piotr Wojtylak, Wojciech Dzik, Roberto Cignoli, Janusz Czelakowski, Kate Pałasińska, Ryszard Wójcicki, Mike Dunn, Massoud Moussavi, Fernando Guzmán, Wolfgang Rautenberg, Sergio Celani, Petr Hájek, Daniele Mundici, James Raftery, Clint van Alten, Jacek Malinowski, Jorge Castro, Petr Cintula, Alessandra Palmigiano, Francesco Paoli, Umberto Rivieccio, Tommaso Moraschini, Hugo Albuquerque, Tomáš Lávička). They have been there, at diferent moments and in different ways: asking questions, answering questions, listening, reading, writing, making resources available, giving advice, or entrusting me with different jobs. I learned something from each one of them.

Although this book is a highly personal project, other people have contributed to shape it as you see it now, and I want to express my debt and warm thanks to all of them: they include students who have worked with preliminary versions of the text and colleagues whose comments on the first serious draft of the complete book have helped to improve it in significant ways. I am particularly indebted to Janusz Czelakowski, José Gil-Férez, Ramon Jansana, Tomáš Lávička, Tommaso Moraschini and Carles Noguera for all their observations. Last, but not least, I want to thank Toffa Evans, Joe Graham and Mike Maudsley for improving my English, and once again, José Gil-Férez for improving my TEX. How I dealt with their recommendations, and the final result, is of course my sole responsibility.

I must also thank Dov Gabbay for including the book in his highly-regarded series Studies in Logic, and Jane Spurr for her editorial assistance.

Teaching at a non-elementary level is intrinsically intertwined with research. Therefore, and also for legal reasons, I must acknowledge (again, with thanks) that my research work over the years has been (partially) supported by grants that the Barcelona Research Group on Algebraic Logic and Non-Classical Logics has received from the governments of Spain and of Catalonia, the latest ones being project MTM2011-25747 (which includes FEDER funds from the European Union) and grant 2014SGR-788, respectively.

I hope you enjoy the book.

Sincerely,

Josep Maria Font
Barcelona, 29 february 2016

Introduction and Reading Guide

If ***algebraic logic*** can be quickly (and roughly) described as *the branch of mathematics that studies the connections between logics and their algebra-based semantics*, then ***abstract algebraic logic*** can be described as its more general and abstract side. Three aspects of the study of these connections deserve a special mention:

- *Describing* them: This was historically the first aspect of the subject to be developed, starting with Boole's pioneering work [44] in 1847. The oldest connections to be described took the form of *completeness theorems*. More recently other, stronger connections (which may be considered refined completeness theorems) have been discovered; these include algebraizability and truth-equationality. The description of the reduced models of a logic and of its algebraic counterpart (in the different senses of this term) also belongs to this aspect of the subject.
- *Exploiting* them: Mostly, this is done by using more powerful and well-developed algebraic theories to prove properties of a logic, but recently some cases of the reverse process have also been worked out. This of course has been done, in different ways, at each stage in the evolution of algebraic logic. One of the distinctive features of abstract algebraic logic, in contrast to more traditional work, is that its results concern not just an individual logic, but classes of logics that are treated uniformly. The general results that support these applications have come to be called *bridge theorems* in the context of the third aspect below; as I say in the general discussion at the beginning of Section 3.6, *bridge theorems are the ultimate justification of abstract algebraic logic*.
- *Explaining* them: This consists of developing *general theories* of these connections and exploring them in many different ways to gain a deeper understanding of how and *why* the different kinds of connections work, what the relations between them are, how large their domain of application is, and similar issues. Ideally, each general theory should support the choice of a specific algebra-based semantics as *the* algebraic counterpart of a given logic, on the basis of general criteria that avoid ad hoc justifications.

Naturally, these three aspects are inextricably intertwined. Moreover, the general notions need to be tested against many *examples*, either natural or ad hoc, not just to obtain properties of particular logics, but in order to gain insights into the general notions themselves, their relations, their applicability, their scope and their limits; this *empirical work*† is an important guide for the more abstract work.

The expression ***algebra-based semantics*** is used in a deliberately ambiguous and informal sense. It is intended to refer to any kind of semantics in which the non-linguistic objects (the domains where the formulas of the language are evaluated) are either just plain algebras, or algebras endowed with some additional structure (a particular element or subset, a family of subsets, an order relation) *and* in which the evaluations are all the homomorphisms from the formula algebra to the algebras underlying the models. Notice that the term *algebraic semantics*, which has been extensively used in a similarly informal way for years, does have a formal meaning in the modern theory (Section 3.1); so I think it is good practice to limit its usage to this strictly technical sense and use "algebra-based semantics" otherwise.

Algebraic logic is an old subject,‡ whose birth is commonly attributed to the work of Boole, De Morgan, Peirce, Schröder and other scholars in the second half of the nineteenth century. Since then the subject has evolved from a bunch of ad-hoc procedures that establish some links between a particular logic (initially classical sentential logic) and certain algebraic structures (initially Boolean algebras) to a full-fledged mathematical theory, with its well-defined object, its typical methods, techniques and constructions, its important general theorems, and its applications to diverse kinds of logics. The development of the subject has been slow, particularly the process of self-structuring as a systematically organized corner of mathematical logic. Without attempting to write a detailed history, one can say that the subject really took off in the 1920s, and began to be something more than a mere *description* of the connections between classical logic and Boolean algebras with the work of Łukasiewicz, Post, Lindenbaum, Tarski and other—mainly Polish—logicians; the notion of *logical matrix* introduced at the time allowed theoretical studies and a potentially universal applicability. Later on, Tarski and his followers focused on the algebraization of (classical) first-order logic. The task of extending the so-called *Lindenbaum-Tarski method* of algebraizing classical logic to other sentential logics generated a host of papers and, ultimately, well-known books by Rasiowa and Sikorski [209, 1963] and by Rasiowa [208, 1974]. Later on, in the 1970s and the early 1980s, a truly general study of the theory of logical matrices was undertaken by several scholars—again, mainly Polish logicians. Czelakowski's theory of *equivalential logics* [67, 1981], for the first time, provided several characterizations, both semantic and syntactic, of a class of logics to which the Lindenbaum-Tarski method, suitably generalized, could

† The importance of dealing with strange, or seemingly pathological examples should not be underestimated. Something similar happens in medicine, where the study of subjects with important neurological disorders provides essential clues to advance in the understanding of the normal behaviour of the human brain and mind.

‡ Some, viewing history of mathematics from a global perspective, will prefer to say that algebraic logic is relatively young.

be applied. This scientific progress is certainly related to the parallel growth of *model theory* and *universal algebra*, the two disciplines that have become the main mathematical tools for algebraic logic. All these trends are well represented in Wójcicki's comprehensive monograph [249, 1988]. Additional historical information can be obtained in [13, 31, 50, 73, 94, 97, 158, 176, 188, 229, 251]; the succint "biased survey of algebraic logic" of [123, pp. 7–9] is also worth reading.

Since the 1980s the subject has seen important changes, the development of new and powerful tools and general frameworks to deal with, classify and compare a much larger number of sentential logics and to establish connections between their metalogical and algebraic properties. This more recent evolution is due mainly to the works of Blok, Pigozzi, Czelakowski, Herrmann, Jansana and Raftery in the more traditional, matrix-based line, and of Bloom, Brown, Suszko,† Font, Jansana, Torrens and Verdú, in the newer line based on abstract logics or generalized matrices. The convergence of these approaches is apparent in [107, 1996], and is also described in [110]. The modern form of the subject started to be called ***abstract algebraic logic*** sometime in the 1990s,‡ and this denomination has been included as entry 03G27 in the 2010 revision of the *Mathematics Subject Classification*. Today, the algebraic study of particular sentential logics is conducted through reference to some general standards that classify logics inside two hierarchies, the Leibniz hierarchy and the Frege hierarchy, according to several criteria. In addition to the results concerning this or that logic or class of algebras, several important general theorems have also been obtained; the building and investigation of these hierarchies is probably the most beautiful and impressive mathematical enterprise to date in the field.

One of the distinctive features of abstract algebraic logic is that it transcends the mere *generalization* of the techniques that have been seen to work in a handful of so-to-speak "classical" (in the sense of "paradigmatic" or "established") cases; for instance, while Rasiowa in her [208] did generalize the Lindenbaum-Tarski process in the sense that she identified a broad class of logics for which this process can be carried out with almost no changes, she did not look for other equivalent characterizations of the logics in this class (which would have allowed her to show which logics *cannot* be treated with her methods). The deeper *abstract framework* underlying Rasiowa's studies was first uncovered by Czelakowski in [67], and more definitively by Blok and Pigozzi in their [30, 32, 34].

As I said before (page xv), abstract algebraic logic is the algebraic logic of the twenty-first century—and of the last quarter of the twentieth century. Hopefully, with the passing of time the word "abstract" currently added to the traditional name will simply be forgotten.§

† Suszko's influence on algebraic logic has been profound, as it has also been on the traditional theory of matrices; this is evident from his early and fundamental [166, 1958].

‡ The term "universal algebraic logic" was used for a short time; see [190, 1998]. As far as I know, the term "abstract algebraic logic" first appears in the literature in the title of Section 5.6 of [138, 1985], which is devoted to a general study of the connections between theories of classical first-order logic and varieties of cylindric algebras. Here I am referring to the later usage of the term in the sentential logic framework.

§ Much as today we all designate simply as "algebra" what was initially called "abstract algebra" or

As a final point on the subject matter of this book, notice that it deals only with sentential logics. So, unless specified otherwise, "logic" always means "sentential logic", either at an informal or at a formal, technical level. This does not mean that *first-order logic* cannot be treated successfully with algebraic tools. Quite to the contrary, for many scholars the expression "algebraic logic" is indeed synonymous with the algebraic study of first-order logic developed either in the style of Tarski and his important school[†] or in the style of Halmos (dealing mainly with cylindric and relation algebras, or with monadic and polyadic algebras, respectively). Indeed, these scholars may be surprised that this study has only a marginal place in this book (as a mater of fact, Section 2.5 briefly reviews the two radically different ways in which predicate logics can be algebraically approached; there you will find a few references). In general, however, these works have a very different character from the algebraic study of sentential logics. Most of their main results, obtained by very sophisticated mathematical techniques, concern the study of particular theories (or particular languages) in *classical* first-order logic. By contrast, a general, abstract theory still has to be built, at least with the degree of systematization and sophistication found nowadays in the algebraic studies of sentential logics; in particular, no hierarchies of logics like those developed for sentential logics in Chapters 6 and 7 have been constructed to classify non-classical first-order logics according to their algebraic behaviour.

Overview of the contents

As explained before, this book adopts a bottom-up approach; that is, it describes the different algebra-based semantics at increasing levels of the mathematical complication of the semantics objects. It starts with the algebraization of implicative logics, where the truth definition is given by a single element of the algebras, and next moves on to algebraizable logics, where it is given by a set of equations. This is followed by an explanation of matrix semantics, where the truth definition is given by an arbitrary set, and it concludes with generalized matrix semantics, where it is given by a family of subsets. The mathematical sophistication does not follow a parallel pattern, though, as the chapter on algebraizability is probably more difficult than the one on matrices. In any event, the last two chapters on the Leibniz and Frege hierarchies are unavoidably the hardest and more complicated in the book. A more detailed description of each chapter's contents follows, with some important reading tips.

As for **Chapter 1**, the name says it all: *Mathematical and logical preliminaries*. Here are most of the general definitions, terminology and notation, and the common mathematical background for the entire book; not everything is needed everywhere, so you may be able to skim the chapter and return to it as and

"modern algebra" in order to emphasize that it differs from the algebra of the nineteenth century. This can be seen from van der Waerden's famous 1930 book *Moderne Algebra*, which has simply been entitled *Algebra* since its 1970 edition.

† The term "Tarski-style algebraic logic" is sometimes used to distinguish its work from the algebraic study of sentential logics, which is then called "Polish-style algebraic logic", or even "Polish logic", a "rather irritating cliché" in Wójcicki's words [249, p. xi].

when needed, depending on your background. Section 1.2 contains a discussion of the general definition of the notion of a sentential logic,[†] which is in some sense the starting point of algebraic logic. Less common is the detailed attention devoted to the theory of closure operators (Sections 1.3–1.5) and to the equational consequence relative to a class of algebras (the final part of Section 1.6). Most of Sections 1.1, 1.2 and 1.6 touch upon very general subjects and are here mainly to establish terminology and notations, but are definitely *not designed for learning the subjects*. You are expected to have a working knowledge of some particular logics and of some classes of algebras (lattices, Boolean algebras), so that the general scene-setting of this chapter looks familiar to you, and builds on a firm footing of concrete examples.

Chapter 2 is mainly (but not exclusively) motivational, and should be easy to read. It starts by reviewing the usual textbook completeness proof of classical two-valued logic, which everyone should be familiar with. Then it shows how the so-called *Lindenbaum-Tarski process* arises naturally from the analysis of this proof, and introduces its more straightforward generalization, the algebraization of *implicative logics*. Nevertheless, *nobody should miss* the first subsection ("The general case", pages 88–91) of Section 2.3, as it introduces *the key notion of filter of a logic*, absolutely indispensable throughout the book. The chapter draws to a close with a discussion (Section 2.4) on how the said process can be extended more or less naturally in several directions, and two digressions (Section 2.5) on the algebraic study of *first-order logic*, the first of which may be viewed as a justification for why sentential logics are studied in a purely algebraic way, as algebraic logic does.

Chapter 3 develops most of the central core of abstract algebraic logic, the theory of *algebraizable logics*, presented as based on the idea of equivalence between the consequence of a sentential logic and the equational consequence relative to a class of algebras, effected by a pair of mutually inverse structural *transformers*. Some proofs are developed as further generalizations of the Lindenbaum-Tarski process (however, connections with the theory of matrices are touched upon only in Section 4.4). Section 3.5, which contains the abstract characterizations of algebraizability known as *Isomorphism Theorems*, is probably the hardest in this chapter. Section 3.6 contains an initial discussion of some central topics to be addressed in later chapters, such as *bridge theorems* and *transfer theorems*, and uses the Deduction-Detachment Theorem as a case study, offering a first taste of what is to come. The chapter ends with a brief account (Section 3.7) of how the basic idea of algebraization can be, and has been, extended and generalized to wider or more abstract notions of logic.

Chapter 4 develops the essentials of the theory of *logical matrices*, the core of the more classical algebraic logic. In particular the chapter contains the first general definition of the notion of *the algebraic counterpart of a logic* and shows how it fits with the constructions done for implicative logics and for algebraizable

[†] Spoiler alert: in abstract algebraic logic a sentential logic is a substitution-invariant consequence relation on the set of formulas of some sentential language. I adhere completely to Wójcicki's words on *the inferential approach to logic* [249, pp. xii–xiii].

logics in previous chapters, and how it fails to give a satisfactory account of the algebraic aspect of some less standard logics. Only notions indispensable for the study of the Leibniz hierarchy are given here; for a more complete view of this well-developed area you should study [70, 249] and other papers.

Chapter 5 contains an exposition of the theory of *generalized matrices*. The material is developed in more detail than that in Chapter 4, because the material in the earlier chapter is better known than the literature on generalized matrices and there are more sources. It is here where the truly general definition of *the algebraic counterpart of a logic* is given, together with that of *the intrinsic variety of a logic*. These classes of algebras are identified in a large number of particular logics, including ones in which the matrix-based definition of the previous chapter seems to fail. I also include a discussion of why I think that this is *the right* definition for this (informal) notion. The most novel idea in this area, the notion of a *full generalized model* of a logic, is dealt with in Section 5.5, where another fundamental *Isomorphism Theorem* is proved. The chapter ends with Section 5.6, which offers a brief glimpse into the usage of generalized matrices as models of Gentzen systems and how this is related to their role as models of sentential logics.

Chapter 6 is really just an *introduction* to the study of the *Leibniz hierarchy*, albeit a detailed one, which should be supplemented by reading and studying [35, 70, 72, 201] and other related works. Things here turn more complicated and advanced, requiring more work on your part; for instance, proofs in the text often rely on results proposed as exercises, with hints when necessary. The most important class in the hierarchy (that of algebraizable logics) is treated in Chapter 3, and it is here only revisited, particularly as far as its relations to other classes in the hierarchy are concerned. Section 6.4 contains some fairly recent results on assertional logics and truth-equational logics.

The final **Chapter 7** on the *Frege hierarchy* is shorter, because this hierarchy is less complicated than the Leibniz one and has been less studied; it also contains some very recent material. With the two hierarchies already studied, this is the natural place to establish the relations between them; inevitably, this means that some of the key results of the final section (Section 7.3) bring together results from distant parts of the book and, therefore, may be harder to follow.

Figure 1 depicts the relations between chapters and sections. From this, you can see several possible partial paths and alternative arrangements of the material, either for reading or for teaching.

An **Appendix** collects, in a summarized way, the features of the best-known non-classical logics (relative to the subject matter of this book), which appear scattered over many numbered examples and other comments. The book concludes with a set of very comprehensive **Indices**.

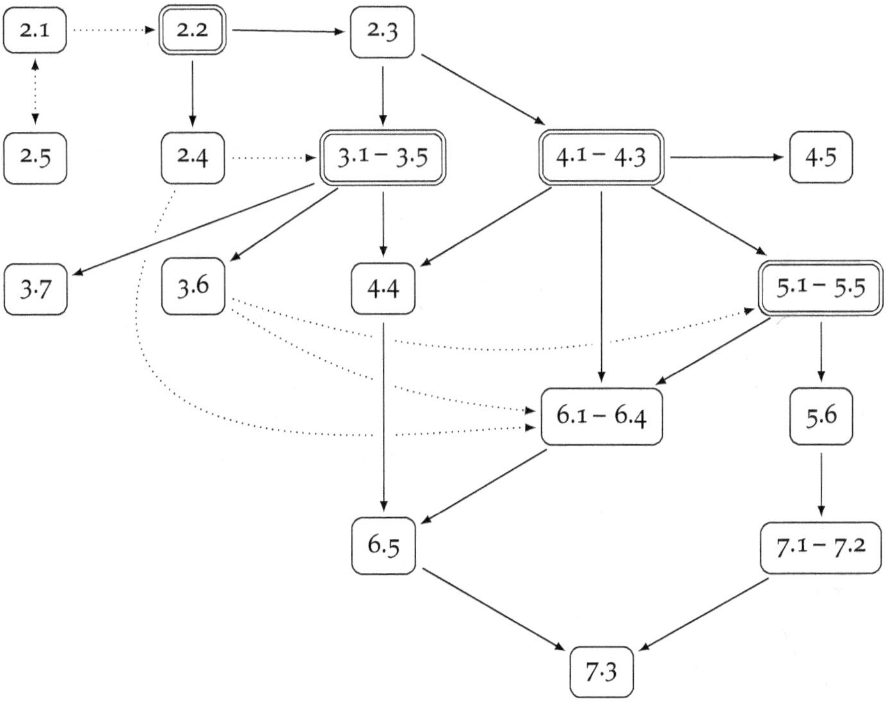

FIGURE 1: A chart of the main dependencies in the book (excluding Chapter 1). Solid arrows indicate mathematical dependency, though this should be taken with a pinch of salt, as not everything in the source of an arrow is needed everywhere in its target. Dotted arrows indicate motivational dependency. The double framed groups of sections mark four possible ways to *start* the study of abstract algebraic logic, with increasing initial level of generality.

Numbers, words, and symbols

- The book is divided into chapters, sections and subsections, but only the first two levels are numbered. All *numbered statements* (definitions, theorems, examples, etc.) share the same two-part number sequence inside each chapter. The exceptions are exercises, which are numbered independently by chapter, and figures, tables and displayed material. Thus, any three-part label refers to a labelled item in a list contained in an already labelled statement. For instance, "Theorem 4.6.3" and "Theorem 4.6(i)" refer, respectively, to item 3 and to item (i) of Theorem 4.6, which is the sixth numbered statement in Chapter 4.
- The symbol ⊠ marks the *end* of some environment, which should be obvious from context: this is most often the end of a proof, but it could be the end of an example, a comment, or a specifically labelled paragraph which might otherwise be difficult to distinguish from surrounding text. It also indicates absence of proof.
- There are three *symbols for equality*: the symbol $=$ is used to express the fact that the mathematical objects represented by the expressions joined by the symbol coincide. The symbol $:=$ means "equal by definition", and is used only when introducing a new symbolic expression, defined in terms of an already meaningful one. Finally, the symbol $\approx$ is used to build formal identities, that is, "equations" between two expressions of some formal language; in fact, an expression like $\alpha \approx \beta$ is just an alternative, more intuitive notation for the pair $\langle \alpha, \beta \rangle$, and it turns into a real equality only when α and β are given values from some mathematical domain (and only then will the equality become either true or false).
- The symbols $\iff$, $\Rightarrow$ and $\Leftarrow$ are used as shorthand for "if and only if", "implies" and "follows from", respectively; the last two are also used to mark the two halves of equivalences, in proofs and comments. The symbol $\overset{\text{def}}{\iff}$ indicates a definition that has the form of an equivalence.
- *The parentheses* "(" and ")" serve several specific purposes in logico-algebraic symbolisms, and I have decided to limit their appearances as much as possible. Specific, technical uses include: expressing the values of functions with more than one argument, as in $f(a_1, \dots, a_n)$; expressing the dependence of a formula on certain variables, as in $\alpha(x)$, and the results of applying a substitution or an evaluation to the formula so denoted, as in $\alpha(\varphi)$ or $\alpha^{\boldsymbol{A}}(a)$; and expressing relation under a binary relation, as in $a \equiv b \ (R)$, which is just shorthand for the assertion that $\langle a, b \rangle \in R$ (technically, a binary relation R on a set A is just an $R \subseteq A \times A$); the *infix notation* aRb is also used. Sequences (in particular, ordered pairs) are denoted by $\langle a_0, \dots, a_{n-1} \rangle$ and so forth, as is popular in set theory; the expression $\vec{a}$ denotes a countable sequence of elements a_i (see Section 1.1).

  In particular, no parentheses are used for functions of one argument, e.g., I have put fa instead of $f(a)$, and $hC_1h^{-1}\bigcup\{C_2hX_i : i \in I\}$ instead of $h\big(C_1(h^{-1}(\bigcup\{C_2(h(X_i)) : i \in I\}))\big)$, and so forth. If you read carefully, no confusion should arise.

In any event, I use parentheses, as any mathematician does, to freely group symbols, in order to either eliminate ambiguity or facilitate legibility.

- The main *typographical conventions* are: $A, B, \dots$ for sets (but C is reserved for closure operators, and L for languages) and $a, b, c, \dots$ for their elements; $\boldsymbol{A}, \boldsymbol{B}, \dots$ for algebras; $\mathscr{B}, \mathscr{C}, \dots$ for families of sets; $\mathcal{L}$ for logics; $\mathcal{M}, \mathcal{N}$ for matrices and g-matrices; $\mathsf{K}, \mathsf{M}, \dots$ for classes (of algebras, matrices or g-matrices); w, x, y, z for sentential variables; $\alpha, \beta, \dots, \varphi, \psi$ for formulas (but $\pi, \rho, \sigma, \tau, \theta$ are used with other purposes); Γ, Δ, Σ for sets of formulas; $f, g, h, \dots$ for functions, etc. No scheme of this kind can be 100% consistent, but I have tried my best.
- *Italic typeface* is used for emphasis, and in particular ***boldface italics*** are used when introducing a new term, either in a formal definition or in the text, while **upright boldface** appears in titles of sectional units and in phrases that perform a similar function in the middle of the text.
- *Universal quantification* in formal statements is sometimes omitted for simplicity or for lack of space, and should be implicitly understood in the natural way, according to the context ("for every logic", "for every algebra", etc.). Obvious *indices* such as "$i \in I$" are also omitted in some symbolic expressions for the same reasons.
- I firmly believe that *all teaching is communication*. This book proposes a *dialogue* between the writer (me) and the reader (you). So when I write "we", I mean "you and me",[†] who are supposed to be travelling together, hand in hand, through the material. This I learned from the late Paul Halmos [134, 135], whose wise counsel I always try to follow (only to find time and again that his standards are too high for me).

Further reading

If you want to go deeper into any of the topics, or seek a different view, here are some recommendations.

To start with, there are a few **big books**, containing lots of information:

- Czelakowski's [70] is the current[‡] encyclopædia of abstract algebraic logic. This book presents in a systematic way most of the published literature, and even previously unpublished work by the author and his co-authors. In Chapters 0 and 1 (including their exercises), it contains general material on algebraic logic, and it brings together many results that are often difficult to find. Beyond this initial, more general part, however, it deals only with the class of protoalgebraic logics and its subclasses, with particular emphasis on the class of algebraizable logics and on the logical theory of quasivarieties.

[†] I presume this is not far from what Wilfrid Hodges meant when, in [144, p. xiii], he wrote the apparently tautological, yet significant sentence: "'I' means I, 'we' means we".

[‡] Given that the book was published in 2001, a number of more recent developments and results have to be looked for elsewhere.

- Rasiowa's classic book [208] established the paradigm of algebraic logic for several generations of logicians. Chapters I to VII examine several classes of algebras, Chapter VIII contains the general theory, and the rest of the book spells out its application to a number of logics whose algebraic counterparts are the classes of algebras dealt with in the first part.
- Wójcicki's [249] is a systematization of "Polish-style" work† in algebraic logic up to the 1980s, and more generally of the study of sentential logics understood as consequence operators. The book also contains some interesting discussions of rather philosophical import.
- The recent *Handbook of Mathematical Fuzzy Logic* [60] contains a chapter by Cintula and Noguera [62] whose aim is, in the authors' own words, "to present a marriage of Mathematical Fuzzy Logic and (Abstract) Algebraic Logic". This 90+-page exposition starts with very general material and then confines itself to the large class of "weakly implicative logics". Although it concentrates on the algebraic study of fuzzy logics, it is written from a very modern perspective and touches on many points of general interest, some of which are not presented here.

There are also a number of **monographs** and **important papers** that are written in a fairly systematic and self-contained way and can be read as introductions to the research lines started in them:

- Blok and Pigozzi's [32] contains pioneering work on the notion of an algebraizable logic, the main ideas and results (for their initial, restricted definition) and an analysis of examples. Blok and Jónsson's [29] is also a pioneering contribution to the building of the most abstract presentation of the idea of equivalence between structural closure operators that underlies the notion of algebraizable logic, as outlined in Section 3.7.
- The short monograph [107] uses the semantics of generalized matrices (here they are called *abstract logics*) to study finitary sentential logics, especially the ones for which neither algebraizability nor the older theory of matrices yields satisfactory enough results, and also to study Gentzen systems. It introduces the key notion of a full generalized model of a sentential logic and a truly general definition of the notion of the algebraic counterpart of a logic. In some sense it is the culmination, at a higher degree of abstraction, of work done on a less general framework by the Barcelona group over the years. It includes a chapter analyzing many examples, although in a rather condensed form.
- The extension of the idea of algebraizability from sentential logics to more general formalisms was started in Rebagliato and Verdú's [212], which deals with Gentzen systems. Raftery's [200] is a more modern and systematic exposition of this idea, and extends it to the equivalence between a Gentzen system and a sentential logic, building also on the ideas of [29].

† See footnote † on page xxii.

Finally, there are quite a few **survey papers** on recent trends in algebraic logic, something natural in a field that has undergone deep changes:

- Blok and Pigozzi's [35] is a combination of a survey of some aspects of abstract algebraic logic and its relation to traditional methods, and a monograph where important results are proved in detail, some for the first time.
- My two papers [94, 97], devoted to the work by Rasiowa, describe her generalization of the Lindenbaum-Tarski method. The first one also touches on the first-order case, and the second one contains an overview of different directions of research sparked by her work. Incidentally, the second one is where the expression "Frege hierarchy" first appears in print.
- Three surveys of abstract algebraic logic to which I have contributed are [96, 102, 110]. Although they are diverse in character and scope, they all try to describe the modern ideas and developments as arising from the essential techniques and results of traditional algebraic logic.
- Raftery's [204] concentrates on algebraizability, mentions some of the latest trends and emphasizes the algebraic side of the topic.
- Jansana's and Pigozzi's (independent) encyclopædia articles [150, 191] are rather condensed but non-trivial, modern expositions of the topic.

The book's **Bibliography** contains far more references than a standard textbook. In addition to the references needed to follow the central contents and the mathematical background, the bibliography also includes specialized works cited in specific places within the book as well as other fundamental works in the area, in order to support the book's somewhat hybrid character, as a cross between a course textbook and a survey monograph, with some historical notes.

Chapter 1

Mathematical and logical preliminaries

1.1 Sets, languages, algebras

A naive theory of sets and classes, including the Axiom of Choice, is sufficient for our purposes. The set of ***natural numbers*** is denoted by ω. A set is ***denumerable*** or *countably infinite* when it can be put in bijective correspondence with ω, and is ***countable*** when it is finite or denumerable. The cardinality of a set A is denoted by $|A|$. The set of ***countable sequences*** of elements of a set A is denoted by $\vec{A}$, and the notation $\vec{a} \in \vec{A}$ indicates that either $\vec{a} = \langle a_1, \ldots, a_n \rangle$ for some $n \geqslant 1$, or $\vec{a} = \langle a_i : i \in \omega \rangle$, with $a_i \in A$ in both cases; context and common sense will make it clear which is the case, and in the finite case which length is intended. A practical convention is to represent the *empty* sequence and the empty set by setting $n = 0$ in expressions of the form $\langle a_1, \ldots, a_n \rangle$ and $\{a_1, \ldots, a_n\}$, respectively.

For a set A, its power set is denoted by $\mathscr{P}A$. A typical feature of algebraic logic is the presence of closure operators, which are (Definition 1.19) certain kind of functions on power sets, $C\colon \mathscr{P}A \to \mathscr{P}A$; according to the notational convention explained on page xxvi, if $X \subseteq A$, then the image of X by C is written as CX rather than as $C(X)$. Additionally, each function $f\colon A \to B$ induces two functions on power sets, which are denoted by f and f^{-1}; namely, $f\colon \mathscr{P}A \to \mathscr{P}B$ given by $fX := \{fa : a \in X\}$ for each $X \subseteq A$, and $f^{-1}\colon \mathscr{P}B \to \mathscr{P}A$ given by $f^{-1}Y := \{a \in A : fa \in Y\}$ for each $Y \subseteq B$.[†] The first of these functions is known as the ***power set extension*** of f. Ordinary functions may be found mixed with their power set extensions and with other functions directly defined on power sets, but I hope no confusion arises, especially with the double usage of the same function symbol (here f) for the original function and its power set extension. It is practical to use a notation for the natural order relation between power set

[†] The two sets are often written in the literature as $f[X]$ and $f^{-1}[Y]$ respectively, but also in other ways, such as $f^*(X)$, $f^{\leftarrow}(Y)$, etc.

functions obtained, for $f,g\colon \mathscr{P}A \to \mathscr{P}B$, by putting $f \leqslant g \overset{\text{def}}{\iff} fX \subseteq gX$ for all $X \subseteq A$. Composition of functions is denoted by juxtaposition unless use of the usual symbol $\circ$ makes the expression more readable. The identity relation on a set A is denoted by Id_A.

Some properties of the power set functions f and f^{-1} relative to the natural order relation $\subseteq$ in power sets are worth recalling for their frequent use: that they are both monotone, that f^{-1} preserves arbitrary intersections and unions while f preserves arbitrary unions, that for all $X \subseteq A$, $X \subseteq f^{-1}fX$, and that for all $Y \subseteq B$, $ff^{-1}Y \subseteq Y$. Moreover, f (the original function, but also the extended one) is injective if and only if $f^{-1}f = \mathrm{Id}_A$, and f is surjective if and only if $ff^{-1} = \mathrm{Id}_B$. Observe that the function f^{-1} is *not* the set-theoretic inverse of f, as this one is not assumed in general to be bijective; strictly speaking, f^{-1} is the ***residual*** of the extended function f, and its main property in this quality is that for all $X \subseteq A$ and all $Y \subseteq B$, $fX \subseteq Y \iff X \subseteq f^{-1}Y$ (for these and related properties, see Exercises 1.1–1.3).

A function $f\colon A \to B$ between ordered sets is ***order-preserving*** or ***monotone*** when $a \leqslant b$ implies $fa \leqslant fb$, for all $a,b \in A$; it is ***order-reflecting*** when the converse implication holds; and it is ***order-reversing*** when $a \leqslant b$ implies $fb \leqslant fa$, for all $a,b \in A$.

An ***(algebraic) language*** or ***similarity type*** is a pair $\boldsymbol{L} = \langle L, \mathrm{ar}\rangle$ where $L \neq \emptyset$ is a set, the set of ***symbols***, and $\mathrm{ar}\colon L \to \omega$ is a function, assigning an ***arity*** to each symbol.

- If $\lambda \in L$ is such that $\mathrm{ar}\lambda = 0$, then λ is a ***constant*** (or ***constant symbol***). The constant symbol that appears more often in this book is $\top$ (called "top"), which in many logics simply represents truth, or some related notion. Another popular constant is $\bot$ (called "bottom"), usually representing falsity. In the literature one also finds 1 and 0 used as constants for the same purposes. However, especially in the literature on substructural logics and linear logics, it is also customary that all the four constants coexist, with varied (sometimes interchanged) interpretations; see for instance foonote ‡ on page 130.
- If $\mathrm{ar}\lambda = n \geqslant 1$, then λ is an ***n-ary connective*** or ***operation*** (or ***operation symbol***). Popular examples of *unary* connectives are $\neg$ ("negation") and $\Box$ and $\Diamond$ ("modalities"); among *binary* connectives, $\wedge, \vee, \to$ and $\leftrightarrow$ (respectively, "conjunction", "disjunction", "implication" and "equivalence") form the traditional core of a vast majority of logics.

While in general there is no restriction on the cardinality of L, a few results may require that the language be countable, or even finite; practically all the examples use a finite language.

In the specification of particular languages, it is customary to describe a language $\boldsymbol{L}$ by a sequence L of the symbols actually used together with the sequence $\langle \mathrm{ar}\lambda : \lambda \in L\rangle$, which is then called its "type"; for instance, one says "let $\langle \wedge, \vee, \bot, \top\rangle$ be a language of type $\langle 2,2,0,0\rangle$".

When an algebraic language $\boldsymbol{L}$ is evaluated in a domain or mathematical universe, the result are the ***algebras of (similarity) type*** $\boldsymbol{L}$, or ***$\boldsymbol{L}$-algebras***; to specify one of them one writes $\boldsymbol{A} = \langle A, \langle \lambda^{\boldsymbol{A}} : \lambda \in L \rangle \rangle$, where $A \neq \emptyset$ is a set, the ***domain*** or ***universe*** of the algebra, and for each $\lambda \in L$ with $n = \operatorname{ar}\lambda$, $\lambda^{\boldsymbol{A}} \colon A^n \to A$ is a function with domain A^n and ranging into A, that is, what is usually called an ***n-ary operation*** or function on A. If $\operatorname{ar}\lambda = 0$ it is customary to identify $\lambda^{\boldsymbol{A}}$ with its image, that is, with an element of A; moreover, it is also traditional to write 0 for $\bot^{\boldsymbol{A}}$ and 1 for $\top^{\boldsymbol{A}}$, especially when these elements are the bounds of an order; but in some contexts this may be different. For other connectives, one usually writes directly λ instead of $\lambda^{\boldsymbol{A}}$ if no confusion is likely, while in particular algebras the operations may have specific definitions and symbols. Thus, one may say "let $\boldsymbol{A} = \langle A, \wedge, \vee, 0, 1 \rangle$ be an algebra of type $\langle 2, 2, 0, 0 \rangle$" to describe an arbitrary algebra of the type, and "for every set X, the power set algebra $\langle \mathscr{P}X, \cap, \cup, \emptyset, X \rangle$ is an algebra of type $\langle 2, 2, 0, 0 \rangle$".

If $\boldsymbol{A}, \boldsymbol{B}$ are algebras of the same similarity type, a function $h \colon A \to B$ is a ***homomorphism*** from $\boldsymbol{A}$ to $\boldsymbol{B}$ when for every $\lambda \in L$, with $n = \operatorname{ar}\lambda$, and every $a_1, \dots, a_n \in A$, $h\lambda^{\boldsymbol{A}}(a_1, \dots, a_n) = \lambda^{\boldsymbol{B}}(ha_1, \dots, ha_n)$; for a constant symbol λ this means that $h\lambda^{\boldsymbol{A}} = \lambda^{\boldsymbol{B}}$. The set of homomorphisms from $\boldsymbol{A}$ to $\boldsymbol{B}$ is denoted by $\operatorname{Hom}(\boldsymbol{A}, \boldsymbol{B})$, and the expression $h \colon \boldsymbol{A} \to \boldsymbol{B}$ means that $h \in \operatorname{Hom}(\boldsymbol{A}, \boldsymbol{B})$. The set of ***endomorphisms*** of an algebra $\boldsymbol{A}$ is $\operatorname{End}\boldsymbol{A} := \operatorname{Hom}(\boldsymbol{A}, \boldsymbol{A})$. Notice that endomorphisms leave the values of constant symbols fixed. Isomorphisms (bijective homomorphisms) are denoted by $\cong$.

If $\boldsymbol{A}$ is an algebra, a binary relation $\theta \subseteq A \times A$ is a ***congruence*** of $\boldsymbol{A}$ when it is an equivalence relation such that for every $\lambda \in L$, with $n = \operatorname{ar}\lambda \geqslant 1$, and every $a_1, \dots, a_n, b_1, \dots, b_n \in A$, if $a_i \equiv b_i \ (\theta)$ for all $i = 1, \dots, n$, then $\lambda^{\boldsymbol{A}}(a_1, \dots, a_n) \equiv \lambda^{\boldsymbol{A}}(b_1, \dots, b_n) \ (\theta)$. The set of congruences of $\boldsymbol{A}$ is denoted by $\operatorname{Con}\boldsymbol{A}$; this set, when ordered under the subset relation, becomes a complete lattice, with $\bigwedge := \bigcap$. If K is a class of algebras (of the same type), then for any algebra $\boldsymbol{A}$, not necessarily in K, the ***K-congruences*** of $\boldsymbol{A}$ are the congruences in the set $\operatorname{Con}_{\mathsf{K}}\boldsymbol{A} := \{\theta \in \operatorname{Con}\boldsymbol{A} : \boldsymbol{A}/\theta \in \mathsf{K}\}$; they are called simply the ***relative congruences*** of $\boldsymbol{A}$ when the class K is clear from context.

A ***trivial*** algebra is one with a one-element universe. All trivial algebras (of the same similarity type) are isomorphic; the symbol $\boldsymbol{A}_t$ denotes any one of them.

Clearly, all these algebraic notions are *relative to the language* $\boldsymbol{L}$. In the general theory one works with an *arbitrary but fixed* language $\boldsymbol{L}$, and so this symbol can be omitted unless it is strictly necessary or there may arise some confusion. Thus, all along the book, *the phrase "for every algebra" always means "for every* $\boldsymbol{L}$*-algebra", and all "arbitrary" classes of algebras are assumed to be of the same type*; and so forth.

The algebra of formulas

This is just what in first-order logic or in universal algebra is called *the algebra of terms*. In this book the words "term" and "formula" are considered in general as synonymous, except in the context of a first-order language.

Fix an arbitrary, countably infinite set V such that $V \cap L = \emptyset$, and such that none of the objects in $V \cup L$ is a finite sequence of other objects in the same set; the members of the set V are called (***sentential***) ***variables***, *atoms*, or *atomic* *formulas*. Then consider the set of *finite* sequences of elements of $V \cup L$ and build the ***set of formulas*** Fm (denoted by $Fm_{\boldsymbol{L}}$ when there is need to emphasize the language) by performing the following construction:

(a) $V \subseteq Fm$,

(b) $L_0 \subseteq Fm$, where $L_0 := \{\lambda \in L : \mathrm{ar}\lambda = 0\}$,

(c) $\lambda\alpha_1 \dots \alpha_n \in Fm$, whenever all $\alpha_i \in Fm$, $\lambda \in L$, $\mathrm{ar}\lambda = n \geqslant 1$. The expression $\lambda\alpha_1 \dots \alpha_n$ represents the sequence produced by *concatenating* the sequence $\langle\lambda\rangle$ and the sequences $\alpha_1, \dots, \alpha_n$. Here, and in (a) and (b), each sequence of length 1 is identified with its only member.

(d) Fm is totally determined by (a), (b) and (c).

This is called *the recursive definition (or construction) of formulas*. In many cases, the only way to prove that a property holds for all formulas is to prove it *inductively*: first, directly, for variables and constants, and then for an arbitrary longer formula of the form $\lambda\alpha_1 \dots \alpha_n$, for $n \geqslant 1$, assuming that each α_i satisfies the property. It can be proved (Exercise 1.5) that the connective λ and the formulas $\alpha_1, \dots, \alpha_n$ used to construct the formula $\lambda\alpha_1 \dots \alpha_n$ are unique; this fact is usually called the *Unique Readability Lemma*, and is needed to prove some of the most basic properties of formulas.

In the (very common) cases in which *all connectives are at most binary* an alternative but *isomorphic* construction using two auxiliary symbols $(\,,\,) \notin V \cup L$, the *parentheses*, is widely used; then rule (c) is replaced by the following two:

(c′) $(\star\alpha) \in Fm$ whenever $\alpha \in Fm$ and $\star \in L$ with $\mathrm{ar}\star = 1$.

(c″) $(\alpha \star \beta) \in Fm$ whenever $\alpha, \beta \in Fm$ and $\star \in L$ with $\mathrm{ar}\star = 2$.

In practice the outermost parentheses are never written, and there may be rules† to omit some of the inner ones without risking ambiguity; thus, for instance, it is customary to write $\neg\alpha \to \beta \wedge \gamma$ instead of $((\neg\alpha) \to (\beta \wedge \gamma))$, but not to write $\alpha \wedge \beta \vee \gamma$ instead of $(\alpha \wedge (\beta \vee \gamma))$ or of $((\alpha \wedge \beta) \vee \gamma)$. In examples I use the version with parentheses.

It is customary to reserve parts of different alphabets for specific purposes. In this book, x, y, z, t, u, v, w denote variables; some lowercase greek letters $\alpha, \beta, \dots, \varphi, \psi$ denote formulas; and uppercase greek letters denote sets of formulas: $\Gamma, \Delta, \dots \subseteq Fm$.

Often a formula α is writen as $\alpha(\vec{x})$ to indicate that the variables occurring in it are *among* those in the sequence $\vec{x}$. Observe that this does not mean that *all* variables in $\vec{x}$ actually occur in α; as a matter of fact, α might not have any variables at all (which is only possible when the language has at least one constant symbol). The set of variables (effectively) occurring in a formula α is denoted by $\mathrm{Var}\,\alpha$, and if $\Gamma \subseteq Fm$, $\mathrm{Var}\,\Gamma := \bigcup\{\mathrm{Var}\,\alpha : \alpha \in \Gamma\}$ is the set of variables occurring in Γ. If you are very fussy, you may find a formal definition of this function Var in Exercise 1.7; and Exercise 1.11 contains an interesting property concerning this notion, which is used later on.

† Usually: the parentheses in (c′) are not written, unary symbols take precedence over binary ones, and $\wedge, \vee$ take precedence over $\to, \leftrightarrow$.

Often unary or binary terms are considered as *derived operations* or *connectives*, and are accordingly denoted by a single symbol: in certain contexts, instead of referring to an arbitrary binary term as $\alpha(x,y)$, it is more practical to say something like "let $\star$ be a binary term", and this means that $x \star y$ is used as an abbreviation of $\alpha(x,y)$, and so forth; but do not forget that this $\star$ need not be among the *primitive* ("real") symbols of the language, which are those in L.

The set of formulas Fm can be turned into an $\boldsymbol{L}$-algebra, denoted by $\boldsymbol{Fm_L}$ or simply by $\boldsymbol{Fm}$ and called ***the formula algebra***, by considering the operations given implicitly by rules (b) and (c) above:

$$\begin{array}{lll} \lambda \in L_0 & \longmapsto & \lambda^{\boldsymbol{Fm}} := \lambda \\ \lambda \in L \smallsetminus L_0 & \longmapsto & \lambda^{\boldsymbol{Fm}}(\alpha_1,\dots,\alpha_n) := \lambda\alpha_1\dots\alpha_n \qquad \text{if } n = \operatorname{ar}\lambda \geqslant 1 \end{array}$$

Exercise 1.5 shows that the functions $\lambda^{\boldsymbol{Fm}}$ are injective. The main property of the algebra of formulas, indeed one that determines it up to isomorphism, is contained in the next result.

LEMMA 1.1
The algebra $\boldsymbol{Fm_L}$ is the ***absolutely[†] free L-algebra generated by*** *V. That is:*

1. *$\boldsymbol{Fm_L}$ is generated by V, in the sense that it is the smallest $\boldsymbol{L}$-algebra containing V.*
2. *It is freely generated, in the absolute sense that every function $h\colon V \to A$ into the domain of an arbitrary $\boldsymbol{L}$-algebra $\boldsymbol{A}$ can be extended in a unique way to a homomorphism (denoted by the same symbol) $h\colon \boldsymbol{Fm_L} \to \boldsymbol{A}$.*

As a consequence:

3. *Every homomorphism $h\colon \boldsymbol{Fm_L} \to \boldsymbol{A}$ is completely determined by its restriction $h{\restriction}V$. That is, two homomorphisms from $\boldsymbol{Fm_L}$ to the same $\boldsymbol{L}$-algebra $\boldsymbol{A}$ are equal if and only if they coincide on variables.*
4. *For each $\varphi \in Fm$, $h\varphi$ depends only on the restriction of h to* $\operatorname{Var}\varphi$.

SKETCH OF THE PROOF: 1. As commented above, clauses (a)–(c) in the construction of the set of formulas imply that $\boldsymbol{Fm_L}$ is an algebra of type $\boldsymbol{L}$ containing V; together with clause (d), they imply that it is the smallest algebra with these properties. This proves 1.

2. Now assume that $h\colon V \to A$ is an arbitrary function, where A is the universe of an arbitrary $\boldsymbol{L}$-algebra $\boldsymbol{A}$. The wanted extension (denoted also by h) is defined recursively as follows:

- hx is the image of x under the original function, for all $x \in V$.
- $h\lambda := \lambda^{\boldsymbol{A}}$, for all $\lambda \in L_0$.
- $h(\lambda\alpha_1\dots\alpha_n) := \lambda^{\boldsymbol{A}}(h\alpha_1,\dots,h\alpha_n)$, for all $\lambda \in L \smallsetminus L_0$,, $n = \operatorname{ar}\lambda \geqslant 1$ and all $\alpha_i \in Fm$.

That this last condition produces a uniquely defined function h depends on the *Unique Readability Lemma* of Exercise 1.5. It is easy to show that this procedure

[†] The "absolutely" emphasizes the difference with the more usual notion of a "free algebra in a class" (free group, free Boolean algebra, free algebra of a variety, etc.), which satisfies similar properties but restricted to algebras in the class.

defines a homomorphism, which clearly extends the original function, and is unique among all the homomorphisms doing so. This proves 2.

3 is just a way of describing the uniqueness of the extension, and 4 can be easily proved, inductively, using that $\mathrm{Var}(\lambda\alpha_1\ldots\alpha_n) = \mathrm{Var}\,\alpha_1 \cup \cdots \cup \mathrm{Var}\,\alpha_n$ (Exercise 1.7). ⊠

Thus, a particular homomorphism is usually defined by just describing its behaviour on variables. The main property of the formula algebra is its free character, and it is used often and without mention. One of its crucial applications is the following, which is also used at several key points in the book.

LEMMA 1.2
If $f\colon \boldsymbol{A} \to \boldsymbol{B}$ is a surjective homomorphism, then for each homomorphism $h\colon \boldsymbol{Fm} \to \boldsymbol{B}$ there is a homomorphism $g\colon \boldsymbol{Fm} \to \boldsymbol{A}$ such that $fg = h$. Graphically:

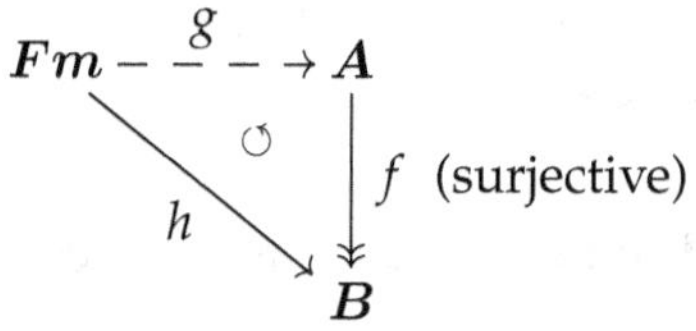

SKETCH OF THE PROOF: The Axiom of Choice allows you to start defining g by choosing, for each $x \in V$, some $gx \in f^{-1}\{hx\} \subseteq A$, so that $fgx = hx$. Then using Lemma 1.1 you will find the wanted $g\colon \boldsymbol{A} \to \boldsymbol{B}$ satisfying $fg\alpha = h\alpha$ for all $\alpha \in Fm$. Exercise 1.8 asks you to do the details. ⊠

Evaluating the language into algebras

An ***evaluation*** (***valuation, assignment***)[†] is any homomorphism from the formula algebra $\boldsymbol{Fm}$ into some $\boldsymbol{L}$-algebra $\boldsymbol{A}$; the set they form is hence denoted by $\mathrm{Hom}(\boldsymbol{Fm}, \boldsymbol{A})$. If $h \in \mathrm{Hom}(\boldsymbol{Fm}, \boldsymbol{A})$, then ***the value of a formula*** α under h is its image $h\alpha \in A$. By the recursive construction of formulas (that is, by the free character of the formula algebra), this value depends only on the variables occurring in α; thus, a popular alternative notation is to write $\alpha^{\boldsymbol{A}}(\vec{a})$ for $h\alpha$, when α is of the form $\alpha(\vec{x})$ and $h\vec{x} = \vec{a}$. This notation is extended in natural ways: for instance, $\alpha^{\boldsymbol{A}}(a, b, \vec{c})$ stands for $h\alpha$, when α is of the form $\alpha(x, y, \vec{z})$ and $hx = a$, $hy = b$ and $h\vec{z} = \vec{c}$; and so forth.

If $\alpha(\vec{x})$ is a formula with n variables, then making $\vec{a}$ range over all finite sequences of length n of elements of A defines a function

$$\begin{aligned} \alpha^{\boldsymbol{A}}\colon A^n &\longrightarrow A \\ \vec{a} &\longmapsto \alpha^{\boldsymbol{A}}(\vec{a}), \end{aligned}$$

that is, an n-ary function on A (by the addition of dummy arguments, it may be viewed as a function of any greater arity, and even as a function on $\vec{A}$). In universal algebra this function is called the ***term function*** associated with α.

[†] It is also very common to use the term "interpretation" for this, but in abstract algebraic logic this term has a technical meaning, which plays a key role in the theory of algebraizability (Chapters 3 and 6; see Definition 3.4).

The relation between a formula α and its associated term function $\alpha^{\boldsymbol{A}}$ is similar to that between polynomials and polynomial functions in ordinary algebra, and this view may be a useful intuition in some situations, but is more general than term functions. More precisely, a function $f\colon A^n \to A$ is a ***polynomial function*** on $\boldsymbol{A}$ when there is a formula $\alpha(\vec{x},\vec{y}) \in Fm$ with $\vec{x}$ of length n and there are $\vec{c} \in \vec{A}$ such that for all $\vec{a} \in A^n$, $f\vec{a} = \alpha^{\boldsymbol{A}}(\vec{a},\vec{c})$. The members of $\vec{y}$, and their values $\vec{c}$, are called ***parameters*** in this context, and have the same role as the coefficients of the polynomials of ordinary algebra. In particular, the term functions as first described above are the polynomial functions defined by formulas without parameters.

Equations and order relations

The next syntactic objects constructed from formulas that have a central role in this book are ***equations***, which are simply *pairs of formulas*. The set of equations of the language $\boldsymbol{L}$ is denoted by $Eq_{\boldsymbol{L}}$ or simply by Eq. Formally it is equal to $Fm_{\boldsymbol{L}} \times Fm_{\boldsymbol{L}}$, but tradition and intuition make it natural to write equations in the form $\alpha \approx \beta$ rather than $\langle \alpha, \beta \rangle$. The set of equations is not treated as an algebra (although, as a power of the universe of an algebra, it could be so). What is important here about equations is that they can be *evaluated* in particular algebras by means of the homomorphisms from the formula algebra into that algebra. An equation $\alpha \approx \beta$ is ***satisfied by the evaluation*** $h \in \mathrm{Hom}(\boldsymbol{Fm}, \boldsymbol{A})$, in symbols

$$\boldsymbol{A} \vDash \alpha \approx \beta \; [\![h]\!] , \tag{1}$$

when $h\alpha = h\beta$; that is, when $\alpha^{\boldsymbol{A}}(\vec{a}) = \beta^{\boldsymbol{A}}(\vec{a})$ if h is identified with $\vec{a}$ (equivalently, when $h\vec{x} = \vec{a}$, for any sequence $\vec{x}$ that includes the variables appearing in the formulas constituting the equation). An equation $\alpha \approx \beta$ ***is valid*** or ***holds*** in $\boldsymbol{A}$ when all evaluations in $\boldsymbol{A}$ satisfy it; in symbols,

$$\boldsymbol{A} \vDash \alpha \approx \beta \iff \boldsymbol{A} \vDash \alpha \approx \beta \; [\![h]\!] \quad \text{for all } h \in \mathrm{Hom}(\boldsymbol{Fm}, \boldsymbol{A}).$$

If K is a class of algebras, then, as is to be expected, an equation $\alpha \approx \beta$ ***is valid*** or ***holds*** in K when it is valid in all the algebras in K; in symbols,

$$\mathsf{K} \vDash \alpha \approx \beta \iff \boldsymbol{A} \vDash \alpha \approx \beta \quad \text{for all } \boldsymbol{A} \in \mathsf{K}.$$

All these relations concerning one equation are extended to sets of equations in the obvious way: they hold for a set when they hold for all equations in the set.

There is another, less common usage of a pair $\langle \alpha , \beta \rangle$ of formulas, which refers to ***order relations***, and is found when all the algebras under consideration have an order $\leqslant$; if in these situations pairs are written as $\alpha \preccurlyeq \beta$, then this indicates that they are intended to denote order relations when evaluated, and one can accordingly define

$$\boldsymbol{A} \vDash \alpha \preccurlyeq \beta \; [\![h]\!]$$

if and only if $h\alpha \leqslant h\beta$, or $\alpha^{\boldsymbol{A}}(\vec{a}) \leqslant \beta^{\boldsymbol{A}}(\vec{a})$; the other derived notions can be defined in the obvious way. Notice that when the order gives the algebra the structure of a (semi)lattice and the infimum operation $\wedge$ already belongs to the type, or is term-definable in it, the formal order $\alpha \preccurlyeq \beta$ is equivalent to the equation $\alpha \wedge \beta \approx \alpha$; this situation is found often in this book.

Sequents, and other wilder creatures

Besides formulas, one can consider other mathematical objects constructed from them. We have just seen equations and "orders", which are the simplest ones. Blok and Pigozzi generalized equations to what they called ***m*-dimensional formulas**, which are finite sequences $\langle \alpha_1, \ldots, \alpha_m \rangle$ of formulas of a fixed length $m \geqslant 1$; for $m = 1$ these are the ordinary formulas, and for $m = 2$ these are the equations. A general theory intended to apply both to formulas and to equations can be developed by working with m-dimensional formulas, for an arbitrary (finite) m; this is enough for the treatment of the notion of algebraizability as a particular case of that of *deductive equivalence* between two "m-dimensional deductive systems", as developed in [33, 35]. But this theory has later on been extended to more general linguistic objects of logical origin, as is briefly explained in Section 3.7.

The other linguistic objects that appear in this book are ***sequents***. In the literature there are several kinds of sequents.† The most common here are pairs $\langle \Gamma, \varphi \rangle$ where Γ is a *finite set* of formulas and φ is a formula; Γ is called the ***antecedent*** and φ the ***succedent***. Such pairs are denoted in the literature in a variety of ways: $\Gamma \rightarrow \varphi$, $\Gamma \Rightarrow \varphi$, $\Gamma : \varphi$, Γ / φ and even $\Gamma \vdash \varphi$, and others. A more neutral notation is always preferable, in order to leave the symbols $\rightarrow, \Rightarrow, \vdash$ free for other, common usages, and to prevent misunderstandings. I use

$$\Gamma \rhd \varphi$$

as suggested recently by several authors. The set of these sequents is denoted by *Seq*, and the set of sequents with non-empty antecedent (i.e., with $\Gamma \neq \emptyset$) by *Seq*°. Sequents are used to describe ***(Hilbert-style) rules*** in a neutral way, without postulating them for any particular logic; for instance, the popular rule of *Modus Ponens* (MP) can be described as the sequent $\{x, x \rightarrow y\} \rhd y$, written simply as $x, x \rightarrow y \rhd y$; and so forth. Additionally, ***generalized sequents*** (***g-sequents***) are obtained in the same way but allowing Γ to be of arbitrary cardinality; they are useful at a few particular points.

Similar sequents popular in the literature are the ones in which Γ represents a *finite multiset*‡ or a *finite sequence* of formulas rather than a finite set. All the sequents mentioned so far are said to be ***two-sided*** and ***single-conclusion***.

Sequents for classical logic as introduced by Gentzen, and also used for many other logics, are of the ***multiple-conclusion*** kind, which are the ones of the form $\Gamma \rhd \Delta$, with Δ a finite set, a multiset, or a sequence. For finitely-valued logics, Rousseau introduced the ***many-sided*** sequents, which have the form

† In some works, such as [8, 62, 216], the term *consecution* has been used instead of *sequent*.

‡ A *multiset* is something in between a set and a sequence. It is easier to understand than to formalize: it is a "list" of objects in which order does not matter but repetition does. It can be formalized as an equivalence class of sequences that are equal up to permutation, or as a set of different objects, each with an attached "multiplicity" (the number of times the object is repeated).

$\Gamma_1 \mid \Gamma_2 \mid \cdots \mid \Gamma_k$ for a fixed $k \geqslant 2$ where, again, the Γ_i can be either finite sets, finite multisets, or finite sequences of formulas. While the length of each sequence (or the cardinal of each finite set) is not bounded, in many-sided sequents the number k of sequences appearing in each sequent is fixed, and corresponds to the number of truth values of the logic to be treated with this formalism.

The literature on proof systems for non-classical logics has created other, more strange objects, such as ***hypersequents***, which combine the above ideas and consist of finite sequences of sequents, denoted by $\Gamma_1 \rhd \Delta_1 \mid \Gamma_2 \rhd \Delta_2 \mid \cdots \mid \Gamma_n \rhd \Delta_n$. Here both the length of the sequences of formulas and the number of sequents (the length of the "outer" sequence) are not bounded. See [17].

When these derived linguistic objects are on stage, several notions developed for formulas can be applied to them as well, just by applying them to each of their constituent formulas. It is clear that an evaluation sends one of these objects to an object with the same "structure" but made of elements of the algebra. Satisfaction of these objects under an evaluation requires having a relation of the same type on the universe of the algebra, just as satisfaction of equations uses the identity relation and satisfaction of pairs $\varphi \preccurlyeq \psi$ uses an order relation.

On variables and substitutions

To call the elements of the set V *variables* is a practice grounded on several reasons.

- Grammatical: The primary role of variables is to act as building blocks of the set of formulas; in this capacity they are also called *atomic formulas*. By Exercise 1.9, they can be replaced by any other set with the same cardinality, and one obtains an isomorphic algebra of formulas. Thus, their proper nature can be disregarded, and only how many of them are there really matters; we never use them explicitly, but only refer to them. In particular, notice that letters such as x, y, z *are not* variables, but merely *denote* them. In this we adhere to the customary abuse of language of ordinary mathematical work, and use "to be" for "to denote" (for example: when we say "let ε, δ be real numbers", we do not imply that these greek letters are real numbers, but simply that they denote them).†
- Semantical: As already seen, variables are used to represent arbitrary elements of the algebras into which the formulas are evaluated. This is because we always consider all possible evaluations, and by Lemma 1.1 these are determined by just giving all possible values to the elements of V. Thus, they (actually, their values) "vary" along all the elements of the algebra. It is the same as when in elementary calculus we speak of a function "of a real variable": we mean that the variable can take any real value.
- Logical: If we are to disregard the exact nature of variables, we want them to work in such a way that, for instance, the formula $x \wedge y \to x$ expresses

† Mathematicians may regard these issues as trivial, but in formal, mathematical logic they may be specially sensible. This is why we have both $=$ and $\approx$; while the expression $x = y$ denotes a *fact* (that the two letters denote the same variable), the expression $x \approx y$ denotes an *object*, the pair $\langle x, y \rangle$.

the same logical law as the formula $z \wedge t \to z$; even more, usually we want it to represent the law $\alpha \wedge \beta \to \alpha$, for arbitrary formulas α and β. To this end, we consider *substitutions*, and we require that logics are invariant under substitutions (Definition 1.5), a property related to the *formal* character of logical consequence.

Actually, these substitutions are just the substitutions of terms for individual variables of first-order languages; but their formal introduction is here simpler:

DEFINITION 1.3
A ***substitution*** *is any endomorphism* $\sigma\colon \boldsymbol{Fm} \to \boldsymbol{Fm}$ *or, equivalently, any function* $\sigma\colon V \to Fm$. *For each* $\alpha \in Fm$, $\sigma\alpha$ *is a* ***substitution instance*** *of* α. *The set of substitutions is denoted by* $\mathrm{End}\boldsymbol{Fm} := \mathrm{Hom}(\boldsymbol{Fm}, \boldsymbol{Fm})$.

Thus substitutions, as endomorphisms, leave the constant symbols of the language fixed and only act on variables and on the formulas that contain variables. Indeed, Lemma 1.1.3 applies, and as a consequence a substitution σ is uniquely determined by the function $\sigma \restriction V$: for each formula α, $\sigma\alpha$ is obtained from $\{\sigma x : x \in \mathrm{Var}\,\alpha\}$ by recursively applying the operations that "construct" α from is constituent variables. Exercise 1.10 draws your attention to a related, stronger fact. The composition of two endormophisms is an endomorphism, and the identity function is obviously an endomorphism. Therefore:

LEMMA 1.4
The set $\mathrm{End}\boldsymbol{Fm}$, *endowed with the operation of composition of functions plus the identity function, is a monoid (a set with a binary operation that is associative and has a unit or neutral element).* ⊠

Notice that a substitution is a particular case of an evaluation, actually an evaluation into the formula algebra itself; you should bear this in mind when certain general definitions are specialized for the formula algebra. In particular, the notations set up above for the value of a formula under a homomorphism can also be used for substitutions, but here one writes simply α rather than $\alpha^{\boldsymbol{Fm}}$. For instance, if $\alpha \in Fm$ and we write $\alpha(x,y)$ with the intended meaning that $x, y \in V$ are such that $\mathrm{Var}\,\alpha \subseteq \{x,y\}$, then we can write $\alpha(\psi_1, \psi_2)$ for $\sigma\alpha$, if σ is any substitution such that $\sigma x = \psi_1$ and $\sigma y = \psi_2$.[†]

This is not a notational shortcut, but an instance of a fact that applies to all homomorphisms (and which is just one of the ways of expressing the property of being a homomorphism; see Exercise 1.12): If $h\colon \boldsymbol{A} \to \boldsymbol{B}$, then for all $\vec{a} \in \vec{A}$ and all $\alpha \in Fm$, $h\alpha^{\boldsymbol{A}}(\vec{a}) = \alpha^{\boldsymbol{B}}(h\vec{a})$. Considering that a substitution σ is an endomorphism of $\boldsymbol{Fm}$, and that in this algebra the superscript is not written, we find that $\sigma\alpha(\vec{x}) = \alpha(\sigma\vec{x})$, which amounts to the expression at the end of the preceding paragraph. These notations are specially useful when mixing substitutions, evaluations and homomorphisms, but at the same time they can be confusing if no attention is paid to them. As an example, you should check

[†] Some authors prefer the notation $\alpha(x/\psi_1, y/\psi_2)$, which certainly helps to avoid misunderstandings in some situations; but beware that other authors prefer to write $\alpha(\psi_1/x_1, \psi_2/x_2)$ to the same effect!

that if $h \in \operatorname{Hom}(\boldsymbol{Fm}, \boldsymbol{A})$, then the obvious equality $h(\sigma\alpha) = (h\sigma)\alpha$ becomes $h(\alpha(\psi_1, \psi_2, \dots)) = \alpha^{\boldsymbol{A}}(h\psi_1, h\psi_2, \dots)$.

Exercises for Section 1.1

Exercise 1.1. Let A and B be sets, and $f\colon \mathscr{P}A \to \mathscr{P}B$ be an arbitrary function. Prove that the following conditions are equivalent:

(i) The function f commutes with arbitrary unions.

(ii) There is a function $g\colon A \to \mathscr{P}B$ such that $fX = \bigcup_{a \in X} ga$ for any $X \subseteq A$.

(iii) The function f has a ***residual***, which is a function $f^{-1}\colon \mathscr{P}B \to \mathscr{P}A$ such that for all $X \subseteq A$ and all $Y \subseteq B$, $X \subseteq f^{-1}Y \iff fX \subseteq Y$. Check that the residual can be defined as $f^{-1}Y := \{a \in A : f\{a\} \subseteq Y\}$ for all $Y \subseteq B$.

Comment. All this is a particular case of a more abstract lattice-theoretic construction, which you can easily imagine. See [123, Section 3.1.1] or [42, Section 1.3]. In some contexts, notations such as f^* or f^+ are preferred to f^{-1}, in order to avoid confusion with the true inverse image construction (but see Exercise 1.3).

Exercise 1.2. Let $f\colon \mathscr{P}A \to \mathscr{P}B$ be a function that commutes with arbitrary unions, and let f^{-1} be its residual (Exercise 1.1). Prove:

1. Both f and f^{-1} are order-preserving (with respect to set inclusion).
2. f^{-1} commutes with arbitrary intersections.
3. The function $X \mapsto f^{-1}fX$, for $X \subseteq A$, is a closure operator on A (see Definition 1.19). In particular $X \subseteq f^{-1}fX$.
4. The function $Y \mapsto ff^{-1}Y$, for $Y \subseteq B$, is an interior operator on B (see Exercise 1.35). In particular $ff^{-1}Y \subseteq Y$.

Exercise 1.3. Let $f\colon A \to B$ be any function, and as usual denote also by f its power set extension; that is, define $fX := \{fa : a \in X\}$ for all $X \subseteq A$, so that $f\colon \mathscr{P}A \to \mathscr{P}B$. Show that this extended f always commutes with arbitrary unions, and that its residual function f^{-1}, as introduced in Exercise 1.1, coincides with the usual inverse image construction (and notation); that is, that $f^{-1}Y = \{a \in A : fa \in Y\}$ for all $Y \subseteq B$ (notice that the f inside the set is the original function).

Exercise 1.4. Find examples of algebraic structures among the mathematical objects that you know well, and describe their similarity type. Find (mathematical) objects that cannot be presented as algebraic structures.

Exercise 1.5. Let $\lambda, \mu \in L$ be connectives of arities $n, m \geqslant 1$ respectively, and $\alpha_1, \dots, \alpha_n$, $\beta_1, \dots, \beta_m \in Fm$. Prove that if $\lambda\alpha_1 \dots \alpha_n = \mu\beta_1 \dots \beta_m$, then $\lambda = \mu$, $n = m$ and $\alpha_i = \beta_i$ for all $i = 1, \dots, n$.

Comment. This result is often called the *Unique Readability Lemma*, and in particular it implies that, as a function, $\lambda^{\boldsymbol{Fm}}$ is injective.

Exercise 1.6. Complete the proof of Lemma 1.1, following the sketch given in the text.

Exercise 1.7. Use Lemma 1.1.2 to prove that there is a function $F\colon Fm \to \mathscr{P}V$ such that $Fx = \{x\}$ for all $x \in V$ and $F(\lambda\alpha_1 \dots \alpha_n) = F\alpha_1 \cup \dots \cup F\alpha_n$ for every formula of the form $\lambda\alpha_1 \dots \alpha_n$, where λ is an n-ary connective of the language with $n \geqslant 1$, and $Fc = \emptyset$ for a constant symbol c. This function F is denoted by Var in the text and other exercises. Confirm that this really corresponds to the intended meaning of "the set of variables occurring in φ", as claimed in the text, by showing that, for any $X \subseteq V$ and any $\alpha \in Fm$, α belongs to the subalgebra of $\boldsymbol{Fm}$ generated by X if and only if $F\alpha \subseteq X$.

EXERCISE 1.8. Complete the proof of Lemma 1.2, following the sketch given in the text. Notice that the found g need not be unique. Try also to prove it for an arbitrary algebra in the place of the formula algebra, and see why you can't; if you have a good algebraic background you may produce a counterexample.

EXERCISE 1.9. Use Lemma 1.1 to prove that two algebras of formulas of the same similarity type generated by sets of variables of the same cardinality are naturally isomorphic (but not, strictly speaking, in a unique way).

EXERCISE 1.10. Let α be a formula and let σ, σ' be two substitutions. Prove that $\sigma\alpha = \sigma'\alpha$ if and only if $\sigma x = \sigma' x$ for all $x \in \operatorname{Var}\alpha$.

HINT. You will need to use Exercise 1.5.

EXERCISE 1.11. Let $\alpha \in Fm$ and σ a substitution. Prove that $\operatorname{Var}\sigma\alpha = \operatorname{Var}\sigma\operatorname{Var}\alpha$, that is, $\operatorname{Var}\sigma\alpha = \bigcup\{\operatorname{Var}\sigma x : x \in \operatorname{Var}\alpha\}$. Deduce from this that for all $\beta \in Fm$, if $\operatorname{Var}\alpha = \operatorname{Var}\beta$, then $\operatorname{Var}\sigma\alpha = \operatorname{Var}\sigma\beta$.

EXERCISE 1.12. Let $\boldsymbol{A}, \boldsymbol{B}$ be algebras (of the same type) and let $h \colon A \to B$ be a function. Prove that h is a homomorphisim from $\boldsymbol{A}$ to $\boldsymbol{B}$ if and only if for every $\alpha(x_1, \ldots, x_n) \in Fm$ and every $a_1, \ldots, a_n \in A$, $h\alpha^{\boldsymbol{A}}(a_1, \ldots, a_n) = \alpha^{\boldsymbol{B}}(ha_1, \ldots, ha_n)$.

1.2 Sentential logics

I hope the Introduction made it clear enough that any general study of logic, as abstract algebraic logic is, must start from a general notion of its own object of study, a logic. When determining a logic, in whatever way, there are usually two components to specify: A *language* (the "formulas" of the logic), which is dealt with in the previous section, and a *relation of consequence* (derivability, inference, entailment), which should be a relation between sets of formulas and formulas. This relation is most often denoted by $\vdash$ or $\vDash$ in the literature, but in this book only $\vdash$ is used to this effect, perhaps sub- or super-scripted. It can be *defined* or *presented* in several ways, but this is not relevant for the general theory. Accordingly, logics are formally defined as pairs of these two components. After the formal definition, and before giving the first examples, there are several comments, on the range of its application or to clarify a few points.

DEFINITION 1.5

A (**sentential**) **logic** (**of type** $\boldsymbol{L}$) *is a pair* $\mathcal{L} = \langle \boldsymbol{L}, \vdash_{\mathcal{L}} \rangle$ *where* $\boldsymbol{L}$ *is an algebraic language and* $\vdash_{\mathcal{L}} \subseteq (\mathscr{P}Fm_{\boldsymbol{L}}) \times Fm_{\boldsymbol{L}}$ *is a relation that satisfies the following properties, for all* $\Gamma \cup \Delta \cup \{\varphi\} \subseteq Fm$.

(I) ***Identity:*** *If* $\varphi \in \Gamma$, *then* $\Gamma \vdash_{\mathcal{L}} \varphi$.

(M) ***Monotonicity:*** *If* $\Gamma \vdash_{\mathcal{L}} \varphi$ *and* $\Gamma \subseteq \Delta$, *then* $\Delta \vdash_{\mathcal{L}} \varphi$.

(C) ***Cut,*** *or* ***Transitivity:*** *If* $\Gamma \vdash_{\mathcal{L}} \varphi$ *and* $\Delta \vdash_{\mathcal{L}} \psi$ *for every* $\psi \in \Gamma$, *then* $\Delta \vdash_{\mathcal{L}} \varphi$.

(S) ***Structurality:*** *If* $\Gamma \vdash_{\mathcal{L}} \varphi$, *then* $\sigma\Gamma \vdash_{\mathcal{L}} \sigma\varphi$ *for every substitution* σ.

The relation $\vdash_{\mathcal{L}}$ *is called the* **derivability relation** *or the* **consequence relation** *of the logic. When* $\Gamma \vdash_{\mathcal{L}} \varphi$ *holds, one says that* φ *is a consequence of* Γ, *or that* φ **follows** *from* Γ.

A **theorem** *of the logic is a formula* φ *such that* $\varnothing \vdash_{\mathcal{L}} \varphi$ *(this is usually written* $\vdash_{\mathcal{L}} \varphi$*).*

A ***theory*** *is a set of formulas closed under consequence, that is, a set $\Gamma \subseteq Fm$ such that $\Gamma \vdash_{\mathcal{L}} \varphi$ implies $\varphi \in \Gamma$. The set of theories of $\mathcal{L}$ is denoted by $\mathcal{T}h\mathcal{L}$.*

The relation $\vdash_{\mathcal{L}}$ is a relation between sets of formulas and formulas; standard infix relational notation is used, so that $\Gamma \vdash_{\mathcal{L}} \varphi$ means $\langle \Gamma, \varphi \rangle \in \vdash_{\mathcal{L}}$. One-element sets are here identified with their member; in particular $\varphi \vdash_{\mathcal{L}} \varphi$ is shorthand for $\{\varphi\} \vdash_{\mathcal{L}} \varphi$. Other self-explanatory list-like shorthand are used for sets as well, such as Γ, Δ for $\Gamma \cup \Delta$, or Γ, φ for $\Gamma \cup \{\varphi\}$. The formal definition of a logic as a pair emphasizes the role of its two components, the language and the consequence relation; however, in practice, and particularly in examples, it is customary to identify a logic with its consequence relation.

It is easy to see (Exercise 1.14) that the set of theorems is the smallest theory of a logic, and that there is always a smallest theory containing a given set Γ of formulas, namely the set $C_{\mathcal{L}}\Gamma := \{\varphi \in Fm : \Gamma \vdash_{\mathcal{L}} \varphi\}$, called ***the theory generated by*** Γ. By definition of the relation, the set Fm of all formulas is always a theory.

COMMENTS ON THE FORMAL DEFINITION OF THE NOTION OF A LOGIC.

- As is explained in the Introduction, this book does not deal with *first-order logics*, except for a brief digression in Section 2.5; thus this definition incorporates only the intuitions about consequence in a sentential environment.
- The above general definition of the notion of a *theorem* may appear slightly awkward to some (what does it mean "to be a consequence of nothing"?); but after looking at what this means in logics defined in a more specific way, either syntactically or semantically, this awkwardness will certainly disappear.
- A logic is defined as a *relation* between sets of formulas and individual formulas, satisfying certain properties. In this aspect, modern (particularly algebraic) studies of sentential logics deviate from *the traditional conception*, coming from Aristotle, that (the science of) logic is a tool or method (the *Organon*) to help the philosopher *discover truth*, by searching for "logical truths", that is, formulas that are "always true". Even in algebraic logic, it is still common to define a logic as *a set of formulas closed under substitutions*; this notion corresponds to the set of theorems of a logic in the present sense. In works of a large scope† this happens typically in studies of the family of all axiomatic extensions of a base logic; in these cases, even if the term "logic" is ascribed to sets of formulas, there is a consequence relation in the background for each of them. However, for really general or abstract studies this is not enough: for instance, it is necessary to treat as different objects the pairs of logics that appear in certain situations having the same theorems and differing in the strength of some proper inference rule (modal logic and the Necessitation Rule is a prototypical case; see Example 1.14). Another reason is that *logics without theorems* are not excluded from the general theory, and in fact we will find several of them, of independent interest (two simple, artificial examples appear in Exercise 1.21, while the first real examples appear in Example 1.13 and in Exercise 1.29). The discussions of these issues in [249, pp. xii–xiii and 43–51] are illustrative.

† Two recent examples are [55] and [123, see note 2 on p. 72].

- Condition (I) needs no comment, as it corresponds to a most basic feature of the idea of consequence; however, it has been contested by some philosophers of the most radical *relevance logic* trends.
- Condition (M) reflects an essential feature of the *deductive* character of the consequences studied in formal logic: addition of more assumptions never invalidates an already done reasoning. The so-called *non-monotone logics* have been studied since some time: these are "logics" in which (M) does not hold in full, but only in some weakened forms (thus, they are not *logics* in the present formal sense of the word). They formalize weaker notions of consequence, which are closer to some kinds of inferences present in everyday life, such as default logic or reasoning with partial or incomplete information. They are not covered by the standard theory of algebraization studied in this book; indeed, they have not been deeply studied from an algebraic point of view. Some useful references may be [167, 223, 227].

  Technically, (M) is in fact superfluous in Definition 1.5, as it follows from (I) and (C), but is included just by tradition and to draw attention to it; it becomes indispensable if (I) is replaced by a weaker form, as shown in Exercise 1.13.
- Condition (C) is also a basic feature of almost any notion of "consequence", and is immediately understood in a proof-theoretic landscape: once something has been proved from certain assumptions, it can be directly included in other proofs without having to prove it again, and it is guaranteed that what is proved with its help actually follows from the original assumptions.
- Condition (S) is also called "invariance under substitutions" or "substitution-invariance". It is the mathematical rendering of the idea that logical consequence is *formal*, which means that, in some sense, it should depend only on the "form" of the sentences, as discussed before. As announced before, it allows elements of the set V to behave as real *variables*, that is, to represent arbitrary formulas in the same sense that variables in real analysis represent arbitrary elements of the domain of a function. For instance, when saying that the consequence $x \wedge y \vdash_{\mathcal{L}} x$ holds for a certain logic $\mathcal{L}$, these "x" and "y" may be viewed as representing arbitrary formulas because, by structurality, this implies that the consequence $\varphi \wedge \psi \vdash_{\mathcal{L}} \varphi$ also holds for any two formulas φ and ψ (just consider any substitution σ such that $\sigma x = \varphi$ and $\sigma y = \psi$).
- Notice that in the context of Gentzen calculi, the term "structural" does not refer to condition (S), but to conditions (I) , (M) and (C), which correspond† to the so-called *structural rules* (Axiom), (Weakening) and (Cut), respectively; possible formulations appear in Table 1.‡

  The term *substructural logic* may refer to a "logic" that does not satisfy some of these rules or conditions, and thus strictly speaking it does not fall under Definition 1.5. Nevertheless, the methods of abstract algebraic logic as

† In the sense that all logics of Definition 1.5 do indeed satisfy these rules (see Definition 5.11).

‡ Moreover, a logic in the present sense also satisfies the rules (Exchange) and (Contraction) simply by the fact that Γ is a *set* of formulas, rather than a *sequence*. There are alternative forms of the rules above, and several subleties arise when playing around with these properties; discussing them belongs more to proof theory than to algebraic logic.

$$\varphi \rhd \varphi \quad (\text{AX}) \qquad \frac{\Gamma \rhd \varphi}{\Gamma, \psi \rhd \varphi} \quad (\text{W}) \qquad \frac{\Gamma \rhd \psi \quad \Gamma, \psi \rhd \varphi}{\Gamma \rhd \varphi} \quad (\text{CUT})$$

TABLE 1: Three of the structural rules of Gentzen calculi

developed in this book can be applied to substructural "logics", in at least three directions: The study of the algebraic models of their sets of theorems; the algebraic study of the sentential-like logics on the set of sequents defined by the corresponding sequent calculi (which do not have all the structural rules); and the algebraic study of the sentential logics associated with the Hilbert-style proof systems that generate their theorems. The "substructural" significance of the third direction is debatable, since the logics defined by these proof systems turn out to be structural, and hence their proof-theoretic properties need not be particularly significant for the "logic", at least for its original meaning or motivation. Anyway, this study is very well developed, as it seems to be a very good tool for investigating their algebraic models, and its results often shed important light on other properties of these "logics". See [123]. ⊠

The notion of a logic with the high degree of generality given in Definition 1.5 goes back basically† to Tarski (1930) for conditions (I) , (M) and (C), completed by Łoś and Suszko (1958) for (S). As a matter of fact, Tarski assumed an additional condition, which has become a *possible* and important, but not compulsory property of logics:

DEFINITION 1.6
A logic $\mathcal{L}$ is ***finitary*** *when the following holds for all $\Gamma \cup \{\varphi\} \subseteq Fm$:*

(F) $\qquad \Gamma \vdash_{\mathcal{L}} \varphi \iff \exists \Delta \subseteq \Gamma,\ \Delta$ *finite, such that* $\Delta \vdash_{\mathcal{L}} \varphi$.

Due to condition (M), the implication ($\Leftarrow$) always holds; therefore, it would be sufficient to put the ($\Rightarrow$) part in the definition.

Most of the examples familiar to you are probably finitary (Example 1.9), but others may be not (Example 1.15). For semantically defined logics this property is not self-evident; for instance, every student of a first course in mathematical logic knows that even for two-valued logic defined through the familiar truth-tables (Example 1.11) one has to *prove* it (and here this amounts to proving the well-known *Compactness Theorem*, either directly or as a corollary to the *Completeness Theorem*). Our general theory needs to start with logics that are not necessarily finitary, so that some beautiful results on this property make sense; but in works of a more limited scope or under other pressupositions, all logics treated may be required to be finitary from the beginning.‡

A few more concepts and notations of a basic character:

† Both Tarski and Suszko used the formalism of closure operators described in Section 1.3 rather than that of closure relations, but as explained there this difference is not relevant. See also the comment in the paragraph at the bottom of page 17.

‡ Finitary logics are called ***deductive systems*** by Blok and Pigozzi and other scholars following their terminology.

DEFINITION 1.7

The second argument of the consequence relation is extended to sets as follows. For all $\Gamma \cup \Delta \subseteq Fm$,

$$\Gamma \vdash_{\mathcal{L}} \Delta \stackrel{\text{def}}{\iff} \Gamma \vdash_{\mathcal{L}} \varphi \text{ holds for all } \varphi \in \Delta.$$

*The consequence or derivability relation induces an **interderivability relation** $\dashv\vdash_{\mathcal{L}}$, which is defined by*

$$\varphi \dashv\vdash_{\mathcal{L}} \psi \stackrel{\text{def}}{\iff} \text{both } \varphi \vdash_{\mathcal{L}} \psi \text{ and } \psi \vdash_{\mathcal{L}} \varphi,$$

for $\varphi, \psi \in Fm$; it is extended to sets $\Gamma, \Delta \subseteq Fm$ by

$$\Gamma \dashv\vdash_{\mathcal{L}} \Delta \stackrel{\text{def}}{\iff} \text{both } \Gamma \vdash_{\mathcal{L}} \Delta \text{ and } \Delta \vdash_{\mathcal{L}} \Gamma.$$

Notice how the definition of the relation $\Gamma \vdash_{\mathcal{L}} \Delta$ *differs* from the usual interpretation of the two-sided sequents $\Gamma \rhd \Delta$ in classical logic, which is (some variant of) "the conjunction of the formulas in Γ implies the disjunction of the formulas in Δ"; neglecting this difference may be a source of misunderstandings.

To finish with the first set of definitions, some logics with a pathological behaviour should be named.

DEFINITION 1.8

*A logic $\mathcal{L}$ is **trivial** when it satisfies $x \vdash_{\mathcal{L}} y$ for two different variables x, y. An **inconsistent** logic is a trivial logic with theorems, and an **almost inconsistent** logic is a trivial logic without theorems. Equivalently, $\mathcal{L}$ is inconsistent when $Th\mathcal{L} = \{Fm\}$, and is almost inconsistent when $Th\mathcal{L} = \{\emptyset, Fm\}$.*

By structurality, a logic is trivial if and only if $\varphi \vdash_{\mathcal{L}} \psi$ for any two formulas φ, ψ, and if and only if Fm is the only non-empty theory (this justifies the "equivalently" in the definition). Observe that on every language there are *exactly two trivial logics*, the inconsistent one and the almost inconsistent one. In many results of a very general character one has to watch whether these logics should be excluded from the assumptions, as they may satisfy certain properties in a trivial, undesired way.

Examples: Syntactically defined logics

Here "syntactically" means "done by purely combinatorial manipulations of formulas, without intervention of elements outside the language".† Just to understand this, consider the rule of *Modus Ponens* $\alpha, \alpha \to \beta \rhd \beta$: In order to use it we need just the ability to "read out" the formulas and recognize their internal structure, thus being able to check that the middle formula is really the result of taking the two extreme ones and plugging a symbol $\to$ between them; we do not need to understand the meaning of the symbols. Most syntactic definitions are *proof-theoretic*, as they rely on some notion of *proof* in some kind of

† Some formalisms, such as *tableaux* or *labelled sequent calculi*, operate on an augmented set of formulas, where some outside objects of a clear semantic origin (such as truth-values) are used to form more complex linguistic objects; but they are manipulated by purely combinatorial rules, and so they should count as syntactic.

calculus, but Exercises 1.21 and 1.33 contain examples that are not. Here I assume you are familiar with the most common proof-theoretic methods: the so-called *Hilbert-style* or *axiomatic* calculi and the *Gentzen-style* calculi.

EXAMPLE 1.9
Let H be a Hilbert-style calculus on a set of formulas Fm of type $\boldsymbol{L}$. For every $\Gamma \subseteq Fm$ and every $\varphi \in Fm$, the relation $\Gamma \vdash_H \varphi$ is defined to hold if and only if there is a proof of φ in H from assumptions in Γ. Then $\langle \boldsymbol{L}, \vdash_H \rangle$ is a finitary logic, and its theorems are the formulas that have a proof in H from no assumptions other than the axioms. See Exercise 1.17. ⊠

In this case the term "theorem" reflects its usage in ordinary mathematics (for instance: the theorems of group theory are the properties that can be proved from the axioms of the theory). Several examples of well-known logics defined in this way, including classical logic and intuitionistic logic, appear on pages 71 and 85.

EXAMPLE 1.10
Let $\mathfrak{G}$ be a Gentzen calculus on the set of sequents of type $\boldsymbol{L}$ such that the structural rules (AX) and (CUT) are admissible† in $\mathfrak{G}$. For $\Gamma \cup \{\varphi\} \subseteq Fm$ the relation $\Gamma \vdash_{\mathfrak{G}} \varphi$ is defined to hold if and only if there is a finite $\Delta \subseteq \Gamma$ such that the sequent $\Delta \rhd \varphi$ is *derivable* in $\mathfrak{G}$. Then $\langle \boldsymbol{L}, \vdash_{\mathfrak{G}} \rangle$ is a finitary logic. See Exercise 1.19. ⊠

The logic defined from a Gentzen calculus $\mathfrak{G}$ in the way just described is called, after [15], ***the internal logic*** of $\mathfrak{G}$ in the literature, in contrast to the one called the ***external logic*** of $\mathfrak{G}$, described in Exercise 1.20.

If one adopts a rather liberal notion of a Hilbert-style calculus, it is easy to prove that a logic is finitary if and only if it can be defined by means of such a calculus (namely, one incorporating all consequences of the logic as rules). This theorem, due to Łoś and Suszko [166, 1958], may appear rather trivial, but it had the merit of highlighting the property of structurality as an essential ingredient of any abstract definition of the notion of a logic; it was not mentioned by Tarski because it is obtained for free for logics with a syntactic or semantic definition (which he assumed). The question of whether a logic is *axiomatizable* by means of such a calculus has its own interest, though, provided "Hilbert-style" is understood in the more restrictive sense postulated by Hilbert: he required (not in these words) that the axioms and rules be *recursive*, that is, that there exists an *algorithm* to decide whether a formula is or is not an (instance of an) axiom and whether a sequent is or is not an (instance of an) inference rule. To be even more restrictive one may require that all these objects be *finite* in number (to keep substitution-invariance, this must refer only to schemata). Under a more general framework, this leads to the questions of whether a logic is *recursively axiomatizable* or whether it is *finitely axiomatizable*. These are interesting questions,

† A rule is *admissible* in a Gentzen calculus $\mathfrak{G}$ when for each substitution instance of the rule, its conclusion is a derivable sequent of $\mathfrak{G}$ whenever its premises are so. The notion can be formulated in full generality, concerning any consequence relation and its theorems; in Exercise 1.18 it is applied to a sentential logic.

and often difficult to answer. The corresponding problem in universal algebra is that of whether a given variety is *finitely based*, that is, of whether its equational theory can be axiomatized by a finite number of equations; this issue has been exhaustively investigated and some deep results have been obtained. Some results in abstract algebraic logic touch upon this problem and find conditions under which the logical problem can be reduced to the algebraic one, or handled by similar techniques; see [30] or [70, p. 161] for instance.

Examples: Semantically defined logics

Generally speaking, semantic definitions of logics imply some kind of *evaluation* of the language into some (mathematical) domain external to the language; the objects of these domains can be of many different kinds: algebraic structures alone, or these augmented by some additional structure, or relational structures in the ordinary sense of model theory, or topological structures, or even more specialized and exotic devices. The logic is then defined with the help of some features of the evaluated universe; in many cases these are read as including some notion of *truth*, but can also be of a more neutral character. While there is a wide consensus about a few general notions of proof and of syntactically defined logic, the landscape of semantics is much richer and there is no comparable consensus as to *what a semantics is* in general. It is better to start by reviewing some common semantic definitions of several logics, of a different character, which appear very often in the book as examples and counterexamples.

EXAMPLE 1.11 (CLASSICAL LOGIC, SEMANTIC PRESENTATION)
Consider the usual two-valued semantics of classical logic $\mathcal{C\ell}$. It is based on the well-known *truth tables* for the logical connectives of *conjunction* ($\wedge$), *disjunction* ($\vee$), *implication* ($\to$) and *negation* ($\neg$):

$\wedge$	0	1
0	0	0
1	0	1

$\vee$	0	1
0	0	1
1	1	1

$\to$	0	1
0	1	1
1	0	1

$\neg$	
0	1
1	0

where 0 represents "false" and 1 represents "true". The connective of *equivalence* is defined by $x \leftrightarrow y := (x \to y) \wedge (y \to x)$.

Viewed algebraically, these tables are just describing operations in the set $2 = \{0,1\}$. If $\boldsymbol{L} = \langle \wedge, \vee, \to, \neg \rangle$ is a language of type $\langle 2,2,2,1 \rangle$, then the tables above define an $\boldsymbol{L}$-algebra $\mathbf{2}$ with universe 2. The class of ***Boolean algebras*** (BA) can be defined as the variety generated by $\mathbf{2}$. If you use these tables to compute the value $v\varphi$ of a formula φ starting from the values $vx \in 2$ for all $x \in \operatorname{Var}\varphi$, this means you are taking v as a *homomorphism* from $\boldsymbol{Fm}$ to $\mathbf{2}$; if you do this in all possible ways, this means you are taking *all homomorphisms* with respect to the operations of $\boldsymbol{L}$. In this semantics the element 1 is *designated* as representing "truth", and from this idea a semantic consequence relation, denoted here provisionally by $\vdash_{\mathbf{2}}$, can be defined in the following, natural way:

$$\Gamma \vdash_{\mathbf{2}} \varphi \overset{\text{def}}{\iff} v\Gamma \subseteq \{1\} \text{ implies } v\varphi = 1, \text{ for all } v \in \operatorname{Hom}(\boldsymbol{Fm}, \mathbf{2}),$$

for every $\Gamma \cup \{\varphi\} \subseteq Fm$. It is easy to check that the relation $\vdash_2$ does indeed satisfy the properties in Definition 1.5. Thus, formally, ***classical logic*** is, for the time being, the logic $\mathcal{C\ell} := \langle \boldsymbol{L}, \vdash_2 \rangle$. After the *Completeness Theorem* (page 72), the consequence relation will be denoted simply by $\vdash_{\mathcal{C\ell}}$.

We will also meet several *fragments*† of $\mathcal{C\ell}$; by Exercise 1.27, they are obtained from the same truth-tables, by just disregarding certain operations. In this book the fragment of $\mathcal{C\ell}$ with just conjunction and disjunction, denoted by $\mathcal{C\ell}_{\wedge\vee}$, plays a prominent role, as it is a paradigmatic and reasonably simple example of several non-standard situations. ⊠

EXAMPLE 1.12 (ŁUKASIEWICZ'S MANY-VALUED LOGICS)
Consider the following operations‡ defined on the real unit interval $[0,1]$ by Łukasiewicz in the 1920s, as an extension of the classical, two-valued truth-tables: For every $a, b \in [0,1]$,

$$\begin{aligned} a \wedge b &:= \min\{a,b\} \\ a \vee b &:= \max\{a,b\} \\ a \to b &:= \min\{1, 1-a+b\} = \begin{cases} 1 & \text{when } a \leqslant b \\ 1-a+b & \text{when } a > b \end{cases} \\ \neg a &:= 1-a. \end{aligned}$$

Observe that $a \to b < 1$ if and only if $a > b$.

Let $[\mathbf{0,1}]$ be the algebra, of the same similarity type $\boldsymbol{L}$ as before, defined on the set $[0,1]$ by these operations. For $n \geqslant 2$ let $[\mathbf{0,1}]_n$ be the finite n-element subalgebra of $[\mathbf{0,1}]$, which has the set $\{0, \frac{1}{n-1}, \frac{2}{n-1}, \dots, \frac{n-2}{n-1}, 1\}$ as universe. Observe that $[\mathbf{0,1}]_2 = \mathbf{2}$. Łukasiewicz considered just the *tautologies* generated by these algebras, starting with the three-valued one, using them as *truth-tables* in an extended sense§ with *designated element* 1; but they can be used to introduce consequences in a way similar to classical logic:

- ***Łukasiewicz's infinitely-valued logic*** is the logic $\text{Ł}_\infty := \langle \boldsymbol{L}, \vdash_\infty \rangle$, where

  $$\Gamma \vdash_\infty \varphi \overset{\text{def}}{\iff} v\Gamma \subseteq \{1\} \text{ implies } v\varphi = 1, \text{ for all } v \in \operatorname{Hom}(\boldsymbol{Fm}, [\mathbf{0,1}]),$$

  for every $\Gamma \cup \{\varphi\} \subseteq Fm$.
- ***Łukasiewicz's finitely-valued logics*** $\text{Ł}_n := \langle \boldsymbol{L}, \vdash_n \rangle$ are defined in the same way, just replacing $[\mathbf{0,1}]$ by the algebra $[\mathbf{0,1}]_n$.

The properties in Definition 1.5 are easily checked. Clearly, $\text{Ł}_2 = \mathcal{C\ell}$. All the Ł_n are finitary (Theorem 4.4 guarantees this); by contrast, Ł_∞ is not finitary, as is shown in Example 1.15. For this reason, the name "Łukasiewicz's infinitely-valued logic" is also applied to the finitary companion $\text{Ł}_{\infty f}$ of Ł_∞ (see Definition 1.47), which is finitely axiomatizable by a Hilbert-style system (this is a difficult result of the theory of many-valued logics).

---

† The notion of a fragment is explained on page 27.

‡ This is not the place to discuss *why* the operations were defined in this way; but observe that on $\{0,1\}$ they agree with the Boolean ones.

§ Since for $[\mathbf{0,1}]$ the number of values is infinite, it is no longer possible to present the operations in a "table"; for $n = 3$, the table for implication is shown in Example 3.20.

The relative positions of all these logics is clarified in a comment on page 186 and in Example 4.10. The class MV of ***MV-algebras***, which turns out to be the algebraic counterpart of this logic (and to be closely related to the algebraic counterpart of $Ł_\infty$), can be defined as the variety generated by $[0,1]$. The subvarieties of MV generated by each of the algebras $[0,1]_n$ play a similar role for the logics $Ł_n$ (Example 5.83).

For more information on many-valued logics, especially from the algebraic side, see [59, 128, 132] and the *Handbook* [60, Chapters 1, 2 and 6]. ⊠

EXAMPLE 1.13 (BELNAP-DUNN'S FOUR-VALUED LOGIC)
Let t and f represent "true" and "false" understood in the *epistemic* and doubly *non-classical* sense that these values reflect only the available information on a situation, and hence are not incompatible and an evaluation may fail to assign any of them to a formula. In this scenario, it may be reasonable to consider that formulas in fact may be evaluated in one of the following possible ways: $\mathsf{t} = \{t\}$, $\mathsf{f} = \{f\}$, $\mathsf{n} = \emptyset$, $\mathsf{b} = \{t,f\}$, and put $M_4 := \mathscr{P}\{t,f\} = \{\mathsf{t},\mathsf{f},\mathsf{n},\mathsf{b}\}$. Consider an order in this set, called "the truth order", correspoding to the idea that, for $a,b \in M_4$, $a \leqslant b$ should mean "b is more true and less false than a". This is implemented by defining $a \leqslant b$ if and only if $\big[(t \in a \text{ implies } t \in b)$ and $(f \in b \text{ implies } f \in a)\big]$; the resulting order is depicted in Figure 2. A similar idea (that the consequences should always be evaluated as "more true and less false" than the premises) leads to a consequence relation $\vdash_{\mathcal{B}}$ defined as follows:

$$\begin{aligned} \Gamma \vdash_{\mathcal{B}} \varphi \overset{\text{def}}{\iff} & \text{ for all } v \colon V \to \mathscr{P}\{t,f\}, \\ & \text{ if } t \in v\gamma \text{ for all } \gamma \in \Gamma, \text{ then } t \in v\varphi, \text{ and} \\ & \text{ if } f \in v\varphi, \text{ then } f \in v\gamma \text{ for some } \gamma \in \Gamma, \end{aligned} \tag{2}$$

where $\Gamma \cup \{\varphi\} \subseteq Fm_{\boldsymbol{L}}$ for the language $\boldsymbol{L} = \langle \wedge, \vee, \neg \rangle$ of type $\langle 2,2,1 \rangle$. The evaluation v, defined in principle on variables, is extended to all formulas by the following stipulations,

$$\begin{aligned} t \in v(\alpha \wedge \beta) &\iff t \in v\alpha \text{ and } t \in v\beta \\ f \in v(\alpha \wedge \beta) &\iff f \in v\alpha \text{ or } f \in v\beta \\[1ex] t \in v(\alpha \vee \beta) &\iff t \in v\alpha \text{ or } t \in v\beta \\ f \in v(\alpha \vee \beta) &\iff f \in v\alpha \text{ and } f \in v\beta \\[1ex] t \in v(\neg\alpha) &\iff f \in v\alpha \\ f \in v(\neg\alpha) &\iff t \in v\alpha, \end{aligned}$$

in which the independent treatment of the two initial epistemic values t and f becomes apparent. These conditions also suggest that they can safely be replaced by the requirement that v is a *homomorphism* with respect to a certain non-standard *algebra of subsets* of the set $\{t,f\}$, that is, with respect to the algebraic

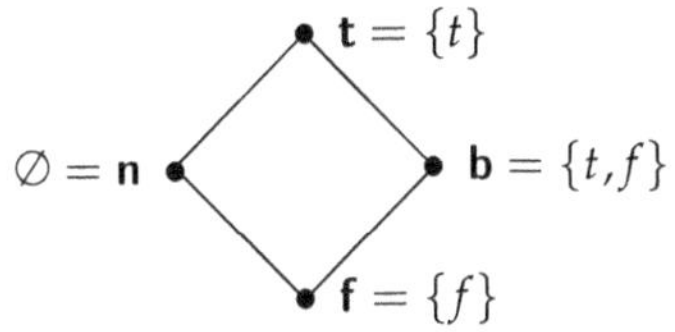

FIGURE 2: The four values of Example 1.13. The "truth order" goes upwards in the graphic, while the subset relation goes rightwards.

structure on M_4 given by the following tables:

$\wedge$	f	n	b	t
f	f	f	f	f
n	f	n	f	n
b	f	f	b	b
t	f	n	b	t

$\vee$	f	n	b	t
f	f	n	b	t
n	n	n	t	t
b	b	t	b	t
t	t	t	t	t

	$\neg$
f	t
n	n
b	b
t	f

We obtain the ordinary lattice operations of the truth order for $\wedge$ and $\vee$, while the negation $\neg$ is not the ordinary, Boolean one, but one of De Morgan character.

The resulting logic $\mathcal{B} := \langle \boldsymbol{L}, \vdash_{\mathcal{B}} \rangle$ is called ***Belnap-Dunn's (four-valued) logic***. The name acknowledges the contributions of both authors in its identification and its popularization; for more information on the origin, motivation and applications of this logic, see Example 4.12, and the references therein.

A perhaps surprising feature of this logic is that *it has no theorems*: it is easy to check that, for $\Gamma = \emptyset$, the definition (2) above amounts to saying that $\vdash_{\mathcal{B}} \varphi$ if and only if $v(\varphi) = \mathbf{t}$ for all $v \colon V \to M_4$; but no formula φ satisfies this, simply because the set $\{\mathbf{b}\}$ is closed under all operations, and hence an evaluation sending all the variables to **b** sends all formulas to **b** (the same holds for **n**).

This definition deviates from the preceding two in that, although still arisen from truth tables, it does not use any "designated element" representing a single truth in the same way. The logic $\mathcal{B}$ appears several times in the book as a particularly interesting example. ⊠

EXAMPLE 1.14 (MODAL LOGICS)
Let F be a class of *frames* in the sense of Kripke-style semantics (sets endowed with a binary relation). The basic normal modal logics arise from them in a well-known way. An *evaluation* v in a frame $\mathcal{W} = \langle W, R \rangle \in \mathsf{F}$ is a function $v \colon V \to \mathscr{P}W$. Assuming a standard modal language $\boldsymbol{L}_{\Box} = \langle \wedge, \vee, \to, \neg, \Box \rangle$, there is a recursive procedure (using the relation R for the $\Box$ case) that defines *truth at a point $p \in W$ under an evaluation v*, a relation sometimes called *forcing* and denoted by $\langle \mathcal{W}, p \rangle \Vdash \alpha \, [\![v]\!]$; I assume you know the details. The set of formulas α that are true at all points and under all evaluations in a certain class F forms a so-called ***normal modal system***. For particular classes of frames, different systems are produced: for instance, the well-known systems $K, S4, S5$ correspond to F being the class of all frames, of all reflexive and transitive frames, of all equivalence relations, respectively; thus, K is the smallest normal modal system.

The forcing relation $\Vdash$ can be used to obtain two consequence relations, and hence two logics, associated with each F. The ***local consequence*** is defined, for all $\Gamma \cup \{\varphi\} \subseteq Fm$, as

$$\begin{aligned} \Gamma \vdash^{\ell}_{\mathsf{F}} \varphi \iff & \text{for all } \mathcal{W} \in \mathsf{F}, \text{ all evaluations } v \text{ and all } p \in W, \\ & \text{if } \langle \mathcal{W}, p \rangle \Vdash \gamma \, [\![v]\!] \text{ for all } \gamma \in \Gamma, \\ & \text{then } \langle \mathcal{W}, p \rangle \Vdash \varphi \, [\![v]\!], \end{aligned}$$

while the ***global consequence*** is

$$\begin{aligned} \Gamma \vdash^{g}_{\mathsf{F}} \varphi \iff & \text{for all } \mathcal{W} \in \mathsf{F} \text{ and all evaluations } v, \\ & \text{if for all } p \in W, \ \langle \mathcal{W}, p \rangle \Vdash \gamma \, [\![v]\!] \text{ for all } \gamma \in \Gamma, \\ & \text{then for all } p \in W, \ \langle \mathcal{W}, p \rangle \Vdash \varphi \, [\![v]\!]. \end{aligned}$$

That these relations satisfy Definition 1.5 is easily checked; but it also follows from a general result (see Exercises 1.22 and 1.23).

It is the different position and scope of the quantifier "for all $p \in W$" that justifies the terms *local* and *global*. If S denotes the modal system defined by the class F, it may be practical to denote the corresponding local and global logics, respectively, by $S^{\ell} := \langle \boldsymbol{L}_{\Box}, \vdash^{\ell}_{S} \rangle$ and $S^{g} := \langle \boldsymbol{L}_{\Box}, \vdash^{g}_{S} \rangle$. Two important (but obvious) features are that the local logic is weaker than the global one (that is, $\Gamma \vdash^{\ell}_{S} \varphi$ implies $\Gamma \vdash^{g}_{S} \varphi$), but the two have the same theorems, which are exactly the formulas of the system S. One of the main differences between the two logics is the *Necessitation Rule* ($x \rhd \Box x$): it is only admissible in the local logics (if $\vdash^{\ell}_{S} \alpha$, then $\vdash^{\ell}_{S} \Box\alpha$, for all $\alpha \in Fm$), while it is derivable in the global logics ($\alpha \vdash^{g}_{S} \Box\alpha$, for all $\alpha \in Fm$). For most of the basic cases, there are Hilbert-style axiomatic presentations of the logics, and the one for the global logic often consists in adding just the Necessitation Rule to the one for the local logic. It is easy to see, directly from the above definitions, that all these logics are expansions of classical logic $\mathcal{C\ell}$.

These two kinds of modal logics appear very often as examples throughout the book; while the two logics of each pair are associated with the same class of algebras, they have quite a different behaviour according to the criteria of abstract algebraic logic. For more information on modal logic, in general, see [26, 55], the *Handbook* [27] and, for Spanish readers, Jansana's excellent survey [151]. ⊠

What is a semantics?

The common features of most of the previous examples lead to setting up a *generic idea* of what a semantics is, which can be applied to a large number of particular kinds of semantics you may have in mind (either algebraic, or relational); we will see that a logic is indeed obtained in all cases falling under this general notion (under reasonable restrictions).

The most popular kinds of semantics are based on some notion of ***truth in a structure***; this is certainly the case of Examples 1.11, 1.12 and 1.14. Their common features include:

- They are based on a class M of "models", also called "structures" or "frames".
- For each $\mathcal{M} \in \mathsf{M}$ there is a set $E_{\mathcal{M}}$ of "evaluations" in $\mathcal{M}$. They need not be exactly functions from Fm to the carrier of $\mathcal{M}$, though; the situation can be more complicated. In certain contexts it is customary to call each of the elements of M "structures" or "frames", and to reserve the term "model" for pairs $\langle \mathcal{M}, v \rangle$ where $\mathcal{M}$ is a structure and $v \in E_{\mathcal{M}}$.
- There is a ternary relation $\mathcal{M} \Vdash \varphi \; [\![v]\!]$, where $\mathcal{M} \in \mathsf{M}$, $\varphi \in Fm$ and $v \in E_{\mathcal{M}}$, implementing the idea that the evaluation v "satisfies" φ in $\mathcal{M}$, or that φ is "true" in $\mathcal{M}$ under the evaluation v. In some contexts this is called the ***truth definition*** or ***truth condition*** of the semantics.

Whenever these elements are given, a ***semantic consequence*** can be defined in the following way:

$$\Gamma \vdash_{\mathsf{M}} \varphi \iff \begin{array}{l} \text{for all } \mathcal{M} \in \mathsf{M} \text{ and all } v \in E_{\mathcal{M}}, \\ \text{if } \; \mathcal{M} \Vdash \gamma \; [\![v]\!] \text{ for all } \gamma \in \Gamma, \text{ then } \; \mathcal{M} \Vdash \varphi \; [\![v]\!] \end{array} \tag{T}$$

If $\mathcal{M}$ is a single model, then $\vdash_{\mathcal{M}}$ is defined as $\vdash_{\{\mathcal{M}\}}$, that is, simply deleting the "for all $\mathcal{M} \in \mathsf{M}$" from (T). It is mostly routine to show that $\vdash_{\mathsf{M}}$ and $\vdash_{\mathcal{M}}$ always satisfy (I), (M) and (C). In order to further obtain (S), and hence a logic, an additional condition is needed; a common and fairly general choice is:

- For each $\mathcal{M} \in \mathsf{M}$, $v \in E_{\mathcal{M}}$ and $\sigma \in \operatorname{End}\boldsymbol{Fm}$ there exist $\mathcal{M}_\sigma \in \mathsf{M}$ and $v_\sigma \in E_{\mathcal{M}_\sigma}$ such that for every $\varphi \in Fm$, $\mathcal{M}_\sigma \Vdash \varphi \; [\![v_\sigma]\!] \iff \mathcal{M} \Vdash \sigma(\varphi) \; [\![v]\!]$.

One can check that with this additional condition $\vdash_{\mathsf{M}}$ is invariant under substitutions, and hence a logic (fixing $\mathcal{M}_\sigma := \mathcal{M}$ makes each $\vdash_{\mathcal{M}}$ invariant under substitutions). See Exercise 1.22.

Logics defined following schemes of this kind can be called ***truth-preserving logics***, especially when only one structure is involved. Condition (T) implements the idea that a consequence holds if and only if

whenever the premisses are *true*, the conclusion is *true* as well,

where the *whenever* corresponds to the first two universal quantifiers appearing in (T). It is common to say that truth is "preserved" under consequence in the sense of being "transmitted" from premisses to conclusion, and actually it is this preservation that characterizes this consequence.

Observe that if you put $\Gamma = \emptyset$ in (T), we obtain that $\vdash_{\mathsf{M}} \varphi$ if and only if $\mathcal{M} \Vdash \varphi \; [\![v]\!]$ for all $\mathcal{M} \in \mathsf{M}$ and all $v \in E_{\mathcal{M}}$; that is, the theorems are the *tautologies*, the formulas that are *always and everywhere true*, a rather natural meaning of the general definition of theorem.

In principle, the definition of Example 1.13 does not conform to this scheme, but to that of ***logics preserving degrees of truth***, which are discussed in Example 4.11 and in Section 5.1. In some sense, these logics can be presented as particular cases of the truth-preserving logics just discussed, but the intended interpretation of their semantic definition is different.

In any event, the development of a general theory of semantics like the one sketched here is not a trivial issue, for it involves some choices that should be motivated by linguistic or philosophical issues.

What is an algebra-based semantics?

In the Introduction, algebraic logic is described as the study of algebra-based semantics for (mostly sentential) logics. Within the rather general setting just outlined, there are some conditions that may be naturally taken to isolate the semantics that are "algebraic" *in a non-technical sense of the word* (a technical one is introduced in Chapter 3), what I prefer to call "algebra-based". Some reasonable proposals might include:

1. Every $\mathcal{M} \in \mathsf{M}$ is of the form $\mathcal{M} = \langle \boldsymbol{A}_{\mathcal{M}}, \dots \rangle$, where $\boldsymbol{A}_{\mathcal{M}}$ is an $\boldsymbol{L}$-algebra, and "…" stands for some (optional) *additional* structure (which can be a special element, a subset, a family of subsets, a binary relation, a topology, or others).

2a. The value of each formula is an *element* of $A_{\mathcal{M}}$: $h\varphi \in A_{\mathcal{M}}$ for each $h \in E_{\mathcal{M}}$. Thus, the evaluations should be functions $h \colon Fm \to A_{\mathcal{M}}$.

2b. The value of a formula should be computed from the values of its variables using exclusively the algebraic structure of $\boldsymbol{A}_{\mathcal{M}}$.

The two last requirements can be summarized in one:

2. $E_{\mathcal{M}} \subseteq \mathrm{Hom}(\boldsymbol{Fm}, \boldsymbol{A}_{\mathcal{M}})$. The $\subseteq$ is actually an $=$ in the vast majority of cases. If it is $\subsetneq$, then $E_{\mathcal{M}}$ has to be closed under the operations $h \longmapsto h_\sigma$ (for each substitution σ) considered above in order to obtain invariance under substitutions (when $E_{\mathcal{M}} = \mathrm{Hom}(\boldsymbol{Fm}, \boldsymbol{A}_{\mathcal{M}})$, defining $h_\sigma := h \circ \sigma$ always works).

3. The relation $\mathcal{M} \Vdash \varphi\ [\![h]\!]$ is defined in terms of $h\varphi$ and a relation on $A_{\mathcal{M}}$, which must be previously, independently defined in $\mathcal{M}$.

In situations where the elements of the algebra $\boldsymbol{A}_{\mathcal{M}}$ can be reasonably understood as *truth-values*, the semantics that satisfy these conditions are commonly called *truth-functional semantics*.†

The most common kinds of algebra-based semantics, which will be treated later on in the book, are the following:

- The closest generalization of Examples 1.11 and 1.12 assumes the algebras $\boldsymbol{A}_{\mathcal{M}}$ have an element 1 which is *designated* as representing truth, so that $\mathcal{M} \Vdash \varphi\ [\![h]\!] \iff h\varphi = 1$. In this case, it is natural to take $\mathcal{M} = \langle \boldsymbol{A}_{\mathcal{M}}, 1 \rangle$. When the element 1 is term-definable in a uniform way in all the $\boldsymbol{A}_{\mathcal{M}}$ (it is an algebraic constant of the class of algebras involved), the logic is called an ***assertional logic*** (see Definition 3.5). A particularly important subclass of these are the *implicative* logics studied in Chapter 2; they constitute the first step in our way to the algebraic study of logics.

---

† Sometimes one can read in the literature that a certain logic is *truth-functional*, or that certain connective is *truth-functional*. This introduces the misunderstanding that truth-functionality is an absolute property (of a logic, or of a connective), while it is clearly relative to a specific semantics.

- A more substantial generalization is, given any set of equations $E(x) \subseteq Eq$ in at most the single variable x, to define $\mathcal{M} \Vdash \varphi \, [\![h]\!] \iff \boldsymbol{A}_{\mathcal{M}} \vDash E(x) \, [\![h\varphi]\!] \iff \varepsilon^{\boldsymbol{A}}(h\varphi) = \delta^{\boldsymbol{A}}(h\varphi)$ for all $\varepsilon(x) \approx \delta(x) \in E(x)$. In this case, the structure $\mathcal{M}$ can be reduced to the algebra $\boldsymbol{A}_{\mathcal{M}}$, and hence the class M can be identified with a class of algebras. This class is then called an ***algebraic semantics*** for the logic, in a technical sense, and it is said that *truth is equationally definable* in M. Chapter 3 is devoted to the study of the logics that can be defined in this way and satisfy another definability condition, which leads to the central notion of an *algebraizable logic*.
- A farther-reaching generalization of 1.11 and 1.12 is *matrix semantics*, based on structures called ***logical matrices***; these are pairs $\langle \boldsymbol{A}, F \rangle$ where $\boldsymbol{A}$ is an algebra and $F \subseteq A$ is a subset *designated* as the *truth set* of the matrix. This motivates defining $\langle \boldsymbol{A}, F \rangle \Vdash \varphi \, [\![h]\!] \iff h\varphi \in F$. Chapter 4 is devoted to the study of this kind of semantics. There we see that *any* logic can be defined in this way by a class of matrices, but this is just a technical result: defining a logic from a matrix or from a class of matrices in this way makes sense only when these matrices are the mathematical counterparts of some intuitive notion of a truth structure.

By contrast, it is clear that the relational semantics of Example 1.14 are not algebra-based.

For semantically defined logics the situation with regard to *finitarity* is different from that of syntactically defined ones: as the next example shows, semantically defined logics need not be finitary. Chapter 4 includes some general results that ensure the finitarity of a large number of semantically defined logics.

EXAMPLE 1.15
Łukasiewicz's infinitely-valued logic $\mathcal{Ł}_\infty$ is not finitary.

PROOF: Consider the defined connective $x \oplus y := \neg x \to y$. According to the definitions of the operations of the algebra $[\mathbf{0,1}]$ given in Example 1.12, it is given, for any $a, b \in [0,1]$, by:

$$a \oplus b = \begin{cases} 1 & \text{when } a + b \geqslant 1 \\ a + b & \text{when } a + b < 1 \end{cases}$$

It is easy to check that, then, for $n \geqslant 2$:

$$a \oplus \cdots^{n} \oplus a = \begin{cases} 1 & \text{when } a \geqslant \frac{1}{n} \\ na & \text{when } a < \frac{1}{n} \end{cases}$$

Now consider two distinct variables x, y, and the following set of formulas $\Sigma := \{(x \oplus \cdots^{n} \oplus x) \to y : n \geqslant 2\} \cup \{\neg x \to y\}$. By doing some elementary real arithmetic, and taking into account the essential fact that, according to the definitions in Example 1.12, $1 \to b = 1$ if and only if $b = 1$, it is not difficult to see (Exercise 1.26) that $\Sigma \vdash_\infty y$ while, for any finite $\Sigma_0 \subseteq \Sigma$, $\Sigma_0 \nvdash_\infty y$. This shows that $\mathcal{Ł}_\infty$ is not finitary. ⊠

Soundness, adequacy, completeness

Just to fix terminology:

DEFINITION 1.16
Let $\mathcal{L}$ be a logic and let M be a semantics defining† *a consequence relation $\vdash_{\mathcal{M}}$ for each $\mathcal{M} \in \mathsf{M}$, and a collective consequence $\vdash_{\mathsf{M}}$.*

1. *An $\mathcal{M} \in \mathsf{M}$ is* ***a model of*** *$\mathcal{L}$ when for every $\Gamma \cup \{\varphi\} \subseteq Fm$,*

$$\Gamma \vdash_{\mathcal{L}} \varphi \implies \Gamma \vdash_{\mathcal{M}} \varphi.$$

2. *The semantics M is* ***sound*** *for $\mathcal{L}$ when for every $\Gamma \cup \{\varphi\} \subseteq Fm$,*

$$\Gamma \vdash_{\mathcal{L}} \varphi \implies \Gamma \vdash_{\mathsf{M}} \varphi.$$

3. *The semantics M is* ***adequate*** *for $\mathcal{L}$ when for every $\Gamma \cup \{\varphi\} \subseteq Fm$,*

$$\Gamma \vdash_{\mathsf{M}} \varphi \implies \Gamma \vdash_{\mathcal{L}} \varphi.$$

4. *The semantics M is (**strongly**) **complete** for $\mathcal{L}$ when it is both sound and adequate, that is, when for every $\Gamma \cup \{\varphi\} \subseteq Fm$,*

$$\Gamma \vdash_{\mathcal{L}} \varphi \iff \Gamma \vdash_{\mathsf{M}} \varphi.$$

5. *The semantics M is* ***weakly complete*** *for $\mathcal{L}$ when for every $\varphi \in Fm$,*

$$\vdash_{\mathcal{L}} \varphi \iff \vdash_{\mathsf{M}} \varphi.$$

The terminology in points 4 and 5 is often reversed; accordingly, one says that ***a logic is strongly/weakly complete with respect to a semantics*** *when the semantics is strongly/weakly complete for the logic.*

*A semantics consisting of a single structure is (**strongly/weakly**) **characteristic** for a logic when it is (strongly/weakly) complete for it.*

Using this terminology, one may say that a logic ***preserves truth*** (***with respect to a semantics***) when this one is *sound* for the logic, and that the logic is *the* truth-preserving logic determined by the semantics when the logic is *complete* with respect to the semantics.

The above notions have many synonyms and variations in the literature. Sometimes "complete" is used with the above meaning of "adequate", in which case "sound and complete" is the present "complete". The qualifying "strongly" and "weakly" are inserted only when there is some risk of confusion; for instance in this book "complete" always means "strongly complete", and only the "weakly" is added, when it is appropriate.

❧ ❧ ❧

† In some way, not necessarily as in (T) on page 23.

Extensions, fragments, expansions, reducts

Again, just to fix terminology:

If $\mathcal{L}$ and $\mathcal{L}'$ are logics on the same language, then $\mathcal{L}$ is an ***extension***† of $\mathcal{L}'$, and this fact is denoted by $\mathcal{L}' \leqslant \mathcal{L}$, when

$$\mathcal{L}' \leqslant \mathcal{L} \overset{\text{def}}{\iff} \vdash_{\mathcal{L}'} \subseteq \vdash_{\mathcal{L}}$$
$$\overset{\text{def}}{\iff} \Gamma \vdash_{\mathcal{L}'} \varphi \text{ implies } \Gamma \vdash_{\mathcal{L}} \varphi, \text{ for all } \Gamma \cup \{\varphi\} \subseteq Fm.$$

In this case, $\mathcal{L}'$ is said to be ***weaker*** than $\mathcal{L}$, and $\mathcal{L}$ to be ***stronger*** than $\mathcal{L}'$. The extension relation is an order relation between logics of the same similarity type.

A ***fragment*** of a language $\boldsymbol{L} = \langle L, \mathrm{ar}\rangle$ is a language $\boldsymbol{L}' = \langle L', \mathrm{ar}'\rangle$ such that $L' \subseteq L$ and $\mathrm{ar}' = \mathrm{ar}\restriction L'$; this fact is denoted by $\boldsymbol{L}' \subseteq \boldsymbol{L}$, and it is also said that $\boldsymbol{L}$ is an ***expansion*** of $\boldsymbol{L}'$. Observe that if $\boldsymbol{L}' \subseteq \boldsymbol{L}$, then $Fm_{\boldsymbol{L}'} \subseteq Fm_{\boldsymbol{L}}$ (for a fixed set of variables).

Let $\mathcal{L}, \mathcal{L}'$ be logics of types $\boldsymbol{L}, \boldsymbol{L}'$, respectively. The logic $\mathcal{L}$ is an ***expansion*** of $\mathcal{L}'$ when $\boldsymbol{L}' \subseteq \boldsymbol{L}$ and $\vdash_{\mathcal{L}'} \subseteq \vdash_{\mathcal{L}}$. Observe that an extension is an expansion with the same language. A ***conservative expansion*** is an expansion such that

$$\text{for all } \Gamma \cup \{\varphi\} \subseteq Fm_{\boldsymbol{L}'},\ \Gamma \vdash_{\mathcal{L}'} \varphi \iff \Gamma \vdash_{\mathcal{L}} \varphi. \tag{3}$$

Given a logic $\mathcal{L}$ of type $\boldsymbol{L}$ and a fragment $\boldsymbol{L}'$ of $\boldsymbol{L}$, ***the $\boldsymbol{L}'$ fragment of $\mathcal{L}$*** is the logic of type $\boldsymbol{L}'$ with consequence relation $\vdash_{\mathcal{L}'} := \vdash_{\mathcal{L}} \cap \big((\mathscr{P}Fm_{\boldsymbol{L}'}) \times Fm_{\boldsymbol{L}'}\big)$, that is, such that (3) holds; it is not difficult to prove that this is always a logic. Thus, $\mathcal{L}'$ is a fragment of $\mathcal{L}$ if and only if $\mathcal{L}$ is a conservative expansion of $\mathcal{L}'$. If necessary the fragment logic may be denoted by $\mathcal{L}\restriction\boldsymbol{L}'$ or, in examples, by $\mathcal{L}_{\boldsymbol{L}'}$ (as an example, $\mathcal{C\ell}_{\to}$ is the $\langle\to\rangle$ fragment of classical logic $\mathcal{C\ell}$).

For syntactically determined logics, finding an explicit presentation of $\mathcal{L}\restriction\boldsymbol{L}'$ from one of $\mathcal{L}$ is often a difficult problem. For Hilbert-style axiomatizations, this leads to the notion of a *separable* axiomatization; for Gentzen-style presentations, the problem is usually addressed through *cut-elimination* theorems. For semantically determined logics the problem is in general easier, at least for algebra-based semantics, as shown in Exercise 1.27.

Similar notions apply to algebras. If $\boldsymbol{A}$ is an algebra of type $\boldsymbol{L}$ and $\boldsymbol{L}' \subseteq \boldsymbol{L}$, then the ***$\boldsymbol{L}'$ reduct*** of $\boldsymbol{A}$ is the algebra *on the same universe* with only the operations of $\boldsymbol{L}'$. A notation like $\boldsymbol{A}\restriction\boldsymbol{L}'$ may be useful, as well as $\mathsf{K}\restriction\boldsymbol{L}' := \{\boldsymbol{A}\restriction\boldsymbol{L}' : \boldsymbol{A} \in \mathsf{K}\}$ for a class K of algebras of type $\boldsymbol{L}$. Observe that the algebra $\boldsymbol{Fm}_{\boldsymbol{L}'}$ is *not* the $\boldsymbol{L}'$ reduct of $\boldsymbol{Fm}_{\boldsymbol{L}}$, but a subalgebra of this reduct.

There is a more general notion of a ***reduct*** of an algebra, which, instead of requiring that $\boldsymbol{L}' \subseteq \boldsymbol{L}$, allows $\boldsymbol{L}'$ to consist of any term-defined operations of $\boldsymbol{L}$ (in this case, the language $\boldsymbol{L}'$ is also called a ***reduct*** of $\boldsymbol{L}$, rather than a fragment); this may occasionally appear in some examples, and plays a key role in Section 7.2.

† The term "strengthening" has also been used in the literature with the same meaning.

Sentential-like notions of a logic on extended formulas

There are other logical systems, besides the ones complying with Definition 1.5, to which the essential intuitions and methods of algebraic logic apply through more or less straightforward analogies. These are the "logics" defined on the sets of "formulas" understood in an extended sense as discussed on pages 8–9 (equations, sequents, etc.). The essential logical characteristics common to all of them are purely set-theoretic, and is thus easily exported to arbitrary sets, so that one might speak about "logics on arbitrary sets":

DEFINITION 1.17
Let A be a set. A ***closure relation on*** *A is a relation $\vdash \subseteq (\mathscr{P}A) \times A$ that satisfies the following properties, for all $a \in A$ and all $X, Y \subseteq A$:*

(I) ***Identity:*** *If $a \in X$, then $X \vdash a$.*

(M) ***Monotonicity:*** *If $X \vdash a$ and $X \subseteq Y$, then $Y \vdash a$.*

(C) ***Cut****, or* ***transitivity:*** *If $X \vdash a$ and $Y \vdash b$ for all $b \in X$, then $Y \vdash a$.*

Thus, the consequence relation $\vdash_{\mathcal{L}}$ of a logic $\mathcal{L}$ is a closure relation on the set of formulas that satisfies the additional condition (S).

The algebraic satisfaction relation (1) displayed on page 7 can be used to define *two kinds of closure relations* on the set of *equations*. The first one has an *absolute* character; it is Birkhoff's ***equational logic*** $\vDash$, which is defined as follows: If $\Theta \cup \{\varepsilon \approx \delta\} \subseteq Eq$, then

$$\Theta \vDash \varepsilon \approx \delta \overset{\text{def}}{\iff} \begin{array}{l}\text{For every algebra (of the similarity type) } \boldsymbol{A}, \\ \text{if } \boldsymbol{A} \vDash \alpha \approx \beta \text{ for all } \alpha \approx \beta \in \Theta, \text{ then } \boldsymbol{A} \vDash \varepsilon \approx \delta\,.\end{array} \tag{4}$$

The second one has a *relative* or *local* character: for each class K one can consider the ***equational consequence relative to*** K, denoted by $\vDash_{\mathsf{K}}$ and defined as follows: If $\Theta \cup \{\varepsilon \approx \delta\} \subseteq Eq$, then

$$\Theta \vDash_{\mathsf{K}} \varepsilon \approx \delta \overset{\text{def}}{\iff} \begin{array}{l}\text{For every } \boldsymbol{A} \in \mathsf{K} \text{ and every } h \in \operatorname{Hom}(\boldsymbol{Fm}, \boldsymbol{A}), \\ \text{if } \boldsymbol{A} \vDash \alpha \approx \beta\, [\![h]\!] \text{ for all } \alpha \approx \beta \in \Theta, \text{ then } \boldsymbol{A} \vDash \varepsilon \approx \delta\, [\![h]\!],\end{array} \tag{5}$$

$$\text{that is, } \overset{\text{def}}{\iff} \begin{array}{l}\text{For every } \boldsymbol{A} \in \mathsf{K} \text{ and every } h \in \operatorname{Hom}(\boldsymbol{Fm}, \boldsymbol{A}), \\ \text{if } h\alpha = h\beta \text{ for all } \alpha \approx \beta \in \Theta, \text{ then } h\varepsilon = h\delta\,,\end{array}$$

$$\text{or } \overset{\text{def}}{\iff} \begin{array}{l}\text{For every } \boldsymbol{A} \in \mathsf{K} \text{ and every } \vec{a} \in \vec{A}, \\ \text{if } \alpha^{\boldsymbol{A}}(\vec{a}) = \beta^{\boldsymbol{A}}(\vec{a}) \text{ for all } \alpha \approx \beta \in \Theta, \text{ then } \varepsilon^{\boldsymbol{A}}(\vec{a}) = \delta^{\boldsymbol{A}}(\vec{a}).\end{array}$$

These relations are easily seen to be closure relations; the second one plays a central role in algebraic logic in general, and in abstract algebraic logic in particular. They are studied in more detail in Section 1.6.

The notion of a closure relation is a very basic, mathematical notion, that supports part of the intuition behind the notion of a "logic", but does not incorporate its "formal" character represented by the substitution-invariance condition (S). The natural notion of a "logic" on one of these extended notions of formula is that of a closure relation on the relevant set of linguistic expressions satisfying some kind of substitution-invariance; this condition should be the generalization of condition (S) in Definition 1.5 to an appropriate notion of a substitution. In order to formulate it, instead of giving each of these sets an algebraic structure and defining substitutions as the corresponding endomorphisms, or considering a brand new notion of a substitution, it is better to *extend the action of ordinary substitutions* to the new sets by extending them to functions on the new universes. Considering the following two examples suffices to grasp the idea; the general, abstract notion is dealt with in Section 1.4.

- A substitution $\sigma \in \mathrm{End}\boldsymbol{Fm}$ can be applied to an equation $\varepsilon \approx \delta$ by defining $\sigma(\varepsilon \approx \delta) := \sigma\varepsilon \approx \sigma\delta$, which is obviously again an equation. You can check (Exercise 1.30) that for any class K, the equational consequence relative to K defined in (5) satisfies:

$$\text{If } \Theta \vDash_{\mathsf{K}} \varepsilon \approx \delta\,, \text{ then } \sigma\Theta \vDash_{\mathsf{K}} \sigma(\varepsilon \approx \delta)$$

  for all $\Theta \cup \{\varepsilon \approx \delta\} \subseteq Eq$. By contrast, the absolute consequence $\vDash$ defined in (4) does not satisfy a property of this kind; see Section 1.6.

- Let Seq be the set of sequents of the first kind mentioned on page 8. Each substitution $\sigma \in \mathrm{End}\boldsymbol{Fm}$ induces a function in Seq, which is denoted in the same way, by defining $\sigma(\Gamma \rhd \varphi) := \sigma\Gamma \rhd \sigma\varphi$, where $\sigma\Gamma := \{\sigma\alpha_1,\ldots,\sigma\alpha_n\}$ for $\Gamma = \{\alpha_1,\ldots,\alpha_n\}$. Clearly the obtained result is again a sequent of the same kind. The function is extended to sets of sequents in the natural way. Then condition (S) for a relation $\vdash_{\mathfrak{G}} \subseteq (\mathscr{P}Seq) \times Seq$ would be

$$\text{If } \boldsymbol{\Gamma} \vdash_{\mathfrak{G}} \varphi\,, \text{ then } \sigma\boldsymbol{\Gamma} \vdash_{\mathfrak{G}} \sigma\varphi$$

  for all $\boldsymbol{\Gamma} \cup \{\varphi\} \subseteq Seq$, and when it holds the relation $\vdash_{\mathfrak{G}}$ is called ***substitution-invariant*** (in the contexts of sequent calculi and the like the term "structural" may be misunderstood). The same applies to a relation on the smaller set of sequents Seq°.

This allows us to formally define a notion we will sometimes find, although it is not one of our central concerns:

DEFINITION 1.18
A ***Gentzen system*** *is a pair $\mathfrak{G} = \langle \boldsymbol{L}, \vdash_{\mathfrak{G}} \rangle$ where $\boldsymbol{L}$ is an algebraic language, and $\vdash_{\mathfrak{G}}$ is a substitution-invariant closure relation on either the set Seq or the set Seq° of sequents built over the formulas of type $\boldsymbol{L}$; this set is denoted by $Seq\,\mathfrak{G}$. The relation $\vdash_{\mathfrak{G}}$ is called the* ***consequence relation*** *of $\mathfrak{G}$. The theorems of this consequence relation are called the* ***derivable sequents*** *of $\mathfrak{G}$.*

The relation between a sequent calculus (or "Gentzen-style calculus") in the ordinary sense and a Gentzen system in the present sense is analogous to that between a Hilbert-style axiomatic system and our general notion of a sentential logic: Every sequent calculus is actually an "axiomatic system" on the set of sequents, and its notion of a proof defines a Gentzen system in the present sense, as long as attention is not limited to the set of *derivable sequents* (as in Example 1.10), but to derivability from premises (arbitrary initial sequents), thus obtaining a *relation of derivability* of a sequent from a set of sequents. Tradition dictates that derivations in Gentzen calculi are displayed in tree-like form, while in Hilbert-style calculi both trees and linearly ordered proofs are equally popular. Notice that *any* sequent calculus defines a logic in this sense, whether it has the structural rules or not; by contrast, in order to use a sequent calculus to define a sentential logic, structural rules certainly play a role, as Example 1.10 shows.

Substitutions are extended to m-dimensional formulas, to either kinds of sequents, to hypersequents, in the obvious way, and a notion of a logic on these other objects would be defined as a substitution-invariant closure relation on the corresponding set. A more abstract notion of structurality, parallel to the abstraction in Definition 1.17, is briefly touched upon in Section 1.4.

Most of the methods, concepts and techniques of algebraic logic, developed originally for sentential logics, can be easily transferred to any of these sentential-like logics. However, this *mutatis mutandis* procedure of adapting things has the drawback that it has to be so-to-speak re-developed once and again every time one wants to do it for different syntactic objects. One of the great insights of Blok and Jónsson's [29] was to recognize the real common setting underlying the framework of "algebraizability" and to start the development of a more abstract theory, a true generalization of the main achievements of abstract algebraic logic. These issues are briefly surveyed in Section 3.7.

Exercises for Section 1.2

EXERCISE 1.13. Concerning Definition 1.5, show that, in the presence of (C), condition (I) implies condition (M), and that in the presence of (M), condition (I) may be replaced by its weaker version:

(I′) $\varphi \vdash_{\mathcal{L}} \varphi$.

Formulate and solve the parallel exercise concerning Definition 1.17, for closure relations on an arbitrary set.

EXERCISE 1.14. Using just the properties in Definition 1.5, prove that the set of theorems of a logic $\mathcal{L}$ is its smallest theory, and that for any $\Gamma \subseteq Fm$, the set $C_{\mathcal{L}}\Gamma := \{\varphi \in Fm : \Gamma \vdash_{\mathcal{L}} \varphi\}$ is the smallest theory of $\mathcal{L}$ containing Γ.

EXERCISE 1.15. Let $\vdash_{\mathcal{L}} \subseteq (\mathscr{P}Fm) \times Fm$ be a relation satisfying conditions (I), (M) and (F). Show that the relation $\vdash_{\mathcal{L}}$ satisfies (C) if and only if it satisfies the (here, only apparently) weaker condition

(C′) If $\Delta \vdash_{\mathcal{L}} \psi$ and $\Gamma, \psi \vdash_{\mathcal{L}} \varphi$, then $\Gamma, \Delta \vdash_{\mathcal{L}} \varphi$,

which more closely resembles the well-known "cut rule" of Gentzen calculi.

EXERCISE 1.16. Investigate whether it is possible to define finitary logics with conditions (I) , (C′) , (S) and (F); that is, either prove or refute this statement.

EXERCISE 1.17. Prove the statement contained in Example 1.9. For this you must first choose a suitable, technical definition of what a Hilbert-style calculus is and of its associated notion of a proof from assumptions.

COMMENT. Although in proof theory you may find calculi with infinitely long proofs and with rules having an infinity of premises, here proofs must be finite and rules must have a finite number of premises.

EXERCISE 1.18. Let $\mathcal{L}$ be a logic in a language $\boldsymbol{L}$, and let $\Gamma \rhd \varphi$ be a generalized sequent. The g-rule $\Gamma \rhd \varphi$ is ***admissible*** in $\mathcal{L}$ when for any $\sigma \in \operatorname{End}\boldsymbol{Fm}$, if $\emptyset \vdash_{\mathcal{L}} \sigma\alpha$ for all $\alpha \in \Gamma$, then $\emptyset \vdash_{\mathcal{L}} \sigma\varphi$. Let the relation $\Gamma \vdash^a_{\mathcal{L}} \varphi$ hold if and only if the rule $\Gamma \rhd \varphi$ is admissible in $\mathcal{L}$. Prove that $\langle \boldsymbol{L}, \vdash^a_{\mathcal{L}} \rangle$ is a logic.

COMMENT. It turns out that, even if $\mathcal{L}$ is finitary, the consequence $\vdash^a_{\mathcal{L}}$ need not be so. This is a non-trivial result; see [207, Example 3.6].

EXERCISE 1.19. Prove the statement contained in Example 1.10. This assumes you have a working knowledge of Gentzen calculi.

EXERCISE 1.20. Let $\mathfrak{G}$ be an arbitrary Gentzen calculus in a language $\boldsymbol{L}$ (not necessarily with the structural rules), and let $\Gamma \cup \{\varphi\} \subseteq Fm_{\boldsymbol{L}}$. Define the relation $\Gamma \vdash^e_{\mathfrak{G}} \varphi$ to hold if and only if the sequent $\emptyset \rhd \varphi$ is derivable from the sequents in the set $\{\emptyset \rhd \gamma : \gamma \in \Gamma\}$ with the rules of the calculus $\mathfrak{G}$. Prove that $\langle \boldsymbol{L}, \vdash^e_{\mathfrak{G}} \rangle$ is a logic.

COMMENT. This logic has been called the *external logic* associated with $\mathfrak{G}$.

EXERCISE 1.21. Let $\boldsymbol{L}$ be an arbitrary algebraic language, and define, for $\Gamma \cup \{\varphi\} \subseteq Fm_{\boldsymbol{L}}$,

$$\Gamma \vdash^0_v \varphi \overset{\text{def}}{\iff} \text{There is some } \alpha \in \Gamma \text{ such that } \operatorname{Var}\alpha = \operatorname{Var}\varphi$$

$$\Gamma \vdash^1_v \varphi \overset{\text{def}}{\iff} \operatorname{Var}\varphi \subseteq \operatorname{Var}\Gamma$$

$$\Gamma \vdash^2_v \varphi \overset{\text{def}}{\iff} \text{There is some } \alpha \in \Gamma \text{ such that } \operatorname{Var}\alpha \subseteq \operatorname{Var}\varphi$$

Put $\mathcal{L}_i := \langle \boldsymbol{L}, \vdash^i_v \rangle$, for $i = 0,1,2$.

1. Prove that the three $\mathcal{L}_i$ are logics, that two of them have no theorems, and that the third has theorems if and only if $\boldsymbol{L}$ has constants.
2. Prove that $\mathcal{L}_1$ and $\mathcal{L}_2$ are extensions of $\mathcal{L}_0$. Thus, $\vdash^0_v \subseteq \vdash^1_v \cap \vdash^2_v$.
3. Prove that $\mathcal{L}_0$ and $\mathcal{L}_2$ are unitary (see Exercise 1.57) and that $\mathcal{L}_1$ is finitary; therefore, the three logics are finitary.
4. Prove that $\vdash^0_v = \vdash^1_v \cap \vdash^2_v$ if and only if $\boldsymbol{L}$ has at most constants and unary operations.
5. Prove that the three logics coincide if and only if $\boldsymbol{L}$ has only unary operations.

COMMENT. In the case where $\boldsymbol{L}$ consists of just a single binary operation, the logics $\mathcal{L}_1$ and $\mathcal{L}_2$ are, respectively, the fragments of classical logic with just conjunction and with just disjunction. See [112] for more details.

EXERCISE 1.22. Let M be a class, and assume that for each $\mathcal{M} \in \mathsf{M}$ there is a set $E_{\mathcal{M}}$ and that there is a ternary relation $\mathcal{M} \Vdash \varphi\, [\![v]\!]$ for $\mathcal{M} \in \mathsf{M}$, $v \in E_{\mathcal{M}}$ and $\varphi \in Fm$.

1. Show that the relations $\vdash_{\mathcal{M}}$ and $\vdash_{\mathsf{M}}$ on the set of formulas Fm, defined by:

$$\Gamma \vdash_{\mathcal{M}} \varphi \overset{\text{def}}{\iff} \text{for all } v \in E_{\mathcal{M}}, \text{ if } \mathcal{M} \Vdash \gamma\, [\![v]\!] \quad \text{for all } \gamma \in \Gamma, \text{ then } \mathcal{M} \Vdash \varphi\, [\![v]\!],$$

$$\Gamma \vdash_{\mathsf{M}} \varphi \overset{\text{def}}{\iff} \text{for all } \mathcal{M} \in \mathsf{M},\ \Gamma \vdash_{\mathcal{M}} \varphi$$

   are closure relations on Fm.
2. Assume that moreover, for each $\mathcal{M} \in \mathsf{M}$, each $v \in E_{\mathcal{M}}$ and each $\sigma \in \operatorname{End}\boldsymbol{Fm}$ there exist $\mathcal{M}_\sigma \in \mathsf{M}$ and $v_\sigma \in E_{\mathcal{M}_\sigma}$ such that for every $\varphi \in Fm$, $\mathcal{M}_\sigma \Vdash \varphi\, [\![v_\sigma]\!] \iff \mathcal{M} \Vdash \sigma(\varphi)\, [\![v]\!]$. Show that then $\vdash_{\mathsf{M}}$ is a logic.
3. Modify the condition in 2 so that each individual $\vdash_{\mathcal{M}}$ is a logic.

EXERCISE 1.23. Discuss how the consequences of normal modal logics (Example 1.14) can be presented as instances of the general semantic scheme (T) on page 23 (see also Exercise 1.22).

HINT. In each case (the local and the global) the class of models and the relation $\Vdash$ appearing in the general scheme have to be instantiated differently, and need not coincide with the ones denoted by the same symbol in 1.14.

EXERCISE 1.24. Review other semantic ways you know of defining a logic and check whether they can be presented as instances of the general semantic scheme (T) on page 23 (see also Exercise 1.22). Think of some that cannot be so.

EXERCISE 1.25. Consider evaluations of the set of formulas of classical logic into the two-element Boolean algebra $\mathbf{2}$. Examine the following relation $\vdash_f$ of *friendliness* between a set of formulas Γ and a formula φ, due to Makinson [168]: $\Gamma \vdash_f \varphi$ if and only if every evaluation $v\colon \operatorname{Var}\Gamma \to \mathbf{2}$ such that $v\gamma = 1$ for all $\gamma \in \Gamma$ can be extended to an evaluation $v^+\colon \operatorname{Var}\big(\Gamma \cup \{\varphi\}\big) \to \mathbf{2}$ such that $v^+\gamma = 1$ for all $\gamma \in \Gamma \cup \{\varphi\}$. Determine which properties does $\vdash_f$ satisfy among (I), (I′), (M), (C), (C′), (F), (S).

EXERCISE 1.26. Complete the proof of non-finitarity of $Ł_\infty$ given in Example 1.15.

HINTS. To see that $\Sigma \vdash_\infty y$, you have to show that for any evaluation v there is some $\alpha \in \Sigma$ such that $v\alpha = 1 \to vy$ (treat the case $vx = 0$ separately). To see the other point, observe first that for each finite $\Sigma_0 \subseteq \Sigma$ there is some $k \geqslant 2$ such that $\Sigma_0 \subseteq \{(x \oplus \cdots^n \oplus x) \to y : 2 \leqslant n \leqslant k\} \cup \{\neg x \to y\}$; then, any evaluation v such that $vx = \frac{1}{k+1}$ and $vy = \frac{k}{k+1}$ should show that $\Sigma_0 \nvdash_\infty y$.

EXERCISE 1.27. Let M be an algebra-based semantics where $E_{\mathcal{M}} = \operatorname{Hom}(\boldsymbol{Fm}, \boldsymbol{A}_{\mathcal{M}})$ for each $\mathcal{M} \in \mathsf{M}$. Let $\boldsymbol{L}$ be the language of $\boldsymbol{Fm}$ and let $\boldsymbol{L}' \subseteq \boldsymbol{L}$ be any fragment. Consider the class $\mathsf{M}{\restriction}\boldsymbol{L}'$ of $\boldsymbol{L}'$ reducts of the structures in M, determined by taking for each $\mathcal{M} \in \mathsf{M}$ a structure $\mathcal{M}{\restriction}\boldsymbol{L}'$ having $\boldsymbol{A}{\restriction}\boldsymbol{L}'$ instead of $\boldsymbol{A}$. Use this class as an algebra-based semantics with $E_{\mathcal{M}\restriction\boldsymbol{L}'} = \operatorname{Hom}(\boldsymbol{Fm}_{\boldsymbol{L}'}, \boldsymbol{A}{\restriction}\boldsymbol{L}')$. Investigate whether $(\vdash_{\mathsf{M}}){\restriction}\boldsymbol{L}' = \vdash_{\mathsf{M}\restriction\boldsymbol{L}'}$.

COMMENT. Given the fuzzyness of the notion of an algebra-based semantics, this may have no general solution. You will certainly be able to *prove* the stated property for the particular cases mentioned on page 24 (assertional logics, logics defined by matrices, etc.). Anyway, you will probably benefit from the investigation.

EXERCISE 1.28. Prove that all fragments of a finitary logic are finitary as well.

EXERCISE 1.29. Following the general scheme of identifying fragments of logics of Exercise 1.27, consider the fragment of classical two-valued logic with only conjunction and disjunction, $\mathcal{C\ell}_{\wedge\vee}$. Realize that this logic is semantically defined by the two-element (distributive) lattice and prove that this logic has no theorems. Do the same for the fragments with just conjunction and with just disjunction, and the two-element semilattice, and observe the difference between the two.

EXERCISE 1.30. Prove that for any class K, the equational consequence relative to K defined in (5) on page 28 satisfies that if $\Theta \vDash_{\mathsf{K}} \varepsilon \approx \delta$, then $\sigma\Theta \vDash_{\mathsf{K}} \sigma(\varepsilon \approx \delta)$, for all $\Theta \cup \{\varepsilon \approx \delta\} \subseteq Eq$.

1.3 Closure operators and closure systems: the basics

Closure operators have already appeared in the context of logics (under the disguise of closure relations). In algebraic logic there are several natural closure operators defined in the algebraic models of the logics, in the algebraic structures under consideration, and in several derived notions. Actually, closure operators are ubiquitous in abstract algebraic logic, which justifies concentrating for a moment on the general notion in order to set up a common ground of terminology, notations, and general properties.

DEFINITION 1.19
A ***closure operator on a set*** *A is a function $C\colon \mathscr{P}A \to \mathscr{P}A$ that satisfies the following conditions:*

(I) *For all $X \subseteq A$, $X \subseteq CX$.*

(M) *For all $X, Y \subseteq A$, if $X \subseteq Y$, then $CX \subseteq CY$.*

(C) *For all $X \subseteq A$, $CCX = CX$.*

The elements of $C\varnothing$ are called the ***theorems*** *of C. A subset $X \subseteq A$ is a* ***closed set*** *or a* ***theory of*** *C when $X = CX$. Subsets $X \subseteq A$ with $CX = A$ and elements $a \in A$ with $C\{a\} = A$ are called* ***inconsistent****.*

The set of closed sets of a closure operator C can be denoted by $\mathcal{T}h C$, but is more commonly denoted by $\mathscr{C}$, following Proposition 1.25 and the typographical convention explained in (9) and (10) on page 36.

Comparing Definitions 1.17 and 1.19, it is easy to see (Exercise 1.31) that closure operators on a set A are in one-to-one correspondence with closure relations on the same set A, through the equivalence

$$a \in CX \iff X \vdash a. \tag{6}$$

If necessary, a more explicit notation for the two correspondences can be used:

$$\vdash \longmapsto C_{\vdash} \qquad \text{and} \qquad C \longmapsto \vdash_C,$$

where, for all $X \cup \{a\} \subseteq A$, $C_{\vdash} X := \{a \in A : X \vdash a\}$ and $X \vdash_C a \overset{\text{def}}{\iff} a \in CX$.

Because of this equivalence, any notion defined for one of the two kinds of objects can be immediately applied to the other (for example, when "finitary" closure operators are introduced in Definition 1.39, the corresponding notion of a "finitary" closure relation is considered automatically defined).

EXAMPLES 1.20

1. If $\mathcal{L} = \langle \boldsymbol{L}, \vdash_{\mathcal{L}} \rangle$ is a logic, and for each $\Gamma \cup \{\varphi\} \subseteq Fm$ we define

   $$\varphi \in C_{\mathcal{L}} \Gamma \overset{\text{def}}{\iff} \Gamma \vdash_{\mathcal{L}} \varphi,$$

   then $C_{\mathcal{L}}$ is a closure operator on Fm.
2. A closure operator on Fm defines a logic through scheme (6) if and only if it satisfies the obvious condition on closure operators parallel to condition (S). You may find several ways of expressing it in Proposition 1.51.
3. Similarly, the other sentential-like logics described at the end of the preceding section define the corresponding closure operators. In particular, the relative equational consequence $\vDash_{\mathsf{K}}$ associated with each class K of algebras defined in (5) on page 28 induces the corresponding closure operator on the set Eq of equations, denoted by C_{K}; they are studied in detail in Section 1.6.
4. The notion of a closure operator is present in many areas of mathematics besides logic, notably in *algebra* and in *topology* (Exercise 1.32); this fact has

had a noticeable influence on the terminology used in the general theory of closure operators. Because of these varied influences, some terminological points, coming from certain intuitions, may conflict with other points in which other intuitions dominate; however, you should not be puzzled by this circumstance, and should not let these intuitions interfere with the abstract definitions and results. ⊠

Notice that a closure operator C is a function on *subsets*: $X \subseteq A \longmapsto CX \subseteq A$. This corresponds to the (reasonable) requirement that a consequence of a set of premises is something that follows from the set *taken as a whole*, something that might not follow from any of its parts or elements. Due to this fact, some notations may seem unnecessarily cumbersome, and some *abbreviations* have become standard, such as $C(X,Y)$ for $C(X \cup Y)$ or $C(X,a)$ for $C(X \cup \{a\})$, when $X,Y \subseteq A$ and $a \in A$. The basic "arithmetical" rules for working with closure operators can be summarized in one property:

PROPOSITION 1.21
Let C be a closure operator on a set A, and let for each $i \in I$, $X_i, Y_i \subseteq A$ be such that $X_i \subseteq Y_i \subseteq CX_i$. Then $C\bigcup_{i\in I} X_i = C\bigcup_{i\in I} Y_i$.

PROOF: From the assumptions and property (M) it immediately follows that $C\bigcup X_i \subseteq C\bigcup Y_i$. To prove the reverse inclusion, observe that for the same reasons $C\bigcup Y_i \subseteq C\bigcup CX_i$. But for each i, $X_i \subseteq \bigcup X_i$, therefore by property (M) and the assumption again, $Y_i \subseteq CX_i \subseteq C\bigcup X_i$, and therefore $\bigcup Y_i \subseteq C\bigcup X_i$. Using (M) and then (C) it follows that $C\bigcup Y_i \subseteq CC\bigcup X_i = C\bigcup X_i$, which establishes the wanted inclusion. ⊠

As particular cases of the above general formula we obtain the most useful forms for its application:

COROLLARY 1.22
Let C be a closure operator on a set A.

1. $C\bigcup_{i\in I} X_i = C\bigcup_{i\in I} CX_i$ *whenever $X_i \subseteq A$ for each $i \in I$.*
2. $C(X,Y) = C(X,Z)$ *whenever $Y,Z \subseteq A$ are such that $Y \subseteq Z \subseteq CY$.*
3. $C(X,Y) = C(X,CY)$ *whenever $X,Y \subseteq A$.* ⊠

These properties are fundamental. They say that when computing a closure, any part of the set to be closed can be replaced by its own closure. For a first example of its use, see Exercise 1.34.

PROPOSITION 1.23
Let C be a closure operator on a set A. A set $X \subseteq A$ is closed if and only if $X = CY$ for some $Y \subseteq A$.

PROOF: If X is closed, then $X = CX$ by definition. If $X = CY$, then by property (C), $CX = CCY = CY = X$. ⊠

This justifies using "theories" for the closed sets (the fixed points) of closure operators, especially in the case of formalized mathematical theories: when logic is understood as a basic notion underlying all mathematical theories, a particular theory can be identified with the set of consequences of some particular set of axioms, that is, a set of the form CY for some Y (it is in this sense that the theory of groups can be identified with the set of consequences of the axioms of group theory, for instance).

Besides closure relations, there is a third presentation of closure operators, which is in many respects *dual* to the other two, and somehow even more intuitive:

DEFINITION 1.24
A ***closure system*** *or* ***closed-set system on*** *a set A is a collection of subsets $\mathscr{C} \subseteq \mathscr{P}A$ that satisfies the following conditions:*

(Cl1) *$A \in \mathscr{C}$.*

(Cl2) *If $\mathscr{B} \subseteq \mathscr{C}$ and $\mathscr{B}$ is non-empty, then $\bigcap \mathscr{B} \in \mathscr{C}$.*

The members of $\mathscr{C}$ are called ***closed sets****.*

Condition (Cl1) might be considered as a particular (degenerate) case of condition (Cl2), if we accept the convention that $\bigcap \emptyset = A$. Notice that the possibility that $\emptyset \in \mathscr{C}$ is not excluded; in fact, there are many natural cases (indeed, logics) where this does happen.

Some simple examples of closure systems are the whole family $\mathscr{P}A$, and the families of the form $\{F, A\}$ for some $F \subseteq A$. As in the case of closure operators, *mathematics is full of closure systems*. They are typically described as families of subsets of some universe closed under arbitrary intersections, for instance the family of all subspaces of a given vector space, or more generally families of the form "all substructures of a given structure", in the algebraic case. The example that had a major influence in the early days, in terminology for example, is the family of closed sets of a topology (which, besides the two properties above, also satisfies closure under finite unions). The family of open sets of a topology inspired the dual notion of an *interior operator*, which has but a marginal presense in this book, and is dealt with in Exercise 1.35. Another big group with a steady presence in algebraic logic are those related to several kinds of "filters" in ordered sets, lattices, and similar structures. However, the requirement of closure under intersections of *any* subfamily may force to include the empty subset in case there is no smallest non-empty "filter", and this may conflict with the natural definitions (for example, in lattice theory a filter is by definition a non-empty subset); in these cases, instead of speaking of "the closure system *of all* the [...]" one should speak of "the closure system *generated by all* the [...]"; see Proposition 1.30 and Exercise 1.42.

The following paradigmatic example is important:

PROPOSITION 1.25
If C is a closure operator on A, then the collection

$$\mathscr{C} := \{X \subseteq A : CX = X\} \tag{7}$$

of its closed sets is a closure system on A.

PROOF: Assume that $F_i \in \mathscr{C}$ for all $i \in I$. By property (I) $\bigcap F_i \subseteq C \bigcap F_i$. Since $\bigcap F_i \subseteq F_j$ for all $j \in I$, by property (M) it follows that $C \bigcap F_i \subseteq CF_j = F_j$ for all

$j \in I$. Therefore, $C \bigcap F_i \subseteq \bigcap F_i$, that is, $C \bigcap F_i = \bigcap F_i$. Finally, $A \in \mathscr{C}$ because $A \subseteq CA$, which holds by (I), implies $CA = A$. ⊠

This $\mathscr{C}$ is called ***the closure system associated with*** C. The first important result of this section establishes that every closure operator is determined by its associated closure system and that all closure systems can be obtained in the preceding way:

THEOREM 1.26
Let $\mathscr{C}$ be a closure system on a set A, and define, for every $X \subseteq A$,

$$CX = \bigcap \{F \in \mathscr{C} : F \supseteq X\}. \tag{8}$$

Then C is a closure operator on A whose closed sets are exactly the members of $\mathscr{C}$.

Conversely, if C is a closure operator on A and $\mathscr{C}$ is its associated closure system as given by (7), *then* (8) *holds, so that CX is the smallest closed set containing X, for every $X \subseteq A$.*

PROOF: The definition already entails that $CX \in \mathscr{C}$ by (Cl2), and that $CX = X$ when $X \in \mathscr{C}$. Therefore, $X \in \mathscr{C}$ if and only if $CX = X$. Thus in particular $CCX = CX$, which is (C). Moreover, the properties (I), that $X \subseteq CX$, and (M), that if $X \subseteq Y$, then $CX \subseteq CY$, also follow directly from (8). For the converse, by Proposition 1.23, $CX \in \mathscr{C}$, and by (I) $CX \supseteq X$, therefore $CX \in \{F \in \mathscr{C} : F \supseteq X\}$. If $F \in \mathscr{C}$ is such that $F \supseteq X$, then $F = CF \supseteq CX$ by (M). Therefore CX is the least element of the family, hence it equals its intersection. This proves (8). ⊠

Accordingly, given a closure system $\mathscr{C}$, the closure operator C defined by (8) is called ***the closure operator associated with*** $\mathscr{C}$. Besides, the theorem establishes that the functions

$$C \longmapsto \mathscr{C} \qquad \text{and} \qquad \mathscr{C} \longmapsto C, \tag{9}$$

given by (7) and (8), respectively, are inverse to one another, and there is a *duality* (which is fully exploited in Theorem 1.33 and following comments):

THEOREM 1.27
For every set A, the functions (9) *are mutually inverse bijections between the set* $\mathrm{CO}(A)$ *of all closure operators on A and the set* $\mathrm{CS}(A)$ *of all closure systems on A.* ⊠

This duality is reflected in the informal typographical rule that represents each pair "closure operator – closure system" by the same uppercase letter (or derived symbol) in two different typefaces:

$$C - \mathscr{C} \quad , \quad C_i - \mathscr{C}_i \quad , \quad D - \mathscr{D} \quad , \quad F - \mathcal{F} \quad , \quad \text{etc.} \tag{10}$$

This rule is used mainly in the general theory, and there are important *exceptions* in cases where one of the two components has a more natural, elaborated or traditional symbol: for instance, as already mentioned, the closure operator of a logic $\mathcal{L}$ is denoted by $C_{\mathcal{L}}$, while its associated closure system is denoted by $\mathcal{T}h\mathcal{L}$; the closure operator corresponding to the closure system $\mathcal{F}i_{\mathcal{L}}\boldsymbol{A}$ of $\mathcal{L}$-filters of an algebra $\boldsymbol{A}$ is denoted by $Fg_{\mathcal{L}}^{\boldsymbol{A}}$, thus facilitating the reading of the symbol as "the filter generated by [...]"; and the closure system of relative congruences $\mathrm{Con}_{\mathsf{K}}\boldsymbol{A}$ induces the closure operator of relative congruence generation denoted by $\Theta_{\mathsf{K}}^{\boldsymbol{A}}$.

Closure systems as ordered sets

As a family of subsets, the members of a closure system are naturally ordered under the subset relation. Then the following property is fundamental in the applications of closure systems to abstract algebraic logic:

PROPOSITION 1.28
If $\mathscr{C}$ is a closure system, then $\langle \mathscr{C}, \subseteq \rangle$ is a partially ordered set, and it is a complete lattice, with operations $\bigwedge F_i = \bigcap F_i$ and $\bigvee F_i = C \bigcup F_i$ for any $\{F_i : i \in I\} \subseteq \mathscr{C}$.

PROOF: The set $\mathscr{C}$ is obviously partially ordered by $\subseteq$. By definition of a closure system, if $\{F_i : i \in I\} \subseteq \mathscr{C}$, then $\bigcap F_i \in \mathscr{C}$, and this intersection is obviously the largest $F \in \mathscr{C}$ such that $F \subseteq F_i$ for all $i \in I$, that is, $\bigcap F_i = \inf\{F_i : i \in I\}$. If $F \in \mathscr{C}$, then $F_i \subseteq F$ for all $i \in I$ if and only if $\bigcup F_i \subseteq F$, and thus if and only if $C \bigcup F_i \subseteq CF = F$. By (8) the smallest F with this property is $C \bigcup F_i$, hence $C \bigcup F_i = \sup\{F_i : i \in I\}$. ⊠

Thus, every closure operator C has an associated ***lattice of closed sets***, which is safe to identify with the associated closure system $\mathscr{C}$, and in particular every logic $\mathcal{L}$ has an associated ***lattice of theories*** $\mathcal{T}h\mathcal{L}$.

Notice that if $\bigcup F_i \in \mathscr{C}$, then $\bigvee F_i = \bigcup F_i$. A good exercise to test understanding of this lattice structure is to prove the following two properties, which are used later on (Exercise 1.36):

PROPOSITION 1.29
Let C be a closure operator on a set A, with associated closure system $\mathscr{C}$.

1. *If $F \in \mathscr{C}$, then* $F = \bigcup\{CY : Y \subseteq F\} = \bigcup\{CY : Y \subseteq F, Y \text{ finite}\} = \bigcup\{C\{a\} : a \in F\}$.

2. *If $X \subseteq A$, then* $CX = \bigcup\{CY : Y \subseteq X\} = \bigvee\{CY : Y \subseteq X\} = \bigvee\{CY : Y \subseteq X, Y \text{ finite}\} = \bigvee\{C\{a\} : a \in X\}$. ⊠

Notice the subtle difference between the two points in this proposition. Observe also that, using Proposition 1.28, the unions in point 1 can be replaced by joins (because $F \in \mathscr{C}$), and that the reverse change is not possible in the two last expressions of point 2, unless C is finitary (Definition 1.39) or unitary (Exercise 1.57), respectively.

In situations in which it is advisable to highlight the lattice where a certain supremum or infimum is taken, self-explaining notations such as $\bigwedge^{\mathscr{C}}$ and $\bigvee^{\mathscr{C}}$ may be used.

Notice that, as every complete lattice, every closure system has a ***minimum***, which is $C\emptyset$, and a ***maximum***, which is A (Exercise 1.37).

Proposition 1.28 can be understood as identifying a particular kind of complete lattices, those arising from closure operators. As a matter of fact, these are prototypical in the sense that every complete lattice is *isomorphic* to the closure system of all closed sets of a closure operator (Exercise 1.39).

Bases of a closure system

Trivially, any family of subsets $\mathscr{B} \subseteq \mathscr{P}A$ is contained in some closure system, namely in the whole family $\mathscr{P}A$; and it is straightforward to prove, in a constructive way, that there is a smallest one:

PROPOSITION 1.30
For any family $\mathscr{B} \subseteq \mathscr{P}A$, the family $\mathscr{C} := \{\bigcap \mathscr{D} : \mathscr{D} \subseteq \mathscr{B}, \mathscr{D} \neq \emptyset\} \cup \{A\}$ is the smallest closure system on A that contains $\mathscr{B}$.

PROOF: From the definition it immediately follows that $\mathscr{B} \subseteq \mathscr{C}$, that $A \in \mathscr{C}$ and that $\mathscr{C}$ is closed under intersections. Thus, $\mathscr{C}$ is a closure system containing $\mathscr{B}$. It is straigthforward to check that if $\mathscr{C}'$ is another closure system with $\mathscr{B} \subseteq \mathscr{C}'$, then also $\mathscr{C} \subseteq \mathscr{C}'$. ⊠

DEFINITION 1.31
Let $\mathscr{C}$ be a closure system. A family $\mathscr{B} \subseteq \mathscr{P}A$ is a **base** *of $\mathscr{C}$ when $\mathscr{C}$ is the smallest closure system that contains $\mathscr{B}$. In this situation, $\mathscr{C}$ is called* ***the closure system generated by*** *$\mathscr{B}$.*

The concept of a base is an abstraction of the notion of a base of closed sets of a topology, and incorporates the feature of "being generated by" but not the feature of "minimality" or "independence", so common in algebra (think of bases in vector spaces) and in logic (the search for independent axiomatizations of logical systems was a trending topic during part of the twentieth century). Every closure system $\mathscr{C}$ has a base; indeed, both $\mathscr{C}$ and $\mathscr{C} \smallsetminus \{A\}$ are bases of $\mathscr{C}$, and notice that this includes the degenerate case of $\emptyset$ being a base of the inconsistent closure system $\{A\}$. Completing Proposition 1.30 (and using it), it is easy to prove the following characterizations (Exercise 1.40).

PROPOSITION 1.32
Let $\mathscr{C}$ be a closure system on A and $\mathscr{B} \subseteq \mathscr{C}$. The following conditions are equivalent:

(i) *$\mathscr{B}$ is a base of $\mathscr{C}$.*

(ii) *For all $F \in \mathscr{C}$, if $F \neq A$, then $F = \bigcap\{B \in \mathscr{B} : B \supseteq F\}$.*

(iii) *For all $X \subseteq A$, either $CX = A$ or $CX = \bigcap\{B \in \mathscr{B} : B \supseteq X\}$.*

(iv) *For all $F \in \mathscr{C}$ and all $a \in A$, if $a \notin F$, then there is some $B \in \mathscr{B}$ such that $F \subseteq B$ and $a \notin B$.* ⊠

Lattice theory is full of examples and applications of this notion. The most noticed one, essentially Stone's "Prime Filter Theorem", can be rephrased as saying that if a lattice is distributive, then the set of prime filters is a base of the closure system generated by its filters;† see more details in Exercise 1.43. Another popular example is that the ultrafilters (maximal filters) form a base of the closure system of all filters of a Boolean algebra.

Bases also appear naturally in algebraic logic, related to completeness theorems. To put just one, simple example, consider completeness of classical logic $\mathcal{C}\ell$

† Actually, this property is *equivalent* to the distributivity of the lattice.

expressed in the following way:

$$\Gamma \vdash_{\mathcal{C}\ell} \varphi \iff v\varphi = 1 \text{ for all } v \in \operatorname{Hom}(\boldsymbol{Fm}, \mathbf{2}) \text{ such that } v\Gamma \subseteq \{1\}.$$

This can be equivalently expressed by saying that either $C_{\mathcal{C}\ell}\Gamma = Fm$ (if the family of homomorphisms described is empty) or

$$C_{\mathcal{C}\ell}\Gamma = \bigcap\{v^{-1}\{1\} : v \in \operatorname{Hom}(\boldsymbol{Fm}, \mathbf{2}) \,,\; \Gamma \subseteq v^{-1}\{1\}\},$$

which says that the family $\{v^{-1}\{1\} : v \in \operatorname{Hom}(\boldsymbol{Fm}, \mathbf{2})\}$ is a base of the closure system $\mathcal{T}h\mathcal{C}\ell$ (compare with point (iii) of Proposition 1.32).

The family of all closure operators on a set

For a set A, consider the set $\mathsf{CO}(A)$ of all closure operators on A. The general order relation $\leqslant$ between power set functions introduced at the top of page 2 makes $\mathsf{CO}(A)$ an ordered set; in this case, if $C, C' \in \mathsf{CO}(A)$, then

$$C \leqslant C' \overset{\text{def}}{\iff} CX \subseteq C'X \quad \text{for all } X \subseteq A.$$

Dually, one can consider the set $\mathsf{CS}(A)$ of all closure systems on A; since closure systems are families of subsets, the subset relation between them also defines an order in $\mathsf{CS}(A)$. It turns out that these two relations correspond to one another in the bijection established in Theorem 1.27:

THEOREM 1.33
The ordered sets $\langle \mathsf{CO}(A), \leqslant \rangle$ and $\langle \mathsf{CS}(A), \subseteq \rangle$ are dually isomorphic under the functions (9) *on page* 36*; that is, it holds that $C \leqslant C' \iff \mathscr{C}' \subseteq \mathscr{C}$.*

PROOF: If $CX \subseteq C'X$ and $X = C'X$, then $X = CX$, because $X \subseteq CX$; hence, $\mathscr{C}' \subseteq \mathscr{C}$. If $\mathscr{C}' \subseteq \mathscr{C}$, then for every $F \supseteq X$, if $F \in \mathscr{C}'$, then also $F \in \mathscr{C}$; therefore, using (8) to express C and C' in terms of $\mathscr{C}$ and $\mathscr{C}'$ respectively, we obtain that $CX \subseteq C'X$. ☒

The order relations between closure operators and between closure systems are described by extending the nomenclature for logics established on page 27. Thus, when $C \leqslant C'$ we say that C' is an ***extension*** of C, and that it is ***stronger*** than C, and that C is ***weaker*** than C'. These terms may also be applied to the corresponding closure systems, but the purely set-theoretic terminology is more expressive; thus, when $\mathscr{C}$ is weaker than $\mathscr{C}'$, that is, when $\mathscr{C}' \subseteq \mathscr{C}$, one simply says that $\mathscr{C}$ is ***larger*** than $\mathscr{C}'$.† To help fix these terms—and to somehow justify them—observe that the stronger closure operator is the one drawing more "consequences" from the same "assumptions". Exercise 1.41 contains another characterization of these order relations.

The ordered set $\mathsf{CO}(A)$ has a minimum, which is the ***identity*** operator, given by: $CX = X$ for all $X \subseteq A$; it corresponds to $\mathscr{C} = \mathscr{P}A$, the maximum of $\mathsf{CS}(A)$. Dually, the maximum of $\mathsf{CO}(A)$ is the ***inconsistent*** operator, defined by $CX = A$ for all $X \subseteq A$, and which corresponds to $\mathscr{C} = \{A\}$, the minimum of $\mathsf{CS}(A)$.

† Using a term of topological origin, it makes sense to say that $\mathscr{C}$ is "finer" than $\mathscr{C}'$ when $\mathscr{C}' \subseteq \mathscr{C}$.

Another prominent member of $\mathsf{CO}(A)$ is the ***almost inconsistent*** operator, which is given by $C\varnothing = \varnothing$ and $CX = A$ for all $X \neq \varnothing$; this corresponds to $\mathscr{C} = \{\varnothing, A\}$. The inconsistent and the almost inconsistent operators are the only closure operators satisfying that $b \in C\{a\}$ for all $a, b \in A$; they are sometimes called ***trivial***, as is done with logics.

The almost inconsistent operator is a maximal[†] element of $\mathsf{CO}(A)$, because its associated closure system $\{\varnothing, A\}$ is clearly minimal in $\mathsf{CS}(A)$. Actually, for each $F \subseteq A$ one can consider the closure system $\mathscr{C} := \{F, A\}$, which is also a minimal element in $\mathsf{CS}(A)$, and yields the operator given by $CX = F$ when $X \subseteq F$, and $CX = A$ otherwise; this operator is thus maximal in $\mathsf{CO}(A)$.

DEFINITION 1.34
An extension C' of a closure operator C is ***axiomatic*** *when there exists a set $F \subseteq A$ such that $C'X = C(X \cup F)$ for all $X \subseteq A$. Then C' is called* ***the axiomatic extension of C determined by*** *F, and it is denoted by C^F. The corresponding closure system is denoted by $\mathscr{C}^F$.*

The last notation is justified by the third of the basic properties contained in the following result.

PROPOSITION 1.35
Let C^F be the axiomatic extension of C determined by $F \subseteq A$.

1. *$C^F\varnothing = CF \in \mathscr{C}$.*
2. *C^F is also the axiomatic extension of C determined by CF.*
3. *$\mathscr{C}^F$ is the closure system $\{G \in \mathscr{C} : G \supseteq F\}$.*

PROOF: Left to you as Exercise 1.44. ⊠

Point 2 establishes that axiomatic extensions can always be considered as determined by closed sets. If $F \in \mathscr{C}$, then the family $\mathscr{C}^F = \{G \in \mathscr{C} : G \supseteq F\}$, as a subset of the ordered set $\mathscr{C}$, is the principal up-set (see Exercise 1.38) generated by the set F.

WARNING.
The term "axiomatic extension" comes obviously from logic, where it is common practice to obtain logics from a given one by just adding some supplementary axioms. However, notice that, in the case of a logic $\mathcal{L}$, not every extension of its associated closure operator $C_{\mathcal{L}}$, in the previous sense, is a logic: for this, the resulting closure operator should be structural. In particular, the axiomatic extensions obtained from an arbitrary closed set (theory) need not be structural; only those determined by substitution-invariant sets are, as Proposition 1.53 shows. In any case, unless explicitly stated, *in this book, when applied to logics, the terms "extension" and "axiomatic extension" always refer to logics.*

[†] The notions of a *minimal* and a *maximal* element of an ordered set are understood to be so after excluding the absolute minimum and the absolute maximum, if they exist. They are also called *atoms* and *dual atoms*, respectively.

The Frege operator

DEFINITION 1.36
*If C is a closure operator, then the **Frege relation** of C is the relation*

$$\boldsymbol{\Lambda} C := \{\langle a,b\rangle \in A \times A : C\{a\} = C\{b\}\},$$

*and the **Frege operator** of C is the function $\boldsymbol{\Lambda}_C \colon \mathscr{P}A \to \mathscr{P}(A \times A)$ defined by*

$$X \subseteq A \quad \longmapsto \quad \boldsymbol{\Lambda}_C X := \{\langle a,b\rangle \in A \times A : C(X,a) = C(X,b)\}.$$

The Frege relation and the Frege operator may also be written as $\boldsymbol{\Lambda}^A C$ and $\boldsymbol{\Lambda}_C^A$ in order to emphasize the universe A on which it is defined. The notations $\boldsymbol{\Lambda}\mathscr{C}$, $\boldsymbol{\Lambda}^A\mathscr{C}$, $\boldsymbol{\Lambda}_{\mathscr{C}}$ and $\boldsymbol{\Lambda}_{\mathscr{C}}^A$ have the obvious meaning (when $\mathscr{C}$ is the closure system associated with C).

NOTATION.
Each logic $\mathcal{L}$ defines a closure operator $C_{\mathcal{L}}$, hence these definitions apply to it, but some simplifications are adopted: the ***Frege relation of*** $\mathcal{L}$, $\boldsymbol{\Lambda}\mathcal{L} := \boldsymbol{\Lambda} C_{\mathcal{L}}$, coincides with the interderivability relation $\dashv\vdash_{\mathcal{L}}$ of $\mathcal{L}$, introduced in Definition 1.7; and the ***the Frege operator of*** $\mathcal{L}$, denoted by $\boldsymbol{\Lambda}_{\mathcal{L}}$, is defined by $\boldsymbol{\Lambda}_{\mathcal{L}} := \boldsymbol{\Lambda}_{C_{\mathcal{L}}}$.

Observe that $\boldsymbol{\Lambda}_C X = \boldsymbol{\Lambda}_C CX$ for all $X \subseteq A$, by the property in Corollary 1.22.3; thus, in practice one may regard the Frege operator as defined over $\mathscr{C}$ rather than over $\mathscr{P}A$.

Straightforward, alternative formulations of the main definitions are as follows:

$$\begin{aligned} a \equiv b\ (\boldsymbol{\Lambda} C) &\iff \text{for all } F \in \mathscr{C},\ [a \in F \iff b \in F] \\ &\iff \text{for all } X \subseteq A,\ C(X,a) = C(X,b) \\ &\iff \text{for all } X \subseteq A,\ C(X,C\{a\}) = C(X,C\{b\}) \\ &\iff \text{for all } X \subseteq A,\ a \equiv b\ (\boldsymbol{\Lambda} C^X). \end{aligned}$$

$$\begin{aligned} a \equiv b\ (\boldsymbol{\Lambda}_C X) &\iff a \equiv b\ (\boldsymbol{\Lambda} C^X) \\ &\iff \text{for all } F \in \mathscr{C},\ \text{if } F \supseteq X, \text{ then } [a \in F \iff b \in F] \\ &\iff \text{for all } Y \supseteq X,\ C(Y,a) = C(Y,b). \end{aligned}$$

Thus, the relation $\boldsymbol{\Lambda}_C X$ is in fact a special case of a Frege relation, and $\boldsymbol{\Lambda}_C \varnothing = \boldsymbol{\Lambda} C$. It is important to become proficient in manipulating all these expressions (Exercise 1.46). Trivially:

PROPOSITION 1.37
If C is a closure operator, then $\boldsymbol{\Lambda} C$ is an equivalence relation. ⊠

Therefore one can always consider the quotient set $A/\boldsymbol{\Lambda} C$. This set is partially ordered by the relation induced by the ***quasi-order*** $\leqslant_C$ ***associated with*** C, which is defined by

$$a \leqslant_C b \overset{\text{def}}{\iff} C\{b\} \subseteq C\{a\} \iff b \in C\{a\}. \tag{11}$$

Actually, $\boldsymbol{\Lambda} C$ is the symmetrization of $\leqslant_C$; this fact automatically implies that $\boldsymbol{\Lambda} C$ is an equivalence relation compatible with $\leqslant_C$, and this one induces an order in the quotient. Moreover, if $C\varnothing \neq \varnothing$, it is easy to show that this set is exactly one of the equivalence classes of $\boldsymbol{\Lambda}C$, and it is the maximum element of this order. See Exercise 1.47.

Notice, however, that for arbitrary closure operators on sets with an algebraic structure, *the relation $\boldsymbol{\Lambda}C$ need not be a congruence*, and so the underlying algebraic structure need not be preserved in the quotient set; all the logics shown in Section 7.2 not to be selfextensional witness this. Actually, this issue provides one of the main classification criteria in abstract algebraic logic, those that will be studied in Chapter 7.

When $\boldsymbol{\Lambda}\mathscr{C} = \mathrm{Id}_A$, one sometimes says that the closure system $\mathscr{C}$ ***separates points***, because this amounts to saying that if $a \neq b$, then $C\{a\} \neq C\{b\}$, or, equivalently, that if $a \neq b$, then there is some $F \in \mathscr{C}$ with either $a \in F$ and $b \notin F$ or $a \notin F$ and $b \in F$ (see Exercise 1.48 for some examples).

LEMMA 1.38
The Frege relation and the Frege operator are monotone (order-preserving) in the following senses:

1. *If $C, C' \in \mathrm{CO}(A)$ and $C \leqslant C'$, then $\boldsymbol{\Lambda}C \subseteq \boldsymbol{\Lambda}C'$.*
2. *If $\mathscr{C}, \mathscr{C}' \in \mathrm{CS}(A)$ and $\mathscr{C}' \subseteq \mathscr{C}$, then $\boldsymbol{\Lambda}\mathscr{C} \subseteq \boldsymbol{\Lambda}\mathscr{C}'$.*
3. *If $X \subseteq X' \subseteq A$, then $\boldsymbol{\Lambda}_C X \subseteq \boldsymbol{\Lambda}_C X'$.*

PROOF: Left to you as Exercise 1.49. ⊠

Before going on with less basic (but important) material on closure operators and closure systems, a section is devoted to the analysis of two properties that have a central position in abstract algebraic logic: finitarity and structurality.

Exercises for Section 1.3

EXERCISE 1.31. Confirm that the correspondence (6) on page 33 makes closure relations correspond to closure operators and conversely, by showing that they preserve conditions (I) and (M) of Definitions 1.17 and 1.19 exactly, while conditions (C) are slightly different, but anyway equivalent in the presence of the other two.

EXERCISE 1.32. Find examples of closure operators and closure systems that you have met before, in other areas of mathematics, especially in algebra and in topology.

EXERCISE 1.33. Let $\boldsymbol{L}$ be an arbitrary algebraic language, and define an operator C on Fm by putting, for any $\Gamma \subseteq Fm$, $C\Gamma := \Gamma$ if there is a substitution σ such that $\sigma\Gamma$ is finite, and $C\Gamma := Fm$ otherwise. Show that C is a structural closure operator on Fm. Compute the associated closure relation $\vdash_C$, and conclude that $\langle \boldsymbol{L}, \vdash_C \rangle$ is a logic.

EXERCISE 1.34. Let C be a closure operator on a set A, and let $f, g \colon \mathscr{P}A \to \mathscr{P}A$ be two functions that commute with unions. Using the fundamental property in Corollary 1.22.1, prove that if $Cf\{a\} = Cg\{a\}$ for all $a \in A$, then $CfX = CgX$ for all $X \subseteq A$.

EXERCISE 1.35. An ***interior operator*** on a set A is a function $I \colon \mathscr{P}A \to \mathscr{P}A$ that satisfies the following properties, for all $X, Y \subseteq A$: $IX \subseteq X$; if $X \subseteq Y$, then $IX \subseteq IY$; and $IIX = IX$. The associated ***open sets*** are the sets in the family $\mathcal{O} := \{X \subseteq A : IX = X\}$. Prove that the family $\mathcal{O}$ is closed under arbitrary unions and contains $\varnothing$; that for all $X \subseteq A$, $IX = \bigcup\{Y \in \mathcal{O} : Y \subseteq X\}$; that given any family $\mathcal{O} \subseteq \mathscr{P}A$ closed under arbitrary unions and containing $\varnothing$, the previous equality defines an interior operator with $\mathcal{O}$ as set of open elements; and that these constructions define a bijective correspondence between interior operators and the corresponding families of open sets, dual to that between closure operators and closure systems worked out in the text.

EXERCISE 1.36. Prove Proposition 1.29. Observe that the proof of the two points is essentially the same, but one is applied set-theoretically and the other is applied lattice-theoretically. Observe also that its point 2 is *not* the particular case of point 1 for $F := CX$, and that the first equality in 2 also holds if the supremum is replaced by a union (like in 1), but this does not happen in the other equalities in 2.

EXERCISE 1.37. Let $\mathscr{C}$ be a closure system on a set A, with associated closure operator C. Prove that $C\varnothing$ is the smallest member of $\mathscr{C}$ and A is the largest one (recall that order is here set inclusion).

EXERCISE 1.38. Let $\langle A, \leqslant \rangle$ be an ordered set (or a quasi-ordered set). An ***up-set*** is a non-empty set $X \subseteq A$ such that if $a, b \in A$ with $a \in X$ and $a \leqslant b$, then $b \in X$. An up-set is ***principal*** when it is of the form $\uparrow a := \{c \in A : a \leqslant c\}$ for some $a \in A$. Prove that the family of all up-sets is a closure system, and that the family of all principal up-sets is a closure system if and only if A is a complete lattice.

EXERCISE 1.39. Let $\langle A, \leqslant \rangle$ be a complete lattice. Define $CX := \{a \in A : a \leqslant \bigvee X\}$, for every $X \subseteq A$. Prove that this function C is a closure operator on A such that $F \in \mathscr{C}$ if and only if either $F = \varnothing$ or $F = \{c \in A : c \leqslant a\}$ for some $a \in A$. Use this to prove that $\langle A, \leqslant \rangle \cong \langle \mathscr{C}, \subseteq \rangle$.

COMMENT. A (***principal***) ***down-set*** is the dual notion to that of (principal) up-set (Exercise 1.38). You can see that this representation uses principal down-sets; clearly, the construction can be dualised, using principal up-sets, but then a *dual* isomorphism is obtained.

EXERCISE 1.40. Prove the equivalence of the conditions that appear in Proposition 1.32.

EXERCISE 1.41. Let C, C' be two closure operators on a set A. Prove, using the equivalence in Theorem 1.33, that $C \leqslant C'$ if and only if $C \circ C' = C'$, and if and only if the function $C \circ C'$ is expansive, that is, it satisfies condition (I) from Definition 1.19.

EXERCISE 1.42. Let $\boldsymbol{A} = \langle A, \wedge, \vee \rangle$ be an arbitrary lattice. A non-empty set $F \subseteq A$ is a ***filter*** of $\boldsymbol{A}$ when for every $a, b \in A$, if $a, b \in F$, then $a \wedge b \in F$, and if $a \leqslant b$ and $a \in F$, then $b \in F$. Denote the family of all filters of $\boldsymbol{A}$ by $\mathcal{F}ilt_\wedge \boldsymbol{A}$, and the closure system generated by this family by $\mathcal{F}ilt^\circ_\wedge \boldsymbol{A}$.

COMMENTS. The above notions make sense, and the results below hold, more generally, for meet-semilattices $\langle A, \wedge \rangle$, as you can check.

The dual notion of an ***ideal*** of a join-semilattice will appear occasionally in the book. Everything can be dualized, with down-sets (Exercise 1.39) instead of up-sets.

1. Show that a filter can be equivalently defined as a non-empty $F \subseteq A$ such that for all $a, b \in A$, $a \wedge b \in F$ if and only if $a \in F$ and $b \in F$.
2. Show that for any $a \in A$, the principal up-set $\uparrow a$ (Exercise 1.38) is a filter of $\boldsymbol{A}$, and indeed the smallest filter of $\boldsymbol{A}$ containing a. Deduce from this that two elements of a lattice are equal if and only if they belong to the same filters.
3. Prove that $\mathcal{F}ilt^\circ_\wedge \boldsymbol{A} = \mathcal{F}ilt_\wedge \boldsymbol{A}$ if and only if A has a maximum element 1, and if and only if there is a least filter of $\boldsymbol{A}$. In this situation, this least filter is $\{1\}$.
4. Suppose that A does not have a maximum. Show that $\mathcal{F}ilt^\circ_\wedge \boldsymbol{A} = \mathcal{F}ilt_\wedge \boldsymbol{A} \cup \{\varnothing\}$.
5. Let F be a filter and $a \notin F$. Prove that the set $\{b \in A : b \geqslant a \wedge c$ for some $c \in F\}$ is the smallest filter containing F and a.

Exercise 1.43. Let $\boldsymbol{A} = \langle A, \wedge, \vee \rangle$ be an arbitrary lattice. A ***prime filter*** is a proper filter ($F \neq A$) satisfying, for every $a, b \in A$, that if $a \vee b \in F$, then $a \in F$ or $b \in F$. Prove that $\boldsymbol{A}$ is distributive if and only if every proper filter is the intersection of all the prime filters that contain it, that is, if and only if the set of prime filters is a base of the closure system generated by $\mathcal{F}ilt_\wedge \boldsymbol{A}$.

> Hint. For the direct implication, assume that F is a filter and there is some $a \notin F$ belonging to every prime filter containing F; then, apply Zorn's Lemma to the family of all filters containing F but not a, in order to reach a contradiction (the only hard step is to prove, using Exercise 1.42.5, that a maximal in this family has to be prime). For the converse, apply Exercise 1.42.2 to show that $a \wedge (b \vee c) = (a \wedge b) \vee (a \wedge c)$.

Exercise 1.44. Let C be a closure operator on a set A and let C' be the axiomatic extension of C determined by some $F \subseteq A$. Prove that $C'\varnothing = CF \in \mathscr{C}$, that C' is also determined from C by CF, and that its associated closure system is $\mathscr{C}' = \{G \in \mathscr{C} : G \supseteq F\}$.

Exercise 1.45. Let C' be an arbitrary extension of a closure operator C. Prove that $C'\varnothing \in \mathscr{C}$ and that $\mathscr{C}' \subseteq \mathscr{C}^{C'\varnothing}$. Prove that C' is an axiomatic extension of C if and only if the equality holds.

Exercise 1.46. Let C be a closure operator on a set A. Check that $\boldsymbol{\Lambda}_C X = \bigcap_{Y \supseteq X} \boldsymbol{\Lambda}_C Y$ for all $X \subseteq A$, that $\boldsymbol{\Lambda} C = \bigcap_{Y \subseteq A} \boldsymbol{\Lambda}_C Y$, and that all the equivalences on the Frege relation and the Frege operator given after Definition 1.36 are correct.

Exercise 1.47. Let C be a closure operator on a set A and consider the relation $\leqslant_C$ defined in (11) on page 41. Prove:

1. The relation $\leqslant_C$ is a quasi-order on A; that is, it is reflexive and transitive.
2. The Frege relation of C, $\boldsymbol{\Lambda} C$, is compatible with the relation $\leqslant_C$; that is, if $\langle a, a' \rangle \in \boldsymbol{\Lambda} C$ and $\langle b, b' \rangle \in \boldsymbol{\Lambda} C$, then $a \leqslant_C b$ implies $a' \leqslant_C b'$.
3. If in $A/\boldsymbol{\Lambda} C$ we define $a/\boldsymbol{\Lambda} C \leqslant b/\boldsymbol{\Lambda} C \iff a \leqslant_C b$, then this correctly defines an order relation on $A/\boldsymbol{\Lambda} C$.
4. Show by an example that this ordered set need not be linearly ordered.
5. If $C\varnothing \neq \varnothing$, then this set constitutes an equivalence class of $\boldsymbol{\Lambda} C$, and this class is the maximum element of $A/\boldsymbol{\Lambda} C$ with respect to the order $\leqslant_C$.
6. Find a necessary and sufficient condition for the existence of a minimum of $\leqslant_C$ in $A/\boldsymbol{\Lambda} C$, and identify it when it exists.

Exercise 1.48. Check that the Frege relation of the closure systems considered in Exercises 1.38 and 1.42 is the identity relation, that is, that they separate points.

Exercise 1.49. Give a proof of the three (anti-)monotonicity properties of the Frege operator stated in Lemma 1.38.

1.4 Finitarity and structurality

These are two properties of logics (the first, an optional one; the second, a compulsory one, actually part of the definition) that can be formulated in more generality for closure operators; while this is straightforward for the first one, the second involves some complication.

Finitarity

DEFINITION 1.39
A closure operator C *on a set* A *is **finitary** when for all* $X \subseteq A$,

$$CX = \bigcup\{CY : Y \subseteq X,\ Y \textit{ finite}\}. \tag{F}$$

In an analogous way, generalizing Definition 1.6, we would obtain the obvious notion of a ***finitary closure relation***. Finitary logics provide examples of finitary closure operators. Notice that all closure operators on a finite set are finitary. Algebraic examples include those related to filter-generation in semilattices, lattices and similar structures (see Exercise 1.53).

It is instructive to compare condition (F) with the properties in Proposition 1.29.1, which hold for an arbitrary closure operator, in order to avoid confusions when applying them.

The following properties are very useful and worth recording, as they are used often without notice:

LEMMA 1.40
Let C *be a finitary closure operator on a set* A*, and let* $X, Z \subseteq A$ *and* $a, b \in A$.

1. *If* $Z \subseteq CX$ *and* Z *is finite, then* $Z \subseteq CY$ *for some finite* $Y \subseteq X$.
2. *If* $C(X,a) = C(X,b)$*, then there is a finite* $Y \subseteq X$ *such that* $C(Y,a) = C(Y,b)$.

PROOF: Left to you as Exercises 1.50 and 1.51. ⊠

Actually, each of these properties is equivalent to finitarity, as pointed out in the exercises.

As a sample of a simple application of finitarity, here is a useful property:

PROPOSITION 1.41
Let C *be a finitary closure operator on a set* A *and let* $X \subseteq A$*. If there is a finite* $Y \subseteq A$ *such that* $CY = CX$*, then there is a finite* $Z \subseteq X$ *such that* $CZ = CX$.

PROOF: If $Y = \emptyset$ we can take $Z = \emptyset$ and everything is trivial. Now assume that $Y = \{a_1, \dots, a_n\}$ for some $n \geqslant 1$. For each i, $a_i \in CX$ and by finitarity there is some finite $Z_i \subseteq X$ such that $a_i \in CZ_i$. Putting $Z := Z_1 \cup \cdots \cup Z_n$ we obtain a finite $Z \subseteq X$ such that $Y \subseteq CZ$, therefore $CX = CY \subseteq CZ$, but $CZ \subseteq CX$ because $Z \subseteq X$. Thus, $CZ = CX$, as desired. ⊠

The logical reading of this property is that if a theory of a finitary logic is finitely axiomatizable (in some, unspecified way), then *each* of its sets of axioms contains a finite subset which also axiomatizes the theory; in this version, it is used at several points in the book.

The closure systems corresponding to finitary closure operators are characterized by the following classical result. A subset $D \subseteq A$ of an ordered set $\langle A, \leqslant \rangle$ is ***upwards directed*** when for each $a, b \in D$ there is a $c \in D$ such that $a \leqslant c$ and $b \leqslant c$; it is equivalent to state this property for any finite number of elements. A non-empty family of sets (in particular, a closure system) is ***inductive***† when

† This notion has nothing to do with the notion of an inductive set in general set theory (a set X satisfying that $\emptyset \in X$ and that if $a \in X$, then $a \cup \{a\} \in X$).

it is closed under unions of non-empty subfamilies that are upwards directed under the subset relation. Recall that by Zorn's Lemma (which is equivalent to the Axiom of Choice on set-theoretic grounds), any inductive family of sets has a maximal element.

THEOREM 1.42
A closure operator C is finitary if and only if its associated closure system $\mathscr{C}$ is inductive.

PROOF: Assume first that C is finitary, and let $\mathscr{D} \subseteq \mathscr{C}$ be an upwards directed non-empty family. In order to show that $\bigcup \mathscr{D} \in \mathscr{C}$ we show that $C \bigcup \mathscr{D} \subseteq \bigcup \mathscr{D}$, so take any $a \in C \bigcup \mathscr{D}$. By finitarity there are[†] $a_1, \dots, a_n \in \bigcup \mathscr{D}$ such that $a \in C\{a_1, \dots, a_n\}$. For each $i = 1, \dots, n$ there is some $F_i \in \mathscr{D}$ such that $a_i \in F_i$. Since $\mathscr{D}$ is upwards directed there is some $F \in \mathscr{D}$ such that all $F_i \subseteq F$. Since $\mathscr{D} \subseteq \mathscr{C}$, $F \in \mathscr{C}$ and it follows that $a \in C(F_1 \cup \dots \cup F_n) \subseteq CF = F$, hence $a \in \bigcup \mathscr{D}$.

For the converse, assume that $\mathscr{C}$ is inductive and for an $X \subseteq A$ consider the family $\mathcal{F} := \{CF : F \subseteq X,\ F \text{ finite}\}$. Since in general $CF_1 \cup CF_2 \subseteq C(F_1 \cup F_2)$ for all $F_1, F_2 \subseteq A$, $\mathcal{F}$ is an upwards directed family, hence $\bigcup \mathcal{F} \in \mathscr{C}$. Now we can use Proposition 1.29.2, and we obtain that

$$CX = \bigvee \mathcal{F} = C \bigcup \mathcal{F} = \bigcup \mathcal{F} = \bigcup \{CF : F \subseteq X,\ F \text{ finite}\},$$

which expresses that C is finitary. ⊠

The preceding result, popularly known as Schmidt's Theorem, can be used to obtain another classical result, an abstract form of the so-called Lindenbaum's Lemma.

LEMMA 1.43
Let C be a finitary closure operator. If $F \in \mathscr{C}$ and $a \notin F$, then there exists $G \in \mathscr{C}$ such that $F \subseteq G$, $a \notin G$, and G is maximal among all the closed sets of $\mathscr{C}$ not containing a.

PROOF: By finitarity and Theorem 1.42, $\mathscr{C}$ is inductive. Consider the family $\mathscr{D} = \{G \in \mathscr{C} : G \supseteq F,\ a \notin G\}$, and show that it is inductive as well. Obviously, $F \in \mathscr{D}$ and so $\mathscr{D} \neq \varnothing$. Let $\{F_i : i \in I\} \subseteq \mathscr{D}$ be upwards directed: $\bigcup F_i \in \mathscr{C}$ because $\mathscr{C}$ is inductive; clearly $a \notin \bigcup F_i$ because $a \notin F_i$ for all $i \in I$; and obviously $F \subseteq \bigcup F_i$. Therefore $\bigcup F_i \in \mathscr{D}$, which shows that $\mathscr{D}$ is inductive. Now, by Zorn's Lemma, the family $\mathscr{D}$ has a maximal element G; since $G \in \mathscr{D}$, it satisfies the first two conditions: $F \subseteq G$ and $a \notin G$. And G is maximal among all the closed sets that do not contain a: if $G \subseteq G' \in \mathscr{C}$ and $a \notin G'$, then $F \subseteq G'$ and so $G' \in \mathscr{D}$, which implies that $G = G'$ because G is maximal in $\mathscr{D}$. ⊠

As a matter of fact, according to Tarski [233, Theorem 56], what Lindenbaum did prove is the following property (for the particular case of a sentential logic); it says, in logical parlance, that under reasonable restrictions every consistent theory can be extended to a maximally consistent one.

COROLLARY 1.44
If C is a finitary closure operator with an inconsistent element, then for all $F \in \mathscr{C}$, $F \neq A$, there is a maximal $G \in \mathscr{C}$ such that $F \subseteq G$.

† In principle, finitarity also includes the case where $a \in C\varnothing$, as $\varnothing$ is a finite subset of $\bigcup \mathscr{D}$. However, since by assumption $\bigcup \mathscr{D} \neq \varnothing$, this case can be reduced to the one above. This subtlety may not be emphasized in some proofs.

Proof: It is easy to see (Exercise 1.52) that if $a \in A$ is inconsistent, then $F \in \mathscr{C}$ is consistent if and only if $a \notin F$. Thus, a maximal closed set not containing a is just a maximal closed set, and applying Lindenbaum's Lemma 1.43 to this element we get the desired result. ⊠

Very often, the most practical way to check finitarity of a closure operator is to use Theorem 1.42 and check inductiveness of the associated closure system; as an example, see Exercise 1.53, and here is another:

Proposition 1.45
If C is a finitary closure operator, then all its axiomatic extensions are finitary.

Proof: It is obvious, by the very definition of inductiveness, that if a closure system $\mathscr{C}$ is inductive, then the closure system $\mathscr{C}^F$ is also inductive, for any $F \in \mathscr{C}$. By Theorem 1.42, this proves the statement. ⊠

The following characterizations of finitarity turn out to be useful in some particular proofs.

Theorem 1.46
For every closure operator C, the following conditions are equivalent:

(i) *C is finitary.*

(ii) *C is **continuous** in the sense that it commutes with unions of upwards directed families; that is, $C \bigcup \mathscr{B} = \bigcup\{CX : X \in \mathscr{B}\}$ for any non-empty upwards directed family $\mathscr{B} \subseteq \mathscr{P}A$.*

(iii) *For every $X \subseteq A$, $\boldsymbol{\Lambda}_C X = \bigcup\{\boldsymbol{\Lambda}_C Y : Y \subseteq X,\ Y \text{ finite}\}$.*

(iv) *$\boldsymbol{\Lambda}_C$ is **continuous** in the sense that it preserves unions of upwards directed families; that is, $\boldsymbol{\Lambda}_C \bigcup \mathscr{B} = \bigcup\{\boldsymbol{\Lambda}_C X : X \in \mathscr{B}\}$ for any non-empty upwards directed family $\mathscr{B} \subseteq \mathscr{P}A$.*

Proof: (i)⇒(ii) By monotonicity $C \bigcup \mathscr{B} \supseteq \bigcup\{CX : X \in \mathscr{B}\}$. Let $a \in C \bigcup \mathscr{B}$. By finitarity, either $a \in C\varnothing$, or there are $a_1, \dots, a_n \in \bigcup \mathscr{B}$ such that $a \in C\{a_1, \dots, a_n\}$. In the first case, $a \in CX$ for any $X \in \mathscr{B}$. In the second case, there are $X_i \in \mathscr{B}$ such that $a_i \in X_i$, and by directedness of $\mathscr{B}$, there is some $X \in \mathscr{B}$ such that $a_i \in X$ for all $i = 1, \dots, n$. Therefore (again, by monotonicity), $a \in CX$.

(ii)⇒(i) Let $X \subseteq A$, and consider the family $\{F \subseteq X : F \text{ finite}\}$. This family is clearly upwards directed. Therefore, applying the assumption to it directly gives finitarity of C.

(i)⇔(iii) This is almost proved in Exercise 1.51: the exercise shows that (i) is equivalent to the inclusion $\boldsymbol{\Lambda}_C X \subseteq \bigcup\{\boldsymbol{\Lambda}_C Y : Y \subseteq X,\ Y \text{ finite}\}$, but the reverse inclusion always holds, by monotonicity of the Frege operator (Lemma 1.38.3).

(iii)⇔(iv) This is proved similarly to the equivalence between (i) and (ii); see Exercise 1.55. ⊠

A similar notion of continuity, related to that in (iv), appears in a different context in Chapter 6.

Now we see that we can always associate a finitary operator with any given closure operator in a natural way.

DEFINITION 1.47
Let C be a closure operator in a set A. The ***finitary companion of*** *C is the operator defined as follows. For all $X \subseteq A$,*

$$C_f X := \bigcup \{CF : F \subseteq X,\ F \text{ finite}\}.$$

Trivially, C is finitary if and only if $C = C_f$. There is more:

THEOREM 1.48
If C is a closure operator, then C_f is a finitary closure operator on A such that $C_f \leqslant C$, and it is the strongest of all finitary closure operators C' on the same set A such that $C' \leqslant C$. Moreover, $C_f X = CX$ whenever $X \subseteq A$ is finite.

PROOF: It is left to you as Exercise 1.58. ⊠

Onviously, this construction applies to logics. The finitary companion of the logic $\mathcal{L}$ is in general denoted by $\mathcal{L}_f$. One example is the finitary version of Łukasiewicz's infinitely-valued logic: it is initially presented as axiomatized in the usual Hilbert-style way (see for instance [132, Chapter 3]), and it turns out to be the finitary companion $Ł_{\infty f}$ of the logic $Ł_\infty$, semantically defined in Example 1.12, which is shown in Example 1.15 to be infinitary (the first fact, known as a *finite completeness theorem*, is not easy to prove).

PROPOSITION 1.49
If C is a closure operator, then $\varLambda C_f = \varLambda C$.

PROOF: It is clear in Definition 1.36 that the Frege relation depends only on the closure of one-element sets. Since by Theorem 1.48, C and C_f coincide on one-element sets, their associated Frege relations coincide. ⊠

Structurality

The final part of Section 1.1 (pages 28ff.) discusses natural notions of "logic" on sets that are not exactly sets of formulas, but sets of syntactic objects constructed from formulas in such a way that they have an "induced" notion of a substitution which makes it reasonably natural to extend condition (S) of substitution-invariance to them. The key idea to formulate this in general† consists in considering the set of substitutions as a monoid that *acts* on the relevant set, be it a set of formulas, of equations, of sequents, and the like.

A ***monoid*** $\boldsymbol{M} = \langle M, \cdot, 1 \rangle$ is a set endowed with an associative binary operation (here denoted by $\cdot$, but the symbol is generally omitted) and a constant 1 which is the unit of this operation. An ***M*-*set*** is a set A endowed with an ***action*** of $\boldsymbol{M}$ on A; this means that for each $\sigma \in M$ there is a function $\sigma\colon A \to A$, denoted in the same way, satisfying

- $(\sigma\sigma')a = \sigma(\sigma' a)$ for all $\sigma, \sigma' \in M$ and all $a \in A$.
- $1a = a$ for all $a \in A$.

(This can be equivalently formulated in terms of a binary operation $M \times A \to A$.)

† The idea was already suggested by Rautenberg in [210], but was fully implemented only 25 years later by Blok and Jónsson in [29].

As with all functions, the extension of σ to the power set of A is denoted in the same way, and its residual is denoted by σ^{-1}; thus, if $X \subseteq A$, then $\sigma X := \{\sigma a : a \in X\}$ and $\sigma^{-1} X := \{a \in A : \sigma a \in X\}$.

The main example is, for any algebra $\boldsymbol{A}$, the set $\langle \mathrm{End} \boldsymbol{A}, \circ, \mathrm{Id}_A \rangle$ of its endomorphisms, with the (associative) binary operation $\circ$ of endomorphism composition, and the identity function Id_A as its unit. Lemma 1.4 shows that the substitutions of the formula algebra form a monoid $\boldsymbol{E} := \langle \mathrm{End} \boldsymbol{Fm}, \circ, \mathrm{Id}_{Fm} \rangle$; on page 29 we see how to extend each substitution to a function on the sets of equations or sequents, and it is straightforward to check that in these ways any of these sets becomes an $\boldsymbol{E}$-set. In a similar way, for any algebra $\boldsymbol{A}$ the set $A \times A$ becomes an $\mathrm{End} \boldsymbol{A}$-set; here the action of an $h \in \mathrm{End} \boldsymbol{A}$ on $A \times A$ is defined by $h\langle a, b \rangle := \langle ha, hb \rangle$. It is easy to check that the power set $\mathscr{P} A$ is also an $\mathrm{End} \boldsymbol{A}$-set, with the usual definition $hX := \{ha : a \in X\}$ for each $X \subseteq A$. And so forth.

With this notion, it is possible to formulate the truly abstract counterpart of the idea of substitution-invariance; since it is no longer related specifically to substitutions or endomorphisms, the old, traditional term "structurality" is generally preferred as a more neutral one:

DEFINITION 1.50
Let $\boldsymbol{M}$ be a monoid and A be an $\boldsymbol{M}$-set. A closure operator C on A is ***structural***† *when for all $\sigma \in M$, $X \subseteq A$ and $a \in A$, if $a \in CX$, then $\sigma a \in C\sigma X$, that is, when $\sigma C X \subseteq C \sigma X$.*

It is not difficult to obtain a few equivalent ways to express structurality, which are worth recalling:

PROPOSITION 1.51
Let $\boldsymbol{M}$ be a monoid, C a closure operator on an $\boldsymbol{M}$-set A, and let $\mathscr{C}$ denote its associated closure system. The following conditions are equivalent:

(i) *C is structural.*
(ii) *For every $\sigma \in M$, $C\sigma C = C\sigma$.*
(iii) *For every $\sigma \in M$ and every $F \subseteq A$, if $F \in \mathscr{C}$, then $\sigma^{-1} F \in \mathscr{C}$.*
(iv) *$\mathscr{C}$ has a base $\mathscr{B}$ such that if $B \in \mathscr{B}$ and $\sigma \in M$, then $\sigma^{-1} B \in \mathscr{B}$.*

PROOF: These equivalences result directly from the more detailed ones of Exercises 1.59 and 1.60. ⊠

Envisaging the application of this notion to real logics, it is easy to see (Exercise 1.61) that the finitary companion of a logic is itself a logic.

It is also interesting to consider axiomatic extensions. Since an arbitrary axiomatic extension of a structural closure operator need not be structural (Exercise 1.62), it is interesting to characterize those that are.

DEFINITION 1.52
Let A be an $\boldsymbol{M}$-set, for some monoid $\boldsymbol{M}$. A set $X \subseteq A$ is ***invariant***† *when $\sigma X \subseteq X$ for every $\sigma \in M$.*

† In some contexts it might be probably better to say "$\boldsymbol{M}$-structural" and "$\boldsymbol{M}$-invariant".

Normally it is enough to say "invariant". We already know that the set of theorems of a logic is invariant under substitutions, that is, it is $\boldsymbol{E}$-invariant. In general, if C is structural, then $\sigma C\emptyset \subseteq C\sigma\emptyset = C\emptyset$, and hence $C\emptyset$ is invariant; more precisely, the following holds:

PROPOSITION 1.53
If C is structural and C' is an axiomatic extension of C, then C' is structural if and only if $C'\emptyset$ is invariant.

PROOF: One implication is mentioned above. As for the converse, suppose that $C'\emptyset$ is invariant and let $X \subseteq A$. Since $C'X = C(X \cup C'\emptyset)$ we get

$$\begin{aligned} \sigma C'X = \sigma C(X \cup C'\emptyset) &\subseteq C(\sigma(X \cup C'\emptyset)) = \\ = C(\sigma X \cup \sigma C'\emptyset) &\subseteq C(\sigma X \cup C'\emptyset) = C'\sigma X \end{aligned}$$

for all $\sigma \in M$. Thus, C' is structural. ☒

A logic is a structural closure operator on the set of formulas. Hence, not all its axiomatic extensions in the technical sense of Definition 1.34 are logics (see the warning on page 40); only those defined by an invariant closed set are. How are invariant closed sets obtained? The next result gives an answer to this question.

LEMMA 1.54
Let C be a structural closure operator. A closed set $F \in \mathscr{C}$ is invariant if and only if $F = CF_0$ for some set $F_0 \subseteq A$ that is invariant.

PROOF: ($\Rightarrow$) is trivial. ($\Leftarrow$) Suppose that $\sigma F_0 \subseteq F_0$ for all $\sigma \in M$. Then whenever $a \in F = CF_0$, we get $\sigma a \in \sigma CF_0 \subseteq C\sigma F_0 \subseteq CF_0 = F$ for every $\sigma \in M$. Thus F is invariant. ☒

This is how *axiomatic extensions of sentential logics* are defined in practice: it is usual to say that they are determined by *axiom schemes*, a term used to refer to formulas intended to represent the set of all their substitution instances, which is of course a set of formulas invariant under substitutions; in general this set is not a closed set, but Proposition 1.35 shows that the axiomatic extension it determines is in fact the one determined by its closure, and Lemma 1.54 guarantees that this closed set is also invariant, therefore the result is indeed a structural closure operator, that is, a logic. Think as an example at what is implied when one says that "classical logic is the axiomatic extension of intuitionistic logic with the axiom $x \vee \neg x$". Without the consideration of axiom schemes, the term "finitely axiomatizable" would be void, because an invariant set is forcefully infinite; however, the term actually refers to axiomatizablility by a finite set of axiom schemes. Although the term "axiomatic extension" is commonly used only for logics given by Hilbert-style axiomatic systems, we see that the present usage in the context of abstract closure operators is consistent with the most common one.

The abstract treatment of structurality hinted at in this section is necessary to obtain abstract versions of other notions in abstract algebraic logic in which substitutions play a central role, such as the isomorphisms involved in the notion of algebraizability (Section 3.5); the present section ends with two technical results that are used exclusively there. They concern the following issue: If C is structural and $F \in \mathscr{C}$, then also $\sigma^{-1}F \in \mathscr{C}$ for every $\sigma \in M$ by Proposition 1.51, but in general it is not the case that $\sigma F \in \mathscr{C}$ (Exercise 1.63); if we want to obtain a closed set after applying an action to a closed set, we have to close the result, that is, take $C\sigma F$. Then:

LEMMA 1.55
Let C be a structural closure operator on an $\boldsymbol{M}$-set A. For each $\sigma \in M$, the function $F \longmapsto C\sigma F$ is a join-preserving operation in the lattice $\mathscr{C}$.

PROOF: Left to you as Exercise 1.64. ⊠

The important result is:

THEOREM 1.56
Let C, D be structural closure operators on $\boldsymbol{M}$-sets A and B, respectively (the actions of the elements of M on A and on B being denoted in the same way). Let Φ be a lattice isomorphism between the respective lattices of closed sets $\mathscr{C}$ and $\mathscr{D}$. For each $\sigma \in M$ the following conditions are equivalent:

(i) $\Phi C\sigma F = D\sigma\Phi F$ *for all* $F \in \mathscr{C}$.
(ii) $\Phi\sigma^{-1}F = \sigma^{-1}\Phi F$ *for all* $F \in \mathscr{C}$.
(iii) $\Phi^{-1}\sigma^{-1}G = \sigma^{-1}\Phi^{-1}G$ *for all* $G \in \mathscr{D}$.
(iv) $\Phi^{-1}D\sigma G = C\sigma\Phi^{-1}G$ *for all* $G \in \mathscr{D}$.

PROOF: (i)⇒(ii) Let $F \in \mathscr{C}$. Since always $\sigma\sigma^{-1}F \subseteq F$, also $C\sigma\sigma^{-1}F \subseteq F$ and hence $\Phi C\sigma\sigma^{-1}F \subseteq \Phi F$. Now, recalling that $\sigma^{-1}F \in \mathscr{C}$ by Proposition 1.51 and using the assumption, we have $\sigma\Phi\sigma^{-1}F \subseteq D\sigma\Phi\sigma^{-1}F = \Phi C\sigma\sigma^{-1} \subseteq \Phi F$, and hence $\Phi\sigma^{-1}F \subseteq \sigma^{-1}\Phi F$. To prove the reverse inclusion, notice that if $F \in \mathscr{C}$, after using Φ, structurality of D and Φ^{-1} back, we obtain that $\Phi^{-1}\sigma^{-1}\Phi F \in \mathscr{C}$, therefore we can apply the assumption to it and obtain $\Phi C\sigma\Phi^{-1}\sigma^{-1}\Phi F = D\sigma\Phi\Phi^{-1}\sigma^{-1}\Phi F = D\sigma\sigma^{-1}\Phi F \subseteq D\Phi F = \Phi F$; the two last steps hold because $\sigma\sigma^{-1} \leqslant \mathrm{Id}_B$ and $\Phi F \in \mathscr{D}$, respectively. Reversing the order isomorphism, this gives $C\sigma\Phi^{-1}\sigma^{-1}\Phi F \subseteq F$ and hence $\sigma\Phi^{-1}\sigma^{-1}\Phi F \subseteq F$, which implies $\Phi^{-1}\sigma^{-1}\Phi F \subseteq \sigma^{-1}F$; using Φ again, we finally obtain $\sigma^{-1}\Phi F \subseteq \Phi\sigma^{-1}F$, which completes the proof of (ii).

(ii)⇒(iii) If $G \in \mathscr{D}$, then applying (ii) to $\Phi^{-1}G \in \mathscr{C}$ gives that $\Phi\sigma^{-1}\Phi^{-1}G = \sigma^{-1}\Phi\Phi^{-1}G = \sigma^{-1}G$. Now applying Φ^{-1} to both sides we obtain $\sigma^{-1}\Phi^{-1}G = \Phi^{-1}\sigma^{-1}G$ as desired.

(iii)⇒(ii) This is the same as the previous implication, after interchanging Φ and Φ^{-1} (which are both isomorphisms).

(ii)⇒(i) Let $F \in \mathscr{C}$. Since $\sigma\Phi F \subseteq D\sigma\Phi F$, we also have $\Phi F \subseteq \sigma^{-1}D\sigma\Phi F$, and hence $F \subseteq \Phi^{-1}\sigma^{-1}D\sigma\Phi F$. Since we already know that (ii) is equivalent to (iii), and $D\sigma\Phi F \in \mathscr{D}$, we can apply (iii) and obtain that $\Phi^{-1}\sigma^{-1}D\sigma\Phi F = \sigma^{-1}\Phi^{-1}D\sigma\Phi F$. Thus, $F \subseteq \sigma^{-1}\Phi^{-1}D\sigma\Phi F$, that is, $\sigma F \subseteq \Phi^{-1}D\sigma\Phi F$. Since this last set is in $\mathscr{C}$, also $C\sigma F \subseteq \Phi^{-1}D\sigma\Phi F$ and by applying Φ we finally obtain $\Phi C\sigma F \subseteq D\sigma\Phi F$, which proves one inclusion in (i). The other inclusion is easier: Since $\sigma F \subseteq C\sigma F$, $F \subseteq \sigma^{-1}C\sigma F$ and then using (ii) $\Phi F \subseteq \Phi\sigma^{-1}C\sigma F = \sigma^{-1}\Phi C\sigma F$, that is, $\sigma\Phi F \subseteq \Phi C\sigma F$. Since this last set is in $\mathscr{D}$, also $D\sigma\Phi F \subseteq \Phi C\sigma F$, which completes the proof of (i).

The equivalence between (iii) and (iv) is the same as that between (i) and (ii), after interchanging the roles of Φ and Φ^{-1}. ⊠

Thus, if according to Lemma 1.55 we regard the function $F \mapsto C\sigma F$ as a unary operation in the closure system $\mathscr{C}$, and the function $G \mapsto D\sigma G$ as a unary operation in $\mathscr{D}$, then the theorem establishes that having (ii) for all $\sigma \in M$ is a necessary and sufficient condition for a lattice isomorphism between the lattices $\mathscr{C}$ and $\mathscr{D}$ to be also a homomorphism with respect to these additional unary operations (one for each $\sigma \in M$), that is, an isomorphism between the ***expanded lattices*** $\langle \mathscr{C}, \langle C\sigma : \sigma \in M \rangle \rangle$ and $\langle \mathscr{D}, \langle D\sigma : \sigma \in M \rangle \rangle$. This is used in Section 3.5 in the case where the monoid $\boldsymbol{M}$ is the monoid of endomorphisms of an algebra, and this equivalent condition is called ***commutativity with endomorphisms***; although the formulation in (ii) suggests we should rather speak of commutativity with *inverse* endomorphisms, this lattice-theoretic view of the unary operations induced by each σ in the lattices of closed sets makes the "direct" terminology also meaningful and it actually reflects a deeper lattice-theoretic fact.

Exercises for Section 1.4

EXERCISE 1.50. Prove that a closure operator C on a set A is finitary if and only if it satisfies the following property: for all $X \subseteq A$, if $Z \subseteq CX$ and Z is finite, then there exists a finite $Y \subseteq X$ such that $Z \subseteq CY$.

EXERCISE 1.51. Prove that a closure operator C on a set A is finitary if and only if for all $X \subseteq A$ and all $a, b \in A$ such that $C(X,a) = C(X,b)$, there exists a finite $Y \subseteq X$ such that $C(Y,a) = C(Y,b)$.

EXERCISE 1.52. Let C be a closure operator on a set A and let $a \in A$ be an inconsistent element. Prove that $F \in \mathscr{C}$ is consistent if and only if $a \notin F$.

EXERCISE 1.53. Let $\boldsymbol{A}$ be a semilattice or a lattice, and denote by $Fg_{\wedge}^{\boldsymbol{A}}$ the closure operator associated with the closure system $\mathcal{F}ilt_{\wedge}^{\circ}\boldsymbol{A}$ generated by the set $\mathcal{F}ilt_{\wedge}\boldsymbol{A}$ of all its filters (Exercise 1.42).

1. Prove that $Fg_{\wedge}^{\boldsymbol{A}}X = \{a \in A : a \geqslant a_1 \wedge \cdots \wedge a_n \text{ for some } a_i \in X, n \geqslant 1\}$ for every non-empty $X \subseteq A$. Use this to show that $Fg_{\wedge}^{\boldsymbol{A}}$ is finitary.
2. Prove the last property in point 1 indirectly by showing that $\mathcal{F}ilt_{\wedge}^{\circ}\boldsymbol{A}$ is inductive.

EXERCISE 1.54. Continue Exercise 1.32 by checking which of the closure operators (resp., systems) you found there are finitary (resp., inductive).

EXERCISE 1.55. Let C be a closure operator on a set A. Prove directly that the following conditions are equivalent:

(i) For every $X \subseteq A$, $\boldsymbol{\Lambda}_C X = \bigcup\{\boldsymbol{\Lambda}_C Y : Y \subseteq X,\ Y \text{ finite}\}$.
(ii) $\boldsymbol{\Lambda}_C \bigcup \mathscr{B} = \bigcup\{\boldsymbol{\Lambda}_C X : X \in \mathscr{B}\}$ for any non-empty upwards directed family $\mathscr{B} \subseteq \mathscr{P}A$.

HINT. Observe that in Exercise 1.51 you proved that (i) is equivalent to finitarity of C. See also Theorem 1.46.

EXERCISE 1.56. Explain why the family $\mathscr{B}$ is required to be non-empty in the statements of Theorem 1.46 and of Exercise 1.55.

EXERCISE 1.57. A closure operator C on a set A is ***unitary*** when $CX = \bigcup\{C\{a\} : a \in X\}$ for all $X \subseteq A$. Clearly unitary implies finitary. Compare this notion with the property in Proposition 1.29.1, which holds for all closure operators, and:

1. Find examples of unitary closure operators, and of finitary closure operators that are not unitary.
   HINT. Thinking about ordered sets, as in Exercises 1.38 and 1.42, may help.
2. Investigate how can Theorems 1.42 and 1.46 be generalized to obtain characterizations of unitary closure operators in terms of unions of arbitrary families of closed sets.

EXERCISE 1.58. Let C be a closure operator on a set A. Prove sucessively:

1. $C_f X = CX$ for all finite $X \subseteq A$; in particular, $C_f\varnothing = C\varnothing$.
2. C_f is a finitary closure operator on A and $C_f \leqslant C$.
3. C_f is the strongest of all finitary closure operators C' on A such that $C' \leqslant C$.

EXERCISE 1.59. Let $\boldsymbol{M}$ be a monoid, C a closure operator on an $\boldsymbol{M}$-set and $\sigma \in M$. Prove that the following conditions are equivalent:

(i) $\sigma C \leqslant C\sigma$.
(ii) $C\sigma C = C\sigma$.
(iii) $\sigma C\sigma^{-1} \leqslant C$.
(iv) $C\sigma^{-1} \leqslant \sigma^{-1}C$.
(v) $C\sigma^{-1}C = \sigma^{-1}C$.

(Here $\leqslant$ is the relation between power set functions introduced at the end of the second paragraph of Section 1.1.)

EXERCISE 1.60. Let $\boldsymbol{M}$ be a monoid, C a closure operator on an $\boldsymbol{M}$-set A, and $\mathscr{C}$ its associated closure system. Prove that C is structural if and only if $\mathscr{C}$ has a base $\mathscr{B}$ such that if $F \in \mathscr{B}$ and $\sigma \in M$, then $\sigma^{-1}F \in \mathscr{B}$.

EXERCISE 1.61. Prove that the finitary companion of a structural closure operator is also structural.

EXERCISE 1.62. Provide an example of an axiomatic extension of a structural closure operator that is not itself structural.

HINT. Of course, logics will be your primary candidates.

EXERCISE 1.63. Show by a simple counterexample that if Γ is a theory of a logic and σ is a substitution, then the set $\sigma\Gamma$ is not necessarily a theory.

HINT. Recall that the set of theorems is closed under substitutions.

EXERCISE 1.64. Let C be a structural closure operator on an $\boldsymbol{M}$-set. Prove that for each $\sigma \in M$, the function $F \mapsto C\sigma F$ is a completely-join-preserving operation in the complete lattice $\mathscr{C}$ of the closed sets of C.

EXERCISE 1.65. Let C be a finitary closure operator on the set of formulas of some algebraic language that satisfies the equivalent properties in Exercise 1.59 for all surjective substitutions σ. Prove that C is structural.

1.5 More on closure operators and closure systems

Lattices of closure operators and lattices of logics

Theorem 1.33 establishes that, for each set A, the sets $\mathsf{CO}(A)$ and $\mathsf{CS}(A)$ are dually isomorphic ordered sets; the first is ordered under the relation $\leqslant$ introduced at the top of page 2 (see also page 39), and the second under the subset relation. But there is more:

THEOREM 1.57
Let A be an arbitrary set.

1. *The set $\mathsf{CO}(A)$ endowed with the relation $\leqslant$ is a complete lattice, with infimum operation defined by $(\bigwedge C_i)X := \bigcap C_i X$ for all $X \subseteq A$.*
2. *The set $\mathsf{CS}(A)$ ordered under $\subseteq$ is a complete lattice, with infimum $\bigwedge \mathscr{C}_i := \bigcap \mathscr{C}_i$.*
3. *These two complete lattices are dually isomorphic.*

PROOF: You probably know that in order to check that a certain ordered set is a complete lattice it is enough to show that it has a maximum and that an arbitrary non-empty family has an infimum (a greatest lower bound). We have already seen that both $\mathsf{CO}(A)$ and $\mathsf{CS}(A)$ have a maximum, and it is a rather simple exercise to check that the given expressions are respectively a closure operator and a closure system, and indeed they are the infimum in the corresponding set. The final point follows from the two previous ones and Theorem 1.33. ⊠

We can also obtain expressions for the supremum operations:

PROPOSITION 1.58
The supremum in $\mathsf{CS}(A)$ is given by the following expression:

$$\bigvee_{i \in I} \mathscr{C}_i := \Big\{ F \subseteq A : F = \bigcap_{i \in I} F_i \text{ with } F_i \in \mathscr{C}_i \text{ for all } i \in I \Big\},$$

and the supremum in $\mathsf{CO}(A)$ is the operator defined, for all $X \subseteq A$, by

$$\Big(\bigvee_{i \in I} C_i\Big) X := \bigcap \Big\{ F \in \bigcap_{i \in I} \mathscr{C}_i : F \supseteq X \Big\}.$$

PROOF: $\bigvee \mathscr{C}_i$ should be the smallest closure system containing all the $\mathscr{C}_i$, that is, containing $\bigcup \mathscr{C}_i$. Applying Proposition 1.30 to this case, and using that each of the $\mathscr{C}_i$ is a closure system, the first wanted expression is easily obtained. Now the dual order isomorphism means that the complete lattice operations are interchanged, therefore the closure system associated with $\bigvee C_i$ is $\bigwedge \mathscr{C}_i = \bigcap \mathscr{C}_i$, and this plus Theorem 1.26 gives the second wanted expression. ⊠

As we see, while the supremum of the closure systems $\mathscr{C}_i$ has a more or less smooth expression directly in terms of the $\mathscr{C}_i$'s, this is not the case for the supremum of the closure operators C_i, which in principle needs to use the associated closure systems $\mathscr{C}_i$.

It is not difficult to show that the set $\mathsf{CO}_{\mathrm{f}}(A)$ of all *finitary* closure operators on a set A is a complete lattice. Interestingly, it is a *sublattice* of $\mathsf{CO}(A)$ (Exercise 1.67), but it *need not be a complete sublattice*. More precisely, both the infimum and the supremum in $\mathsf{CO}(A)$ of two (or a finite number of) finitary closure operators are finitary, and even the supremum in $\mathrm{CO}(A)$ of an infinite number of them is finitary, hence these operations in $\mathrm{CO_f}(A)$ coincide with those in $\mathrm{CO}(A)$; but the infimum in $\mathrm{CO}(A)$ of an infinite number of them may not be finitary, and thus it may not coincide with the infimum in $\mathrm{CO_f}(A)$ (see Example 4.10). Expressions for the lattice operations of $\mathrm{CO_f}(A)$ are suggested in Exercise 1.68.

If A is an $\boldsymbol{M}$-set, it is easy to see that the property of being *structural* is preserved by any of the suprema and infima just considered; thus, one might as well consider the lattice of all structural closure operators on a fixed $\boldsymbol{M}$-set and the lattice of all the finitary and structural ones. One might also define the "structural companion" of a given closure operator (the strongest of all structural closure operators that are weaker than it), and so forth. Anyway, these constructions are not of special interest to us, except in the special case of the formula algebra, that is, of logics.

The above constructions and properties apply in particular to the set of logics on a fixed language, as they are closure operators on a fixed $\boldsymbol{M}$-set, in this case the $\boldsymbol{E}$-set of all formulas of the language. Hence we obtain ***the lattice of logics*** (***over a fixed language***) and ***the lattice of finitary logics*** (***over a fixed language***). Some details are worked out in Chapter 1 of [249]; see also Exercise 1.70. Observe that the order relation between logics introduced on page 27 (the relative strength relation) coincides with the order relation between the associated closure operators as considered here ($\mathcal{L} \leqslant \mathcal{L}'$ if and only if $C_{\mathcal{L}} \leqslant C_{\mathcal{L}'}$); this relation appears often in this book, while the lattice structure of the sets seldom does, and no particular notation for these lattices and their operations is necessary.

Notice that in terms of the consequence relations $\vdash_{\mathcal{L}}$, the operation corresponding to the infimum in the lattice of all logics is also intersection: If $\{\mathcal{L}_i : i \in I\}$ is a family of logics on the same language, and $\Gamma \cup \{\varphi\} \subseteq Fm$, then $\Gamma \vdash_{\bigwedge \mathcal{L}_i} \varphi$ if and only if $\varphi \in C_{\bigwedge \mathcal{L}_i}\Gamma = \bigcap C_{\mathcal{L}_i}\Gamma$ if and only if $\varphi \in C_{\mathcal{L}_i}\Gamma$ for all $i \in I$ if and only if $\Gamma \vdash_{\mathcal{L}_i} \varphi$ for all $i \in I$. Thus, $\vdash_{\bigwedge \mathcal{L}_i} = \bigcap_{i \in I} \vdash_{\mathcal{L}_i}$. By contrast, as remarked above, the infimum of an infinite number of finitary logics in the lattice of all the finitary ones need not be their intersection (while the infimum of a finite number of finitary logics is always finitary).

Irreducible sets and saturated sets

Definition 1.59

A closed set F of a closure system $\mathscr{C}$ is ***irreducible*** *(also* ***completely irreducible*** *or* ***C-irreducible****) when $F \neq A$ and for every family $\{F_i : i \in I\} \subseteq \mathscr{C}$, if $F = \bigcap F_i$, then $F = F_k$ for some $k \in I$. The word* ***finitely*** *is prepended to define the parallel notion with the property limited to finite I.*

F is ***saturated*** *(or* ***maximal****)* ***relatively to*** *$a \in A$ when $a \notin F$ and for all $G \in \mathscr{C}$, if $F \subsetneq G$, then $a \in G$.*

Finally, F is ***saturated*** *or* ***relatively maximal*** *when it is so relatively to some $a \in A$.*

The irreducible closed sets are the closed sets that cannot be expressed as intersections of *other* closed sets. Obviously every irreducible element must belong to every base of $\mathscr{C}$. The following characterization is easy to see and very practical for applications:

PROPOSITION 1.60
F is saturated relatively to some $a \in A$ if and only if $a \notin F$ and for each $b \in A$, if $b \notin F$, then $a \in C(F, b)$.

PROOF: Left to you as Exercise 1.71. ⊠

The proofs of the following two interesting results show typical features of how to work with closure operators and closure systems.

PROPOSITION 1.61
If $F \in \mathscr{C}$, then F is irreducible if and only if F is saturated.

PROOF: ($\Rightarrow$) Clearly $F \subseteq \bigcap\{C(F,b) : b \notin F\}$, and $F \subsetneq C(F,b)$ for each $b \notin F$. Thus, if F is irreducible, the first inclusion cannot be an equality; that is, there must be some $a \notin F$ such that $a \in C(F,b)$ for each $b \notin F$. Hence F is saturated.

($\Leftarrow$) If F is not irreducible, then there is some family $\{F_i : i \in I\} \subseteq \mathscr{C}$ with $F \subsetneq F_i$ for every i, such that $F = \bigcap F_i$. So, for each $i \in I$ there is a $b_i \in F_i$ such that $b_i \notin F$. Then $C(F, b_i) \subseteq F_i$ and hence $F \subsetneq C(F, b_i) \subseteq F_i$, which implies that $F = \bigcap\{C(F, b_i) : i \in I\}$; from this it follows that $F = \bigcap\{C(F,b) : b \notin F\}$, against the definition of being saturated, modulo Proposition 1.60. ⊠

PROPOSITION 1.62
If $F \in \mathscr{C}$, then F is maximal if and only if it is proper ($F \neq A$) and it is saturated relatively to every $a \notin F$.

PROOF: Clearly, if F is maximal and $a, b \notin F$, then necessarily $F \subsetneq C(F,b)$ implies $C(F,b) = A$ and so $a \in C(F,b)$. By Proposition 1.60 this shows that F is saturated relatively to any $a \notin F$. Conversely, assume that F is saturated relatively to every $a \notin F$ and that $F \subsetneq G \in \mathscr{C}$. By definition of saturation, this implies that $a \in G$ for all $a \notin F$, but since $F \subseteq G$ as well, it results that $G = A$, which shows that F is maximal. ⊠

Lindenbaum's Lemma 1.43 can be rephrased using saturated sets as follows: If C is a finitary closure operator, $F \in \mathscr{C}$ and $a \notin F$, then there exists $G \in \mathscr{C}$ saturated relatively to a such that $F \subseteq G$.

There is a relation between irreducible sets and bases. Clearly, every irreducible set must belong to every base. Therefore, if the set of irreducibles is a base of $\mathscr{C}$, then it must be the smallest base (Exercise 1.72). A situation where this happens is when the closure operator is finitary; this is proved by a typical application of Lindenbaum's Lemma.

PROPOSITION 1.63
If C is a finitary closure operator, then $CX = \bigcap\{F \in \mathscr{C} : F \text{ irreducible and } F \supseteq X\}$ for all $X \subseteq A$. Hence the collection of all irreducible sets of $\mathscr{C}$ is a base of $\mathscr{C}$.

PROOF: The inclusion $\subseteq$ in the first statement holds in general. To see the reverse one, let $a \notin CX$. By Lemma 1.43 there exists $G \in \mathscr{C}$, $CX \subseteq G$, such that $a \notin G$ and G is saturated relatively to a, and hence G is irreducible. Therefore

$a \notin \bigcap\{F \in \mathscr{C} : F \text{ irreducible and } F \supseteq X\}$. The second statement is a consequence of the first. ⊠

The converse to this proposition does not hold (Exercise 1.73).

Finitarity and compactness

Next we explore the relations between finitarity and compactness.[†] First of all, a general property:

A closed set $F \in \mathscr{C}$ is ***finitely generated*** when $F = CF_0$ for some finite $F_0 \subseteq A$, and is ***principal*** when $F = C\{a\}$ for some $a \in A$. Notice that if $C\varnothing \neq \varnothing$, then $C\varnothing$ is principal, because $C\varnothing = C\{a\}$ for any $a \in C\varnothing$. It is interesting to observe that, by Proposition 1.29, in all closure systems every closed set is a supremum of finitely generated closed sets, and also of principal ones. As a consequence, the next property, which is used in Chapter 6, follows immediately:

PROPOSITION 1.64
Let C be a closure operator, with associated closure system $\mathscr{C}$. The family of its finitely-generated closed sets $\mathscr{C}^{\omega}$ is a join-sub-semilattice of $\mathscr{C}$, which is generated as a semilattice by the set of principal closed sets $\{C\{a\} : a \in A\}$. ⊠

An element a of a lattice $\boldsymbol{A}$ is ***compact*** when for all $D \subseteq A$ such that $\bigvee D$ exists, if $a \leqslant \bigvee D$, then $a \leqslant \bigvee D'$ for a *finite* $D' \subseteq D$. The parallelism with the topological notion is obvious; for a useful equivalent definition see Exercise 1.75. Observe that in a power set lattice $\mathscr{P}A$, an $X \in \mathscr{P}A$ is compact if and only if it is finite. More generally, Exercise 1.76 establishes a useful (and natural) characterization of compactness for the lattices associated with closure systems: If $\mathscr{C}$ is a closure system on A, then a closed set $F \in \mathscr{C}$ is compact in $\mathscr{C}$ if and only if for any $\{X_i : i \in I\} \subseteq \mathscr{P}A$, if $F \subseteq C\bigcup_{i \in I} X_i$, then there is a finite $J \subseteq I$ such that $F \subseteq C\bigcup_{i \in J} X_i$.

PROPOSITION 1.65
Let C be a closure operator, and let $F \in \mathscr{C}$.

1. *If F is compact in $\mathscr{C}$, then F is finitely generated.*
2. *If C is finitary, then F is compact in $\mathscr{C}$ if and only if F is finitely generated.*

PROOF: See Exercise 1.77. This is a typical application of finitarity and a good exercise. Alternatively, see [51, Theorem I.5.5]. ⊠

This result is needed in Section 3.5, together with the next one.

PROPOSITION 1.66
Compactness is a lattice-theoretic property preserved by isomorphisms between complete lattices.

PROOF: Left to you as Exercise 1.78. ⊠

[†] *Compactness* is taken here in the lattice-theoretic sense. The term is also used in logic: a theorem stating that a certain logic is finitary is often called a "Compactness Theorem". The name is also applied to the property that a set of formulas is consistent if and only if all its finite subsets are consistent; the two properties are related but in general they are not equivalent.

A lattice is ***compactly generated*** when each of its elements is the join of compact elements (hence, it is the join of all compact elements below itself), and is ***algebraic*** when it is complete and compactly generated. In the case of closure systems only the second condition matters, because they are always complete as lattices. Then:

PROPOSITION 1.67
If C is a finitary closure operator, then its associated closure system $\mathscr{C}$ is an algebraic lattice. In other words: every inductive closure system is algebraic as a lattice.

PROOF: This follows immediately from Definition 1.39, after considering Proposition 1.65.2. ⊠

The converse of this result is not true: not every closure system that is algebraic as a lattice is inductive; in particular, a non-finitary closure operator can have an algebraic but non-inductive closure system, as Exercise 1.79 shows. The very close result that does hold here is that a closure system that is algebraic as a lattice is isomorphic to an inductive closure system, namely the closure system of a certain finitary closure operator (Exercise 1.80). But *inductiveness is a set-theoretic property* rather than an algebraic or order-theoretic one, and it need not be preserved by order isomorphisms.

Exercises for Section 1.5

EXERCISE 1.66. Complete the details in the proofs of Theorem 1.57 and of Proposition 1.58.

EXERCISE 1.67. Let $\mathrm{CO_f}(A)$ denote the set of finitary closure operators on a set A. Prove, without using Exercise 1.68, that $\mathrm{CO_f}(A)$ is a sublattice of the lattice $\mathrm{CO}(A)$ of all closure operators on A.

EXERCISE 1.68. Let $\{C_i : i \in I\} \subseteq \mathrm{CO_f}(A)$, $I \neq \varnothing$, and let $\bigvee C_i \in \mathrm{CO}(A)$ be the supremum of the family in the lattice $\mathrm{CO}(A)$. Prove that $\bigvee C_i$ is finitary, and that it can be described by the following expression: for all $X \subseteq A$ and all $a \in A$, $a \in (\bigvee C_i)X$ if and only if there exist $i_1, \dots, i_n \in I$ (for some $n \geqslant 1$) such that $a \in C_{i_1} C_{i_2} \dots C_{i_n} X$. Conclude that the set $\mathrm{CO_f}(A)$ is a complete lattice, and observe that its infimum operation is given by $\bigwedge^{\mathrm{f}} C_i := (\bigwedge C_i)_{\mathrm{f}}$, where $\bigwedge$ is the infimum in the bigger lattice CO(A).

EXERCISE 1.69. Let $\{C_i : i \in I\} \subseteq \mathrm{CO_f}(A)$ be a non-empty upwards directed family. Define $CX := \bigcup_{i \in I} C_i X$ for all $X \subseteq A$. Prove that C is a finitary closure operator on A, and that it is the supremum of the family both in $\mathrm{CO}(A)$ and in $\mathrm{CO_f}(A)$.

EXERCISE 1.70. Let A be an $\boldsymbol{M}$-set. Prove that the set $\mathrm{CO_s}(A)$ of all structural closure operators on A is a complete sublattice of $\mathrm{CO}(A)$. Use Exercise 1.68 to prove that the set $\mathrm{CO_{fs}}(A)$ of all finitary and structural closure operators on A is a complete sublattice of $\mathrm{CO_f}(A)$. Particularize this and the preceding exercises to the case of logics.

EXERCISE 1.71. Let $\mathscr{C}$ be a closure system, and $F \in \mathscr{C}$. Prove that F is saturated relatively to some $a \in A$ if and only if $a \notin F$ and for all $b \notin F$, $a \in C(F, b)$.

EXERCISE 1.72. Let $\mathscr{B}$ be a base of a closure system $\mathscr{C}$. Show that $\mathscr{B}$ is the smallest base of $\mathscr{C}$ if and only if $\mathscr{B}$ is the collection of all irreducible sets of $\mathscr{C}$.

COMMENT. It may be the case that none of the two things happen, i.e., that the closure system does not have a smallest base and that the family of all irreducible closed sets does not form a base (for instance, this family may be empty).

EXERCISE 1.73. Give an example of a non-finitary closure system in which every closed set is irreducible (and hence the set of irreducibles is a base).

HINT. Work on some infinite, linearly ordered set.

EXERCISE 1.74. Let $\mathscr{C}$ be a closure system, and $F \in \mathscr{C}$, $F \neq \emptyset$. Prove that F is finitely generated if and only if $F = C\{a_1, \ldots, a_n\}$ for some $a_1, \ldots, a_n \in F$. Compare with the definition given on page 57 and see why this reformulation does not work for the empty set (in case it is a closed set).

EXERCISE 1.75. Let $\boldsymbol{A}$ be a lattice, and $a \in A$. Prove that a is compact in $\boldsymbol{A}$ if and only if for any non-empty upwards directed $D \subseteq A$ such that $\bigvee D$ exists, if $a \leqslant \bigvee D$, then $a \leqslant d$ for some $d \in D$.

EXERCISE 1.76. Let $\mathscr{C}$ be a closure system, and $F \in \mathscr{C}$. Prove that F is compact in $\mathscr{C}$ (with respect to its lattice structure) if and only if for any family $\{X_i : i \in I\} \subseteq \mathscr{P}A$, if $F \subseteq C \bigcup_{i\in I} X_i$, then there is a finite $J \subseteq I$ such that $F \subseteq C \bigcup_{i \in J} X_i$.

EXERCISE 1.77. Let C be a closure operator on a set A and $F \in \mathscr{C}$. Prove that if F is compact in $\mathscr{C}$, then it is finitely generated, and that if C is finitary the converse also holds.

EXERCISE 1.78. Prove that an isomorphism between two complete lattices puts the compact elements of one of them in correspondence with the compact elements of the other.

EXERCISE 1.79. Let $A := \omega \cup \{\infty\}$, where ω is the set of natural numbers and ∞ is a new element, and consider the linear order on A obtained by putting ∞ on top of ω. Check that A becomes a complete lattice. For each $a \in A$ define ${\downarrow}a := \{b \in A : b \leqslant a\}$ and for each $X \subseteq A$ define $CX := {\downarrow} \bigvee X$. Prove:

1. The function C is a closure operator on the power set of A.
2. Its associated closure system is $\mathscr{C} := \{\emptyset\} \cup \{{\downarrow}a : a \in A\}$, and as a lattice it is algebraic.
3. The closure operator C is not finitary.
   HINT. Show that $\infty \in C\omega$ but $\infty \notin CF$ for all finite $F \subseteq \omega$.
4. Give an example witnessing that $\mathscr{C}$ is not inductive.

COMMENT. The construction in this exercise is just the set-theoretic representation of complete lattices in terms of down-sets worked out in Exercise 1.39.

EXERCISE 1.80. Let $\boldsymbol{A}$ be an algebraic lattice. Prove that the construction in Exercise 1.39, restricted to the set X of the compact elements of $\boldsymbol{A}$, defines an isomorphism between $\boldsymbol{A}$ and the algebraic lattice of the closed sets of a finitary closure operator on X.

1.6 Consequences associated with a class of algebras

The material in the first subsection should be covered in any standard introductory course on universal algebra, such as [24] or [51]; depending on your background, you may safely skip it and return to it to check some result, or you can review it now only to fix notations, definitions and terminology, or you can study it in detail.

Recall that all algebras are assumed to have the same fixed, arbitrary similarity type, and that equations are identified with pairs of formulas.

The equational consequence, and varieties

- Recall the definitions of satisfaction and validity of equations (page 7):

$$A \vDash \alpha \approx \beta \,[\![h]\!] \iff h\alpha = h\beta$$

$$A \vDash \alpha \approx \beta \iff A \vDash \alpha \approx \beta \,[\![h]\!] \text{ for all } h \in \mathrm{Hom}(\boldsymbol{Fm}, A)$$

$$\mathsf{K} \vDash \alpha \approx \beta \iff A \vDash \alpha \approx \beta \text{ for all } A \in \mathsf{K}$$

Thus, in ordinary model-theoretic terms, $A \vDash \alpha \approx \beta$ is shorthand for the first-order condition that $A \vDash \forall \vec{x}(\alpha \approx \beta)$, where $\vec{x}$ is the list of variables occurring in the equation.

- A class K is an ***equational class*** or ***variety*** when there is a set $\Theta \subseteq Eq$ such that $\mathsf{K} = \{A : A \vDash \Theta\}$.

- The "absolute" equational consequence, denoted also† by $\vDash$, is defined on page 28 as: For all $\Theta \subseteq Eq$ and all $\varepsilon \approx \delta \in Eq$,

$$\Theta \vDash \varepsilon \approx \delta \iff \begin{array}{l}\text{For every algebra } A, \\ \text{if } A \vDash \alpha \approx \beta \text{ for all } \alpha \approx \beta \in \Theta, \text{ then } A \vDash \varepsilon \approx \delta.\end{array} \tag{4}$$

This relation is sometimes called ***Birkhoff's equational consequence*** (see why below) or simply ***the equational consequence***, or even ***equational logic***, but in our context these other terms are prone to confusion.

If you look at the occurrences of the quantifier "for all $h \in \mathrm{Hom}(\boldsymbol{Fm}, A)$" hidden in (4), you will realize that this consequence reflects the transmission of *global* satisfaction of equations, that is, of *validity*: $\Theta \vDash \varepsilon \approx \delta$ means that "$\varepsilon \approx \delta$ is *valid* in all algebras where the equations in Θ are *valid*". Expressed in first-order terms, we would say that this is the consequence between universally quantified sentences: $\Theta \vDash \varepsilon \approx \delta$ means that every "model" of Θ is a "model" of $\varepsilon \approx \delta$.

As an example, consider what is meant when one says that the equation $\neg(x \vee y) \approx \neg x \wedge \neg y$ is a *consequence* of the set Θ of equations that axiomatize Heyting algebras: one actually means that the former equation is valid in all algebras where the equations in Θ are valid, that is, in all Heyting algebras. By contrast, one says that the equation $\neg(x \wedge y) \approx \neg x \vee \neg y$ is not a consequence of Θ, because it fails in some Heyting algebras; however, this equation is a consequence of $\Theta \cup \{x \approx \neg\neg x\}$, because the equation is valid in the algebras where this set is valid (the Boolean algebras).

- It is easy to see that $\vDash$ is a closure relation on the set Eq of equations; its closed sets are called ***equational theories***, and they correspond bijectively with varieties: The class of *all* algebras satisfying an equational theory is by definition a variety, and the set of *all* the equations satisfied by a variety is an equational theory. This correspondence works as that between theories and axiomatizable classes in ordinary model theory.

† The double usage of the symbol $\vDash$, for the satisfaction and validity relations in an algebra, and for the consequence relation between equations, should lead to no confusion: the object appearing to its left clarifies its meaning.

• Birkhoff's Completeness Theorem asserts that the consequence $\vDash$ can be axiomatized by means of the so-called ***Birkhoff's rules*** (the first one is actually an *axiom*):

(Reflexivity) $\alpha \approx \alpha$

(Symmetry) $\dfrac{\alpha \approx \beta}{\beta \approx \alpha}$ (Transitivity) $\dfrac{\alpha \approx \beta \qquad \beta \approx \gamma}{\alpha \approx \gamma}$

(Replacement) $\dfrac{\alpha \approx \beta}{\varphi(\alpha,\vec{z}) \approx \varphi(\beta,\vec{z})}$ for each formula $\varphi(z,\vec{z})$ with at least one variable z

(Substitution) $\dfrac{\alpha \approx \beta}{\sigma\alpha \approx \sigma\beta}$ for each $\sigma \in \operatorname{End}\boldsymbol{Fm}$

The set of rules (Replacement) is equivalent to the following set of rules:

(Congruence) $\dfrac{\alpha_1 \approx \beta_1 \,, \;\dots\;, \alpha_n \approx \beta_n}{\lambda\alpha_1 \dots \alpha_n \;\approx\; \lambda\beta_1 \dots \beta_n}$ for each $\lambda \in L$, $n = \operatorname{ar}\lambda \geqslant 1$

The set of rules (Substitution) is equivalent to the following set:

(Substitution) $\dfrac{\alpha(x,\vec{z}) \approx \beta(x,\vec{z})}{\alpha(\varphi,\vec{z}) \approx \beta(\varphi,\vec{z})}$ for each formula φ

The alternative formulations facilitate the comparison between (Replacement) and (Substitution).†

• Observe that for a closure operator on the set Eq, to be structural (substitution-invariant) is not the same as to satisfy the (Substitution) rule. Actually, the latter rule is not invariant under substitutions, and therefore the consequence $\vDash$ is not structural as a closure operator; what this rule says is that its theories are indeed closed under substitutions. For an example, see Exercise 1.81.

• The class operators $\mathbb{H}, \mathbb{I}, \mathbb{S}, \mathbb{P}, \mathbb{P}_{\mathrm{U}}, \mathbb{P}_{\mathrm{R}}, \mathbb{P}_{\mathrm{SD}}$ of collecting, respectively, all homomorphic images, isomorphic images, subalgebras, products (of non-empty families), ultraproducts, reduced products and subdirect products of algebras in a given class, are defined as usual.‡

Since $\mathbb{H}$ includes $\mathbb{I}$, every class closed under $\mathbb{H}$ is also closed under $\mathbb{I}$. The classes of algebras closed under isomorphisms are sometimes called ***abstract classes***.

† Some confusion may arise, because both rules involve operations of "replacement": In (Replacement) a variable z is replaced by formulas α and β in different places of the equation $\varphi(z,\vec{z}) \approx \varphi(z,\vec{z})$, while in (Substitution) each variable x in α and β is uniformly replaced by σx. But observe that only (Substitution) asserts that the result of a replacement follows from the original, non-replaced equation. Besides, notice that the replacement that takes place in (Replacement) is not *uniform*, that is, it is not a substitution.

‡ Some authors prefer the product-like operators to be applied also to the empty family, producing a trivial algebra; then it is not necessary to add a trivial algebra in a number of places. Notice that the operator $\mathbb{S}$ and the product-like ones do not include closure under isomorphisms; some textbooks do, which dispenses them from using $\mathbb{I}$ in a number of places. In these issues I follow [51].

The operator $\mathbb{H}$ includes the formation of quotients, therefore for classes closed under $\mathbb{I}$, to be closed under $\mathbb{H}$ is the same as to be closed under formation of quotients (this follows from the First Isomorphism Theorem of universal algebra). Finally, notice that any function onto a trivial algebra is an epimorphism, hence any class closed under $\mathbb{H}$ contains all trivial algebras.

• By an important theorem of Birkhoff, varieties are the classes of algebras that are closed under $\mathbb{H}, \mathbb{S}$ and $\mathbb{P}$. And by another important theorem of Tarski, the smallest variety containing a given class K of algebras is the class $\mathbb{V}\mathsf{K} := \mathbb{HSP}\mathsf{K}$. This class is axiomatized by the set of equations $\{\alpha \approx \beta \in Eq : \mathsf{K} \vDash \alpha \approx \beta\}$.

• Observe that a congruence of the formula algebra $\boldsymbol{Fm}$ is just a set of equations (pairs of formulas) closed under the rules (Reflexivity), (Symmetry), (Transitivity) and (Replacement) or (Congruence). A congruence $\theta \in \mathrm{Con}\boldsymbol{Fm}$ is ***fully invariant*** when it is closed under the rule (Substitution); that is, when $\alpha \equiv \beta\ (\theta)$ implies $\sigma\alpha \equiv \sigma\beta\ (\theta)$ for all substitutions σ. Hence, a set of equations is an equational theory if and only if it is a fully invariant congruence of the algebra of formulas $\boldsymbol{Fm}$. The lattice of fully invariant congruences of $\boldsymbol{Fm}$ is thus dually isomorphic to the lattice of all varieties of the similarity type. Fully invariant congruences have the following important property; its proof is left to you as Exercise 1.82.

LEMMA 1.68
If θ is a fully invariant congruence of $\boldsymbol{Fm}$, then $\boldsymbol{Fm}/\theta \vDash \alpha \approx \beta$ if and only if $\alpha \equiv \beta\ (\theta)$. As a consequence, the variety generated by $\boldsymbol{Fm}/\theta$ is axiomatized by θ (considered as a set of equations), and $\boldsymbol{Fm}/\theta$ is the free algebra in this variety with denumerably many generators; one set of generators is $\{x/\theta : x \in V\}$. ⊠

The relative equational consequence

While some of the material in this and the next subsections is covered in courses in universal algebra that deal with quasivarieties, some is not; the properties that are not straightforward are included as exercises for readers seriously interested in the universal algebra of algebraic logic; some clues may be found in [29], [144, Chapter 9] and [200].

DEFINITION 1.69
*The **relative equational consequence** associated with a class K of algebras is the relation $\vDash_{\mathsf{K}}$ between sets of equations and equations defined by formula (5) on page 28, which is repeated here:*

$$\begin{aligned} \Theta \vDash_{\mathsf{K}} \varepsilon \approx \delta \iff & \textit{For every } \boldsymbol{A} \in \mathsf{K} \textit{ and for every } h \in \mathrm{Hom}(\boldsymbol{Fm}, \boldsymbol{A}), \\ & \textit{if } \boldsymbol{A} \vDash \alpha \approx \beta\ [\![h]\!] \textit{ for all } \alpha \approx \beta \in \Theta, \textit{ then } \boldsymbol{A} \vDash \varepsilon \approx \delta\ [\![h]\!]. \end{aligned} \tag{5}$$

The associated closure operator is denoted by $\mathcal{C}_{\mathsf{K}}$.
For each algebra $\boldsymbol{A}$, $\vDash_{\boldsymbol{A}}$ is defined as $\vDash_{\{\boldsymbol{A}\}}$, hence $\vDash_{\mathsf{K}} = \bigcap\{\vDash_{\boldsymbol{A}} : \boldsymbol{A} \in \mathsf{K}\}$.

This relation reflects the transmission of *local* satisfaction of equations in algebras, and moreover *relatively* to the class K, that is, satisfaction relative to evaluations in algebras in K: $\Theta \vDash_{\mathsf{K}} \varepsilon \approx \delta$ means that "$\varepsilon \approx \delta$ is *satisfied* by an evaluation in an algebra in K whenever this same evaluation (hence, in the same algebra) *satisfies* all the equations in Θ".

The consequences of this kind are usually formulated as concerning arbitrary elements of an algebra of the class, with a universal quantifier encompassing the whole implication. Here are two examples:

- The *cancellative law* of a binary operation $\cdot$ states that "$a{\cdot}b = a{\cdot}c$ implies $b = c$, for all a, b, c". This can be expressed as "$x{\cdot}y \approx x{\cdot}z \vDash_{\mathsf{K}} y \approx z$"; this more formal expression allows one to make distinctions, for instance, to say that this holds if K is a class of groups, while it does not hold if K is a class of rings.
- If K is a class of residuated lattices, one half of the *residuation law* "$a \cdot b \leqslant c$ implies $b \leqslant a \backslash c$ for all a, b, c" can be formally expressed as the consequence $(x \cdot y) \vee z \approx z \vDash_{\mathsf{K}} y \wedge (x \backslash z) \approx y$. The other half is expressed by the reverse consequence.

PROPOSITION 1.70

For any class K (even when it is empty and when it is the class of all algebras of the similarity type), $\mathsf{C_K}$ is a structural closure operator. Therefore the set $\mathsf{C_K}\varnothing$ is invariant, that is, it is an equational theory. More precisely, the theorems of $\mathsf{C_K}$ are the equations that are valid in all algebras in K, that is, they form the equational theory of K. As a consequence, there is no class K such that $\vDash = \vDash_{\mathsf{K}}$. ⊠

On the other hand, $\mathsf{C_K}$ does *not* satisfy the rule (Substitution), except when K is empty or when it consists only of trivial algebras, in which case $\mathsf{C_K}$ is the inconsistent consequence. This leads to an alternative proof of the previous fact. See Exercise 1.83 and, for a less artificial example, Exercise 1.84.

Moreover, $\mathsf{C_K}$ satisfies the first four Birkhoff's rules, and hence its theories are congruences of the algebra $\boldsymbol{Fm}$. Actually, this property *characterizes* the closure operators on $\boldsymbol{Fm}$ of the form $\mathsf{C_K}$ for some K:

THEOREM 1.71

Let C be a structural closure operator on Eq. The following conditions are equivalent:

(i) $\mathsf{C} = \mathsf{C_K}$ *for some class K of algebras.*

(ii) *C satisfies the first four Birkhoff's rules.*

(iii) *Every closed set or theory of C is a congruence of $\boldsymbol{Fm}$.*

PROOF: Left as Exercise 1.85. ⊠

Thus, one of the main differences between $\vDash$ and $\vDash_{\mathsf{K}}$ lie in the properties they satisfy concerning substitutions: the former satisfies the rule (Substitution) but is not substitution-invariant (structural), while the latter is substitution-invariant (structural) but does not satisfy the rule (except in trivial cases).

If $\mathsf{K}_1 \subseteq \mathsf{K}_2$, then obviously $\vDash_{\mathsf{K}_2} \subseteq \vDash_{\mathsf{K}_1}$. In particular, if A is the class of all algebras of a certain type, then $\vDash_{\mathsf{A}} \subseteq \vDash_{\mathsf{K}}$ for all classes K. By Theorem 1.71, the consequence $\vDash_{\mathsf{A}}$ can be axiomatized by the first four Birkhoff's rules, and its closed sets are exactly *all* congruences on $\boldsymbol{Fm}$. As a consequence, $\vDash_{\mathsf{A}} \subsetneq \vDash$. Notice that for arbitrary classes K there need not be a relation between $\vDash_{\mathsf{K}}$ and $\vDash$.

Observe that, while $\vDash$ is "absolute" and by contrast $\vDash_{\mathsf{K}}$ is "relative to K", the latter is *not* the "relativization" of the former to K; Exercise 1.86 shows that considering such a relativization is nonsense.

Quasivarieties and generalized quasivarieties

A ***quasi-equation*** is a first-order formula of the form $\left(\bigwedge_{i<n} \alpha_i \approx \beta_i\right) \to \varepsilon \approx \delta$ for some $n \in \omega$; it is understood that for $n = 0$ this expression denotes the equation $\varepsilon \approx \delta$; thus, equations are considered as particular cases of quasi-equations. Here the symbols $\wedge$ and $\to$ are the conjunction and implication of a first-order language, disjoint from $\boldsymbol{L}$, and satisfaction and validity of a quasi-equation in an algebra or class of algebras are defined in the usual model-theoretic sense (universal quantification is implicit), so that

$$\mathsf{K} \vDash \left(\bigwedge_{i<n} \alpha_i \approx \beta_i\right) \to \varepsilon \approx \delta \iff \{\alpha_i \approx \beta_i : i < n\} \vDash_{\mathsf{K}} \varepsilon \approx \delta.$$

A ***quasivariety*** can be defined as a class of algebras definable by a set of quasi-equations. An important theorem of Mal'cev shows that quasivarieties are the classes that are closed under $\mathbb{I}, \mathbb{S}, \mathbb{P}$ and $\mathbb{P}_{\mathrm{U}}$ and contain a trivial algebra (see Exercise 1.99 for a particularly simple proof); closure under $\mathbb{P}$ and $\mathbb{P}_{\mathrm{U}}$ can be jointly replaced by closure under $\mathbb{P}_{\mathrm{R}}$.

Thus, a quasivariety is a variety if and only if it is closed under $\mathbb{H}$, or equivalently if and only if it is closed under the construction of quotients; the usual way to show that a certain quasivariety is not a variety is to find an algebra in the class having a quotient outside the class. Quasivarieties that are not varieties are called ***proper***.

Obviously, for a class K, the least quasivariety containing K is the class $\mathbb{Q}\mathsf{K}$ of all algebras where all the quasi-equations that hold in K are valid. Mal'cev also proved that $\mathbb{Q}\mathsf{K} = \mathbb{ISPP}_{\mathrm{U}}(\mathsf{K} \cup \{\boldsymbol{A}_t\})$. If K is a finite set of finite algebras, then $\mathbb{Q}\mathsf{K} = \mathbb{ISP}(\mathsf{K} \cup \{\boldsymbol{A}_t\})$.

If K is a quasivariety presented by a set *Id* of equations and a set *QId* of quasi-equations, then a Hilbert-style presentation of $\vDash_{\mathsf{K}}$ is obtained by taking as *axioms* all equations in *Id* plus $x \approx x$, and as *rules of inference* those of (Symmetry), (Transitivity) and (Congruence), plus the quasi-equations in *QId* formulated as rules of inference between a finite set of equations and an equation.

If $\mathsf{K}' \subseteq \mathsf{K}$, then K' is called a ***relative sub(quasi)variety*** of K when the algebras in K' are the algebras in K where a certain set of (quasi-)equations hold. This notion is most usually applied when K is itself a quasivariety or a variety; then K' can be presented by adding (quasi-)equations to a presentation of K.

If one restricts attention to finitary logics, then quasivarieties are the natural framework to work with. Our more liberal definition of a logic needs the following infinitary extension of the above notions.

DEFINITION 1.72
A ***generalized quasi-equation*** *is an infinitary formula* $\left(\bigwedge_{i\in I} \alpha_i \approx \beta_i\right) \to \varepsilon \approx \delta$, *for some set* I. *Satisfaction and validity of infinitary formulas is defined in the obvious way,*

so that

$$\mathsf{K} \vDash \big(\bigwedge_{i\in I} \alpha_i \approx \beta_i\big) \to \varepsilon \approx \delta \iff \{\alpha_i \approx \beta_i : i \in I\} \vDash_{\mathsf{K}} \varepsilon \approx \delta .$$

A ***generalized quasivariety*** *is a class defined by generalized quasi-equations.*

For a class K, *the class* $\mathbb{GQ}\mathsf{K}$ *is the class of all algebras where all the generalized quasi-equations that hold in* K *are valid.*

Obviously, $\mathbb{GQ}\mathsf{K}$ is a generalized quasivariety, and indeed the least generalized quasivariety containing K. Moreover, by definition $\vDash_{\mathbb{GQ}\mathsf{K}} = \vDash_{\mathsf{K}}$. This class plays a key role in Chapter 3; its only properties we really need are:

THEOREM 1.73

1. *For any two classes* K *and* K', $\vDash_{\mathsf{K}'} = \vDash_{\mathsf{K}}$ *if and only if* $\mathbb{GQ}\mathsf{K} = \mathbb{GQ}\mathsf{K}'$.
2. *For each class* K, *the class* $\mathbb{GQ}\mathsf{K}$ *is the largest class* K', *and the only generalized quasivariety, such that* $\vDash_{\mathsf{K}'} = \vDash_{\mathsf{K}}$.
3. *Generalized quasivarieties are closed under the operators* $\mathbb{I}, \mathbb{S}$ *and* $\mathbb{P}$, *and contain all trivial algebras.*

PROOF: 1 is a direct consequence of Definition 1.72, and 2 follows directly from it. 3 is easily checked. ⊠

Since quasi-equations are particular cases of generalized quasi-equations, the quasi-equations that hold in K also hold in $\mathbb{GQ}\mathsf{K}$; thus, $\mathbb{GQ}\mathsf{K} \subseteq \mathbb{Q}\mathsf{K} \subseteq \mathbb{V}\mathsf{K}$. It is easy to show that for any class K, $\vDash_{\mathsf{K}}$ is finitary if and only if $\mathbb{GQ}\mathsf{K} = \mathbb{Q}\mathsf{K}$ (Exercise 1.87). Therefore, when $\vDash_{\mathsf{K}}$ is finitary, $\mathbb{Q}\mathsf{K}$ is the largest class K' such that $\vDash_{\mathsf{K}'} = \vDash_{\mathsf{K}}$. For an arbitrary class K, the operator $C_{\mathbb{Q}\mathsf{K}}$ is the finitary companion of C_{K}. As a consequence of Exercise 1.87, it is not difficult to show:

PROPOSITION 1.74

Let K *be a generalized quasivariety.*† *The following conditions are equivalent:*

(i) C_{K} *is finitary.*

(ii) K *is a quasivariety.*

(iii) K *is closed under* $\mathbb{P}_{\mathrm{U}}$. ⊠

Examples of generalized quasivarieties that are not closed under ultraproducts are found in Section 3.4, when dealing with finitarity issues around algebraizability; there, this result is important.

Notice that $\mathbb{V}\mathsf{K} = \mathbb{Q}(\mathbb{V}\mathsf{K})$, because every variety is closed under $\mathbb{I}, \mathbb{S}, \mathbb{P}$ and $\mathbb{P}_{\mathrm{U}}$. This, however, does not mean that $\mathbb{V}\mathsf{K} = \mathbb{Q}\mathsf{K}$, which turns out to be a quite different, and far less trivial issue.

Relative congruences

Recall (page 3) that a congruence $\theta \in \mathrm{Con}\boldsymbol{A}$ is a ***congruence of*** $\boldsymbol{A}$ ***relative to the class*** K, or K-***congruence*** for short, when $\boldsymbol{A}/\theta \in \mathsf{K}$; the set $\mathrm{Con}_{\mathsf{K}}\boldsymbol{A}$ of all K-congruences of $\boldsymbol{A}$ is ordered under the subset relation, but in general it need not be a sublattice of $\mathrm{Con}\boldsymbol{A}$. However:

† Actually, it is enough to assume that K is closed under $\mathbb{I}, \mathbb{S}$ and $\mathbb{P}$.

PROPOSITION 1.75

Let K *be a class of algebras closed under* $\mathbb{I}, \mathbb{S}$ *and* $\mathbb{P}$*, and containing a trivial algebra.*

1. *For every algebra* $\boldsymbol{A}$*, the set* $\mathrm{Con}_{\mathsf{K}}\boldsymbol{A}$ *is a closure system, and hence a complete lattice.*
2. *The lattice* $\mathrm{Con}_{\mathsf{K}}\boldsymbol{Fm}$ *is the lattice of theories of the operator* C_{K}.

PROOF: Left as Exercise 1.88. ⊠

The most crucial consequence of this result is that, under its assumptions, the smallest K-congruence of an algebra $\boldsymbol{A}$ containing a specified subset $X \subseteq A \times A$ always exists; it is denoted by $\boldsymbol{\Theta}_{\mathsf{K}}^{\boldsymbol{A}} X$. This result is relevant for us when applied to quasivarieties and generalized quasivarieties (as they satisfy the assumption). For a refinement of point 1, see Exercise 1.89. A generalization of point 2 to arbitrary algebras is the following important result, which is used later on:

THEOREM 1.76

Let K *be a generalized quasivariety, let* $\boldsymbol{A}$ *be any algebra, and let* $R \subseteq A \times A$*. Then* $R \in \mathrm{Con}_{\mathsf{K}}\boldsymbol{A}$ *if and only if it is a "closed set" under the consequence* $\vDash_{\mathsf{K}}$ *in the following sense: For all* $\Theta \cup \{\varepsilon \approx \delta\} \subseteq Eq$ *such that* $\Theta \vDash_{\mathsf{K}} \varepsilon \approx \delta$*, if* $h \in \mathrm{Hom}(\boldsymbol{Fm}, \boldsymbol{A})$ *is such that* $h\Theta \subseteq R$*, then* $\langle h\varepsilon, h\delta \rangle \in R$*.*

PROOF: Left as Exercise 1.90; it does not use any real universal algebra. ⊠

Exercise 1.92 contains a couple of general properties of relative congruence-generation, which are used at some point in the book.

The operator $\mathbb{U}$

This subsection is mainly informative, as its results are seldom used; however, doing the exercises may be instructive.

Like quasivarieties, the class $\mathbb{GQ}\mathsf{K}$ can be described as the result of applying certain class operators to K, such that generalized quasivarieties are precisely the classes closed under them. In this matter, generalized quasivarieties are sometimes confused† with ***prevarieties***, which are the classes defined by generalized quasi-equations with an unbounded number of variables (for this, a *proper class* of variables is needed, but still the conjunctions of the involved generalized quasi-equations should be made over infinite *sets* of equations; see [144, Chapter 9]). Prevarieties are the classes that are closed under $\mathbb{I}, \mathbb{S}$ and $\mathbb{P}$ (and contain a trivial algebra), and accordinginly they are also called "S-P-classes". Since our construction of the formula algebra assumes a *fixed set of variables* V, it is necessary to consider one more‡ operator: for any class K of algebras,

$$\mathbb{U}(\mathsf{K}) := \{\boldsymbol{A} : \text{If } \boldsymbol{B} \leqslant \boldsymbol{A} \text{ and } \boldsymbol{B} \text{ is generated by a set of cardinality} \leqslant |V|, \\ \text{then } \boldsymbol{B} \in \mathsf{K}\}.$$

† Generalized quasivarieties are also called "σ-quasivarieties" in the literature, while the term "implicational" (or "implicative") class has been used for ordinary quasivarieties, for generalized quasivarieties, and for prevarieties.

‡ The operator $\mathbb{U}$ was introduced in [29]. An alternative strategy to characterize σ-quasivarieties, mentioned without proof by Czelakowski in [70, p. 188], is to use σ-filtered products.

The real reason why this construction arises here is that whenever we consider an evaluation $h \in \mathrm{Hom}(\boldsymbol{Fm}, \boldsymbol{A})$, its range is a subalgebra of $\boldsymbol{A}$ generated by a set of cardinality $\leqslant |V|$. As a consequence, the operator $\mathbb{U}$ is easily seen to preserve the validity of generalized quasi-equations (Exercise 1.93).

We will see that $\mathbb{GQ}\mathsf{K} = \mathbb{UISP}(\mathsf{K} \cup \{\boldsymbol{A}_t\})$, and that generalized quasivarieties are the classes closed under these operators and containing a trivial algebra. Observe that $\mathbb{U}$ is not an operator of the "construction" kind, and the fact that $\mathbb{ISP}(\mathsf{K} \cup \{\boldsymbol{A}_t\}) \subseteq \mathbb{UISP}(\mathsf{K} \cup \{\boldsymbol{A}_t\})$, which is indeed true (Exercise 1.94), is not a direct consequence of "closing under one more construction" (because it is not true in general that $\mathsf{K} \subseteq \mathbb{U}\mathsf{K}$). Anyway: that $\boldsymbol{A} \in \mathbb{UISP}(\mathsf{K} \cup \{\boldsymbol{A}_t\})$ means that every countably[†] generated subalgebra of $\boldsymbol{A}$ can be embedded into a product of a family of algebras from K, or else it is the trivial algebra. The main properties of this construction are summarized in the following result:

THEOREM 1.77

For every class K *and every algebra* $\boldsymbol{A}$:

1. *If* $\boldsymbol{A} \in \mathbb{UISP}(\mathsf{K} \cup \{\boldsymbol{A}_t\})$, *then* $\vDash_{\mathsf{K}} \subseteq \vDash_{\boldsymbol{A}}$.
2. *If* $\mathsf{K} \subseteq \mathsf{K}' \subseteq \mathbb{UISP}(\mathsf{K} \cup \{\boldsymbol{A}_t\})$, *then* $\vDash_{\mathsf{K}'} = \vDash_{\mathsf{K}}$.
3. *If* θ *is a theory of* C_{K}, *then* $\boldsymbol{Fm}/\theta \in \mathbb{ISP}(\mathsf{K} \cup \{\boldsymbol{A}_t\})$.
4. $\mathbb{GQ}\mathsf{K} = \mathbb{UISP}(\mathsf{K} \cup \{\boldsymbol{A}_t\})$. *That is,* $\mathbb{UISP}(\mathsf{K} \cup \{\boldsymbol{A}_t\})$ *is the class axiomatized by the generalized quasi-equations valid in* K *(hence, it is a generalized quasivariety). In other words,* $\boldsymbol{A} \in \mathbb{UISP}(\mathsf{K} \cup \{\boldsymbol{A}_t\})$ *if and only if* $\vDash_{\mathsf{K}} \subseteq \vDash_{\boldsymbol{A}}$.
5. K *is a generalized quasivariety if and only if* $\mathbb{UISP}(\mathsf{K} \cup \{\boldsymbol{A}_t\}) = \mathsf{K}$.
6. K *is a generalized quasivariety if and only if it is closed under* $\mathbb{U}, \mathbb{I}, \mathbb{S}$ *and* $\mathbb{P}$ *and contains a trivial algebra.*

PROOF: 1 is left as Exercise 1.96, and 2 follows trivially from it. 3 is left as Exercise 1.97, and 4 as Exercise 1.98. 5 follows from 4, because a class K is a generalized quasivariety if and only if $\mathsf{K} = \mathbb{GQ}\mathsf{K}$. Finally, the direct implication in 6 is proved by checking that generalized quasi-equations are preserved by these operators; after Theorem 1.73, only the case of $\mathbb{U}$ remains to be checked, and it is left to you as Exercise 1.93. The converse implication is a trivial consequence of 5. ⊠

For an interesting application of this result, see Exercise 1.99. The fact that generalized quasivarieties are closed under $\mathbb{U}$ is used in Exercise 3.7.

Exercises for Section 1.6

EXERCISE 1.81. Show that $x \approx y \vDash x \approx z$ while $x \approx x \nvDash x \approx z$, and use this fact to show that $\vDash$ is not structural.

[†] Recall that we assume that V is denumerable. Disregarding the above comment, all the contents of this subsection holds for a set V of arbitrary infinite cardinality, in which case the above comment would read "every κ-generated subalgebra, for some cardinal $\kappa \leqslant |V|$, can be [...]".

Exercise 1.82. Use Lemma 1.2 to show that if θ is a fully invariant congruence of $\boldsymbol{Fm}$, then $\boldsymbol{Fm}/\theta \vDash \alpha \approx \beta$ if and only if $\alpha \equiv \beta\ (\theta)$, and deduce from this that the variety generated by $\boldsymbol{Fm}/\theta$ is axiomatized by the equations in θ, and $\boldsymbol{Fm}/\theta$ is a free algebra in this variety, a set of generators being $\{x/\theta : x \in V\}$.

Exercise 1.83.

1. Prove that if K is empty or contains only trivial algebras, then $\vDash_{\mathsf{K}}$ is inconsistent and hence satisfies the (Substitution) rule.
2. Assume that K contains at least one non-trivial algebra. Show that then $x \approx y \nvDash_{\mathsf{K}} x \approx z$, and use this fact to show that $\vDash_{\mathsf{K}}$ does not satisfy the (Substitution) rule.

Conclude that $\vDash \neq \vDash_{\mathsf{K}}$ for all classes K.

Exercise 1.84. Assume that the similarity type contains a constant $\top$ and a unary connective $\Box$. Prove that $x \approx \top \vDash \Box x \approx \top$ and characterize the classes K such that $x \approx \top \vDash_{\mathsf{K}} \Box x \approx \top$.

Comment. If you know something about algebraic models of normal modal logics, you can connect this issue with the Necessitation Rule $\varphi \vdash \Box\varphi$ of modal logic.

Exercise 1.85. Let C be a structural closure operator on Eq. Prove that the following conditions are equivalent:

(i) $\mathsf{C} = \mathsf{C}_{\mathsf{K}}$ for some class K of algebras.

(ii) C satisfies the first four Birkhoff's rules.

(iii) Every closed set or theory of C is a congruence of $\boldsymbol{Fm}$: $\mathscr{C} \subseteq \mathrm{Con}\boldsymbol{Fm}$.

Hint. The only non-trivial step is (iii)$\Rightarrow$(i); for this, take $\mathsf{K} := \{\boldsymbol{Fm}/\theta : \theta \in \mathscr{C}\}$.

Exercise 1.86. For any class K of algebras, consider the relation obtained by replacing the quantifier "for every algebra $\boldsymbol{A}$" in (4) on page 28 by "for every $\boldsymbol{A} \in \mathsf{K}$". Show that the resulting relation is a closure relation that is an extension both of $\vDash$ and of $\vDash_{\mathsf{K}}$, and that its theorems are the equations valid in K.

Exercise 1.87. Prove that for any class of algebras K, the following conditions are equivalent:

(i) C_{K} is finitary.

(ii) $\mathbb{GQ}\mathsf{K} = \mathbb{Q}\mathsf{K}$.

(iii) $\mathbb{GQ}\mathsf{K}$ is actually a quasivariety.

Exercise 1.88. Let K be a class of algebras closed under $\mathbb{I}, \mathbb{S}$ and $\mathbb{P}$ and containing a trivial algebra. Prove that for every algebra $\boldsymbol{A}$, the ordered set $\mathrm{Con}_{\mathsf{K}}\boldsymbol{A}$ is a closure system, and hence a complete lattice, and that in the case of the formula algebra, the lattice $\mathrm{Con}_{\mathsf{K}}\boldsymbol{Fm}$ is the lattice of theories of the operator C_{K}.

Exercise 1.89. Check that the first statement of Exercise 1.88 can be proved by just requiring that K be closed under $\mathbb{P}_{\mathrm{SD}}$ (subdirect products) and contain a trivial algebra. Prove that this condition is almost necessary in the following sense: If K is a class closed under isomorphisms and with a trivial algebra, then K is closed under $\mathbb{P}_{\mathrm{SD}}$ if and only if for every algebra $\boldsymbol{A}$, the ordered set $\mathrm{Con}_{\mathsf{K}}\boldsymbol{A}$ is a complete lattice.

Exercise 1.90. Let K be a generalized quasivariety, $\boldsymbol{A}$ any algebra, and $R \subseteq A \times A$. Prove that $R \in \mathrm{Con}_{\mathsf{K}}\boldsymbol{A}$ if and only if for all $\Theta \cup \{\varepsilon \approx \delta\} \subseteq Eq$ such that $\Theta \vDash_{\mathsf{K}} \varepsilon \approx \delta$ and all $h \in \mathrm{Hom}(\boldsymbol{Fm}, \boldsymbol{A})$ such that $h\Theta \subseteq R$, then $\langle h\varepsilon, h\delta \rangle \in R$.

Comment. Observe that when $\boldsymbol{A}$ is $\boldsymbol{Fm}$ this exercise states the same as the last part of Exercise 1.88, but there the assumption on the class K is weaker.

Exercise 1.91. Use Exercise 1.90 to show that if K is a quasivariety, then for every algebra $\boldsymbol{A}$, the set $\mathrm{Con}_{\mathsf{K}}\boldsymbol{A}$ is an inductive closure system, and hence an algebraic lattice.

Exercise 1.92. Let K be a generalized quasivariety, let $\boldsymbol{A}, \boldsymbol{B}$ be arbitrary algebras, and let $h\colon \boldsymbol{A} \to \boldsymbol{B}$. Prove:

1. If $\theta \in \mathrm{Con}_{\mathsf{K}}\boldsymbol{B}$, then $h^{-1}\theta \in \mathrm{Con}_{\mathsf{K}}\boldsymbol{A}$.
2. For any $\theta \subseteq A \times A$, $\boldsymbol{\Theta}^{\boldsymbol{B}}_{\mathsf{K}} h\theta = \boldsymbol{\Theta}^{\boldsymbol{B}}_{\mathsf{K}} h \boldsymbol{\Theta}^{\boldsymbol{A}}_{\mathsf{K}} \theta$.

Exercise 1.93. Prove that if a generalized quasi-equation holds in a class K, then it holds in the class $\mathbb{U}\mathsf{K}$.

Exercise 1.94. Observe that if K is closed under $\mathbb{S}$, then $\mathsf{K} \subseteq \mathbb{U}\mathsf{K}$, but in general this is not the case. Use the first property to prove that $\mathbb{ISP}(\mathsf{K} \cup \{\boldsymbol{A}_t\}) \subseteq \mathbb{GQ}\mathsf{K}$ for any class K.

Exercise 1.95. Prove that if K is a class of algebras closed under $\mathbb{I}, \mathbb{S}$ and $\mathbb{P}_{\mathrm{U}}$, then it is also closed under $\mathbb{U}$.

Hint. You will need Theorem v.2.14 of [51].

Exercise 1.96. Let $\boldsymbol{A} \in \mathbb{UISP}(\mathsf{K} \cup \{\boldsymbol{A}_t\})$. Prove that $\vDash_{\mathsf{K}} \subseteq \vDash_{\boldsymbol{A}}$.

Exercise 1.97. Let θ be a theory of C_{K}. By Exercise 1.85, $\theta \in \mathrm{Con}\boldsymbol{Fm}$. Prove that $\boldsymbol{Fm}/\theta \in \mathbb{ISP}(\mathsf{K} \cup \{\boldsymbol{A}_t\})$.

Exercise 1.98. Use the result in Exercise 1.97 to show that $\boldsymbol{A} \in \mathbb{UISP}(\mathsf{K} \cup \{\boldsymbol{A}_t\})$ whenever $\vDash_{\mathsf{K}} \subseteq \vDash_{\boldsymbol{A}}$.

Exercise 1.99. Use the characterization of generalized quasivarieties in Theorem 1.77.6 to give a proof of Mal'cev's Theorem: a class of algebras is a quasivariety if and only if it closed under $\mathbb{I}, \mathbb{S}, \mathbb{P}$ and $\mathbb{P}_{\mathrm{U}}$ and contains a trivial algebra.

Hint. ($\Rightarrow$) is shown by routinely checking that quasi-equations are preserved by all these operators and hold in all trivial algebras. To prove ($\Leftarrow$) use Exercise 1.95, and then Proposition 1.74.

Chapter 2

The first steps in the algebraic study of a logic

Section 2.1 describes, in a rather loose way, the process that leads to establishing the links between classical logic and the class of Boolean algebras. It is historically the first one of its kind, and moreover it can be done in a way that shows the possibilities for its generalization; these lead naturally to the notion of an *implicative logic* due to Rasiowa, which is presented and its algebraization studied in Section 2.2. As a natural offspring of this discussion the general notion of a *filter* of a logic emerges; its basic properties are studied in Section 2.3, a section you should not miss, even if your background allows you to skip the discussion around classical logic, Boolean algebras, and even that on implicative logics. Section 2.4 describes, at an informal level, several directions along which the standard procedures of this chapter can be generalized; this constitutes, to some extent, a first overview of the contents of some of the more advanced chapters in the book. Finally, Section 2.5 gathers two digressions concerning first-order logic; the first one might even be taken as a motivating start to the chapter.

2.1 From two-valued truth tables to Boolean algebras: the Lindenbaum-Tarski process for classical logic

Example 1.11 presented classical logic $\mathcal{Cl}$ from a semantic standpoint. You know that this logic can be presented proof-theoretically in several ways, among them a Hilbert-style calculus or axiomatic system. Let us denote the resulting syntactic consequence (Example 1.9) by $\vdash_{\mathcal{Cl}}$. Since this is not a specific book on this logic, we are not concerned by any particular presentation; all we need to know here is that the following formulas are theorems of $\vdash_{\mathcal{Cl}}$:

(Ax1)	$x \to (y \to x)$
(Ax2)	$(x \to (y \to z)) \to ((x \to y) \to (x \to z))$
(Ax3)	$x \to x \vee y$
(Ax4)	$y \to x \vee y$
(Ax5)	$(x \to z) \to ((y \to z) \to (x \vee y \to z))$
(Ax6)	$x \wedge y \to x$
(Ax7)	$x \wedge y \to y$
(Ax8)	$(x \to y) \to ((x \to z) \to (x \to y \wedge z))$
(Ax9)	$(x \to \neg y) \to (y \to \neg x)$
(Ax10)	$\neg(x \to x) \to y$
(Ax11)	$x \vee \neg x$

I assume the chosen axiomatic system has the rule of *Modus Ponens* as its only proper inference rule:

$$x \,,\, x \to y \;\rhd\; y. \tag{MP}$$

The fact that the above list of formulas (Axn), together with the rule (MP), actually constitute an axiomatization† of the logic $\mathcal{C\ell}$ is the so-called COMPLETENESS THEOREM, that is, the statement that for all $\Gamma \cup \{\varphi\} \subseteq Fm$,

$$\Gamma \vdash_{\mathcal{C\ell}} \varphi \iff \Gamma \vdash_{\mathbf{2}} \varphi. \tag{12}$$

Let us review the *standard completeness proof* found in (almost) every elementary logic textbook:

($\Rightarrow$) This part, the *Soundness Theorem*, is proved by what is normally qualified as *routine checking*; that is, by just checking, for an arbitrary $h \in \mathrm{Hom}(\boldsymbol{Fm}, \mathbf{2})$, that $h\varphi = 1$ for all axioms φ of your chosen axiomatization, and that (MP) preserves the property of "being equal to 1": If $h\varphi = 1$ and $h(\varphi \to \psi) = 1$, then $h\psi = 1$; this amounts to checking that in $\mathbf{2}$, if $1 \to a = 1$, then $a = 1$, for all $a \in 2$, and this is obvious from the truth-table of $\to$. Then, induction on the length of proofs completes the proof.

($\Leftarrow$) This is proved by contraposition: One assumes that $\Gamma \nvdash_{\mathcal{C\ell}} \varphi$, and one shows that $\Gamma \nvdash_{\mathbf{2}} \varphi$ by constructing a particular $h \in \mathrm{Hom}(\boldsymbol{Fm}, \mathbf{2})$ that satisfies $h\Gamma \subseteq \{1\}$ and $h\varphi \neq 1$, that is, $h\varphi = 0$. This is done in a two-step process:

(a) Starting from the assumption that $\Gamma \nvdash_{\mathcal{C\ell}} \varphi$ one shows that there exists a maximally consistent theory Γ' of $\mathcal{C\ell}$ such that $\Gamma \subseteq \Gamma'$ and $\varphi \notin \Gamma'$. This is traditionally called the *Lindenbaum's Lemma*; you can check that it is a particular case of Corollary 1.44. See also Exercise 2.1.

(b) Given this Γ', one defines a function $h \colon Fm \to 2$ by putting, for any $\gamma \in Fm$, $h\gamma = 1$ if and only if $\gamma \in \Gamma'$ (and hence $h\gamma = 0$ if and only if $\gamma \notin \Gamma'$). Then

† The reason for choosing precisely this axiomatization here becomes clear on pages 84–85, in relation with Table 2.

the important task is to show that this is actually a *homomorphism*; to this end, in contrast to the previous step, many particular properties of maximally consistent theories of classical logic have to be used in an essential way.†

By its own definition and the construction of Γ', it follows that $h\Gamma \subseteq \{1\}$ while $h\varphi = 0$, as demanded.

In this proof, the logical apparatus seems somehow superfluous; one might delete the terminology of homomorphisms and the like, and speak only of evaluations into the set 2 equipped with its truth-tables. However, a closer and non-standard view of step (b) reveals it hides more algebraic facts than it shows; this more algebrac view facilitates the introduction of a variant of step (a) that leads directly to the *completeness with respect to the class of all Boolean algebras*.

First, once Γ' is obtained, instead of considering directly the definition of the evaluation h into **2**, consider the following binary relation between formulas, which is denoted by $\boldsymbol{\Omega}\Gamma'$ for reasons that become clear later on:

$$\alpha \equiv \beta \ (\boldsymbol{\Omega}\Gamma') \overset{\text{def}}{\iff} \text{ either } \alpha,\beta \in \Gamma' \text{ or } \alpha,\beta \notin \Gamma'. \tag{13}$$

This is the equivalence relation that partitions the set of formulas in two classes, Γ' and $Fm \smallsetminus \Gamma'$. Therefore, the quotient set can be identified with $2 = \{1,0\}$, so that the function h in (b) above becomes the canonical projection from Fm onto 2 (and $\boldsymbol{\Omega}\Gamma'$ is its kernel). The properties of maximally consistent sets in $\mathcal{Cl}$ can be used to show that $\boldsymbol{\Omega}\Gamma'$ is a congruence of the formula algebra, so that the quotient algebra can be identified with **2**, and that h becomes a homomorphism. This is still a too particular construction, but it already uncovers *the core construction* under the completeness proof, namely, that of *factoring out the formula algebra by a suitable congruence*.

There is an alternative definition of $\boldsymbol{\Omega}\Gamma'$ of a higher interest:

$$\begin{aligned} \alpha \equiv \beta \ (\boldsymbol{\Omega}\Gamma') &\overset{\text{def}}{\iff} \alpha \leftrightarrow \beta \in \Gamma' \\ &\overset{\text{def}}{\iff} \alpha \to \beta, \beta \to \alpha \in \Gamma'. \end{aligned} \tag{14}$$

The second equivalence follows from elementary properties of conjunction, and holds for an arbitrary theory Γ'. It is not difficult to see (Exercise 2.2) that the definitions in (13) and (14) are actually equivalent; the equivalence again requires properties of maximally consistent theories of $\mathcal{Cl}$. The interesting thing is that the proof that $\boldsymbol{\Omega}\Gamma'$ is a congruence of $\boldsymbol{Fm}$, starting from (14), requires significantly fewer particular properties of $\mathcal{Cl}$; as detailed in Section 2.2, it requires a few properties of $\leftrightarrow$ (or of $\to$) that are present in a large number of logics, and moreover it can be done for arbitrary theories. So, (14) can be used in many other cases, and still a quotient algebra $\boldsymbol{Fm}/\boldsymbol{\Omega}\Gamma'$ is obtained such that the canonical projection h is a homomorphism.

---

† For instance: If Γ' is a maximally consistent theory, then $\alpha \vee \beta \in \Gamma'$ if and only if $\alpha \in \Gamma'$ or $\beta \in \Gamma'$. Given the definition of h, this is to say that $h(\alpha \vee \beta) = 1$ if and only if $h\alpha = 1$ or $h\beta = 1$, and given the truth-table of $\vee$, this implies that $h(\alpha \vee \beta) = h\alpha \vee h\beta$; that is, this makes h a homomorphism with respect to disjunction. And so on.

Now observe what happens when the extension performed from Γ to Γ' in step (a) above is replaced by the most trivial one, that is, when Γ' is taken to be just the theory of $\mathcal{C}\ell$ generated by the set of formulas Γ. This extension is enough to make the so-called ***Lindenbaum-Tarski process*** possible.[†] This is what we see if we review its steps more closely:

(LT1) One starts from the assumption that $\Gamma \nvdash_{\mathcal{C}\ell} \varphi$, and one considers the theory Γ' of $\mathcal{C}\ell$ generated by Γ. Thus, $\Gamma \subseteq \Gamma'$ while $\varphi \notin \Gamma'$.

(LT2) One defines the relation $\boldsymbol{\Omega}\Gamma'$ as in (14).

(LT3) One shows that $\boldsymbol{\Omega}\Gamma'$ is a congruence of $\boldsymbol{Fm}$. This requires using specific axioms of $\mathcal{C}\ell$ such as the ones in the above list and other theorems, plus (MP); for instance, that $\vdash_{\mathcal{C}\ell} (x \to y) \to (\neg y \to \neg x)$ is used to show that $\boldsymbol{\Omega}\Gamma'$ is a congruence with respect to $\neg$.

(LT4) One shows that $\boldsymbol{Fm}/\boldsymbol{\Omega}\Gamma'$ is a Boolean algebra. There are several ways of doing this; one would be to use previous knowledge that BA is a variety, and that $\vdash_{\mathcal{C}\ell} \alpha \leftrightarrow \beta$ for all equations $\alpha \approx \beta$ in one of its equational presentations (some of the axioms and theorems highlighted before give some of these).

(LT5) One shows that for any $\alpha \in Fm$, $\alpha \in \Gamma'$ if and only if $\alpha/\boldsymbol{\Omega}\Gamma' \in \Gamma'/\boldsymbol{\Omega}\Gamma'$. It is easy to see that what is required for this to hold is just that if $\beta \in \Gamma'$ and $\beta \equiv \alpha\ (\boldsymbol{\Omega}\Gamma')$, then $\alpha \in \Gamma'$, and this in turn follows from (MP), given the definition of $\boldsymbol{\Omega}\Gamma'$ in (LT2).

(LT6) Finally, one shows that all formulas in Γ' are mutually equivalent, therefore Γ' constitutes a single element in the quotient algebra, *and* this element is the top element 1 in the Boolean algebra $\boldsymbol{Fm}/\boldsymbol{\Omega}\Gamma'$. Both facts follow by applying (MP) to the formula $\alpha \to (\beta \to \alpha)$, a substitution instance of (Ax1), and the fact that in a Boolean algebra, $a \leqslant b \iff a \to b = 1$.

After this, the property in (LT5) becomes: For any $\alpha \in Fm$, $\alpha \in \Gamma'$ if and only if $\alpha/\boldsymbol{\Omega}\Gamma' = 1$.

The "process" is now finished. The quotient algebra $\boldsymbol{Fm}/\boldsymbol{\Omega}\Gamma'$ is a Boolean algebra, by (LT4), and the canonical projection $\pi \in \mathrm{Hom}(\boldsymbol{Fm}, \boldsymbol{Fm}/\boldsymbol{\Omega}\Gamma')$ defined by $\pi(\alpha) := \alpha/\boldsymbol{\Omega}\Gamma'$ for all $\alpha \in Fm$ is a homomorphism such that, by (LT1) and (LT6), $\pi(\Gamma) \subseteq \pi(\Gamma') = \{1\}$ while $\pi(\varphi) \neq 1$.

We can use BA, the class of Boolean algebras, to set up a semantic consequence relation $\vDash_{\mathsf{BA}}$ after *designating* 1 as representing *truth* in each Boolean algebra:

$$\Gamma \vDash_{\mathsf{BA}} \varphi \overset{\text{def}}{\iff} h\Gamma \subseteq \{1\} \text{ implies } h\varphi = 1, \\ \text{for all } h \in \mathrm{Hom}(\boldsymbol{Fm}, \boldsymbol{A}) \text{ and all } \boldsymbol{A} \in \mathsf{BA}.$$

[†] Its original formulation by Lindenbaum referred to the particular case in which $\Gamma = \emptyset$, that is, to the proof of the *weak* completeness theorem. Moreover, there has been a certain debate in the literature about whether the step of factoring the formula algebra and obtaining a Boolean algebra in the quotient should be attributed only to Tarski, or to both scholars; what seems clear is that the idea of considering the set of formulas as an abstract algebra is Lindenbaum's. See [94].

Then the above process shows by contraposition that $\vdash_{\mathsf{BA}} \subseteq \vdash_{\mathcal{C}\ell}$. This provides a proof of part ($\Leftarrow$) of the following ALGEBRAIC COMPLETENESS THEOREM:

$$\Gamma \vdash_{\mathcal{C}\ell} \varphi \iff \Gamma \vdash_{\mathsf{BA}} \varphi. \tag{15}$$

As in the proof of (12), that of the ($\Rightarrow$) half would be done by checking that for an arbitrary $\boldsymbol{A} \in \mathsf{BA}$ and an arbitrary $h \in \mathrm{Hom}(\boldsymbol{Fm}, \boldsymbol{A})$, $h\varphi = 1$ for every axiom φ, and that *Modus Ponens* preserves the property of "being equal to 1". This part (also called the *Soundness Theorem*) requires using that $\boldsymbol{A}$ is a Boolean algebra.

As becomes clear in Section 2.2, the Lindenbaum-Tarski process can be applied to a very large number of logics; there is a general definition of the class of algebras that plays the role of BA, but in each case the properties of these algebras produce properties of the logic, among them *enhanced completeness theorems*; that is, completeness theorems with respect to restricted classes of algebras having a more definite structure such as algebras of sets or algebras of a specific kind of sets in a topological space.

Still in the case of $\mathcal{C}\ell$, some observations or consequences of the above process can be highlighted.

1. Using algebraic properties of Boolean algebras, the standard completeness (12) with respect to $\vdash_{\mathbf{2}}$ can be obtained from the Algebraic Completeness Theorem (15). One needs just to show that $\vdash_{\mathbf{2}} = \vdash_{\mathsf{BA}}$. Since $\mathbf{2} \in \mathsf{BA}$, it is clear that $\vdash_{\mathsf{BA}} \subseteq \vdash_{\mathbf{2}}$. The reverse inclusion is proved by contraposition: If $\Gamma \nvdash_{\mathsf{BA}} \varphi$, there is some $\boldsymbol{A} \in \mathsf{BA}$ and some $h \in \mathrm{Hom}(\boldsymbol{Fm}, \boldsymbol{A})$ such that $h\Gamma \subseteq \{1\}$ but $h\varphi \neq 1$. It is well known that if $\boldsymbol{A} \in \mathsf{BA}$ and $a \in A$ with $a \neq 1$, then there is an ultrafilter U of $\boldsymbol{A}$ such that $a \notin U$; and it is also well known that $\boldsymbol{A}/U \cong \mathbf{2}$. Applying this to $a := h\varphi$ and denoting by π the canonical projection from $\boldsymbol{A}$ onto $\boldsymbol{A}/U \cong \mathbf{2}$, then πh is the required homomorphism that witnesses that $\Gamma \nvdash_{\mathbf{2}} \varphi$. See Exercise 2.3.
2. Any *representation theorem* of Boolean algebras in terms of algebras in a particular subclass yields a corresponding completeness theorem for the smaller class (for instance, fields of sets, or algebras of clopen sets of Stone spaces). If $\mathsf{K} \subseteq \mathsf{BA}$, then $\vdash_{\mathcal{C}\ell} = \vdash_{\mathsf{BA}} \subseteq \vdash_{\mathsf{K}}$. By a representation theorem I mean that for each $\boldsymbol{A} \in \mathsf{BA}$ there is some $\boldsymbol{A}' \in \mathsf{K}$ such that $\boldsymbol{A}$ can be embedded in $\boldsymbol{A}'$. Using this one proves that if $\Gamma \nvdash_{\mathsf{BA}} \varphi$, then also $\Gamma \nvdash_{\mathsf{K}} \varphi$, which completes the proof that $\vdash_{\mathcal{C}\ell} = \vdash_{\mathsf{K}}$. There are more complicated representations (such as embeddings into products of algebras), and they would produce other completeness theorems.

   Traditional algebraic logic is particularly rich in representation theorems; Makinson argues in [169] that their relevance is at the same level as that of completeness theorems.
3. Observe that step (a) above can be rephrased as saying that the family of all maximally consistent theories is *a base* of the closure system of the theories of $\mathcal{C}\ell$. Thus, a similar extension $\Gamma \mapsto \Gamma'$ can be performed each time a special base of the closure system of the theories of a particular logic is identified,

so that the theory Γ' has some special properties, and the quotient algebra $\boldsymbol{Fm}/\Omega\Gamma'$ is of a particular kind. The combination of this extension process with the Lindenbaum-Tarski process enables to obtain *completeness theorems with respect to restricted classes of algebras*, relevant for the logic under study. Examples of this strategy are the completeness of the basic fuzzy logic $\mathcal{BL}$ with respect to linear BL-algebras in [132], and its generalization to the so-called *implicational semilinear logics* in [61, Section 4], which include all fuzzy logics† in the ordinary sense.

The conclusion of these examples is that the *logical work* in the algebra of formulas required for step (a) above may sometimes be replaced, to a certain extent, by *purely algebraic work* in Boolean algebras. In this case both are very similar, in particular because there is a completeness with respect to the single algebra **2**; but the analogous situation in other logics may make the algebraic work more intuitive than the logical one. The study of Boolean algebras has progressed in parallel to that of classical logic, and many known facts in one of these theories have a parallel in the other. For other logics, however, their algebraic study has begun much later than the study of the algebras themselves (which may have arisen for other reasons), and hence working with the algebras as a semantics of a given logic may provide more sophisticated tools.

Exercises for Section 2.1

EXERCISE 2.1. Find a proof in some logic textbook of the fact that if $\Gamma \nvdash_{\mathcal{C}\ell} \varphi$, then there exists a maximally consistent theory of $\mathcal{C}\ell$, Γ', such that $\Gamma \subseteq \Gamma'$ while $\varphi \notin \Gamma'$. Compare it with the proof obtained using Corollary 1.44 (or using Lemma 1.43).

EXERCISE 2.2. Let Γ' be any maximally consistent theory of $\mathcal{C}\ell$. Show that the two definitions of the relation $\Omega\Gamma'$ given in (13) and (14) on page 73 are in fact equivalent. Identify the properties (axioms, theorems) of $\mathcal{C}\ell$ used in the proof. Observe that one of the implications between the two definitions actually holds for an arbitrary theory.

EXERCISE 2.3. If you are proficient with Boolean algebras, complete the details of the argument in point 1 on page 75 and provide examples of the completeness theorems mentioned in point 2.

EXERCISE 2.4. Prove the points in the steps (LT3)–(LT6) of the Lindenbaum-Tarski process that you do not find obvious. If you are acquainted with lattices and similar structures, try to use as few properties of negation as possible.

2.2 Implicative logics

This section shows how the Lindenbaum-Tarski process described in the previous section can be based entirely on very general properties of implication—except, of course, step (LT4) which concerns the particular kind of algebras to be obtained.

† As argued by Běhounek and Cintula in [20], completeness with respect to a class of chains (linearly ordered algebras) seems to be a key feature of fuzzy logics.

This leads naturally to the introduction of *implicative logics*, which were defined and studied by Rasiowa in her famous book [208] as the culmination of 25 years of studies† of the algebras that appear as the algebraic counterpart of a family of logics conforming to the same algebraization pattern. The idea of this section is to review what she did, adding some results and modern readings of facts actually known to her (or almost so).

Actually, the exposition slightly deviates from Rasiowa's setting in two inessential points, highlighted at the beginning of Section 2.4. There, it is also explored what happens when admitting some more substantial deviations but still yielding logics with a similar algebraic character.

Throughout this section it is assumed that $\boldsymbol{L}$ is an algebraic language that has $\to$ as a binary connective, either among its primitive ones, or as a defined one by an algebraic term in exactly two variables.

DEFINITION 2.1

Let $\mathcal{L}$ be a logic of type $\boldsymbol{L}$, and let Γ be a theory of $\mathcal{L}$. The binary relations $\leqslant_\Gamma$ and $\boldsymbol{\Omega}\Gamma$ on the set Fm are defined, for any $\alpha, \beta \in Fm$, as follows:

$$\alpha \leqslant_\Gamma \beta \overset{\text{def}}{\iff} \alpha \to \beta \in \Gamma$$

$$\alpha \equiv \beta \; (\boldsymbol{\Omega}\Gamma) \overset{\text{def}}{\iff} \alpha \to \beta, \beta \to \alpha \in \Gamma .$$

The relation $\boldsymbol{\Omega}\Gamma$ is called the ***equivalence modulo the theory*** *Γ.*

This coincides with the definition of $\boldsymbol{\Omega}\Gamma'$ for $\mathcal{C}\ell$ used in (14). Notice that in particular, if $\vdash_{\mathcal{L}} \alpha \to \beta$, then $\alpha \leqslant_\Gamma \beta$ for any $\Gamma \in \mathcal{T}h\mathcal{L}$.

Now the properties of the logic required for each of the steps of the Lindenbaum-Tarski process to work are identified.

THEOREM 2.2

Let $\mathcal{L}$ be a logic.

1. *The relation $\leqslant_\Gamma$ is a quasi-order on Fm for all $\Gamma \in \mathcal{T}h\mathcal{L}$ if and only if:*

   (IL1) $\vdash_{\mathcal{L}} x \to x$

   (IL2) $x \to y, \; y \to z \vdash_{\mathcal{L}} x \to z$

   In this case, the relation $\boldsymbol{\Omega}\Gamma$ is an equivalence relation on Fm that is compatible with the quasi-order $\leqslant_\Gamma$, so that the relation $\leqslant_\Gamma$ defines an order $\leqslant$ in the quotient set $Fm/\boldsymbol{\Omega}\Gamma$.

2. *The relation $\boldsymbol{\Omega}\Gamma$ is a congruence of the formula algebra $\boldsymbol{Fm}$, for all $\Gamma \in \mathcal{T}h\mathcal{L}$, if and only if in addition:*

   (IL3) $\left\{ \begin{array}{l} x_1 \to y_1, \dots, x_n \to y_n \\ y_1 \to x_1, \dots, y_n \to x_n \end{array} \right\} \vdash_{\mathcal{L}} \lambda x_1 \dots x_n \to \lambda y_1 \dots y_n,$

   for each $\lambda \in L$, with arity $n \geqslant 1$.

† Historically speaking, the names of Henkin and Sikorski have to be associated with her in this respect, in particular the book [209] with Sikorski. For more historical information on Rasiowa's contributions see [94, 97].

In this case the quotient set becomes an algebra $\boldsymbol{Fm}/\boldsymbol{\Omega}\Gamma$ *of type* $\boldsymbol{L}$.

3. *Every theory* Γ *of* $\mathcal{L}$ *is an up-set of its associated quasi-order* $\leqslant_\Gamma$ *(that is,* $\alpha \in \Gamma$ *and* $\alpha \leqslant_\Gamma \beta$ *imply* $\beta \in \Gamma$*), if and only if:*

   (IL4) $x, x \to y \vdash_{\mathcal{L}} y$.

   Moreover, in this case the canonical projection of $\boldsymbol{Fm}$ *onto* $\boldsymbol{Fm}/\boldsymbol{\Omega}\Gamma$ *preserves the theory* Γ *in the sense that for all* $\alpha \in Fm$, $\alpha \in \Gamma$ *if and only if* $\alpha/\boldsymbol{\Omega}\Gamma \in \Gamma/\boldsymbol{\Omega}\Gamma$*; and the set* $\Gamma/\boldsymbol{\Omega}\Gamma$ *is an up-set of the order* $\leqslant$ *in* $Fm/\boldsymbol{\Omega}\Gamma$.

4. *Every theory* Γ *of* $\mathcal{L}$ *constitutes a single equivalence class under* $\boldsymbol{\Omega}\Gamma$ *(that is, if* $\alpha \in \Gamma$*, then* $\alpha \equiv \beta \ (\boldsymbol{\Omega}\Gamma)$ *if and only if* $\beta \in \Gamma$*) if and only if:*

   (G) $x, y \vdash_{\mathcal{L}} y \to x$.

5. *This equivalence class is the top element of the order in the algebra* $\boldsymbol{Fm}/\boldsymbol{\Omega}\Gamma$ *(that is, for all* $\alpha, \beta \in Fm$, $\alpha \in \Gamma$ *implies* $\beta \leqslant_\Gamma \alpha$*), if and only if in addition:*

   (IL5) $x \vdash_{\mathcal{L}} y \to x$.

(In each step the properties of the preceding steps are assumed to hold.)

Proof: Notice that each of the syntactic conditions† is stated to be equivalent to the fact that the corresponding algebraic or order-theoretic property holds *for every theory*; and that structurality is essential to make the conditions involve only a few variables.

1 and 2 are fairly straightforward: For instance, by structurality (IL1) is equivalent to saying that for all α, $\alpha \to \alpha$ is a theorem of $\mathcal{L}$, and this is equivalent to saying that $\alpha \to \alpha \in \Gamma$ for all theories Γ of $\mathcal{L}$, which gives reflexivity of $\leqslant_\Gamma$. Similarly (IL2) amounts to its transitivity, and (IL3) to its compatibility with the operations. Recall also that the symmetrization of a quasi-order is always an equivalence relation, which is automatically compatible with the quasi-order and defines an order relation in the quotient. The main part of 3 is easy. Then, in the next equivalence, notice that $\alpha \in \Gamma$ implies $\alpha/\boldsymbol{\Omega}\Gamma \in \Gamma/\boldsymbol{\Omega}\Gamma$ by definition, and that the converse is *equivalent* to saying that if $\alpha \equiv \beta \ (\boldsymbol{\Omega}\Gamma)$ and $\beta \in \Gamma$, then $\alpha \in \Gamma$; to see that this is guaranteed by (IL4) is again easy. 4 is trivial; given the property in 3, the condition that Γ constitutes a single equivalence class reduces to the simpler property that if $\alpha, \beta \in \Gamma$, then $\alpha \equiv \beta \ (\boldsymbol{\Omega}\Gamma)$. Notice that (G) could be written as $x, y \vdash_{\mathcal{L}} \{x \to y, y \to x\}$, but this amounts to the same by symmetry. 5 is also straightforward. The details are left to you as Exercise 2.5. ⊠

After this "investigation", the following definition appears as a completely natural one:

Definition 2.3

An ***implicative logic*** *is a logic* $\mathcal{L}$ *in a language* $\boldsymbol{L}$ *with a binary term* $\to$ *such that the following conditions are satisfied:*

(IL1) $\vdash_{\mathcal{L}} x \to x$.

(IL2) $x \to y, \ y \to z \vdash_{\mathcal{L}} x \to z$.

† Why condition (G) is labelled in this way is explained in footnote † on page 141.

(IL3) $\left\{ \begin{array}{l} x_1 \to y_1, \ldots, x_n \to y_n \\ y_1 \to x_1, \ldots, y_n \to x_n \end{array} \right\} \vdash_{\mathcal{L}} \lambda x_1 \ldots x_n \to \lambda y_1 \ldots y_n,$

for each $\lambda \in L$, *with arity* $n \geqslant 1$.

(IL4) $x, x \to y \vdash_{\mathcal{L}} y$.

(IL5) $x \vdash_{\mathcal{L}} y \to x$.

The examples of implicative logics treated by Rasiowa, and a few more, are reviewed at the end of this section. It is not difficult to see that the infinitary logic $Ł_\infty$ of Example 1.15 is implicative, using how the operations in $[0,1]$ are defined; notice that an infinitary logic is implicative if and only if its finitary companion is implicative, because the conditions in the definition involve only consequences with a finite number of premises.

Observe that, by (IL1), all implicative logics have theorems, and that (IL5) implies (G). Therefore, all theorems (actually, all formulas of Γ) are identified as a single element in the quotient $\boldsymbol{Fm}/\boldsymbol{\Omega}\Gamma$. Sometimes it is assumed that the language includes a constant $\top$ as a particular theorem, and which is thus evaluated as this single element of the quotient; this makes some mathematical details simpler, but is not strictly necessary, as it is possible to consider $\top$ as a derived connective, in the way explained after Lemma 2.6.

It is clear that the Lindenbaum-Tarski process can be mimicked for implicative logics. (LT1) starts the process in the same way; Definition 2.1 gives (LT2), and Theorem 2.2 gives (LT3), (LT5) and (LT6). As to (LT4), it just identifies a class of algebras to which the quotients belong, and is not essential at this point (but see below). To summarize:

COROLLARY 2.4
If Γ *is a theory of an implicative logic* $\mathcal{L}$, *then* $\leqslant_\Gamma$ *is a quasi-order and* $\boldsymbol{\Omega}\Gamma$ *is a congruence of* $\boldsymbol{Fm}$ *compatible with* $\leqslant_\Gamma$; *this relation defines an order* $\leqslant$ *in* $\boldsymbol{Fm}/\boldsymbol{\Omega}\Gamma$, *which has a top element* 1 *constituted by the set* Γ, *which is a single equivalence class, and for all* $\alpha \in Fm$ *it holds that* $\alpha \in \Gamma$ *if and only if* $\alpha/\boldsymbol{\Omega}\Gamma = 1$. ⊠

The quotient algebra $\boldsymbol{Fm}/\boldsymbol{\Omega}\Gamma$ is popularly called ***the Lindenbaum-Tarski algebra of the theory*** Γ. In the particular case of $\Gamma := C_{\mathcal{L}}\varnothing$, the quotient algebra is called the ***pure Lindenbaum-Tarski algebra of the logic*** $\mathcal{L}$. It is easy to show just now (Exercise 2.6) that this algebra is weakly characteristic for $\mathcal{L}$, but fails to be characteristic for it, simply because different logics with the same theorems produce the same pure Lindenbaum-Tarski algebra. See also Theorem 2.10 for a fundamental property of this algebra. This issue is retaken in Section 3.3 in the context of algebraizable logics (in particular, see Exercise 3.21).

With the constructions made up to this point, it is clear that the Lindenbaum-Tarski process already provides a completeness theorem for any implicative logic, namely a *completeness with respect to the class of all the Lindenbaum-Tarski algebras of its theories*. However, this class is not particularly interesting; for instance, all these algebras have at most the cardinality of the set of formulas (hence, in most particular cases, they are countable), and they are indeed constructed from the formula algebra and the theories of the logic, and up to now it has no characterization independent from the logic.

The only step lacking a generalization up to now is (LT4), and this is to be expected, as it is the only step that is strongly dependent on the particular logic under scrutiny, since, obviously, in general the quotient algebra is not a Boolean algebra. We need to identify a class of algebras that includes the Lindenbaum-Tarski algebras, in order for step (LT4) to work and give us part ($\Leftarrow$) of the Algebraic Completeness Theorem; but at the same time part ($\Rightarrow$) has to hold, and this restricts the candidate classes of algebras in a decisive way. The solution is obtained so-to-speak by *cheating (but not absolutely)*: We take the class of all algebras for which part ($\Rightarrow$) *does* hold, *and* which satisfy an additional property, which is later described as "to be reduced". It turns out that this class includes all Lindenbaum-Tarski algebras, and hence it also makes part ($\Leftarrow$) work.

Definition 2.5

*Let $\mathcal{L}$ be an implicative logic in the language $\boldsymbol{L}$. An **$\mathcal{L}$-algebra** is an algebra $\boldsymbol{A}$ of similarity type $\boldsymbol{L}$ that has an element 1 with the following properties:*

(LALG1) *For all $\Gamma \cup \{\varphi\} \subseteq Fm$ and all $h \in \mathrm{Hom}(\boldsymbol{Fm}, \boldsymbol{A})$, if $\Gamma \vdash_{\mathcal{L}} \varphi$ and $h\Gamma \subseteq \{1\}$, then $h\varphi = 1$.*

(LALG2) *For all $a, b \in A$, if $a \to b = 1$ and $b \to a = 1$, then $a = b$.*

The class of $\mathcal{L}$-algebras is denoted by $\mathsf{Alg}^*\mathcal{L}$.

Condition (LALG1) may appear as a bit daunting as part of a definition of a class of algebras, since it involves *all* the consequences of $\vdash_{\mathcal{L}}$; however, Proposition 2.7 shows that for a finitary logic with a Hilbert-style presentation it is enough to restrict it to the axioms and rules of the presentation.

Lemma 2.6

Let $\mathcal{L}$ be an implicative logic.

1. *If $\vdash_{\mathcal{L}} \varphi$ and $\boldsymbol{A} \in \mathsf{Alg}^*\mathcal{L}$, then $h\varphi = 1$ for all $h \in \mathrm{Hom}(\boldsymbol{Fm}, \boldsymbol{A})$.*
2. *If $\vdash_{\mathcal{L}} \varphi$ and $\vdash_{\mathcal{L}} \psi$, then $\mathsf{Alg}^*\mathcal{L} \vDash \varphi \approx \psi$.*
3. *$\mathsf{Alg}^*\mathcal{L} \vDash \varphi \approx x \to x$ for all $\varphi \in Fm$ such that $\vdash_{\mathcal{L}} \varphi$.*
4. *$\mathsf{Alg}^*\mathcal{L} \vDash x \to x \approx y \to y$ for all $x, y \in V$.*

Proof: The first point is a consequence of (LALG1), and the rest follow easily in succession (Exercise 2.7). ⊠

Thus, the term $x \to x$ is an *algebraic constant* in the class $\mathsf{Alg}^*\mathcal{L}$, and its value in all these algebras equals the value of all the theorems of $\mathcal{L}$. It is good to keep things so that $\mathsf{Alg}^*\mathcal{L}$ is a class of algebras of the same similarity type as the logic; hence if $\boldsymbol{L}$ itself does not contain a primitive constant $\top$ that is a theorem of $\mathcal{L}$, it can be *defined* as an *abbreviation* of $x \to x$ for a fixed variable x (in this way it is not a constant of the language, but only an algebraic constant of the class of algebras). Using it, Lemma 2.6.3 would be expressed as:

$$\text{If } \vdash_{\mathcal{L}} \varphi, \text{ then } \mathsf{Alg}^*\mathcal{L} \vDash \varphi \approx \top \tag{16}$$

(It is also clear that one could take any theorem of $\mathcal{L}$ to be defined as $\top$.) The above property suggests using the "transformation" $\varphi \longmapsto \varphi \approx \top$ to turn every formula into an equation, and every (generalized) sequent or rule into a (generalized) quasi-equation. For instance, (MP) becomes the quasi-equation $(x \approx \top) \wedge (x \to y \approx \top) \to (y \approx \top)$. With this trick, it is not difficult to prove:

PROPOSITION 2.7
For any implicative logic $\mathcal{L}$, the class $\mathsf{Alg}^*\mathcal{L}$ is a generalized quasivariety. If moreover $\mathcal{L}$ is finitary, then $\mathsf{Alg}^*\mathcal{L}$ is a quasivariety, and it can be presented by the equations and the quasi-equations that result by applying the transformation $\varphi \longmapsto \varphi \approx \top$ to the axioms and rules of any Hilbert-style presentation of $\mathcal{L}$.

PROOF: Left to you as Exercise 2.8. ⊠

THEOREM 2.8
If Δ is a theory of an implicative logic $\mathcal{L}$, then $\boldsymbol{Fm}/\boldsymbol{\Omega}\Delta \in \mathsf{Alg}^*\mathcal{L}$. Moreover, Δ is consistent if and only if the algebra $\boldsymbol{Fm}/\boldsymbol{\Omega}\Delta$ is non-trivial.

PROOF: Recall that, after Corollary 2.4, $\Delta \in \boldsymbol{Fm}/\boldsymbol{\Omega}\Delta$, therefore we can take $1 := \Delta$. In order to check that $\boldsymbol{Fm}/\boldsymbol{\Omega}\Delta$ is an $\mathcal{L}$-algebra, the two properties in Definition 2.5 have to be checked.

(LALG1) Assume that $\Gamma \vdash_{\mathcal{L}} \varphi$ and $h\Gamma \subseteq \{1\}$ for some $\Gamma \cup \{\varphi\} \subseteq Fm$ and some $h \in \mathrm{Hom}(\boldsymbol{Fm}, \boldsymbol{Fm}/\boldsymbol{\Omega}\Delta)$. Since the canonical projection $\pi\colon \boldsymbol{Fm} \to \boldsymbol{Fm}/\boldsymbol{\Omega}\Delta$ is surjective, by Lemma 1.2 there is a homomorphism $\sigma\colon \boldsymbol{Fm} \to \boldsymbol{Fm}$ (i.e., a substitution) such that $h = \pi\sigma$, that is, $h\alpha = \sigma\alpha/\boldsymbol{\Omega}\Delta$ for all $\alpha \in Fm$. In particular, if $\alpha \in \Gamma$, then $\sigma\alpha/\boldsymbol{\Omega}\Delta = h\alpha = 1$, that is, $\sigma\alpha \in \Delta$. By structurality, $\sigma\Gamma \vdash_{\mathcal{L}} \sigma\varphi$. Since Δ is a theory of $\mathcal{L}$ and we have just seen that $\sigma\Gamma \subseteq \Delta$, we conclude that $\sigma\varphi \in \Delta$, which implies that $h\varphi = \sigma\varphi/\boldsymbol{\Omega}\Delta = 1$. Thus (LALG1) holds.

(LALG2) Assume that $\alpha/\boldsymbol{\Omega}\Delta \to \beta/\boldsymbol{\Omega}\Delta = 1$ and $\beta/\boldsymbol{\Omega}\Delta \to \alpha/\boldsymbol{\Omega}\Delta = 1$, for some $\alpha, \beta \in Fm$. This means that $(\alpha \to \beta)/\boldsymbol{\Omega}\Delta = 1$ and that $(\beta \to \alpha)/\boldsymbol{\Omega}\Delta = 1$; that is, that $\alpha \to \beta \in \Delta$ and that $\beta \to \alpha \in \Delta$. By definition, this means that $\alpha \equiv \beta \ (\boldsymbol{\Omega}\Delta)$, and this implies that $\alpha/\boldsymbol{\Omega}\Delta = \beta/\boldsymbol{\Omega}\Delta$. Thus (LALG2) also holds.

The proof of the second statement is left to you as Exercise 2.9. ⊠

We now have all the machinery to prove the main general relations linking each implicative logic with its class of algebras:

THEOREM 2.9 (ALGEBRAIC COMPLETENESS THEOREM)
If $\mathcal{L}$ is an implicative logic, then $\mathcal{L}$ is complete with respect to the class $\mathsf{Alg}^*\mathcal{L}$ in the following sense: For all $\Gamma \cup \{\varphi\} \subseteq Fm$,

$$\begin{aligned} \Gamma \vdash_{\mathcal{L}} \varphi \iff & h\Gamma \subseteq \{1\} \textit{ implies } h\varphi = 1 \\ & \textit{for all } h \in \mathrm{Hom}(\boldsymbol{Fm}, \boldsymbol{A}) \textit{ and all } \boldsymbol{A} \in \mathsf{Alg}^*\mathcal{L}. \end{aligned} \tag{17}$$

Proof: ($\Rightarrow$) This is exactly condition (LALG1) in Definition 2.5.

($\Leftarrow$) This is obtained by the generalization of the Lindenbaum-Tarski process, using the properties collected in Corollary 2.4. Assume that $\Gamma \nvdash_{\mathcal{L}} \varphi$ for some $\Gamma \cup \{\varphi\} \subseteq Fm$. Consider the theory Δ of $\mathcal{L}$ generated by Γ; we know that $\varphi \notin \Delta$. Now consider $\boldsymbol{A} := \boldsymbol{Fm}/\boldsymbol{\Omega}\Delta$ and the canonical projection $\pi \in \mathrm{Hom}(\boldsymbol{Fm}, \boldsymbol{A})$, defined by $\pi\alpha := \alpha/\boldsymbol{\Omega}\Delta$ for all $\alpha \in Fm$. It satisfies that $\pi\Gamma \subseteq \pi\Delta = \{1\}$ while $\pi\varphi \neq 1$. But, by Theorem 2.8, $\boldsymbol{A} \in \mathsf{Alg}^*\mathcal{L}$. Therefore, the right-hand side of (17) fails. By contraposition, this establishes the wanted implication. ⊠

This says that the algebra-based semantics constituted by the class of structures

$$\{\langle \boldsymbol{A}, \{1\}\rangle : \boldsymbol{A} \in \mathsf{Alg}^*\mathcal{L}\}$$

is complete for $\mathcal{L}$. In particular, (16) can now be enhanced:

$$\vdash_{\mathcal{L}} \varphi \ \text{ if and only if } \ \mathsf{Alg}^*\mathcal{L} \vDash \varphi \approx \top \tag{18}$$

Other traditional consequences of the same basic facts are:

Theorem 2.10

Let $\mathcal{L}$ be a consistent implicative logic, and let $\Gamma_0 = C_{\mathcal{L}}\varnothing$ be the set of theorems of $\mathcal{L}$. The quotient $\boldsymbol{Fm}/\boldsymbol{\Omega}\Gamma_0$ is the free algebra in $\mathsf{Alg}^*\mathcal{L}$ with denumerably many generators. The set $Var/\boldsymbol{\Omega}\Gamma_0 = \{x/\boldsymbol{\Omega}\Gamma_0 : x \in V\}$ can be taken as a set of generators of $\boldsymbol{Fm}/\boldsymbol{\Omega}\Gamma_0$, and if $x, y \in V$ with $x \neq y$, then $x/\boldsymbol{\Omega}\Gamma_0 \neq y/\boldsymbol{\Omega}\Gamma_0$.

Proof: Left to you as Exercise 2.10. ⊠

Theorem 2.11

Let $\mathcal{L}$ be an implicative logic. If $\boldsymbol{A} \in \mathsf{Alg}^*\mathcal{L}$ is countably generated, then there exists a theory Γ of $\mathcal{L}$ such that $\boldsymbol{Fm}/\boldsymbol{\Omega}\Gamma \cong \boldsymbol{A}$.

Proof: Since $\boldsymbol{A}$ is countably generated, there exists a surjective function from the set V onto the generating set of $\boldsymbol{A}$, and this function extends to a homomorphism h from $\boldsymbol{Fm}$ onto $\boldsymbol{A}$, which will be obviously surjective; therefore, $\boldsymbol{Fm}/\ker h \cong \boldsymbol{A}$. Define $\Gamma := h^{-1}\{1\}$. Since $\boldsymbol{A} \in \mathsf{Alg}^*\mathcal{L}$, it satisfies condition (LALG1); if you rewrite it using $h^{-1}\{1\}$, you will observe that it says that Γ is a closed set of the closure relation $\vdash_{\mathcal{L}}$, that is, it is a theory of $\mathcal{L}$. So to prove the theorem it only remains to show that $\boldsymbol{\Omega}\Gamma = \ker h$. That is, that $\alpha \equiv \beta\ (\boldsymbol{\Omega}\Gamma)$ if and only if $h\alpha = h\beta$. Equivalently, that $\alpha \to \beta, \beta \to \alpha \in \Gamma$ if and only if $h\alpha = h\beta$. Given the definition of Γ, this is a consequence of (LALG2). ⊠

The relations between an implicative logic $\mathcal{L}$ and its class of $\mathcal{L}$-algebras go much beyond the Algebraic Completeness Theorem and the other results above. This was the great discovery of Blok and Pigozzi, published in [32]. These relations can be smoothly expressed here, extending (18), with the help of the "transformation" $\varphi \longmapsto \varphi \approx \top$ and the relative equational consequence $\vDash_{\mathsf{K}}$ (Definition 1.69):

THEOREM 2.12 (ALGEBRAIZABILITY THEOREM)
Let $\mathcal{L}$ be an implicative logic, and put $\mathsf{K} := \mathsf{Alg}^*\mathcal{L}$ to simplify notation. For any $\Gamma \cup \{\varphi\} \subseteq Fm$ and any $\Theta \cup \{\varepsilon \approx \delta\} \subseteq Eq$, the following properties hold:

(ALG1) $\Gamma \vdash_{\mathcal{L}} \varphi \iff \{\gamma \approx \top : \gamma \in \Gamma\} \vDash_{\mathsf{K}} \varphi \approx \top$.

(ALG2) $\Theta \vDash_{\mathsf{K}} \varepsilon \approx \delta \iff \{\alpha \to \beta,\ \beta \to \alpha : \alpha \approx \beta \in \Theta\} \vdash_{\mathcal{L}} \{\varepsilon \to \delta,\ \delta \to \varepsilon\}$.

(ALG3) $\varphi \dashv\vdash_{\mathcal{L}} \{\varphi \to \top,\ \top \to \varphi\}$.

(ALG4) $\varepsilon \approx \delta \mathrel{=\!\!\mid\!\!\models}_{\mathsf{K}} \{\varepsilon \to \delta \approx \top,\ \delta \to \varepsilon \approx \top\}$.

PROOF: Observe that property (ALG1) is just Theorem 2.9. Property (ALG3) is easy to prove from the definition of implicative logic. To show property (ALG4), recall that by Lemma 2.6, the equation $y \to y \approx \top$ is valid in every algebra in $\mathsf{Alg}^*\mathcal{L}$; from this the entailment $\varepsilon \approx \delta \vDash_{\mathsf{K}} \{\varepsilon \to \delta \approx \top,\ \delta \to \varepsilon \approx \top\}$ follows. The converse entailment amounts to (LALG2). Finally, it is straightforward to see that (ALG2) follows mechanically by applying (ALG1) to the right-hand side of (ALG2) and then using (ALG4). ⊠

Actually, it is easy to see that also (ALG3) follows mechanically from (ALG1) and (ALG4), and conversely that both (ALG1) and (ALG4) follow, again mechanically, from (ALG2) and (ALG3). Section 3.2 confirms that this is a more general feature; there, the name "Algebraizability Theorem" given above to this result is dully justified.

This section ends by introducing a class of algebras whose algebraic properties are those shared by all the $\mathcal{L}$-algebras of all implicative logics.

DEFINITION 2.13
*An **implicative algebra** is an algebra $\boldsymbol{A} = \langle A, \to \rangle$ of type $\langle 2 \rangle$ where the following equations and quasi-equations hold; in them, $\top$ is shorthand for the term $x \to x$ (here x is a specific variable):*

(IA1) $y \to y \approx \top$

(IA2) $(x \to y \approx \top) \wedge (y \to z \approx \top) \to (x \to z \approx \top)$

(IA3) $x \to \top \approx \top$

(IA4) $(x \to y \approx \top) \wedge (y \to x \approx \top) \to x \approx y$

The class of all implicative algebras is denoted by IA.

As in the case of implicative logics, after (IA1) the symbol $\top$ can be treated as a real constant of the language, as far as work in implicative algebras is concerned, and its value in any of them is denoted by 1. Thus one can say that an implicative algebra is an algebra with a binary operation $\to$ and an element 1 such that, for all $a, b, c \in A$,

(IA1) $a \to a = 1$;

(IA2) if $a \to b = 1$ and $b \to c = 1$, then $a \to c = 1$;

(IA3) $a \to 1 = 1$; and

(IA4) if $a \to b = 1$ and $b \to a = 1$, then $a = b$.

The following result contains a basic property of implicative algebras, so basic that it can actually replace the definition:

Proposition 2.14
An algebra $\boldsymbol{A} = \langle A, \to \rangle$ *of type* $\langle 2 \rangle$ *is an implicative algebra if and only if it has an element* 1 *such that the relation defined by*

$$a \leqslant b \overset{\text{def}}{\iff} a \to b = 1\,, \textit{ for all } a, b \in A, \tag{19}$$

is an order relation on A *and* 1 *is its maximum.*

Proof: Just realize that after (19), conditions (IA1)–(IA4) say exactly that $\leqslant$ is an order relation and that 1 is its maximum. ⊠

Thus, the only places of the table of the operation $\to$ that matter for an algebra to be an implicative algebra are those where the element 1 appears; this fact may be of some interest for computational issues, and in general for the search of suitable counterexamples.

Now, it is straightforward to prove:

Proposition 2.15
If $\mathcal{L}$ *is an implicative logic, then* $(\mathsf{Alg}^*\mathcal{L}) \restriction \langle \to \rangle \subseteq \mathsf{IA}$.

One can loosely paraphrase this result by saying that "the $\mathcal{L}$-algebras of implicative logics *are* implicative algebras". Exercise 2.13 shows that implicative algebras are the $\mathcal{L}$-algebras of *the weakest implicative logic*. This explains the interest in implicative algebras in the literature. From Definition 2.13 it follows that IA is a quasivariety; Exercise 2.14 shows that it is actually a proper quasivariety (that is, it is not a variety).

Examples of classes of algebras whose implication reduct is an implicative algebra include Boolean algebras, Heyting algebras, and Hilbert algebras, to name but a few (see Example 3.64 for a less standard one, in which the implication connective is not primitive and not unique). As a matter of fact, in her book [208], Rasiowa, besides developing the general theory, examined in detail a number of implicative logics and determined their classes of $\mathcal{L}$-algebras; the two most prominent topics she treated are notions of filter and their relations to congruences, and representation theorems, mainly of the topological kind. Table 2 summarizes these cases.† The axiomatization of $\mathcal{C}\ell$ on page 71 is the one she used, and it is designed to facilitate the incremental presentation of most of these logics (*Modus Ponens* is the only rule in all of them). Here are some:

$\mathcal{I}\ell_{\to}$	(Ax1) + (Ax2)
$\mathcal{C}\ell_{\to}$	(Ax1) + (Ax2) + $\big((x \to y) \to x\big) \to x$
$\mathcal{I}\ell^{+}$	(Ax1) $+ \cdots +$ (Ax8)
Minimal	(Ax1) $+ \cdots +$ (Ax9)

† These are the ones that she treated in the main body of the book; other logics were briefly sketched in exercises.

$\mathcal{I}\ell$	(Ax1) $+ \cdots +$ (Ax10)
$\mathcal{C}\ell$	(Ax1) $+ \cdots +$ (Ax11)

The axiom particular to $\mathcal{C}\ell_{\to}$ is usually called "Peirce's axiom". In the next section we see how to show that, indeed, $\mathsf{Alg}^*\mathcal{I}\ell_{\to} = \mathsf{HiA}$, $\mathsf{Alg}^*\mathcal{I}\ell = \mathsf{HA}$, and so forth (Exercises 2.28 and 2.31). To prove that some of the logics, so axiomatized, are indeed the stated fragments of $\mathcal{C}\ell$ or of $\mathcal{I}\ell$ is quite a different issue, not treated in this book. The class of implicative logics is much larger:

THEOREM 2.16

1. *A logic is implicative if and only if its finitary companion is implicative.*
2. *All extensions of an implicative logic are implicative.*
3. *All fragments of an implicative logic whose language contains $\to$ are implicative.*
4. *An expansion of an implicative logic is implicative if and only if the expanded logic satisfies condition* (IL3) *for the new connectives.*

PROOF: Just considering Definition 2.3, the first three points are obvious, and the fourth one is easy (Exercise 2.16). ⊠

L	$\mathcal{L}$	name of the logic	$\mathsf{Alg}^*\mathcal{L}$
$\to$	$\mathcal{I}\ell_{\to}$	positive implicative logic [implication fragment of $\mathcal{I}\ell$]	HiA: positive implication algebras [Hilbert algebras]
$\to$	$\mathcal{C}\ell_{\to}$	classical implicative logic [implication fragment of $\mathcal{C}\ell$]	implication algebras
$\to, \vee, \wedge$	$\mathcal{I}\ell^{+}$	positive logic [negation-less fragment of $\mathcal{I}\ell$]	relatively pseudo-complemented lattices [generalized Heyting algebras]
$\to, \vee, \wedge, \neg$		minimal logic [Johansson's logic]	contrapositionally complemented lattices
$\to, \vee, \wedge, \neg$		positive logic with semi-negation	semi-complemented lattices
$\to, \vee, \wedge, \neg$	$\mathcal{I}\ell$	intuitionistic logic	HA: pseudo-Boolean algebras [Heyting algebras]
$\to, \vee, \wedge, \neg, \Rightarrow, \sim$		constructive logic with strong negation [Nelson's logic]	quasi-pseudo-Boolean algebras [Nelson algebras]
$\to, \vee, \wedge, \neg$	$\mathcal{C}\ell$	classical logic	BA: Boolean algebras
$\to, \vee, \wedge, \neg, \Box$	$S4^{g}$	modal logic [global logic of system $S4$]	topological Boolean algebras [closure algebras]
$\to, \vee, \wedge, \neg, \{e_i : i<m\}$		Post's m-valued logic [Rousseau's version]	Post algebras of order m

TABLE 2: The logics in Rasiowa's book, and their algebras. The names of logics and algebras are those she used [with alternative or descriptive names]. The symbols are the ones used here.

EXAMPLES 2.17

1. The logics associated with normal modal systems (Example 1.14) are expansions of classical logic, but only the global ones fall under Theorem 2.16.4. Condition (IL3) for the primitive connective $\Box$ becomes $x \to y, y \to x \vdash_{\mathcal{L}} \Box x \to \Box y$. All those logics have the formula $\Box(x \to y) \to (\Box x \to \Box y)$ among their theorems, and when using it to obtain (IL3), the Necessitation Rule $\alpha \vdash_{\mathcal{L}} \Box\alpha$ is also used; this is why only the global logics associated with normal modal systems are implicative, while the local logics (which have only the weaker (admissible) rule "if $\vdash_{\mathcal{L}} \alpha$, then $\vdash_{\mathcal{L}} \Box\alpha$") are not. Thus, $S4^g$ is not the only modal logic that can be treated with Rasiowa's methods; clearly K^g is one as well, and then by Theorem 2.16.2, the global logics associated with all the normal modal systems ($K4, T, B, S5$, etc.) are implicative as well. It is easy, using Kripke semantics for instance, to show that the standard implication connective does not satisfy the properties needed to make the local logics implicative; in principle there might be some other binary connective satisfying them, but this is not the case, though the proof is not so direct: in Example 3.61 it is shown that this is not possible using tools from the theory of algebraizable logics, thus confirming that the local logics associated with normal modal systems are not implicative.
2. Another large group of implicative logics whose relation with a class of algebras has been studied in the literature are the fuzzy logics of Hájek's well-known book [132], starting with his basic logic $\mathcal{BL}$; in its Section 2.3 he proves its completeness with respect to the class of BL-algebras by following the standard Lindenbaum-Tarski method. Well-known logics in this group are Łukasiewicz's many-valued logics (Example 1.12). They form a significant subclass of the logics associated with varieties of residuated lattices studied in [123]; actually, all those for integral varieties are implicative. ⊠

Exercises for Section 2.2

EXERCISE 2.5. Complete the proof of Theorem 2.2, following the steps indicated there. Recall that in each step the properties of the preceding steps are assumed to hold.

EXERCISE 2.6. Let $\mathcal{L}$ be an implicative logic, and let $\Gamma_0 := C_{\mathcal{L}}\varnothing$ be the set of its theorems. Prove, without using the notion of an $\mathcal{L}$-algebra and its properties, that $\boldsymbol{\Omega}\Gamma_0$ is a fully invariant congruence of $\boldsymbol{Fm}$ and that $\vdash_{\mathcal{L}} \varphi$ if and only if $\boldsymbol{Fm}/\boldsymbol{\Omega}\Gamma_0 \vDash \varphi \approx \top$ for any formula φ, where $\top := x \to x$ for a fixed variable x.

EXERCISE 2.7. Check that all points of Lemma 2.6 follow from just the concept of an $\mathcal{L}$-algebra (Definition 2.5).

EXERCISE 2.8. Let $\mathcal{L}$ be an implicative logic.

1. Prove that the class $\mathsf{Alg}^*\mathcal{L}$ is a generalized quasivariety, by applying the transformation $\varphi \mapsto \varphi \approx \top$ to conditions (LALG1) and (LALG2).
2. Prove that if $\mathcal{L}$ is finitary, then $\mathsf{Alg}^*\mathcal{L}$ is a quasivariety.
3. Assume that $\mathcal{L}$ is a finitary logic that has a Hilbert-style presentation. Prove that condition (LALG1) can be limited to the axioms and rules of the presentation, so that

this supplies a (hopefuly small) set of equations and quasi-equations constituting a presentation of $\mathsf{Alg}^*\mathcal{L}$.

Exercise 2.9. Let Δ be a theory of an implicative logic $\mathcal{L}$. Prove that Δ is consistent if and only if the algebra $\boldsymbol{Fm}/\Omega\Delta$ is non-trivial.

Exercise 2.10. Prove Theorem 2.10. Show first the second part, using the second part of Theorem 2.8.

Exercise 2.11. Explain why axiom (IA1) in Definition 2.13 cannot be formulated as "$x \to x \approx \top$".

Exercise 2.12. Let $\langle A, \leqslant \rangle$ be an arbitrary ordered set with a maximum element 1, and define a binary operation $\to$ on A as follows: for each $a, b \in A$, $a \to b := 1$ when $a \leqslant b$, and $a \to b := b$ otherwise. Prove that then $\langle A, \to \rangle$ is an implicative algebra. This operation is sometimes called ***the canonical implication***.

Exercise 2.13. Show (by intersection considerations) that the weakest implicative logic in the language $\langle \to \rangle$ of type $\langle 2 \rangle$ exists. Denote it by $\mathcal{I}mp$. Give a Hilbert-style axiomatization of this logic, prove that $\mathsf{Alg}^*\mathcal{I}mp = \mathsf{IA}$, and that the theorems of this logic are all formulas of the form $\alpha_1 \to (\alpha_2 \to \dots (\alpha_n \to (\beta \to \beta)))$, where $\alpha_1, \dots, \alpha_n, \beta$ are arbitrary formulas ($n \geqslant 0$).

Exercise 2.14. Consider the two algebras $\boldsymbol{A}_1 = \langle \{a, b, 1\}, \to \rangle$ and $\boldsymbol{A}_2 = \langle \{c, 1\}, \to' \rangle$ given by the following operations:

$\to$	a	b	1
a	1	a	1
b	a	1	1
1	a	a	1

$\to'$	c	1
c	1	1
1	1	1

Check that $\boldsymbol{A}_1$ is an implicative algebra while $\boldsymbol{A}_2$ is not, and that $\boldsymbol{A}_2$ is a homomorphic image of $\boldsymbol{A}_1$. Conclude that the class IA is not a variety.

Exercise 2.15. Let $\mathcal{I}mp'$ be the logic obtained by deleting a rule corresponding to (IL3) from the presentation of $\mathcal{I}mp$ obtained in Exercise 2.13 (assuming you *did* include such a rule). Show that $\mathcal{I}mp'$ is not an implicative logic. However, you can still apply Definition 2.5 to $\mathcal{I}mp'$ and prove that $\mathsf{Alg}^*\mathcal{I}mp' = \mathsf{IA}$. Conclude that $\mathcal{I}mp'$ does not satisfy the same completeness theorem as $\mathcal{I}mp$ does, with respect to the class of implicative algebras.

Exercise 2.16. Let $\mathcal{L}$ be an implicative logic in a language $\boldsymbol{L}$ and let $\mathcal{L}'$ be an expansion of $\mathcal{L}$ in a language $\boldsymbol{L}' \supseteq \boldsymbol{L}$. Prove that $\mathcal{L}'$ is implicative if and only if it satisfies condition (IL3) for the connectives in $\boldsymbol{L}' \setminus \boldsymbol{L}$.

Exercise 2.17. Compare the definition of implicative logics (2.3) with that of implicative algebras (2.13), under the light of Proposition 2.15. It seems as if there is a correspondence between the properties in each definition: (IL1) and (IL2) clearly correspond to (IA1) and (IA2) respectively. Show that (IL5) corresponds to (IA3). However, observe that there are no algebraic analogues of (IL3) and (IL4), and there is no logical analogue of (IA4). Try to explain why.

Comment. This exercise has no "formally correct solution", but the reflection it will lead you to do will be beneficial in order to deeply understand some key ideas of the algebraic logic setting.

2.3 Filters

The general case

This subsection contains some general notions and results, which apply to every logic and which play *a central, fundamental role* in this book. The proofs are easy, but illustrate basic techniques to be retained for forthcoming chapters.

DEFINITION 2.18
Let $\mathcal{L}$ be a logic, and let $\boldsymbol{A}$ be an algebra. A subset $F \subseteq A$ is a ***filter*** *of the logic $\mathcal{L}$, or* ***$\mathcal{L}$-filter****, when the following condition holds for all $\Gamma \cup \{\varphi\} \subseteq Fm$:*

$$\text{If } \Gamma \vdash_{\mathcal{L}} \varphi \text{ and } h \in \operatorname{Hom}(\boldsymbol{Fm}, \boldsymbol{A}) \text{ is such that } h\Gamma \subseteq F, \text{ then } h\varphi \in F. \tag{20}$$

The set of $\mathcal{L}$-filters of the algebra $\boldsymbol{A}$ is denoted by $\mathcal{F}i_{\mathcal{L}}\boldsymbol{A}$.

Sometimes the terms ***deductive filter*** or ***logical filter*** are also used for emphasis.† An $\mathcal{L}$-filter is ***proper*** when it is not the whole universe A of the algebra (which is always, obviously, a filter).

The definition of $\mathcal{L}$-filter can be considerably simplified for logics with a finite Hilbert-style presentation: it is enough to limit the condition that F be "closed under all consequences" of the logic to hold just for its axioms and rules; and instead of speaking of "all evaluations" it is simpler to speak of "all elements of the algebra" needed to evaluate the axioms and rules. An example makes this clearer: F is an $\mathcal{IL}_{\to}$-filter of an (arbitrary) algebra $\boldsymbol{A}$ if and only if for all $a, b, c \in A$, $a \to (b \to a) \in F$;

$$(a \to (b \to c)) \to ((a \to b) \to (a \to c)) \in F; \text{ and}$$
$$\text{if } a \in F \text{ and } a \to b \in F, \text{ then } b \in F.$$

Exercise 2.18 contains a couple of simple examples. Proposition 2.28 and Exercise 2.31 contain descriptions of the filters of some well-known logics for selected algebras (namely, for the algebras in $\mathsf{Alg}^*\mathcal{L}$).

$\mathcal{L}$-filters are in some sense like the theories of $\mathcal{L}$: they are sets that contain the theorems and are closed under $\mathcal{L}$-consequence, but, since the theorems and the consequence live in the formula algebra, they have to be evaluated; thus, more precisely, $\mathcal{L}$-filters are the sets that *contain all values of the $\mathcal{L}$-theorems and are closed under all evaluations of the $\mathcal{L}$-consequences.* The following (trivial) fact reinforces this parallelism:

PROPOSITION 2.19
Let $\mathcal{L}$ be a logic, and let $\Delta \subseteq Fm$. The following conditions are equivalent:

(i) *Δ is a theory of $\mathcal{L}$. That is, if $\Delta \vdash_{\mathcal{L}} \varphi$, then $\varphi \in \Delta$.*

(ii) *If $\Gamma \vdash_{\mathcal{L}} \varphi$ and $\Gamma \subseteq \Delta$, then $\varphi \in \Delta$.*

† The term *deductive system* has also been used with the same meaning by Monteiro's school [178], extending Tarski's usage of this term for "theory" in the English translation of his early works; see [236, pp. 33,185,etc.]. However, in modern algebraic logic literature, this term is often taken as a synonym for "finitary logic".

(iii) *If* $\Gamma \vdash_{\mathcal{L}} \varphi$ *and* $\sigma \in \mathrm{End}\boldsymbol{Fm}$ *is such that* $\sigma\Gamma \subseteq \Delta$, *then* $\sigma\varphi \in \Delta$. ⊠

Notice that 2.19(iii) says exactly that "Δ is an $\mathcal{L}$-filter of $\boldsymbol{Fm}$"; comparison with (20) makes the mentioned parallelism between theories and filters clear. Morover, condition (20) can be rewritten using the residual function h^{-1}. Thus:

COROLLARY 2.20

1. *For any logic* $\mathcal{L}$, *the* $\mathcal{L}$*-filters of the formula algebra are the theories of* $\mathcal{L}$; *in symbols,* $\mathcal{F}i_{\mathcal{L}}\boldsymbol{Fm} = \mathcal{T}h\mathcal{L}$.
2. *For all* $\boldsymbol{A}$ *and all* $F \subseteq A$, $F \in \mathcal{F}i_{\mathcal{L}}\boldsymbol{A}$ *if and only if for every* $h \in \mathrm{Hom}(\boldsymbol{Fm}, \boldsymbol{A})$, $h^{-1}F \in \mathcal{T}h\mathcal{L}$. ⊠

In particular, if $\Delta \subseteq Fm$, then $\Delta \in \mathcal{T}h\mathcal{L}$ if and only if for every $\sigma \in \mathrm{End}\boldsymbol{Fm}$, $\sigma^{-1}\Delta \in \mathcal{T}h\mathcal{L}$; this result is already obtained in Section 1.4 as a consequence of the structurality of the logic.

The first important property of the notion of a filter is that *every* logic is *complete* with respect to the algebra-based semantics whose structures are the pairs $\langle \boldsymbol{A}, F\rangle$ with $F \in \mathcal{F}i_{\mathcal{L}}\boldsymbol{A}$, when $\boldsymbol{A}$ ranges over all possible algebras, and in each case the evaluations are all homomorphisms; this semantics is the *semantic of matrices*, and it is dealt with in depth in Chapter 4.

THEOREM 2.21

For any logic $\mathcal{L}$ *and any* $\Gamma \cup \{\varphi\} \subseteq Fm$,

$$\Gamma \vdash_{\mathcal{L}} \varphi \iff \text{for all algebras } \boldsymbol{A}, \text{ all } F \in \mathcal{F}i_{\mathcal{L}}\boldsymbol{A} \text{ and all } h \in \mathrm{Hom}(\boldsymbol{Fm}, \boldsymbol{A}), \text{ if } h\Gamma \subseteq F, \text{ then } h\varphi \in F.$$

PROOF: ($\Rightarrow$) This is a direct consequence of the definition of $\mathcal{L}$-filter.
($\Leftarrow$) In order to prove this by contraposition, assume that $\Gamma \nvdash_{\mathcal{L}} \varphi$, and let Δ be the $\mathcal{L}$-theory generated by Γ. Thus, $\Gamma \subseteq \Delta$ but $\varphi \notin \Delta$; but this contradicts the assumption, because the identity function is an endomorphism of $\boldsymbol{Fm}$ and, by Corollary 2.20.1, Δ is an $\mathcal{L}$-filter. ⊠

The first basic facts on $\mathcal{L}$-filters are straightforward to check:

PROPOSITION 2.22

Let $\mathcal{L}$ *be a logic and* $\boldsymbol{A}$ *an algebra.*

1. *The set* $\mathcal{F}i_{\mathcal{L}}\boldsymbol{A}$ *is a closure system.*
2. φ *is a theorem of* $\mathcal{L}$ *if and only if* $h\varphi \in F$ *for all* $F \in \mathcal{F}i_{\mathcal{L}}\boldsymbol{A}$, *all* $h \in \mathrm{Hom}(\boldsymbol{Fm}, \boldsymbol{A})$ *and all* $\boldsymbol{A}$, *and if and only if* $h\varphi \in \bigcap \mathcal{F}i_{\mathcal{L}}\boldsymbol{A}$ *for all* $h \in \mathrm{Hom}(\boldsymbol{Fm}, \boldsymbol{A})$ *and all* $\boldsymbol{A}$.
3. $\mathcal{L}$ *has theorems if and only if every* $\mathcal{L}$*-filter is non-empty.*
4. $\mathcal{L}$ *does not have theorems if and only if* $\emptyset \in \mathcal{F}i_{\mathcal{L}}\boldsymbol{A}$ *for all* $\boldsymbol{A}$. ⊠

Some additional useful information appears in Exercise 2.19. Observe that the statement in point 4 is not the logical negation of that in point 3 (which would be that $\mathcal{L}$ does not have theorems if and only if $\emptyset \in \mathcal{F}i_{\mathcal{L}}\boldsymbol{A}$ for some $\boldsymbol{A}$).

Notice that $\bigcap \mathcal{F}i_{\mathcal{L}}\boldsymbol{A}$ is the smallest $\mathcal{L}$-filter of $\boldsymbol{A}$; we have just seen that it contains all the values of all theorems. However, this does not imply that the set of all such values *is* the smallest $\mathcal{L}$-filter; this is a common mistake, but Exercise 2.21 provides a counterexample (the problem is that this set may not be an $\mathcal{L}$-filter itself; of course, in case it is, it is certainly the smallest one). The curious property of Exercise 2.22 is also related to these issues.

To check that a given subset F of an algebra $\boldsymbol{A}$ is an $\mathcal{L}$-filter involves in principle considering all cases of $\Gamma \vdash_{\mathcal{L}} \varphi$ and of $h \in \operatorname{Hom}(\boldsymbol{Fm}, \boldsymbol{A})$ such that $h\Gamma \subseteq F$, and proving that in each case $h\varphi \in F$. Exercise 2.23 shows that in this task one may assume *without loss of generality* that $hz = a$ for a chosen variable z and a chosen point $a \in A$; this is a *trick* that helps in some proofs later on.

Since $\mathcal{F}i_{\mathcal{L}}\boldsymbol{A}$ is a closure system, all the machinery of Section 1.3 can be used. For instance, $\langle \mathcal{F}i_{\mathcal{L}}\boldsymbol{A}, \subseteq \rangle$ is a complete lattice, and it has an associated closure operator, which is denoted by $Fg_{\mathcal{L}}^{\boldsymbol{A}}$; thus, for $X \subseteq A$, $Fg_{\mathcal{L}}^{\boldsymbol{A}} X$ denotes the least $\mathcal{L}$-filter of $\boldsymbol{A}$ containing X. In the finitary case there is a very useful characterization of this operator:

Theorem 2.23

Let $\mathcal{L}$ be a finitary logic and let $\boldsymbol{A}$ be an arbitrary algebra.

1. *The set $\mathcal{F}i_{\mathcal{L}}\boldsymbol{A}$ is inductive. That is, the operator $Fg_{\mathcal{L}}^{\boldsymbol{A}}$ is finitary.*
2. *For every $X \subseteq A$, $Fg_{\mathcal{L}}^{\boldsymbol{A}} X = \bigcup_{n \geqslant 0} X_n$, where the sets X_n are defined inductively as follows:*

$$X_0 := X$$

$$X_{n+1} := \{a \in A : \text{ there is } \Gamma \cup \{\varphi\} \subseteq Fm,\ \Gamma \text{ finite, such that } \Gamma \vdash_{\mathcal{L}} \varphi, \text{ and there is } h \in \operatorname{Hom}(\boldsymbol{Fm}, \boldsymbol{A}) \text{ with } h\Gamma \subseteq X_n \text{ and } h\varphi = a\}$$

Proof: 1. Let $\{F_i\}_{i \in I} \subseteq \mathcal{F}i_{\mathcal{L}}\boldsymbol{A}$ be an upwards directed family; we have to see that $\bigcup F_i \in \mathcal{F}i_{\mathcal{L}}\boldsymbol{A}$. Assume that $\Gamma \vdash_{\mathcal{L}} \varphi$ and $h \in \operatorname{Hom}(\boldsymbol{Fm}, \boldsymbol{A})$ such that $h\Gamma \subseteq \bigcup F_i$. Since $\mathcal{L}$ is finitary, there is a finite $\Gamma_0 \subseteq \Gamma$ such that $\Gamma_0 \vdash_{\mathcal{L}} \varphi$. Then $h\Gamma_0 \subseteq \bigcup F_i$, and the upwards-directedness of the family implies that there is some $i_0 \in I$ such that $h\Gamma_0 \subseteq F_{i_0}$. Since F_{i_0} is an $\mathcal{L}$-filter this implies that $h\varphi \in F_{i_0} \subseteq \bigcup F_i$.

2. Put $Z = \bigcup_{n \geqslant 0} X_n$. Thus, $X \subseteq Z$. It is easy to see that the family $\{X_n\}_{n \geqslant 0}$ is increasing (using that $x \vdash_{\mathcal{L}} x$). Let us see that Z is an $\mathcal{L}$-filter: If $\Gamma \vdash_{\mathcal{L}} \varphi$ and $h\Gamma \subseteq Z$, by finitarity there is a finite $\Gamma_0 \subseteq \Gamma$ such that $\Gamma_0 \vdash_{\mathcal{L}} \varphi$. Then $h\Gamma_0 \subseteq Z$, which by the increasing character of the family implies that there is some $n \geqslant 0$ such that $h\Gamma_0 \subseteq X_n$. Then by the definition of the X_n, $h\varphi \in X_{n+1} \subseteq Z$. Finally, Z is the smallest $\mathcal{L}$-filter of $\boldsymbol{A}$ that contains X: If $F \in \mathcal{F}i_{\mathcal{L}}\boldsymbol{A}$ with $X \subseteq F$, then $X_n \subseteq F$ for all $n \geqslant 0$ is shown by induction: $X_0 = X \subseteq F$ by assumption, and assuming that $X_n \subseteq F$ and taking $a \in X_{n+1}$, by construction $a = h\varphi$ for some $h \in \operatorname{Hom}(\boldsymbol{Fm}, \boldsymbol{A})$ and some finite $\Gamma \cup \{\varphi\} \subseteq Fm$ such that $h\Gamma \subseteq X_n \subseteq F$ and $\Gamma \vdash_{\mathcal{L}} \varphi$. Since F is an $\mathcal{L}$-filter, it follows that $a = h\varphi \in F$. This shows that $X_{n+1} \subseteq F$ and thus that $Z \subseteq F$. ☒

In Exercise 2.26 the previous characterization is presented in an apparently more "constructive" way; it may be more practical in some situations. For a characterization in the general case, see Exercise 2.25.

The definition of $\mathcal{L}$-filter, in the expression of Corollary 2.20.2, immediately yields the following:

PROPOSITION 2.24
Let $\mathcal{L}$ be a logic, let $\boldsymbol{A}, \boldsymbol{B}$ be algebras and let $h \in \mathrm{Hom}(\boldsymbol{A}, \boldsymbol{B})$. If $G \in \mathcal{F}i_{\mathcal{L}}\boldsymbol{B}$, then $h^{-1}G \in \mathcal{F}i_{\mathcal{L}}\boldsymbol{A}$. ⊠

Thus, the inverse image of a filter by a homomorphism is always a filter as well; in general, however, it is not true that the direct image of a filter is also a filter (Exercise 2.27).

Not much more can be said about the relations between $\mathcal{L}$-filters and homomorphisms without putting restrictions on the logic $\mathcal{L}$ or on the homomorphisms themselves.† As an important instance, the following properties of *surjective* homomorphisms should be retained, as they are profusely used:

PROPOSITION 2.25
Let $\mathcal{L}$ be a logic, let $\boldsymbol{A}, \boldsymbol{B}$ be algebras and let $h \in \mathrm{Hom}(\boldsymbol{A}, \boldsymbol{B})$ be surjective.

1. *If $G \subseteq B$ is such that $h^{-1}G \in \mathcal{F}i_{\mathcal{L}}\boldsymbol{A}$, then $G \in \mathcal{F}i_{\mathcal{L}}\boldsymbol{B}$.*
2. *If $F \subseteq A$ is such that $F = h^{-1}hF$, then $F \in \mathcal{F}i_{\mathcal{L}}\boldsymbol{A}$ if and only if $hF \in \mathcal{F}i_{\mathcal{L}}\boldsymbol{B}$.*

PROOF: 1. Use Definition 2.18: assume that $\Gamma \vdash_{\mathcal{L}} \varphi$ and that $g \in \mathrm{Hom}(\boldsymbol{Fm}, \boldsymbol{B})$ is such that $g\Gamma \subseteq G$. Since h is surjective, Lemma 1.2 implies that there is a $g' \in \mathrm{Hom}(\boldsymbol{Fm}, \boldsymbol{A})$ such that $g = hg'$. Thus, $hg'\Gamma = g\Gamma \subseteq G$, so $g'\Gamma \subseteq h^{-1}G$. Since $h^{-1}G \in \mathcal{F}i_{\mathcal{L}}\boldsymbol{A}$, this implies that $g'\varphi \in h^{-1}G$, that is, $g\varphi = hg'\varphi \in G$.

2. The implication from left to right is a particular case of the property in 1, while the converse is a particular case of Proposition 2.24. ⊠

Concerning the additional assumption of point 2, it is interesting to recall that $G = hh^{-1}G$ for any $G \in \mathcal{F}i_{\mathcal{L}}\boldsymbol{B}$, because h is surjective. The correspondences between $\mathcal{L}$-filters across surjective homomorphisms are further exploited in Lemma 4.36.

The implicative case

Now we return to the restricted world of implicative logics. To start with, observe that condition (LALG1) of Definition 2.5 just says that $\{1\}$ is an $\mathcal{L}$-filter; this provides an alternative definition of $\mathcal{L}$-algebras, and some more information.

COROLLARY 2.26
Let $\mathcal{L}$ be an implicative logic. An algebra $\boldsymbol{A}$ is an $\mathcal{L}$-algebra if and only if there is $1 \in A$ such that the set $\{1\}$ is an $\mathcal{L}$-filter and for all $a, b \in A$, if $a \to b = 1$ and $b \to a = 1$, then $a = b$. Moreover, the set $\{1\}$ is the least $\mathcal{L}$-filter in any $\mathcal{L}$-algebra, and the logic

† This is a verbatim quotation of [70, p. 58].

$\mathcal{L}$ is complete with respect to the semantics $\{\langle \boldsymbol{A}, F\rangle : \boldsymbol{A} \in \mathsf{Alg}^*\mathcal{L}, F \in \mathcal{F}i_{\mathcal{L}}\boldsymbol{A}\}$, an algebra-based one.

PROOF: As said before, the first statement is just a rephrasing of Definition 2.5 using the notion of a filter. Since $1 = h\varphi$ for any $h \in \mathrm{Hom}(\boldsymbol{Fm}, \boldsymbol{A})$ and any theorem φ, $1 \in F$ for any $F \in \mathcal{F}i_{\mathcal{L}}\boldsymbol{A}$, hence $\{1\}$ is the least $\mathcal{L}$-filter of $\boldsymbol{A}$ (this is a general fact, see Exercise 2.20). It also follows that the class of structures mentioned in the statement contains the class $\{\langle \boldsymbol{A}, \{1\}\rangle : \boldsymbol{A} \in \mathsf{Alg}^*\mathcal{L}\}$; moreover, it is trivially contained in the class $\{\langle \boldsymbol{A}, F\rangle : \boldsymbol{A} \text{ any algebra }, F \in \mathcal{F}i_{\mathcal{L}}\boldsymbol{A}\}$. Since by Theorems 2.9 and 2.21, respectively, $\mathcal{L}$ is complete with respect to these two other classes, it follows that $\mathcal{L}$ is complete with respect to the class mentioned in the statement as well. ⊠

See Exercise 2.28 for a simple application of this characterization, allowing you to find the $\mathcal{I}\ell_{\rightarrow}$-algebras. For other logics the process, though similar, may require more technical work at the level of axiomatics of the classes of algebras; see for instance Exercise 2.31. In general, when one works with a particular implicative logic, after identifying its algebras one wants to identify the notion of a filter of these algebras, and one finds that it usually coincides with some particular kind of subset naturally arising in the purely algebraic theory of the corresponding classes of algebras. This confirms a relation between logical inference from assumptions and algebraic generation of certain subsets, very often called "filters" as well, with perhaps some adjective. This constitutes an important theme in this book, and one can even view it as the base of the general theory of algebraizability. For now:

DEFINITION 2.27
Let $\boldsymbol{A}$ be an implicative algebra. A subset $F \subseteq A$ is an ***implicative filter*** *of $\boldsymbol{A}$ when $1 \in F$, and for all $a, b \in A$, if $a \in F$ and $a \rightarrow b \in F$, then $b \in F$. The set of implicative filters of an implicative algebra $\boldsymbol{A}$ is denoted by $\mathcal{F}ilt_{\rightarrow}\boldsymbol{A}$.*

The notion of an implicative filter is modelled after the Hilbert-style presentations of many logics based on the rule of *Modus Ponens*. Actually:

PROPOSITION 2.28
Let $\mathcal{L}$ be an implicative logic, and $\boldsymbol{A} \in \mathsf{Alg}^*\mathcal{L}$. Then $\mathcal{F}i_{\mathcal{L}}\boldsymbol{A} \subseteq \mathcal{F}ilt_{\rightarrow}\boldsymbol{A}$. If moreover $\mathcal{L}$ can be presented by a Hilbert-style calculus with some axioms and the only rule of Modus Ponens, then $\mathcal{F}i_{\mathcal{L}}\boldsymbol{A} = \mathcal{F}ilt_{\rightarrow}\boldsymbol{A}$; that is, the $\mathcal{L}$-filters of the $\mathcal{L}$-algebras coincide with their implicative filters.

PROOF: Let $\boldsymbol{A} \in \mathsf{Alg}^*\mathcal{L}$ and $F \in \mathcal{F}i_{\mathcal{L}}\boldsymbol{A}$. By Corollary 2.26, $\{1\}$ is the least $\mathcal{L}$-filter of $\boldsymbol{A}$, hence $1 \in F$. Now assume that $a, a \rightarrow b \in F$ for some $a, b \in A$. Then take $x, y \in V$ with $x \neq y$ and define an evaluation h in $\boldsymbol{A}$ such that $hx = a$ and $hy = b$. By (IL4), $x, x \rightarrow y \vdash_{\mathcal{L}} y$, therefore, using Definition 2.18, one concludes that $b \in F$. Thus, F is an implicative filter. For the converse, assume that F is an implicative filter, that $\Gamma \vdash_{\mathcal{L}} \varphi$, and that $h \in \mathrm{Hom}(\boldsymbol{Fm}, \boldsymbol{A})$ is such that $h\Gamma \subseteq F$. If $\Gamma = \emptyset$ this means that φ is a theorem of $\mathcal{L}$ and by Lemma 2.6 $h\varphi = 1 \in F$.

Otherwise, one considers a proof of φ from Γ in the given axiomatic presentation, and one shows by induction that $h\psi \in F$ for each step ψ in this proof, using that all axioms are theorems and that F is closed under *Modus Ponens*. This shows that $h\varphi \in F$ and thus that F is an $\mathcal{L}$-filter. ⊠

Notice that the second part of the property includes an implicit assumption of finitarity. By (IL4), *Modus Ponens* is a rule of all implicative logics; what is essential in the second part, as Exercise 2.29 shows, is that it is *the only* rule in some axiomatization of the logic. This applies to many well-known logics. Thus, in particular cases some algebraic work may be done to describe the implicative filters in more typical algebraic terms; for instance, in Heyting algebras and in Boolean algebras they coincide with the lattice filters (Exercise 2.31).

Filters and congruences are related in a way that generalizes the constructions in Section 2.2, and goes much further. As a matter of fact, these relations constitute one of the cornerstones of the theory of algebraizable logics. The versions that hold in the case of implicative logics are easy to prove.

DEFINITION 2.29
Let $\mathcal{L}$ be an implicative logic and let F be an $\mathcal{L}$-filter of an arbitrary algebra $\boldsymbol{A}$. The relations $\leqslant_F^{\boldsymbol{A}}$ and $\boldsymbol{\Omega}^{\boldsymbol{A}}F$ on A are defined as follows, for all $a, b \in A$:

$$a \leqslant_F^{\boldsymbol{A}} b \overset{\text{def}}{\iff} a \to b \in F,$$

$$a \equiv b\ (\boldsymbol{\Omega}^{\boldsymbol{A}}F) \overset{\text{def}}{\iff} a \to b, b \to a \in F.$$

Since the $\mathcal{L}$-filters of $\boldsymbol{Fm}$ are the theories of $\mathcal{L}$, this generalizes Definition 2.1, and the notation $\boldsymbol{\Omega}\Gamma$ used there becomes shorthand for the present one $\boldsymbol{\Omega}^{\boldsymbol{Fm}}\Gamma$. Then, Corollary 2.26 can be expressed in a more compact way:

COROLLARY 2.30
Let $\mathcal{L}$ be an implicative logic. An algebra $\boldsymbol{A}$ is an $\mathcal{L}$-algebra if and only if there is $1 \in A$ such that $\{1\}$ is an $\mathcal{L}$-filter and $\boldsymbol{\Omega}^{\boldsymbol{A}}\{1\}$ is the identity relation on A. ⊠

The first basic properties of $\boldsymbol{\Omega}^{\boldsymbol{A}}F$ are those that generalize the ones included in Corollary 2.4; Exercise 2.30 asks you to check that everything works by just generalizing its proof as well:

PROPOSITION 2.31
Let $\mathcal{L}$ be an implicative logic and let F be an $\mathcal{L}$-filter of an arbitrary algebra $\boldsymbol{A}$. The relation $\leqslant_F^{\boldsymbol{A}}$ is a quasi-order on A, the relation $\boldsymbol{\Omega}^{\boldsymbol{A}}F$ is a congruence of $\boldsymbol{A}$, the quotient algebra $\boldsymbol{A}/\boldsymbol{\Omega}^{\boldsymbol{A}}F$ is ordered under the relation $a/\boldsymbol{\Omega}^{\boldsymbol{A}}F \leqslant b/\boldsymbol{\Omega}^{\boldsymbol{A}}F \overset{\text{def}}{\iff} a \leqslant_F^{\boldsymbol{A}} b$, and has a maximum 1 such that $a \in F$ if and only if $a/\boldsymbol{\Omega}^{\boldsymbol{A}}F = 1$ for all $a \in A$, that is, $F = \pi^{-1}\{1\}$, where $\pi\colon \boldsymbol{A} \to \boldsymbol{A}/\boldsymbol{\Omega}^{\boldsymbol{A}}F$ is the canonical projection. Moreover, $\boldsymbol{A}/\boldsymbol{\Omega}^{\boldsymbol{A}}F$ is an $\mathcal{L}$-algebra. Finally, the relation $\leqslant_F^{\boldsymbol{A}}$ is an order if and only if the relation $\boldsymbol{\Omega}^{\boldsymbol{A}}F$ is the identity, and if and only if $\boldsymbol{A}$ is an $\mathcal{L}$-algebra and $F = \{1\}$. ⊠

SKETCH OF THE PROOF: Observe that the $\mathcal{L}$-filters have all the homologous properties used to prove Corollary 2.4 for theories. It is not difficult to see that $\boldsymbol{A}/\boldsymbol{\Omega}^{\boldsymbol{A}}F$

satisfies (LALG2), and by Corollary 2.26, in order to prove (LALG1) it is enough to prove that $\{1\}$ is an $\mathcal{L}$-filter; but this follows from Proposition 2.25 because $F = \pi^{-1}\{1\}$ and π is surjective. This shows the second part. The final part is easily shown using (ALG3). ⊠

The construction in Proposition 2.31 can be applied to obtain two (mutually inverse) isomorphisms:

THEOREM 2.32
Let $\mathcal{L}$ be an implicative logic and let θ be a congruence of an arbitrary algebra $\boldsymbol{A}$, such that $\boldsymbol{A}/\theta$ is an $\mathcal{L}$-algebra. The set

$$F_\theta := \{a \in A : a \equiv a \to a \ (\theta)\} \tag{21}$$

is an $\mathcal{L}$-filter of $\boldsymbol{A}$ such that $\boldsymbol{\Omega}^{\boldsymbol{A}}(F_\theta) = \theta$. Moreover, the functions $F \mapsto \boldsymbol{\Omega}^{\boldsymbol{A}}F$ and $\theta \mapsto F_\theta$ are mutually inverse order isomorphisms between the ordered set of the $\mathcal{L}$-filters of $\boldsymbol{A}$ and that of the congruences θ of $\boldsymbol{A}$ such that $\boldsymbol{A}/\theta$ is an $\mathcal{L}$-algebra; that is, these functions establish that $\mathcal{F}i_{\mathcal{L}}\boldsymbol{A} \cong \mathrm{Con}_{\mathsf{Alg}^*\mathcal{L}}\boldsymbol{A}$.

PROOF: The equation $x \to x \approx \top$ holds in $\boldsymbol{A}/\theta$ by the assumption on it, therefore actually $F_\theta = \pi_\theta^{-1}\{1\}$, where $\pi_\theta \colon \boldsymbol{A} \to \boldsymbol{A}/\theta$ is the canonical projection and 1 is the algebraic constant of $\boldsymbol{A}/\theta$ evaluating $\top$. Then $\{1\} \in \mathcal{F}i_{\mathcal{L}}(\boldsymbol{A}/\theta)$, by Corollary 2.26, and $F_\theta \in \mathcal{F}i_{\mathcal{L}}\boldsymbol{A}$, by Proposition 2.24. Since the algebra $\boldsymbol{A}/\theta$ is assumed to be an $\mathcal{L}$-algebra, it satisfies condition (LALG2), and it is easy to see that this implies the equality $\boldsymbol{\Omega}^{\boldsymbol{A}}(F_\theta) = \theta$. This shows that one of the compositions of the two functions is the identity. As for the other composition, we have to show that $F = F_{\boldsymbol{\Omega}^{\boldsymbol{A}}F}$ for all $F \in \mathcal{F}i_{\mathcal{L}}\boldsymbol{A}$; but this is the same as showing that $F = \pi_{\boldsymbol{\Omega}^{\boldsymbol{A}}F}^{-1}\{1\}$, by the observation at the beginning, and this is proved in Proposition 2.31. Thus, the compositions of the two functions in the two orders are the respective identity relations, therefore the two functions are mutually inverse bijections; since by their definitions they are order-preserving, they are order isomorphisms, which ends the proof. ⊠

Notice that the trick of using $\top$ as shorthand for $x \to x$ can be made to work only for the algebras in $\mathsf{Alg}^*\mathcal{L}$; since the algebra in the statement of the theorem is arbitrary, it would be incorrect to write $a \equiv 1 \ (\theta)$ in (21). However, this can be done when the similarity type explicitly includes a constant symbol $\top$ such that $\vdash_{\mathcal{L}} \top$; then $\top^{\boldsymbol{A}}$ is a constant in every algebra $\boldsymbol{A}$, and after defining $1 := \top^{\boldsymbol{A}}$ we can also write $1/\theta$ instead of F_θ, thus obtaining a more intuitive description of the second function.

Notice that, since $\mathsf{Alg}^*\mathcal{L}$ is a generalized quasivariety, by Proposition 1.75 the set of relative congruences of any algebra $\boldsymbol{A}$, $\mathrm{Con}_{\mathsf{Alg}^*\mathcal{L}}\boldsymbol{A}$, is a complete lattice. Recall that an order isomorphism between two complete lattices is always a complete isomorphism.

Since 1986, the function $\boldsymbol{\Omega}^{\boldsymbol{A}}$ considered in Theorem 2.32 has been called ***the Leibniz operator***, and this result is informally called *the Isomorphism Theorem*:

$$\boldsymbol{\Omega}^{\boldsymbol{A}} \colon \mathcal{F}i_{\mathcal{L}}\boldsymbol{A} \cong \mathrm{Con}_{\mathsf{Alg}^*\mathcal{L}}\boldsymbol{A}.$$

What may be surprising is that this holds *for any algebra* $\boldsymbol{A}$. These isomorphisms appear, though rather implicitly, in the book by Rasiowa; anyway, its particular cases had been known since many years, for instance in the case of classical logic $\mathcal{C}\ell$, for the class BA of Boolean algebras; see Exercise 2.31. A more general version of this isomorphism is proved, for a larger class of logics, in Section 3.2, and the Leibniz operator is studied in full generality in Section 4.2, where the name is justified.

THEOREM 2.33

Let $\mathcal{L}$ be an implicative logic and let F be an $\mathcal{L}$-filter of an arbitrary algebra $\boldsymbol{A}$. The function $G \mapsto \pi G := G/\boldsymbol{\Omega}^{\boldsymbol{A}}F$ is an isomorphism between the complete lattices $(\mathcal{F}i_{\mathcal{L}}\boldsymbol{A})^F$ and $\mathcal{F}i_{\mathcal{L}}(\boldsymbol{A}/\boldsymbol{\Omega}^{\boldsymbol{A}}F)$.

PROOF: Recall from Definition 1.34 that $(\mathcal{F}i_{\mathcal{L}}\boldsymbol{A})^F := \{G \in \mathcal{F}i_{\mathcal{L}}\boldsymbol{A} : F \subseteq G\}$. Observe that the assumptions are those of Proposition 2.31. Therefore, $\boldsymbol{A}/\boldsymbol{\Omega}^{\boldsymbol{A}}F$ is an $\mathcal{L}$-algebra and $\pi F = \{1\}$, which is the least $\mathcal{L}$-filter of $\boldsymbol{A}/\boldsymbol{\Omega}^{\boldsymbol{A}}F$. Now, let $H \in \mathcal{F}i_{\mathcal{L}}(\boldsymbol{A}/\boldsymbol{\Omega}^{\boldsymbol{A}}F)$. Then $\pi^{-1}H \in \mathcal{F}i_{\mathcal{L}}\boldsymbol{A}$ by Proposition 2.24, and $F \subseteq \pi^{-1}H$ because $\{1\} \subseteq H$, and $\pi\pi^{-1}H = H$ by the surjectivity of π. The key point is to show that if $G \in \mathcal{F}i_{\mathcal{L}}\boldsymbol{A}$ with $F \subseteq G$, then $G = \pi^{-1}\pi G$. That $G \subseteq \pi^{-1}\pi G$ is always true. Now let $a \in \pi^{-1}\pi G$. Thus, $\pi a = \pi b$ for some $b \in G$, and then $a \equiv b\ (\boldsymbol{\Omega}^{\boldsymbol{A}}F)$, that is, $a \to b, b \to a \in F$. Since $F \subseteq G$, $b \in G$ and G is an $\mathcal{L}$-filter, by *Modus Ponens* it follows that $a \in G$. Now, since π is surjective, we can use Proposition 2.25 and conclude that $\pi G \in \mathcal{F}i_{\mathcal{L}}(\boldsymbol{A}/\boldsymbol{\Omega}^{\boldsymbol{A}}F)$. Obviously π and π^{-1} are both order-preserving, therefore they are mutually inverse lattice isomorphisms, as desired. ⊠

This result can be compared to the Correspondence Theorem of universal algebra; see Exercise 2.32. Moreover, as a particular case, a classical result of traditional algebraic logic is recovered:

COROLLARY 2.34

If $\mathcal{L}$ is an implicative logic, then its lattice of theories is isomorphic to the lattice of $\mathcal{L}$-filters of the pure Lindenbaum-Tarski algebra of $\mathcal{L}$; that is, $\mathcal{T}h\mathcal{L} \cong \mathcal{F}i_{\mathcal{L}}(\boldsymbol{Fm}/\boldsymbol{\Omega}\Gamma_0)$, where Γ_0 is the set of theorems of $\mathcal{L}$. The isomorphism is established by the function $\Gamma \mapsto \Gamma/\boldsymbol{\Omega}\Gamma_0$. ⊠

For particular logics this means that the lattice of theories of the logic is isomorphic to a certain well-determined lattice of subsets of an algebra of a specific kind (for example, the lattice filters of a Boolean algebra, for $\mathcal{C}\ell$; or of a Heyting algebra, for $\mathcal{I}\ell$; see Exercise 2.31), and therefore the properties of these (better studied) lattices "transfer" to the lattice of theories of the logic.

Exercises for Section 2.3

EXERCISE 2.18. Check that the set $\{1\}$ is not an $\mathcal{I}\ell_{\to}$-filter of the algebra $\boldsymbol{A}_1$ of Exercise 2.14. Check that the set $\{1\}$ is an $\mathcal{I}\ell$-filter of the five-element Heyting algebra shown at the right, but not a $\mathcal{C}\ell$-filter.

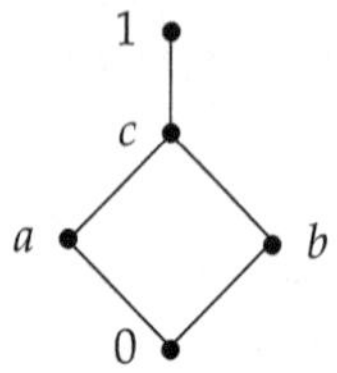

HINT. You can check the second part using axiomatic presentations of the two logics, or using the results in the first two points of Exercise 2.31.

EXERCISE 2.19. Check all the statements in Proposition 2.22, and moreover prove the following:

1. $\mathcal{L}$ is the inconsistent logic if and only if for any $\boldsymbol{A}$, $\mathcal{F}i_{\mathcal{L}}\boldsymbol{A} = \{A\}$.
2. $\mathcal{L}$ is the almost inconsistent logic if and only if for any $\boldsymbol{A}$, $\mathcal{F}i_{\mathcal{L}}\boldsymbol{A} = \{\emptyset, A\}$.

EXERCISE 2.20. Let $\mathcal{L}$ be a logic with theorems and $\boldsymbol{A}$ any algebra. Prove that if for some $a \in A$ the set $\{a\}$ is an $\mathcal{L}$-filter, then it is the smallest $\mathcal{L}$-filter of $\boldsymbol{A}$.

EXERCISE 2.21. Let $\langle \to, \top \rangle$ be a language of type $\langle 2, 0 \rangle$, and let $\mathcal{L}$ be the logic in this language having the constant $\top$ as its single axiom, and *Modus Ponens* $(x, x \to y \rhd y)$ as its single inference rule. Find an algebra of this similarity type that provides a counterexample showing that, in general, the set $\{h\varphi : \vdash_{\mathcal{L}} \varphi,\ h \in \mathrm{Hom}(\boldsymbol{Fm}, \boldsymbol{A})\}$ need not be an $\mathcal{L}$-filter of $\boldsymbol{A}$.

HINT. Start by finding the theorems of this logic.

EXERCISE 2.22. Let $\mathcal{L}$ be a logic with theorems and $\boldsymbol{A}$ any algebra. Prove that if for some $a \in A$ the set $\{a\}$ is a subalgebra of $\boldsymbol{A}$, then $\{a\} \subseteq F$ for any $\mathcal{L}$-filter F of $\boldsymbol{A}$. If the example you constructed in Exercise 2.21 is simple enough, it will show that the set $\{a\}$ need not be itself an $\mathcal{L}$-filter.

EXERCISE 2.23. Let $\mathcal{L}$ be a logic, $\boldsymbol{A}$ an algebra, $h \in \mathrm{Hom}(\boldsymbol{Fm}, \boldsymbol{A})$ and $\Gamma \vdash_{\mathcal{L}} \varphi$. Prove that for each $z \in V$ and each $a \in A$ there are $\Gamma' \cup \{\varphi'\} \subseteq Fm$ and $h' \in \mathrm{Hom}(\boldsymbol{Fm}, \boldsymbol{A})$ such that z does not occur in $\Gamma' \cup \{\varphi'\}$, $\Gamma' \vdash_{\mathcal{L}} \varphi'$, $h'\Gamma' = h\Gamma$, $h'\varphi' = h\varphi$, and $h'z = a$. Show that a similar property holds for any countable set of variables with a denumerable complement, instead of a single variable z.

HINT. Consider a bijection $\sigma\colon V \to V \smallsetminus \{z\}$, use structurality, and define h' accordingly.

EXERCISE 2.24. Let $\boldsymbol{B}$ be a subalgebra of the formula algebra $\boldsymbol{Fm}$ (of some similarity type), $\Gamma \subseteq F$, and $\mathcal{L}$ a logic of that type. Prove that $Fg_{\mathcal{L}}^{\boldsymbol{B}}\Gamma = (C_{\mathcal{L}}\Gamma) \cap B$.

EXERCISE 2.25. Let $\mathcal{L}$ be a logic, $\boldsymbol{A}$ an algebra, and $X \subseteq A$. Generalize the result in Theorem 2.23.2 to not necessarily finitary logics, by defining a family $\{X_\alpha\}_{\alpha<\kappa^+}$, where κ is the cardinality of the set of formulas, such that $Fg_{\mathcal{L}}^{\boldsymbol{A}}X = \bigcup_{\alpha<\kappa^+} X_\alpha$.

EXERCISE 2.26. Let $\mathcal{L}$ be a finitary logic and $\boldsymbol{A}$ an algebra. For every $X \cup \{a\} \subseteq A$, prove that $a \in Fg_{\mathcal{L}}^{\boldsymbol{A}}X$ if and only for some $n \geqslant 1$ there are $a_1, \dots, a_n \in A$ such that $a_n = a$ and for each $i = 1, \dots, n$ either $a_i \in X$ or there is some finite $\Gamma \cup \{\varphi\} \subseteq Fm$ with $\Gamma \vdash_{\mathcal{L}} \varphi$ and some $h \in \mathrm{Hom}(\boldsymbol{Fm}, \boldsymbol{A})$ such that $h\Gamma \subseteq \{a_1, \dots, a_{i-1}\}$ and $h\varphi = a_i$.

COMMENT. The sequence $\langle a_1, \dots, a_n \rangle$ is called an "$\mathcal{L}$-proof in A" in [70] because, to a certain extent, it mimics the structure of an ordinary proof in an axiomatic system, i.e., it is a finite sequence of the results of the application of rules of $\mathcal{L}$, used with elements of an algebra (which are the values of formulas); however, notice that in each step the evaluation can change, and this is important. If there is a Hilbert-style presentation of $\mathcal{L}$, then in each step one can use one rule of the presentation (instead of a generic finite consequence of $\mathcal{L}$). A totally general version of this property (i.e., for not necessarily finitary logics) can be obtained if the proof system admits proofs of arbitrary length, in the spirit of Exercise 2.25.

EXERCISE 2.27. Consider the natural embedding of $\mathbf{2}$ into $\mathbf{2} \times \mathbf{2}$, and show that while the set $\{0,1\}$ is obviously a $\mathcal{C\ell}$-filter of $\mathbf{2}$, its image is not a $\mathcal{C\ell}$-filter of $\mathbf{2} \times \mathbf{2}$. This shows that the direct image of an $\mathcal{L}$-filter by a homomorphism need not be an $\mathcal{L}$-filter. Check that Exercise 1.63 provides another counterexample of this fact.

EXERCISE 2.28. Use Corollary 2.26 and the comment that preceds it to show that $\mathsf{Alg}^*\mathcal{I\ell}_{\rightarrow}$ is the class of algebras $\boldsymbol{A} = \langle A, \rightarrow \rangle$ of type $\langle 2 \rangle$ in which the following equations and quasi-equation hold, with $\top$ being shorthand for $x \rightarrow x$ (which by the first equation is an algebraic constant):

$$\begin{aligned}
& y \rightarrow y \approx \top \\
& x \rightarrow (y \rightarrow x) \approx \top \\
& (x \rightarrow (y \rightarrow z)) \rightarrow ((x \rightarrow y) \rightarrow (x \rightarrow z)) \approx \top \\
& (x \rightarrow y \approx \top \;\wedge\; y \rightarrow x \approx \top) \boldsymbol{\rightarrow} x \approx y
\end{aligned}$$

The algebras in this class are usually called ***Hilbert algebras***, and are denoted by HiA in Table 2 on page 85.

COMMENT. By definition this class is a quasivariety. It is well known that it is actually a variety; an explicit equational presentation is displayed on page 441, but the quasi-equational one just given is more convenient for logical purposes. In some works, such as [208], they are presented in a language of type $\langle 2,0 \rangle$, incorporating $\top$ as a primitive constant symbol.

EXERCISE 2.29. Let $\mathcal{I}mp$ be the weakest implicative logic (Exercise 2.13). Find an implicative algebra $\boldsymbol{A}$ such that the $\mathcal{I}mp$-filters of $\boldsymbol{A}$ do not coincide with its implicative filters. Use this fact to prove that $\mathcal{I}mp$ cannot be axiomatized by a Hilbert style system whose only proper inference rule is *Modus Ponens*.

EXERCISE 2.30. Prove Proposition 2.31 in the way indicated in the text.

EXERCISE 2.31.

1. Prove that $\mathsf{Alg}^*\mathcal{I\ell} = \mathsf{HA}$, the class of Heyting algebras, and that on these algebras the notion of an implicative filter coincides with that of a lattice filter. Then, use Proposition 2.28 to show that on a Heyting algebra the notion of an $\mathcal{I\ell}$-filter and the notion of a lattice filter coincide.
2. Prove that $\mathsf{Alg}^*\mathcal{C\ell} = \mathsf{BA}$, and use point 1 to conclude that on Boolean algebras the notion of a $\mathcal{C\ell}$-filter and that of lattice filter coincide.
3. In the two cases, check that the isomorphism in Theorem 2.32 amounts to the usual isomorphism between lattice filters and congruences in the involved algebras.

   HINT. The procedure for characterizing the algebras is similar to that in Exercise 2.28, but the technicalities may be more involved, depending on your knowledge of Heyting and Boolean algebras and on how much does the axiomatization chosen for the logics differ from that chosen for the algebras.

EXERCISE 2.32. Let $\mathcal{L}$ be an implicative logic such that $\mathsf{Alg}^*\mathcal{L}$ is a variety. Obtain Theorem 2.33 as a consequence of the Correspondence Theorem of universal algebra (see [51, Theorem II.6.20], for instance).

EXERCISE 2.33. Fact VIII.13.1 of [208] contains the following result: If Γ is a theory of an implicative logic $\mathcal{L}$, then for each $\Gamma' \subseteq Fm$ the set $\Gamma'_\Gamma := \{\alpha/\Omega\Gamma : \Gamma \cup \Gamma' \vdash_\mathcal{L} \alpha\}$ is an $\mathcal{L}$-filter of the implicative algebra $\boldsymbol{Fm}/\Omega\Gamma$, and this filter is generated by the set $\Gamma'/\Omega\Gamma = \{\gamma/\Omega\Gamma : \gamma \in \Gamma'\}$. Prove it, using as many of the results in the present section as possible.

2.4 Extensions of the Lindenbaum-Tarski process

There are two ways to generalize the Lindenbaum-Tarski process in an inessential way, thus obtaining larger classes of logics to which the process can be applied *mutatis mutandis*. The difference lies in whether the process is viewed mainly as *implication-based*, or mainly as *equivalence-based*. Some aspects of the process are essentially equivalence-based, as they focus on the process of factoring out the formula algebra in order to obtain a similar algebra which belongs to a class that can serve as a semantics for the logic. However, others are essentially implication-based: we want the quotient algebras to be ordered and have a maximum, and we saw that this is obtained by using a quasi-order defined in an implicative way. In the Rasiowa approach, the implicational aspect takes the lead, as the equivalence is obtained by using the symmetrized formula $(x \to y) \wedge (y \to x)$ or the symmetrized set $\{x \to y, y \to x\}$, so what is required of it is formulated through properties of $\to$.

Implication-based extensions

1. The first inessential generalization has been already incorporated into this exposition, and as a matter of fact it was already suggested by Wójcicki in his seminal paper [246, 1973], when Rasiowa's book was still to appear: it is to allow an arbitrary formula in two variables to play the role of $x \to y$ instead of requiring (as Rasiowa did) that $\to$ is effectively among the primitive connectives of the language.
2. The other generalization, also already incorporated, is that the similarity type $\boldsymbol{L}$ may contain connectives of arbitrary finite arity (Rasiowa assumed that there are no connectives of arity greater than 2).
3. The next one would be to replace the formula $x \to y$ by *a set of formulas* in two variables $I(x,y) \subseteq Fm$ and restate the conditions in the definition in the obvious way:

   (IL1′) $\vdash_\mathcal{L} I(x,x)$.

   (IL2′) $I(x,y) \cup I(y,z) \vdash_\mathcal{L} I(x,z)$.

   (IL3′) $\bigcup_{i=1}^n \big(I(x_i,y_i) \cup I(y_i,x_i)\big) \vdash_\mathcal{L} I(\lambda x_1 \dots x_n, \lambda y_1 \dots y_n)$, for each $\lambda \in L$ of arity $n \geqslant 1$.

   (IL4′) $x,\ I(x,y) \vdash_\mathcal{L} y$.

   (IL5′) $x \vdash_\mathcal{L} I(y,x)$.

It is an easy (but instructive) task to check that everything works, *without any further assumption on the set* $I(x,y)$. The quasi-order associated with a theory Γ and its Leibniz congruence afe now defined by:

$$\begin{aligned} \alpha \leqslant_\Gamma \beta &\stackrel{\text{def}}{\iff} I(\alpha,\beta) \subseteq \Gamma \\ \alpha \equiv \beta \ (\boldsymbol{\Omega}\Gamma) &\stackrel{\text{def}}{\iff} I(\alpha,\beta) \cup I(\beta,\alpha) \subseteq \Gamma . \end{aligned} \tag{22}$$

It is clear that that by (IL4′) and (IL5′) all formulas in each theory become identified under $\boldsymbol{\Omega}\Gamma$ and form the top element 1 in the ordered quotient. In particular all theorems are identified, and any of them can play the role of $x \to x$, that is, it can act as $\top$ and be evaluated as the top element 1 of all $\mathcal{L}$-algebras; these are defined as those having an element 1 and satisfying the following properties:

(LALG1) For all $\Gamma \cup \{\varphi\} \subseteq Fm$ and all $h \in \operatorname{Hom}(\boldsymbol{Fm}, \boldsymbol{A})$, if $\Gamma \vdash_{\mathcal{L}} \varphi$ and $h\Gamma \subseteq \{1\}$, then $h\varphi = 1$.

(LALG2′) For all $a, b \in A$, if $I^{\boldsymbol{A}}(a,b) = \{1\}$ and $I^{\boldsymbol{A}}(b,a) = \{1\}$, then $a = b$.

If the set $I(x,y)$ is *finite*, then the variations are nil, and basically consist in complicating some mathematical expressions. If it is not, then one so-to-speak important (though not essential) variation is that the class $\mathsf{Alg}^*\mathcal{L}$ need not be a quasivariety even when $\mathcal{L}$ is finitary, because for instance condition (LALG2′) need not be expressible by a quasi-equation.

This extension has been considered in the literature only recently, and for theoretical reasons, in [61, 2010], where the logics satisfying it are called *Rasiowa-implicational*, while our implicative logics are called "Rasiowa-implicative". As far as I know, there is no natural example requiring more than one formula to construct the implication set; in [61] there is an ad hoc example of a logic with two implication connectives $\to_1$ and $\to_2$ such that the set $\{x \to_1 y, x \to_2 y\}$ satisfies conditions (IL1′)–(IL5′) but where no single binary connective can exist that satisfies (IL1)–(IL5).

4. The first substantial extension corresponds to not requiring that all formulas in a theory (neither all theorems) are identified in the quotient, so that $\top$ no longer makes sense as a general shorthand for any of them; this implies throwing (IL5) or (G) away. If there is a way to describe *equationally* the equivalence class $\Gamma/\boldsymbol{\Omega}\Gamma$, this produces sensible versions of the Algebraic Completeness Theorem and of the Algebraizability Theorem, after replacing the *semantic equation* $x \approx \top$ that is implicit in 2.5, 2.9 and 2.12 by another one. The most general way of doing this, added to the extension considered in the previous point, leads simply to the class of *algebraizable logics*, which is studied in Chapter 3; the additional assumption makes them more properly "equivalence-based" than "implication-based". However, there is a large number of logics in the literature that do so in a more "implicative" way. The most typical is to have a constant term $\top$ such that $\vdash_{\mathcal{L}} \top$, and require a particular, weaker version of (IL5):

(IL5w) $x \vdash_{\mathcal{L}} \top \to x$

This weaker condition no longer implies (G), so members of Γ are not identified under $\boldsymbol{\Omega}\Gamma$, but together with (IL1)–(IL4), it implies that for every theory Γ and every $\alpha \in Fm$, $\alpha \in \Gamma$ if and only if $\top \leqslant_{\Gamma} \alpha$ if and only if $\top/\boldsymbol{\Omega}\Gamma \leqslant \alpha/\boldsymbol{\Omega}\Gamma$; lacking (IL5), the element $\top/\boldsymbol{\Omega}\Gamma$ is not the top of the quotient algebra. Then, the semantic equation $x \approx \top$ is replaced by the inequation $\top \preccurlyeq x$ and the definition of $\mathcal{L}$-algebras would become:

(LALG1″) For all $\Gamma \cup \{\varphi\} \subseteq Fm$ and all $h \in \mathrm{Hom}(\boldsymbol{Fm}, \boldsymbol{A})$, if $\Gamma \vdash_{\mathcal{L}} \varphi$ and $\top^{\boldsymbol{A}} \leqslant h\gamma$ for all $\gamma \in \Gamma$, then $\top^{\boldsymbol{A}} \leqslant h\varphi$.

(LALG2″) For all $a, b \in A$, if $\top^{\boldsymbol{A}} \leqslant a \to b$ and $\top^{\boldsymbol{A}} \leqslant b \to a$, then $a = b$.

Thus, $\mathsf{Alg}^*\mathcal{L}$ would have to be considered specifically as a class of *ordered algebras*, and a Completeness Theorem would be obtained, but in general an Algebraizability Theorem would not, unless the order relation can be expressed equationally (as in lattices, where $a \leqslant b$ is equivalent to $a \wedge b = a$).

Examples of logics where this happens are several substructural logics related to Lambek's sequent calculus FL, and their extensions, studied in [123]; their $\mathcal{L}$-algebras are certain varieties of residuated lattices, as reviewed in Example 3.34.2.

Of course, this extension can also be performed for a set $I(x,y)$ of formulas instead of a single formula $x \to y$, and everything would be changed *mutatis mutandis*, considering the condition

(IL5w′) $x \vdash_{\mathcal{L}} I(\top, x)$.

5. A very systematic investigation of a hierarchy of logics subject to algebraizability processes like that for implicative logics, based on a very general notion of what a set of implication formulas is or may be, has been proposed by Cintula and Noguera in [61]; the resulting *implicational hierarchy* encompasses the Leibniz hierarchy to be discussed in Chapter 6 together with classes not definable under the Leibniz paradigm.

Equivalence-based extensions

6. Let us examine what happens if condition (IL5) is *replaced* by the weaker condition (G); or, in the extended case outlined in paragraph 3 above, if condition (IL5′) is replaced by

   (G′) $x, y \vdash_{\mathcal{L}} I(y, x)$.

   It is clear from Theorem 2.2 that then all formulas in a theory Γ would still be identified in the quotient $\boldsymbol{Fm}/\boldsymbol{\Omega}\Gamma$. Thus, the shorthand $\top$ representing *any* of the theorems would still make sense, and the associated algebras would have an algebraic constant $\top^{\boldsymbol{A}}$. The difference is that it would not necessarily be the top element; instead, it would simply be a *designated* element, which would nevertheless enable the Algebraic Completeness Theorem 2.9 and the stronger Algebraizability Theorem 2.12 to work without any essential changes; the Definition 2.5 of $\mathcal{L}$-algebras is either the same or as in paragraph 3 above.

The resulting logics are called *regularly algebraizable logics*, and appear in Sections 3.4, 4.4 and 6.5 of this book. Typical non-implicative examples are the equivalence fragments of classical and of intuitionistic logics.

However, lack of condition (IL5) or (IL5′) entails that, in some sense, in this move *the implicative character is somehow lost*: By symmetry, from condition (G′) it follows that

$$x, y \vdash_{\mathcal{L}} I(x,y) \cup I(y,x).$$

Then, if we define $\Delta(x,y) := I(x,y) \cup I(y,x)$, it is easy to see that this set also satisfies the same conditions (IL1′)–(IL4′)[†] plus (G′), and also that

$$\Delta(x,y) \vdash_{\mathcal{L}} \Delta(y,x), \tag{23}$$

while it does not satisfy (IL5); and notice that this set is the one used to define the congruence (22), and condition (LALG2′) in the definition of $\mathcal{L}$-algebras can be equivalently expressed by

$$\text{For all } a, b \in A, \text{ if } \Delta^{\boldsymbol{A}}(a,b) = \{\top^{\boldsymbol{A}}\}, \text{ then } a = b.$$

Hence we might start using some $\Delta(x,y)$ subject to similar conditions plus symmetry,[‡] and we realize that we are actually formalizing *a notion of an equivalence* rather than one of an implication.

7. If we go one step further, and simply delete (IL5) or (IL5′) with no replacement for them, we are left with just conditions (IL1′)–(IL4′). The resulting logics are the so-called *equivalential logics*, historically the first real generalization of the Lindenbaum-Tarski process, due to Czelakowski in 1981 and here studied in Section 6.3. It is easy to see in Theorem 2.2 and in the proof of Theorem 2.8 that the process can still be carried on, but there is no workable description of the $\mathcal{L}$-algebras and of the quotient set $\Gamma/\Omega\Gamma$. Thus, this would be a truly essential extension, and this explains why Czelakowski, in the second part of [67], also studied the equivalential logics that moreover satisfy condition (G′).

Conclusion

It seems clear that the essential points for the Lindenbaum-Tarski process to work in *all* its aspects, though reasonably relaxed, can be abstracted to the following two *definability conditions*:

- One needs to be able to factor out the formula algebra by a congruence $\Omega\Gamma$ that is definable from Γ through some set of formulas in two variables in the form of (22).

[†] With (IL3′) slightly reformulated as it would not need to make the symmetrization explicitly.

[‡] To make things even clearer, in Theorem 6.60 we see that if a set $\Delta(x,y)$ satisfies conditions (IL1′), (IL3′) and (IL4′), then it also satisfies conditions (IL2′) and (23).

- One needs to be able to define the set $\Gamma/\Omega\Gamma$ in the quotient algebra $\boldsymbol{Fm}/\Omega\Gamma$ (be it a single class or not) by some condition that amounts to an equation or a set of equations, and which can then be used in the definition of $\mathcal{L}$-algebras, in the Algebraic Completeness Theorem, and in the Algebraizability Theorem.

Imposing these conditions in varying degrees of generality deploys the so-called *Leibniz hierarchy* that gives structure to a *very large* class of logics as studied with the methods of abstract algebraic logic. The core classes in this hierarchy, the *algebraizable logics* and some of its subclasses, are studied in Chapter 3, while the larger classes are dealt with in Chapter 6.

2.5 Two digressions on first-order logic

This section gathers two unrelated digressions having in common that they refer to first-order logic. The first one aims to somehow justify that the algebraic framework already discussed in this chapter is the *natural* way to study of the logic of the sentential connectives of *ordinary first-order logic*. The second one aims to give just a glimpse into the (much more intrincate) world of the algebraization of first-order logic in full.

The logic of the sentential connectives of first-order logic

The study of the logical properties of certain sentential connectives can be approached from a first-order perspective, as the sentential connectives are part of the construction of the formulas of a first-order language. In such a perspective, one may look at the restriction of the consequence relation of the first-order logic to the quantifier-free and equality-free formulas, that is, disregarding everything that is not intrinsically "sentential". This digression intends to show that, in the case of classical first-order logic, this boils down to the ordinary, purely algebraic framework of sentential two-valued logic, and one can dispense with atomic formulas of relational kind, relational structures, assignments, and the like.

Assume an arbitrary first-order language $\mathscr{L}$ and a set V of individual variables as given, and consider:

$$\begin{aligned} T(\mathscr{L}) &:= \text{the set of terms of } \mathscr{L}\text{, constructed from } V \\ At^{\circ}(\mathscr{L}) &:= \text{the set of atomic equality-free formulas of } \mathscr{L} \\ Fm^{\circ}(\mathscr{L}) &:= \text{the set of quantifier-free equality-free formulas of } \mathscr{L} \end{aligned}$$

The formulas in $Fm^{\circ}(\mathscr{L})$ are built from those in $At^{\circ}(\mathscr{L})$ with the help of the sentential connectives $\boldsymbol{\wedge},\boldsymbol{\vee},\boldsymbol{\rightarrow},\boldsymbol{\neg}$ of classical, first-order logic;[†] we can thus consider the algebra $\boldsymbol{F} := \langle Fm^{\circ}(\mathscr{L}),\boldsymbol{\wedge},\boldsymbol{\vee},\boldsymbol{\rightarrow},\boldsymbol{\neg}\rangle$ of the same similarity type as

[†] This presentation, and the construction of Theorem 2.35, can be done in any similarity type appropriate for classical logic and Boolean algebras. Writing the connectives in boldface emphasizes that here we are dealing with first-order formulas, and is coherent with the way quasi-equations are written (Section 1.6) and with the main construction in Section 4.5.

the Boolean algebra $\mathbf{2} = \langle 2, \wedge, \vee, \rightarrow, \neg \rangle$. Therefore, $\boldsymbol{F}$ is (isomorphic to) the free algebra, of the similarity type given by the classical sentential connectives, generated by the set $At^{\circ}(\mathscr{L})$. The symbol $\vDash$ represents, as in ordinary first-order logic, both the satisfaction relation in an $\mathscr{L}$-structure by an assignment ($\mathscr{A} \vDash \alpha \, [\![\overline{a}]\!]$), and the resulting semantic consequence between first-order formulas (not necessarily sentences). Then:

THEOREM 2.35
If $\Gamma \cup \{\varphi\} \subseteq Fm^{\circ}(\mathscr{L})$, then $\Gamma \vDash \varphi$ if and only if for every $h \in \mathrm{Hom}(\boldsymbol{F}, \mathbf{2})$, if $h\gamma = 1$ for all $\gamma \in \Gamma$, then $h\varphi = 1$.

PROOF: ($\Rightarrow$) Assume that $\Gamma \vDash \varphi$, and let $h \in \mathrm{Hom}(\boldsymbol{F}, \mathbf{2})$ be such that $h\gamma = 1$ for all $\gamma \in \Gamma$. We construct a particular $\mathscr{L}$-structure $\mathscr{A}$, sometimes called *the Herbrand universe*: Take the set $T(\mathscr{L})$ of terms as its universe, evaluate the constants and function symbols as themselves,

$$c^{\mathscr{A}} := c \qquad \text{for each constant symbol } c$$

$$f^{\mathscr{A}}(t_1, \ldots, t_n) := f t_1 \ldots t_n \qquad \text{for each function symbol } f$$

and evaluate the relation symbols as follows: if $t_1, \ldots, t_n \in T(\mathscr{L})$, then

$$\langle t_1, \ldots, t_n \rangle \in r^{\mathscr{A}} \overset{\text{def}}{\iff} h(r t_1 \ldots t_n) = 1 \qquad \text{for each relation symbol } r.$$

Consider the "identity assignment" $\overline{a}(x) := x$ for all $x \in V$. A straightforward induction on the construction of terms readily shows that $t^{\mathscr{A}}(\overline{a}) = t$ for all $t \in T(\mathscr{L})$. Taking this last fact into account, the definition of $r^{\mathscr{A}}$ can be rewritten in the form

$$\mathscr{A} \vDash r t_1 \ldots t_n \, [\![\overline{a}]\!] \iff h(r t_1 \ldots t_n) = 1.$$

because the atomic equality-free formulas are those of the form $r t_1 \ldots t_n$. Now, by induction on the construction of the formulas in $Fm^{\circ}(\mathscr{L})$, it is easy to extend this property to all of them, so that for all $\alpha \in Fm^{\circ}(\mathscr{L})$,

$$\mathscr{A} \vDash \alpha \, [\![\overline{a}]\!] \iff h\alpha = 1. \tag{24}$$

It is in this induction proof where the fact that h is a homomorphism of the sentential connectives is used. To check just one case, observe that $\mathscr{A} \vDash \alpha \boldsymbol{\wedge} \beta \, [\![\overline{a}]\!]$ if and only if both $\mathscr{A} \vDash \alpha \, [\![\overline{a}]\!]$ and $\mathscr{A} \vDash \beta \, [\![\overline{a}]\!]$, which by the inductive hypothesis holds if and only if $h\alpha = 1$ and $h\beta = 1$. Now, looking at the table of the operation $\wedge$ in $\mathbf{2}$ (Example 1.11), we see that this holds if and only if $h\alpha \wedge h\beta = 1$; and using that h is a homomorphism with respect to $\boldsymbol{\wedge}$ and $\wedge$, this holds if and only if $h(\alpha \boldsymbol{\wedge} \beta) = 1$, as is to be proved. The other connectives are handled similarly. Now, applying (24) to the assumption on h, it follows that $\mathscr{A} \vDash \gamma \, [\![\overline{a}]\!]$ for all $\gamma \in \Gamma$. Since $\Gamma \vDash \varphi$, it follows that $\mathscr{A} \vDash \varphi \, [\![\overline{a}]\!]$, which by (24) again gives $h\varphi = 1$, as is to be proved.

($\Leftarrow$) Assume that the condition holds, and let $\mathscr{A}$ be any $\mathscr{L}$-structure and $\overline{a}$ any assignment in it such that $\mathscr{A} \vDash \gamma \, [\![\overline{a}]\!]$ for all $\gamma \in \Gamma$. Define a function

$h\colon Fm^\circ(\mathscr{L}) \to 2$ by putting, for all $\alpha \in Fm^\circ(\mathscr{L})$,

$$h\alpha = 1 \overset{\text{def}}{\iff} \mathscr{A} \vDash \alpha \; [\![\bar{a}]\!]. \tag{25}$$

Then I claim that $h \in \mathrm{Hom}(\boldsymbol{F}, \mathbf{2})$. This is a consequence of the fact that the rules for extending the relation of satisfaction $\vDash$ under a fixed assignment $\bar{a}$ from atomic formulas to quantifier-free formulas actually use the conditions on the sentential connectives that correspond to their truth-tables in the algebraic structure of **2**; again, a simple example will suffice to convince you of this: We know that $\mathscr{A} \vDash \alpha \boldsymbol{\wedge} \beta \; [\![\bar{a}]\!]$ if and only if $\mathscr{A} \vDash \alpha \; [\![\bar{a}]\!]$ and $\mathscr{A} \vDash \beta \; [\![\bar{a}]\!]$. After (25), this says that $h(\alpha \boldsymbol{\wedge} \beta) = 1$ if and only if $h\alpha = 1$ and $h\beta = 1$. Since 2 is a two-element set, this implies that $h(\alpha \boldsymbol{\wedge} \beta) = 0$ if and only if $h\alpha = 0$ or $h\beta = 0$. Looking at the table of the operation $\wedge$ in **2**, we realize that this is equivalent to saying that $h(\alpha \boldsymbol{\wedge} \beta) = (h\alpha) \wedge (h\beta)$, that is, h is a homomorphism with respect to $\boldsymbol{\wedge}$ and $\wedge$. A similar reasoning applies to all the other sentential connectives. Thus the claim is proved. Since by (25) the initial assumption on $\mathscr{A}$ and $\bar{a}$ says that $h\gamma = 1$ for all $\gamma \in \Gamma$, we can conclude that $h\varphi = 1$, which by (25) again, means that $\mathscr{A} \vDash \varphi \; [\![\bar{a}]\!]$. This proves that $\Gamma \vDash \varphi$. ⊠

Homomorphisms into **2** have appeared because of the way the satisfaction relation $\mathscr{A} \vDash \varphi \; [\![\bar{a}]\!]$ is defined. In the proof we recognize that the inductive clauses that tell us how to extend this relation from atomic equality-free formulas to complex formulas are actually describing the usual algebraic structure of **2**. The internal construction of the atomic equality-free formulas $rt_1 \dots t_n$ has not played any role in the argument. So it seems reasonable to replace them by an unspecified set V of "sentential variables". This leads us to the ordinary construction of the algebra of sentential formulas and the algebraic/semantic presentation of the consequence of classical sentential logic from the two-valued truth-tables arising from **2**, as is done in Example 1.11.

Although this is a very simple and particular example, this might convince some that the study of the properties of the sentential connectives in first-order logic can be reduced to that of the same connectives in the typical, algebraic framework of sentential logic.

The algebraic study of first-order logics

The algebraization (in a broad sense of the term) of first-order logic is an independent branch of algebraic logic. Probably, the most basic question when facing it is to decide *what to do with quantifiers*. In the completeness proof of a sentential logic by the Lindenbaum-Tarski process, the algebras arise by factoring out the algebra of formulas, and their algebraic operations arise from the sentential connectives of the logic under study. Now, in a first-order language, quantifiers can also be viewed as "operations" on formulas, in the sense that when applied to a formula they produce other formulas. Thus, the key point is how these operations are handled. In the literature we can find two radically different ways to do so.

- The "soft" way originates with Mostowski [181], and its central idea is to evaluate quantifiers as *infinite lattice operations* (namely infinite meet for the universal quantifier and infinite join for the existential one) on the algebras of truth values (whose type corresponds to the sentential connectives). This is clear in the case of classical logic: its satisfaction relation is defined as: $\mathscr{A} \vDash \forall x \alpha \, [\![\overline{a}]\!]$ if and only if $\mathscr{A} \vDash \alpha \, [\![\overline{a}(\frac{c}{x})]\!]$ for all $c \in A$. Under the constructions in the proof of Theorem 2.35, this would be to say that $h(\forall x \alpha) = 1$ if and only if $h_x^c(\alpha) = 1$ for all $c \in A$, where h_x^c is the homomorphism corresponding to the modified assignment $\overline{a}(\frac{c}{x})$. Now, considering that the infimum of a set in $\mathbf{2}$ equals 1 if and only if all the values in the set are equal to 1, this can be alternatively described as

  $$h(\forall x \alpha) = \inf\{h_x^c(\alpha) : c \in A\}.$$

  When the logic is based on a richer set of truth values, this expression becomes more interesting.

  This approach fits well with the standard algebraic studies of sentential logics, and the algebras have a much clearer role as *algebras of truth-values*, on which certain first-order structures are built. This idea, developed for implicative logics by Rasiowa [208, Supplement, pp. 347–379], fits perfectly with first-order logics based on many-valued logics [132, Chapter 5], and has been recently extended to arbitrary algebraizable logics [64]. In this approach, the most significant change concerns the relations between the logic and the algebras, which are limited to the Completeness Theorem and its consequences; there is no analogue of the Algebraizability Theorem, essentially because in first-order logic the semantic consequence is not the relative equational consequence of the associated classes of algebras (more on this below).

- The "hard" way is to make quantifiers correspond to a family of *unary primitive operations* in the algebras. This idea was exploited by Halmos [133], and independently in the celebrated works by Tarski and his school [137, 138]; the two approaches have produced *polyadic algebras* and *cylindric algebras*, respectively. These works have developed rather ad-hoc techniques, which are mainly suited to treat first-order theories on different languages based on *classical first-order logic*. An enormous amount of work, often of great difficulty, has been produced over the last half of a century, and the links with classical problems and techniques of model theory have been more explicit, touching, among others, on issues of representability, (finite) axiomatizability and decidability.

  To a certain extent, this "hard" approach is closer in spirit to the view of algebraic logic of this book than the "soft" one: if we associate a class of algebras with a logic, then the algebras should have the same similarity type as the language of that logic, a fact that does not happen in Mostowski's approach (unless one does not count the quantifiers among the true "operations" of the algebra of formulas). The first eight sections of [222] contain a short introduction to the Tarskian approach and a quick summary of the process of how cylindric algebras of various kinds are obtained from the algebra of formulas by a generalization of the Lindenbaum-Tarksi process. An introduction to the algebraic theory of cylindric algebras and their relation with classical first-order logic can be found in Monk's singular textbook [175, Chapter 12]; however, it does not even reach the completeness theorem, as it requires a much more complicated development, as is found in Section 4.3 of [138].

The main difference between the Halmos approach and the Tarskian one is the way each handles substitutions (of terms for individual variables), an essential ingredient of first-order languages. Halmos incorporates them into the algebraic language (after all, they also define unary operations on the algebra of formulas), while Tarski simply "eliminates" them, thanks to his insightful observation that if $\varphi(x)$ is a formula where the individual variable x is free and t is a term where x does not appear, then the substituted formula $\varphi(t)$ is logically equivalent to the formula $\exists x\big(x \approx t \wedge \varphi(x)\big)$.

However, in both approaches *structurality is lost* for the resulting semantic consequence: the role of sentential variables is played by atomic formulas, and the concept of substitution itself becomes quite different.† As a consequence, no substitution-invariance (structurality) holds. And this seems to be the crucial point in abstract algebraic logic, for the (relative) equational consequence is always structural. So, to a certain extent and up to the present state of the art, the algebraization of first-order logic seems to significantly diverge from the theory of abstract algebraic logic treated in this book.

I should mention that there is a way of *presenting first-order logic in disguise as a sentential logic*, in particular as a structural consequence, so that it can be treated with the techniques here exposed; see Appendix C of [32]. The result of this effort is nevertheless remarkably artificial.

There have been some attempts, mainly by Andréka, Németi and Sain [10, 11, 12] to build a general theory of the algebraization of logic in the Tarskian approach and under the influence of abstract model theory. However, even the "general" paper [182] does not account for any general theory, which is assumed as "given", and concentrates on particular problems of both algebraic and logical interest. It is noticeable that over time these authors have evolved towards a sentential-like presentation of first-order logic and its models (and in particular towards modal-like versions of certain fragments of first-order logic). As a matter of fact, one can establish a correspondence between these proposals and the approach presented in this book; see [105] and [110, Section 6.1].

† In first-order languages we replace an individual (free) variable by a term; we do not replace an atomic formula, inside a more complex one, by an arbitrary formula, which is what is done in a sentential language.

Chapter 3

The semantics of algebras

The algebraic study of implicative logics shows that their $\mathcal{L}$-algebras constitute a semantics in which *no further structure is necessary* in order to evaluate the language into the algebras and to obtain a Completeness Theorem. This appears most strikingly in the Algebraizability Theorem 2.12, where the stronger relations between the logic $\mathcal{L}$ and the class $\mathsf{Alg}^*\mathcal{L}$ are expressed using the relative equational consequence $\vDash_{\mathsf{Alg}^*\mathcal{L}}$ and a particular element 1 of the algebras; both tools (the consequence and this element) are *intrinsic* to the algebras.

This chapter deals with the notions of *algebraic semantics* and *algebraizability*, which implement the above ideas at a higher level of generality. It was the genius of Blok and Pigozzi that uncovered a powerful way of generalizing the situation of implicative logics and raising it to a mathematically more powerful level: that the relation between the logic and the class of algebras, the latter represented by its relative equational consequence, is effected by means of two *transformers*, a generalization of the functions

$$\varphi \longmapsto \varphi \approx \top \qquad\qquad \alpha \approx \beta \longmapsto \{\alpha \to \beta, \beta \to \alpha\}$$

that transform a formula into an equation and an equation into a set of formulas; these transformers are implicit in Theorem 2.12, particularly in conditions (ALG1) and (ALG2), but if you look at conditions (ALG3) and (ALG4) you will realize that their right-hand halves are the results of performing the two transformations in succession.

The first stage of the study concerns the simpler situation in which there is only one of the transformers.

3.1 Transformers, algebraic semantics, and assertional logics

DEFINITION 3.1
*A **transformer from formulas to (sets of) equations** is any function $\boldsymbol{\tau}\colon Fm \to \mathscr{P}Eq$. It is extended to a function $\boldsymbol{\tau}\colon \mathscr{P}Fm \to \mathscr{P}Eq$ by setting, for any $\Gamma \subseteq Fm$,*

$$\boldsymbol{\tau}\Gamma := \bigcup_{\gamma \in \Gamma} \boldsymbol{\tau}\gamma.$$

*A transformer $\boldsymbol{\tau}$ is **finitary** when the set $\boldsymbol{\tau}\varphi$ is finite, for all $\varphi \in Fm$.*

Although for simplicity you may often read "a transformer from formulas to equations", $\boldsymbol{\tau}\varphi$ is always a *set* of equations for each formula φ; this is why the union symbol appears in the definition of $\boldsymbol{\tau}\Gamma$.

By its definition, any transformer commutes with arbitrary unions, that is: $\boldsymbol{\tau} \bigcup_{i \in I} \Gamma_i = \bigcup_{i \in I} \boldsymbol{\tau}\Gamma_i$. Alternatively, one can define a transformer as a function $\boldsymbol{\tau}\colon \mathscr{P}Fm \to \mathscr{P}Eq$ that commutes with arbitrary unions; see Exercise 1.1.

DEFINITION 3.2
*A transformer $\boldsymbol{\tau}$ is **structural** when it commutes with substitutions, in the sense that $\boldsymbol{\tau}\sigma = \sigma\boldsymbol{\tau}$ for every substitution σ.*

The σ in the right-hand side of the preceding equality refers to the action of σ on (sets of) equations as described on page 29; thus, the condition means that for any $\varphi \in Fm$, $\boldsymbol{\tau}\sigma\varphi = \sigma\boldsymbol{\tau}\varphi := \{\sigma\alpha \approx \sigma\beta : \alpha \approx \beta \in \boldsymbol{\tau}\varphi\}$.

The main feature of structural transformers is that they are determined by their output on a single variable:

PROPOSITION 3.3
A transformer from formulas to equations $\boldsymbol{\tau}$ is structural if and only if there is a set of equations $E(x) \subseteq Fm$ in at most the variable x such that $\boldsymbol{\tau}\varphi = E(\varphi)$ for all $\varphi \in Fm$.

PROOF: Recall that for any ψ, $E(\psi)$ is shorthand for $\sigma_\psi E(x)$ where σ_ψ is a substitution such that $\sigma_\psi x = \psi$.

($\Rightarrow$) Assume that $\boldsymbol{\tau}$ is a structural transformer, and define $E(x) := \boldsymbol{\tau}x \subseteq Eq$. Now, to show that $\boldsymbol{\tau}\varphi = E(\varphi)$ is to show that $\boldsymbol{\tau}\sigma_\varphi x = \sigma_\varphi\boldsymbol{\tau}x$, which is a particular case of $\boldsymbol{\tau}$ being structural. It remains only to prove that the equations in $\boldsymbol{\tau}x$ contain at most the variable x. For this, consider another variable $y \neq x$ and the substitution $\sigma z := x$ for all $z \in V$. It is easy to see (by induction on the length of φ) that for any φ, $\sigma\varphi$ contains at most the variable x, and the same applies to equations. Then, by structurality of $\boldsymbol{\tau}$, $\boldsymbol{\tau}x = \boldsymbol{\tau}\sigma y = \sigma\boldsymbol{\tau}y$, which as just seen contains at most x.

($\Leftarrow$) If $\boldsymbol{\tau}\varphi = E(\varphi) = \sigma_\varphi E(x)$ for some $E(x) \subseteq Fm$ in at most the variable x, then for any substitution σ we have that $\sigma\boldsymbol{\tau}\varphi = \sigma\sigma_\varphi E(x) = \sigma_{\sigma\varphi}E(x) = \boldsymbol{\tau}\sigma\varphi$, because $(\sigma\sigma_\varphi)x = \sigma(\sigma_\varphi x) = \sigma\varphi$, taking the notational shorthand recalled above into account. ⊠

Thus, a structural transformer is specified by just giving its image for a single variable (an arbitrarily chosen one, it does not matter which) and then it is "extended by substitutions" to arbitrary formulas, and finally it is "extended by unions" to arbitrary sets of formulas:

$$\begin{aligned}
&\boldsymbol{\tau} x := E(x)\\
&\boldsymbol{\tau} \varphi := E(\varphi) = \{\alpha(\varphi) \approx \beta(\varphi) : \alpha(x) \approx \beta(x) \in E(x)\}\\
&\boldsymbol{\tau} \Gamma := \bigcup_{\gamma\in\Gamma} \boldsymbol{\tau}\gamma = \bigcup_{\gamma\in\Gamma} E(\gamma) = \{\alpha(\gamma) \approx \beta(\gamma) : \alpha(x) \approx \beta(x) \in E(x), \gamma \in \Gamma\}
\end{aligned}$$

Notice that a structural transformer is finitary if and only if the set $E(x)$ is finite.

Transformers have also been called *translations* in the literature, but it is preferable to avoid this term, because it has been extensively used to refer to functions from the formulas of a language into formulas of another language (or the same) that are defined recursively and intend to express relations between two different logics; as such they have been used since 1925 by Kolmogorov, Gödel and others, and in general these translations are not structural.

Definition 3.4
Let $\mathcal{L}$ be a logic, K a class of algebras, and $\boldsymbol{\tau}$ a structural transformer. The class K is a ***$\boldsymbol{\tau}$-algebraic semantics*** *for $\mathcal{L}$ when the following condition is satisfied, for all $\Gamma \cup \{\varphi\} \subseteq Fm$:*

$$\Gamma \vdash_{\mathcal{L}} \varphi \iff \boldsymbol{\tau}\Gamma \vDash_{\mathsf{K}} \boldsymbol{\tau}\varphi \tag{ALG1}$$

The set $E(x)$ corresponding to the transformer $\boldsymbol{\tau}$ is called the set of ***defining equations****. An* ***algebraic semantics*** *for a logic is a $\boldsymbol{\tau}$-algebraic semantics for some $\boldsymbol{\tau}$.*

An alternative terminology is to say that $\boldsymbol{\tau}$ is an ***interpretation***† *of the logic $\mathcal{L}$ into the consequence $\vDash_{\mathsf{K}}$ when the equivalence in* (ALG1) *holds; then K is an algebraic semantics for $\mathcal{L}$ when there is a* ***structural interpretation*** *of $\mathcal{L}$ into $\vDash_{\mathsf{K}}$.*

Definition 3.4 was originally introduced by Blok and Pigozzi in [32], in the restricted case in which the logic and the transformer are finitary and the class of algebras is a quasivariety. The following three comments around the term "algebraic semantics" may be of interest:

- If K is an algebraic semantics for $\mathcal{L}$ through transformer $\boldsymbol{\tau}$, then K constitutes an algebra-based semantics (in the sense of the general discussion on pages 22–25) where the truth condition can be defined through the transformer, in the following way (for $\boldsymbol{A} \in \mathsf{K}$): $\boldsymbol{A} \Vdash \varphi\ [\![h]\!] \overset{\text{def}}{\iff} \boldsymbol{A} \vDash E(x)\ [\![h\varphi]\!]$. Thus, only purely algebraic features of the algebras $\boldsymbol{A} \in \mathsf{K}$ need to be used to set up this semantics. The condition that K is an algebraic semantics for $\mathcal{L}$ amounts to requiring that this semantics is *complete* for $\mathcal{L}$.

† There is little terminological agreement on these and similar notions in the literature. As a relation between two logics, what is called here an "interpretation" is also called a "faithful interpretation" or a "conservative translation", while the terms "semi-interpretation", "representation", "translation" or even "interpreation" as well, have been used when the left-to-right implication in (ALG1) holds. See Exercise 6.37.

- Using a rather algebraic terminology, one can say that $\boldsymbol{A} \Vdash \varphi\, [\![h]\!]$ if and only if $h\varphi$ belongs to the set of "solutions" of the equations in $E(x)$. Then, recalling that in geometry *algebraic sets* are the sets of roots of polynomials, and realizing that being a root of a polynomial is a particular case of being a solution of an equation, one can say that an algebraic semantics is a semantics where the *truth set* is an "algebraic set" in this generalized sense. This observation is due to Cignoli [personal communication].
- In the 1970s, Suszko (in unpublished lectures) proposed the term "algebraic semantics" to denote a matrix semantics in which the filters are one-element sets, and this terminology was adopted by Czelakowski in the context of equivalential logics [67]. The "equivalential logics with an algebraic semantics" he introduced are nowadays called *regularly algebraizable*, and are treated in Section 3.4; other important kinds of logics having an algebraic semantics are treated in Section 6.4.

Notice that the set $E(x)$, that is, the set $\boldsymbol{\tau} x$, must be non-empty, except for the inconsistent logic. Actually, any class of algebras is an algebraic semantics for the inconsistent logic, with defining equation $x \approx x$. See Exercise 3.1.

Given a class K of algebras and a transformer $\boldsymbol{\tau}$, there is always a logic having K as its $\boldsymbol{\tau}$-algebraic semantics, namely the logic *defined* by condition (ALG1). A particular, very large and very important class of logics with an algebraic semantics arises from pointed classes of algebras. A class K of algebras is ***pointed*** when each algebra in K has a designated element that is the value of an algebraic constant common to the class; the constant term, which is conventionally denoted by the symbol $\top$, need not be a primitive constant of the language, and may have variables, but one can always take one with at most one variable, hence it may be denoted by $\top(x)$ if necessary. As an example, the class of implicative algebras, defined in the language $\langle \to \rangle$ of type $\langle 2 \rangle$, is pointed, with $\top := x \to x$; more in general, Lemma 2.6 shows that the class $\mathsf{Alg}^*\mathcal{L}$ is pointed for any implicative logic $\mathcal{L}$ and that any theorem of $\mathcal{L}$ can be taken as the constant term.

Definition 3.5

*Let K be a pointed class of algebras of a similarity type $\boldsymbol{L}$. The **assertional logic of** K is the logic $\mathcal{L}^{\top}_{\mathsf{K}} := \langle \boldsymbol{L}, \vdash^{\top}_{\mathsf{K}} \rangle$, where, for all $\Gamma \cup \{\varphi\} \subseteq Fm_{\boldsymbol{L}}$,*

$$\Gamma \vdash^{\top}_{\mathsf{K}} \varphi \stackrel{\text{def}}{\iff} \{\gamma \approx \top : \gamma \in \Gamma\} \vDash_{\mathsf{K}} \varphi \approx \top .$$

*A logic is **an assertional logic** when it is the assertional logic of some pointed class of algebras.*†

By definition the logic $\mathcal{L}^{\top}_{\mathsf{K}}$ has K as an algebraic semantics and $x \approx \top$ as a defining equation. It is easy to see (Exercise 3.3) that then $\top$ is a theorem of the logic. Assertional logics are probably the most common kind of logics having an

† For emphasis, or to exclude any ambiguity (K might be doubly pointed), some authors specifiy "1-assertional" or "$\top$-assertional". More in general, the logic satisfying (ALG1) for some transformer $\boldsymbol{\tau}$ and some class K can be called "the $\boldsymbol{\tau}$-assertional logic of K".

algebraic semantics, and they encompass the following very large and important group of examples.

PROPOSITION 3.6
A logic is implicative if and only if it is the assertional logic of a class of algebras K *such that for some binary term* $\to$, $\mathsf{K}{\restriction}\langle\to\rangle \subseteq \mathsf{IA}$.

In particular, every implicative logic is an assertional logic. Therefore, all implicative logics have an algebraic semantics.

PROOF: Notice that the requirement that $\mathsf{K}{\restriction}\langle\to\rangle \subseteq \mathsf{IA}$ already entails that K is pointed, with $\top := x \to x$, and hence it makes sense to consider its assertional logic. Then on all algebras of K the relation $a \leqslant b \overset{\text{def}}{\iff} a \to b = 1 := \top^{\boldsymbol{A}}$ is an order such that 1 is its maximum (Proposition 2.14). This property is just what is needed in order to show that the assertional logic of K satisfies the conditions (IL1)–(IL5) of Definition 2.3 and is hence an implicative logic. Conversely, if $\mathcal{L}$ is an implicative logic, by Lemma 2.6 the class $\mathsf{Alg}^*\mathcal{L}$ is pointed with $\top := x \to x$, by Theorem 2.9 $\mathcal{L}$ is the assertional logic of $\mathsf{Alg}^*\mathcal{L}$, and by Proposition 2.15 $(\mathsf{Alg}^*\mathcal{L}){\restriction}\langle\to\rangle \subseteq \mathsf{IA}$. ⊠

For instance, BA is an algebraic semantics for $\mathcal{C}\ell$, HA is an algebraic semantics for $\mathcal{I}\ell$, and so forth (Table 2 on page 85 shows other examples). Assertional logics are given a more in-depth treatment in Section 6.4 (pages 383ff.).

An algebraic semantics for a logic is seldom unique. The definition itself makes it clear that if K is a $\boldsymbol{\tau}$-algebraic semantics for $\mathcal{L}$, then any other class K' such that $\vDash_{\mathsf{K}'} = \vDash_{\mathsf{K}}$ is also a $\boldsymbol{\tau}$-algebraic semantics for $\mathcal{L}$. In particular, if K is the variety or quasivariety generated by a single algebra $\boldsymbol{A}$, then K is an algebraic semantics for $\mathcal{L}$ if and only if $\{\boldsymbol{A}\}$ alone is. For instance, $\{\boldsymbol{2}\}$ is an algebraic semantics for $\mathcal{C}\ell$, therefore so is any class of Boolean algebras containing it, such as the class of all finite Boolean algebras. In general, if a logic has an algebraic semantics, then there is a largest one among all its algebraic semantics with the same transformer:

DEFINITION 3.7
Let $\mathcal{L}$ *be a logic and let* $\boldsymbol{\tau}$ *be a structural transformer. Define*

$$\mathsf{K}(\mathcal{L},\boldsymbol{\tau}) := \{\boldsymbol{A} : \boldsymbol{\tau}\Gamma \vDash_{\boldsymbol{A}} \boldsymbol{\tau}\varphi \text{ for all } \Gamma \cup \{\varphi\} \subseteq Fm \text{ such that } \Gamma \vdash_{\mathcal{L}} \varphi\}.$$

In some contexts the algebras in the class $\mathsf{K}(\mathcal{L},\boldsymbol{\tau})$ *are called the* ***τ-models*** *of* $\mathcal{L}$.

Observe that the class $\mathsf{K}(\mathcal{L},\boldsymbol{\tau})$ can always be defined, for any logic and any transformer, and is never empty, for it contains at least all trivial algebras (check this). Depending on the logic and on the transformer, it may not contain other algebras, and it need not be an algebraic semantics for $\mathcal{L}$. But:

PROPOSITION 3.8
Let $\mathcal{L}$ *be a logic, and* $\boldsymbol{\tau}$ *a structural transformer. If* $\mathcal{L}$ *has a* $\boldsymbol{\tau}$*-algebraic semantics, then the class* $\mathsf{K}(\mathcal{L},\boldsymbol{\tau})$ *is its largest* $\boldsymbol{\tau}$*-algebraic semantics.*

PROOF: Pretty obvious, but left to you as Exercise 3.4. ⊠

The class $\mathsf{K}(\mathcal{L},\boldsymbol{\tau})$ is a generalized quasivariety, as it is defined by a set of generalized quasi-equations, because the condition "$\boldsymbol{\tau}\Gamma \vDash_{\boldsymbol{A}} \boldsymbol{\tau}\varphi$" is the same as "$\boldsymbol{A} \vDash \bigwedge \boldsymbol{\tau}\Gamma \to \alpha \approx \beta$, for all $\alpha \approx \beta \in \boldsymbol{\tau}\varphi$". Of course, a smaller (perhaps finite) equivalent set can be obtained from any axiomatic system defining the logic. By Proposition 3.8, $\mathsf{K} \subseteq \mathsf{K}(\mathcal{L},\boldsymbol{\tau})$ for any $\boldsymbol{\tau}$-algebraic semantics K for $\mathcal{L}$, hence $\mathbb{GQ}\mathsf{K} \subseteq \mathsf{K}(\mathcal{L},\boldsymbol{\tau})$ as well. However, this inclusion may be strict (see Exercise 3.5 for an example), which means that the class $\mathsf{K}(\mathcal{L},\boldsymbol{\tau})$ may not generate the same relative equational consequence as K; that the two classes are both $\boldsymbol{\tau}$-algebraic semantics for $\mathcal{L}$ just means that their relative equational consequences coincide on their "$\boldsymbol{\tau}$-transformed" fragments.

Despite its natural character, there may be no "canonical" algebraic semantics for a given logic, and algebraic semantics can be rather weird, as the following examples and observations show.

- If $\mathcal{L}$ and $\boldsymbol{\tau}$ are finitary, then $\mathsf{K}(\mathcal{L},\boldsymbol{\tau})$ can be defined by a set of quasi-equations, and hence it is a quasivariety. However, the same logic may have other $\boldsymbol{\tau}$-algebraic semantics with a non-finitary relative equational consequence. An example of this situation is Herrmann's logic $\mathcal{LJ}$ described in Example 3.42.
- One and the same class of algebras can be an algebraic semantics for a logic with two different transformers. For instance, since $x \approx \neg\neg x$ holds in BA, this class of algebras is an algebraic semantics for $\mathcal{C}\ell$ with the transformer $\boldsymbol{\tau}'x := \{\neg\neg x \approx \top\}$.
- A relatively natural idea, to a certain extent, may be that different algebraic semantics for a logic, for the same transformer, should generate the same quasivariety or the same generalized quasivariety (as happens in the Boolean examples for $\mathcal{C}\ell$ mentioned before), but this is not true. More strongly, the different algebraic semantics may even be quasivarieties themselves: It turns out that the class HA of Heyting algebras is an algebraic semantics for $\mathcal{C}\ell$ with the transformer $\boldsymbol{\tau}'$ just introduced (Exercise 3.5). This also shows that the class $\mathsf{K}(\mathcal{L},\boldsymbol{\tau})$ need not coincide with the class $\mathsf{K}(\mathcal{L},\boldsymbol{\tau}')$, and that algebraic semantics and the associated transformers can be quite unnatural (at least from the point of view of the logic). Another example of an unexpected algebraic semantics for $\mathcal{C}\ell$ appears in Example 4.79.
- The previous points also showcase the (expected) facts that different transformers can produce different algebraic semantics for the same logic, and that different logics can have the same algebraic semantics (obviously, for different transformers); see also Example 3.20.
- Not every logic has an algebraic semantics, although the counterexamples are rare: The logic $\mathcal{I}$ in the language $\langle \to \rangle$ whose only axiom is $x \to x$ and whose only rule is *Modus Ponens* for $\to$ does not have an algebraic semantics; this logic is treated with more detail in Example 6.78. Other, less artificial examples are the negation fragments of $\mathcal{C}\ell$ and of $\mathcal{I}\ell$ [38], and the logic known as $\mathcal{PW}$ (among several names) in relevance logic literature [201, Example 4].

- By contrast, it is also proved in [38] that all fragments of $\mathcal{Cl}$ and of $\mathcal{Il}$ having either conjunction or disjunction do have an algebraic semantics (with a weird defining equation, $x \wedge x \approx x$ or $x \vee x \approx x$), and that all local logics associated with normal modal logics also have an algebraic semantics with a similar defining equation; however, the corresponding classes of algebras are rather wild and artificial constructs, with no known intuitive meaning.
- Finally, a logic can have the class of all algebras of its type as an algebraic semantics. The following example is a logic formulated in the so-called "mono-unary" language (having only one operation, a unary one). In general, languages with only unary operations provide interesting counterexamples in abstract algebraic logic, and logics in mono-unary languages are particularly easy to handle; therefore it is worth to devote some attention to them now, and to give something more than a single example:

EXAMPLE 3.9
Let $\mathcal{L}$ be a finitary logic in the language $\boldsymbol{L} = \langle \Box \rangle$ of type $\langle 1 \rangle$. The logic $\mathcal{L}$ has an algebraic semantics if and only if $x \vdash_{\mathcal{L}} \Box x$; in case it has, the equation $x \approx \Box x$ can be taken as defining equation. The logic axiomatized by just the rule $x \rhd \Box$ has the class of all algebras of the type as algebraic semantics.

PROOF: Assume that K is an algebraic semantics for $\mathcal{L}$ with defining equations the set $E(x)$. Since the language has only a single unary operation, all terms are unary and have the form $\delta(x) = \Box^n x$ for some $x \in V$ and some $n \in \omega$, therefore $\delta(\Box x) = \Box^{n+1} x = \Box \delta(x)$. As a consequence, the implication $E(x) \vDash_{\mathsf{K}} E(\Box x)$ holds trivially. Now, by the assumption, this means that $x \vdash_{\mathcal{L}} \Box x$. The converse and the rest of the statement are proved jointly thanks to a non-trivial general result from [38]: Any extension of a logic with an algebraic semantics has an algebraic semantics as well, and with the same set of defining equations. Therefore, it is enough to show that the logic $\mathcal{L}$ defined by a Hilbert system with no axioms and the single rule $x \rhd \Box x$ has an algebraic semantics with $x \approx \Box x$ as defining equation. By Proposition 3.8, if $\mathcal{L}$ has some algebraic semantics with transformer $\boldsymbol{\tau}$, then the class $\mathsf{K} := \mathsf{K}(\mathcal{L}, \boldsymbol{\tau})$ is also a $\boldsymbol{\tau}$-algebraic semantics for $\mathcal{L}$ (indeed, the largest one); but since $\boldsymbol{\tau}$ is given by $x \approx \Box x$, this K is the class of all algebras of the type, because the translation of the rule $x \rhd \Box x$ is the condition $x \approx \Box x \vDash_{\mathsf{K}} \Box x \approx \Box\Box x$, which holds trivially in all algebras. Since this class by definition satisfies that $\Gamma \vdash_{\mathcal{L}} \varphi$ implies $\boldsymbol{\tau}\Gamma \vDash_{\mathsf{K}} \boldsymbol{\tau}\varphi$, in order to show that it is really an algebraic semantics for $\mathcal{L}$ it only remains to prove that $\Gamma \nvdash_{\mathcal{L}} \varphi$ implies that $\boldsymbol{\tau}\Gamma \nvDash_{\mathsf{K}} \boldsymbol{\tau}\varphi$. If $\varphi = \Box^n x$, then clearly $\Box^i x \notin \Gamma$ for $i \leqslant n$. Then consider $A := \{0, \ldots, n{+}1\}$, turn it into an $\boldsymbol{L}$-algebra $\boldsymbol{A}$ by defining $\Box j := j{+}1$ for $j \leqslant n$ and $\Box(n{+}1) := n{+}1$, and take the evaluation given by $hx := 0$ and $hy := n{+}1$ for all $y \in V, y \neq x$. We know that all $\gamma \in \Gamma$ are either of the form $\Box^i x$ with $i \geqslant n{+}1$ or of the form $\Box^k y$ for any k and $y \neq x$. Then, $h\gamma = h\Box\gamma = n{+}1$ for all $\gamma \in \Gamma$, while $h\varphi = n \neq n{+}1 = h\Box\varphi$. This shows that $\boldsymbol{\tau}\Gamma \nvDash_{\boldsymbol{A}} \boldsymbol{\tau}\varphi$ and, therefore, finishes the proof. ⊠

The bare concept of algebraic semantics has not been extensively studied. Exercise 3.46 contains one of the few characterizations of this notion, of a lattice-theoretic character. At the time of writing, [38] and [201] are probably the only papers that contain significant results about it. One of the most interesting and difficult results in [38], which corrects a too pessimistic statement in [32], is that if a logic has an algebraic semantics, then all its extensions have one, with the same transformer (this result is used in the preceding proof). The same happens to fragments, with a limitation, and this is easy to show:

PROPOSITION 3.10
Let $\mathcal{L}$ be a logic with a $\boldsymbol{\tau}$-algebraic semantics K and let $\boldsymbol{L}'$ be a fragment of the language of $\mathcal{L}$ such that $E(x) \subseteq Eq_{\boldsymbol{L}'}$. The class $\mathsf{K}{\restriction}\boldsymbol{L}'$ of all $\boldsymbol{L}'$ reducts of algebras in K is a $\boldsymbol{\tau}$-algebraic semantics for $\mathcal{L}{\restriction}\boldsymbol{L}'$, the $\boldsymbol{L}'$ fragment of $\mathcal{L}$. Moreover, if K is a generalized quasivariety, then $\mathsf{K}' := \mathbb{S}(\mathsf{K}{\restriction}\boldsymbol{L}')$, which is a generalized quasivariety as well, is also a $\boldsymbol{\tau}$-algebraic semantics for $\mathcal{L}{\restriction}\boldsymbol{L}'$.

PROOF: It is straightforward to see that $\vDash_{\mathsf{K}{\restriction}\boldsymbol{L}'} = (\vDash_{\mathsf{K}}){\restriction}\boldsymbol{L}'$ (actually, this is a particular case of Exercise 1.27). Thus, condition (ALG1) is preserved by a restriction on the language provided it makes sense (that is, that the transformer stays inside $\boldsymbol{L}'$). Therefore the first part is trivial. The second part follows from the fact (Exercise 3.7) that if K is a generalized quasivariety, then the generalized quasivariety generated by the class $\mathsf{K}{\restriction}\boldsymbol{L}'$ is the class $\mathbb{S}(\mathsf{K}{\restriction}\boldsymbol{L}')$. This implies that they define the same relative equational consequence, and therefore $\mathbb{S}(\mathsf{K}{\restriction}\boldsymbol{L}')$ is also a $\boldsymbol{\tau}$-algebraic semantics for $\mathcal{L}{\restriction}\boldsymbol{L}'$. ⊠

It is important to realize that even if K is closed under $\mathbb{S}$, the class of reducts $\mathsf{K}{\restriction}\boldsymbol{L}'$ need not be closed under $\mathbb{S}$ as well (Exercise 3.8). Hence, consideration of the algebras in $\mathbb{S}(\mathsf{K}{\restriction}\boldsymbol{L}')$ is unavoidable if one wants to obtain a (generalized) quasivariety; the class $\mathbb{S}(\mathsf{K}{\restriction}\boldsymbol{L}')$ is called the class of ($\boldsymbol{L}'$) ***subreducts*** of the algebras in K; subreducts appear frequently in works on particular non-classical logics and their algebraization.

Thus, all fragments of $\mathcal{C}\ell$ or of $\mathcal{I}\ell$ containing $\top$, or where $\top$ is term-definable, have an algebraic semantics, which is a quasivariety; in each case, the transformer is written with the connectives of the fragment. The algebraic work is to determine and characterize the quasivariety.

The conclusion of this section is that, while the notion of an algebraic semantics seems to formalize an intuitively natural kind of algebraic completeness, it is not enough by itself alone to reflect the really meaningful links between a logic and a class of algebras, when they exist, and it needs to be further restricted or refined in order to do so.

Exercises for Section 3.1

EXERCISE 3.1. Prove that any class of algebras is an algebraic semantics for the inconsistent logic, both with an empty set of defining equations, and with $x \approx x$ as defining equation, and that it is the only logic to have an algebraic semantics with these transformers.

EXERCISE 3.2. Explore how an algebraic semantics for the almost inconsistent logic in a language with at least one non-constant symbol should look like. You can even construct one, for each such language.

EXERCISE 3.3. Let $\mathcal{L}$ be the assertional logic of a pointed class of algebras K, with constant term $\top$. Prove that $\top$ is a theorem of $\mathcal{L}$.

HINT. Write $\top(x)$, because in principle the variable x may appear in $\top$. You have to prove that $\mathsf{K} \vDash \top(x) \approx \top(\top(x))$, using that $\top$ is constant over K.

EXERCISE 3.4. Let K be a $\boldsymbol{\tau}$-algebraic semantics for $\mathcal{L}$. Prove that then $\mathsf{K} \subseteq \mathsf{K}(\mathcal{L},\boldsymbol{\tau})$, and that if K' is such that $\mathsf{K} \subseteq \mathsf{K}' \subseteq \mathsf{K}(\mathcal{L},\boldsymbol{\tau})$, then K' is a $\boldsymbol{\tau}$-algebraic semantics for $\mathcal{L}$. Conclude that, if $\mathcal{L}$ has a $\boldsymbol{\tau}$-algebraic semantics, then the class $\mathsf{K}(\mathcal{L},\boldsymbol{\tau})$ is its largest $\boldsymbol{\tau}$-algebraic semantics.

EXERCISE 3.5. Glivenko's Theorem is a classic result (1929) which, in modern terms, asserts that for any $\Gamma \cup \{\varphi\} \subseteq Fm$, $\Gamma \vdash_{\mathcal{C}\ell} \varphi \iff \{\neg\neg\gamma : \gamma \in \Gamma\} \vdash_{\mathcal{I}\ell} \neg\neg\varphi$. Use it to prove that HA is an algebraic semantics for $\mathcal{C}\ell$ with defining equation $\neg\neg x \approx \top$, and deduce from this that $\mathsf{BA} \subsetneq \mathsf{K}(\mathcal{C}\ell,\boldsymbol{\tau}')$, and that $\mathsf{K}(\mathcal{C}\ell,\boldsymbol{\tau}) \neq \mathsf{K}(\mathcal{C}\ell,\boldsymbol{\tau}')$, where $\boldsymbol{\tau}x := \{x \approx \top\}$ and $\boldsymbol{\tau}'x := \{\neg\neg x \approx \top\}$.

HINT. Use that the quasi-equation $\neg\neg x \approx \top \rightarrow x \approx \top$ does not hold in HA.

EXERCISE 3.6. Prove that if $\mathcal{L}$ is a logic with an algebraic semantics and defining equations $E(x)$, then for each equation $\varepsilon(x) \approx \delta(x) \in E(x)$, each $\alpha(w,\vec{z}) \in Fm$ and each $\vec{\beta} \in \vec{Fm}$, it holds that $x, \alpha(\varepsilon(x),\vec{\beta}) \vdash_{\mathcal{L}} \alpha(\delta(x),\vec{\beta})$.

COMMENT. In [38] it is shown that the negation fragment of $\mathcal{C}\ell$ and the negation fragment of $\mathcal{I}\ell$ satisfy the conclusion but do not have an algebraic semantics. Thus, the converse of the property in the exercise does not hold.

EXERCISE 3.7. Let K be a generalized quasivariety of type $\boldsymbol{L}$, and let $\boldsymbol{L}'$ be a fragment of $\boldsymbol{L}$. Show that the class $\mathbb{S}(\mathsf{K}{\restriction}\boldsymbol{L}')$ of the $\boldsymbol{L}'$-algebras that can be embedded into the $\boldsymbol{L}'$ reduct of some algebra of K, is a generalized quasivariety as well. Deduce from this that the generalized quasivariety generated by the class $\mathsf{K}{\restriction}\boldsymbol{L}'$ is the class $\mathbb{S}(\mathsf{K}{\restriction}\boldsymbol{L}')$.

HINT. Use the easy half of Theorem 1.77.6. When proving that the class $\mathbb{S}(\mathsf{K}{\restriction}\boldsymbol{L}')$ is closed under $\mathbb{U}$ you will have to use the standard universal algebraic technique of embedding any algebra into a reduced product (actually, an ultraproduct) of the family of its finitely generated subalgebras [51, Theorem V.2.14]. For quasivarieties, the result in the exercise is well known; see [171, p. 216] for instance.

EXERCISE 3.8. Find an example of a class of algebras K of some type $\boldsymbol{L}$ such that K is closed under subalgebras but for some fragment $\boldsymbol{L}'$ of $\boldsymbol{L}$, the class $\mathsf{K}{\restriction}\boldsymbol{L}'$ is not closed under subalgebras.

HINT. If you are a mathematician, think about groups, and about the set of natural numbers as included in the additive group of the integers.

3.2 Algebraizable logics

In a completely parallel way to the beginning of Section 3.1, one can define ***transformers of equations into (sets of) formulas*** as functions $\boldsymbol{\rho}\colon Eq \to \mathscr{P}Fm$ that extend to sets of equations by taking unions, or as functions from $\mathscr{P}Eq$ to $\mathscr{P}Fm$ that commute with unions, and consider the ***structural*** ones, which are those determined by a set $\Delta(x,y)$ of formulas in at most the two variables x,y by setting $\rho(\varphi \approx \psi) := \Delta(\varphi,\psi)$. We have already met examples of these functions in Chapter 2: it is not difficult to recognize the role of the transformer $\rho(x \approx y) := \{x \to y, y \to x\}$ in the Algebraizability Theorem for implicative logics (Theorem 2.12).

The notations $\boldsymbol{\tau}$ and $\boldsymbol{\rho}$ are used to focus on the properties of the functions, while it is better to use $E(x)$ and $\Delta(x,y)$ to focus on the sets of formulas that define these functions. The latter notations can be extended like the transformers: If $\Gamma \subseteq Fm$, then $E(\Gamma) := \bigcup_{\gamma \in \Gamma} E(\gamma)$, therefore $\boldsymbol{\tau}\Gamma = E(\Gamma)$; likewise, if $\Theta \subseteq Eq$, then $\Delta(\Theta) := \bigcup_{\varepsilon \approx \delta \in \Theta} \Delta(\varepsilon,\delta)$, and then $\boldsymbol{\rho}\Theta = \Delta(\Theta)$.

DEFINITION 3.11
*A logic $\mathcal{L}$ is **algebraizable** when there is a class K of algebras and there are structural transformers $\boldsymbol{\tau}, \boldsymbol{\rho}$ (from formulas into equations and from equations into formulas, respectively) such that, for all $\Gamma \cup \{\varphi\} \subseteq Fm$ and all $\Theta \cup \{\varepsilon \approx \delta\} \subseteq Eq$, the following conditions are satisfied:*

(ALG1) $\Gamma \vdash_{\mathcal{L}} \varphi \iff \boldsymbol{\tau}\Gamma \vDash_{\mathsf{K}} \boldsymbol{\tau}\varphi$ *i.e.* $\iff E(\Gamma) \vDash_{\mathsf{K}} E(\varphi)$

(ALG2) $\Theta \vDash_{\mathsf{K}} \varepsilon \approx \delta \iff \boldsymbol{\rho}\Theta \vdash_{\mathcal{L}} \boldsymbol{\rho}(\varepsilon \approx \delta)$ *i.e.* $\iff \Delta(\Theta) \vdash_{\mathcal{L}} \Delta(\varepsilon,\delta)$

(ALG3) $\varphi \dashv\vdash_{\mathcal{L}} \boldsymbol{\rho\tau}\varphi$ *i.e.* $\varphi \dashv\vdash_{\mathcal{L}} \Delta(E(\varphi))$

(ALG4) $\varepsilon \approx \delta \mathrel{=\!\!|\!\vDash_{\mathsf{K}}} \boldsymbol{\tau\rho}(\varepsilon \approx \delta)$ *i.e.* $\varepsilon \approx \delta \mathrel{=\!\!|\!\vDash_{\mathsf{K}}} E(\Delta(\varepsilon,\delta))$

*As in Definition 3.4, the equations in $E(x)$ are called the **defining equations**. Moreover, the formulas in $\Delta(x,y)$ are called the **equivalence formulas**. The transformers $\boldsymbol{\tau}$ and $\boldsymbol{\rho}$ are said to **witness** the algebraizability of $\mathcal{L}$ **with respect to** the class K. The consequences $\vDash_{\mathsf{K}}$ and $\vdash_{\mathcal{L}}$ are said to be **(deductively) equivalent**.*

Using the alternative terminology set up after Definition 3.4, $\mathcal{L}$ is algebraizable with respect to K if and only if there are ***two structural interpretations*** $\boldsymbol{\tau}, \boldsymbol{\rho}$, of $\mathcal{L}$ into $\vDash_{\mathsf{K}}$ and of $\vDash_{\mathsf{K}}$ into $\mathcal{L}$ respectively, which are ***mutually inverse*** in the sense that (ALG3) and (ALG4) are satisfied.

We see in this section that the equivalence formulas really deserve this name, as they implement a generalization of the Lindenbaum-Tarski procedure of factoring the formula algebra through a congruence defined by a set of formulas.

Consider the new conditions (ALG3) and (ALG4) for a moment. First, observe that, due to structurality, they can be equivalently put in a simpler way:

$$\begin{array}{llll} \text{(ALG3)} & x \dashv\vdash_{\mathcal{L}} \boldsymbol{\rho\tau} x & \text{i.e.} & x \dashv\vdash_{\mathcal{L}} \Delta(E(x)) \\ \text{(ALG4)} & x \approx y \mathrel{=\!\!|\!\vDash_{\mathsf{K}}} \boldsymbol{\tau\rho}(x \approx y) & \text{i.e.} & x \approx y \mathrel{=\!\!|\!\vDash_{\mathsf{K}}} E(\Delta(x,y)) \end{array} \tag{26}$$

Second, observe also that they can be equivalently stated for *sets* of formulas and equations respectively: for all $\Gamma \subseteq Fm$ and all $\Theta \subseteq Eq$,

$$\begin{array}{llll} \text{(ALG3)} & \Gamma \dashv\vdash_{\mathcal{L}} \boldsymbol{\rho\tau}\Gamma & \text{i.e.} & \Gamma \dashv\vdash_{\mathcal{L}} \Delta(E(\Gamma)) \\ \text{(ALG4)} & \Theta \mathrel{=\!\!|\!\vDash_{\mathsf{K}}} \boldsymbol{\tau\rho}\Theta & \text{i.e.} & \Theta \mathrel{=\!\!|\!\vDash_{\mathsf{K}}} E(\Delta(\Theta)) \end{array} \tag{27}$$

(see Exercise 3.9 for the exact formulation of these "equivalently"). Finally, notice that (ALG4) is a property that actually holds for each algebra in K individually, that is, it amounts to requiring that for each $\boldsymbol{A} \in \mathsf{K}$, $x \approx y \dashv\models_{\boldsymbol{A}} E(\Delta(x,y))$. Lemma 3.22 expresses this in a different way.

The first important thing to say concerning this definition is that it is redundant; two of the four conditions suffice for it:

PROPOSITION 3.12
A logic $\mathcal{L}$ is algebraizable if and only if there is a class K of algebras and there are structural transformers $\boldsymbol{\tau}, \boldsymbol{\rho}$ such that conditions (ALG1) *and* (ALG4) *are satisfied; or, equivalently, conditions* (ALG2) *and* (ALG3).

PROOF: Straightforward (Exercise 3.10). ⊠

The class K is, thus, an algebraic semantics for $\mathcal{L}$ with the transformer $\boldsymbol{\tau}$ from formulas into equations that in addition satisfies condition (ALG4) for the same $\boldsymbol{\tau}$ and the transformer $\boldsymbol{\rho}$ of equations into formulas. Actually, in particular examples where one wants to show algebraizability of a logic by using the definition, the most usual way is to show (ALG1), a typical Algebraic Completeness Theorem, and (ALG4), which requires some work just inside the class of algebras.

As a first sample of what one can do with algebraizable logics, the main properties of the set of equivalence formulas are obtained. These are used in an essential way later on in this section; besides, they somehow justify the name "equivalence formulas" given to the formulas in $\Delta(x,y)$, and moreover we see that this set also satisfies a generalized version of the rule of *Modus Ponens*.

PROPOSITION 3.13
Let $\mathcal{L}$ be an algebraizable logic, with equivalence formulas $\Delta(x,y) \subseteq Fm$. The following conditions are satisfied:

(R) $\vdash_{\mathcal{L}} \Delta(x,x)$

(Sym) $\Delta(x,y) \vdash_{\mathcal{L}} \Delta(y,x)$

(Trans) $\Delta(x,y) \cup \Delta(y,z) \vdash_{\mathcal{L}} \Delta(x,z)$

(Re) $\bigcup_{i=1}^{n} \Delta(x_i,y_i) \vdash_{\mathcal{L}} \Delta(\lambda x_1 \dots x_n, \lambda y_1 \dots y_n)$ *for each* $\lambda \in L$, $n = \operatorname{ar}\lambda \geqslant 1$

(MP) $x\,,\Delta(x,y) \vdash_{\mathcal{L}} y$

PROOF: The first four conditions are exactly the $\boldsymbol{\rho}$-transforms of the first four of Birkhoff's rules (page 61), the fourth one in the form (Congruence). These are satisfied by the relative equational consequence $\models_{\mathsf{K}}$, therefore the conditions hold by (ALG2). As to (MP), it follows from the trivial fact that $E(x)\,, x \approx y \models_{\mathsf{K}} E(y)$, after applying (ALG2) to it and then using (ALG3). ⊠

The meaning of the acronyms is self-explanatory: Reflexivity, Symmetry, Transitivity, Replacement, *Modus Ponens*.

The sets $E(x)$ and $\Delta(x,y)$ must be non-empty except in the case of the inconsistent logic, which is trivially algebraizable as shown in Exercise 3.11. Then:

COROLLARY 3.14
If $\mathcal{L}$ is an algebraizable logic, with equivalence formulas $\Delta(x,y)$, then $\mathcal{L}$ has theorems, among them the formulas in $\Delta(x,x)$ when $\Delta \neq \emptyset$.

PROOF: If $\Delta \neq \emptyset$, this is (R) from Proposition 3.13; if $\Delta = \emptyset$, $\mathcal{L}$ must be the inconsistent logic (by Exercise 3.11), for which the property is trivially true. ⊠

Thus, all logics without theorems are non-algebraizable; you will find some non-trivial ones along this book, for instance, the fragments of classical logic with only conjunction $\mathcal{C\ell}_\wedge$ (Example 4.46) and with only conjunction and disjunction $\mathcal{C\ell}_{\wedge\vee}$ (Example 4.47), and Belnap-Dunn's logic $\mathcal{B}$ (Example 4.12); later on some criteria to determine that some other logics are not algebraizable will be given. Usually, showing that a specific logic *is* algebraizable is easier; anyway, Theorem 2.12 has already provided a first, large, important class of examples:

PROPOSITION 3.15
Every implicative logic $\mathcal{L}$ is algebraizable with respect to the class $\mathsf{Alg}^*\mathcal{L}$, and the algebraizability is witnessed by the defining equation $x \approx x \to x$ and the equivalence formulas $\{x \to y, y \to x\}$. ⊠

For instance, $\mathcal{C\ell}$, $\mathcal{I\ell}$ and $\mathcal{I\ell}_\to$ are algebraizable, with respect to the classes BA, HA and HiA respectively, through the natural transformers, and so are all logics appearing in Table 2 on page 85. As with any general mathematical notion, considering as many examples as possible right from the start is essential to understand it. The class of implicative logics is already very large. If you want to meet other, non-implicative algebraizable logics, you may jump to Section 3.4 just to see them; however, the proof that (some of) these logics are indeed algebraizable requires results to be given later on in this and next sections.

In particular, we can examine the situation of the algebraizability of $\mathcal{C\ell}$. We saw in Section 3 that it has both the classes BA and HA as algebraic semantics. It is natural to think that it is algebraizable with respect to BA but *not* with respect to HA. We know that $\mathcal{C\ell}$ satisfies (ALG1) relatively to HA with $\tau' x = \{\neg\neg x \approx \top\}$, and it is already easy to check (Exercise 3.13) that it does not satisfy (ALG4) with this τ' and the usual $\rho x = \{x \to y, y \to x\}$. However, how do we see that it does not satisfy this condition, or both (relatively to HA) for other, perhaps obscure transformers? There are two answers for this: first, the *uniqueness theorem* that follows, and second, the *Isomorphism Theorems* of Section 3.5.

Uniqueness of the algebraization: the equivalent algebraic semantics

Besides particular examples such as the one just discussed, it is true in general that a class of algebras satisfying conditions (ALG1)–(ALG4) for a logic need not be unique, because these conditions are formulated in terms of the consequences of the logic and of the class of algebras, and thus any other class with the same relative equational consequence satisfies these conditions as well. This also happens with the notion of an algebraic semantics, but in contrast to it, here there is a "canonical" or distinguished class among them, namely the absolute largest one, which is *unique* and *independent of the transformers*.

Proposition 3.13 and the following enhancement are used in a tricky way in the proof of the main result:

LEMMA 3.16
If $\mathcal{L}$ is an algebraizable logic, with equivalence formulas $\Delta(x,y)$, then

$$\Delta(x,y) \vdash_{\mathcal{L}} \Delta(\alpha(x,\vec{z}),\alpha(y,\vec{z}))$$

for any $\alpha(w,\vec{z}) \in Fm$, where $x,y,w \in V$ with $w \neq x,y$.

PROOF: This is clearly a generalization of condition (Re) in Proposition 3.13, and can be derived from it, using also (R) and (Trans), by an inductive argument on the length of α. Alternatively, you may view it as the ρ-transform of Birkhoff's rule (Replacement) on page 61, which holds for $\vDash_{\mathsf{K}}$, and use (ALG2). ⊠

THEOREM 3.17
Let $\mathcal{L}$ be an algebraizable logic with respect to a class K, with defining equations $E(x)$ and equivalence formulas $\Delta(x,y)$. The logic $\mathcal{L}$ is algebraizable with respect to a class K', with defining equations $E'(x)$ and equivalence formulas $\Delta'(x,y)$ if and only if the following conditions are satisfied:

1. $\vDash_{\mathsf{K}'} = \vDash_{\mathsf{K}}$.
2. $\Delta(x,y) \dashv\vdash_{\mathcal{L}} \Delta'(x,y)$.
3. $E(x) \mathrel{=\!\mid\!\vDash}_{\mathsf{K}} E'(x)$.

Conditions 2 and 3, in terms of the associated transformers $\boldsymbol{\tau},\boldsymbol{\rho}$ and $\boldsymbol{\tau}',\boldsymbol{\rho}'$, are:

2′. $C_{\mathcal{L}}\boldsymbol{\rho}\Theta = C_{\mathcal{L}}\boldsymbol{\rho}'\Theta$ *for all $\Theta \subseteq Eq$.*

3′. $C_{\mathsf{K}}\boldsymbol{\tau}\Gamma = C_{\mathsf{K}}\boldsymbol{\tau}'\Gamma$ *for all $\Gamma \subseteq Fm$.*

PROOF: Assume first that $\mathcal{L}$ is also algebraizable with respect to K', with defining equations $E'(x)$ and equivalence formulas $\Delta'(x,y)$. Observe that then $\mathcal{L}$ satisfies Proposition 3.13 and Lemma 3.16 both for Δ and for Δ'. We start by proving condition 2: If we apply Lemma 3.16 to each of the formulas $\alpha(w,\vec{z})$ in the set $\Delta'(x,w)$, by considering $\vec{z} := \langle x \rangle$, we obtain

$$\Delta(x,y) \vdash_{\mathcal{L}} \Delta(\Delta'(x,x),\Delta'(x,y)). \tag{28}$$

On the other hand, by structurality, from (MP) for Δ it follows that

$$\Delta'(x,x)\,,\ \Delta(\Delta'(x,x),\Delta'(x,y)) \vdash_{\mathcal{L}} \Delta'(x,y) \tag{29}$$

and since, by (R), $\vdash_{\mathcal{L}} \Delta'(x,x)$, from (28) and (29) we obtain $\Delta(x,y) \vdash_{\mathcal{L}} \Delta'(x,y)$. By symmetry we finally obtain $\Delta(x,y) \dashv\vdash_{\mathcal{L}} \Delta'(x,y)$. Now we show condition 1: By structurality, the property in 2 implies that $\Delta(\alpha,\beta) \dashv\vdash_{\mathcal{L}} \Delta'(\alpha,\beta)$ for any $\alpha,\beta \in Fm$. Then, using this together with (ALG2) for both K and K' we obtain

$$\begin{aligned}\Theta \vDash_{\mathsf{K}} \varepsilon \approx \delta &\iff \Delta(\Theta) \vdash_{\mathcal{L}} \Delta(\varepsilon,\delta)\\ &\iff \Delta'(\Theta) \vdash_{\mathcal{L}} \Delta'(\varepsilon,\delta) \iff \Theta \vDash_{\mathsf{K}'} \varepsilon \approx \delta,\end{aligned}$$

which proves that $\vDash_{\mathsf{K}} = \vDash_{\mathsf{K}'}$. Finally, to prove condition 3, we use (ALG3) for both pairs of transformers and condition 2, and we have

$$\Delta(E(x)) \dashv\vdash_{\mathcal{L}} x \dashv\vdash_{\mathcal{L}} \Delta'(E'(x)) \dashv\vdash_{\mathcal{L}} \Delta(E'(x)),$$

but by (ALG2) this is equivalent to $E(x) \Dashv\vDash_{\mathsf{K}} E'(x)$.

The converse is trivial: If the three conditions are satisfied, then properties (ALG1)–(ALG4) for $\mathsf{K}, \boldsymbol{\tau}, \boldsymbol{\rho}$ imply the same ones for $\mathsf{K}', \boldsymbol{\tau}', \boldsymbol{\rho}'$.

Finally, the alternative forms of 2 and 3 follow from the original ones after using structurality and taking unions, as is done in Exercise 3.9. ⊠

This theorem establishes that *a logic cannot have two essentially different algebraizations*: If it has two, the relative equational consequences must coincide, and the only differences may appear in the sets of defining equations and equivalence formulas, which can be *formally* different, but *not essentially* as the two forms must be (deductively) equivalent. Moreover, from Theorems 1.73 and 3.17 it follows:

COROLLARY 3.18
If $\mathcal{L}$ is algebraizable with respect to K, then the class $\mathbb{GQ}\mathsf{K}$ is the largest class of algebras, and the only generalized quasivariety, with respect to which $\mathcal{L}$ is algebraizable. These facts are independent of the transformers that witness the algebraizability. ⊠

Therefore, the following definition makes sense:

DEFINITION 3.19
Let $\mathcal{L}$ be an algebraizable logic. Its **equivalent algebraic semantics**† *is the largest class of algebras K such that $\mathcal{L}$ is algebraizable with respect to K.*

Thus, if $\mathcal{L}$ is algebraizable with respect to a class K and this class is a generalized quasivariety, then K is directly the equivalent algebraic semantics for $\mathcal{L}$. For instance, $\mathsf{Alg}^*\mathcal{L}$ is the equivalent algebraic semantics of any implicative logic $\mathcal{L}$. Moreover, if one happens to know that $\vDash_{\mathsf{K}}$ is finitary, then one also knows that $\mathbb{GQ}\mathsf{K} = \mathbb{Q}\mathsf{K}$, a quasivariety; in particular, when K is a variety or a quasivariety, it is the largest equivalent algebraic semantics, and it is sometimes referred to as ***the equivalent (quasi-)variety of*** $\mathcal{L}$. Thus, one says, for instance, that BA, HA and HiA are, respectively, *the equivalent varieties* of $\mathcal{C\ell}$, of $\mathcal{I\ell}$ and of $\mathcal{I\ell}_{\rightarrow}$; and that IA is *the equivalent quasivariety* of $\mathcal{I}mp$, the weakest implicative logic (see Exercises 2.13 and 2.14).

In general, the class $\mathbb{GQ}\mathsf{K}$ can be characterized by a set of generalized quasi-equations (those that hold in K), but in the case where it is the equivalent algebraic semantics of an algebraizable logic, it can also be characterized by a set of quasi-equations derived from the logic through the transformers; see Theorem 3.30, and Proposition 3.44 for the totally finitary case.

† In the early versions of the theory this term was applied to *any* class K with respect to which $\mathcal{L}$ is algebraizable, and in Definition 3.11 one said "K is *an* equivalent algebraic semantics for $\mathcal{L}$". Raftery started in [200] the wise practice of reserving this term for the largest of these classes, so that it makes sense to speak of *the* equivalent algebraic semantics of an algebraizable logic.

Observe that an algebraizable logic can have algebraic semantics other than its equivalent one, and that it need not be algebraizable with respect to them. This certainly happens when $\mathbb{GQ}\mathsf{K} \subsetneq \mathsf{K}(\mathcal{L},\boldsymbol{\tau})$, that is, when the largest $\boldsymbol{\tau}$-algebraic semantics is larger than the equivalent algebraic semantics; this is the case for instance of $\mathcal{Cl}$, as shown in Exercises 3.5 and 3.13.

In the opposite direction, an algebraizable logic can have algebraic semantics which are smaller than, or even disjoint from the equivalent one; it may be algebraizable with respect to some of them, as happens for instance when a small class generates the same relative equational consequence as the whole equivalent algebraic semantics, but it may also be not, as in the ones for $\mathcal{Cl}$ presented in Exercise 4.25 and in Example 4.79. Exercise 3.15 is also relevant to this discussion.

Finally, concerning uniqueness, a surprising fact is that the dual situation to Theorem 3.17 does not hold: While a logic cannot have two essentially different algebraizations, a class of algebras can be the equivalent algebraic semantics of two different logics (of course, through different transformers), as the next example shows. The logics become "deductively equivalent" in a certain, technical sense (see Section 3.7), but their consequence relations are not identical (in contrast to condition 1 of Theorem 3.17).

EXAMPLE 3.20

Consider the three-element Łukasiewicz algebra† $\boldsymbol{A}_3 = \langle \{0, 1/2, 1\}, \neg, \rightarrow \rangle$ given by the tables below. Besides $\rightarrow$ and $\neg$, consider two defined modal operations $\Diamond x := \neg x \rightarrow x$ (this definition is due to Tarski) and $\Box x := \neg \Diamond \neg x$:

$\rightarrow$	0	½	1
0	1	1	1
½	½	1	1
1	0	½	1

	$\neg$	$\Box$	$\Diamond$
0	1	0	0
½	½	0	1
1	0	1	1

The table of $\rightarrow$ shows that $x \rightarrow x$ is an algebraic constant with value 1; it is represented as $\top$. Condition (ALG1) for this algebra, used with two different transformers, *defines* two different logics:

- $Ł_3$, Łukasiewicz's three-valued logic, with $\tau_1 x := x \approx \top$.
- $\mathcal{J}_3$, the paraconsistent logic of Da Costa and D'Ottaviano,‡ with the transformer $\tau_2 x := \Diamond x \approx \top$.

Equivalently, you can check (Exercise 3.16) that these are the logics defined by the matrices $\langle \boldsymbol{A}_3, \{1\} \rangle$ and $\langle \boldsymbol{A}_3, \{1/2, 1\} \rangle$, respectively. That the two logics are different is witnessed by the facts that $\Diamond x \vdash_{\mathcal{J}_3} x$ while $\Diamond x \nvdash_{Ł_3} x$. Actually, neither logic is an extension of the other, because moreover $x \nvdash_{\mathcal{J}_3} \Box x$ while $x \vdash_{Ł_3} \Box x$. All this is easy to check.

† This is just a reduct of the algebra denoted by $[\mathbf{0,1}]_3$ in Example 1.12.

‡ The logic $\mathcal{J}_3$ has been denoted in various ways in the literature, such as dC or **LFI1**, and has been presented in various similarity types, very often with $\neg, \vee, \Diamond$ as primitive connectives (sometimes an "inconsistency connective" is also included). The definitionally equivalent presentation choosen here makes the direct comparison with $Ł_3$ easier. See [52, Section 4.4.3] and [53, Example 18].

With these definitions, $\{\boldsymbol{A}_3\}$ constitutes an algebraic semantics for each of the logics, and you can check that this algebra satisfies condition (ALG4) with different transformers:

- For $Ł_3$, with $\rho_1(x \approx y) := \{x \to y\,, y \to x\}$.
- For $\mathcal{J}_3$, with $\rho_2(x \approx y) := \{x \rightsquigarrow y\,, y \rightsquigarrow x\,, \neg x \rightsquigarrow \neg y\,, \neg y \rightsquigarrow \neg x\}$, the derived implication being defined as $x \rightsquigarrow y := (\neg \Diamond x \to y) \to y$.

Thus, the two logics are algebraizable with respect to $\{\boldsymbol{A}_3\}$. Since this is a finite algebra, $\vDash_{\boldsymbol{A}_3}$ is finitary, therefore the common equivalent algebraic semantics is the quasivariety $\mathbb{Q}\{\boldsymbol{A}_3\} = \mathbb{ISP}\{\boldsymbol{A}_3\}$; it turns out that this class is actually a variety. Then, by using condition (ALG1) for one logic and condition (ALG2) for the other, and after some computations (Exercise 3.16 again), one finds:

$$\begin{aligned} &\Gamma \vdash_{Ł_3} \varphi \iff \rho_2\tau_1\Gamma \vdash_{\mathcal{J}_3} \rho_2\tau_1\varphi \iff [\ldots] \iff \Box\Gamma \vdash_{\mathcal{J}_3} \Box\varphi \\ &\Gamma \vdash_{\mathcal{J}_3} \varphi \iff \rho_1\tau_2\Gamma \vdash_{Ł_3} \rho_1\tau_2\varphi \iff [\ldots] \iff \Diamond\Gamma \vdash_{Ł_3} \Diamond\varphi \\ &\varphi \dashv\vdash_{Ł_3} \Diamond\Box\varphi \\ &\varphi \dashv\vdash_{\mathcal{J}_3} \Box\Diamond\varphi \end{aligned} \tag{30}$$

These expressions establish the "deductive equivalence" of the two logics, witnessed by the "transformers" $x \mapsto \Box x$ and $x \mapsto \Diamond x$. ⊠

Other examples of pairs of different algebraizable logics with the same equivalent algebraic semantics are found in [32, Section 5.2.4] and in [141].

Exercises for Section 3.2

EXERCISE 3.9. Prove that the formulations of conditions (ALG3) and (ALG4) given in Definition 3.11 and in the displays (26) and (27) on page 116 are equivalent, in the sense that a logic $\mathcal{L}$ satisfies any of these three forms of (ALG3) if and only if it satisfies the others, and that for any class K the consequence $\vDash_{\mathsf{K}}$ satisfies any of these three forms of (ALG4) if and only if it satisfies the others. You only need to assume that the transformers are structural and commute with unions, but neither condition (ALG1) nor (ALG2). Expressing everything in terms of the associated closure operators $C_{\mathcal{L}}$ and C_{K} and recalling Exercise 1.34 may help.

EXERCISE 3.10. Prove the equivalence of the two pairs of conditions (ALG1)+(ALG4) and (ALG2)+(ALG3) as stated in Proposition 3.12.

EXERCISE 3.11. Investigate the algebraizability of logics with empty transformers, and the corresponding equivalent algebraic semantics.

EXERCISE 3.12. Show that the identity logic (the logic whose consequence operator is the identity function on the power set of formulas) is not algebraizable.

EXERCISE 3.13. Show that, in contrast to the result in Exercise 3.5, classical logic $\mathcal{Cl}$ is not algebraizable with respect to the class of Heyting algebras HA with defining equation $\neg\neg x \approx \top$ and equivalence formulas $\{x \to y\,, y \to x\}$.

EXERCISE 3.14. Does the converse of Proposition 3.15 hold? That is: Is it true that a logic is algebraizable with defining equation $x \approx x \to x$ and equivalence formulas the set

$\{x \to y, y \to x\}$, for some binary term $\to$, if and only if it is an implicative logic? What if we add condition (G), or its strengthening (IL5)?

HINT. If you cannot decide now, look into Section 3.4.

EXERCISE 3.15. Let $\mathcal{L}$ be an algebraizable logic with equivalent algebraic semantics K, and let K' be any algebraic semantics for $\mathcal{L}$. Show that then $\mathsf{K} \subseteq \mathbb{V}\mathsf{K}'$. Notice that it is not required that the defining equations are the same for K and for K'.

EXERCISE 3.16. Work out all the missing details (that is, all the unproven assertions) in Example 3.20.

3.3 A syntactic characterization, and the Lindenbaum-Tarski process again

The definition of the algebraizability of a logic allows us to establish this property for a given logic only with prior knowledge of the class K and of the transformers. Now we will find a *purely syntactic criterion* that allows us to check whether a given pair of transformers bears witness to the algebraizability of a given logic just by looking at their behaviour with regard to the consequence relation of the logic. Such a criterion is sometimes qualified as an *intrinsic characterization* of algebraizability, but in fact this is only partially the case, as it still depends on knowledge of the transformers.

The goal is to prove the following result:

THEOREM 3.21
A logic $\mathcal{L}$ is algebraizable if and only if there are equations $E(x) \subseteq Eq$ and formulas $\Delta(x,y) \subseteq Fm$ such that $\mathcal{L}$ satisfies the following five conditions:

(R) $\vdash_{\mathcal{L}} \Delta(x,x)$

(Sym) $\Delta(x,y) \vdash_{\mathcal{L}} \Delta(y,x)$

(Trans) $\Delta(x,y) \cup \Delta(y,z) \vdash_{\mathcal{L}} \Delta(x,z)$

(Re) $\bigcup_{i=1}^{n} \Delta(x_i,y_i) \vdash_{\mathcal{L}} \Delta(\lambda x_1 \dots x_n, \lambda y_1 \dots y_n)$ *for each* $\lambda \in L$, $n = \operatorname{ar}\lambda \geqslant 1$

(ALG3) $x \dashv\vdash_{\mathcal{L}} \Delta(E(x))$

Then $E(x)$ is the set of defining equations and $\Delta(x,y)$ is the set of equivalence formulas.

It is possible to show that (MP) is a consequence of these conditions (Exercise 3.17). It is possible as well to show that conditions (Sym) and (Trans) can be replaced by (MP) in this theorem, but this requires a different theoretical context; see Proposition 6.130.

The direct part of the theorem is proved in Proposition 3.13. In order to prove the converse, we need to exhibit an appropriate class of algebras K such that conditions (ALG1)–(ALG4) are satisfied. We find it by *a generalization of the Lindenbaum-Tarski process*, which yields a minimal† class with respect to which the logic is algebraizable, and at the same time we find a description of the largest

† By this it is not meant that such a class is minimal in a technical, set-theoretic sense, but rather that it contains just the algebras needed for the proof to work.

one (which exists by Corollary 3.18), that is, of its equivalent algebraic semantics. There is an obvious parallelism between conditions (R), (Trans) and (Re) and conditions (IL1)–(IL3) of the definition of implicative logics; the proofs reveal that the role played there by conditions (IL4) and (IL5) is played here by condition (ALG3), and this is somehow confirmed by Exercise 3.20.

The following observation, although trivial, is nevertheless useful:

LEMMA 3.22
Let $E(x) \subseteq Eq$ and $\Delta(x,y) \subseteq Fm$. For any algebra $\boldsymbol{A}$, the condition (ALG4) *is equivalent to the following pair of conditions:*

(R) $\boldsymbol{A} \vDash E(\Delta(x,x))$

(Red) $\boldsymbol{A} \vDash \bigwedge E(\Delta(x,y)) \rightarrow x \approx y$ *or, equivalently,* $E(\Delta(x,y)) \vDash_{\boldsymbol{A}} x \approx y$. ⊠

A class of algebras is said to satisfy the condition (R) or (Red) when all its members satisfy it; that is, when the shown equations or quasi-equation hold in the class. Notice that in Theorem 3.21 the label "(R)" is also used for a related property; since there it refers to a logic, confusion is unlikely. The label "(Red)" means "reduced". Observe that while (R) corresponds to a (possibly infinite) set of ordinary equations, (Red) corresponds to a single generalized quasi-equation (which is an ordinary quasi-equation only when both Δ and E are finite).

The equivalent algebraic semantics we are after is the following class.

DEFINITION 3.23
For any logic $\mathcal{L}$ and structural transformers $\boldsymbol{\tau}, \boldsymbol{\rho}$, the class $\mathsf{K}(\mathcal{L}, \boldsymbol{\tau}, \boldsymbol{\rho})$ is defined as

$$\mathsf{K}(\mathcal{L}, \boldsymbol{\tau}, \boldsymbol{\rho}) := \{\boldsymbol{A} : \boldsymbol{\tau}\Gamma \vDash_{\boldsymbol{A}} \boldsymbol{\tau}\varphi \text{ whenever } \Gamma \vdash_{\mathcal{L}} \varphi, \text{ and } E(\Delta(x,y)) \vDash_{\boldsymbol{A}} x \approx y\}.$$

This is the subclass of algebras in $\mathsf{K}(\mathcal{L}, \boldsymbol{\tau})$ that satisfy (Red). Notice how this generalizes the definition of the class $\mathsf{Alg}^*\mathcal{L}$ for implicative logics (Definition 2.5). The first condition is exactly (LALG1), while the second one is a generalized form of (LALG2). We can see that this is the natural choice: Given E and Δ, we want a class K such that (ALG1) and (ALG4) are satisfied. To satisfy (ALG1) is to say that K is an algebraic semantics for $\mathcal{L}$, hence forcefully we should end with a subclass of $\mathsf{K}(\mathcal{L}, \boldsymbol{\tau})$. And we also need that at least one half of (ALG4) is satisfied, which is the second condition (Red). Thus we decide to take all algebras satisfying these two conditions. But the logic satisfies (R) by assumption, therefore if we prove (ALG1), then condition (R) of Lemma 3.22 is satisfied in K, and we have (ALG4). So, the real challenge is to prove (ALG1), that is, that the chosen class is indeed an algebraic semantics for $\mathcal{L}$; this is a *completeness theorem*, precisely what the Lindenbaum-Tarski process is able to prove. This discussion also clarifies why (LALG2) is included in Definition 2.5: we want not just algebraic completeness, but algebraizability!

Until Proposition 3.29 we keep assuming that $\mathcal{L}$ satisfies the five conditions in Theorem 3.21 for certain $E(x) \subseteq Eq$ and $\Delta(x,y) \subseteq Fm$; the corresponding structural transformers are denoted by $\boldsymbol{\tau}$ and $\boldsymbol{\rho}$, respectively.

Definition 3.24
For each $T \in \mathcal{T}h\mathcal{L}$, define $\boldsymbol{\Omega} T$ as: $\alpha \equiv \beta\ (\boldsymbol{\Omega} T)$ if and only if $\Delta(\alpha,\beta) \subseteq T$, for all $\alpha,\beta \in Fm$.

From the four conditions (R) , (Sym) , (Trans) , (Re) it trivially follows:

Lemma 3.25
For every $T \in \mathcal{T}h\mathcal{L}$, $\boldsymbol{\Omega} T \in \mathrm{Con}\boldsymbol{Fm}$. ⊠

Thus, the following (provisional) definition makes sense:

Definition 3.26
*The **Lindenbaum-Tarski algebras** of $\mathcal{L}$ are the algebras in the class*

$$\mathsf{LTAlg}^*\mathcal{L} := \{\boldsymbol{Fm}/\boldsymbol{\Omega} T : T \in \mathcal{T}h\mathcal{L}\}.$$

Although it is not reflected in the notation, in principle the relation $\boldsymbol{\Omega} T$ and this class of algebras depend on the formulas $\Delta(x,y)$; however, once algebraizability of $\mathcal{L}$ is proved with $\Delta(x,y)$ as the set of equivalence formulas, one can use Theorem 3.17 to see that $\boldsymbol{\Omega} T$ does not depend on a particular set of equivalence formulas (provided T is a theory of $\mathcal{L}$). Moreover, a class of Lindenbaum-Tarski algebras can be associated with any logic (Definition 4.42). Corollary 4.55 contains an intrinsic characterization of the congruence $\boldsymbol{\Omega} T$, independent of any set of equivalence formulas, and confirms that the two definitions of the class $\mathsf{LTAlg}^*\mathcal{L}$ coincide for algebraizable logics.

The core of *the generalized Lindenbaum-Tarski process* is the following:

Theorem 3.27
For all $\Gamma \cup \{\varphi\} \subseteq Fm$, if $\Gamma \nvdash_{\mathcal{L}} \varphi$, then $\boldsymbol{\tau}\Gamma \nvDash_{\mathsf{LTAlg}^*\mathcal{L}} \boldsymbol{\tau}\varphi$.

Proof: This tries to follow the steps of the process explained in Section 2.1 as closely as possible.

(LT1) Assume that $\Gamma \nvdash_{\mathcal{L}} \varphi$ and consider the $\mathcal{L}$-theory T generated by Γ. Thus, $\Gamma \subseteq T$ but $\varphi \notin T$.

(LT2) Consider the relation $\boldsymbol{\Omega} T$ obtained by applying Definition 3.24 to T.

(LT3) By Lemma 3.25 this relation is a congruence of $\boldsymbol{Fm}$. Consider the canonical projection $\pi\colon \boldsymbol{Fm} \to \boldsymbol{Fm}/\boldsymbol{\Omega} T$.

(LT4) $\boldsymbol{Fm}/\boldsymbol{\Omega} T \in \mathsf{LTAlg}^*\mathcal{L}$ by Definition 3.26.

(LT5) For any $\alpha \in Fm$, $\alpha \in T$ if and only if $\boldsymbol{Fm}/\boldsymbol{\Omega} T \vDash \boldsymbol{\tau}\alpha\ [\![\pi]\!]$. This is left to you as Exercise 3.18; notice that this is where condition (ALG3) is used in an essential way.

The process stops here. Applying (LT5) to (LT1) we see that $\boldsymbol{Fm}/\boldsymbol{\Omega} T \vDash \boldsymbol{\tau}\Gamma\ [\![\pi]\!]$ while $\boldsymbol{Fm}/\boldsymbol{\Omega} T \nvDash \boldsymbol{\tau}\varphi\ [\![\pi]\!]$. By (LT4), this establishes that $\boldsymbol{\tau}\Gamma \nvDash_{\mathsf{LTAlg}^*\mathcal{L}} \boldsymbol{\tau}\varphi$. ⊠

There is no analogue of step (LT6) here, as the properties that all formulas in a theory become equivalent and that their equivalence class is the maximum of the quotient algebra need not hold in an arbitrary algebraizable logic, while they do hold in all the implicative ones (see Theorem 2.2); as a matter of fact, the quotient algebra need not be ordered, and certainly our equivalence relation is not obtained as the symmetrization of a quasi-order, as in the implicative case.

The proof of Theorem 3.21 is broken into smaller, technical steps, so as to make its global structure more visible; as a consequence, apparently impressive statements have rather short proofs:

THEOREM 3.28
If K *is any class of algebras satisfying both that* $\mathsf{LTAlg}^*\mathcal{L} \subseteq \mathsf{K}$ *and that* $\mathsf{K} \subseteq \mathsf{K}(\mathcal{L}, \boldsymbol{\tau}, \boldsymbol{\rho})$, *then* $\mathcal{L}$ *is algebraizable with respect to* K *with transformers* $\boldsymbol{\tau}$ *and* $\boldsymbol{\rho}$.

PROOF: We prove that $\mathcal{L}, \mathsf{K}, \boldsymbol{\tau}$ and $\boldsymbol{\rho}$ satisfy conditions (ALG1) and (ALG4). If $\Gamma \vdash_{\mathcal{L}} \varphi$, then $\boldsymbol{\tau}\Gamma \vDash_{\mathsf{K}} \boldsymbol{\tau}\varphi$, because $\mathsf{K} \subseteq \mathsf{K}(\mathcal{L}, \boldsymbol{\tau}, \boldsymbol{\rho})$. If $\Gamma \nvdash_{\mathcal{L}} \varphi$, then $\boldsymbol{\tau}\Gamma \nvDash_{\mathsf{LTAlg}^*\mathcal{L}} \boldsymbol{\tau}\varphi$ by Theorem 3.27; and since $\mathsf{LTAlg}^*\mathcal{L} \subseteq \mathsf{K}$, it also follows that $\boldsymbol{\tau}\Gamma \nvDash_{\mathsf{K}} \boldsymbol{\tau}\varphi$. This establishes (ALG1). Now we see that all algebras in K satisfy the two conditions in Lemma 3.22: (R) because by assumption $\emptyset \vdash_{\mathcal{L}} \Delta(x, x)$ and $\mathsf{K} \subseteq \mathsf{K}(\mathcal{L}, \boldsymbol{\tau}, \boldsymbol{\rho})$, and (Red) by the second part in the definition of $\mathsf{K}(\mathcal{L}, \boldsymbol{\tau}, \boldsymbol{\rho})$. Therefore, by the lemma, condition (ALG4) is satisfied, and this concludes the proof of the theorem. ⊠

Observe that this does *not* prove that $\mathcal{L}$ is algebraizable, unless the *existence* of a class of algebras K satisfying the two assumptions of the theorem is proved. Here two classes with these properties are found at the same time:

PROPOSITION 3.29
$\mathsf{LTAlg}^*\mathcal{L} \subseteq \mathsf{K}(\mathcal{L}, \boldsymbol{\tau}, \boldsymbol{\rho})$.

PROOF: Let $T \in \mathit{Th}\mathcal{L}$. We have to show that $\boldsymbol{Fm}/\boldsymbol{\Omega}T$ satisfies the two conditions in Definition 3.23. Assume that $\Gamma \vdash_{\mathcal{L}} \varphi$ and take any $h \in \mathrm{Hom}(\boldsymbol{Fm}, \boldsymbol{Fm}/\boldsymbol{\Omega}T)$ such that $\boldsymbol{Fm}/\boldsymbol{\Omega}T \vDash \boldsymbol{\tau}\Gamma\, [\![h]\!]$. Consider $\pi \colon \boldsymbol{Fm} \to \boldsymbol{Fm}/\boldsymbol{\Omega}T$; since π is surjective, by Lemma 1.2 there is a substitution σ such that $h = \pi\sigma$. So $\boldsymbol{Fm}/\boldsymbol{\Omega}T \vDash \boldsymbol{\tau}\Gamma\, [\![\pi\sigma]\!]$. Using that $\boldsymbol{\tau}$ is structural, this is to say that $\boldsymbol{Fm}/\boldsymbol{\Omega}T \vDash \boldsymbol{\tau}\sigma\Gamma\, [\![\pi]\!]$. By the result obtained in step (LT5) of the proof of Theorem 3.27 (or by Exercise 3.18), this is to say that $\sigma\Gamma \subseteq T$. By structurality of the logic and the fact that T is a theory, it follows that $\sigma\varphi \in T$, and, after reversing the process, we obtain that $\boldsymbol{Fm}/\boldsymbol{\Omega}T \vDash \boldsymbol{\tau}\varphi\, [\![h]\!]$. This shows that $\boldsymbol{\tau}\Gamma \vDash_{\boldsymbol{Fm}/\boldsymbol{\Omega}T} \boldsymbol{\tau}\varphi$, which is the first condition in Definition 3.23. The proof of the second condition uses the same trick, and is left to you as Exercise 3.19. ⊠

If one wants the Lindenbaum-Tarski process to directly prove algebraizability with respect to $\mathsf{K}(\mathcal{L}, \boldsymbol{\tau}, \boldsymbol{\rho})$, one simply plugs this proof in† as step (LT4) of the proof of Theorem 3.27; notice, however, that the order of the steps in the process has been slightly altered, as (LT5) has been used to prove Proposition 3.29.

The conclusion of Proposition 3.29 is that both classes $\mathsf{LTAlg}^*\mathcal{L}$ and $\mathsf{K}(\mathcal{L}, \boldsymbol{\tau}, \boldsymbol{\rho})$ satisfy the assumptions in Theorem 3.28, and therefore $\mathcal{L}$ is algebraizable with respect to each of them, with the transformers $\boldsymbol{\tau}$ and $\boldsymbol{\rho}$.

† After renaming the set Γ.

This completes the proof of Theorem 3.21; but in passing we have proved something more, which is worth recording:

THEOREM 3.30

Let $\mathcal{L}$ be an algebraizable logic, with transformers $\boldsymbol{\tau}, \boldsymbol{\rho}$. The logic $\mathcal{L}$ is algebraizable with respect to the classes $\mathsf{LTAlg}^*\mathcal{L}$ *and* $\mathsf{K}(\mathcal{L}, \boldsymbol{\tau}, \boldsymbol{\rho})$ *and also with respect to any class of algebras* K *such that* $\mathsf{LTAlg}^*\mathcal{L} \subseteq \mathsf{K} \subseteq \mathsf{K}(\mathcal{L}, \boldsymbol{\tau}, \boldsymbol{\rho})$. *Moreover the class* $\mathsf{K}(\mathcal{L}, \boldsymbol{\tau}, \boldsymbol{\rho})$ *is the equivalent algebraic semantics of $\mathcal{L}$, and* $\mathsf{K}(\mathcal{L}, \boldsymbol{\tau}, \boldsymbol{\rho}) = \mathbb{GQ}\mathsf{LTAlg}^*\mathcal{L}$. ⊠

Thus, the class of all Lindenbaum-Tarski algebras of an algebraizable logic generates, as a generalized quasivariety, its equivalent algebraic semantics. As already observed, by Theorem 3.17 the class $\mathsf{K}(\mathcal{L}, \boldsymbol{\tau}, \boldsymbol{\rho})$ does not depend on the particular $\boldsymbol{\tau}, \boldsymbol{\rho}$ taken to define it. Notice also that *all* Lindenbaum-Tarski algebras are needed in order to generate the equivalent algebraic semantics. That this may be achieved with just the pure Lindenbaum-Tarski algebra $\boldsymbol{Fm}/\Omega C_{\mathcal{L}}\emptyset$ is in general unlikely, because this algebra characterizes only the theorems of the logic; Exercise 3.21 describes a situation where this indeed happens.

The algebraizability of a large number of logics has been shown in the literature by using Theorem 3.21, either directly (see Section 3.4) or through the next straightforward application, which settles the issue of the algebraizability of extensions, fragments and expansions:

PROPOSITION 3.31

Let $\mathcal{L}$ be an algebraizable logic with respect to a class K *and with transformers $\boldsymbol{\tau}, \boldsymbol{\rho}$.*

1. *Every extension $\mathcal{L}'$ of $\mathcal{L}$ is algebraizable as well, with respect to a subclass* K' *of* K *and with the same transformers.*
2. *If $\boldsymbol{L}'$ is a fragment of the language $\boldsymbol{L}$ of $\mathcal{L}$ satisfying that $\boldsymbol{\tau} x \subseteq Eq_{\boldsymbol{L}'}$ and that $\boldsymbol{\rho}(x \approx y) \subseteq Fm_{\boldsymbol{L}'}$, then $\mathcal{L}' := \mathcal{L} \restriction \boldsymbol{L}'$, the $\boldsymbol{L}'$ fragment of $\mathcal{L}$, is algebraizable with respect to the class* $\mathsf{K} \restriction \boldsymbol{L}'$ *and with the same transformers.*
3. *If $\mathcal{L}'$ is an expansion of $\mathcal{L}$ such that $\vdash_{\mathcal{L}'}$ satisfies condition* (Re) *for the additional connectives, then $\mathcal{L}'$ is algebraizable, with the same transformers.*

PROOF: This is a straightforward application of Theorem 3.21, and is left to you as Exercise 3.22. It is interesting to notice that in the case of axiomatic extensions, the property in 1 can be proved directly from the definitions (Exercise 3.23). ⊠

Notice that, as a particular case of point 3, all expansions by constants of an algebraizable logic are algebraizable.

The above result does not claim that the classes of algebras mentioned there are the *equivalent* algebraic semantics of the logics $\mathcal{L}'$, even if K is the equivalent algebraic semantics of $\mathcal{L}$. But some relations can be established. For instance, in point 1, it is easy to see that if there is a presentation of $\mathcal{L}'$ relatively to $\mathcal{L}$ by an axiomatic system, then the equivalent algebraic semantics $\mathsf{K}(\mathcal{L}', \boldsymbol{\tau}, \boldsymbol{\rho})$ of $\mathcal{L}'$ is obtained from that of $\mathcal{L}$ by adding the generalized quasi-equations that result from the additional axioms and rules. In point 2 the situation is similar to that for algebraic semantics (Proposition 3.10 and Exercise 3.7):

PROPOSITION 3.32
Let $\mathcal{L}$ be an algebraizable logic with K as equivalent algebraic semantics, and with transformers τ, ρ. Let $\boldsymbol{L}'$ be a fragment of the language of $\mathcal{L}$ satisfying that $\tau x \subseteq Eq_{\boldsymbol{L}'}$ and that $\rho(x \approx y) \subseteq Fm_{\boldsymbol{L}'}$. The class $\mathbb{S}(\mathsf{K} \restriction \boldsymbol{L}')$ is the equivalent algebraic semantics of $\mathcal{L}' := \mathcal{L} \restriction \boldsymbol{L}'$. ⊠

This result has been used in practice to find the equivalent algebraic semantics of many fragments of a known algebraizable logic such as $\mathcal{C\ell}$, $\mathcal{I\ell}$, or many substructural logics.

As to point 1, it can be enhanced in the following sense:

THEOREM 3.33
Let $\mathcal{L}$ be an algebraizable logic, with equivalent algebraic semantics the generalized quasivariety K. Associating with every algebraizable logic its equivalent algebraic semantics establishes a dual isomorphism between the lattice of all extensions of $\mathcal{L}$ and the lattice of all generalized quasivarieties contained in K.

PROOF: Clearly, if $\mathcal{L} \leqslant \mathcal{L}'$ then $\mathsf{K}(\mathcal{L}', \tau, \rho) \subseteq \mathsf{K}(\mathcal{L}, \tau, \rho) = \mathsf{K}$, and both are generalized quasivarieties. If $\mathcal{L}$ is algebraizable, then by Proposition 3.31 $\mathcal{L}'$ is also algebraizable with the same transformers, and the function $\mathcal{L}' \longmapsto \mathsf{K}(\mathcal{L}', \tau, \rho)$ is order-reversing. Now let $\mathsf{K}' \subseteq \mathsf{K}$ be a generalized quasivariety. Since K satisfies (R) and (Red), K' also satisfies them for the same transformers, that is, it satisfies (ALG4). Therefore, the logic $\mathcal{L}'$ defined by K' thorugh condition (ALG1) with the same transformers is algebraizable with respect to K', and its equivalent algebraic semantics is forcefully K' because it is already a generalized quasivariety. Finally, that $\mathcal{L} \leqslant \mathcal{L}'$ is shown by using (ALG1) twice, first for K and then, after using that $\mathsf{K}' \subseteq \mathsf{K}$, for K'. Thus the function $\mathsf{K}' \longmapsto \mathcal{L}'$ also reverses order. The two functions are clearly inverses to one another. ⊠

Again, this isomorphism does not depend on the particular transformers taken to define the classes $\mathsf{K}(\mathcal{L}, \tau, \rho)$ for each $\mathcal{L}$. A consequence of this isomorphism is that if two algebraizable logics have the same equivalent algebraic semantics (obviously, with at least one different transformer), then neither can be an extension of the other, that is, they must be incomparable; this is what happens in Example 3.20.

Exercises for Section 3.3

EXERCISE 3.17. Let $\mathcal{L}$ be a logic satisfying the five conditions in Theorem 3.21. Show in a purely syntactic way (not using algebraizability) that $\mathcal{L}$ satisfies (MP): $x, \Delta(x,y) \vdash_{\mathcal{L}} y$.

HINT. You should first prove that, under these assumptions, the property in Lemma 3.16 holds as well.

EXERCISE 3.18. Let $\mathcal{L}$ be a logic satisfying the first four conditions (R)–(Re) in Theorem 3.21. Prove that the logic satisfies (ALG3) if and only if for all $T \in \mathcal{Th}\mathcal{L}$ and all $\alpha \in Fm$, $\alpha \in T$ if and only if $\boldsymbol{Fm}/\Omega T \vDash \tau\alpha\, [\![\pi]\!]$, where $\pi \colon \boldsymbol{Fm} \to \boldsymbol{Fm}/\Omega T$ is the canonical projection.

EXERCISE 3.19. Let $\mathcal{L}$ be a logic satisfying the five conditions in Theorem 3.21, and let $T \in \mathcal{T}h\mathcal{L}$. Prove that the quotient algebra $\boldsymbol{Fm}/\Omega T$ satisfies the condition (Red) of Lemma 3.22.

HINT. Use the result in Exercise 3.18.

EXERCISE 3.20. Let $\mathcal{L}$ be a logic with theorems and with a binary connective $\to$ satisfying (IL2). Prove that it satisfies conditions (IL4) and (IL5) if and only if it satisfies $x \dashv\vdash_{\mathcal{L}} \top \to x$ and $\vdash_{\mathcal{L}} x \to \top$ for some theorem $\top$ of $\mathcal{L}$.

EXERCISE 3.21. Let $\mathcal{L}$ be an algebraizable logic, and put $\Omega_0 := \Omega C_{\mathcal{L}}\emptyset$.

1. Show that a (g-)rule $\Gamma \rhd \varphi$ is admissible in $\mathcal{L}$ if and only if for all $\alpha \approx \beta \in \tau\varphi$, where τ is the transformer witnessing algebraizability, the (generalized) quasi-equation $\bigwedge \tau\Gamma \to \alpha \approx \beta$ is valid in $\boldsymbol{Fm}/\Omega_0$.
2. Conclude that $\mathcal{L}$ is structurally complete if and only if the class $\mathbb{GQ}\{\boldsymbol{Fm}/\Omega_0\}$ coincides with the equivalent algebraic semantics of $\mathcal{L}$.

COMMENT. Admissible (g-)rules are defined in Exercise 1.18. A logic is ***structurally complete*** when all its admissible rules are derivable in it: if $\Gamma \rhd \varphi$ is admissible in $\mathcal{L}$, then $\Gamma \vdash_{\mathcal{L}} \varphi$. In a few works the notion refers to g-rules.

EXERCISE 3.22. Prove Proposition 3.31, using Theorem 3.21.

EXERCISE 3.23. Prove directly from the definition of algebraizability that an axiomatic extension of an algebraizable logic is also algebraizable, with the same transformers.

EXERCISE 3.24. Complete the details skipped in the proof of Theorem 3.33.

EXERCISE 3.25. Explore the following alternative (and, in some sense, dual) proof of Theorem 3.21. Assume that $\mathcal{L}$ satisfies the five conditions in it, with respect to transformers τ and ρ; define a structural closure operator C on the set *Eq* in the appropriate way; use Exercise 1.85 to obtain a class of algebras; prove that the logic is algebraizable with respect to it. Prove that the class of algebras found (more indirectly) in this way also coincides with the class of Lindenbaum-Tarski algebras of the logic (Definition 3.26).

3.4 More examples, and special kinds of algebraizable logics

We know (Proposition 3.15) that all implicative logics are algebraizable. A very close, but larger class of algebraizable logics (Exercise 3.14) is obtained by conditions (ALG1) and (ALG4) for the particular transformers $x \approx \top$ and $\{x \to y, y \to x\}$ and some class K. In many of the cases where there is a reasonable candidate for K these two conditions are the method of choice to show algebraizability: (ALG1) is a typical Algebraic Completeness Theorem, and (ALG4) involves exclusively the class K, and may thus be established by purely algebraic work inside K; indeed, in each algebra in K individually (Lemma 3.22). Examples of this procedure are the algebraizability of $\L_\infty$ with respect to $[0,1]$ and of $\L_n$ with respect to $[0,1]_n$ (see Example 3.41 for more details).

Many other cases are shown to be algebraizable by using Proposition 3.31: any extension of an algebraizable logic (in the same language) is also algebraizable with the same transformers, and any fragment of an algebraizable logic that contains the language of the transformers is so as well; of course both procedures can be combined, and this produces many examples out from a single one. As for expansions, Proposition 3.31 establishes under which conditions an expansion of an algebraizable logic is algebraizable.

Finally, in some cases, Theorem 3.21 has been used as well: it establishes a set of purely syntactic conditions under which a logic is algebraizable; the main interest of this characterization is that it does not require to guess a candidate for the equivalent algebraic semantics in advance.

EXAMPLES 3.34

1. As a first application of these results, from Proposition 3.15 it follows that all the *superintuitionistic* or *intermediate logics* (the extensions† of intuitionistic logic) are also algebraizable with the same transformers.

2. A large class of algebraizable logics, not all of them implicative, is that of the so-called *substructural logics*, understood in [123] as the axiomatic extensions of a very weak logic denoted by $\mathcal{FL}$ and related to the (associative) *Full Lambek calculus*; in [123, Theorem 2.27] it is shown that $\mathcal{FL}$, which is presented axiomatically, is algebraizable and its equivalent algebraic semantics is the variety of *residuated lattices*.‡ The transformers are

$$x \longmapsto x \wedge \top \approx \top \qquad\qquad x \approx y \longmapsto \{x\backslash y, y\backslash x\}$$

where $\backslash$ is the left residual (or left division) connective; notice that the defining equation can be read as $\top \preccurlyeq x$. By Proposition 3.31 all the logics in the family are also algebraizable, with the same transformers, and their equivalent algebraic semantics is a variety of residuated lattices. The transformers may be replaced by simpler ones thanks to the properties of a specific logic or a specific variety. For instance, in the commutative cases the residual connective $\backslash$ becomes the more familiar implication $\rightarrow$, and in the integral cases the defining equation is equivalent to $x \approx \top$. The fragments containing at least $\wedge, \backslash, \top$ also fall in the same category, with the corresponding class of subreducts (Proposition 3.32).

The class of algebraizable logics obtained in the way just explained is surprisingly enormous; it is described in Section 2.3 of [123]. It includes:

3. Some *weak logics of implication* whose introduction was motivated by the study of *combinatory logic*. Their language contains just $\rightarrow$, but they need not be implicative in the technical sense of Definition 2.3. The best known in

† In some works the terms "intermediate" and "superintuitionistic" are applied only to axiomatic extensions of $\mathcal{I}\ell$; often in these works a logic is not defined as a consequence relation but as a set of theorems.

‡ A ***residuated lattice*** can be defined as an algebraic structure in a language $\langle \wedge, \vee, \cdot, \backslash, /, \top \rangle$ of type $\langle 2,2,2,2,2,0 \rangle$ such that its $\langle \wedge, \vee \rangle$ reduct is a lattice, its $\langle \cdot, \top \rangle$ reduct is a monoid, and for all $a, b, c \in A$, $a \cdot b \leqslant c \Leftrightarrow b \leqslant a\backslash c \Leftrightarrow a \leqslant c/b$. It is ***commutative*** when the monoid operation is so (this implies that the two residuals $\backslash$ and $/$ coincide, being denoted by $\rightarrow$), and it is ***integral*** when $\top^{\boldsymbol{A}}$, the unit of the monoid, is at the same time the maximum of the order. Notice that in the literature a different constant symbol, 1, is often used for our $\top$, and both can even coexist, with different roles; see [123, p. 111].

this group is the logic $\mathcal{BCK}$, which is algebraizable with defining equation $x \approx x \to x$ and equivalence formulas $\{x \to y, y \to x\}$, and its equivalent algebraic semantics is the quasivariety of BCK algebras; this is well known as one of the few examples of a proper quasivariety naturally arising from logic. By contrast, the related, weaker logic $\mathcal{BCI}$, also a well known one, is shown in [32] not to be algebraizable.

4. Some *relevance logics* with a *truth constant*; see the final part of point 7 below.
5. Several versions and fragments of *linear logic* (classical, intuitionistic, multiplicative, additive, etc.), all without exponentials, can be presented as definitionally equivalent to an extension of, or an expansion of some extension of $\mathcal{FL}$, and are hence algebraizable.
6. Hájek's *basic logic* $\mathcal{BL}$, and all the *fuzzy logics* presented as its axiomatic extensions in [132]. Their equivalent algebraic semantics are subvarieties of the variety of ***BL-algebras***, which are the commutative integral residuated lattices with minimum in which the equations $x \wedge y \approx x \cdot (x \to y)$ and $(x \to y) \vee (y \to x) \approx \top$ hold; it is integrality that makes all these cases implicative. Therefore, in these cases the defining equation is simply $x \approx \top$. In this very large group we find Łukasiewicz's many-valued logics $\L_{\infty f}$ and the $\L_n$, for $n > 2$ (see Examples 3.41 and 5.83 for their equivalent algebraic semantics); *product logic* Π; and the Gödel-Dummett logic, the extension of $\mathcal{I}\ell$ with the *prelinearity axiom* $(x \to y) \vee (y \to x)$.

Further examples of algebraizable logics, not belonging to the previous group:

7. Several logics from the field of *relevance logic*. One of the best known is $\mathcal{R}$, a version of Anderson-Belnap's system R of relevance logic, which was proved to be algebraizable in [32] by purely syntactic means (using Theorem 3.21); its equivalent algebraic semantics (the class of *R-algebras*) was identified in [114]. The language of $\mathcal{R}$ is $\langle \wedge, \vee, \to, \neg \rangle$, and in it one does not consider a constant $\top$ as shorthand for some theorem;[†] the defining equation is $x \wedge (x \to x) \approx x \to x$. The well-known logic $\mathcal{RM}$ of *relevance with mingle* is the extension of $\mathcal{R}$ with the so-called "mingle axiom" $x \to (x \to x)$, and so it is algebraizable as well, with the simpler defining equation $x \approx x \to x$; its equivalent algebraic semantics is the variety of *Sugihara algebras*. These cases are also recorded in [123], because their conservative expansions obtained by addition of a truth constant $\top$ (with the appropriate axioms) do belong to the class of substructural logics, and are hence algebraizable in the standard way described in point 2 above.
8. $\mathcal{C}\ell_{\leftrightarrow}$, the *equivalence fragment of* $\mathcal{C}\ell$, is algebraizable and its equivalent algebraic semantics is the variety of ***Boolean groups***, the Abelian groups where every non-zero element has order 2 (Exercise 3.26). $\mathcal{C}\ell_{\leftrightarrow}$ was axiomatized

[†] Mainly for the philosophical motivations around relevance logic; technically, because in this logic two theorems need not be equivalent, and thus privileging one among them seems to make little sense.

by Łukasiewicz (1939) with the single axiom $(x \leftrightarrow y) \leftrightarrow ((z \leftrightarrow y) \leftrightarrow (z \leftrightarrow x))$ and the rule of *Modus Ponens* for $\leftrightarrow$, that is, $x, x \leftrightarrow y \rhd y$.

9. $\mathcal{I}\ell_{\leftrightarrow}$, the *equivalence fragment of* $\mathcal{I}\ell$, is also algebraizable, its equivalent algebraic semantics being the variety of *intuitionistic equivalence algebras*. A famous axiomatization of $\mathcal{I}\ell_{\leftrightarrow}$ was given by Tax (1973), and contains four axioms and the so-called *Tax Rule* $x \rhd y \leftrightarrow (y \leftrightarrow x)$, besides the rule of *Modus Ponens* (for $\leftrightarrow$). For more details on $\mathcal{C}\ell_{\leftrightarrow}$ and $\mathcal{I}\ell_{\leftrightarrow}$, see [70, Section 6.6] or [123, p. 118].

10. Among the *logics of formal inconsistency* (a subgroup of *paraconsistent logics*) there is a very large (more than eight thousand members) family of three-valued logics, defined like the logic $\mathcal{J}_3$ of Example 3.20, by taking $\{1/2, 1\}$ as designated subset and varying the truth-tables. All these logics are algebraizable [53, Theorem 135], with defining equation $(x \to x) \to x \approx x \to x$ and with equivalence formula either $(x \leftrightarrow y) \wedge (\circ x \leftrightarrow \circ y)$ or $\sim\sim((x \leftrightarrow y) \wedge (\circ x \leftrightarrow \circ y))$, where $\circ$ is an "inconsistency" unary operator, and $\sim$ is a classical negation operator; these connectives are definable, in different ways, depending on the logic. In this group we find $\mathcal{J}_3$ itself, which has already appeared in Example 3.20, and Sette's logic P^1, whose algebraizability had been independently studied earlier [162, 164]. As Avron argues in [16], the logic $\mathcal{J}_3$ has several properties that justify its role as one of the most "natural" three-valued paraconsistent logics. ⊠

Finitarity issues

Now several issues around *finitarity* are considered, in two senses: finitarity of the two consequences involved ($\vdash_{\mathcal{L}}$ and $\vDash_{\mathsf{K}}$), and finitarity of the two structural transformers involved, which means finiteness of the sets $E(x)$ and $\Delta(x,y)$. To start with, the syntactic characterization of algebraizability in Theorem 3.21 trivially gives:

PROPOSITION 3.35
Let $\mathcal{L}$ be a logic, and let $\mathcal{L}_{\mathrm{f}}$ be its finitary companion.

1. *If $\mathcal{L}_{\mathrm{f}}$ is algebraizable, then $\mathcal{L}$ is algebraizable as well, with the same transformers.*
2. *If $\mathcal{L}$ is algebraizable with finitary transformers, then $\mathcal{L}_{\mathrm{f}}$ is algebraizable as well, with the same transformers.* ⊠

We are going to find all implications between the four "finitarity conditions" mentioned above; they are summarized in Figure 3 on page 137. Recall from Proposition 1.74 that $\vDash_{\mathsf{K}}$ is finitary if and only if K is a quasivariety (because we assume that K is the equivalent algebraic semantics of the algebraizable logic, which is a generalized quasivariety). The first property, already an interesting one, requires only a very general argument (Exercise 3.27):

LEMMA 3.36

Let $\mathcal{L}$ be algebraizable with equivalent algebraic semantics K.

1. *If $\mathcal{L}$ is finitary and has a finite set $\Delta(x,y)$ of equivalence formulas, then every set of equivalence formulas for $\mathcal{L}$ contains a finite subset which is also a set of equivalence formulas for $\mathcal{L}$.*
2. *If $\vDash_{\mathsf{K}}$ is finitary and $\mathcal{L}$ has a finite set $E(x)$ of defining equations, then every set of defining equations for $\mathcal{L}$ contains a finite subset which is also a set of defining equations for $\mathcal{L}$.* ⊠

Observe the assumptions of existence of finite sets. By contrast, the next results, which may seem similar, really prove this existence.

THEOREM 3.37

Let $\mathcal{L}$ be algebraizable with equivalent algebraic semantics K.

1. *If $\mathcal{L}$ is finitary, then every set $E(x)$ of defining equations for $\mathcal{L}$ contains a finite subset which is also a set of defining equations for $\mathcal{L}$.*
2. *If $\vDash_{\mathsf{K}}$ is finitary, then every set $\Delta(x,y)$ of equivalence formulas for $\mathcal{L}$ contains a finite subset which is also a set of equivalence formulas for $\mathcal{L}$.*
3. *If $\mathcal{L}$ is finitary and has a finite set $\Delta(x,y)$ of equivalence formulas, then $\vDash_{\mathsf{K}}$ is finitary.*
4. *If $\vDash_{\mathsf{K}}$ is finitary and has a finite set $E(x)$ of defining equations, then $\mathcal{L}$ is finitary.*

SKETCH OF THE PROOF: For 1, apply finitarity of $\mathcal{L}$ to $\Delta(E(x)) \vdash_{\mathcal{L}} x$ to find that $\Delta_0(E_0(x)) \vdash_{\mathcal{L}} x$ for a finite $\Delta_0 \subseteq \Delta$ and a finite $E_0 \subseteq E$. From this it follows that (ALG3) is satisfied with Δ and E_0, which together with (ALG2) for Δ establishes algebraizability. For 2, act similarly applying finitarity of $\vDash_{\mathsf{K}}$ to one half of (ALG4). The other two properties are almost trivial. Complete the details as asked in Exercise 3.28. ⊠

Thus, under the assumption of finitarity of the corresponding logic, in each case where *some* finite transformer exists, one gets more: *every* transformer can be "finitized", in the obvious sense. You may observe that each point of Lemma 3.36 also follows by combining two points of Theorem 3.37; nevertheless, it was interesting to see that the lemma had a quite direct proof.

COROLLARY 3.38

Let $\mathcal{L}$ be algebraizable with equivalent algebraic semantics K, with sets $E(x)$ of defining equations and $\Delta(x,y)$ of equivalence formulas.

1. *If both $E(x)$ and $\Delta(x,y)$ are finite, then $\mathcal{L}$ is finitary if and only if $\vDash_{\mathsf{K}}$ is finitary.*
2. *If $\mathcal{L}$ is finitary, then $\vDash_{\mathsf{K}}$ is finitary if and only if $\Delta(x,y)$ can be taken finite.*
3. *If $\vDash_{\mathsf{K}}$ is finitary, then $\mathcal{L}$ is finitary if and only if $E(x)$ can be taken finite.*

In each of the cases, if the two equivalent conditions hold, then all the four "finitarity conditions" hold. ⊠

In particular, observe that when working with finitary logics, the set of defining equations $E(x)$ can always be taken finite, and then just assuming that the equivalence set $\Delta(x,y)$ is finite implies that the relative equational consequence is finitary as well. This was the scenario of the original notion of algebraizability introduced by Blok and Pigozzi in their seminal monograph [32], therefore the following makes sense:

DEFINITION 3.39
*Let $\mathcal{L}$ be an algebraizable logic. $\mathcal{L}$ is **finitely algebraizable** when it has a finite set of equivalence formulas. $\mathcal{L}$ is **BP-algebraizable** when it is both finitary and finitely algebraizable.*

Thus, for finitary logics, finite algebraizability and BP-algebraizability is the same property. There is a certain lack of symmetry in calling "finitely algebraizable" the algebraizable logics whose set of equivalence formulas is finite, even when the set of defining equations need not be finite; this is a consequence of the history of the subject and particularly of the previous existence of the notion of a "finitely equivalential" logic (Section 6.3) and the relations between the two concepts (Section 6.5).

From Definition 3.23 it follows that *the equivalent algebraic semantics of a BP-algebraizable logic is a quasivariety*; Proposition 3.44 shows how to obtain a set of quasi-equations axiomatizing it from any Hilbert-style presentation of the logic.

Theorem 3.33 specializes in a straightforward way to the following results, much better known and of greater applicability, particularly in contexts in which all logics are assumed to be finitary (recall that by Proposition 1.45 an axiomatic extension of a finitary logic is always finitary):

COROLLARY 3.40
1. *Let $\mathcal{L}$ be a BP-algebraizable logic, with equivalent algebraic semantics the quasivariety K. There is a dual isomorphism between the lattice of all finitary extensions of $\mathcal{L}$ and the lattice of all subquasivarieties of K.*

   This isomorphism restricts to a dual isomorphism between the lattice of all the axiomatic extensions of $\mathcal{L}$ and the lattice of all relative subvarieties of K (that is, the subquasivarieties of K that can be defined by adding only equations to a set of quasi-equations defining K).
2. *Let $\mathcal{L}$ be a BP-algebraizable logic whose equivalent algebraic semantics K is a variety.*[†] *There is a dual isomorphism between the lattice of all axiomatic extensions of $\mathcal{L}$ and the lattice of all subvarieties of K.* ⊠

These facts are interesting because there is a significant amount of work in the universal algebra literature on lattices of subvarieties of a given variety and related topics; for instance, the lattices of varieties corresponding to modal logics and to intermediate logics have been extensively studied.

[†] The logics satisfying these conditions have sometimes been called ***strongly algebraizable***.

All algebraizable logics of Example 3.34 are in fact finitely algebraizable. To meet logics that are algebraizable but not finitely, and to see that no further implications hold between the four finitarity conditions, one must consider some less standard examples.

EXAMPLE 3.41
The infinitely-valued logic Ł_∞ of Łukasiewicz, introduced in Example 1.12 is shown not to be finitary in Example 1.15. From the definition itself it follows that the algebra $[\mathbf{0,1}]$ is an algebraic semantics for Ł_∞ with defining equation $x \approx \top$, thus (ALG1) holds. In this algebra, $a \leqslant b$ if and only if $a \to b = 1$, therefore condition (ALG4) is satisfied with equivalence formulas $\{x \to y, y \to x\}$. Thus, the logic is finitely algebraizable with respect to the single algebra $[\mathbf{0,1}]$ and with both transformers finite. Therefore its equivalent algebraic semantics is the class $\mathbb{GQ}\{[\mathbf{0,1}]\}$, a subclass of MV, the variety of MV-algebras. By Theorem 3.37.4, the consequence $\vDash_{[\mathbf{0,1}]}$ cannot be finitary, and this might be algebraically confirmed by showing that $\mathbb{GQ}\{[\mathbf{0,1}]\}$ is not closed under $\mathbb{P}_U$ (thus, this class is not a quasivariety). This is an example of an algebraizable logic with both transformers finite, but where this does not imply finitarity of any of the two consequences. Actually, the logic is implicative: it is easy to check that the conditions in the definition are satisfied.

As to the finitary companion $\text{Ł}_{\infty f}$, the ordinary (syntactically defined) version of Łukasiewicz's infinitely-valued logic, by Theorem 2.16.2 it is also implicative, and by Proposition 3.35 it is algebraizable with the same transformers are Ł_∞; its equivalent algebraic semantics is MV; a direct proof of its algebraizability with respect to this class of algebras (in the term-definable equivalent form of Wajsberg algebras) appeared in [218].

As to the finitely-valued logics Ł_n, their algebraizability is determined in exactly the same way as that of Ł_∞ above, as they are defined from the algebra $[\mathbf{0,1}]_n$ with the condition (ALG1) for the same defining equation, and since $[\mathbf{0,1}]_n$ is a subalgebra of $[\mathbf{0,1}]$, it also satisfies conditioin (ALG4) for the same equivalence formulas. Thus, each of the Ł_n is algebraizable with respect to $[\mathbf{0,1}]_n$, for $n \geqslant 2$. Their equivalent algebraic semantics is $\mathbb{V}\{[\mathbf{0,1}]_n\}$ (Example 5.83). ⊠

EXAMPLE 3.42
Herrmann's "Last Judgement" logic $\mathcal{LJ}$, introduced in [141]. The name comes from a Kripke-style semantics used to motivate the logic, but it is actually defined axiomatically (the relational semantics is not complete). The language is $\langle \wedge, \vee, \to, \neg, \Box \rangle$. Its axioms are:

- All instances of $\mathcal{C}\ell$-tautologies in this language.
- The formulas $\Box^n \varphi$ for every instance φ of an $\mathcal{I}\ell$-theorem and every $n \geqslant 0$.
- The formulas $\Box^n(\Box(x \to y) \to (\Box x \to \Box y))$, for all $n \geqslant 0$.
- The formulas $(x \to y) \to \Box^n(\neg y \to \neg x)$, for all $n \geqslant 0$.

The only inference rule is (MP) for $\to$.

$\mathcal{LJ}$ is very similar to K^ℓ, the local logic associated with the least normal modal system K, but $\mathcal{LJ}$ does not satisfy the weak version of the Necessitation Rule ("if φ is a theorem, then $\Box\varphi$ is a theorem"). By definition $\mathcal{LJ}$ is a finitary logic, and one can prove that it is algebraizable, with the single defining equation $\neg x \approx \neg(x \to x)$, and the infinite set $\Delta(x,y) = \{\Box^n(x \to y), \Box^n(y \to x) : n \geqslant 0\}$ as set of equivalence formulas. Using either the original Kripke semantics or a suitable algebraic model it is not difficult to show that it is not possible to find an equivalent finite subset of the given set $\Delta(x,y)$; by Theorem 3.17 and Lemma 3.36, this implies that no set of equivalence formulas can be finite, and hence the logic is not finitely algebraizable. In turn, this implies that its equivalent consequence $\vDash_{\mathsf{K}}$ cannot be finitary, by Theorem 3.37. This example shows that the finitarity of $\mathcal{L}$ does not imply that of $\vDash_{\mathsf{K}}$ and does not imply the finiteness of Δ, and that this one is not implied by the finiteness of E either. Finally, observe that the largest τ-algebraic semantics for $\mathcal{LJ}$ (Definition 3.7) is a quasivariety, by the finitarity of $\mathcal{LJ}$ and of τ; hence this class is forcefuly larger than the equivalent algebraic semantics. See [70, pp. 329ff.] for more details. ☒

Example 3.43
Quite recently, Raftery [202] has found a (rather artificial) example of a non-finitary logic that is finitely algebraizable, but is not BP-algebraizable. This logic is defined in a language of type $\langle 2,1,1,1\rangle$ by condition (ALG1) from a suitable finitely based variety and an *infinite* family of defining equations; it is shown to be algebraizable with respect to the given variety, with that infinite set of defining equations and a single equivalence formula, and to be non-finitary, and it is also shown that no finite subset of defining equations is sufficient. Hence, the finitarity of $\vDash_{\mathsf{K}}$ and the finiteness of Δ, even together, imply neither the finitarity of $\mathcal{L}$ nor the finiteness of E. This settles an open problem posed for instance in [70, p. 314], namely the existence of an infinitary logic that is equivalent to a finitary equational consequence, hence algebraizable with respect to a quasivariety; in this case, the equivalent algebraic semantics is actually a variety. ☒

It is now straightforward to check that these examples, together with the already shown implications, complete the analysis: no other implications hold between the four finitarity conditions, nor between different pairs of them. Clearly, any three of them imply the fourth, and there are some pairs that also imply the other two conditions, namely:

- Both $\mathcal{L}$ and $\vDash_{\mathsf{K}}$ are finitary.
- $\mathcal{L}$ is finitary and Δ is finite. This is the original definition of BP-algebraizability.
- $\vDash_{\mathsf{K}}$ is finitary and E is finite. This is dual to the previous one.

All these cases amount to the same: BP-algebraizability. The situation is expressed graphically in Figure 3.

A related *open problem* is the following: At the time of writing, no example is known of an algebraizable logic that has a finite set of defining equations, but

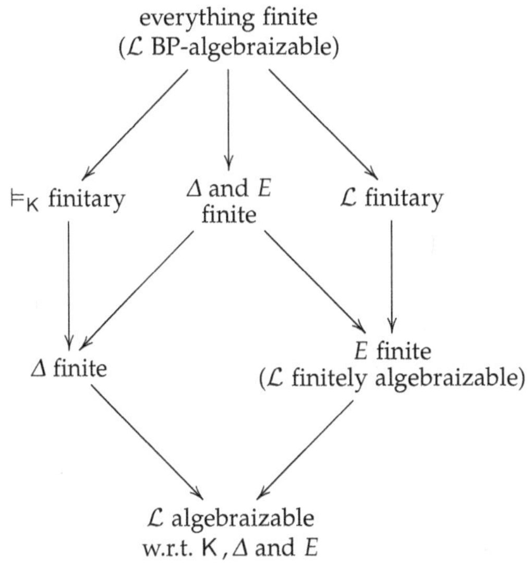

FIGURE 3: Implications (represented as arrows) between the four finitarity conditions and their combinations. Joins in the graph represent intersections of classes, corresponding to the combination of two or more conditions.

such that this set cannot be a one-element set; that is, that more than one defining equations are required for its algebraization, but finitely many are sufficient.

To close this subsection, observe that the following situation is in principle possible: that neither $\mathcal{L}$ nor $\vDash_{\mathsf{K}}$ are finitary, and that there are two sets of equivalence formulas, a finite one and an infinite one, but the latter does not contain a finite equivalent subset. Again, I do not know of any example in the literature. Connected with this, Exercise 3.29 shows that any (finite) set of equivalence formulas can be enlarged to one that is, in general, infinite. Of course, the situation for sets of defining equations is the same.

Axiomatization

In the case of BP-algebraizability, in which everything is finitary, it makes sense to consider the issue of axiomatizability of one of the consequences in terms of the other via the transformers.

PROPOSITION 3.44
Let $\mathcal{L}$ be a BP-algebraizable logic, with equivalence formulas $\Delta(x,y)$ and defining equations $E(x)$, and let it be axiomatized by a set Ax of axioms and a set $InfR$ of proper inference rules. Its equivalent algebraic semantics is the quasivariety presented by the sets of equations

$(\boldsymbol{\tau}\text{-}Ax)$ $\quad E(\varphi)$ $\qquad$ *for each* $\varphi \in Ax$

and the sets of quasi-equations

$$(\tau\text{-}InfR) \qquad \Big(\bigwedge \bigcup_{i=1}^{n} E(\alpha_i)\Big) \to \varepsilon \approx \delta \qquad \textit{for each } \alpha_1,\dots,\alpha_n \rhd \beta \in InfR \textit{ and each } \varepsilon \approx \delta \in E(\beta)$$

$$(\text{Red}) \qquad \Big(\bigwedge E(\Delta(x,y))\Big) \to x \approx y.$$

Notice that the last condition is a single quasi-equation.

PROOF: Left to you as Exercise 3.30. ⊠

The advantage of this presentation of the equivalent algebraic semantics is that it is finite whenever the axiomatization of the logic is so; see Corollary 3.48. Observe that it is not enough to transform an axiomatization of the logic in order to obtain an axiomatization of the relative equational consequence (and conversely): To conditions (τ-Ax) and (τ-$InfR$), which are the τ-transforms of the axiomatization of $\mathcal{L}$, one has to add condition (Red). This is indeed necessary, otherwise we are axiomatizing only the class $\mathsf{K}(\mathcal{L},\tau)$, the largest τ-algebraic semantics for $\mathcal{L}$, which need not coincide with its equivalent algebraic semantics, as is commented before (an example appears in Exercise 4.25).

Notice that the equations in the set $E(\Delta(x,x))$, that is, (R) of Lemma 3.22, might, in principle, appear in the given presentation of K, in case the formulas in $\Delta(x,x)$ happen to be among the axioms of the logic; but anyway they hold in K, because these formulas must be theorems of $\mathcal{L}$. By contrast, the quasi-equation (Red) has to be added: the particular form of its tail guarantees that it is never the τ-transform of an inference rule.

Satisfying (R) and (Red) appears to be characteristic of the quasivarieties obtained in this way; this has suggested the following notion and characterization.

DEFINITION 3.45
A quasivariety Q *is a* ***quasivariety of logic*** *when it is the equivalent algebraic semantics of a BP-algebraizable logic.*

PROPOSITION 3.46
A quasivariety Q *is a quasivariety of logic if and only if there are finite sets* $\Delta(x,y) \subseteq Fm$ *and* $E(x) \subseteq Eq$ *such that* Q *satisfies* (R) *and* (Red)*; i.e., such that all equations in* $E(\Delta(x,x))$ *and the quasi-equation* $\Big(\bigwedge E(\Delta(x,y))\Big) \to x \approx y$ *hold in* Q.

PROOF: Just take condition (ALG1) to *define* a logic $\mathcal{L}$ from the class Q with the help of the equations in $E(x)$. Then the assumptions and Lemma 3.22 guarantee that condition (ALG4) is satisfied. Since the two sets are finite and $\vDash_{\mathsf{Q}}$ is fintiary, this means that this $\mathcal{L}$ is BP-algebraizable, and its equivalent algebraic semantics is the generalized quasivariety generated by Q, which is obviously Q itself; thus, Q is a quasivariety of logic. ⊠

This is actually a way to *find an algebraizable logic associated with a given quasivariety*, when one conjectures that such a logic exists and how the transformers look like. The conditions are purely algebraic, that is, they refer only to validity of certain equations and quasi-equations in the algebras of the quasivariety. The main clue arises when a quasi-equation similar to (Red), i.e., one whose tail is $x \approx y$, holds in the quasivariety (or variety, of course). If its head has the form $\bigwedge E(\Delta(x,y))$ for finite sets $\Delta(x,y) \subseteq Fm$ and $E(x) \subseteq Eq$, and the equations in $E(\Delta(x,x))$ hold in the class, then (ALG4) holds, and we are done, because then (ALG1) *defines* a logic so that the two conditions for algebraizability are automatically satisfied.

The dual process is also possible: from a presentation of the quasivariety, to obtain a Hilbert-style axiomatization of the logic:

PROPOSITION 3.47
Let $\mathcal{L}$ be a BP-algebraizable logic, with equivalence formulas $\Delta(x,y)$ and defining equations $E(x)$, and let its equivalent quasivariety K be presented by a set Id of equations and a set QId of proper quasi-equations. The logic $\mathcal{L}$ can be axiomatized by the Hilbert-style system with the (sets of) axioms

$$(\rho\text{-}Id) \qquad \Delta(\varepsilon,\delta) \qquad \text{for each } \varepsilon \approx \delta \in Id$$

$$(\mathrm{R}) \qquad \Delta(x,x)$$

and the (sets of) inference rules

$$(\rho\text{-}QId) \qquad \Big(\bigcup_{i=1}^{n} \Delta(\alpha_i,\beta_i)\Big) \rhd \Delta(\varepsilon,\delta) \qquad \text{for each } \bigwedge_{i=1}^{n} \alpha_i \approx \beta_i \to \varepsilon \approx \delta \in QId$$

$$(\mathrm{Sym}) \qquad \Delta(x,y) \rhd \Delta(y,x)$$

$$(\mathrm{Trans}) \qquad \Delta(x,y) \cup \Delta(y,z) \rhd \Delta(x,z)$$

$$(\mathrm{Re}) \qquad \bigcup_{i=1}^{n} \Delta(x_i,y_i) \rhd \Delta(\lambda x_1 \dots x_n, \lambda y_1 \dots y_n) \qquad \text{for each } \lambda \in L, \text{ with } n = \operatorname{ar}\lambda \geqslant 1$$

$$(\mathrm{ALG3\text{-}a}) \qquad x \rhd \Delta(E(x))$$

$$(\mathrm{ALG3\text{-}b}) \qquad \Delta(E(x)) \rhd x$$

where the abbreviation $X \rhd Y := \{X \rhd y : y \in Y\}$, for $X, Y \subseteq Fm$, is used.

PROOF: Denote by $\mathcal{L}'$ the logic axiomatized by the Hilbert-style system of the statement. The proof consists in showing that $\mathcal{L}'$ is algebraizable with respect to K with the same $E(x)$ and $\Delta(x,y)$; then by (ALG1) applied to the two logics one concludes that $\mathcal{L} = \mathcal{L}'$. The (groups of) rules (ALG3-a,b) show that $\mathcal{L}'$ satisfies (ALG3), therefore we need only prove (ALG2); that is, that for all $\Theta \cup \{\varepsilon \approx \delta\} \subseteq Eq$,

$$\Theta \vDash_{\mathsf{K}} \varepsilon \approx \delta \iff \Delta(\Theta) \vdash_{\mathcal{L}'} \Delta(\varepsilon,\delta). \tag{31}$$

To start with, since $\mathcal{L}$ is algebraizable with respect to K with these $E(x)$ and $\Delta(x,y)$, all axioms and rules that define $\mathcal{L}'$ hold in $\mathcal{L}$, and therefore $\mathcal{L}' \leqslant \mathcal{L}$. As a consequence, $\Delta(\Theta) \vdash_{\mathcal{L}'} \Delta(\varepsilon,\delta)$ implies $\Delta(\Theta) \vdash_{\mathcal{L}} \Delta(\varepsilon,\delta)$, which by algebraizability of $\mathcal{L}$ implies that $\Theta \vDash_{\mathsf{K}} \varepsilon \approx \delta$. This proves the ($\Leftarrow$) half of (31). To prove the ($\Rightarrow$) half, consider the set $\theta := \{\langle \alpha,\beta \rangle \in Fm \times Fm : \Delta(\Theta) \vdash_{\mathcal{L}'} \Delta(\alpha,\beta)\}$. It is left to you as Exercise 3.31 to prove that $\theta \in \mathrm{Con}_{\mathsf{K}}\boldsymbol{Fm}$. Since obviously $\Theta \subseteq \theta$, we have that $\boldsymbol{Fm}/\theta \vDash \Theta\ [\![\pi]\!]$, and that the canonical projection $\pi\colon \boldsymbol{Fm} \to \boldsymbol{Fm}/\theta$ is an evaluation into an algebra of K. Therefore, from the assumption that $\Theta \vDash_{\mathsf{K}} \varepsilon \approx \delta$, it follows that also $\boldsymbol{Fm}/\theta \vDash \varepsilon \approx \delta\ [\![\pi]\!]$, that is, $\langle \varepsilon, \delta \rangle \in \theta$; but by definition of θ this means that $\Delta(\Theta) \vdash_{\mathcal{L}'} \Delta(\varepsilon, \delta)$, thus completing the proof. ⊠

If you compare Propositions 3.44 and 3.47 you will notice a certain *lack of symmetry*. In each case, the axiomatization contains the transforms of the presentation of the logic or of the class of algebras, respectively, plus some condition(s) directly concerning algebraizability: condition (Red) in the first case, and the two dual ones (ALG3-a,b) in the second. However, in the second case also the ρ-transforms of the first four Birkhof rules appear, while nothing parallel appears in the first case. The reason is apparent if you review the step of the proof where these additional rules are used, namely that the logic $\mathcal{L}'$ defined by the given axiomatic system satisfies part ($\Rightarrow$) of (31): Besides the entailments coming from the particular class K, the consequence $\vDash_{\mathsf{K}}$ is not an arbitrary closure relation on the set of equations, but one that should satisfy the four Birkhoff rules of equational logic (recall Exercise 1.85), therefore $\mathcal{L}'$ has to satisfy their transforms; it is not difficult to realize (Exercise 3.32) that they need not be contained in, or entailed by, the rest of the axiomatization.

This axiomatization has the finiteness advantage already commented after Proposition 3.44. Therefore:

COROLLARY 3.48
Let $\mathcal{L}$ be a BP-algebraizable logic with the quasivariety K as equivalent algebraic semantics. The logic $\mathcal{L}$ is finitely axiomatizable if and only if K can be finitely presented. ⊠

Regularly algebraizable logics

If we compare Definition 2.3 with Proposition 3.13 (under the light of Theorem 3.21), we can observe the qualitative differences between implicative logics and algebraizable logics: while conditions (R), (Trans), (Re) and (MP) are weakened versions of conditions (IL1)–(IL4), nothing parallel to (IL5) holds in algebraizable logics (this issue is relevant to Exercise 3.14). In Theorem 2.2.4 we see that a consequence of (IL5), denoted there by (G), is to identify all formulas of the same theory in the same equivalence class. Accordingly:

DEFINITION 3.49
A logic $\mathcal{L}$ is ***regularly algebraizable*** *when it is algebraizable and satisfies*

$$x, y \vdash_{\mathcal{L}} \Delta(x, y) \tag{G}$$

for some (equivalently: for any) non-empty set $\Delta(x, y)$ of equivalence formulas.

A logic is ***finitely regularly algebraizable*** *when it is both finitely algebraizable and regularly algebraizable.*

A logic is ***regularly BP-algebraizable*** *when it is both BP-algebraizable and regularly algebraizable; that is, when it is finitary and finitely regularly algebraizable.*

The above condition (G) is sometimes called in the literature, in this context, the "G-rule"; hence the label. Surprisingly, the "G" is for Gödel, and its usage in this sense seems to originate with Suszko.†

EXAMPLES 3.50
Obviously, *every implicative logic is finitely regularly algebraizable*; this provides a large class of examples. Condition (G) can be viewed as the symmetrization of (IL5), and it is to be expected that it is weaker. That's right: $\mathcal{C}\ell_{\leftrightarrow}$, $\mathcal{I}\ell_{\leftrightarrow}$ and $\mathcal{I}\ell_{\neg,\leftrightarrow}$ (the $\langle\neg,\leftrightarrow\rangle$ fragment of $\mathcal{I}\ell$) are examples of finitely regularly algebraizable logics that are not implicative,‡ as shown in [61, Example 1.1]. The logics $\mathcal{FL}$, $\mathcal{R}$ and $\mathcal{RM}$ are BP-algebraizable, but not regularly; the same happens to all substructural logics whose equivalent quasivariety includes some non-integral residuated lattice (see [123] for examples). Of the two logics shown in Example 3.20 to be BP-algebraizable with respect to the same quasivariety, $Ł_3$ is regularly algebraizable while $\mathcal{J}_3$ is not. The affirmative facts are easy to check, and the negative ones are shown in Examples 6.144. ⊠

PROPOSITION 3.51
Let $\mathcal{L}$ be regularly algebraizable, with equivalent algebraic semantics the class K.

1. *If φ,ψ are any two theorems of $\mathcal{L}$, then $\mathsf{K}\vDash\varphi\approx\psi$.*
2. *If φ is any theorem of $\mathcal{L}$, then φ is an algebraic constant of the class* K.

PROOF: 1. If $\vdash_{\mathcal{L}}\varphi$ and $\vdash_{\mathcal{L}}\psi$, then by (G) also $\vdash_{\mathcal{L}}\Delta(\varphi,\psi)$, and by (ALG2) this implies that $\mathsf{K}\vDash\varphi\approx\psi$, as desired.

2. Let φ be a theorem of $\mathcal{L}$, let $\boldsymbol{A}\in\mathsf{K}$ and let $h,h'\in\operatorname{Hom}(\boldsymbol{Fm},\boldsymbol{A})$. Since $\operatorname{Var}\varphi$ is finite, there is a bijective substitution σ sending all the variables in the set $\operatorname{Var}\varphi$ to variables outside this set. Then it is possible to consider any $h''\in\operatorname{Hom}(\boldsymbol{Fm},\boldsymbol{A})$ such that $h''x:=hx$ and $h''\sigma x:=h'x$, for all $x\in\operatorname{Var}\varphi$. Obviously, we have that $h''\varphi=h\varphi$ and that $h''\sigma\varphi=h'\varphi$. Now, by structurality, $\sigma\varphi$ is also a theorem of $\mathcal{L}$, and by 1 it follows that $h\varphi=h''\varphi=h''\sigma\varphi=h'\varphi$. This shows that φ is an algebraic constant of K. ⊠

Next we see that thanks to these algebraic constants the defining equations can be reduced to a single one, indeed one of a very simple form, and that this kind of algebraizability can be characterized syntactically by using exclusively the equivalence formulas; this contrasts with Theorem 3.21, which characterizes plain algebraizability in a purely syntactic way, but using both the equivalence formulas and the defining equations.

† Suszko discussed this rule in the context of his study of non-Fregean logics. Gödel was the first (1933) to suggest axiomatizing a modal logic (S4) with the Necessitation Rule $x\rhd\Box x$, and in Suszko's context this rule is equivalent to (G). See [230, p. 34] and [231, pp. 190–193]; for more details see [189, p. 483].

‡ Showing that a logic is not implicative is in principle harder than showing, for instance, that it is algebraizable but not regularly so. Since no equivalent characterizations are available, and there is no analogue of Theorem 3.17.2 for the implication connective of implicative logics, one has to show that the logic cannot satisfy the conditions (IL1–IL5) for any conceivable binary term $\to$.

THEOREM 3.52
For a logic $\mathcal{L}$ the following conditions are equivalent:

(i) *$\mathcal{L}$ is regularly algebraizable.*

(ii) *There is a set $\Delta(x,y) \subseteq Fm$ such that $\mathcal{L}$ satisfies conditions* (R), (Sym), (Trans), (Re), (MP) *and* (G) *with respect to it.*

(iii) *$\mathcal{L}$ is algebraizable with respect to some class K and defining equation $x \approx \top$, where $\top$ is an algebraic constant of K with at most the variable x.*

Moreover, the term $\top$ is a theorem of $\mathcal{L}$.

PROOF: (i)$\Rightarrow$(ii) follows from Theorem 3.21. In Exercise 3.33 you are asked to prove that the conditions in (ii) imply that $x \dashv\vdash_{\mathcal{L}} \Delta(x,\top)$ for any theorem $\top$ of $\mathcal{L}$. In particular, if $\top$ is a theorem in at most the variable x, this is (ALG3) for $E(x) := x \approx \top$ and the same Δ, therefore together with Theorem 3.21 again this establishes (i). But this also shows that the defining equation can be taken to be $x \approx \top$, and by Proposition 3.51 $\top$ will be a constant of the equivalent algebraic semantics. This proves that (i)$\Rightarrow$(iii). Finally, to show that (iii)$\Rightarrow$(i) let $\Delta(x,y)$ be a set of equivalence formulas for $\mathcal{L}$, and write $\top(x)$ for clarity. By assumption condition (ALG3) is satisfied with $\Delta(x,y)$ and the defining equation $x \approx \top(x)$. Therefore, and using structurality, $x,y \vdash_{\mathcal{L}} \Delta(x,\top(x)) \cup \Delta(y,\top(y))$. Since $\top$ is constant over K, the equation $\top(x) \approx \top(y)$ holds in K, therefore by (ALG2) $\vdash_{\mathcal{L}} \Delta(\top(x),\top(y))$. From all this, and using (Sym) and (Trans), which hold by algebraizability, it follows that $x,y \vdash_{\mathcal{L}} \Delta(x,y)$, which is condition (G), thus proving regular algebraizability. ☒

The characterization in (ii) is provisional, and is improved in Corollary 6.61, where it is shown that it is actually possible to delete conditions (Sym) and (Trans), because in the presence of (MP) they follow from (R) and (Re).

Obviously, Theorem 3.52 facilitates the proof that certain logics are regularly algebraizable. Notice that the examples previously mentioned are finitely algebraizable. However, there is no relation between these two modes of algebraizability, the finitely and the regularly ones, even for finitary logics; some examples of finitely algebraizable logics that are not regularly algebraizable have been mentioned before, and here is one, due to Dellunde [79], that is regularly algebraizable but not finitely so:

EXAMPLE 3.53
Define a logic $\mathcal{L}_D$ in the language $\langle \leftrightarrow, \Box \rangle$ of type $\langle 2,1 \rangle$ by the following axiomatic system:

Axiom: $x \leftrightarrow x$

Rules: $x, x \leftrightarrow y \rhd y$ (this is (MP) for $\leftrightarrow$)

$x, y \rhd \Box^n x \leftrightarrow \Box^n y$, for all $n \geqslant 0$

$x_1 \leftrightarrow y_1,, x_2 \leftrightarrow y_2 \rhd \Box^n (x_1 \leftrightarrow x_2) \leftrightarrow \Box^n (y_1 \leftrightarrow y_2)$, for all $n \geqslant 0$

By definition this logic is finitary, and it is not difficult to see that it satisfies the conditions of Theorem 3.52 for the set $\Delta(x,y) := \{\Box^n x \leftrightarrow \Box^n y : n \geqslant 0\}$;

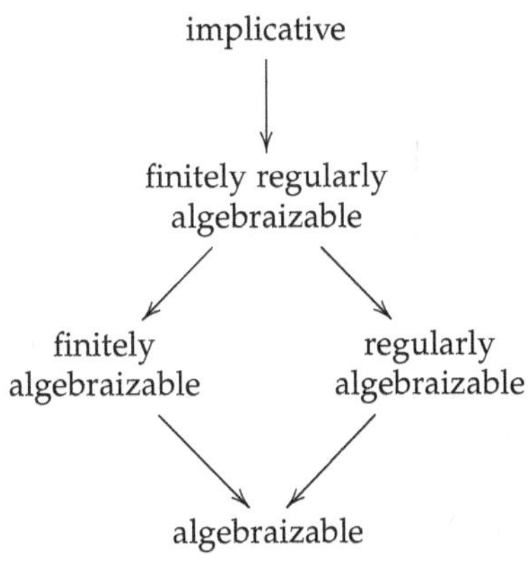

FIGURE 4: Inclusions (represented as arrows) between some classes of logics. Classes in lower positions are larger, and classes in upper positions have better properties.

notice that some are already in the presentation, and that checking (Sym) and (Trans) can be dispensed with if you rely on Theorem 6.60. Thus $\mathcal{L}_D$ is regularly algebraizable. However, it is not finitely algebraizable, because it is not possible to extract a finite subset of equivalence formulas from Δ (Exercise 3.35). This example solved the open Problem 3.18 in [140]. ⊠

The logic $\mathcal{LJ}$ of Example 3.42 is finitary and algebraizable, but is neither finitely nor regularly algebraizable (the latter fact is shown in [141]). The relations between the modes of algebraizability considered up to now can be graphically expressed as in Figure 4; recall that in the finitary case finite algebraizability is the same as BP-algebraizability. The definitions ensure that the graphical join represents intersection of the two classes, and the examples given here and before show that the graphical meet does not represent their union, that the inclusions are proper, and that no other inclusions hold.

Regularly algebraizable logics were first considered in the literature by Czelakowski in his [67]; he defined them by the conditions appearing in Theorem 3.52(ii), under the name *equivalential logics with algebraic semantics* (thus using "algebraic semantics" in a more restricted sense). Later on, Herrmann in his [140, 141, 142] called them 1-*equivalential* or 1-*algebraizable*. The origin of the term "regularly" is related to the behaviour of congruences in the equivalent algebraic semantics, as is explained next to Theorem 3.63, but has acquired a meaning of its own in abstract algebraic logic, explained after Proposition 4.58; see also the comment after Theorem 6.142.

Exercises for Section 3.4

EXERCISE 3.26. Show that the logic $\mathcal{C}\ell_{\leftrightarrow}$, the equivalence fragment of $\mathcal{C}\ell$, is finitely regularly algebraizable, with equivalence formula $x \leftrightarrow y$ and defining equation $x \approx x \leftrightarrow x$.

Show that its equivalent algebraic semantics is the variety of Boolean groups (see Example 3.34.8).

HINT. Observe that $\mathbf{2}_{\leftrightarrow}$, the $\langle \leftrightarrow \rangle$ reduct of the two-element Boolean algebra, is a Boolean group, with unit 1.

EXERCISE 3.27. Prove Lemma 3.36, using Proposition 1.41, together with Theorem 3.17.2.

EXERCISE 3.28. Complete the proof of Theorem 3.37 sketched in the text. Realize that in principle it is clear that in each point you cannot prove more than stated, and locate an example that confirms that this is really the case.

HINT. The reformulations of conditions (ALG3) and (ALG4) with only variables in (26) on page 116 is key here.

EXERCISE 3.29. Let $\Delta(x,y)$ be a set of equivalence formulas for an algebraizable logic $\mathcal{L}$. Consider a substitution σ that leaves x and y fixed and sends all other variables either to x or to y, and define $\Delta'(x,y) := \sigma C_{\mathcal{L}}\Delta(x,y)$. Prove that Δ' is an infinite set of equivalence formulas for $\mathcal{L}$.

HINT. Show that the set Δ' may be finite only when the language has just constants; but in this case the logic would not be algebraizable, as there would be no formulas in two variables.

EXERCISE 3.30. Prove Proposition 3.44, recalling that by Theorem 3.30, the equivalent algebraic semantics of an algebraizable logic can always be described as the class $\mathsf{K}(\mathcal{L},\boldsymbol{\tau},\boldsymbol{\rho})$, for any pair of witnessing transformers $\boldsymbol{\tau}$ and $\boldsymbol{\rho}$.

EXERCISE 3.31. Prove the claim made in the middle of the proof of Proposition 3.47.

HINT. You have to use in an essential way how the logic $\mathcal{L}'$ is axiomatized.

EXERCISE 3.32. After Proposition 3.44 it is explained that the quasi-equation (Red) appears explicitly there because it is never the $\boldsymbol{\tau}$-transform of a rule of the logic. Examine the parallel issue in the case of Proposition 3.47: Check that the rules (ALG3-a,b) are never the $\boldsymbol{\rho}$-transforms of any quasi-equations, but that the rules (R), (Sym), (Trans) and (Re) might in principle be the $\boldsymbol{\rho}$-transforms of some quasi-equations in the presentation of the quasivariety K, thus making this axiomatization of the logic redundant. However, this last situation is highly unlikely; argue why.

EXERCISE 3.33. Show that a logic $\mathcal{L}$ satisfying conditions (Sym), (MP) and (G) for some set $\Delta(x,y) \subseteq Fm$ also satisfies $x \dashv\vdash_{\mathcal{L}} \Delta(x,\top)$, where $\top$ represents any theorem of $\mathcal{L}$.

EXERCISE 3.34. Establish the following kind of converse to Exercise 3.33: If a logic $\mathcal{L}$ satisfies conditions (Sym) and (Trans) for some set $\Delta(x,y) \subseteq Fm$, and satisfies that $x \dashv\vdash_{\mathcal{L}} \Delta(x,\top)$ for a formula $\top$ where the variable x does not appear, then it also satisfies conditions (MP) and (G) for the same $\Delta(x,y)$. Observe that if in addition (R) holds, then this $\top$ is a theorem of the logic.

EXERCISE 3.35. Consider the logic $\mathcal{L}_D$ of Example 3.53.

1. Show that if it were possible to extract a finite set of equivalence formulas from $\Delta(x,y) := \{\Box^n x \leftrightarrow \Box^n y : n \geqslant 0\}$, then there would be some $m \geqslant 0$ such that $\{\Box^n x \leftrightarrow \Box^n y : n \leqslant m\} \vdash_{\mathcal{L}_D} \Box^{m+1} x \leftrightarrow \Box^{m+1} y$.
2. Show that such an m cannot exist by considering the following algebraic model: Take $A := \omega \times \omega$ and define the operations $\Box$ and $\leftrightarrow$ on it as follows:

$$\Box\langle i,j\rangle := \langle i{+}1,j\rangle \quad \text{for all } \langle i,j\rangle \in A,$$

and for all $\langle i,j\rangle, \langle k,l\rangle \in A$,

$$\langle i,j\rangle \leftrightarrow \langle k,l\rangle := \begin{cases} \langle 1,0\rangle & \text{when } i=k \text{ and either } j=l, \text{ or } i<j,l; \text{ and} \\ \langle 0,0\rangle & \text{otherwise.} \end{cases}$$

Put $F := \{\langle 1,0\rangle\}$, $a := \langle 0,m{+}1\rangle$ and $b := \langle 0,m{+}2\rangle$. Show that F is an $\mathcal{L}_D$-filter such that $\Box^n a \leftrightarrow \Box^n b \in F$ for all $n \leqslant m$ while $\Box^{m+1} a \leftrightarrow \Box^{m+1} b \notin F$.

3. Conclude that $\mathcal{L}_D$ is not finitely algebraizable.

3.5 The Isomorphism Theorems

Existential definitions are good for positive proofs, and universal ones are good for refutations. The definition of an algebraizable logic, and its intrinsic characterization in Theorem 3.21, have an existential character, thus providing a means to show that a given logic *is* algebraizable; but in principle they do not provide a way to show that a given logic *is not* algebraizable (except for the logics without theorems): since both contain properties of certain structural transformers, in order to show that a certain logic is not algebraizable we would have to show that no such transformers can exist.

The main tool to overcome this is the so-called Isomorphism Theorem, which does have a universal character (it states that something should hold in every algebra). It is based on the study of the residuals of the transformer functions $\boldsymbol{\tau}\colon \mathscr{P}Fm \to \mathscr{P}Eq$ and $\boldsymbol{\rho}\colon \mathscr{P}Eq \to \mathscr{P}Fm$, and of their extensions to arbitrary algebras. Thus, we define them and study their main, basic properties.

Since $\boldsymbol{\tau}$ and $\boldsymbol{\rho}$ are functions defined on power sets which by their definition commute with arbitrary unions, they have the ***residuals***, $\boldsymbol{\tau}^{-1}\colon \mathscr{P}Eq \to \mathscr{P}Fm$ and $\boldsymbol{\rho}^{-1}\colon \mathscr{P}Fm \to \mathscr{P}Eq$, defined as follows: If $\Theta \subseteq Eq$ and $\Gamma \subseteq Fm$, then

$$\boldsymbol{\tau}^{-1}\Theta := \{\varphi \in Fm : \boldsymbol{\tau}\varphi \subseteq \Theta\} \qquad \boldsymbol{\rho}^{-1}\Gamma := \{\varepsilon \approx \delta \in Eq : \boldsymbol{\rho}(\varepsilon \approx \delta) \subseteq \Gamma\}$$

Notice that, despite the notation, these are *not* the set-theoretic inverses of $\boldsymbol{\tau}$ and $\boldsymbol{\rho}$. The key properties that characterize them are the equivalences:

$$\Gamma \subseteq \boldsymbol{\tau}^{-1}\Theta \iff \boldsymbol{\tau}\Gamma \subseteq \Theta \qquad \Theta \subseteq \boldsymbol{\rho}^{-1}\Gamma \iff \boldsymbol{\rho}\Theta \subseteq \Gamma$$

and the fact that they commute with arbitrary intersections (while $\boldsymbol{\tau}$ and $\boldsymbol{\rho}$ commute with arbitrary unions). The residual construction is a key tool in this section, and it is useful to review Exercises 1.1–1.3.

The first application of the residuals is a useful reformulation of the four conditions defining algebraizability in terms of the closure operators; it is relatively straightforward to check (Exercise 3.36) that, individually, each condition can be rewritten in the following forms:

$$\begin{array}{lllll} \text{(ALG1)} & C_{\mathcal{L}} = \boldsymbol{\tau}^{-1} C_{\mathsf{K}} \boldsymbol{\tau} & & & \\ \text{(ALG2)} & C_{\mathsf{K}} = \boldsymbol{\rho}^{-1} C_{\mathcal{L}} \boldsymbol{\rho} & & & \\ \text{(ALG3)} & C_{\mathcal{L}} = C_{\mathcal{L}} \boldsymbol{\rho} \boldsymbol{\tau} & \text{or} & \boldsymbol{\tau}^{-1}\boldsymbol{\rho}^{-1} C_{\mathcal{L}} = C_{\mathcal{L}} & \\ \text{(ALG4)} & C_{\mathsf{K}} = C_{\mathsf{K}} \boldsymbol{\tau} \boldsymbol{\rho} & \text{or} & \boldsymbol{\rho}^{-1}\boldsymbol{\tau}^{-1} C_{\mathsf{K}} = C_{\mathsf{K}} & \end{array} \tag{32}$$

As a second application, from the preceding formulations of the conditions, the following interesting, and useful reformulation follows (same exercise):

PROPOSITION 3.54
Let $\mathcal{L}$ be a logic and K a class of algebras and let $\boldsymbol{\tau}, \boldsymbol{\rho}$ be transformers such that conditions (ALG3) *and* (ALG4) *are satisfied. Then*

$$\text{(ALG1)} \iff \boldsymbol{\rho}^{-1}C_{\mathcal{L}} = C_{\mathsf{K}}\boldsymbol{\tau}, \text{ and}$$
$$\text{(ALG2)} \iff \boldsymbol{\tau}^{-1}C_{\mathsf{K}} = C_{\mathcal{L}}\boldsymbol{\rho}. \qquad \boxtimes$$

From now on, the labels "(ALGi)" for $i \in \{1,2,3,4\}$ refer indistinctly to the original forms of these conditions in Definition 3.11, to the reformulations given after it, or to the ones given here.

The evaluated transformers and their residuals

Let $\boldsymbol{\tau}$ be a structural transformer of formulas into equations, defined by a set $E(x) \subseteq Eq$. For each algebra $\boldsymbol{A}$ the function $\boldsymbol{\tau}^{\boldsymbol{A}}\colon A \to \mathscr{P}(A \times A)$ is defined by setting, for each $a \in A$,

$$\boldsymbol{\tau}^{\boldsymbol{A}}a := E^{\boldsymbol{A}}(a) = \{\langle \varepsilon^{\boldsymbol{A}}(a), \delta^{\boldsymbol{A}}(a)\rangle : \varepsilon \approx \delta \in E(x)\} \subseteq A \times A;$$

recall that $E^{\boldsymbol{A}}(a)$ is shorthand notation for $hE(x)$ where $h \in \mathrm{Hom}(\boldsymbol{Fm}, \boldsymbol{A})$ is such that $hx = a$. This function can be extended to subsets of A by taking unions,

$$\boldsymbol{\tau}^{\boldsymbol{A}}F := \bigcup\{\boldsymbol{\tau}^{\boldsymbol{A}}a : a \in F\} \qquad \text{for all } F \subseteq A,$$

so that a function $\boldsymbol{\tau}^{\boldsymbol{A}}\colon \mathscr{P}A \to \mathscr{P}(A \times A)$ is obtained. By this definition, this function commutes with arbitrary unions, and therefore (Exercise 1.1) it has a residual function $\boldsymbol{\tau}^{\boldsymbol{A}^{-1}}\colon \mathscr{P}(A \times A) \to \mathscr{P}A$, defined, for $\theta \subseteq A \times A$, by

$$\boldsymbol{\tau}^{\boldsymbol{A}^{-1}}\theta := \{a \in A : \boldsymbol{\tau}^{\boldsymbol{A}}a \subseteq \theta\} = \{a \in A : E^{\boldsymbol{A}}(a) \subseteq \theta\} \subseteq A.$$

Symmetrically, for a structural transformer $\boldsymbol{\rho}$ of equations into formulas defined by some $\Delta(x,y) \subseteq Fm$, a function $\boldsymbol{\rho}^{\boldsymbol{A}}\colon A \times A \to \mathscr{P}A$ can be defined in each algebra $\boldsymbol{A}$ by putting, for all $a, b \in A$,

$$\boldsymbol{\rho}^{\boldsymbol{A}}\langle a,b\rangle := \Delta^{\boldsymbol{A}}(a,b) \subseteq A.$$

By taking unions, this function is extended to $\boldsymbol{\rho}^{\boldsymbol{A}}\colon \mathscr{P}(A \times A) \to \mathscr{P}A$, which has a residual function $\boldsymbol{\rho}^{\boldsymbol{A}^{-1}}\colon \mathscr{P}A \to \mathscr{P}(A \times A)$, given, for a subset $F \subseteq A$, by

$$\begin{aligned} \boldsymbol{\rho}^{\boldsymbol{A}^{-1}}F &:= \{\langle a,b\rangle \in A \times A : \boldsymbol{\rho}^{\boldsymbol{A}}\langle a,b\rangle \subseteq F\} \\ &= \{\langle a,b\rangle \in A \times A : \Delta^{\boldsymbol{A}}(a,b) \subseteq F\} \subseteq A \times A. \end{aligned}$$

These constructions should look familiar to you, from Theorem 2.32. There, $E(x) = \{x \approx x \to x\}$; thus, if $a \in A$, then $\boldsymbol{\tau}^{\boldsymbol{A}}a = \{\langle a, a \to a\rangle\}$, with residual the function $\boldsymbol{\tau}^{\boldsymbol{A}^{-1}}\theta = \{a \in A : \langle a, a \to a\rangle \in \theta\}$, denoted there by F_θ, for $\theta \subseteq A \times A$.

Dually, there we had $\Delta(x,y) = \{x \to y, y \to x\}$, and the resulting function was $\boldsymbol{\rho}^{\boldsymbol{A}^{-1}} F = \{\langle a,b\rangle \in A \times A : a \to b, b \to a \in F\}$, denoted by $\boldsymbol{\Omega}^{\boldsymbol{A}} F$.

Informally, the functions $\boldsymbol{\tau}^{\boldsymbol{A}}$ and $\boldsymbol{\rho}^{\boldsymbol{A}}$ are called the ***evaluated transformers***, and their residuals $\boldsymbol{\tau}^{\boldsymbol{A}^{-1}}$ and $\boldsymbol{\rho}^{\boldsymbol{A}^{-1}}$ the ***evaluated residuals***. It is worth to recall here the residuation properties

$$F \subseteq \boldsymbol{\tau}^{\boldsymbol{A}^{-1}} \theta \iff \boldsymbol{\tau}^{\boldsymbol{A}} F \subseteq \theta$$

$$\theta \subseteq \boldsymbol{\rho}^{\boldsymbol{A}^{-1}} F \iff \boldsymbol{\rho}^{\boldsymbol{A}} \theta \subseteq F$$

for all $F \subseteq A$ and all $\theta \subseteq A \times A$, and that they commute with intersections; again, see Exercises 1.1–1.3 for other properties of residual functions. Next, the fact that the transformers are structural implies that the evaluated ones commute with homomorphisms, and their residuals commute with inverse homomorphisms, in the following very general senses:

PROPOSITION 3.55
Let $\boldsymbol{\tau}, \boldsymbol{\rho}$ be structural transformers from formulas to equations and conversely, and let $\boldsymbol{A}, \boldsymbol{B}$ be algebras. For every $h \in \mathrm{Hom}(\boldsymbol{A}, \boldsymbol{B})$, every $F \subseteq A$ and every $\theta \subseteq A \times A$,

$$\boldsymbol{\tau}^{\boldsymbol{B}} hF = h\boldsymbol{\tau}^{\boldsymbol{A}} F \quad \textit{and} \quad \boldsymbol{\rho}^{\boldsymbol{B}} h\theta = h\boldsymbol{\rho}^{\boldsymbol{A}} \theta, \tag{33}$$

and for every $G \subseteq B$ and every $\vartheta \subseteq B \times B$,

$$\boldsymbol{\tau}^{\boldsymbol{A}^{-1}} h^{-1} \vartheta = h^{-1} \boldsymbol{\tau}^{\boldsymbol{B}^{-1}} \vartheta \quad \textit{and} \quad \boldsymbol{\rho}^{\boldsymbol{A}^{-1}} h^{-1} G = h^{-1} \boldsymbol{\rho}^{\boldsymbol{B}^{-1}} G. \tag{34}$$

PROOF: The first point derives from of the property of arbitrary homomorphisms contained in Exercise 1.12, and is left to you as Exercise 3.37; notice that, since all the functions involved in (33) commute with arbitrary unions, it is enough to check the property for one-element sets, by Exercise 1.34. Then (34) follows from (33) directly by residuation. ⊠

Several particular cases are used in key points in the sequel: when the two algebras coincide, and when the first one is the formula algebra and the second is an arbitrary algebra. Properties (34) are more interesting than (33); they are usually described by saying that the residuals of the evaluated transformers ***commute with homomorphisms*** (resp., ***with endomorphisms*** when the two algebras coincide, and ***with substitutions*** when both are the formula algebra).†

The theorems, in many versions

The first version of the Isomorphism Theorem is a weak‡ one, in which the isomorphisms are stated only as properties of algebraizability but still not as equivalent to it; for this reason, its proof at this point of the theory is almost an exercise, and is given only in sketched form.

† It would be more natural to say "commute *with the residuals* of homo/endomorphisms/substitutions"; but see the explanation after Theorem 3.58.

‡ Notice, however, that this form is already sufficient to show that some logics are not algebraizable.

THEOREM 3.56 (THE ISOMORPHIM THEOREM, FIRST VERSION)
Let $\mathcal{L}$ be an algebraizable logic with equivalent algebraic semantics K *and transformers* $\boldsymbol{\tau}, \boldsymbol{\rho}$, *and let* $\boldsymbol{A}$ *be an arbitrary algebra.*

1. *If* $\theta \in \mathrm{Con}_{\mathsf{K}}\boldsymbol{A}$, *then* $\boldsymbol{\tau}^{\boldsymbol{A}^{-1}}\theta \in \mathcal{F}i_{\mathcal{L}}\boldsymbol{A}$.
2. *If* $F \in \mathcal{F}i_{\mathcal{L}}\boldsymbol{A}$, *then* $\boldsymbol{\rho}^{\boldsymbol{A}^{-1}}F \in \mathrm{Con}_{\mathsf{K}}\boldsymbol{A}$.
3. *The functions* $\boldsymbol{\rho}^{\boldsymbol{A}^{-1}}$ *and* $\boldsymbol{\tau}^{\boldsymbol{A}^{-1}}$ *are mutually inverse lattice isomorphisms between the lattices* $\mathcal{F}i_{\mathcal{L}}\boldsymbol{A}$ *and* $\mathrm{Con}_{\mathsf{K}}\boldsymbol{A}$. *In symbols,* $\boldsymbol{\rho}^{\boldsymbol{A}^{-1}} : \mathcal{F}i_{\mathcal{L}}\boldsymbol{A} \cong \mathrm{Con}_{\mathsf{K}}\boldsymbol{A} : \boldsymbol{\tau}^{\boldsymbol{A}^{-1}}$.

These isomorphisms commute with endomorphisms in the sense of (34).

SKETCH OF THE PROOF: A nice, and at this point most probably a simple exercise for you (Exercise 3.38). Points 1 and 2 are proved by checking (with the help of the first two algebraizability conditions) that the relevant sets satisfy the definition of $\mathcal{L}$-filter and the characterization of relative congruences in Theorem 1.76, respectively. Point 3 is straightforward, because by definition the functions $\boldsymbol{\rho}^{\boldsymbol{A}^{-1}}$ and $\boldsymbol{\tau}^{\boldsymbol{A}^{-1}}$ preserve order, and the simplified form of condition (ALG3) in (26) on page 116 readily implies that $\boldsymbol{\tau}^{\boldsymbol{A}^{-1}}\boldsymbol{\rho}^{\boldsymbol{A}^{-1}}F = F$ for any $F \in \mathcal{F}i_{\mathcal{L}}\boldsymbol{A}$; and symmetrically, (ALG4) implies that $\boldsymbol{\rho}^{\boldsymbol{A}^{-1}}\boldsymbol{\tau}^{\boldsymbol{A}^{-1}}\theta = \theta$ for any $\theta \in \mathrm{Con}_{\mathsf{K}}\boldsymbol{A}$. Finally, the commutativity with endomorphisms is obtained here by Proposition 3.55, that is, for free, simply because the isomorphisms are given by the residuals of structural transformers. ⊠

See Exercise 3.39 for an alternative proof of the particular case where $\boldsymbol{A}$ is the formula algebra.

This result generalizes Theorem 2.32, given for implicative logics with the notations $\boldsymbol{\Omega}^{\boldsymbol{A}}F$ for $\boldsymbol{\rho}^{\boldsymbol{A}^{-1}}F$ and F_θ for $\boldsymbol{\tau}^{\boldsymbol{A}^{-1}}\theta$; these notations suggest that *the isomorphisms are independent of the particular transformers* under consideration, and this is indeed true in general, as Exercise 3.40 shows. Actually, in Section 4.4 we find a way to describe these unique isomorphisms without using any of the transformers, thus giving this uniqueness a deeper significance.

For now, we can get a clearer picture of the isomorphisms still using the transformers. By Theorem 3.56, the functions $\boldsymbol{\rho}^{\boldsymbol{A}^{-1}}$ and $\boldsymbol{\tau}^{\boldsymbol{A}^{-1}}$, which are introduced as the residuals of $\boldsymbol{\rho}^{\boldsymbol{A}}$ and $\boldsymbol{\tau}^{\boldsymbol{A}}$, when restricted to filters and congruences are bijective and mutually inverse; therefore we can write that $\boldsymbol{\rho}^{\boldsymbol{A}^{-1}} = \left(\boldsymbol{\tau}^{\boldsymbol{A}^{-1}}\right)^{-1}$ on $\mathcal{F}i_{\mathcal{L}}\boldsymbol{A}$, and dually $\boldsymbol{\tau}^{\boldsymbol{A}^{-1}} = \left(\boldsymbol{\rho}^{\boldsymbol{A}^{-1}}\right)^{-1}$ on $\mathrm{Con}_{\mathsf{K}}\boldsymbol{A}$, and in both equalities the last, outermost "$^{-1}$" indicates the real inverse function. One might even be tempted to cancel one "$^{-1}$" symbol on each side, and conclude that actually $\boldsymbol{\rho}^{\boldsymbol{A}} = \boldsymbol{\tau}^{\boldsymbol{A}^{-1}}$ and $\boldsymbol{\tau}^{\boldsymbol{A}} = \boldsymbol{\rho}^{\boldsymbol{A}^{-1}}$ on the specified domains; but this would be wrong: even if F is an $\mathcal{L}$-filter, its image $\boldsymbol{\tau}^{\boldsymbol{A}}F$ need not be a relative congruence, and dually. But from the fact, proved above, that $F \subseteq \boldsymbol{\tau}^{\boldsymbol{A}^{-1}}\boldsymbol{\rho}^{\boldsymbol{A}^{-1}}F$, it follows by residuation that $\boldsymbol{\tau}^{\boldsymbol{A}}F \subseteq \boldsymbol{\rho}^{\boldsymbol{A}^{-1}}F$; thus, $\boldsymbol{\rho}^{\boldsymbol{A}^{-1}}F$ is a K-congruence containing $\boldsymbol{\tau}^{\boldsymbol{A}}F$, and it turns out that the closest possible approximation is indeed true:

COROLLARY 3.57
Let $\mathcal{L}$ be an algebraizable logic with equivalent algebraic semantics K and transformers $\boldsymbol{\tau}, \boldsymbol{\rho}$, and let $\boldsymbol{A}$ be an arbitrary algebra. For every $F \subseteq A$ and every $\theta \subseteq A \times A$,

$$\boldsymbol{\rho}^{\boldsymbol{A}^{-1}} Fg_{\mathcal{L}}^{\boldsymbol{A}} F = \Theta_{\mathsf{K}}^{\boldsymbol{A}} \boldsymbol{\tau}^{\boldsymbol{A}} F \qquad \textit{and} \qquad \boldsymbol{\tau}^{\boldsymbol{A}^{-1}} \Theta_{\mathsf{K}}^{\boldsymbol{A}} \theta = Fg_{\mathcal{L}}^{\boldsymbol{A}} \boldsymbol{\rho}^{\boldsymbol{A}} \theta .$$

In particular, $\boldsymbol{\rho}^{\boldsymbol{A}^{-1}} = \Theta_{\mathsf{K}}^{\boldsymbol{A}} \boldsymbol{\tau}^{\boldsymbol{A}}$ on $\mathcal{F}i_{\mathcal{L}} \boldsymbol{A}$, and $\boldsymbol{\tau}^{\boldsymbol{A}^{-1}} = Fg_{\mathcal{L}}^{\boldsymbol{A}} \boldsymbol{\rho}^{\boldsymbol{A}}$ on $\mathrm{Con}_{\mathsf{K}} \boldsymbol{A}$.

PROOF: Given $F \subseteq A$, apply the reasonings made in the previous paragraph to $Fg_{\mathcal{L}}^{\boldsymbol{A}} F$ and use residuation, and you will find the first displayed equality. If $F \in \mathcal{F}i_{\mathcal{L}} \boldsymbol{A}$, then $Fg_{\mathcal{L}}^{\boldsymbol{A}} F = F$ and the first equality in the next line is obtained. The other two are just dual to these. Exercise 3.41 gives more details. ⊠

These properties, in the particular case of the formula algebra, are those found in Proposition 3.54 from simply the definition of algebraizability. These descriptions of the isomorphisms serve as a guide in the proof of the next result, one of the greatest in abstract algebraic logic.

THEOREM 3.58 (THE ISOMORPHISM THEOREM, SECOND VERSION)
Let $\mathcal{L}$ be a logic, and let K be a generalized quasivariety. The following conditions are equivalent:

(i) *$\mathcal{L}$ is algebraizable with equivalent algebraic semantics the class K.*

(ii) *For every algebra $\boldsymbol{A}$ there is an isomorphism $\Phi^{\boldsymbol{A}} \colon \mathcal{F}i_{\mathcal{L}} \boldsymbol{A} \cong \mathrm{Con}_{\mathsf{K}} \boldsymbol{A}$ that commutes with endomorphisms in the sense that $\Phi^{\boldsymbol{A}} h^{-1} F = h^{-1} \Phi^{\boldsymbol{A}} F$ for every $F \in \mathcal{F}i_{\mathcal{L}} \boldsymbol{A}$ and every $h \in \mathrm{End} \boldsymbol{A}$.*

(iii) *There is an isomorphim $\Phi \colon \mathcal{T}h\mathcal{L} \cong \mathcal{T}h\mathsf{C}_{\mathsf{K}} = \mathrm{Con}_{\mathsf{K}} \boldsymbol{Fm}$ that commutes with substitutions in the sense that $\Phi \sigma^{-1} T = \sigma^{-1} \Phi T$ for every $T \in \mathcal{T}h\mathcal{L}$ and every $\sigma \in \mathrm{End} \boldsymbol{Fm}$.*

Moreover, every isomorphism as in (iii) is of the form $\boldsymbol{\rho}^{-1}$ and its inverse of the form $\boldsymbol{\tau}^{-1}$, for a pair of structural transformers $\boldsymbol{\rho}$ and $\boldsymbol{\tau}$ establishing algebraizability of the logic. As a consequence, such an isomorphism is unique.

PROOF: Theorem 3.56 proves that (i) implies (ii) for $\Phi^{\boldsymbol{A}} := \boldsymbol{\rho}^{\boldsymbol{A}^{-1}}$, and (iii) is a particular case of (ii). Now assume (iii) and let us prove (i). Since Φ is an isomorphism between the lattices of closed sets of two closure operators, we can apply Theorem 1.56, which tells us that if Φ commutes with substitutions in the stated sense, then also Φ^{-1} commutes with substitutions, and these commutativities can be equivalently expressed by the properties:

$$\Phi \mathsf{C}_{\mathcal{L}} \sigma T = \mathsf{C}_{\mathsf{K}} \sigma \Phi T \qquad \text{and} \qquad \Phi^{-1} \mathsf{C}_{\mathsf{K}} \sigma \theta = \mathsf{C}_{\mathcal{L}} \sigma \Phi^{-1} \theta \tag{35}$$

for all $T \in \mathcal{T}h\mathcal{L}$, all $\theta \in \mathcal{T}h\mathsf{C}_{\mathsf{K}}$ and all $\sigma \in \mathrm{End} \boldsymbol{Fm}$. Moreover, by Proposition 1.51, structurality of the operators $\mathsf{C}_{\mathcal{L}}$ and C_{K} can be equivalently expressed by the properties:

$$\mathsf{C}_{\mathcal{L}} \sigma \Gamma = \mathsf{C}_{\mathcal{L}} \sigma \mathsf{C}_{\mathcal{L}} \Gamma \qquad \text{and} \qquad \mathsf{C}_{\mathsf{K}} \sigma \Theta = \mathsf{C}_{\mathsf{K}} \sigma \mathsf{C}_{\mathsf{K}} \Theta \tag{36}$$

for all $\Gamma \subseteq Fm$, all $\Theta \subseteq Eq$ and all $\sigma \in \mathrm{End} \boldsymbol{Fm}$.

How should the transformers $\boldsymbol{\tau}$ and $\boldsymbol{\rho}$ be defined so that they witness algebraizability of $\mathcal{L}$ with respect to K? The leading idea is that once they are found and algebraizability is proved, we want the induced isomorphisms $\boldsymbol{\rho}^{-1}$ and $\boldsymbol{\tau}^{-1}$ resulting from Theorem 3.56 to coincide with the given isomorphisms Φ and Φ^{-1} respectively; according to Proposition 3.54, we should then have

$$\Phi C_{\mathcal{L}}\Gamma = C_{\mathsf{K}}\boldsymbol{\tau}\Gamma \qquad \text{and} \qquad \Phi^{-1}C_{\mathsf{K}}\Theta = C_{\mathcal{L}}\boldsymbol{\rho}\Theta \tag{37}$$

for all $\Gamma \subseteq Fm$ and all $\Theta \subseteq Eq$. Thus, we want to define $\boldsymbol{\tau}$ and $\boldsymbol{\rho}$ so that these equalities hold. This is done step by step, starting with the simplest cases where Γ is a single variable (in the case of $\boldsymbol{\tau}$) and where Θ is a single equation (for $\boldsymbol{\rho}$).

Take an arbitrary $x \in V$ and consider $C_{\mathcal{L}}\{x\} \in \mathcal{T}h\mathcal{L}$; then $\Phi C_{\mathcal{L}}\{x\} \in \mathcal{T}hC_{\mathsf{K}}$. Let $\Theta_0 \subseteq Eq$ be a set of generators[†] of this relative equational theory, that is, such that $\Phi C_{\mathcal{L}}\{x\} = C_{\mathsf{K}}\Theta_0$. Consider the substitution $\sigma_0 \in \operatorname{End}\boldsymbol{Fm}$ such that $\sigma_0 z := x$ for all $z \in V$, and define $\boldsymbol{\tau}x := \sigma_0\Theta_0$. By construction this is a set of equations in at most the variable x, and hence it defines a structural transformer $\boldsymbol{\tau}$ from formulas into equations, extended to sets by taking unions as usual.

Symmetrically, starting from the equation $x \approx y$ where x, y are different variables, and using any substitution σ_1 such that $\sigma_1 x := x$, $\sigma_1 y := y$ and $\sigma_1 z \in \{x, y\}$ for all other $z \in V$, we get a structural transformer $\boldsymbol{\rho}$ from equations into formulas by putting $\boldsymbol{\rho}(x \approx y) := \sigma_1\Gamma_1$, where $\Gamma_1 \subseteq Fm$ is such that $C_{\mathcal{L}}\Gamma_1 = \Phi^{-1}C_{\mathsf{K}}\{x \approx y\} \in \mathcal{T}h\mathcal{L}$. Clearly, $\sigma_1\Gamma_1$ is a set of formulas in at most the variables x and y.

To start with,

$$\begin{aligned} \Phi C_{\mathcal{L}}\{x\} = \Phi C_{\mathcal{L}}\sigma_0\{x\} &\underset{(36)}{=} \Phi C_{\mathcal{L}}\sigma_0 C_{\mathcal{L}}\{x\} \underset{(35)}{=} C_{\mathsf{K}}\sigma_0\Phi C_{\mathcal{L}}\{x\} \\ &= C_{\mathsf{K}}\sigma_0 C_{\mathsf{K}}\Theta_0 \underset{(36)}{=} C_{\mathsf{K}}\sigma_0\Theta_0 = C_{\mathsf{K}}\boldsymbol{\tau}x. \end{aligned} \tag{38}$$

Now let $\varphi \in Fm$ and consider any substitution $\sigma_2 \in \operatorname{End}\boldsymbol{Fm}$ such that $\sigma_2 x := \varphi$. Then

$$\begin{aligned} \Phi C_{\mathcal{L}}\{\varphi\} = \Phi C_{\mathcal{L}}\sigma_2\{x\} &\underset{(36)}{=} \Phi C_{\mathcal{L}}\sigma_2 C_{\mathcal{L}}\{x\} \underset{(35)}{=} C_{\mathsf{K}}\sigma_2\Phi C_{\mathcal{L}}\{x\} \\ &\underset{(38)}{=} C_{\mathsf{K}}\sigma_2 C_{\mathsf{K}}\boldsymbol{\tau}x \underset{(36)}{=} C_{\mathsf{K}}\sigma_2\boldsymbol{\tau}x = C_{\mathsf{K}}\boldsymbol{\tau}\sigma_2 x = C_{\mathsf{K}}\boldsymbol{\tau}\varphi. \end{aligned} \tag{39}$$

This property is extended to sets by using some general properties of closure operators of Section 1.3 and that Φ is an isomorphism between the complete lattices $\mathcal{T}h\mathcal{L}$ and $\mathcal{T}hC_{\mathsf{K}}$; so, for all $\Gamma \subseteq Fm$,

$$\begin{aligned} \Phi C_{\mathcal{L}}\Gamma \underset{1.29}{=} \Phi\Big(\bigvee_{\varphi\in\Gamma} C_{\mathcal{L}}\{\varphi\}\Big) &= \bigvee_{\varphi\in\Gamma} \Phi C_{\mathcal{L}}\{\varphi\} \underset{(39)}{=} \bigvee_{\varphi\in\Gamma} C_{\mathsf{K}}\boldsymbol{\tau}\varphi \\ &= C_{\mathsf{K}}\Big(\bigcup_{\varphi\in\Gamma} C_{\mathsf{K}}\boldsymbol{\tau}\varphi\Big) \underset{1.22}{=} C_{\mathsf{K}}\Big(\bigcup_{\varphi\in\Gamma} \boldsymbol{\tau}\varphi\Big) = C_{\mathsf{K}}\boldsymbol{\tau}\Gamma, \end{aligned} \tag{40}$$

[†] This is not essential to the proof of the present statement as it is done here. One might take $\Theta_0 := \Phi C_{\mathcal{L}}\{x\}$ as well, but this would not save any work. Moreover, the present choice has other theoretical advantages, as witnessed by Exercise 3.43.

where each supremum is taken in the corresponding lattice of theories. Observe that this is the left equation in (37); so, we have reached our first goal. Now algebraizability follows in a straightforward way: if $\Gamma \cup \{\varphi\} \subseteq Fm$, then

$$\begin{aligned} \Gamma \vdash_{\mathcal{L}} \varphi &\iff C_{\mathcal{L}}\{\varphi\} \subseteq C_{\mathcal{L}}\Gamma \\ &\iff \Phi C_{\mathcal{L}}\{\varphi\} \subseteq \Phi C_{\mathcal{L}}\Gamma && \text{because } \Phi \text{ is an isomorphism} \\ &\iff C_{\mathsf{K}}\boldsymbol{\tau}\varphi \subseteq C_{\mathsf{K}}\boldsymbol{\tau}\Gamma && \text{by (39) and (40)} \\ &\iff \boldsymbol{\tau}\Gamma \vDash_{\mathsf{K}} \boldsymbol{\tau}\varphi, \end{aligned}$$

that is, we obtain (ALG1). Using the definition of $\boldsymbol{\rho}$ and again that Φ is an isomorphism, we can now prove (ALG4):

$$\begin{aligned} C_{\mathsf{K}}\{x \approx y\} &= C_{\mathsf{K}}\sigma_1\{x \approx y\} \underset{(36)}{=} C_{\mathsf{K}}\sigma_1 C_{\mathsf{K}}\{x \approx y\} \\ &= \Phi\Phi^{-1}C_{\mathsf{K}}\sigma_1 C_{\mathsf{K}}\{x \approx y\} \underset{(35)}{=} \Phi C_{\mathcal{L}}\sigma_1\Phi^{-1}C_{\mathsf{K}}\{x \approx y\} \\ &= \Phi C_{\mathcal{L}}\sigma_1 C_{\mathcal{L}}\Gamma_1 \underset{(36)}{=} \Phi C_{\mathcal{L}}\sigma_1\Gamma_1 = \Phi C_{\mathcal{L}}\boldsymbol{\rho}\{x \approx y\} \underset{(40)}{=} C_{\mathsf{K}}\boldsymbol{\tau}\boldsymbol{\rho}\{x \approx y\}. \end{aligned}$$

With (ALG1) and (ALG4) we have shown that $\mathcal{L}$ is algebraizable with respect to K with the transformers $\boldsymbol{\tau}$ and $\boldsymbol{\rho}$. Since K is assumed to be a generalized quasivariety, it is then the equivalent algebraic semantics for $\mathcal{L}$.

Finally, to show uniqueness for (iii), observe that the proof of the implication (iii)⇒(i) shows, as a particular case of (40), that any isomorphism Φ satisfying (iii) satisfies $\Phi\Gamma = C_{\mathsf{K}}\boldsymbol{\tau}\Gamma$ for all $\Gamma \in \mathcal{T}h\mathcal{L}$, for some structural transformer $\boldsymbol{\tau}$ witnessing algebraizability; but by Theorem 3.17 any two of these transformers are equivalent modulo C_{K}, therefore only one isomorphism can exist. ⊠

This proof follows rather closely Blok and Pigozzi's original proof. However, their original setting assumed that $\mathcal{L}$ is finitary and that K is a quasivariety, and their notion of algebraizability demanded that the transformers are finitary (see the comments before Definition 3.39). From the very definitions it follows (Theorem 3.37) that once we have proved algebraizability, there are such finitary transformers, but it is possible to see (Exercise 3.43) that one can obtain them directly by just inserting an additional step in the middle of the very same proof. Exercise 3.44 proposes an alternative proof, still under the same methodology.

Using the terminology of the abstract setting of structurality developed in the final part of Section 1.4, the equivalent expression of commutativity with homomorphisms given in Theorem 1.56 allows us to view the isomorphisms as being isomorphisms between *the expanded lattices* $\langle \mathcal{F}i_{\mathcal{L}}\boldsymbol{A}, \langle h_{\mathcal{L}} : h \in \operatorname{End}\boldsymbol{A}\rangle\rangle$ and $\langle \operatorname{Con}_{\mathsf{K}}\boldsymbol{A}, \langle h_{\mathsf{K}} : h \in \operatorname{End}\boldsymbol{A}\rangle\rangle$: each $h \in \operatorname{End}\boldsymbol{A}$ induces an operation in each of the specified lattices: If $F \in \mathcal{F}i_{\mathcal{L}}\boldsymbol{A}$, then hF may not be an $\mathcal{L}$-filter, but closing it under the corresponding operator produces an $\mathcal{L}$-filter, so $F \mapsto h_{\mathcal{L}}F := Fg^{\boldsymbol{A}}_{\mathcal{L}}hF$ is a unary operation in $\mathcal{F}i_{\mathcal{L}}\boldsymbol{A}$. And dually for relative congruences, for which the operation is $\theta \mapsto h_{\mathsf{K}}\theta := \boldsymbol{\Theta}^{\boldsymbol{A}}_{\mathsf{K}}h\theta$. Then, particularizing Theorem 1.56 to the present situation, a lattice isomorphism $\Phi^{\boldsymbol{A}} : \mathcal{F}i_{\mathcal{L}}\boldsymbol{A} \cong \operatorname{Con}_{\mathsf{K}}\boldsymbol{A}$ satisfies the property of Theorem 3.58(ii) if and only if $\Phi^{\boldsymbol{A}} h_{\mathcal{L}} F = h_{\mathsf{K}} \Phi^{\boldsymbol{A}} F$ for all $F \in \mathcal{F}i_{\mathcal{L}} \boldsymbol{A}$, which is to say that $\Phi^{\boldsymbol{A}}$ commutes with the corresponding operations induced by each endomorphism in the lattices of $\mathcal{L}$-filters and of relative congruences. This fully justifies calling this property *commutativity with endomorphisms* (or simply *structurality*) rather than "commutativity with *the residuals* of endomorphisms" or "with *inverse* endomorphisms", as is also found in the literature.†

This commutativity property is extended in Exercise 3.45 to homomorphisms between two algebras. The situation begins to acquire a definite categorial aspect, which has been exploited in a more abstract setting only recently.

Thus, algebraizability is characterized by the mere *existence* of *some* lattice isomorphism between theories and relative congruences on the formula algebra that commutes with substitutions; if such an isomorphism exists, then the proof has shown that it is determined by a structural transformer, and is hence *unique*, by Theorem 3.17. These isomorphism theorems admit several variations and formulations, and only the main ones are labelled as "Isomorphism Theorem". Here are two more. The first, of an intermediate strength, follows from the fact that, by Proposition 3.55, the residuals of structural transformers always commute with endomorphisms:

COROLLARY 3.59
Let $\mathcal{L}$ be a logic, and let K be a generalized quasivariety. The following conditions are equivalent:

(i) *$\mathcal{L}$ is algebraizable with equivalent algebraic semantics the class K.*
(ii) *There is a structural transformer ρ from equations into formulas such that for every algebra $\boldsymbol{A}$, the evaluated residual $\rho^{\boldsymbol{A}^{-1}}$ is a lattice isomorphism between $\mathcal{F}i_{\mathcal{L}} \boldsymbol{A}$ and $\mathrm{Con}_{\mathsf{K}} \boldsymbol{A}$.*
(iii) *There is a structural transformer ρ from equations into formulas such that its residual ρ^{-1} is a lattice isomorphim between $\mathcal{T}h\mathcal{L}$ and $\mathcal{T}h\mathrm{C}_{\mathsf{K}} = \mathrm{Con}_{\mathsf{K}} \boldsymbol{Fm}$.* ⊠

Still, these characterizations of algebraizability require knowledge of a candidate class K to play the role of equivalent algebraic semantics. Now, we can mix several of the approaches used so far and obtain an "intrinsic" characterization requiring knowledge of only one of the transformers, of a partly syntactic and partly lattice-theoretic character:

THEOREM 3.60
A logic $\mathcal{L}$ is algebraizable if and only if there is a structural transformer ρ from equations into formulas, determined by a set of equivalence formulas Δ that satisfies conditions (R), (Sym), (Trans) *and* (Re) *of Theorem* 3.21 *and such that its residual function ρ^{-1} is injective over $\mathcal{T}h\mathcal{L}$.*

PROOF: Only the converse needs to be proved, and in view of Theorem 3.21 we need only find a transformer $\boldsymbol{\tau}$ from formulas into equations such that condition

† To cite two of the most authoritative sources: this convention already appears in Czelakowski's [70], and is also used by Raftery in [200].

(ALG3) is satisfied. This is done in a very similar (but much simpler) manner to the proof of Theorem 3.58. First, observe that $C_{\mathcal{L}}\rho\rho^{-1}T = T$ for any theory T. To see this, put $T' := C_{\mathcal{L}}\rho\rho^{-1}T$. We have that $T, T' \in \mathcal{T}h\mathcal{L}$, and $T' \subseteq T$ because $\rho\rho^{-1}T \subseteq T$ by residuation (Exercise 1.2). But by definition $\rho\rho^{-1}T \subseteq T'$, that is, $\rho^{-1}T \subseteq \rho^{-1}T'$. Since ρ^{-1} is injective by assumption, and always preserves meets, it follows (Exercise 3.48) that it reflects order on $\mathcal{T}h\mathcal{L}$, therefore $T \subseteq T'$. In conclusion, $T = T'$ or, equivalently, $C_{\mathcal{L}}\rho\rho^{-1}T = T$ as claimed. Now, this claim, for $T := C_{\mathcal{L}}\{x\}$, where $x \in Var$, says that $x \dashv\vdash_{\mathcal{L}} \rho\rho^{-1}C_{\mathcal{L}}\{x\}$. Consider the substitution $\sigma_x \in \mathrm{End}\boldsymbol{Fm}$ such that $\sigma_x z := x$ for all $z \in V$, and define $\boldsymbol{\tau} x := \sigma_x\rho^{-1}C_{\mathcal{L}}\{x\}$. By construction this is a set of equations in at most the variable x, and hence it defines a structural transformer $\boldsymbol{\tau}$ from formulas into equations, extended to sets by taking unions as usual. Now we need only apply structurality of both $\mathcal{L}$ and ρ and we find that

$$x = \sigma_x x \dashv\vdash_{\mathcal{L}} \sigma_x\rho\rho^{-1}C_{\mathcal{L}}\{x\} = \rho\sigma_x\rho^{-1}C_{\mathcal{L}}\{x\} = \rho\boldsymbol{\tau} x,$$

which is (ALG3) in the single-variable form (26) on page 116. ⊠

For a dual property see Exercise 3.49.

The Isomorphism Theorems provide tools for showing both algebraizability and non-algebraizability. For instance, the method of Corollary 3.59 was used in [218] to show algebraizability of the finitary version $Ł_{\infty f}$ of Łukasiewicz's infinitely-valued logic $Ł_{\infty f}$ with respect to MV, the variety of MV-algebras (there presented as the term-equivalent variety of Wajsberg algebras): Taking the transformer $\rho(x \approx y) := \{x \to y, y \to x\}$, it is shown that on any $\boldsymbol{A} \in \mathsf{MV}$, its evaluated residual, which is the function $F \mapsto \{\langle a,b\rangle \in A \times A : a \to b, b \to a \in F\}$, is the mentioned lattice isomorphism. In principle the reverse transformer is not needed, but in practice surjectivity of the function is established by considering the inverse function $\theta \mapsto 1/\theta$; in this case, this happens to be the evaluated residual of the transformer $\boldsymbol{\tau} x := \{x \approx \top\}$, but in general any function that helps to prove surjectivity would do. A similar situation can be found in other cases.

Theorem 3.56.3 provides the main tool to show *non-algebraizability*; if one can show that for some particular, hopefully small finite algebra such an isomorphism is impossible (due to the algebraic structure itself), then the logic cannot be algebraizable. Here are a couple of examples of this procedure:

Example 3.61 (Local modal logics)
The local logic $S5^{\ell}$ associated with the modal system $S5$ is not algebraizable. Let $\boldsymbol{B}_4$ be the four-element Boolean algebra, where $B_4 = \{0, a, b, 1\}$ with $b = \neg a$, expanded with the "simple" modal operator given by $\Box a = \Box b = \Box 0 := 0$ and $\Box 1 := 1$. It is not difficult to show that this algebra is a simple algebra (while the non-expanded Boolean algebra is not), that is, $\mathrm{Con}\boldsymbol{B}_4$ has exactly *two* elements. Therefore, all lattices of the form $\mathrm{Con}_{\mathsf{K}}\boldsymbol{B}_4$ for some K have *at most two* elements. However, all *four* distinct lattice filters of this algebra are filters of $S5^{\ell}$. Thus, no isomorphism between the lattice of filters of $S5^{\ell}$ and a lattice of the form $\mathrm{Con}_{\mathsf{K}}\boldsymbol{B}_4$ for some K is possible (and observe that we need not specify K). Exercise 3.50

asks you to provide the missing details. By Proposition 3.31, this implies that neither of the local logics associated with weaker modal systems ($K, S4$, etc.) is algebraizable, and a fortiori not implicative; this confirms a claim made in Example 2.17.1. By contrast, the corresponding global logics are implicative, as shown there, and hence algebraizable. A different argument showing that the local logics are not algebraizable precisely because of their relations with the global logics is given in Example 7.47.1. ☒

EXAMPLE 3.62 (SOME RELEVANCE LOGICS)
The logics $\mathcal{E}$ of *Entailment* and $\mathcal{T}$ of *Ticket Entailment*, and the logic $\mathcal{R}_{\rightarrow}$, the implication fragment of relevance logic $\mathcal{R}$, are not algebraizable. As in Example 3.61, this is shown by exhibiting a simple algebra that has more than two filters of the logic. Here are the algebras: the one for $\mathcal{E}$, which is formulated in the language $\langle \wedge, \vee, \rightarrow, \neg \rangle$, is based on the three-element chain with universe $\{0, ½, 1\}$, with the two tables on the left below plus the lattice operations $\wedge$ and $\vee$ of the order $0 < ½ < 1$; the one for $\mathcal{R}_{\rightarrow}$, which has only implication, is given by the four-element table on the right.

$\rightarrow$	0	½	1
0	1	1	1
½	0	1	1
1	0	0	1

$\neg$	
0	1
½	½
1	0

$\rightarrow$	0	a	b	1
0	1	1	1	1
a	0	b	0	1
b	0	a	b	1
1	0	0	0	1

In the first case, the filters of $\mathcal{E}$ are the subsets $\{1\}$, $\{1, ½\}$ and $\{1, ½, 0\}$. In the second case, the filters of $\mathcal{R}_{\rightarrow}$ are the subsets $\{1, b\}$, $\{1, a, b\}$ and $\{1, a, b, 0\}$. It is easy to check, just by inspecting the tables of the operations, that in both cases the algebras are simple, which means that they have exactly two congruences. As a consequence, no isomorphism between the lattices of filters of the logic and of relative congruences is possible. Since $\mathcal{E}$ is an axiomatic extension of $\mathcal{T}$, it follows that the latter is not algebraizable either. Again, Exercise 3.50 asks you to provide the missing details.

The case of $\mathcal{R}_{\rightarrow}$ is interesting, for it shows that the presence of conjunction in the defining equation for $\mathcal{R}$ (Example 3.34.7) is essential; by contrast, the absence of conjunction in the implication fragment of $\mathcal{RM}$ is not a problem, because $\mathcal{RM}$ is algebraizable with defining equation $x \approx x \rightarrow x$, and therefore its implication fragment is algebraizable with this defining equation as well. All this was essentially noted by Blok and Pigozzi in their [32] for an axiomatically defined logic, which was proved to coincide with the implication fragment of $\mathcal{R}$ only in [146, Theorem 8.4]. ☒

The Isomorphism Theorem (in its third version still to come, Theorem 4.56) can also be used (in a more tricky way) to show that a given variety *cannot* be the equivalent algebraic semantics of *any* algebraizable logic; the case of the variety of distributive lattices is detailed in Example 4.61.

Regularity

As a final application of the Isomorphism Theorem we see how *regularly* algebraizable logics got their name. It came from a property called *congruence regularity* in universal algebra; it is most often considered for varieties, and for quasivarieties in its relativized form, but its scope is potentially larger. A class of algebras K is ***congruence-regular*** when for every $\boldsymbol{A} \in \mathsf{K}$, if $\theta, \theta' \in \operatorname{Con}\boldsymbol{A}$ are such that $a/\theta = a/\theta'$ for some $a \in A$, then $\theta = \theta'$; that is, congruences are determined by any of their equivalence classes. A restricted variant of the property makes sense in pointed classes of algebras: a pointed class K is ***point-regular*** when for every $\boldsymbol{A} \in \mathsf{K}$, if $\theta, \theta' \in \operatorname{Con}\boldsymbol{A}$ are such that $\top^{\boldsymbol{A}}/\theta = \top^{\boldsymbol{A}}/\theta'$, then $\theta = \theta'$; that is, the equivalence classes of the designated element determine the congruences. If these properties hold only for relative congruences, that is, for congruences in $\operatorname{Con}_{\mathsf{K}}\boldsymbol{A}$, then K is called ***relatively congruence-regular*** and ***relatively point-regular***, respectively; of course these relative notions coincide with the plain ones when K is a variety.

THEOREM 3.63
Let $\mathcal{L}$ be an algebraizable logic, and let K be its equivalent algebraic semantics. If $\mathcal{L}$ is regularly algebraizable, then K is pointed and relatively point-regular.

PROOF: Let $\boldsymbol{A} \in \mathsf{K}$ and let $\top$ be any theorem of $\mathcal{L}$ with at most the variable x. Proposition 3.51 shows that $\top$ is an algebraic constant of K, hence this is a pointed class. We know by Theorem 3.52 that $x \approx \top$ can be taken as the defining equation for $\mathcal{L}$, and so $\boldsymbol{\tau}^{\boldsymbol{A}^{-1}}\theta = \{a \in A : \boldsymbol{\tau}^{\boldsymbol{A}}a \subseteq \theta\} = \{a \in A : \langle a, \top^{\boldsymbol{A}}\rangle \in \theta\} = \top^{\boldsymbol{A}}/\theta$. By Theorem 3.56 the function $\boldsymbol{\tau}^{\boldsymbol{A}^{-1}}$ is one-to-one over relative congruences. Thus, if $\theta, \theta' \in \operatorname{Con}_{\mathsf{K}}\boldsymbol{A}$ are such that $\top^{\boldsymbol{A}}/\theta = \top^{\boldsymbol{A}}/\theta'$, it follows that $\theta = \theta'$. That is, K is relatively point-regular. ⊠

Together with Theorem 3.52, this says that every regularly algebraizable logic is the assertional logic of a pointed, relatively point-regular generalized quasivariety. Point-regular varieties and relatively point-regular quasivarieties have been extensively studied in universal algebra and in abstract algebraic logic, and have provided some difficult results. One of them is the following enhanced converse of Theorem 3.63, in the all-finitary case: *A logic is regularly BP-algebraizable if and only if it is the assertional logic of a pointed, relatively point-regular quasivariety* (Theorem 6.146). By the Isomorphism Theorem, the filters of these logics on these algebras have the form $1/\theta$ for some $\theta \in \operatorname{Con}_{\mathsf{K}}\boldsymbol{A}$. Thus, in relatively point-regular quasivarieties the filters of the corresponding assertional logics can be defined in a purely algebraic way, and can be studied independently of the logical motivation and the logical apparatus; sometimes they are called "filters" or "ideals" of the algebra. Classes of algebras with similar properties are said to have "ideal-determined congruences". There is an important amount of universal algebraic work done with these and similar logical ideas in the background, as witnessed for instance by [3, 36, 237].

Moreover, there is a bijection (actually, a dual isomorphism) between the families of regularly BP-algebraizable logics and of pointed, relatively point-regular quasivarieties; see Theorem 6.146 again. Thus, these quasivarieties can be the equivalent algebraic semantics of several logics, but there is one (and only one) "canonical" logic (the assertional one) that is "best" associated with it, in the sense that its defining equation $x \approx \top$ is determined "intrinsically" by the quasivariety, while others depend on a more or less arbitrary transformer.

A prototypical case of this situation appears in Example 3.20. There the two BP-algebraizable logics $Ł_3$ and $\mathcal{J}_3$ are shown to have the same equivalent algebraic semantics $\mathbb{Q}\{\boldsymbol{A}_3\}$ (in this case, this class turns out to be a variety). Since $Ł_3$ is regularly algebraizable (Example 3.50), we know that $\mathbb{Q}\{\boldsymbol{A}_3\}$ is (relatively) point-regular; by contrast, $\mathcal{J}_3$ is not regularly algebraizable (Example 6.144.3), and is not the assertional logic of its equivalent variety. Moreover, this provides a counterexample to a naïve converse of Theorem 3.63 you might be wondering about (that if $\mathcal{L}$ is BP-algebraizable and its equivalent quasivariety is a pointed and relatively point-regular quasivariety, then $\mathcal{L}$ is regularly algebraizable); actually, this is almost true, but an additional condition is needed, as shown in Theorem 6.145. However, while in a technical sense $Ł_3$ is better associated with $\mathbb{Q}\{\boldsymbol{A}_3\}$ than $\mathcal{J}_3$, the interest of the logician in studying one or the other most probably depends on extra-logical considerations, and the algebraic logician can just report on the "degree" of algebraizability of each of the alternatives, and draw as many consequences from this fact as possible.

One case where this approach is useful is the following.

EXAMPLE 3.64

The "canonical" *quantum logic* is the assertional logic of the variety OML of all orthomodular lattices (it is formally defined in Example 6.122.5). The language of orthomodular lattices includes a constant $\top$, but does not include an implication; in fact, there are several (at least five) choices in the literature for an implication operation $x \to y$ defining order in them (that is, such that $a \leqslant b$ if and only if $a \to b = 1$); perhaps the best known is Sasaki's implication $x \to y := \neg x \vee (x \wedge y)$. If one shows that OML is a point-regular variety, then by Theorem 6.146 its assertional logic is regularly BP-algebraizable; however, Exercise 3.51 contains a more direct way to prove this, and also that any of these implications is suitable to define the set of equivalence formulas $\Delta(x,y) = \{x \to y, y \to x\}$. Thus, this is the only BP-algebraizable logic satisfying the G-rule and having OML as equivalent algebraic semantics. This provides a criterion that selects it among other logics associated with OML. This is one of the few historical situations of algebraic logic in which the class of algebras has preceded the logic in the landscape: it seems there has been a wide consensus since long ago that the right algebraic structures to describe the logic of quantum mechanics are orthomodular lattices, and the problem is to choose among several logics that satisfy some kind of completeness theorem with respect to OML. The theory of algebraizability has provided a sound criterion to do so. ⊠

Exercises for Section 3.5

EXERCISE 3.36. Consider the forms of conditions (ALG1–4) postulated in (32) on page 145. Check that the ones for (ALG1) and (ALG2) are really equivalent to the original ones in Definition 3.11, and that the ones for (ALG3) and (ALG4) are just a re-writing of the forms of (27) on page 116. Still on the same issue, prove Proposition 3.54; notice that this proposition needs an assumption that is not necessary for the other equivalences.

EXERCISE 3.37. Prove Proposition 3.55. Moreover, write out explicity the forms of the properties (33) and (34) appearing there, in the particular cases where $h \in \mathrm{Hom}(\boldsymbol{Fm}, \boldsymbol{A})$ and where h is an endormorphism of $\boldsymbol{Fm}$, that is, a substitution.

EXERCISE 3.38. Complete all the details in the proof of Theorem 3.56, following the indications therein. Beware that when using the properties in (33) you have to change the names of the algebras, because you have to apply them to an $h \in \mathrm{Hom}(\boldsymbol{Fm}, \boldsymbol{A})$, as you did in Exercise 3.37.

EXERCISE 3.39. Use the result in Proposition 3.54 to obtain an alternative and very direct proof of Theorem 3.56 in the particular case of the formula algebra.

EXERCISE 3.40. Let $\mathcal{L}$ be an algebraizable logic, with equivalent algebraic semantics K, and let τ_1, ρ_1 and τ_2, ρ_2 be two pairs of structural transformers, each pair witnessing the algebraizability of $\mathcal{L}$ with respect to K. Show that for each algebra $\boldsymbol{A}$:

1. If $F \in \mathcal{F}i_{\mathcal{L}}\boldsymbol{A}$, then $\rho_1^{\boldsymbol{A}^{-1}}F = \rho_2^{\boldsymbol{A}^{-1}}F$.
2. If $\theta \in \mathrm{Con}_{\mathsf{K}}\boldsymbol{A}$, then $\tau_1^{\boldsymbol{A}^{-1}}\theta = \tau_2^{\boldsymbol{A}^{-1}}\theta$.

Conclude that the isomorphisms in Theorems 3.56 and 3.58(ii) are independent of the particular transformers taken to witness algebraizability.

COMMENT. Observe that in the case of the formula algebra this already entails that there can be only one isomorphism satisfying (iii) of Theorem 3.58; but this does *not* ensure the same for (ii).

EXERCISE 3.41. Let $\mathcal{L}$ be an algebraizable logic with equivalent algebraic semantics K and transformers τ, ρ, and let $\boldsymbol{A}$ be an arbitrary algebra. Use Theorem 3.56 to prove that $\rho^{\boldsymbol{A}^{-1}}\mathcal{F}g_{\mathcal{L}}^{\boldsymbol{A}}F = \Theta_{\mathsf{K}}^{\boldsymbol{A}}\tau^{\boldsymbol{A}}F$ for every $F \subseteq A$.

HINTS. You have to see, first, that $\tau^{\boldsymbol{A}}F \subseteq \rho^{\boldsymbol{A}^{-1}}\mathcal{F}g_{\mathcal{L}}^{\boldsymbol{A}}F \in \mathrm{Con}_{\mathsf{K}}\boldsymbol{A}$ in order to obtain the inclusion $\Theta_{\mathsf{K}}^{\boldsymbol{A}}\tau^{\boldsymbol{A}}F \subseteq \rho^{\boldsymbol{A}^{-1}}\mathcal{F}g_{\mathcal{L}}^{\boldsymbol{A}}F$; and, second, that $F \subseteq \tau^{\boldsymbol{A}^{-1}}\Theta_{\mathsf{K}}^{\boldsymbol{A}}\tau^{\boldsymbol{A}}F \in \mathcal{F}i_{\mathcal{L}}\boldsymbol{A}$, in order to obtain the reverse inclusion.

EXERCISE 3.42. Let $\mathcal{L}$ be an algebraizable logic with equivalent algebraic semantics K and transformers τ, ρ, and let $\boldsymbol{A}$ be an arbitrary algebra. Given that, by Corollary 3.57, or Exercise 3.41, we know that the isomorphisms between $\mathcal{F}i_{\mathcal{L}}\boldsymbol{A}$ and $\mathrm{Con}_{\mathsf{K}}\boldsymbol{A}$ are given by the functions $F \mapsto \Theta_{\mathsf{K}}^{\boldsymbol{A}}\tau^{\boldsymbol{A}}F$ and $\theta \mapsto \mathcal{F}g_{\mathcal{L}}^{\boldsymbol{A}}\rho^{\boldsymbol{A}}\theta$, you may try to prove this fact directly, without using the residuals framework at all. If you succeed, compare with the proof of Theorem 3.56 suggested in the text (including Exercise 3.38).

EXERCISE 3.43. By careful inspection of the proof of Theorem 3.58, show that if we restrict its assumptions by adding that $\mathcal{L}$ is finitary, then the very same proof can show directly that algebraizability of $\mathcal{L}$ is witnessed by a finite set of defining equations. Dually, show that if we assume that K is a quasivariety, then that proof can provide a finite set of equivalence formulas. Check that these facts are coherent with Theorem 3.37.

HINT. Go back and revise Propositions 1.65 and 1.66.

Exercise 3.44. Provide an alternative proof of the implication (iii)⇒(i) of Theorem 3.58, showing algebraizability by establishing (ALG2) and (ALG3).

Exercise 3.45. Let $\mathcal{L}$ be an algebraizable logic, with transformers $\boldsymbol{\tau}, \boldsymbol{\rho}$ and equivalent algebraic semantics K. Let $\boldsymbol{A}, \boldsymbol{B}$ be algebras and $h \in \mathrm{Hom}(\boldsymbol{A}, \boldsymbol{B})$. Prove that for any $F \in \mathcal{F}i_{\mathcal{L}}\boldsymbol{A}$, $\boldsymbol{\Theta}_{\mathsf{K}}^{\boldsymbol{B}} \boldsymbol{\tau}^{\boldsymbol{B}} Fg_{\mathcal{L}}^{\boldsymbol{B}} hF = \boldsymbol{\Theta}_{\mathsf{K}}^{\boldsymbol{B}} h \boldsymbol{\Theta}_{\mathsf{K}}^{\boldsymbol{A}} \boldsymbol{\tau}^{\boldsymbol{A}} F$. Use this to show the commutativity of the following diagram

$$
\begin{array}{ccc}
\mathcal{F}i_{\mathcal{L}}\boldsymbol{A} & \xrightarrow{\ \ \Phi^{\boldsymbol{A}}\ \ } & \mathrm{Con}_{\mathsf{K}}\boldsymbol{A} \\
{\scriptstyle h_{\mathcal{L}}}\big\downarrow & & \big\downarrow{\scriptstyle h_{\mathsf{K}}} \\
\mathcal{F}i_{\mathcal{L}}\boldsymbol{B} & \xrightarrow[\ \ \Phi^{\boldsymbol{B}}\ \ ]{} & \mathrm{Con}_{\mathsf{K}}\boldsymbol{B}
\end{array}
$$

where here $h_{\mathcal{L}}F := \mathcal{F}i_{\mathcal{L}}^{\boldsymbol{B}} hF$ for any $F \subseteq A$, and $h_{\mathsf{K}}\theta := \boldsymbol{\Theta}_{\mathsf{K}}^{\boldsymbol{B}} h\theta$ for any $\theta \subseteq A \times A$. Confirm that this generalizes the commutativity of the isomorphisms with endomorphisms as explained after Theorem 3.58.

Hint. You will find the properties in Exercise 1.92 useful.

Exercise 3.46. Prove the following result, which is an enhanced version of Theorem 3.7(i) of [32]: A class K is an algebraic semantics for a logic $\mathcal{L}$ if and only if there is an order isomorphism Φ between $\mathcal{T}h\mathcal{L}$ and a join-complete sub-semilattice of $\mathrm{Con}_{\mathsf{K}}\boldsymbol{Fm}$ such that Φ commutes with substitutions in the sense of the left-hand half equation of (35).

Hint. You will get the necessary inspiration, and most of the proof of the reverse direction, in Theorem 3.58. Lemma 1.55, Proposition 3.55 and Corollary 3.57 will also be useful.

Exercise 3.47. Show that the assumption of commutativity with substitutions of Theorem 3.58(iii) and of Exercise 3.46 can be weakened to require commutativity with just surjective substitutions, if $\mathcal{L}$ is finitary and K is a quasivariety.

Comment. This observation corresponds to a line of research, started by Czelakowski and Jansana, on limiting other properties where substitutions appear (such as structurality of non-finitary consequences, or of the transformers), and also homomorphisms, to the surjective ones; see [70, 72]. Exercise 1.65 is relevant to this issue.

Exercise 3.48. Let $\boldsymbol{A} = \langle A, \leqslant \rangle$ and $\boldsymbol{A}' = \langle A', \leqslant' \rangle$ be lattices, and $f \colon A \to A'$ an injective function that preserves meets. Prove that f reflects order; that is, that for all $a, b \in A$, if $fa \leqslant' fb$, then $a \leqslant b$.

Exercise 3.49. Let K be a generalized quasivariety. Prove that K is the equivalent algebraic semantics of an algebraizable logic if and only if there is a set $E(x) \subseteq Eq$ such that for each algebra $\boldsymbol{A}$ the function $\theta \longmapsto \{a \in A : \varepsilon^{\boldsymbol{A}}(a) \equiv \delta^{\boldsymbol{A}}(a)\ (\theta) \text{ for all } \varepsilon \approx \delta \in E(x)\}$ is injective over $\mathrm{Con}_{\mathsf{K}}\boldsymbol{A}$.

Exercise 3.50. Prove all the statements made in Examples 3.61 and 3.62. Notice that in particular this means using Hilbert-style axiomatizations, in order to check that the exhibited sets are filters of the corresponding logic. In the second case, you may find some useful information in [123, pp. 104–107].

Exercise 3.51. Consider the class OML of orthomodular lattices mentioned in Example 3.64, and let $\to$ be any binary operation behaving as an "implication" in the sense that in any $\boldsymbol{A} \in \mathsf{OML}$ and for any $a, b \in A$, $a \leqslant b$ if and only if $a \to b = 1$. Show that the assertional logic of OML is regularly algebraizable, with the set $\Delta(x, y) = \{x \to y, y \to x\}$

is a set of equivalence formulas, directly through the definition. You need not enter into many details of orthomodular lattices. If you connect this issue with Proposition 3.6, then you will realize that this is a much more general, and essentially known, result.

3.6 Bridge theorems and transfer theorems

Suppose that a framework has been established in which a general procedure or criterion has been adopted to associate with each logic $\mathcal{L}$ (perhaps only for logics of a certain kind) a class K of algebras, or algebra-based models, as its algebraic counterpart (algebraizability is one such framework; but other, more general ones appear in Definitions 4.37 and 5.67). In such a framework, a ***bridge theorem*** is a mathematical result stating that for a certain property **P** concerning a logic and a certain property **P′** concerning a class of algebras,[†] if $\mathcal{L}$ is any logic (in the class, if applicable) and K is the corresponding algebraic counterpart, then

$$\mathcal{L} \text{ satisfies } \mathbf{P} \quad \text{if and only if} \quad \mathsf{K} \text{ satisfies } \mathbf{P'}.$$

The term "bridge theorem" was coined a long time ago by Andréka, Németi and Sain (see [12, pp. 135–136 and 186–188]); according to them, results of this kind establish a bridge between two diferent lands, that of logic and that of algebra, and allow one to transform problems about a logic into problems about a class of algebras, so that one can use the powerful tools of algebra, a much more intensively studied discipline, to solve the problem, and then go back (crossing the bridge again) and obtain a solution to the original logical problem. In particular, they emphasize that this methodology makes it possible to approach the ever growing forest of new logics from a more uniform and better known territory, that of algebra.

In the bridge theorems that created the name, **P** is a typically logical property (such as interpolation) and **P′** is typically algebraic, that is, it is a global property of a class of algebras (such as amalgamation). In other bridge theorems the property **P′** has the form "every algebra in K satisfies property **Q**", where **Q** is a property concerning a single algebra (such as having a distributive congruence lattice); besides, in many of these cases it is the property **Q** that refers to the class K, while the restriction "in K" in the theorem can be deleted, and a stronger result is obtained.

A particular kind of bridge theorems comprises the ones in which both **P** and **P′** (**P** and **Q**, in the second case) are essentially the same property, suitably interpreted on each side (such as "to be finitely axiomatizable"). These results have been called ***transfer theorems***, as they can be phrased as "the property **P** transfers to K" (respectively, "to all algebras in K", or even "to all algebras", in the stronger cases).

[†] In other frameworks, bridge and transfer theorems concern classes of more general algebra-based models, such as classes of matrices, or classes of generalized matrices (and some examples will be given later on in this section); for simplicity, only classes of algebras are mentioned in this introductory discussion.

An important subclass of the transfer theorems in which **Q** is just "**P** interpreted in an algebra" are the ones in which **P** is a property of a closure operator or of a closure system. Such a property clearly applies to a logic $\mathcal{L}$ through its closure system of theories $\mathcal{T}h\mathcal{L}$ or its associated closure operator $C_{\mathcal{L}}$; and it applies to an algebra $\boldsymbol{A}$ through the closure system $\mathcal{F}i_{\mathcal{L}}\boldsymbol{A}$ of all the $\mathcal{L}$-filters of $\boldsymbol{A}$ or its associated closure operator $Fg_{\mathcal{L}}^{\boldsymbol{A}}$. Thus, transfer theorems of this kind state that

$$\begin{gathered} \mathcal{T}h\mathcal{L} \ \text{ (or } C_{\mathcal{L}}\text{) satisfies } \mathbf{P} \\ \text{if and only if} \\ \text{for each } \boldsymbol{A}(\in \mathsf{K}),\ \mathcal{F}i_{\mathcal{L}}\boldsymbol{A} \ \text{ (or } Fg_{\mathcal{L}}^{\boldsymbol{A}}\text{) satisfies } \mathbf{P}. \end{gathered}$$

When this holds one says that "**P** *transfers* to all algebras (in K)". Some transfer theorems are unconditional, that is, they hold for all logics and for all algebras; others have restrictions, either to the logics to which they apply, or to the algebras to which the property is transferred.

Establishing a procedure for associating classes of algebras with logics would not be that interesting without the bridge and transfer theorems obtained in recent decades. Thus, one is tempted to say:

> Bridge theorems and transfer theorems are the ultimate justification of abstract algebraic logic.

It is thus natural that bridge theorems and transfer theorems appear in various places in the book, particularly in Chapter 6. There are bridge theorems and transfer theorems of widely diverse degrees of difficulty, and some of them need extensive development of the general theory, or complicated algebraic technicalities (often both). Thus, this section is just an informative appetizer, concentrating on the algebraizability framework; but first the simplest examples of truly general transfer theorems will be considered. In order to formulate the results below in their most general form, the following terminology (partly advanced on page 8) may be useful.

DEFINITION 3.65

*A **generalized sequent** (**g-sequent** for short) is any pair* $\langle \Gamma, \varphi \rangle \in (\mathscr{P}Fm) \times Fm$, *which is denoted by* $\Gamma \rhd \varphi$.

A g-sequent $\Gamma \rhd \varphi$ *is a* (***Hilbert-style***) ***generalized rule*** (***g-rule***) ***of a logic*** $\mathcal{L}$ *when* $\Gamma \vdash_{\mathcal{L}} \varphi$.

A closure operator C on the universe of an algebra $\boldsymbol{A}$ ***satisfies a g-sequent*** $\Gamma \rhd \varphi$ *when, for all* $h \in \mathrm{Hom}(\boldsymbol{Fm}, \boldsymbol{A})$, $h\varphi \in Ch\Gamma$.

When Γ *is finite the "generalized" (the "g-") are deleted.*

By structurality, the consequence operator $C_{\mathcal{L}}$ of a logic $\mathcal{L}$ satisfies a g-sequent if and only if this one is a Hilbert-style g-rule of $\mathcal{L}$ in the above sense. Then:

Proposition 3.66
The following properties transfer from a logic $\mathcal{L}$ to the closure operator $Fg_{\mathcal{L}}^{\boldsymbol{A}}$ on an arbitrary algebra $\boldsymbol{A}$:

1. *To satisfy a g-sequent.*
2. *To have theorems.*
3. *To have no theorems.*
4. *To be finitary.*

Proof: You can check that point 1 is actually implicit in the definition of $\mathcal{L}$-filter (Exercise 3.53), and that points 2–4 appear in (or easily follow from) Proposition 2.22 and Theorem 2.23. ⊠

The closure operators that satisfy all the Hilbert-style g-rules of a logic are easily characterized:

Lemma 3.67
Let $\mathcal{L}$ be a logic, and let C be a closure operator on an arbitrary algebra $\boldsymbol{A}$, with associated closure system $\mathscr{C}$. Then C satisfies all Hilbert-style g-rules of $\mathcal{L}$ if and only if $\mathscr{C} \subseteq \mathcal{F}i_{\mathcal{L}}\boldsymbol{A}$.

Proof: Just use that $h\varphi \in \mathsf{C}h\Gamma$ if and only if for every $F \in \mathscr{C}$, if $h\Gamma \subseteq F$, then $h\varphi \in F$, and interchange two universal quantifiers. ⊠

A more thorough treatment of these issues appears in Chapter 5, where closure operators on arbitrary algebras are treated as models of a logic in a certain, technical sense. Later on in this section we will see how to proceed in the case of a very well-known property of a logic, the Deduction Theorem, as a first "taste" of the method.

The theory of algebraizability is one of the frameworks mentioned at the beginning of this section: it associates a class of algebras K with each algebraizable logic $\mathcal{L}$, namely its equivalent algebraic semantics. The link between the two is actually a link between two closure operators, $\mathsf{C}_{\mathcal{L}}$ and C_{K}, and this link is effected by means of two structural transformers, that is, by means of two sets of equations and formulas. Thus, it is natural to expect that many of the properties of a closure operator that can be syntactically expressed will transfer between them. We have already seen one: Corollary 3.38.1 states exactly a transfer theorem (point 1 in the result below), which can moreover be reformulated as a bridge theorem (point 2):

Proposition 3.68
Let $\mathcal{L}$ be an algebraizable logic with finitary transformers, and let K be its equivalent algebraic semantics.

1. *$\mathcal{L}$ is finitary if and only if C_{K} is finitary.*
2. *$\mathcal{L}$ is finitary if and only if K is a quasivariety.* ⊠

The following transfer theorems are also predictable:

PROPOSITION 3.69
Let $\mathcal{L}$ be a BP-algebraizable logic, with equivalent quasivariety K.

1. *$\mathcal{L}$ is finitely axiomatizable if and only if* K *has a finite presentation.*
2. *$\mathcal{L}$ is decidable if and only if the equational theory of* K *is decidable.*

PROOF: 1 has already appeared as Corollary 3.48. As to 2, we know that $\vdash_{\mathcal{L}} \varphi$ if and only if $\vDash_{\mathsf{K}} E(\varphi)$, that is, if and only if all equations in $E(\varphi)$ hold in K. Since E is finite, a decision procedure for the equational theory of K easily yields one for theoremhood of $\mathcal{L}$. And conversely, using that $\mathsf{K} \vDash \varphi \approx \psi$ if and only if $\vdash_{\mathcal{L}} \Delta(\varphi, \psi)$. ⊠

In this result, decidability is understood, as usual, as decidability of the theoremhood property; obviously, decidability of consequences from a finite number of assumptions is also equivalent to decidability of quasi-equations.

The most famous bridge theorems, which boosted research in abstract algebraic logic in the last twenty years of the twentieth century and were key steps in its development, can be *informally* described as the following correspondences between a property of a BP-algebraizable logic and a property of its equivalent algebraic semantics:

Deduction Theorem	$\Longleftrightarrow$	Equationally definable principal relative congruences
Local Deduction Theorem	$\Longleftrightarrow$	Relative congruence extension
Craig's Interpolation Theorem	$\Longleftrightarrow$	Amalgamation
Beth's Definability Theorem	$\Longleftrightarrow$	Epimorphisms are surjective

Some of these had been known for some time before abstract algebraic logic entered into the picture, and had been proved in an ad hoc way for restricted families of logics. For instance, since the 1970s, Maksimova has studied Beth-like properties and interpolation properties in large families of normal modal and intermediate logics [120, 121, 170], and Andréka, Németi and Sain have studied the last displayed correspondence in the framework of the algebraization of first-order logic with cylindric algebras [12, 221]. For a study of bridge theorems concerning the Beth property in the modern framework of abstract algebraic logic, see [28, 145], and for those concerning interpolation properties, see [74]. The systematic study of these correspondences with the tools of abstract algebraic logic has revealed that the first two bridge theorems mentioned work generally for algebraizable logics, while the last two work in the larger class of *equivalential logics* (when reformulated as concerning the semantics of matrices). Moreover, the two variants of the Deduction Theorem mentioned above satisfy other bridge theorems[†] in the even larger class of *protoalgebraic logics*.

[†] There are bridge and transfer theorems for other variants of the Deduction Theorem, such as the Contextual Deduction Theorem considered on page 174. Recently Raftery [205, 206] has obtained similar results for properties called "Inconsistency Lemma" and "Classical Inconsistency Lemma",

The rest of this section is devoted to the description of some aspects of the first correspondence stated above.

The classical Deduction Theorem

A logic $\mathcal{L}$ satisfies the ***classical Deduction-Detachment Theorem***, DDT, when for all $\Gamma \cup \{\varphi, \psi\} \subseteq Fm$,

$$\Gamma, \varphi \vdash_{\mathcal{L}} \psi \iff \Gamma \vdash_{\mathcal{L}} \varphi \to \psi,$$

where $\to$ is a binary connecive of "implication" belonging to the language of $\mathcal{L}$.[‡] Informally one speaks of "the Deduction Theorem" to refer to the whole equivalence, but the DDT denomination, coined by Blok and Pigozzi, emphasizes that the two halves of the theorem have a much distinct character: The DT (***Deduction Theorem***) is the $\Rightarrow$ half, while ***Detachment*** is the $\Leftarrow$ half. Symbolically one can write DDT $=$ DT $+$ (MP). While by Lemma 3.70 (MP) amounts to a Hilbert-style rule, by Theorem 3.75 the (DT) is not equivalent to any Hilbert-style rule, not even to a set of them.[§]

LEMMA 3.70
A logic $\mathcal{L}$ satisfies the property that for all $\Gamma \cup \{\varphi, \psi\} \subseteq Fm$, $\Gamma \vdash_{\mathcal{L}} \varphi \to \psi$ implies $\Gamma, \varphi \vdash_{\mathcal{L}} \psi$, if and only if it satisfies the rule (MP) $x, x \to y \rhd y$.

PROOF: For ($\Rightarrow$) take $\Gamma = \varphi \to \psi$, and for ($\Rightarrow$) use (C). ⊠

That $\mathcal{C\ell}$ satisfies the DT was known to Tarski as early as 1921, and was independently published by Herbrand in 1928. Tarski considered it as a kind of "metalogical axiom" characterizing classical logic (together with certain properties of negation, discussed on page 246).

How is it proved? For $\mathcal{C\ell}$, a *semantic proof* is trivial, because both the two-valued truth-table of $\to$ and the definition of $\vdash_{\mathcal{C\ell}}$ from the two-valued semantics implement essentially the same idea: that (for a given valuation) "either the antecedent is false or the succedent is true". However, a direct *syntactic proof* is usually given in standard introductory courses of mathematical logic, because it greatly simplifies the necessary work with the syntactic consequence in the way towards the Completeness Theorem. The proof is well known: starting from the assumption that $\Gamma, \varphi \vdash_{\mathcal{L}} \psi$, it works by induction on the length of a proof of ψ from $\Gamma \cup \{\varphi\}$, showing that for each step φ_i in such a proof, there is a proof of $\varphi \to \varphi_i$ from Γ. For this, one only uses the following theorems of $\mathcal{C\ell}$,

$$\text{(Ax1)} \qquad x \to (y \to x)$$

---

which are natural generalizations of the Properties of *Reductio ad Absurdum* of intuitionistic and classical logic, called PIRA and PRA in Chapter 5, respectively.

[‡] The same property, with $\to$ representing a binary term of the language, is included in the more general notion of a *uniterm* DDT; see Definition 3.76 and the comment after it.

[§] Some may take this to show that DT is a *truly metalogical* property. The "meta" refers here to the (perhaps old-fashioned) practice of calling the theorems *about* a logic $\mathcal{L}$ its *metatheorems*, in order to distinguish them from the theorems *of* $\mathcal{L}$ (the formulas φ such that $\vdash_{\mathcal{L}} \varphi$). In this sense, the DDT is clearly a metatheorem; but observe that the assertion "φ is a theorem of $\mathcal{L}$" is a metatheorem of $\mathcal{L}$!

$$\text{(Ax2)}\quad (x \to (y \to z)) \to ((x \to y) \to (x \to z)),$$

which are often included among its axioms (see page 71), and the rule (MP). A close look at the proof makes it clear that the same proof works for any axiomatically presented logic having these formulas as theorems and having (MP) as *the only proper rule*. Conversely, it is straightforward to show:

LEMMA 3.71
If $\mathcal{L}$ satisfies the classical DDT, *then* (Ax1) *and* (Ax2) *are theorems of $\mathcal{L}$, and $\mathcal{L}$ satisfies the rule* (MP). ⊠

THEOREM 3.72
A finitary logic $\mathcal{L}$ in a language containing $\to$ satisfies the classical DDT *if and only if $\mathcal{L}$ is an axiomatic extension of the logic defined in the language of $\mathcal{L}$ by the axioms* (Ax1) *and* (Ax2) *and the rule* (MP).

PROOF: Denote by $\mathcal{L}_0$ the logic defined in the statement; the classical proof outlined above shows that $\mathcal{L}_0$ satisfies the DDT. Now assume that $\mathcal{L}$ satisfies the DDT and take $\Gamma_1 := C_{\mathcal{L}}\varnothing$, the set of theorems of $\mathcal{L}$: we see that $\mathcal{L}$ is the axiomatic extension of $\mathcal{L}_0$ by Γ_1. By Lemma 3.71, $C_{\mathcal{L}_0}\Gamma \subseteq C_{\mathcal{L}}\Gamma$ for all $\Gamma \subseteq Fm$, so in particular $C_{\mathcal{L}_0}(\Gamma \cup \Gamma_1) \subseteq C_{\mathcal{L}}(\Gamma \cup \Gamma_1) = C_{\mathcal{L}}\Gamma$. For the reverse inclusion, take any $\varphi \in C_{\mathcal{L}}\Gamma$. If $\Gamma = \varnothing$, then $\varphi \in C_{\mathcal{L}}\varnothing = \Gamma_1 \subseteq C_{\mathcal{L}_0}(\Gamma \cup \Gamma_1)$. If $\Gamma \neq \varnothing$, by finitarity there are $\psi_1, \ldots, \psi_n \in \Gamma$, for some $n \geqslant 1$, such that $\varphi \in C_{\mathcal{L}}(\psi_1, \ldots, \psi_n)$. Using the DDT n times we obtain that $\psi_1 \to (\ldots \to (\psi_n \to \varphi)) \in C_{\mathcal{L}}\varnothing = \Gamma_1 \subseteq C_{\mathcal{L}_0}(\Gamma \cup \Gamma_1)$. Since the $\psi_i \in \Gamma$ and $\mathcal{L}_0$ satisfies (MP), using it n times we finally obtain that $\varphi \in C_{\mathcal{L}_0}(\Gamma \cup \Gamma_1)$. Thus, $C_{\mathcal{L}}\Gamma \subseteq C_{\mathcal{L}_0}(\Gamma \cup \Gamma_1)$ for all $\Gamma \subseteq Fm$, which completes the proof. Conversely, if $\mathcal{L}$ is an axiomatic extension of $\mathcal{L}_0$, there is a set $\Gamma_1 \subseteq Fm$ such that $C_{\mathcal{L}}\Gamma = C_{\mathcal{L}_0}(\Gamma \cup \Gamma_1)$ for all $\Gamma \subseteq Fm$. Using this, the DDT for $\mathcal{L}$ is a straightforward consequence of the DDT for $\mathcal{L}_0$. ⊠

COROLLARY 3.73
1. *The logic $\mathcal{IL}_{\to}$, the implication fragment of $\mathcal{IL}$, is the weakest logic in the language $\langle\to\rangle$ satisfying the classical* DDT.
2. *A finitary logic in a language containing $\to$ satisfies the classical* DDT *if and only if it has* (Ax1) *and* (Ax2) *among its theorems and can be axiomatized with* (MP) *as the only proper rule.* ⊠

Thus, the same, classical form of the DDT holds for a huge number of intermediate logics, including $\mathcal{IL}$, and even for weaker ones: actually, it does hold for all axiomatic expansions of $\mathcal{IL}_{\to}$, such as the local logics associated with normal modal logics, for instance, because they can be formulated with (MP) as their only rule of inference, in contrast to the global logics, which need the Necessitation Rule, or equivalent ones.

In order to show that the DDT (specifically: the DT) is not "equivalent" to any set of Hilbert-style rules, one needs to formulate the DDT as a possible property of a closure operator:

DEFINITION 3.74
Let C *be a closure operator on the universe of an algebra* $\boldsymbol{A}$ *whose type includes a binary operation* $\rightarrow$. C ***satisfies the classical Deduction-Detachment Theorem***, DDT, *when for any* $X \cup \{a,b\} \subseteq A$, $b \in C(X,a)$ *if and only if* $a \rightarrow b \in CX$.

THEOREM 3.75
There is no set of Hilbert-style g-rules such that for all closure operators C *on arbitrary algebras* $\boldsymbol{A}$, *satisfying that set is equivalent to satisfying the classical* DDT.

PROOF: Assume that such a set of rules exists. Then these rules are satisfied by any logic $\mathcal{L}$ with the DDT, and by Lemma 3.67 they are satisfied by all closure operators C on an algebra $\boldsymbol{A}$ such that $\mathscr{C} \subseteq \mathcal{F}i_{\mathcal{L}}\boldsymbol{A}$. But then, the assumption would imply that these closure operators would satisfy the DDT in the sense of Definition 3.74. An example shows that this is not the case. Take $\mathcal{I}\ell_{\rightarrow}$ as the logic, and the three-element algebra $\boldsymbol{A}$ where $A = \{0, 1/2, 1\}$ and the operation $\rightarrow$ is given by $a \rightarrow b := 1$ when $a \leqslant b$, and $a \rightarrow b := b$ otherwise. By Exercise 2.12, this turns $\boldsymbol{A}$ into an implicative algebra, and since $\mathcal{I}\ell_{\rightarrow}$ is an implicative logic, $\{1\} \in \mathcal{F}i_{\mathcal{I}\ell_{\rightarrow}}\boldsymbol{A}$. Therefore, we can consider $\mathscr{C} = \{\{1\}, A\} \subseteq \mathcal{F}i_{\mathcal{I}\ell_{\rightarrow}}\boldsymbol{A}$, and by the assumption the associated closure operator C should satisfy the DDT. But it is easy to check that $0 \in C\{1/2, 1\} = A$ while $1/2 \rightarrow 0 = 0 \notin C\{1\} = \{1\}$; this shows that C does not satisfy the DDT. ⊠

In Chapter 5 closure operators are treated as models of a logic in their own right; Exercise 5.7 asks you to show that, for finitary logics, satisfying the classical DDT is indeed equivalent to being a model of two natural Gentzen-style rules.

Other logics exhibit a DDT-like property similar to the classical one, but with a different, non-primitive term playing the role of $x \rightarrow y$:

- The global logic $S4^g$ associated with modal system $S4$ satisfies the following modified form of the DDT: $\Gamma, \varphi \vdash^g_{S4} \psi \iff \Gamma \vdash^g_{S4} \Box\varphi \rightarrow \psi$; this is easy to show semantically. The same holds for $S5$.
- Fuzzy logics with the so-called "Baaz-Monteiro Delta" connective $\triangle$ satisfy the same form, with $\triangle$ in the place of $\Box$ [21, Theorem 2.2.1, Example 3.2.7].
- Łukasiewicz's finitely-valued logics $Ł_n$ $(n \geqslant 2)$ satisfy

$$\Gamma, \varphi \vdash_n \psi \iff \Gamma \vdash_n \varphi \rightarrow^{n-1} \psi,$$

  where the terms $\rightarrow^n$ are defined recursively as follows:

$$x \rightarrow^0 y := y$$
$$x \rightarrow^{n+1} y := x \rightarrow (x \rightarrow^n y) \quad (n \geqslant 0)$$

  Recall that $Ł_2 = \mathcal{C}\ell$, and notice how for $n = 2$ the classical DDT is obtained. See [249, p. 278].
- The relevance logic $\mathcal{RM}$ satisfies: $\Gamma, \varphi \vdash_{\mathcal{RM}} \psi \iff \Gamma \vdash_{\mathcal{RM}} (\neg(\varphi \rightarrow \neg\psi) \vee (\varphi \rightarrow \psi)) \wedge (\neg\varphi \vee \psi)$. This was proved by Tokarz, see [70, p. 126].

It is thus natural to consider a more general formulation of this property.

The general Deduction Theorem and its transfer

DEFINITION 3.76
A set of formulas $I(x,y) \subseteq Fm$ in at most two variables is a **Deduction-Detachment (DD) set** *for a logic $\mathcal{L}$ when, for all $\Gamma \cup \{\varphi,\psi\} \subseteq Fm$,*

$$\Gamma, \varphi \vdash_{\mathcal{L}} \psi \iff \Gamma \vdash_{\mathcal{L}} I(\varphi,\psi).$$

A logic **satisfies the Deduction-Detachment Theorem**, DDT, *when it has a DD set.*

As with the classical form, the real DT is the $\Rightarrow$ half, while the $\Leftarrow$ half amounts to (MP) for $I(x,y)$, that is, to the Hilbert-style (g-)rule $x, I(x,y) \rhd y$.

All the natural logics in the literature known to satisfy the DDT do so for a DD set containing just one formula; when this happens and one wants to emphasize this fact, one speaks of the ***uniterm*** DDT. However, the abstract theory has raised the case in which the role of the "implication" connective is played collectively by a set of formulas. This is the general DDT appearing in some bridge theorems of abstract algebraic logic (and in the above definition); when one wants to emphasize this fact one speaks of a ***multiterm*** DDT.

One should not forget that, being a generalization, the multiterm DDT is a *weaker* property than the uniterm DDT. But observe that its weakness is in the (MP) half, not in the DT half. For, if $\mathcal{L}$ has the DDT for a set $I(x,y)$, then for each $\delta(x,y) \in I(x,y)$ one has that $\Gamma, \varphi \vdash_{\mathcal{L}} \psi$ implies $\Gamma \vdash_{\mathcal{L}} \delta(\varphi,\psi)$; that is, each formula in $I(x,y)$ satisfies the DT individually. If one of these formulas satisfies the (MP), then it already constitutes a uniterm DD set for $\mathcal{L}$. Thus, a truly multiterm DDT is one where the whole set $I(x,y)$ is needed in order to obtain (MP): $x, I(x,y) \vdash_{\mathcal{L}} y$; a (rather artificial) example where this is so can be found in [70, p. 170].

The key of the DDT is that it allows one to take a formula in the set of assumptions and, so to speak, make it "jump over the symbol $\vdash_{\mathcal{L}}$" and become part of the conclusion, though combined inside a more complex formula. Of course, if the set Γ is non-empty, this process can be iterated. Starting from a set $I(x,y)$ in two variables, the following family of sets, each with an increasing number of variables, can be inductively defined ($n \geqslant 1$):

$$\begin{aligned} I_0(x_0,y) &:= I(x_0,y) \\ I_n(x_0,\dots,x_n,y) &:= \bigcup\{I(x_0,\varphi) : \varphi \in I_{n-1}(x_1,\dots,x_n,y)\} \end{aligned} \tag{41}$$

If $I(x,y)$ is a single formula, say $x \to y$, then also $I_n(x_0,\dots,x_n,y)$ is a single formula, say $x_0 \to (\dots \to (x_n \to y)\dots)$. One routinely shows:

PROPOSITION 3.77
If $I(x,y)$ is a DD set for $\mathcal{L}$, then for all $n \geqslant 0$ and all $\Gamma \cup \{\varphi_0,\dots,\varphi_n,\psi\} \subseteq Fm$,

$$\Gamma, \varphi_0,\dots,\varphi_n \vdash_{\mathcal{L}} \psi \iff \Gamma \vdash_{\mathcal{L}} I_n(\varphi_0,\dots,\varphi_n,\psi). \qquad \boxtimes$$

In particular, putting $\Gamma = \emptyset$, we obtain that for all $\varphi_0,\dots,\varphi_n,\psi \in Fm$,

$$\varphi_0,\dots,\varphi_n \vdash_{\mathcal{L}} \psi \iff \vdash_{\mathcal{L}} I_n(\varphi_0,\dots,\varphi_n,\psi).$$

This emphasizes the main metalogical aspect of the DDT: It *reduces the consequence from assumptions in $\mathcal{L}$ to theoremhood in $\mathcal{L}$*.

Can a logic have several DD sets? Yes, but it is easy to see (Exercise 3.56):

LEMMA 3.78
Assume that $\mathcal{L}$ satisfies the DDT *with respect to a DD set $I(x,y)$, and let $J(x,y) \subseteq Fm$. Then $J(x,y)$ is a DD set for $\mathcal{L}$ if and only if $I(x,y) \dashv\vdash_{\mathcal{L}} J(x,y)$.* ⊠

Concerning the cardinality of the DD set, we have:

PROPOSITION 3.79
If $\mathcal{L}$ is finitary and satisfies the DDT, *then any DD set contains a finite DD set.*

PROOF: Left to you as Exercise 3.57. ⊠

The following characterization is essentially due to Czelakowski [68], and has its roots in Hilbert-Bernays' *Praemissentheorem* (1939).

PROPOSITION 3.80
A logic $\mathcal{L}$ satisfies the DDT *with respect to the DD set $I(x,y)$ if and only if the following conditions are satisfied:*

1. $\vdash_{\mathcal{L}} I(x,x)$.
2. $y \vdash_{\mathcal{L}} I(x,y)$.
3. $x, I(x,y) \vdash_{\mathcal{L}} y$.
4. *For all $\Gamma \cup \{\varphi, \psi\} \subseteq Fm$, if $\Gamma \vdash_{\mathcal{L}} \varphi$, then $\bigcup\{I(\psi,\gamma) : \gamma \in \Gamma\} \vdash_{\mathcal{L}} I(\psi,\varphi)$.*

PROOF: Left to you as Exercise 3.58. You will see that no induction on proofs is required (in particular, no assumption of a Hilbert-style presentation is needed); thus, it actually holds in the more abstract form given in the exercise. ⊠

In case $\mathcal{L}$ is finitary and presented by a Hilbert-style axiomatization, then this characterization can be strengthened by requiring the property in point 4 to hold just for each of the proper rules of the axiomatization; this is probably its most useful version. See Exercise 3.59.

The previous characterization is tailored so as to obtain a smooth proof that the DDT *transfers to every algebra*, in the following sense.

THEOREM 3.81
A logic $\mathcal{L}$ satisfies the DDT *with DD set $I(x,y)$ if and only if for every algebra $\boldsymbol{A}$ and all $X \cup \{a,b\} \subseteq A$,*

$$b \in Fg^{\boldsymbol{A}}_{\mathcal{L}}(X,a) \iff I^{\boldsymbol{A}}(a,b) \subseteq Fg^{\boldsymbol{A}}_{\mathcal{L}} X. \tag{42}$$

SKETCH OF THE PROOF: The particular case of (42) when $\boldsymbol{A}$ is the formula algebra is just the DDT, so in one direction there is nothing to prove. Now assume that $\mathcal{L}$ satisfies the said DDT, let $\boldsymbol{A}$ be an algebra, and fix some $X \cup \{a,b\} \subseteq A$. If we define $G := \{b \in A : I^{\boldsymbol{A}}(a,b) \subseteq Fg^{\boldsymbol{A}}_{\mathcal{L}} X\}$, then to prove (42) is to prove that $G = Fg^{\boldsymbol{A}}_{\mathcal{L}}(X,a)$, that is, that G is the least $\mathcal{L}$-filter of $\boldsymbol{A}$ containing $X \cup \{a\}$. The proof goes roughly as follows; the details are left to you as Exercise 3.60. That

$a \in G$ is a consequence of 3.80.1; that $X \subseteq G$ follows from 3.80.2; and that if $H \in \mathcal{F}i_{\mathcal{L}}\boldsymbol{A}$ is such that $X \cup \{a\} \subseteq H$, then $G \subseteq H$ follows from 3.80.3. The only involved part is to show that G is an $\mathcal{L}$-filter. For this, assume that $\Gamma \vdash_{\mathcal{L}} \varphi$ and let $h \in \mathrm{Hom}(\boldsymbol{Fm}, \boldsymbol{A})$ be such that $h\Gamma \subseteq G$. By Exercise 2.23 we can assume, without loss of generality, that there is $z \in V$ such that $hz = a$. From 3.80.4 it follows that $\bigcup\{I(z,\gamma) : \gamma \in \Gamma\} \vdash_{\mathcal{L}} I(z,\varphi)$. But $h\gamma \in G$ for each $\gamma \in \Gamma$, and this means that $hI(z,\gamma) = I^{\boldsymbol{A}}(a,h\gamma) \subseteq Fg_{\mathcal{L}}^{\boldsymbol{A}}X$, and therefore also $hI(z,\varphi) = I^{\boldsymbol{A}}(a,h\varphi) \subseteq Fg_{\mathcal{L}}^{\boldsymbol{A}}X$. This proves that $h\varphi \in G$ and hence that G is an $\mathcal{L}$-filter. ⊠

The DDT has been the object of extensive research in the literature. Porte's paper [197] is a survey of the work previous to 1980, and states some open problems. One of them is the lack of absolute negative results (i.e., stating that a certain logic does not satisfy *any form* of the DDT; of course, that some logics do not satisfy the classical DDT has always been known). These have finally arrived thanks to the bridge theorems obtained by Czelakowski in [68] and by Blok and Pigozzi in [33, 35].† We will see some of them.

The Deduction Theorem in algebraizable logics and its applications

The idea is simple: when two consequences are deductively equivalent through structural transformers, each can be expressed in terms of the other by means of certain syntactic transformations, and this makes the DDT to be preserved by these transformations in an almost mechanical way. Here only the case of equivalence between a logic and a consequence of the form C_{K} for some class K of algebras (that is, the algebraizability of Section 3.2) is treated.

It may be useful to denote equations by a single symbol in boldface, in a uniform manner, as in $\boldsymbol{x} := x_0 \approx x_1$,, $\boldsymbol{\alpha} := \alpha_0 \approx \alpha_1$, and so on. Privileging the view of equations as pairs of formulas, we realize that we can substitute two equations for the variables of a formula with *four* variables, and we can write $\varphi(\boldsymbol{x},\boldsymbol{y})$ instead of $\varphi(x_0,x_1,y_0,y_1)$, $\varphi(\boldsymbol{\alpha},\boldsymbol{\beta})$ instead of $\varphi(\alpha_0,\alpha_1,\beta_0,\beta_1)$, and the same in equations with four variables.

Definition 3.82
Let $P(\boldsymbol{x},\boldsymbol{y}) \subseteq Eq$ be a set of equations in four variables, and let C be a consequence (a structural closure operator) on the set of equations. The set P is a **DD set** *for C when for all $\Theta \cup \{\boldsymbol{\alpha},\boldsymbol{\beta}\} \subseteq Eq$,*

$$\Theta, \boldsymbol{\alpha} \vdash_C \boldsymbol{\beta} \iff \Theta \vdash_C P(\boldsymbol{\alpha},\boldsymbol{\beta}).$$

The consequence C ***satisfies the*** DDT *when it has a DD set.*

The simplified notation tries to make the formulation of the DDT more transparent, and its transfer more likely, but notice that syntactically this is

† Actually, it seems that Blok and Pigozzi's basic results had already been obtained around 1978, but they began to circulate some time later (some were published *much* later, and some are still *unpublished*; see footnote † on page 171), which explains that Porte didn't know of them in 1982.

more complicated: The elements of $P(x,y)$ are equations in the four variables x_0, x_1, y_0, y_1, and these may appear in either of the two formulas that constitute each equation. That is, $P(x,y) = \{\eta_j(x,y) : j \in J\}$, so that for each $j \in J$, $\eta_j(x,y)$ is an equation of the form $\eta_{j0}(x_0,x_1,y_0,y_1) \approx \eta_{j1}(x_0,x_1,y_0,y_1)$, and hence $\eta_j(\alpha,\beta) = \eta_{j0}(\alpha_0,\alpha_1,\beta_0,\beta_1) \approx \eta_{j1}(\alpha_0,\alpha_1,\beta_0,\beta_1)$.

The main *transfer theorem* concerning the DDT is the following.

THEOREM 3.83
Let $\mathcal{L}$ be an algebraizable logic with equivalent algebraic semantics K *and finitary structural transformers τ, ρ. The logic $\mathcal{L}$ satisfies the* DDT *if and only if* C_{K} *satisfies the* DDT.

PROOF: Let us put $\rho(x \approx y) = \Delta(x,y) = \{\delta_1(x,y),\ldots,\delta_m(x,y)\}$, consider a DD set $I(x,y) \subseteq Fm$ for $\mathcal{L}$, and let $\{I_n : n \geqslant 0\}$ be the sets constructed from $I(x,y)$ as in (41). Observe that for all $\varphi \in Eq$, $\rho\varphi = \{\delta_1(\varphi_0,\varphi_1),\ldots,\delta_m(\varphi_0,\varphi_1)\}$ is a set of m formulas, $m \geqslant 1$ since Δ is non-empty. Then, for all $\Theta \cup \{\alpha,\beta\} \subseteq Eq$,

$$\begin{aligned}
\Theta,\alpha \vDash_{\mathsf{K}} \beta &\underset{(1)}{\iff} \rho\Theta, \rho\alpha \vdash_{\mathcal{L}} \rho\beta \\
&\iff \rho\Theta, \{\delta_1(\alpha_0,\alpha_1),\ldots,\delta_m(\alpha_0,\alpha_1)\} \vdash_{\mathcal{L}} \delta_j(\beta_0,\beta_1) \ \text{ for all } j \\
&\underset{(2)}{\iff} \rho\Theta \vdash_{\mathcal{L}} I_{m-1}(\delta_1(\alpha_0,\alpha_1),\ldots,\delta_m(\alpha_0,\alpha_1),\delta_j(\beta_0,\beta_1)) \ \text{ for all } j \\
&\iff \rho\Theta \vdash_{\mathcal{L}} \bigcup_{j=1}^{m} I_{m-1}(\delta_1(\alpha_0,\alpha_1),\ldots,\delta_m(\alpha_0,\alpha_1),\delta_j(\beta_0,\beta_1)) \\
&\underset{(3)}{\iff} \tau\rho\Theta \vDash_{\mathsf{K}} \tau \bigcup_{j=1}^{m} I_{m-1}(\delta_1(\alpha_0,\alpha_1),\ldots,\delta_m(\alpha_0,\alpha_1),\delta_j(\beta_0,\beta_1)) \\
&\underset{(4)}{\iff} \Theta \vDash_{\mathsf{K}} P(\alpha,\beta)
\end{aligned}$$

where $j \in \{1,\ldots,m\}$. Step (1) holds by (ALG2), step (2) by Proposition 3.77, step (3) by (ALG1) and step (4) by (ALG4), in the last case by taking

$$P(x_0,x_1,y_0,y_1) := \tau \bigcup_{j=1}^{m} I_{m-1}(\delta_1(x_0,x_1),\ldots,\delta_m(x_0,x_1),\delta_j(y_0,y_1)).$$

This is certainly a set of equations in four variables, because τ is a structural transformer of formulas into (sets of) equations; hence the final equations will have the variables of the formulas that replace the variable x in the set τx.

This proves that if $\mathcal{L}$ satisfies the DDT, then so does C_{K}. The converse is proved in a similar way. ⊠

The interesting—and important—fact is that the DDT for a consequence of the form C_{K}, when K is a variety or a quasivariety, is equivalent to a purely algebraic property of the class K, thus turning this transfer theorem into a true bridge theorem, the most famous one in the recent history of abstract algebraic logic.

Recall from Section 1.6 that if a class K of algebras is closed under $\mathbb{I}$ and $\mathbb{P}_{\mathsf{SD}}$, then for any algebra $\boldsymbol{A}$ the set $\mathrm{Con}_{\mathsf{K}}\boldsymbol{A}$ is a closure system, with associated closure operator $\Theta_{\mathsf{K}}^{\boldsymbol{A}}$: if $X \subseteq A \times A$, then $\Theta_{\mathsf{K}}^{\boldsymbol{A}} X$ is the smallest K-congruence of $\boldsymbol{A}$ that contains X, that is, that identifies all the pairs in X. To follow tradition and to simplify notation, given $a, b \in A$, the congruence $\Theta_{\mathsf{K}}^{\boldsymbol{A}}\{\langle a, b\rangle\}$ is written as $\Theta_{\mathsf{K}}^{\boldsymbol{A}}(a, b)$; using the general terminology for closure systems (page 57), this is called the ***principal*** relative congruence generated by the pair $\langle a, b\rangle$, and is the smallest K-congruence of $\boldsymbol{A}$ that identifies a and b. If K is a variety and $\boldsymbol{A} \in \mathsf{K}$, then $\operatorname{Con}_{\mathsf{K}} \boldsymbol{A} = \operatorname{Con} \boldsymbol{A}$, because K is closed under quotients, and one can write $\Theta_{\mathsf{K}}^{\boldsymbol{A}}$ simply as $\Theta^{\boldsymbol{A}}$; if the algebra $\boldsymbol{A}$ is clear from the context, even the superscript may be omitted, particularly in examples.

DEFINITION 3.84
A quasivariety K *has* ***equationally definable principal relative congruences****,* EDPRC*, when there is a finite, non-empty set of equations in four variables* $\{\eta_i(x_0, x_1, y_0, y_1) : i = 1, \dots, m\}$ *such that for all* $\boldsymbol{A} \in \mathsf{K}$ *and all* $a, b, c, d \in A$*,*

$$\begin{aligned} c \equiv d \ \big(\Theta_{\mathsf{K}}^{\boldsymbol{A}}(a,b)\big) &\iff \boldsymbol{A} \vDash \eta_i(x_0, x_1, y_0, y_1)\, [\![a,b,c,d]\!] \ \text{for all } i \\ &\iff \eta_{i0}^{\boldsymbol{A}}(a,b,c,d) = \eta_{i1}^{\boldsymbol{A}}(a,b,c,d) \ \text{for all } i, \end{aligned} \tag{43}$$

where $i \in \{1, \dots, m\}$.

If K is a variety, then the word "relative" in the name of the property, the letter "R" in the acronym, and the subindex "K" in the formula, are dropped.

The EDPRC property has an undeniable "DDT flavour", if one thinks of $\Theta_{\mathsf{K}}^{\boldsymbol{A}}$ as a closure operator: If $\boldsymbol{A} \in \mathsf{K}$, then the relative congruence $\Theta_{\mathsf{K}}^{\boldsymbol{A}} \varnothing$ is the identity, and the equivalence (43) can be expressed as

$$c \equiv d \ \big(\Theta_{\mathsf{K}}^{\boldsymbol{A}}(a,b)\big) \iff \{\eta_i^{\boldsymbol{A}}(a,b,c,d) : i = 1, \dots, m\} \subseteq \Theta_{\mathsf{K}}^{\boldsymbol{A}} \varnothing, \tag{44}$$

which is a particular case of a DDT for the operator $\Theta_{\mathsf{K}}^{\boldsymbol{A}}$ with empty set of "side assumptions". It is not difficult to show that this is a consequence of the DDT for C_{K}, by proving an analogue of Theorem 3.81. That the converse entailment also holds is undoubtedly more difficult, and needs further universal algebra and abstract algebraic logic techniques. Anyway, it is important to record here that it is possible to prove the fundamental:

THEOREM 3.85
A quasivariety K *has* EDPRC *if and only if the consequence* C_{K} *satisfies the* DDT *for a finite DD set.* ⊠

The proof, due to Blok and Pigozzi, has not appeared in print.† Nevertheless, what is important here is the following bridge theorem that results as a consequence of Theorems 3.83 and 3.85:

COROLLARY 3.86
Let $\mathcal{L}$ *be a BP-algebraizable logic with equivalent algebraic semantics a quasivariety* K*. The logic* $\mathcal{L}$ *satisfies the* DDT *if and only if* K *has* EDPRC. ⊠

† It was to be published in [35], but Blok's untimely death deferred its publication indefinitely; it is possible to find a draft version in the internet. Some of the results were already reviewed in [33].

The interest of this result is three-fold:

- It allows one to use algebraic properties to show that a certain algebraizable logic has some DDT.
- More often, and more importantly, it allows one to use algebraic properties to show that a certain algebraizable logic does *not* have *any* DDT.
- Historically, it was the search for a proper context in which this equivalence between DDT and EDPRC could be properly formulated and proved that originated the notion of an algebraizable logic, as Pigozzi explains in [190].

It is nevertheless interesting to review some examples of how this result is applied. An important advantage of the bridge theorem is that EDPC is an intensively studied property in universal algebra literature, and is related to other properties which come into play in examples.

Some varieties with EDPC:

- In Boolean and Heyting algebras, principal congruences are characterized in this way: $c \equiv d\ (\Theta(a,b))$ if and only if $(a \leftrightarrow b) \wedge c = (a \leftrightarrow b) \wedge d$. This reflects the fact that $\mathcal{Cl}$ and $\mathcal{Il}$ have DDT. Actually, it is possible (Exercise 3.63) to "find" this expression by applying the algorithm implicit in the proof of Theorem 3.83 and then using (44).
- ***Closure algebras*** are Boolean algebras expanded with a unary operator $\Box$ such that the equations $\Box(x \wedge y) \approx \Box x \wedge \Box y$, $\Box\top \approx \top$, $\Box x \preccurlyeq x$ and $\Box^2 x \approx \Box x$ hold. They are the equivalent algebraic semantics of the global logic $S4^g$ associated with the normal modal system $S4$, which is algebraizable. Principal congruences are characterized in this way: $c \equiv d\ (\Theta(a,b))$ if and only if $\Box(a \leftrightarrow b) \wedge c = \Box(a \leftrightarrow b) \wedge d$. The logic satisfies the DDT (page 165). The same is true for ***monadic algebras***, the variety of closure algebras in which the equation $\Box(x \vee \Box y) \approx \Box x \vee \Box y$ holds, and which is the equivalent algebraic semantics of $S5^g$, an axiomatic extension of $S4^g$.
- The variety DL of distributive lattices has EDPC: $c \equiv d\ (\Theta(a,b))$ if and only if $(a \wedge b) \wedge c = (a \wedge b) \wedge d$ and $(a \vee b) \wedge c = (a \vee b) \wedge d$. This is an example where EDPC seems to require more than one equation. DL is not the equivalent algebraic semantics of any algebraizable logic (this is proved in Example 4.61); therefore, the fact that it has EDPC does not imply that some related logic has the DDT. However, in Example 3.87 we see that DL is the equivalent algebraic semantics of an algebraizable Gentzen system $\mathfrak{G}_{\wedge\vee}$. *Mutatis mutandis*, there are results parallel to Theorem 3.83, and hence to Corollary 3.86, for algebraizable Gentzen systems, therefore $\mathfrak{G}_{\wedge\vee}$ satisfies a form of the DDT for the consequence relation between sequents. This means that the derivability of sequents from other sequents can be reduced to the absolute derivability of sequents in this calculus.

If K is a discriminator variety, then K has EDPC; therefore, algebraizable logics whose equivalent algebraic semantics is a discriminator variety satisfy the DDT.

In general, the ***ternary discriminator function*** on an algebra $\boldsymbol{A}$ is the function $t\colon A \times A \times A \to A$ defined as follows:

$$t(a,b,c) := \begin{cases} c & \text{if } a = b, \\ a & \text{if } a \neq b. \end{cases}$$

A variety K is a ***discriminator variety*** when $\mathsf{K} = \mathbb{V}\mathsf{K}'$ for a class K' where the ternary discriminator function is (uniformly) a term function: $t = \alpha^{\boldsymbol{A}}$ for some $\alpha(x,y,z)$. One can show that if $\boldsymbol{A} \in \mathsf{K}$, then $c \equiv d \;(\boldsymbol{\Theta}(a,b))$ if and only if $\alpha^{\boldsymbol{A}}(a,b,c) = \alpha^{\boldsymbol{A}}(a,b,d)$ for all $a,b,c,d \in A$.

Besides Boolean algebras, also monadic algebras, relation algebras, the varieties $\mathbb{V}\{[\mathbf{0,1}]_n\}$ are discriminator varieties. Since $Ł_n$ is algebraizable with equivalent algebraic semantics the variety $\mathbb{V}\{[\mathbf{0,1}]_n\}$ (this is shown in Example 5.83), this explains that $Ł_n$ has the DDT; its form is given on page 165.

The negative results are most often established using algebraic consequences of EDP(R)C. Notice that a negative result means that the logic in question does not satisfy the DDT for *any* DD set.

- EDP(R)C is a particular case of *having first-order definable principal (relative) congruences*. One consequence of this is that the classes of the (relatively) simple and of the (relatively) subdirectly irreducible algebras in K are closed under ultraproducts. This is a property that can be tested in particular examples:
  - $\mathcal{BCK}$ does not have any DDT, because one can find a family of relatively simple BCK-algebras (recall that these form a quasivariety, not a variety) with a non-relatively simple ultraproduct.
  - The global logics associated with the normal modal systems K and T do not have any DDT, because one can find a family of simple modal algebras having a non-simple ultraproduct.
- EDP(R)C implies (R)CEP, the (*relative*) *congruence extension property*, which says that if $\boldsymbol{A}, \boldsymbol{B} \in \mathsf{K}$ with $\boldsymbol{B} \leqslant \boldsymbol{A}$ and $\theta \in \operatorname{Con}\boldsymbol{B}$ (resp. $\operatorname{Con}_{\mathsf{K}}\boldsymbol{B}$), then $\theta = \theta' \cap (B \times B)$ for some $\theta' \in \operatorname{Con}\boldsymbol{A}$ (resp. $\operatorname{Con}_{\mathsf{K}}\boldsymbol{A}$); that is, any (relative) congruence on a subalgebra can be "extended" to a (relative) congruence of the whole algebra. In particular, this implies that all subalgebras of a (relatively) simple algebra in K should be (relatively) simple.

  Using this, one can show that quantum logic (Example 3.64) does not have any DDT at all: one exhibits a simple orthomodular lattice with a non-simple subalgebra, which makes CEP, and hence EDPC, impossible.
- EDP(R)C implies (*relative*) *congruence distributivity*, the property that for any $\boldsymbol{A} \in \mathsf{K}$, the lattice $\operatorname{Con}_{\mathsf{K}}\boldsymbol{A}$ is distributive. This implication (in the case of varieties) had been an important open problem in universal algebra for some time, and was solved by Köhler and Pigozzi in 1980 thanks to the logical intuitions connecting EDPC with the DDT. The distributivity of lattices of congruences and related lattices is an intensively studied property in universal algebra. Already in his [68], Czelakowski found connections between the DDT

and the distributivity of certain lattices of theories or of $\mathcal{L}$-filters, in a wider context than algebraizability; see in Corollary 6.30.

This property has been used to show that $\mathcal{C\ell}_{\leftrightarrow}$, the equivalence fragment of $\mathcal{C\ell}$ (Example 3.34.8), does not have any DDT. This logic is algebraizable, with equivalent algebraic semantics the class of Boolean groups, and this variety is known not to be congruence distributive. Hence, it cannot have EDPC, and the logic cannot have the DDT.

You will find more details in [35, 70].

Weak versions of the Deduction Theorem

In the literature some truly weakened versions of the DDT have been investigated, in connection with bridge theorems of several kinds. Most of the logics mentioned above as not having the DDT satisfy one of these weaker forms.

- The oldest idea is to make it *local* in the sense that a family $\{I_j(x,y) : j \in J\}$ of sets is involved (instead of a single DD set) such that in each instance of the theorem only one of the sets is used: For all $\Gamma \cup \{\varphi, \psi\} \subseteq Fm$,

$$\Gamma, \varphi \vdash_{\mathcal{L}} \psi \iff \Gamma \vdash_{\mathcal{L}} I_j(\varphi, \psi) \text{ for some } j \in J.$$

  There is a large group of paradigmatic examples: Almost all substructural logics studied in [123, Corollary 2.15] (namely, all axiomatic extensions of $\mathcal{FL}_e$, Lambek's logic with exchange) have a Local DDT with respect to the same family of (one-element) DD sets. In general the expression of these sets is complicated, but for logics that are reasonably behaved it may reduce to the following: For all $\Gamma \cup \{\varphi, \psi\} \subseteq Fm$,

$$\Gamma, \varphi \vdash_{\mathcal{L}} \psi \iff \Gamma \vdash_{\mathcal{L}} \varphi \to^n \psi \text{ for some } n \geqslant 0$$

  where $x \to^n y$ is the term introduced on page 165. The two logics in this group for which this form of the Local DDT had been early identified are $\mathcal{BCK}$ and the finitary version $\text{Ł}_{\infty f}$ of Łukasiewicz's Ł_∞.

  Another well-known logic with a Local DDT is the global logic K^g associated with the normal modal system K: For all $\Gamma \cup \{\varphi, \psi\} \subseteq Fm$,

$$\Gamma, \varphi \vdash^g_K \psi \iff \Gamma \vdash^g_K (\varphi \wedge \Box\varphi \wedge \ldots \Box^n \varphi) \to \psi \text{ for some } n \geqslant 0.$$

  This kind of DDT was first investigated in general by Czelakowski in his [69], obtaining a first bridge theorem; later on also Blok and Pigozzi studied it in their [33], and they proved the bridge theorem mentioned on page 162 connecting the Local DDT to the property RCEP described above. Using this theorem, and some algebraic properties of orthomodular lattices, they showed that quantum logic, already mentioned as not having a DDT, has no Local DDT either.

- One can *add parameters* to the formulas in the (single) DD set in such a way that each instance of the theorem uses particular values of these parameters. The DD set takes the form $I(x, y, \vec{z})$, and the property becomes: For all

$\Gamma \cup \{\varphi, \psi\} \subseteq Fm$,

$$\Gamma, \varphi \vdash_{\mathcal{L}} \psi \iff \Gamma \vdash_{\mathcal{L}} I(\varphi, \psi, \vec{\delta}) \text{ for some } \vec{\delta} \in \vec{Fm}.$$

- Of course, one can combine the two previous weakenings and obtain the notion of a *Parametrised Local* DDT; this is an important property in abstract algebraic logic, for theoretical reasons, and is studied in detail in Chapter 6, where several examples are given.
- The most recent addition to this family is the *Contextual* DDT (CDDT), in which the DDT-like property is not applied uniformly to the relevant formulas, but in a way that *depends on the variables* occurring in them. Here one assumes the existence of a family of sets $\{I_n(x_0, \dots, x_{n+1}) : n \geqslant 1\}$, such that for all $n \geqslant 1$ and all $\Gamma \cup \{\varphi, \psi\} \subseteq Fm$ with $\mathrm{Var}(\Gamma, \varphi, \psi) \subseteq \{x_0, \dots, x_{n-1}\}$,

$$\Gamma, \varphi \vdash_{\mathcal{L}} \psi \iff \Gamma \vdash_{\mathcal{L}} I_n(x_0, \dots, x_{n-1}, \varphi, \psi).$$

  This property has been introduced and studied by Raftery in [203]. The CDDT, which has also a local version, enjoys a natural transfer theorem and several bridge theorems: The one for BP-algebraizable logics is similar to Corollary 3.86, with EDPRC replaced by the property of having "equationally semi-definable" principal relative congruences, and the one for finitary protoalgebraic logics is similar to (part of) Theorem 6.28, with the same algebraic property but limited to finitely generated algebras.
- In all the previous variants of the DDT, a fixed set of *side assumptions* (the "Γ") is present and remains in place in each application of the theorem. A further weakened form of the DDT is one in which all side assumptions are necessarily moved to the other side of the symbol $\vdash_{\mathcal{L}}$. Here one assumes directly the existence of a family of sets $\{I_n(x_0, \dots, x_n, y) : n \geqslant 0\}$, not necessarily constructed from a basic set $I(x, y)$ as in (41), such that for all $n \geqslant 0$ and all $\varphi_0, \dots, \varphi_n, \psi \in Fm$,

$$\varphi_0, \dots, \varphi_n \vdash_{\mathcal{L}} \psi \iff \vdash_{\mathcal{L}} I_n(\varphi_0, \dots, \varphi_n, \psi).$$

  This property has been called the *Graded* DDT in [109], and the *General* DT in [220]; it turns out to have surprising connections with several issues in abstract algebraic logic, but which are not exactly bridge theorems.
- The last weakening to consider goes in quite a different direction: what it does is to *limit the cardinality* of the set Γ of side assumptions. One would have DDTs of this form: For all $\Gamma \cup \{\varphi, \psi\} \subseteq Fm$ such that $|\Gamma| \leqslant n$,

$$\Gamma, \varphi \vdash_{\mathcal{L}} \psi \iff \Gamma \vdash_{\mathcal{L}} \varphi \to \psi,$$

  for a fixed $n \in \omega$ associated with the logic. This has been algebraically studied in [47] in connection with strengthened versions of *Modus Ponens*; one of the logics discussed in this paper in relation with the case where $n = 1$ appears in Example 6.78.2. The extreme case where $n = 0$, that is, the property that for all $\varphi, \psi \in Fm$,

$$\varphi \vdash_{\mathcal{L}} \psi \iff \vdash_{\mathcal{L}} \varphi \to \psi,$$

has been studied by Wójcicki under the name *Weak Deduction Theorem*, WDT, see [249, p. 47].

Exercises for Section 3.6

EXERCISE 3.52. Prove Lemmas 3.70 and 3.71.

EXERCISE 3.53. Prove that for any logic $\mathcal{L}$ and any $\Gamma \cup \{\varphi\} \subseteq Fm$, $\Gamma \vdash_{\mathcal{L}} \varphi$ if and only if $h\varphi \in Fg_{\mathcal{L}}^{\boldsymbol{A}} h\Gamma$ for all $h \in \mathrm{Hom}(\boldsymbol{Fm}, \boldsymbol{A})$ and all algebras $\boldsymbol{A}$.

EXERCISE 3.54. If you have never worked out in detail a proof of the ordinary DDT for a syntactic formulation of $\mathcal{C}\ell$, do it, and realize that this amounts to proving the right-to-left half of Theorem 3.72 (see the comments before Lemma 3.71).

EXERCISE 3.55. Prove the following "finitized" version of Theorem 3.75: There is no set of Hilbert-style rules such that for all finitary closure operators C on arbitrary algebras $\boldsymbol{A}$, satisfying that set is equivalent to satisfying the DDT.

EXERCISE 3.56. Assume that a logic $\mathcal{L}$ satisfies the DDT with respect to a DD set $I(x,y)$, and let $J(x,y) \subseteq Fm$. Show that $\mathcal{L}$ satisfies the DDT with respect to the set $J(x,y)$ if and only if $I(x,y) \dashv\vdash_{\mathcal{L}} J(x,y)$.

EXERCISE 3.57. Show that if $\mathcal{L}$ is a finitary logic with the DDT, and $I(x,y)$ is a DD set for $\mathcal{L}$, then there is a finite $I_0(x,y) \subseteq I(x,y)$ such that $I_0(x,y)$ is a DD set for $\mathcal{L}$.

HINT. Apply finitarity to (MP).

EXERCISE 3.58. Let C be a closure operator on an algebra $\boldsymbol{A}$ and $I(x,y) \subseteq Fm$. Prove that the following conditions are equivalent:

(i) C satisfies the DDT with respect to I, in the sense that for all $X \cup \{a,b\} \subseteq A$, $b \in C(X,a)$ if and only if $I^{\boldsymbol{A}}(a,b) \subseteq CX$.

(ii) The following conditions are satisfied, for all $X \cup \{a,b\} \subseteq A$:

1. $I^{\boldsymbol{A}}(a,a) \subseteq C\varnothing$.
2. $I^{\boldsymbol{A}}(a,b) \subseteq C\{b\}$.
3. $b \in C(I^{\boldsymbol{A}}(a,b), a)$.
4. If $b \in CX$, then $I^{\boldsymbol{A}}(a,b) \subseteq C\bigcup\{I^{\boldsymbol{A}}(a,c) : c \in X\}$.

EXERCISE 3.59. Confirm that the comment after Proposition 3.80 is right.

EXERCISE 3.60. Supply the details in the proof of Theorem 3.81, following the sketch given in the text. In particular, check carefully that there is no cheating in the application of Exercise 2.23, and that in case of doubt, instead of just saying "we can assume, without loss of generality, that…", it is possible to apply the exercise in its full formulation and carry on with the proof.

EXERCISE 3.61. Give an alternative proof of Theorem 3.81 for a finitary $\mathcal{L}$, where the two halves of (42) are shown separately, and ($\Rightarrow$) is shown inductively, from the description of $Fg_{\mathcal{L}}^{\boldsymbol{A}}(X,a)$ given in Theorem 2.23.

EXERCISE 3.62. Justify the necessity of assuming the finiteness of the transformers in Theorem 3.83. Observe that by contrast no finitarity of the logics is assumed: check that this is right.

EXERCISE 3.63. Using algebraizabilit and the DDTs of the corresponding logics, and following the indications in the text (page 171), prove the following characterizations of principal congruences:

1. $c \equiv d \ (\boldsymbol{\Theta}(a,b))$ if and only if $(a \leftrightarrow b) \wedge c = (a \leftrightarrow b) \wedge d$ in Heyting algebras and also in Boolean algebras.
2. $c \equiv d \ (\boldsymbol{\Theta}(a,b))$ if and only if $\Box(a \leftrightarrow b) \wedge c = \Box(a \leftrightarrow b) \wedge d$ in monadic algebras and in closure algebras.

3.7 Generalizations and abstractions of algebraizability

The tools and perspective taken in Section 3.5 to handle the Isomorphism Theorems in a highly lattice-theoretic way show that algebraizability is a notion with a clear potential to be generalized beyond its original framework of sentential logics. There is no space in this book to go into the details; to give a summary, the process of generalization and abstraction can be described as a progression through several steps.

Step 1: Algebraization of other sentential-like logical systems

Soon after the appearance of [32] it was clear to some algebraic logicians that algebraizability was essentially a property relating two consequence relations (namely, that of a sentential logic and the relative equational consequence of certain class of algebras) by means of a pair of transformers. It is not surprising that by 1991 the idea of algebraizing Gentzen systems had been introduced (by several members of the Barcelona school†). The core idea is to transform sequents into (sets of) equations and conversely. This has been particularly successful for logics without an implication connective in their language. Here is the first, simplest example.

EXAMPLE 3.87 (THE CONJUNCTION AND DISJUNCTION FRAGMENT OF $\mathcal{C}\ell$)
Let $\boldsymbol{L} = \langle \wedge, \vee \rangle$ be a language of type $\langle 2,2 \rangle$. From Exercise 1.27 it follows that $\mathcal{C}\ell_{\wedge\vee}$, the fragment of classical logic with only conjunction and disjunction, is determined by the two-valued truth-tables of Example 1.11, considering the reduct $\mathbf{2}_{\wedge\vee}$ of $\mathbf{2}$ with only conjunction and disjunction; thus, checking (non-)consequence relations in this logic is trivial.

Now recall from page 8 that Seq° denotes the set of sequents (here, on this language) of the form $\Gamma \rhd \varphi$ where $\Gamma \subseteq Fm$ is *finite* and *nonempty*. Consider the Gentzen system $\mathfrak{G}_{\wedge\vee}$ defined on Seq° by the calculus that has the usual structural

† The idea was first presented in a *Semester on Algebraic Logic* held at the Banach Centre of the Polish Academy of Sciences in the Fall/Winter of 1991; the first publication on this trend is [212].

rules plus the following ones:

$$\frac{\Gamma \rhd \varphi \qquad \Gamma \rhd \psi}{\Gamma \rhd \varphi \wedge \psi} \qquad\qquad \frac{\Gamma, \alpha, \beta \rhd \varphi}{\Gamma, \alpha \wedge \beta \rhd \varphi}$$

$$\frac{\Gamma \rhd \varphi}{\Gamma \rhd \varphi \vee \psi} \qquad \frac{\Gamma \rhd \psi}{\Gamma \rhd \varphi \vee \psi} \qquad \frac{\Gamma, \alpha \rhd \varphi \qquad \Gamma, \beta \rhd \varphi}{\Gamma, \alpha \vee \beta \rhd \varphi}$$

where Γ is any finite set of formulas, and $\alpha, \beta, \varphi, \psi$ are arbitrary formulas. Denote by $\vdash_{\mathfrak{G}_{\wedge\vee}}$ the consequence relation induced on the set of sequents by the relation "a sequent is derivable from a set of sequents using these rules". This is a structural closure relation on Seq°. It was introduced and studied, from the present point of view, in [117]. It is related to $\mathcal{C\ell}_{\wedge\vee}$ in several ways; notably, that for any $\Gamma \cup \{\varphi\} \subseteq Fm$ with Γ finite and non-empty,

$$\emptyset \vdash_{\mathfrak{G}_{\wedge\vee}} \Gamma \rhd \varphi \iff \Gamma \vdash_{\mathcal{C\ell}_{\wedge\vee}} \varphi. \tag{45}$$

Since $\mathcal{C\ell}_{\wedge\vee}$ is finitary and has no theorems, this equivalence completely recovers its consequence relation from the Gentzen system. Another relation is that $\mathcal{C\ell}_{\wedge\vee}$ is characterized by the above set of rules in the sense that it is the weakest logic satisfying them.†

We know (page 118) that the logic $\mathcal{C\ell}_{\wedge\vee}$ is not algebraizable. But the Gentzen system $\mathfrak{G}_{\wedge\vee}$ is so, in the following sense. Consider the transformers

$$\begin{array}{llcll} \tau: & x_1,\dots,x_n \rhd y & \longmapsto & x_1 \wedge \cdots \wedge x_n \preccurlyeq y & \\ & \{x \rhd y\,,\ y \rhd x\} & \longleftarrow & x \approx y & : \rho \end{array} \tag{46}$$

where, in general, $\alpha \preccurlyeq \beta$ is shorthand for the equation $\alpha \approx \alpha \wedge \beta$, and some association convention is fixed for the binary connective $\wedge$. Clearly these transformers are structural, and transform every sequent in Seq° into an equation, and hence every set of such sequents into a set of equations; and every equation and set of equations into a set of sequents in Seq°.

Let DL denote the class of distributive lattices (it is a variety). It is not difficult† to prove that the relations between the consequences $\vdash_{\mathfrak{G}_{\wedge\vee}}$ and $\vDash_{\mathsf{DL}}$ satisfy the properties parallel to (ALG1)–(ALG4) but formulated for these two transformers, *mutatis mutandis*. It is in this sense that one can say that the Gentzen system $\mathfrak{G}_{\wedge\vee}$ is "algebraizable" and that DL is its "equivalent algebraic semantics". ⊠

† This property, which is essentially equivalent to (45), is obtained in Example 5.120.3 using typical tools of abstract algebraic logic. The same property holds for intuitionistic logic (therefore, the conjunction-disjunction fragments of the two logics coincide), as was first proved in [195] using relational semantics.

† Proving part ($\Rightarrow$) of (ALG1), and (ALG4), involves only elementary properties of order in distributive lattices. Checking this will allow you to realize what is going on with this kind of "algebraizability". The distributive lattices needed in order to prove part ($\Leftarrow$) of (ALG1) by a Lindenbaum-Tarski-style process (where a class of algebras has to be produced) are quotients of the formula algebra (not of the set of sequents!) by a congruence associated with each theory of the sequent system in a way similar to Definition 3.24. An alternative proof is to use generalized matrices (Chapter 5) and apply Proposition 5.111.

The development of a general theory of the algebraization of Gentzen systems is parallel to that of the algebraization of sentential logics, and only notationally more complicated. Perhaps the main difference is the additional complication of the homologue of Proposition 3.3: A structural transformer from sequents into (sets of) equations is not determined by a single set of equations, but by an infinite family of such sets, one for each value of n in (46), with an increasing number of variables. In order for the Isomorphism Theorem to make sense, an appropriate notion of a "filter" of the Gentzen system on an arbitrary algebra has to be introduced; the idea, similar to that used in Theorem 1.76 to characterize relative congruences as the "filters" of $\vDash_{\mathsf{K}}$, basically amounts to the notion of a "finite closure relation" briefly discussed on page 311. This general theory was started in the unpublished [214], and has been pursued further, and in a more modern presentation, in [200].

In a completely parallel way, one can define and study the notion of algebraizability of the other sentential-like logical systems that are reviewed at the end of Section 1.2, such as m-sided Gentzen systems or hypersequent systems. One just needs to consider the appropriate transformers of the relevant syntactic objects into sets of equations, and conversely, and the corresponding notion of a "filter". In all these cases, results analogous to most of the general theorems obtained in this chapter, including the Isomorphism Theorems, are obtained.

Step 2: The notion of deductive equivalence

The next step in this route of progressive generalization and abstraction of the notion of algebraizability is the recognition that the four conditions (ALG1)–(ALG4) need not involve a sentential-like logic on one side and a relative equational consequence on the other side: If C and D are two structural closure operators on different sets of linguistic expressions (formulas, equations, sequents, hypersequents, etc.), and $\boldsymbol{\tau}$ and $\boldsymbol{\rho}$ are structural transformers between their respective languages, then one can say that C and D are ***deductively equivalent*** when the analogues of those four conditions hold. Thus, in this step the focus is still on the transformers, that is, it is "language based". This idea was first considered by Blok and Pigozzi themselves, starting in [33], in the more restricted context of equivalence between two finite-dimensional deductive systems, possibly of different dimension,[†] and was further extended to systems that are not necessarily formulated in the same similarity type by Gyuris in his unpublished [131].

Algebraizability becomes the particular case of deductive equivalence between some (sentential-like) logical system and an equational consequence relative to a class of algebras, and can be mixed with the other cases. The following interesting

[†] The 1-dimensional systems are the sentential logics, whose language is Fm, and the 2-dimensional systems are the consequences of the form $\vDash_{\mathsf{K}}$, whose language is $Fm \times Fm$. A unified treatment of the two kinds of logics was achieved by considering n-dimensional systems, that is, structural closure relations on the set Fm^n for an arbitrary, fixed n.

observation appears: since the new notion of deductive equivalence is obviously transitive, if two closure operators are deductively equivalent in this sense, then one of them is algebraizable if and only if the other one is.

Two particular cases deserve to be highlighted.

- **The deductive equivalence between two sentential logics.** This notion is stronger than other common notions of equivalence between logics, notably that of *definitional equivalence*, which usually applies to sentential logics formulated in different languages. A typical example at hand is the relation between the logics $\mathcal{J}_3$ and $Ł_3$ of Example 3.20: As seen there, both are algebraizable, through different transformers, with respect to the same class of algebras, and the composition of two of the transformers produces transformers between their respective formula algebras that also satisfy the properties (ALG1)–(ALG4) as shown in (30) on page 122. That is, the two sentential logics are deductively equivalent in the above sense.
- **The deductive equivalence between a Gentzen system and a sentential logic.** This provides a novel way of expressing relations between a Gentzen system and a sentential logic, going further than the fact that the derivable sequents correspond to consequences of the logic as in (45). For instance, this notion has helped in the clarification of the landscape of relevance logics and substructural logics in general, in which it is fairly usual that the same logical system is formalized both as an axiomatically presented sentential logic and as a Gentzen calculus. Here the transformers would send formulas to (sets of) sequents and conversely; one of the more common pairs of transformers is ($n \geqslant 0$):

$$\begin{array}{lcccl} \tau: & x_1,\dots,x_n \rhd y & \longmapsto & x_1 \to (x_2 \to \dots \to (x_n \to y)\dots) & \\ & \varnothing \rhd x & \longleftarrow & x & : \rho \end{array} \tag{47}$$

  Since here sequents with an empty antecedent are admitted, the definition of τ should be understood as including the inverse of ρ as a degenerate case. In [200], these transformers are used to establish the deductive equivalence between the sentential and the sequential versions of the logic $\mathcal{BCI}$; as mentioned in Example 3.34.3, the former is not algebraizable, hence its sequent version is not algebraizable either, but still the two logical systems are deductively equivalent in this sense.

The described situation had been (partially) considered earlier in the literature, when dealing with the issue of how a (sentential) logic should be associated with a Gentzen system, without a relation to algebraizability. If $\mathfrak{G}$ is a Gentzen system and $\mathcal{L}$ is the logic such that for all $\Gamma \cup \{\varphi\} \subseteq Fm$,

$$\Gamma \vdash_{\mathcal{L}} \varphi \iff \rho\Gamma \vdash_{\mathfrak{G}} \rho\varphi,$$

where ρ is the transformer in (47), then $\mathcal{L}$ is called the *external logic* of $\mathfrak{G}$. By contrast, when the relation follows the pattern of (45), that is, when for finite $\Gamma \cup \{\varphi\} \subseteq Fm$,

$$\Gamma \vdash_{\mathcal{L}} \varphi \iff \varnothing \vdash_{\mathfrak{G}} \Gamma \rhd \varphi,$$

then $\mathcal{L}$ is called the *internal logic* of $\mathfrak{G}$. See Example 1.10 and the comment that follows it.

While in this step the possible applications of the notion of deductive equivalence to many specific cases are clearly recognized, its level of generality does not help to develop a whole theory parallel to that of algebraizability (for instance, including characterizations via Isomorphism Theorems). Formally speaking, the different syntactic structure of the languages involved requires a different parallel formalization in each case; for instance, [200] develops it in the case of equivalence between Gentzen systems, which is general enough for its purposes because sentential logics and relative equational consequences can be viewed as particularly simple cases of Gentzen systems. But its results do not encompass, for instance, the algebraizability of hypersequents. A fully general setting requires us to move up to a further level of abstraction.†

Step 3: Equivalence of structural closure operators

The first *truly abstract* study of the notion of algebraizability arose from a reflection on *the Isomorphism Theorems* (Theorem 3.58 and related ones) and on the role of substitutions. In the ordinary setting, the logical consequence relations are structural and the transformers are structural, in both cases in the sense of commuting with substitutions; in the case of the formula algebra, the substitutions are the endomorphisms, but in the case of equations, Gentzen systems and other syntactic objects derived from formulas, the "substitutions" are the actions on these objects induced by the endomorphisms of the formula algebra (see pages 28 and 49). But substitutions also act on the two lattices of theories when coupled with the closure operators, inducing internal operations on these lattices, and therefore the isomorphisms commute with respect to these operations, in the sense expressed in (35) on page 149 and in Theorem 1.56; that is, they are isomorphisms of the expanded lattices of theories.

Blok and Jónsson realized that the key point in all this is the fact that there is a monoid acting on a set, and that this monoid expands the lattice of closed sets of a closure operator by additional unary operations whenever the closure operator is "structural" in the sense that it commutes with the action of the monoid; this is how structurality was discussed in Section 1.4. Thus the setting can be recreated without assuming any further (linguistic) structure in the two base sets on which two closure operators are considered. Moreover, they thought that it is *the Isomorphism Theorems that capture the abstract essence of algebraizability*; accordingly, in [29] the focus of the general theory is reversed: if two closure operators on two different sets are structural with respect to the actions of a common monoid on each set, then they are *defined* to be ***equivalent*** when there exists an isomorphism between the respective lattices of closed sets that commutes with the actions that

† EXERCISE in philosophy of mathematics: can you perceive a difference between *generalization* and *abstraction*?

the monoid induces in these lattices (notice that Theorem 1.56 still holds in this setting, thus giving two different ways of expressing this commutativity).

In this framework, then, something that in the initial theory is an important result (the Isomorphism Theorem) becomes the definition of the abstract notion. Then, a new and challenging problem emerges: how to formulate the additional abstract conditions under which one can recover (an abstract version of) the initial definition, namely, that the isomorphism is *induced by certain transformers* (between the universes of the closure operators) that commute with the monoid actions (the "structural" transformers).

If you are interested in the history of mathematics, you will recognize this is a relative common phenomenon in the evolution of a subject; its emergence here indicates the relevance of these abstract notions.

Step 4: Getting rid of points

This is also a typical jump in the abstraction process of other areas in mathematics. In this case, it concerns the fact that, while closure operators are functions directly defined on the power sets of some base sets, the substitutions are functions on these base sets, and the transformers are functions from the base sets to the power sets, but all are extended to the power sets by taking unions (hence becoming residuated functions); the conditions defining equivalence, analogous to (ALG1)–(ALG4), can be stated simply for the extended functions, as at the beginning of Section 3.5. Hence, everything can be done at the level of the power sets; more precisely, of the complete lattices of theories.[†] In [124] Galatos and Tsinakis recognized the key role of the complete lattice structures[‡] of both the power sets and the families of closed sets, and that it was their *residuated* character that characterized the behaviour of substitutions and transformers on these lattices; focusing on this character allowed them to disregard the fact that these functions on the power sets are actually extensions of point functions on the base sets.

This view originated a far-reaching reformulation of the whole setting in the context of *modules over a complete residuated lattice*.[†] The notion of a *module* makes the action of the monoid (the complete residuated lattice) become an integral part of the algebraic structure, so that structurality and commutativity properties become algebraic properties; for instance, structural transformers become module homomorphisms, the lattice of fixed points (the theories) of a structural closure operator on one such module also has a module structure, and in fact a closure operator is just a module homomorphism. As in the previous step, equivalence of two structural closure operators on two of these modules is defined as the

[†] Notice that a closure operator is determined by its lattice of theories, that structurality can be characterized by a condition on this lattice alone (Proposition 1.51), etc.

[‡] You may have noticed that the operation of set complement (which is the negation operation in the Boolean structure of power sets) has not been used at all.

[†] Or "modules over a quantale", because complete residuated lattices are also called "quantales" in the literature.

existence of a *module isomorphism* between the associated modules of fixed points. If you are beginning to sense that the framework is moving to category theory, then you are right.

An interesting problem, again, is that of identifying when such an isomorphism, when it exists, is *induced* by transformers, as happens in the original setting; that is, the problem of identifying the abstract correlate of the properties of the algebra of formulas that make the equivalence in Theorem 3.58 possible. This has come to be known as *the isomorphism problem*. Some sufficient, but not necessary, conditions had already been found by [29]; more complete solutions, obtained at different levels in [124, 113], are formulated in categorial terms, and involve the notion of projectivity.

These lines of research seem to be expanding in several directions. Both the idea of equivalence between deductive systems and the isomorphism problem can be studied in other contexts. Let me just mention that [122] moves up again, to the categories of *modules over a quantaloid*, which facilitates the merging of this line of research with the already very fruitful one of *categorical abstract algebraic logic* pursued by Voutsadakis (see [241] and many others), covering many-sorted logics; that [219] addresses the issue of considering the equivalence between deductive systems in different languages; and finally that [179] generalizes, at an abstract level, the semantic side of the Isomorphism Theorem 3.58 (the equivalence between its parts (i) and (ii)); this is in contrast to all the previously mentioned studies, which deal exclusively with abstract versions of its syntactic side (the equivalence between its parts (i) and (iii)).

But that is another story.

Chapter 4

The semantics of matrices

Unlike algebraic semantics, matrix semantics is truly general. As is proved in this chapter, *every logic has a (complete) matrix semantics*. Actually, it has several of them, which arise from the logic in some fairly natural ways, and the corresponding completeness theorems are presented as generalizations of the Lindenbaum-Tarski process. As is to be expected, the factorization process is done through a congruence defined in a very abstract way, because in general it is not possible to rely on a set of formulas or equations with particular properties.

This chapter presents only the essentials of the theory of logical matrices, and the particular points needed for later chapters; good references to start further reading on this extensively studied area are [62, 70, 249].

4.1 Logical matrices: basic concepts

Logical matrices are briefly introduced in Section 1.2 as constituting a very general *algebra-based semantics* for a logic. This section contains the main definitions and gives their most elementary properties.

DEFINITION 4.1
Let $\boldsymbol{L}$ *be an algebraic type. An* ***L-matrix*** *is a pair* $\mathcal{M} = \langle \boldsymbol{A}, F \rangle$ *where* $\boldsymbol{A}$ *is an algebra of type* $\boldsymbol{L}$ *and* $F \subseteq A$*. The algebra* $\boldsymbol{A}$ *is called the* ***algebraic reduct*** *of the matrix, and the set* F *is called the* ***filter*** *of the matrix.*

A matrix is ***finite*** *when its algebraic reduct is so. A matrix* $\langle \boldsymbol{A}, F \rangle$ *is* ***proper*** *when* $F \neq A$*, and* ***almost trivial*** *when* $F = \varnothing$*. A* ***trivial*** *matrix is a non-proper matrix on a trivial algebra. If* M *is a class of matrices, then* $\mathsf{Alg}\,\mathsf{M}$ *denotes the class of the algebraic reducts of the matrices in* M*.*

When no confusion is likely one simply says ***matrix***, or sometimes *logical matrix*. The set F is also called the *designated subset* or the *truth filter* or the *truth set* because, recalling the general discussion on algebra-based semantics in Section 1.2, this set represents *truth* when the matrix is used to define a logic or when modelling one.

Observe that on each algebra $\boldsymbol{A}$ there is a non-proper matrix $\langle \boldsymbol{A}, A \rangle$ and an almost trivial one $\langle \boldsymbol{A}, \varnothing \rangle$. On a trivial algebra $\boldsymbol{A}_t$ with universe $\{a\}$ there are two (different) matrices, the non-proper, trivial one $\langle \boldsymbol{A}_t, \{a\} \rangle$ and the proper, almost trivial one $\langle \boldsymbol{A}_t, \varnothing \rangle$; in spite of its aspect, the latter is non-trivial. Actually, all non-proper matrices may be regarded as trivial in some sense, as the observation after Definition 4.7 confirms, but the term is technically used only when the algebraic reduct is trivial as well.

Logics defined by matrices

DEFINITION 4.2

If $\mathcal{M} = \langle \boldsymbol{A}, F \rangle$ *is an* $\boldsymbol{L}$*-matrix, then* ***the logic defined by*** $\mathcal{M}$ *is* $\mathcal{L}_{\mathcal{M}} := \langle \boldsymbol{L}, \vdash_{\mathcal{M}} \rangle$, *where for all* $\Gamma \cup \{\varphi\} \subseteq Fm$,

$$\Gamma \vdash_{\mathcal{M}} \varphi \overset{\text{def}}{\iff} \textit{for all } h \in \operatorname{Hom}(\boldsymbol{Fm}, \boldsymbol{A}), \textit{ if } h\Gamma \subseteq F, \textit{ then } h\varphi \in F.$$

It is straightforward to see that *this indeed defines a logic*, that is, a structural closure relation (this is done in more generality in Exercise 1.22; structurality is due to having used all the homomorphisms and to the fact that the composition of a substitution and a homomorphism is also a homomorphism). Examples of logics defined by a matrix should be familiar to you: $\mathcal{C\ell}$ is the logic defined by the matrix $\langle \mathbf{2}, \{1\} \rangle$, where $\mathbf{2}$ is the two-element Boolean algebra in the appropriate similarity type; its fragments are defined by the appropriate reduct. The presentations of Łukasiewicz's many-valued logics $Ł_\infty$ and $Ł_n$ in Example 1.12 can be rephrased as definitions of these logics from the matrices $\langle [\mathbf{0},\mathbf{1}], \{1\} \rangle$ and $\langle [\mathbf{0},\mathbf{1}]_n, \{1\} \rangle$, respectively. The two three-valued logics presented in Example 3.20 are the logics defined by the matrices $\langle \boldsymbol{A}_3, \{1\} \rangle$ and $\langle \boldsymbol{A}_3, \{1/2, 1\} \rangle$ respectively. And so forth. Actually, all logics with an algebraic semantics (in the technical sense) are defined by a class of matrices, indeed by one of a particular kind (Proposition 4.8).

Some more or less obvious facts (Exercise 4.1) concerning non-proper and trivial matrices are:

LEMMA 4.3

1. *For every* $\boldsymbol{A}$, *the non-proper matrix* $\langle \boldsymbol{A}, A \rangle$ *defines the inconsistent logic, and the almost trivial matrix* $\langle \boldsymbol{A}, \varnothing \rangle$ *defines the almost inconsistent logic.*
2. *If* $\mathcal{M} = \langle \boldsymbol{A}, F \rangle$ *is a proper matrix, then* $\Gamma \subseteq Fm$ *is* $\mathcal{L}_{\mathcal{M}}$*-inconsistent if and only if there is no* $h \in \operatorname{Hom}(\boldsymbol{Fm}, \boldsymbol{A})$ *such that* $h\Gamma \subseteq F$. ⊠

While $\mathcal{L}_{\mathcal{M}}$ is always a logic, it need not be finitary, as shown in Example 1.15. However, there is the following important property:

THEOREM 4.4
Any logic defined by a finite matrix is finitary.

PROOF: This theorem has been proved in the literature by a variety of methods (see [249, 4.1.8] for a survey). I give here the topological proof due to Łoś and Suszko [166], probably the first published one. Another, model-theoretic proof is presented in Corollary 4.69.

Let $\langle \boldsymbol{A}, F\rangle$ be a finite matrix. Consider the set A as a topological space, with the discrete topology: every subset is open (hence every subset is closed). Since A is finite, it is a compact space. The set $\mathrm{Hom}(\boldsymbol{Fm}, \boldsymbol{A})$ can be identified with the set of functions A^V and hence with the infinite product $\prod_V A$. Thus, this set can be considered as a topological space with the product topology of the discrete topology on A. Recall that the (infinite) product topology has as a base of open sets the sets that are the product of open sets of each of the factors but with only a finite number of these open sets being proper (not equal to A).

One can see (Exercise 4.2) that, for each formula ψ, the set $Val\,\psi := \{h \in \mathrm{Hom}(\boldsymbol{Fm}, \boldsymbol{A}) : h\psi \in F\}$ is the union of a finite number of open sets in the just mentioned base. Therefore $Val\,\psi$ is an open set in the product topology. Similarly, its complement $(Val\,\psi)^c = \{h \in \mathrm{Hom}(\boldsymbol{Fm}, \boldsymbol{A}) : h\psi \notin F\}$ is of the same form, and hence an open set as well; here $(\cdot)^c$ indicates the set-theoretic complement with respect to $\mathrm{Hom}(\boldsymbol{Fm}, \boldsymbol{A})$. Then Definition 4.2 can be expressed as:

$$\Gamma \vdash_{\mathcal{M}} \varphi \iff \bigcap_{\gamma\in\Gamma} Val\,\gamma \subseteq Val\,\varphi \iff \big(\bigcup_{\gamma\in\Gamma}(Val\,\gamma)^c\big) \cup Val\,\varphi = \mathrm{Hom}(\boldsymbol{Fm}, \boldsymbol{A}).$$

Thus, we have an open covering of the space. Since by Tychonoff's theorem this product space is compact, this covering has a finite sub-covering: there is a finite $\Gamma_0 \subseteq \Gamma$ such that

$$\big(\bigcup_{\gamma\in\Gamma_0}(Val\,\gamma)^c\big) \cup Val\,\varphi = \mathrm{Hom}(\boldsymbol{Fm}, \boldsymbol{A}) \iff \bigcap_{\gamma\in\Gamma_0} Val\,\gamma \subseteq Val\,\varphi \iff \Gamma_0 \vdash_{\mathcal{M}} \varphi. \boxtimes$$

In order to deal with examples, it is interesting to record here two elementary matrix constructions; recall what a fragment is from page 27.

DEFINITION 4.5
*Let $\mathcal{M} = \langle \boldsymbol{A}, F\rangle$ be an $\boldsymbol{L}$-matrix, for some similarity type $\boldsymbol{L}$. A **submatrix** of $\mathcal{M}$ is an $\boldsymbol{L}$-matrix $\langle \boldsymbol{B}, G\rangle$ such that $\boldsymbol{B}$ is a subalgebra of $\boldsymbol{A}$, with universe $B \subseteq A$, and $G = F \cap B$. For each fragment $\boldsymbol{L}'$ of $\boldsymbol{L}$, the $\boldsymbol{L}'$ **reduct** of $\mathcal{M}$ is the $\boldsymbol{L}'$-matrix $\mathcal{M}{\restriction}\boldsymbol{L}' := \langle \boldsymbol{A}{\restriction}\boldsymbol{L}', F\rangle$. If M is a class of $\boldsymbol{L}$-matrices, then $\mathsf{M}{\restriction}\boldsymbol{L}' := \{\mathcal{M}{\restriction}\boldsymbol{L}' : \mathcal{M} \in \mathsf{M}\}$.*

PROPOSITION 4.6
Let $\mathcal{L}$ be the logic defined by an $\boldsymbol{L}$-matrix $\mathcal{M}$.

1. *If $\mathcal{L}'$ is defined by a submatrix of $\mathcal{M}$, then $\mathcal{L} \leqslant \mathcal{L}'$.*
2. *If $\boldsymbol{L}'$ is a fragment of $\boldsymbol{L}$, then the fragment $\mathcal{L}{\restriction}\boldsymbol{L}'$ is the logic defined by $\mathcal{M}{\restriction}\boldsymbol{L}'$.*

PROOF: See Exercises 4.3 and 4.4. ⊠

You will find an extension of the second point in Exercise 4.6.

One of the most usual applications of the submatrix construction is when $F = \{1\}$ for some algebraic constant 1 of $\boldsymbol{A}$, which guarantees that $1 \in \boldsymbol{B}$ for any subalgebra $\boldsymbol{B}$ of $\boldsymbol{A}$; in this case submatrices can be identified with subalgebras. For instance, Łukasiewicz's finitely-valued logic $Ł_n$ is defined by the matrix $\langle [\mathbf{0},\mathbf{1}]_n, \{1\} \rangle$, which is clearly a submatrix of the matrix $\langle [\mathbf{0},\mathbf{1}], \{1\} \rangle$ that defines the infinitely-valued $Ł_\infty$; therefore, $Ł_\infty \leqslant Ł_n$ for all $n \geqslant 2$. In a similar way you can determine some relations between different finitely-valued logics (Exercise 4.5).

It makes sense to study logics defined not just by a single matrix, but by a set or a class of matrices. This is not only a natural generalization; there may be sound philosophical or applied motivations for doing so: for instance, we may want to study the logic that preserves truth in a certain kind of truth structure, but we may not have a fixed, complete description of our model of truth and thus several of its realizations must be considered (for example: we may believe that truth is many-valued, but be unsure about how many truth values are there; or believe that in temporal logic time structures are linearly ordered, but be undecided about whether they are continuous or discrete). Another reason appears in logics preserving degrees of truth (Example 4.11): we may believe that truth does not reside in a single matrix, but in the interactions inside a family of matrices, understood as a single truth structure when taken as a whole.

Definition 4.7

If M *is a class of matrices of type* $\boldsymbol{L}$, ***the logic defined by*** M *is* $\mathcal{L}_{\mathsf{M}} := \langle \boldsymbol{L}, \vdash_{\mathsf{M}} \rangle$, *where* $\vdash_{\mathsf{M}} := \bigcap\limits_{\mathcal{M} \in \mathsf{M}} \vdash_{\mathcal{M}}$; *that is, for all* $\Gamma \cup \{\varphi\} \subseteq Fm$,

$$\Gamma \vdash_{\mathsf{M}} \varphi \overset{\text{def}}{\iff} \text{for all } \langle \boldsymbol{A}, F \rangle \in \mathsf{M} \text{ and all } h \in \mathrm{Hom}(\boldsymbol{Fm}, \boldsymbol{A}), \text{ if } h\Gamma \subseteq F, \text{ then } h\varphi \in F.$$

That $\mathcal{L}_{\mathsf{M}}$ is a logic is a consequence of the fact that the intersection of a family of structural closure relations is always a structural closure relation (Theorem 1.57 and Exercise 1.70). It is also easy to see that for any class M of matrices and any non-proper matrix $\mathcal{M}$, the logics defined by the classes M and $\mathsf{M} \cup \{\mathcal{M}\}$ coincide; therefore, when considering the logic defined by some class of matrices, it is safe to assume that the matrices are all proper. It is in this sense that non-proper matrices may be regarded as "trivial".

In fact, we have already met many logics defined by a class of matrices, namely the ones with filters "equationally definable" in the following sense:

Proposition 4.8

Let K *be a class of algebras and* $\boldsymbol{\tau}$ *a structural transformer defined by* $\mathrm{E}(x) \subseteq Eq$. *The class* K *is a* $\boldsymbol{\tau}$*-algebraic semantics for a logic* $\mathcal{L}$ *if and only if* $\mathcal{L}$ *is the logic defined by the class of matrices* $\{\langle \boldsymbol{A}, Sol_{\boldsymbol{\tau}}^{\boldsymbol{A}} \rangle : \boldsymbol{A} \in \mathsf{K}\}$, *where* $Sol_{\boldsymbol{\tau}}^{\boldsymbol{A}} := \{a \in A : \boldsymbol{A} \vDash \mathrm{E}(x)\, [\![a]\!]\}$.

Proof: This is just an easy application of the definitions involved, and is left to you as Exercise 4.7. ⊠

Thus, in particular, all algebraizable logics are logics defined by a class of matrices. The issue of equational definability of filters is seriously studied in Section 6.4, where some details on the notation $Sol^{\boldsymbol{A}}_{\tau}$ are given.

By Exercise 1.67, when all the logics in a set are finitary and they are finite in number, their intersection is also finitary; thus, from Theorem 4.4:

COROLLARY 4.9
Any logic defined by a finite set of finite matrices is finitary. ⊠

Theorem 4.68 characterizes, in a model-theoretic way, more general classes of matrices that define finitary logics; Corollary 4.9 can then be obtained as a consequence, but the proof given here, using Theorem 4.4, does not need the model-theoretic framework of Section 4.5.

In the literature, logics of the form $\mathcal{L}_{\mathsf{M}}$ are sometimes called ***strongly finite*** when M is a finite set of finite matrices, and ***finitely approximable*** when M is any class of finite matrices. Corollary 4.9 states that strongly finite logics are finitary; by contrast, finitely approximable logics are the infimum (intersection) of a (possibly infinite) family of strongly finite (hence finitary) logics, but may not be finitary themselves. Here is one case.

EXAMPLE 4.10 (ŁUKASIEWICZ'S MANY-VALUED LOGICS)
Recall (from Example 1.12 and the remark on page 184) that each Łukasiewicz's finitely-valued logic $Ł_n$ is the logic defined by the matrix $\langle [\mathbf{0,1}]_n, \{1\}\rangle$. Consider the infimum of all these logics in the lattice of all logics, $Ł^{\infty} := \bigwedge\{Ł_n : n \geqslant 2\}$. By Definition 4.7, $Ł^{\infty}$ is the logic defined by the (denumerable) set of finite matrices $\{\langle [\mathbf{0,1}]_n, \{1\}\rangle : n \geqslant 2\}$; thus, this logic is finitely approximable. However, $Ł^{\infty}$ is not finitary. To see this, consider the set Σ used in Example 1.15 to show that $Ł_{\infty}$ is not finitary. There it is proved that $\Sigma \vdash_{\infty} y$; since $Ł_{\infty} \leqslant Ł_n$ (page 186), it follows that $Ł_{\infty} \leqslant Ł^{\infty}$, hence $\Sigma \vdash^{\infty} y$ as well. Now let $\Sigma_0 \subseteq \Sigma$ be finite, and let $k \geqslant 2$ be such that $\Sigma_0 \subseteq \{(x \oplus \cdots^{n} \oplus x) \to y : 2 \leqslant n \leqslant k\} \cup \{\neg x \to y\}$. The evaluation suggested in Exercise 1.26 shows that $\Sigma_0 \nvdash_{k+2} y$, and therefore $\Sigma_0 \nvdash^{\infty} y$.

Other interesting properties to notice are that $Ł_{\infty} \neq Ł^{\infty}$ (Exercise 4.8) and that the infimum of the family $\{Ł_n : n \geqslant 2\}$ in the lattice of all finitary logics is $Ł_{\infty f}$ (Exercise 4.9). Therefore, $Ł_{\infty f} < Ł_{\infty} < Ł^{\infty} < Ł_n$ for every $n \geqslant 2$. This example confirms the claim made on page 54 that the lattice of finitary logics (in a fixed language) need not be a complete sublattice of the lattice of all logics. ⊠

Among the classes of matrices that are commonly used to define logics there are the so-called ***bundles of matrices***: sets of matrices with the same algebraic reduct, that is, sets† of the form $\{\langle \boldsymbol{A}, F\rangle : F \in \mathscr{D}\}$ for some family $\mathscr{D} \subseteq \mathscr{P}A$. A very large class of logics that can be viewed as defined by a particular kind of bundles of matrices is the following:

† Since all filters in the bundle belong to the power set of the same set, these classes are always sets. However, most of the classes of matrices that appear in general results are indeed proper classes.

EXAMPLE 4.11 (LOGICS PRESERVING DEGREES OF TRUTH)
Let $\boldsymbol{A}$ be an *ordered algebra*, that is, an algebra with an order relation $\leqslant$ on its universe, without assuming any further property linking the order to the algebraic operations.† If the elements of $\boldsymbol{A}$ are to be understood as *truth values* in a certain sense, then one can consider the principal up-sets $\uparrow a := \{b \in A : a \leqslant b\}$ as representing the *truth degree* "to have at least truth value a". The consequence relation defined by the bundle of matrices $\{\langle \boldsymbol{A}, \uparrow a\rangle : a \in A\}$, according to Definition 4.7, is given by:

$$\begin{aligned} \Gamma \vdash^{\leqslant}_{\boldsymbol{A}} \varphi \overset{\text{def}}{\iff} & \text{ for all } a \in A \text{ and all } h \in \operatorname{Hom}(\boldsymbol{Fm}, \boldsymbol{A}), \\ & \text{if } a \leqslant h\gamma \text{ for all } \gamma \in \Gamma, \text{ then } a \leqslant h\varphi. \end{aligned} \tag{48}$$

The associated logic deserves to be called ***the logic that preserves degrees of truth*** with respect to the structure $\langle \boldsymbol{A}, \leqslant\rangle$. This name was first used in algebraic logic by Nowak in [183, 184]; the discussion in Scott's well-known [224] can be considered as an early proposal to adopt this scheme as an alternative to the matrix-based one for many-valued logics.

It is not difficult to realize (Exercise 4.10) that, in general, $\vdash^{\leqslant}_{\boldsymbol{A}}$ has theorems if and only if the order in $\boldsymbol{A}$ has a term-definable maximum, that is, a maximum that is an algebraic constant of the algebra.

When the ordered algebra is a complete meet-semilattice‡ (in particular, when it is a complete lattice), the definition (48) can be equivalently written as

$$\Gamma \vdash^{\leqslant}_{\boldsymbol{A}} \varphi \iff \text{for all } h \in \operatorname{Hom}(\boldsymbol{Fm}, \boldsymbol{A}),\ \bigwedge h\Gamma \leqslant h\varphi. \tag{49}$$

This expression is particularly meaningful when $\Gamma \neq \emptyset$, but notice that it actually covers also the case $\Gamma = \emptyset$, because formally $\bigwedge \emptyset = \max A$. If the algebra $\boldsymbol{A}$ is finite, then the bundle is finite as well, and hence by Corollary 4.9 the consequence $\vdash^{\leqslant}_{\boldsymbol{A}}$ is finitary; if moreover $\boldsymbol{A}$ has a maximum definable by a term $\top$, then the definition can be reduced to the two clauses

$$\begin{aligned} \varphi_1, \dots, \varphi_n \vdash^{\leqslant}_{\boldsymbol{A}} \varphi &\iff \boldsymbol{A} \vDash \varphi_1 \wedge \dots \wedge \varphi_n \preccurlyeq \varphi, \\ \emptyset \vdash^{\leqslant}_{\boldsymbol{A}} \varphi &\iff \boldsymbol{A} \vDash \varphi \approx \top, \end{aligned}$$

where $n \geqslant 1$, and, recall, $\boldsymbol{A} \vDash \alpha \preccurlyeq \beta$ is shorthand for $\boldsymbol{A} \vDash \alpha \approx \alpha \wedge \beta$; thus, these logics appear as completely determined by the equational theory of the algebra $\boldsymbol{A}$. A large class of logics of this kind, called *semilattice-based*, is studied in Section 7.2. ⊠

Bundles of matrices of these and similar forms can be understood as *truth structures* representing *a collective notion of truth* that is not represented by any

† The term "ordered algebra" has been often used in the literature to mean that the operations are monotone with respect to the order. In algebraic logic one often finds anti-monotone operations (the typical example is negation). In the present general discussion nothing of either kind is assumed.

‡ An ordered set $\langle A, \leqslant\rangle$ is a ***complete meet-semilattice*** when any subset $X \subseteq A$ has a meet or infimum, denoted by $\bigwedge X$. This definition includes the extreme cases $\bigwedge A = \min A$ and $\bigwedge \emptyset = \max A$; thus, all complete meet-semilattices are bounded, but notice that the bounds need not be term-definable.

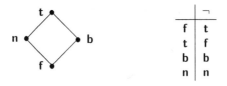

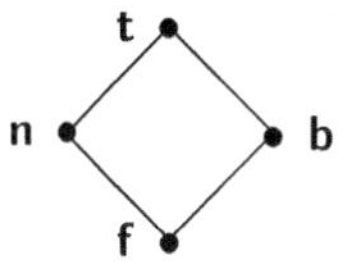

	$\neg$
f	t
t	f
b	b
n	n

FIGURE 5: The De Morgan lattice M_4, described by the Hasse diagram of its lattice structure, and the table of its negation operation.

of the individual matrices $\langle \boldsymbol{A}, F\rangle$, but rather by the whole family; in the case of ordered algebras, in the ordered algebraic structure $\langle \boldsymbol{A}, \leqslant\rangle$. Proposals to define logics in ways similar to these are not uncommon in the literature; one of the most famous is the following.

EXAMPLE 4.12 (BELNAP-DUNN'S FOUR-VALUED LOGIC)
The logic $\mathcal{B}$, in the language $\langle \wedge, \vee, \neg\rangle$ of type $\langle 2,2,1\rangle$, is described in Example 1.13. It is not difficult to see (Exercise 4.26) that the definition given there amounts exactly to saying that $\mathcal{B}$ is the logic defined by the bundle of two matrices $\{\langle \boldsymbol{M}_4, \{\mathbf{t},\mathbf{b}\}\rangle, \langle \boldsymbol{M}_4, \{\mathbf{t},\mathbf{n}\}\rangle\}$, where $\boldsymbol{M}_4 := \langle M_4, \wedge, \vee, \neg\rangle$ is the four-element De Morgan† lattice, that is, the lattice described in Figure 5 endowed with the negation operation $\neg$ also shown there. Ironically, it turns out (Exercise 4.26 again) that $\mathcal{B}$ is actually the logic defined by each of these two matrices alone; but this is a purely technical fact with no logical interpretation.

Using that the two sets $\{\mathbf{t},\mathbf{b}\}$ and $\{\mathbf{t},\mathbf{n}\}$ form a basis of the closure system of lattice filters of $\boldsymbol{M}_4$, that the maximum of this order is not term-definable (because each of the subsets $\{\mathbf{n}\}$ and $\{\mathbf{b}\}$ is a subalgebra) and Corollary 4.9, it is easy to see that $\vdash_{\mathcal{B}} = \vdash^{\leqslant}_{\boldsymbol{M}_4}$, and that $\mathcal{B}$ can be more compactly described as the logic defined by the following clauses (where $n \geqslant 1$):

$$\varphi_1, \dots, \varphi_n \vdash_{\mathcal{B}} \varphi \overset{\text{def}}{\iff} \boldsymbol{M}_4 \vDash \varphi_1 \wedge \dots \wedge \varphi_n \preccurlyeq \varphi,$$

$\mathcal{B}$ has no theorems,

$\mathcal{B}$ is finitary.

This logic appeared, essentially, in Dunn's 1966‡ algebraic completeness proof of the system of so-called *first degree entailments*, a fragment common to the logics $\mathcal{E}$ of entailment and $\mathcal{R}$ of relevance. It was extraordinarily popularized by its presentation in Belnap's [23] as a logic convenient for what are now called *deductive data-bases*, formalizing situations in which one has to handle perhaps incomplete and perhaps inconsistent information (this may happen, for instance, when the information is collected from several independent sources, as is common nowadays when navigating through the Web). Belnap and Dunn suggest to admit two primary *epistemic* values t, f (standing for "told true" and

† See Example 5.82 for algebraic details on De Morgan lattices.
‡ In Dunn's dissertation, whose advisor was Belnap; the Introduction to [85] gives precise details. Their main results on these logics appear in different sections of [8, 9].

"told false", respectively) which are not considered as incompatible, and to admit undefined situations; this leads to the definition of the associated logic as described in Example 1.13 and above. The logic $\mathcal{B}$ has become very popular in applications; moreover, it provides interesting examples in abstract algebraic logic, which are reviewed in Exercise 4.36 and in Examples 5.82 and 5.92.

Recognition of the existence of another lattice structure in the same set, given by the "knowledge order" or "information order" (going from left to right in Figure 5, which corresponds to the subset relation in $\mathscr{P}\{t,f\}$ as shown in Figure 2 on page 21) and of the relations between the two orders, engendered the also popular notion of a *bilattice*, due to Ginsberg [126] and extensively studied by Fitting and others [91, 14]; recently some logics defined by blattices have been studied with the methods of abstract algebraic logic in [48], as they turn out to provide some interesting examples of non-standard situations (see Example 7.6.7). ⊠

Matrices as models of a logic

Instantiating the terminology set up in Definition 1.16 for a generic semantics, we obtain:

DEFINITION 4.13

Let $\mathcal{L}$ be a logic. A matrix $\mathcal{M}$ is a ***model*** *of $\mathcal{L}$ when $\mathcal{L} \leqslant \mathcal{L}_{\mathcal{M}}$; that is, when for all $\Gamma \cup \{\varphi\} \subseteq Fm$,*

$$\text{if } \Gamma \vdash_{\mathcal{L}} \varphi \text{ and } h \in \operatorname{Hom}(\boldsymbol{Fm}, \boldsymbol{A}) \text{ is such that } h\Gamma \subseteq F, \text{ then } h\varphi \in F.$$

The class of all models of $\mathcal{L}$ is denoted by $\mathsf{Mod}\mathcal{L}$.

*$\mathcal{L}$ is (**strongly**) **complete**† with respect to a class of matrices M when $\mathcal{L} = \mathcal{L}_{\mathsf{M}}$, that is, when $\vdash_{\mathcal{L}} = \vdash_{\mathsf{M}}$; and is **weakly complete** with respect to M when $\mathcal{L}$ and $\vdash_{\mathsf{M}}$ have the same theorems.*

*When the class M consists of a single matrix, this one is called **strongly / weakly characteristic** for $\mathcal{L}$ when $\mathcal{L}$ is strongly / weakly complete with respect to it.*

Thus, $\mathsf{M} \subseteq \mathsf{Mod}\mathcal{L}$ if and only if $\mathcal{L} \leqslant \mathcal{L}_{\mathsf{M}}$. Recalling that trivial matrices define the inconsistent logic, it follows that the class $\mathsf{Mod}\mathcal{L}$ always contains all trivial matrices. Some earlier results in this book can be rephrased in terms of completeness with respect to classes of matrices. Let me highlight three:

PROPOSITION 4.14

1. *Let $\mathcal{L}$ be an implicative logic. Then $\mathcal{L}$ is complete with respect to the class of matrices $\{\langle \boldsymbol{A}, \{1\}\rangle : \boldsymbol{A} \in \mathsf{Alg}^*\mathcal{L}\}$.*
2. *A logic $\mathcal{L}$ is the assertional logic of a pointed class of algebras K if and only if $\mathcal{L}$ is complete with respect to the class of matrices $\{\langle \boldsymbol{A}, \{1\}\rangle : \boldsymbol{A} \in \mathsf{K}\}$.*

---

† When $\mathcal{L}$ is complete with respect to M, some authors say that "M is *a matrix semantics for* $\mathcal{L}$", extending to matrices the notion of a class of algebras being an *algebraic semantics for* $\mathcal{L}$ of Definition 3.4. This terminology is not used in this book.

3. *A class* K *of algebras is a* $\boldsymbol{\tau}$*-algebraic semantics for a logic* $\mathcal{L}$ *if and only if* $\mathcal{L}$ *is complete with respect to the class of matrices* $\{\langle \boldsymbol{A}, Sol^{\boldsymbol{A}}_{\boldsymbol{\tau}}\rangle : \boldsymbol{A} \in \mathsf{K}\}$ *defined in Proposition* 4.8. ⊠

The first statement is Theorem 2.9, the second is a rephrasing of Definition 3.5, and the third, of Proposition 4.8. Putting Definitions 2.18 and 4.13 together, the following is obvious.

Lemma 4.15
A matrix $\langle \boldsymbol{A}, F\rangle$ *is a model of a logic* $\mathcal{L}$ *if and only if* $F \in \mathcal{F}i_{\mathcal{L}}\boldsymbol{A}$. ⊠

In particular, by Corollary 2.20 the models of a logic on the formula algebra are given by the theories of the logic. This simple observation leads to a trivial solution to the completeness problem of a logic. The matrices of the form $\langle \boldsymbol{Fm}, \Gamma\rangle$ with $\Gamma \in \mathcal{T}h\mathcal{L}$ are called the ***Lindenbaum models*** of $\mathcal{L}$; by Lemma 4.15 they are indeed models of $\mathcal{L}$. The idea of using the theories of the logic as models of the logic itself, due to Lindenbaum, was a fundamental advance in the mathematical approach to semantics of sentential logics, and one can even say that, in some sense, it marked a turning point in the evolution of algebraic logic. The following enhanced version of his result is due to Wójcicki [244, 245].

Theorem 4.16 (Completeness)
Every logic is complete with respect to the class of all its Lindenbaum models. More generally, if L *denotes the class of all Lindenbaum models of* $\mathcal{L}$*, then every logic* $\mathcal{L}$ *is complete with respect to any class* M *of matrices such that* $\mathsf{L} \subseteq \mathsf{M} \subseteq \mathsf{Mod}\mathcal{L}$*. In particular every logic is complete with respect to the class of all its models.*

Proof: If $\mathsf{M} \subseteq \mathsf{Mod}\mathcal{L}$, then by assumption $\vdash_{\mathcal{L}} \subseteq \vdash_{\langle \boldsymbol{A},F\rangle}$ for each $\langle \boldsymbol{A}, F\rangle \in \mathsf{M}$, so $\vdash_{\mathcal{L}} \subseteq \vdash_{\mathsf{M}}$. The converse is proved by contradiction, namely by the most radical simplification of the Lindenbaum-Tarski process: Assume that $\Gamma \nvdash_{\mathcal{L}} \varphi$, and do only step (LT1), obtaining the theory Γ' generated by Γ. Then $\langle \boldsymbol{Fm}, \Gamma'\rangle \in \mathsf{L}$, therefore $\langle \boldsymbol{Fm}, \Gamma'\rangle \in \mathsf{M}$. Since the identity function is an evaluation from $\boldsymbol{Fm}$ into itself, this shows that $\Gamma \nvdash_{\mathsf{M}} \varphi$.

An interesting, more advanced topic is the investigation of characteristic or weakly characteristic matrices for a given logic. Here are some miscellaneous basic facts.

- Lindenbaum's original theorem (ca 1920) was that for every logic $\mathcal{L}$, the matrix $\langle \boldsymbol{Fm}, C_{\mathcal{L}}\emptyset\rangle$ is weakly characteristic for $\mathcal{L}$. It may appear as trivial to our eyes, as does Theorem 4.16 in its modern formulation; however, its truth depends on recognizing the role of structurality: while the fact that $C_{\mathcal{L}}\emptyset$ is closed under substitutions was recognized early, structurality of the consequence relation had to wait until Łoś and Suszko's 1958 paper [166], and only Wójcicki in 1969 realized the full form of the Completeness Theorem as given above.
- By definition, $\mathcal{C}\ell$ and all its fragments have a characteristic matrix: $\langle \mathbf{2}, \{1\}\rangle$, where $\mathbf{2}$ is the algebraic reduct of the two-element Boolean algebra given by the operations corresponding to the connectives of the fragment.

- Trivially, every logic defined by a single matrix has it as characteristic; if the matrix is finite, then by Theorem 4.4 the logic is finitary. Then an interesting problem for these logics is that of whether they have *a finite Hilbert-style presentation* (this is sometimes called to be "finitely H-based", or simply "finitely axiomatizable"). This can be a difficult problem. Even very simple and small finite matrices can define logics that cannot be finitely presented; a famous example, probably the simplest one, was found by Wroński [250]: On the three-element set $M = \{0, 1/2, 1\}$ consider the binary operation $\star$ given by $0 \star 1/2 := 0$ and $a \star b := 1$ otherwise; then the logic defined by the matrix $\langle \boldsymbol{M}, \{1\} \rangle$ is finitary but not finitely axiomatizable.
- While every logic has a weakly characteristic matrix $\langle \boldsymbol{Fm}, C_{\mathcal{L}}\varnothing \rangle$, not every logic has a characteristic matrix. In [166], as corrected in [245], the following result is proved: A finitary logic $\mathcal{L}$ has a characteristic matrix if and only if $\mathcal{L}$ is *uniform*: If $\Gamma \cup \Delta \vdash_{\mathcal{L}} \varphi$, Δ is consistent, and $\operatorname{Var}(\Gamma, \varphi) \cap \operatorname{Var} \Delta = \varnothing$, then $\Gamma \vdash_{\mathcal{L}} \varphi$. If the assumption of finitarity is deleted, then besides uniformity one has to require *co-uniformity*: If $\mathscr{E} \subseteq \mathscr{P}Fm$ is a family such that any two of its distinct members have no variables in common, $\operatorname{Var}(\bigcup \mathscr{E}) \neq V$ and $\bigcup \mathscr{E}$ is inconsistent, then there is some inconsistent set in $\mathscr{E}$.

  A sufficient condition for a logic to be both uniform and co-uniform is that for each consistent Γ there is a substitution σ such that $\sigma\Gamma \subseteq C_{\mathcal{L}}\varnothing$. Using it one can prove that all intermediate logics (including $\mathcal{I}\ell$) and all their fragments do have a characteristic matrix; however, the one for $\mathcal{I}\ell$ cannot be finite (Gödel, 1932; see Exercise 4.11) nor denumerable (Wroński, 1974).
- A logic can have a weakly characteristic matrix that is not characteristic. Exercise 4.12 details the example of a three-valued one for $\mathcal{C}\ell$ appearing in [166], as corrected in [251]; Exercise 4.13 shows a similar one. Another, noteworthy example is $\mathcal{I}\ell$, which in spite of the just mentioned facts, does have a weakly characteristic denumerable matrix (the "limit" of the denumerable sequence of "Jaśkowski's matrices"; see [84]).
- If there is no characteristic matrix for a logic, then it makes sense to define its *ramification number* (Wójcicki) as the smallest number of matrices needed to achieve completeness, and to (try to) compute its value.

All these issues belong to an established direction in the theory of matrices for sentential logics; Wójcicki's classical monograph [249] contains a good, detailed sample of these topics, and many references.

Exercises for Section 4.1

Exercise 4.1. Let $\mathcal{M} = \langle \boldsymbol{A}, F \rangle$ be a proper matrix. Prove that a set $\Gamma \subseteq Fm$ is $\mathcal{L}_{\mathcal{M}}$-inconsistent if and only if there is no $h \in \operatorname{Hom}(\boldsymbol{Fm}, \boldsymbol{A})$ such that $h\Gamma \subseteq F$.

Exercise 4.2. Show in detail that if $\langle \boldsymbol{A}, F \rangle$ is a finite matrix, then for each formula ψ, the set $Val\,\psi := \{h \in \operatorname{Hom}(\boldsymbol{Fm}, \boldsymbol{A}) : h\psi \in F\}$ and its complement can each be written as the union of a finite number of basic open sets in the product topology of $\mathrm{Hom}(\boldsymbol{Fm}, \boldsymbol{A})$ as described in the proof of Theorem 4.4, and hence are open sets of this topology.

HINT. Assume for simplicity that $\mathrm{Var}\,\psi = \{x_1, \dots, x_n\}$ with $n \geqslant 1$. Start by showing that for any $\langle a_1, \dots, a_n \rangle \in A^n$, the set $\{h \in \mathrm{Hom}(\boldsymbol{Fm}, \boldsymbol{A}) : hx_i = a_i \text{ for } i = 1, \dots, n\}$ is a basic open set in the product topology. In this exercise, it may be helpful to view a homomorphism h as the tuple $\langle hx : x \in V \rangle$.

EXERCISE 4.3. Let $\mathcal{M} = \langle \boldsymbol{A}, F \rangle$ be a matrix, let $\boldsymbol{B}$ be a subalgebra of $\boldsymbol{A}$, and consider the matrix $\mathcal{N} = \langle \boldsymbol{B}, F \cap B \rangle$, the submatrix of $\mathcal{M}$ determined by the subalgebra $\boldsymbol{B}$. Show that $\mathcal{L}_{\mathcal{M}} \leqslant \mathcal{L}_{\mathcal{N}}$. Observe that this implies that if $\mathcal{M}$ is a model of a logic $\mathcal{L}$, then $\mathcal{N}$ is a model of $\mathcal{L}$ as well; or, equivalently, that if $F \in \mathcal{F}i_{\mathcal{L}}\boldsymbol{A}$ and $\boldsymbol{B}$ is a subalgebra of $\boldsymbol{A}$, then $F \cap B \in \mathcal{F}i_{\mathcal{L}}\boldsymbol{B}$.

COMMENT. Notice that the converse property is not true; that is, not all $\mathcal{L}$-filters of a subalgebra $\boldsymbol{B}$ of $\boldsymbol{A}$ are of the form $F \cap B$ for some $F \in \mathcal{F}i_{\mathcal{L}}\boldsymbol{A}$. When this holds it is said that the logic has the "Filter Extension Property", which under some assumptions characterizes the Local Deduction Theorem; see Theorem 6.24.

EXERCISE 4.4. Let $\mathcal{L}$ be the logic defined by the $\boldsymbol{L}$-matrix $\langle \boldsymbol{A}, F \rangle$. Prove that for each fragment $\boldsymbol{L}'$ of $\boldsymbol{L}$, the $\boldsymbol{L}'$ fragment of the logic $\mathcal{L}$ is the logic defined by the $\boldsymbol{L}'$ reduct of $\langle \boldsymbol{A}, F \rangle$. Compare with Exercise 1.27.

EXERCISE 4.5. Consider Łukasiewciz's finitely-valued logics $Ł_n$ of Example 1.12. Prove that if $2 \leqslant n, m < \omega$ and $m-1$ divides $n-1$, then $Ł_n \leqslant Ł_m$.

HINT and COMMENT. Prove that $[\mathbf{0,1}]_m$ is a subalgebra of $[\mathbf{0,1}]_n$ if and only if $m-1$ divides $n-1$, and use Proposition 4.6.1. The converse of the stated implication is also true, but this is more tricky to prove; see [21, Theorem 2.4.2].

EXERCISE 4.6. Let $\boldsymbol{L}_1, \boldsymbol{L}_2$ be two similarity types with $\boldsymbol{L}_1 \subseteq \boldsymbol{L}_2$, and let $\mathcal{L}_i$ be a logic of type $\boldsymbol{L}_i$ defined by a class M_i of $\boldsymbol{L}_i$-matrices, for $i = 1, 2$. Prove that if $\mathsf{M}_2 {\restriction} \boldsymbol{L}_1 \subseteq \mathsf{M}_1$, then $\mathcal{L}_2$ is an expansion of $\mathcal{L}_1$.

COMMENT. This property is useful in many particular cases; see Exercise 4.36 for an example of its application.

EXERCISE 4.7. Prove Proposition 4.8, using just Definitions 3.4 and 4.7.

EXERCISE 4.8. Consider the logics $Ł_\infty$ and $Ł^\infty$ defined in Examples 1.12 and 4.10 respectively. Define $\Sigma := \{x_k \leftrightarrow (x_{k+1} \oplus x_{k+1}) : k \geqslant 1\} \subseteq Fm$, where $\oplus$ is the connective defined in Example 1.15, and $\langle x_k : k \geqslant 1 \rangle$ is a denumerable sequence of distinct variables. Prove that $\Sigma \nvdash_\infty x_1 \vee \neg x_1$, and that $\Sigma \vdash_n x_1 \vee \neg x_1$ for every $n \geqslant 2$. Conclude that $Ł_\infty \neq Ł^\infty$.

HINTS. Begin by checking that, in $[\mathbf{0,1}]$, $a \leftrightarrow (b \oplus b) = 1$ if and only if $a = 2b$ (for $b \leqslant \frac{1}{2}$), and that $a \vee \neg a = 1$ if and only if $a = 0$ or $a = 1$. Then, define $h \in \mathrm{Hom}(\boldsymbol{Fm}, [\mathbf{0,1}])$ by $hx_k := \frac{1}{2^k}$, for all $k \geqslant 1$; this evaluation should show the first point. For the second, observe that if $h \in \mathrm{Hom}(\boldsymbol{Fm}, [\mathbf{0,1}]_n)$, then the only way to obtain that $h\alpha = 1$ for all $\alpha \in \Sigma$ is that $hx_1 = 0$ or $hx_1 = 1$.

EXERCISE 4.9. Prove that $Ł_{\infty f}$ is the infimum of the family $\{Ł_n : n \geqslant 2\}$ in the lattice of all finitary logics.

HINT. Notice that the definition of $Ł_n$ from the matrix $\langle [\mathbf{0,1}]_n, \{1\} \rangle$ can be expressed in terms of validity in $[\mathbf{0,1}]_n$ of a (transformed) quasi-equation and that the same holds for $Ł_{\infty f}$ with respect to $[\mathbf{0,1}]$. Use that $[\mathbf{0,1}] \in \mathsf{MV}$, and the known fact [127, Corollary 7.2] that $\mathbb{Q}\{[\mathbf{0,1}]_n : n \geqslant 2\} = \mathsf{MV}$. For an alternative proof with more information, see [82, Theorem 6.4.12].

EXERCISE 4.10. Let $\langle \boldsymbol{A}, \leqslant \rangle$ be an algebra endowed with an order. Show that the consequence $\vdash_{\boldsymbol{A}}^{\leqslant}$ has theorems if and only if this order has a maximum element and this element is an algebraic constant of $\boldsymbol{A}$ (it is definable by a term of the language).

EXERCISE 4.11. Consider the formula $\bigvee_{0 \leqslant i < j \leqslant n} (x_i \to x_j)$ of the language of $\mathcal{IL}$, where $x_0, \dots x_n$ are different variables, and $n \geqslant 2$. Prove that this formula holds in all Heyting algebras with at most n elements, while it fails in chains of at least $n+1$ elements. Conclude that no finite Heyting algebra can provide a weakly characteristic matrix for $\mathcal{IL}$ (with its top element as designated).

HINT. As an implicative logic, $\mathcal{IL}$ is complete with respect to $\mathsf{Alg}^*\mathcal{IL}$, which is the class of all Heyting algebras (Exercise 2.31). Besides this general fact, you need only use that in a Heyting algebra $a \to a = 1$, and that if the algebra is a chain, then $a \to b = b$ when $b < a$.

EXERCISE 4.12. Consider the three-element algebra $\boldsymbol{A}_3$ on $\{0, 1/2, 1\}$ with the operations given by:

$\to$	0	½	1
0	1	1	1
½	0	1	1
1	0	1	1

$\wedge$	0	½	1
0	0	0	0
½	0	1	1
1	0	1	1

$\vee$	0	½	1
0	0	1	1
½	1	1	1
1	1	1	1

	$\neg$
0	1
½	0
1	0

That is, ½ has been added as a "copy" of 1 to the ordinary two-element algebra **2**. Denote by $\mathcal{L}_3$ the logic defined by the matrix $\langle \boldsymbol{A}_3, \{1\} \rangle$. Prove:

1. $\mathcal{L}_3 \leqslant \mathcal{C\ell}$.
2. $\mathcal{L}_3$ has exactly the same theorems as $\mathcal{C\ell}$.
3. $\mathcal{L}_3 \neq \mathcal{C\ell}$.

Conclude that the above matrix (a three-valued one with a single designated element) is weakly characteristic but not characteristic for $\mathcal{C\ell}$.

HINTS. For 1, use Exercise 4.3. For 2, observe that a formula that is not a variable is never evaluated to ½ in $\boldsymbol{A}_3$. For 3, show that $\{1\}$ is not closed under (MP).

EXERCISE 4.13. Do the same as Exercise 4.12 but for the matrix $\langle \boldsymbol{A}_3', \{1/2, 1\} \rangle$, where $\boldsymbol{A}_3'$ is the algebra on the same set $\{0, 1/2, 1\}$ with the operations given by:

$\to$	0	½	1
0	1	1	1
½	1	1	1
1	0	0	1

$\wedge$	0	½	1
0	0	0	0
½	0	0	0
1	0	0	1

$\vee$	0	½	1
0	0	0	1
½	0	0	1
1	1	1	1

	$\neg$
0	1
½	1
1	0

This time ½ is a "copy" of 0. Conclude that there are weakly characteristic, but not characteristic, matrices for $\mathcal{C\ell}$ in which the designated set is not $\{1\}$ but $\{1/2, 1\}$.

4.2 The Leibniz operator

The main construction with matrices of interest for us is factoring them out under appropriate congruences. These congruences, though, should "preserve" the matrix as a relational structure, and not only the algebraic structure.

Let $\boldsymbol{A}$ be an algebra and let $\theta \in \mathrm{Con}\boldsymbol{A}$. The canonical projection $\pi\colon \boldsymbol{A} \to \boldsymbol{A}/\theta$ is a surjective homomorphism; it is denoted by π_θ if necessary. Both notations

πa and a/θ, for $a \in A$, are used. Accordingly one can define, for each $F \subseteq A$, the set $\pi F := F/\theta := \{a/\theta : a \in F\} \subseteq A/\theta$. Thus, if $b \in F$, then $b/\theta \in F/\theta$; however, the converse implication fails, for it is possible that $b/\theta \in F/\theta$ because $b/\theta = a/\theta$ for some $a \in F$ while $b \notin F$. The interesting situation is when the equivalence holds, hence it is useful to give it a name:

DEFINITION 4.17
A congruence θ of an algebra $\boldsymbol{A}$ is **compatible** *with a set $F \subseteq A$ when any of the following properties holds:*

(i) *For any $a, b \in A$, if $a \in F$ and $a \equiv b$ (θ), then $b \in F$.*

(ii) *For any $a \in A$, $a \in F$ if and only if $a/\theta \in F/\theta$.*

(iii) $F = \pi^{-1}\pi F$.

(iv) $F = \bigcup_{a \in F} a/\theta$.

(v) *F is a union of blocks of θ.*

(These properties are trivially equivalent.)

Thus, a congruence is compatible with its saturated sets (the unions of some of its equivalence classes or blocks). Conversely, the congruences compatible with a given set are those with respect to which the set is saturated. Notice that this is a purely algebraic notion: given the algebraic structure it is possible to know whether a given congruence is compatible or not with a given subset. Exercise 4.14 contains yet another, often useful description of the notion of compatibility.

As mentioned in Definition 4.17, compatibility of a congruence with a set F can be expressed in terms of the canonical projection π by the condition that $F = \pi^{-1}\pi F$. This can be generalized to a useful property. Since compatibility is a novel notion and you need to get some practice in dealing with it, it is good to see its (easy!) proof in detail. Recall that the kernel of a homomorphism $h\colon \boldsymbol{A} \to \boldsymbol{B}$ is $\ker h := \{\langle a, b\rangle \in A \times A : ha = hb\}$ and that $\ker h \in \mathrm{Con}\boldsymbol{A}$.

LEMMA 4.18
Let $h\colon \boldsymbol{A} \to \boldsymbol{B}$ and $F \subseteq A$.

1. *$F = h^{-1}hF$ if and only if $\ker h$ is compatible with F.*
2. *If moreover h is surjective, then it induces an order isomorphism between the sets $\{F \subseteq A : \ker h$ is compatible with $F\}$ and $\mathscr{P}B$, with inverse given by the residual function h^{-1}.*

   In particular, for each $\theta \in \mathrm{Con}\boldsymbol{A}$, the canonical projection induces an order isomorphism between the sets $\{F \subseteq A : \theta$ is compatible with $F\}$ and $\mathscr{P}(A/\theta)$.

PROOF: 1. ($\Rightarrow$) Assume that $a \equiv b$ $(\ker h)$ and $a \in F$. Then $ha = hb$, hence $b \in h^{-1}hF$, and then $b \in F$.

($\Leftarrow$) $F \subseteq h^{-1}hF$ always holds, so assume that $a \in h^{-1}hF$; this actually means that $ha = hb$ for some $b \in F$. But then $a \equiv b$ $(\ker h)$ and by the compatibility assumption $a \in F$. Thus, $F = h^{-1}hF$.

2. Given any $G \subseteq B$, trivially $\ker h$ is compatible with the set $h^{-1}G \subseteq A$. Then the first part, plus the fact that $hh^{-1} = \mathrm{Id}_B$ by the surjectivity of h, implies that h and h^{-1} are mutually inverse bijections between the two sets involved, namely the set $\{F \subseteq A : \ker h \text{ is compatible with } F\}$ and the whole power set $\mathscr{P}B$. These two sets are ordered under the set inclusion relation, and the two functions are order-preserving, therefore they are order isomorphisms. ⊠

Actually, the lemma holds for an ordinary function between two sets, as the proof makes apparent (obviously, the notion of compatibility applies to arbitrary equivalence relations without any changes); but in practice it is used only in the algebraic context.

The order isomorphism of Lemma 4.18.2 is restricted to models of a logic in Lemma 4.36, and becomes an important fact in the theory of matrices.

Definition 4.19
*The **strict congruences** of a matrix $\langle \boldsymbol{A}, F\rangle$, or **matrix congruences**, are the congruences of $\boldsymbol{A}$ that are compatible with F; the set they form is denoted by $\mathrm{Con}\langle \boldsymbol{A}, F\rangle$.*
*If $\theta \in \mathrm{Con}\langle \boldsymbol{A}, F\rangle$, then the **quotient matrix** is $\langle \boldsymbol{A}, F\rangle/\theta := \langle \boldsymbol{A}/\theta, F/\theta\rangle$.*

Theorem 4.20
The set $\mathrm{Con}\langle \boldsymbol{A}, F\rangle$ is a complete sublattice of $\mathrm{Con}\boldsymbol{A}$, and its least member is the identity relation. Moreover it is a down-set of $\mathrm{Con}\boldsymbol{A}$, that is, if $\theta, \theta' \in \mathrm{Con}\boldsymbol{A}$ are such that $\theta \in \mathrm{Con}\langle \boldsymbol{A}, F\rangle$ and $\theta' \subseteq \theta$, then $\theta' \in \mathrm{Con}\langle \boldsymbol{A}, F\rangle$, and hence it is an ideal of $\mathrm{Con}\boldsymbol{A}$.

Proof: I assume you know that the set of equivalence relations on a set A is a complete lattice where infimum is intersection and supremum is given by the following expression

$$\bigvee_{i\in I} \theta_i = \bigcup\{\theta_{i_1} \circ \cdots \circ \theta_{i_n} : i_j \in I, n \geqslant 0\},$$

(where $\circ$ is relational composition, or product) and that the lattice $\mathrm{Con}\boldsymbol{A}$ of all congruences of an algebra is a complete sublatttice of this lattice, that is, its infimum and supremum are the same. First, we check that the compatible congruences form a down-set: if $\theta' \subseteq \theta$, $a \in F$ and $a \equiv b \ (\theta')$, then also $a \equiv b \ (\theta)$ and the compatibility of θ implies that $b \in F$, thus proving the compatibility of θ'. As a particular case, the intersection of a family of compatible congruences is also compatible. It is easy to check that the union of a family of compatible congruences is also compatible, and that the composition of two compatible congruences is compatible as well; then from the above display it follows that the supremum of a family of compatible congruences is compatible as well. Therefore, the set $\mathrm{Con}\langle \boldsymbol{A}, F\rangle$ is a complete sublattice of $\mathrm{Con}\boldsymbol{A}$, and since it is a down-set, it is also an ideal. Obviously, the identity relation, denoted by Id_A, is a compatible congruence, and hence it is the least one. ⊠

Any complete lattice has a maximum, hence it makes sense to single out the maximum of the set of strict congruences of a matrix and give it a name:

DEFINITION 4.21
*The **Leibniz congruence** of a matrix $\langle \boldsymbol{A}, F\rangle$ is the largest congruence of the algebra $\boldsymbol{A}$ compatible with F:*

$$\boldsymbol{\Omega}^{\boldsymbol{A}}F := \max \mathrm{Con}\langle \boldsymbol{A}, F\rangle = \max\{\theta \in \mathrm{Con}\boldsymbol{A} : \theta \textit{ is compatible with } F\}.$$

In the case of the formula algebra the superscript is omitted: if $\Gamma \subseteq Fm$, then its Leibniz congruence is denoted by $\boldsymbol{\Omega}\Gamma$.

The notations $\boldsymbol{\Omega}\Gamma$ and $\boldsymbol{\Omega}^{\boldsymbol{A}}F$ are used in Chapters 2 and 3, namely in the analysis of the Lindenbaum-Tarski process for $\mathcal{C\ell}$, for implicative logics and for algebraizable logics; Corollary 4.55 confirms the coherence with the present general usage. From the second property in Theorem 4.20 and Definition 4.21 it follows that the condition "θ is compatible with F" can now be rewritten as "$\theta \subseteq \boldsymbol{\Omega}^{\boldsymbol{A}}F$"; that is:

COROLLARY 4.22
$\mathrm{Con}\langle \boldsymbol{A}, F\rangle = \{\theta \in \mathrm{Con}\boldsymbol{A} : \theta \subseteq \boldsymbol{\Omega}^{\boldsymbol{A}}F\}$. ⊠

The main result on the Leibniz congruence is the following characterization, due independently to Shoesmith and Smiley [225, 1978] and to Czelakowski [66, 1980], and which has its roots in the work of Łoś [165, 1949] for the case of the formula algebra.

THEOREM 4.23
Let $\boldsymbol{A}$ be any algebra, let $F \subseteq A$, and let $a, b \in A$. The following conditions are equivalent:

(i) $a \equiv b\ (\boldsymbol{\Omega}^{\boldsymbol{A}}F)$.

(ii) *For all unary polynomial functions f on $\boldsymbol{A}$, $[fa \in F \Leftrightarrow fb \in F]$.*

(iii) *For all $\delta(x, \vec{z}) \in Fm$ and all $\vec{c} \in \vec{A}$, $[\delta^{\boldsymbol{A}}(a, \vec{c}) \in F \Leftrightarrow \delta^{\boldsymbol{A}}(b, \vec{c}) \in F]$.*

PROOF: Clearly (ii) and (iii) are two ways of expressing the same property.

(i)⇒(iii) If $a \equiv b\ (\boldsymbol{\Omega}^{\boldsymbol{A}}F)$, a straightforward inductive argument using that $\boldsymbol{\Omega}^{\boldsymbol{A}}F$ is a congruence shows that $\delta^{\boldsymbol{A}}(a, \vec{c}) \equiv \delta^{\boldsymbol{A}}(b, \vec{c})\ (\boldsymbol{\Omega}^{\boldsymbol{A}}F)$. Then the compatibility of $\boldsymbol{\Omega}^{\boldsymbol{A}}F$ with F plus symmetry of $\boldsymbol{\Omega}^{\boldsymbol{A}}F$ show that $\delta^{\boldsymbol{A}}(a, \vec{c}) \in F$ if and only if $\delta^{\boldsymbol{A}}(b, \vec{c}) \in F$.

(ii)⇒(i) Define the relation $a \equiv b\ (\theta)$ if and only if a and b satisfy the condition in (ii). Observe that the previous part of the proof actually shows that $\boldsymbol{\Omega}^{\boldsymbol{A}}F \subseteq \theta$. Thus, in order to show that $\boldsymbol{\Omega}^{\boldsymbol{A}}F = \theta$ it is enough to show that $\theta \in \mathrm{Con}\langle \boldsymbol{A}, F\rangle$. It is trivial that θ is an equivalence relation. Now let $\lambda \in L$ with $\mathrm{ar}\lambda = n \geqslant 1$ and take $a_1, \dots, a_n, b_1, \dots, b_n \in A$ such that $a_i \equiv b_i\ (\theta)$ for $i = 1, \dots, n$. Let f be any unary polynomial function on $\boldsymbol{A}$ and define the following auxiliary functions on A: For any $c \in A$,

$$g_1 c := f(\lambda^{\boldsymbol{A}}(c, a_2, \dots, a_n))$$
$$g_2 c := f(\lambda^{\boldsymbol{A}}(b_1, c, a_3, \dots, a_n))$$

$$g_3 c := f(\lambda^{\boldsymbol{A}}(b_1, b_2, c, \dots, a_n))$$
$$\vdots$$
$$g_n c := f(\lambda^{\boldsymbol{A}}(b_1, b_2, \dots, b_{n-1}, c))$$

Clearly, these are all unary polynomial functions, therefore the assumption that $a_i \equiv b_i \ (\theta)$ implies that

$$\begin{aligned} f(\lambda^{\boldsymbol{A}}(a_1, a_2, \dots, a_n)) = g_1 a_1 \in F &\iff g_1 b_1 = g_2 a_2 \in F \\ &\iff g_2 b_2 = g_3 a_3 \in F \\ &\vdots \\ &\iff g_n b_n = f(\lambda^{\boldsymbol{A}}(b_1, b_2, \dots, b_n)) \in F, \end{aligned}$$

and therefore $\lambda^{\boldsymbol{A}}(a_1, \dots, a_n) \equiv \lambda^{\boldsymbol{A}}(b_1, \dots, b_n) \ (\theta)$. Thus, $\theta \in \mathrm{Con}\boldsymbol{A}$. Finally, θ is compatible with F, because the identity function is a unary polynomial function (corresponding to the formula $x \in V$), therefore if $a \equiv b \ (\theta)$ and $a \in F$, then by definition of θ, $b \in F$. ☒

In the formula algebra, to evaluate the variables in a formula by other formulas is just to apply a substitution, and produces further formulas; therefore in this case the second expression in the theorem can be simplified, as is easy to see:

COROLLARY 4.24
If $\Gamma \cup \{\alpha, \beta\} \subseteq Fm$, then $\alpha \equiv \beta \ (\boldsymbol{\Omega}\Gamma)$ if and only if for all $\delta(x, \vec{z}) \in Fm$,

$$\delta(\alpha, \vec{z}) \in \Gamma \iff \delta(\beta, \vec{z}) \in \Gamma .$$ ☒

As a first example of the application of this characterization, Exercise 4.16 establishes a result that is used elsewhere in the book.

The above characterizations may be taken to justify some interpretations of the Leibniz congruence. As a relation between formulas, that is, objects of a formal language, $\boldsymbol{\Omega}\Gamma$ has been called *the synonymity relation* (*relative to* Γ) in linguistics and in computer science: Corollary 4.24 can be read as saying that two formulas are equivalent modulo $\boldsymbol{\Omega}\Gamma$ when each can be replaced by the other inside any linguistic context without altering Γ's "view" of the result. Blok and Pigozzi (who, in their 1989 monograph [32] did christen the notion after Leibniz) say that for a matrix $\langle \boldsymbol{A}, F \rangle$,

> "the relation $\boldsymbol{\Omega}^{\boldsymbol{A}}F$ is the first-order analogue of second-order identity".

In second-order logic, the identity relation can be *defined* by postulating that $x \approx y \leftrightarrow \forall P(Px \leftrightarrow Py)$, with P ranging over all unary predicates. Two objects are called *indiscernible* when they satisfy the same properties; the philosophical principle that "indiscernibility implies identity" is considered as one of the cornerstones of the metaphysics of Leibniz. This interpretation, which supports also calling $\boldsymbol{\Omega}^{\boldsymbol{A}}F$ ***the indiscernibility relation***, is better understood in the setting of Section 4.5; the precise result appears in Proposition 4.62.

The ***Leibniz operator*** on the algebra $\boldsymbol{A}$ is the function

$$\begin{aligned} \boldsymbol{\Omega}^{\boldsymbol{A}} : \mathscr{P}A &\longrightarrow \operatorname{Con}\boldsymbol{A} \\ F &\longmapsto \boldsymbol{\Omega}^{\boldsymbol{A}}F, \end{aligned}$$

and $\boldsymbol{\Omega}$ denotes the Leibniz operator on the formula algebra. Notice that this function is a purely algebraic object, independent of any logic. However, the properties of this function when its domain is restricted to the family $\mathcal{F}i_{\mathcal{L}}\boldsymbol{A}$ (or to $\mathcal{T}h\mathcal{L}$ for the formula algebra) turn out to be very relevant to describe (some aspects of) the algebraic and metalogical behaviour of the logic $\mathcal{L}$ (and Lemma 4.41 shows that its target set is also restricted in this case). In particular, this function produces ***the Leibniz hierarchy***, a classification of logics according to several properties of this operator. In this chapter I just show that the Leibniz congruence makes it possible to generalize the Lindenbaum-Tarski process to a completely arbitrary logic (Definition 4.42 and Theorem 4.43), and that in the case of algebraizable logics this function, when restricted to $\mathcal{F}i_{\mathcal{L}}\boldsymbol{A}$, actually coincides with the isomorphism between filters and relative congruences given by the evaluated residuals of the transformers (Theorem 4.56). Chapter 6 contains an introduction to the study of this hierarchy, one of the cornerstones of abstract algebraic logic.

Strict homomorphisms and the reduction process

In universal algebra there is a correspondence between congruences (that is, quotients) and surjective homomorphisms in algebras. In the theory of matrices a similar correspondence exists for strict congruences; the corresponding kind of homomorphisms are the strict surjective homomorphisms. The next property says that *the Leibniz operator commutes with (inverse) surjective homomorphisms*, in a sense parallel to that of Theorem 3.58.

PROPOSITION 4.25
If $h \in \operatorname{Hom}(\boldsymbol{A},\boldsymbol{B})$ and $G \subseteq B$, then $h^{-1}\boldsymbol{\Omega}^{\boldsymbol{B}}G \subseteq \boldsymbol{\Omega}^{\boldsymbol{A}}h^{-1}G$. If moreover h is surjective, then $h^{-1}\boldsymbol{\Omega}^{\boldsymbol{B}}G = \boldsymbol{\Omega}^{\boldsymbol{A}}h^{-1}G$.

PROOF: I asume you know that the inverse image of a congruence by any homomorphism is a congruence (Exercise 4.15). Then $h^{-1}\boldsymbol{\Omega}^{\boldsymbol{B}}G \subseteq \boldsymbol{\Omega}^{\boldsymbol{A}}h^{-1}G$ holds if and only if $h^{-1}\boldsymbol{\Omega}^{\boldsymbol{B}}G$ is compatible with $h^{-1}G$, and this is straightforward to check. Now assume that h is surjective. Then, to show that $\boldsymbol{\Omega}^{\boldsymbol{A}}h^{-1}G \subseteq h^{-1}\boldsymbol{\Omega}^{\boldsymbol{B}}G$ we have to show that if $a \equiv b \ (\boldsymbol{\Omega}^{\boldsymbol{A}}h^{-1}G)$, then $ha \equiv hb \ (\boldsymbol{\Omega}^{\boldsymbol{B}}G)$. To this end Theorem 4.23 can be used, so let $\delta(x,\vec{z}) \in Fm$ and $\vec{c} \in \vec{B}$. Surjectivity of h implies that there are $\vec{d} \in \vec{A}$ such that $h\vec{d} = \vec{c}$. Then

$$\begin{aligned} \delta^{\boldsymbol{B}}(ha,\vec{c}) = \delta^{\boldsymbol{B}}(ha,h\vec{d}) = h\delta^{\boldsymbol{A}}(a,\vec{d}) \in G &\iff \delta^{\boldsymbol{A}}(a,\vec{d}) \in h^{-1}G \\ \iff \delta^{\boldsymbol{A}}(b,\vec{d}) \in h^{-1}G \iff \delta^{\boldsymbol{B}}(hb,\vec{c}) &= \delta^{\boldsymbol{B}}(hb,h\vec{d}) = h\delta^{\boldsymbol{A}}(b,\vec{d}) \in G. \quad \boxtimes \end{aligned}$$

In particular this holds for endomorphisms of a single algebra, and more in particular for substitutions of the formula algebra: If σ is a surjective substitution and $\Gamma \subseteq Fm$, then $\sigma^{-1}\boldsymbol{\Omega}\Gamma = \boldsymbol{\Omega}\sigma^{-1}\Gamma$; this is particularly appealing because if $\Gamma \in \mathcal{T}h\mathcal{L}$, then $\sigma^{-1}\Gamma \in \mathcal{T}h\mathcal{L}$ as well.

DEFINITION 4.26
*A **strict homomorphism from** $\langle \boldsymbol{A}, F\rangle$ **to** $\langle \boldsymbol{B}, G\rangle$ is an $h \in \mathrm{Hom}(\boldsymbol{A},\boldsymbol{B})$ such that $F = h^{-1}G$. A **matrix isomorphism** is a bijective strict homomorphism (that is, a strict algebraic isomorphism). Matrix isomorphisms wil be denoted by $\cong$.*

Hence, given $G \subseteq B$, any $h\colon \boldsymbol{A} \to \boldsymbol{B}$ is by definition strict from $\langle \boldsymbol{A}, h^{-1}G\rangle$ to $\langle \boldsymbol{B}, G\rangle$. The reverse view is more delicate: re-stating Lemma 4.18 in the present terminology, and particularizing it for the case of the canonical projection associated with a congruence, we obtain:

PROPOSITION 4.27
1. *If $h\colon \boldsymbol{A} \to \boldsymbol{B}$ and $F \subseteq A$, then h is strict from $\langle \boldsymbol{A}, F\rangle$ to $\langle \boldsymbol{B}, hF\rangle$, that is, $F = h^{-1}hF$, if and only if $\ker h \subseteq \boldsymbol{\Omega}^{\boldsymbol{A}}F$; that is, if and only if $\ker h$ is a strict congruence of $\langle \boldsymbol{A}, F\rangle$.*
2. *Let $\theta \in \mathrm{Con}\boldsymbol{A}$. The canonical projection $\pi\colon \langle \boldsymbol{A}, F\rangle \to \langle \boldsymbol{A}/\theta, F/\theta\rangle$ is a strict homomorphism if and only if θ is a strict congruence of $\langle \boldsymbol{A}, F\rangle$, or, equivalently, if and only if $\theta \subseteq \boldsymbol{\Omega}^{\boldsymbol{A}}F$.* ⊠

Recall that if $h\colon A \to B$ is a surjective function, then $hh^{-1}G = G$ for all $G \subseteq B$, and similarly for subsets of $B \times B$, therefore in particular $hh^{-1}\boldsymbol{\Omega}^{\boldsymbol{B}}G = \boldsymbol{\Omega}^{\boldsymbol{B}}G$. Thus, using Proposition 4.25:

COROLLARY 4.28
Let h be a strict and surjective homomorphism from $\langle \boldsymbol{A}, F\rangle$ to $\langle \boldsymbol{B}, G\rangle$.
1. *$F = h^{-1}G$ and $G = hF$. As a consequence, $F = h^{-1}hF$ and $\ker h$ is a strict congruence of $\langle \boldsymbol{A}, F\rangle$.*
2. *$\boldsymbol{\Omega}^{\boldsymbol{A}}F = h^{-1}\boldsymbol{\Omega}^{\boldsymbol{B}}G$ and $\boldsymbol{\Omega}^{\boldsymbol{B}}G = h\boldsymbol{\Omega}^{\boldsymbol{A}}F$.* ⊠

Obviously, all properties of surjective homomorphisms have their particular case for the canonical projections associated with congruences; you should not be distracted by the coexistence of the notations F/θ and πF, and so forth. For instance, after taking Proposition 4.27.2 into account, from Corollary 4.28.2 it immediately follows:

LEMMA 4.29
If $\theta \in \mathrm{Con}\boldsymbol{A}$ and $\theta \subseteq \boldsymbol{\Omega}^{\boldsymbol{A}}F$, then $\boldsymbol{\Omega}^{\boldsymbol{A}/\theta}(F/\theta) = (\boldsymbol{\Omega}^{\boldsymbol{A}}F)/\theta$. ⊠

It is possible, and not particularly difficult, to state and prove Homomorphism, Isomorphism and Correspondence Theorems for strict congruences and strict homomorphisms, analogous to the typical ones found in any first course in universal algebra (see for instance [51, Section II.6]); the following results are seldom used.

THEOREM 4.30

1. *If* $h\colon \langle \boldsymbol{A},F\rangle \to \langle \boldsymbol{B},G\rangle$ *is strict and surjective, then* $\ker h \in \operatorname{Con}\langle \boldsymbol{A},F\rangle$ *and* $\langle \boldsymbol{A},F\rangle/\ker h \cong \langle \boldsymbol{B},G\rangle$.
2. *If* $\theta,\theta' \in \operatorname{Con}\langle \boldsymbol{A},F\rangle$ *and* $\theta \subseteq \theta'$, *then* $\theta'/\theta \in \operatorname{Con}\langle \boldsymbol{A}/\theta,F/\theta\rangle$, *and moreover* $\langle \boldsymbol{A}/\theta,F/\theta\rangle/(\theta'/\theta) \cong \langle \boldsymbol{A}/\theta',F/\theta'\rangle$.
3. *For each* $\theta \in \operatorname{Con}\langle \boldsymbol{A},F\rangle$ *the function* $\theta' \mapsto \theta'/\theta$ *establishes an order isomorphism between the segment* $[\theta, \boldsymbol{\Omega}^{\boldsymbol{A}}F]$ *of* $\operatorname{Con}\langle \boldsymbol{A},F\rangle$ *and the lattice* $\operatorname{Con}\langle \boldsymbol{A}/\theta,F/\theta\rangle$.

PROOF: The first fact in point 1 appears in Corollary 4.28.1. The rest of the proofs actually consist in "adding strictness" to the homologous theorems of universal algebra. Since this is more or less straightforward, it is left to you as Exercises 4.17–4.19; doing this is recommended only if you have a universal-algebraic inclination. ⊠

Putting Proposition 4.27.2 and Theorem 4.30.1 together, we see that strict congruences are exactly the kernels of strict surjective homomorphisms, and that all strict images of a matrix are obtained, modulo isomorphism, as its quotients under strict congruences.

As a particular case of factorization under a strict congruence, we see that for each $F \subseteq A$, factoring under the Leibniz congruence of F is always strict from $\langle \boldsymbol{A},F\rangle$ to $\langle \boldsymbol{A}/\boldsymbol{\Omega}^{\boldsymbol{A}}F, F/\boldsymbol{\Omega}^{\boldsymbol{A}}F\rangle$. This process is called ***reduction*** because of the following definition and properties:

DEFINITION 4.31

A matrix $\langle \boldsymbol{A},F\rangle$ *is* ***reduced*** *when* $\boldsymbol{\Omega}^{\boldsymbol{A}}F = \mathrm{Id}_A$, *or, equivalently, when the identity is its only strict congruence. For every matrix* $\mathcal{M} = \langle \boldsymbol{A},F\rangle$, *its* ***reduction*** *is the matrix* $\mathcal{M}^* = \langle \boldsymbol{A},F\rangle^* := \langle \boldsymbol{A}/\boldsymbol{\Omega}^{\boldsymbol{A}}F, F/\boldsymbol{\Omega}^{\boldsymbol{A}}F\rangle$. *If* M *is a class of matrices, then we put* $\mathsf{M}^* := \{\mathcal{M}^* : \mathcal{M} \in \mathsf{M}\}$.

To consider some trivial situations, notice that in general $\boldsymbol{\Omega}^{\boldsymbol{A}}\varnothing = \boldsymbol{\Omega}^{\boldsymbol{A}}A = A \times A$. As a consequence (Exercise 4.21), all matrices on a trivial algebra are reduced, the reduction of a non-proper matrix is a trivial matrix, and a non-proper matrix is reduced if and only if it is in fact trivial.

For each matrix $\mathcal{M}$, the canonical projection $\mathcal{M} \to \mathcal{M}^*$ is a strict and surjective homomorphism. Moreover:

PROPOSITION 4.32

1. *For every matrix* $\mathcal{M}$, *the matrix* $\mathcal{M}^*$ *is reduced.*
2. *If* $\mathcal{M} \cong \mathcal{N}$, *then* $\mathcal{M}$ *is reduced if and only if* $\mathcal{N}$ *is.*
3. *If there is a strict and surjective homomorphism from* $\mathcal{M}$ *onto* $\mathcal{N}$, *then* $\mathcal{M}^* \cong \mathcal{N}^*$.
4. *If* $\mathcal{M} \cong \mathcal{N}$, *then* $\mathcal{M}^* \cong \mathcal{N}^*$.
5. *If* $\theta \in \operatorname{Con}\langle \boldsymbol{A},F\rangle$, *then* $\langle \boldsymbol{A},F\rangle^* \cong \langle \boldsymbol{A}/\theta,F/\theta\rangle^*$.

PROOF: To prove 1, put $\mathcal{M} = \langle \boldsymbol{A},F\rangle$ and use Lemma 4.29 with $\theta = \boldsymbol{\Omega}^{\boldsymbol{A}}F$. Then $\boldsymbol{\Omega}^{\boldsymbol{A}/\boldsymbol{\Omega}^{\boldsymbol{A}}F}(F/\boldsymbol{\Omega}^{\boldsymbol{A}}F) = (\boldsymbol{\Omega}^{\boldsymbol{A}}F)/\boldsymbol{\Omega}^{\boldsymbol{A}}F = \mathrm{Id}_{A/\boldsymbol{\Omega}^{\boldsymbol{A}}F}$. To prove 2, apply Corollary 4.28.2 to the isomorphism, which in particular is a surjective strict homomorphism, and to its inverse. To prove 3, assume that $h\colon \langle \boldsymbol{A},F\rangle \to \langle \boldsymbol{B},G\rangle$ is strict and surjective and $\pi\colon \langle \boldsymbol{B}, G\rangle \to \langle \boldsymbol{B}, G\rangle^*$ performs the reduction process, that is, the quotient under $\boldsymbol{\Omega}^{\boldsymbol{B}}G$. Then the composition $\pi h\colon \langle \boldsymbol{A}, F\rangle \to \langle \boldsymbol{B}, G\rangle^*$ is also strict and surjective (Exercise 4.22), and as a consequence previous results can be applied. A straightforward computation, together with Corollary 4.28.2, shows that $\ker(\pi h) = h^{-1}\boldsymbol{\Omega}^{\boldsymbol{B}}G = \boldsymbol{\Omega}^{\boldsymbol{A}}F$, and then Theorem 4.30.1 tells us that $\langle \boldsymbol{A}, F\rangle^* \cong \langle \boldsymbol{B}, G\rangle^*$. Finally, 4 and 5 are particular cases of 3. ⊠

Thus, in some sense, a strict and surjective homomorphism is somehow like a "partial reduction" that goes halfway towards the true reduction; this one, as is to be expected, cannot be further reduced, as the next result shows.

COROLLARY 4.33
For any matrix $\mathcal{M}$ the following conditions are equivalent:

(i) *$\mathcal{M}$ is reduced.*

(ii) $\mathcal{M} \cong \mathcal{M}^*$.

(iii) *$\mathcal{M}$ is isomorphic to some reduced matrix.*

(iv) *Any strict and surjective homomorphism from $\mathcal{M}$ onto any other matrix is in fact an isomorphism.*

PROOF: Just combine previous results (Exercise 4.23). ⊠

That is, a reduced matrix is one that cannot be further reduced.

As a final point on the reduction construction, notice that the converse of Proposition 4.32.3 cannot be true in general (just for cardinality considerations). The result in Exercise 4.24 helps in showing that the relation of having isomorphic reductions is the symmetric and transitive closure of the relation of one matrix being a strict and surjective image of another; this relation plays an important role in some model-theoretic investigations on the theory of matrices along the lines sketched in Section 4.5.

Exercises for Section 4.2

EXERCISE 4.14. Let $\boldsymbol{A}$ be any algebra, $\theta \in \operatorname{Con}\boldsymbol{A}$ and $F \subseteq A$. Prove that θ is compatible with F if and only if for all $G \subseteq A$, $G \subseteq F$ if and only if $G/\theta \subseteq F/\theta$.

EXERCISE 4.15. Recall that $h^{-1}\theta := \{\langle a,b\rangle \in A \times A : \langle ha, hb\rangle \in \theta\}$ for any function $h\colon A \to B$ and any $\theta \subseteq B \times B$. Prove that if $h\colon \boldsymbol{A} \to \boldsymbol{B}$ and $\theta \in \operatorname{Con}\boldsymbol{B}$, then $h^{-1}\theta \in \operatorname{Con}\boldsymbol{A}$.

EXERCISE 4.16. Let $\mathcal{L}$ be a logic having an algebraic semantics, with $E(x)$ as set of defining equations. Use the result in Exercise 3.6 to prove that for each $\varepsilon \approx \delta \in E(x)$ and each $\langle \boldsymbol{A}, F\rangle \in \mathsf{Mod}\mathcal{L}$, if $a \in F$, then $\varepsilon^{\boldsymbol{A}}(a) \equiv \delta^{\boldsymbol{A}}(a)\ (\boldsymbol{\Omega}^{\boldsymbol{A}}F)$. Deduce from this that for each $\varepsilon \approx \delta \in E(x)$, $\varepsilon \equiv \delta\ (\boldsymbol{\Omega} C_{\mathcal{L}}\{x\})$.

EXERCISE 4.17. Prove that if $h\colon \langle \boldsymbol{A}, F\rangle \to \langle \boldsymbol{B}, G\rangle$ is a strict and surjective homomorphism, then $\langle \boldsymbol{A}, F\rangle / \ker h \cong \langle \boldsymbol{B}, G\rangle$.

HINT. Adapt Theorem II.6.12 of [51].

EXERCISE 4.18. Prove that if $\theta, \theta' \in \operatorname{Con}\langle \boldsymbol{A}, F\rangle$ and $\theta \subseteq \theta'$, then $\theta'/\theta \in \operatorname{Con}\langle \boldsymbol{A}/\theta, F/\theta\rangle$ and $\langle \boldsymbol{A}/\theta, F/\theta\rangle/(\theta'/\theta) \cong \langle \boldsymbol{A}/\theta', F/\theta'\rangle$.

HINT. Adapt Theorem II.6.15 of [51].

EXERCISE 4.19. Let $\theta \in \mathrm{Con}\langle \boldsymbol{A}, F\rangle$ and consider the function $\theta' \mapsto \theta'/\theta$. Prove that this function establishes an order isomorphism between the segment $[\theta, \boldsymbol{\Omega}^{\boldsymbol{A}} F]$ of $\mathrm{Con}\langle \boldsymbol{A}, F\rangle$ and the lattice $\mathrm{Con}\langle \boldsymbol{A}/\theta, F/\theta\rangle$.

HINT. Adapt Theorem II.6.20 of [51].

EXERCISE 4.20. Let $\boldsymbol{A}$ be an arbitrary algebra, and let $F, G \subseteq A$. Show that the matrix $\langle \boldsymbol{A}/(\boldsymbol{\Omega}^{\boldsymbol{A}} F \cap \boldsymbol{\Omega}^{\boldsymbol{A}} G), (F \cap G)/(\boldsymbol{\Omega}^{\boldsymbol{A}} F \cap \boldsymbol{\Omega}^{\boldsymbol{A}} G)\rangle$ is isomorphic to a subdirect product of the reductions of the matrices $\langle \boldsymbol{A}, F\rangle$ and $\langle \boldsymbol{A}, G\rangle$. This construction is used in the proof of Theorem 6.17.

HINT. Show that the natural homomorphism $ha := \langle a/\boldsymbol{\Omega}^{\boldsymbol{A}} F,\ a/\boldsymbol{\Omega}^{\boldsymbol{A}} G\rangle$ induces a strict homomorphism from the matrix $\langle \boldsymbol{A}, F \cap G\rangle$ to the product of the mentioned reduced matrices. Then restrict h to its image to make it strict and surjective, determine $\ker h$, and use Theorem 4.30.1.

EXERCISE 4.21. Check that in any algebra $\boldsymbol{A}$ any congruence is compatible with the empty subset and with the whole universe. Deduce that $\boldsymbol{\Omega}^{\boldsymbol{A}} \emptyset = \boldsymbol{\Omega}^{\boldsymbol{A}} A = A \times A$. Use these properties to show that all matrices on a trivial algebra are reduced, that the reduction of a non-proper matrix is a trivial matrix, and that a non-proper matrix is reduced if and only if it is in fact a trivial matrix.

EXERCISE 4.22. Prove that the composition of two strict and surjective homomorphisms between matrices is also strict and surjective.

EXERCISE 4.23. Prove that a matrix is reduced if and only if it is isomorphic to its own reduction, and if and only if every strict and surjective homomorphism from it to another matrix is actually an isomorphism.

EXERCISE 4.24. Let $\mathcal{M}_1,, \mathcal{M}_2$ be two matrices. Prove that $\mathcal{M}_1^* \cong \mathcal{M}_2^*$ if and only if for some matrix $\mathcal{M}_3$, there are strict and surjective homomorphisms from $\mathcal{M}_1$ onto $\mathcal{M}_3$ and from $\mathcal{M}_2$ onto $\mathcal{M}_3$.

4.3 Reduced models and Leibniz-reduced algebras

The preceding section contains no references to a logic; its constructions and results are purely mathematical. Now their effects upon matrix models of a logic will be examined. The key point is the following.

THEOREM 4.34

If there is a strict surjective homomorphism between two matrices, then these matrices define the same logic. A matrix and its reduction define the same logic. Isomorphic matrices define the same logic.

PROOF: Let $h \colon \langle \boldsymbol{A}, F\rangle \to \langle \boldsymbol{B}, G\rangle$ be the homomorphism assumed in the first statement. Take any $\Gamma \cup \{\varphi\} \in Fm$ and assume first that $\Gamma \vdash_{\langle \boldsymbol{B}, G\rangle} \varphi$. In order to show that $\Gamma \vdash_{\langle \boldsymbol{A}, F\rangle} \varphi$ take any $g \colon \boldsymbol{Fm} \to \boldsymbol{A}$ such that $g\Gamma \subseteq F$. Then $hg \in \mathrm{Hom}(\boldsymbol{Fm}, \boldsymbol{B})$ is such that $hg\Gamma \subseteq hF = G$ and hence by assumption $hg\varphi \in G$, that is, $g\varphi \in h^{-1}G = F$. For the converse, assume that $\Gamma \vdash_{\langle \boldsymbol{A}, F\rangle} \varphi$ and that $f \colon \boldsymbol{Fm} \to \boldsymbol{B}$ is such that $f\Gamma \subseteq G$. By surjectivity of h, you can apply Lemma 1.2 and find a $g \in \mathrm{Hom}(\boldsymbol{Fm}, \boldsymbol{A})$ such that $hg = f$. Hence

$hg\Gamma \subseteq G$, that is, $g\Gamma \subseteq h^{-1}G = F$. The assumption implies that $g\varphi \in F$, so $f\varphi = hg\varphi \in hF = G$. This shows that $\Gamma \vdash_{\langle \boldsymbol{B},G\rangle} \varphi$. The remaining two statements are particular cases of the first one. ⊠

The application of the first property produces many non-standard definitions of well-known logics; see for instance Exercise 4.25, where this is used to confirm some statements on algebraic semantics (of $\mathcal{C\ell}$, in this case) made on page 121. The last property also helps in proving that the logic $\mathcal{B}$ (Example 4.12) can actually be defined by a single matrix (Exercise 4.26).

Many properties of models are just reformulations of properties of filters, because of Lemma 4.15: $\langle \boldsymbol{A}, F\rangle \in \mathsf{Mod}\,\mathcal{L}$ if and only if $F \in \mathcal{F}i_{\mathcal{L}}\boldsymbol{A}$; this provides two ways of phrasing the same property: if focusing on a fixed algebra it is more usual to speak of filters, while in more global views one usually speaks of models. You are particularly advised to review Propositions 2.24 and 2.25, which are specially relevant here.

PROPOSITION 4.35
Let $\mathcal{L}$ be a logic.

1. *If there is a strict surjective homomorphism between the matrices $\langle \boldsymbol{A}, F\rangle$ and $\langle \boldsymbol{B}, G\rangle$, then $F \in \mathcal{F}i_{\mathcal{L}}\boldsymbol{A}$ if and only if $G \in \mathcal{F}i_{\mathcal{L}}\boldsymbol{B}$.*
2. *For any matrix $\langle \boldsymbol{A}, F\rangle$ and any $\theta \in \mathrm{Con}\langle \boldsymbol{A}, F\rangle$, $F \in \mathcal{F}i_{\mathcal{L}}\boldsymbol{A}$ if and only if $F/\theta \in \mathcal{F}i_{\mathcal{L}}(\boldsymbol{A}/\theta)$.*
3. *For any matrix $\langle \boldsymbol{A}, F\rangle$, $F \in \mathcal{F}i_{\mathcal{L}}\boldsymbol{A}$ if and only if $F/\boldsymbol{\Omega}^{\boldsymbol{A}}F \in \mathcal{F}i_{\mathcal{L}}(\boldsymbol{A}/\boldsymbol{\Omega}^{\boldsymbol{A}}F)$.*
4. *The class $\mathsf{Mod}\,\mathcal{L}$ is closed under images and inverse images by strict homomorphisms (in particular, under reductions) and under isomorphisms.*

PROOF: 1 is a re-statement of the results proved in the propositions mentioned above (see also Exercise 4.27). 2 and 3 are particular cases of 1. The first fact in 4 is a re-statement of 1 in terms of the class $\mathsf{Mod}\,\mathcal{L}$, and the second one is a particular case of the first. An alternative proof of 1 is to use Theorem 4.34 and that $F \in \mathcal{F}i_{\mathcal{L}}\boldsymbol{A}$ if and only if $\vdash_{\mathcal{L}} \subseteq \vdash_{\langle \boldsymbol{A},F\rangle}$. ⊠

LEMMA 4.36
Let $\mathcal{L}$ be a logic, and let $h\colon \boldsymbol{A} \to \boldsymbol{B}$ be a surjective homomorphism. Then the functions h and h^{-1} induce mutually inverse order isomorphisms between the set $\{F \in \mathcal{F}i_{\mathcal{L}}\boldsymbol{A} : \ker h \subseteq \boldsymbol{\Omega}^{\boldsymbol{A}}F\}$ and the set $\mathcal{F}i_{\mathcal{L}}\boldsymbol{B}$. As a consequence, $h^{-1}\mathcal{F}i_{\mathcal{L}}\boldsymbol{B} = \{F \in \mathcal{F}i_{\mathcal{L}}\boldsymbol{A} : \ker h \subseteq \boldsymbol{\Omega}^{\boldsymbol{A}}F\}$ and $\mathcal{F}i_{\mathcal{L}}\boldsymbol{B} = \{hF : F \in \mathcal{F}i_{\mathcal{L}}\boldsymbol{A}, \ker h \subseteq \boldsymbol{\Omega}^{\boldsymbol{A}}F\}$.

In particular, if $\theta \in \mathrm{Con}\boldsymbol{A}$ and π is the associated canonical projection, then the functions π and π^{-1} induce mutually inverse order isomorphisms between the set $\{F \in \mathcal{F}i_{\mathcal{L}}\boldsymbol{A} : \theta \subseteq \boldsymbol{\Omega}^{\boldsymbol{A}}F\}$ and the set $\mathcal{F}i_{\mathcal{L}}(\boldsymbol{A}/\theta)$, and therefore $\pi^{-1}\mathcal{F}i_{\mathcal{L}}(\boldsymbol{A}/\theta) = \{F \in \mathcal{F}i_{\mathcal{L}}\boldsymbol{A} : \theta \subseteq \boldsymbol{\Omega}^{\boldsymbol{A}}F\}$ and $\mathcal{F}i_{\mathcal{L}}(\boldsymbol{A}/\theta) = \{F/\theta : F \in \mathcal{F}i_{\mathcal{L}}\boldsymbol{A}, \theta \subseteq \boldsymbol{\Omega}^{\boldsymbol{A}}F\}$.

PROOF: This is the restriction to $\mathcal{L}$-filters of the order isomorphism of Lemma 4.18, so we need only check that h and h^{-1} send $\mathcal{L}$-filters to $\mathcal{L}$-filters; but this is what Proposition 4.35.1 asserts. The condition "$\ker h$ compatible with F" of Lemma 4.18 can be rewritten here as "$\ker h \subseteq \boldsymbol{\Omega}^{\boldsymbol{A}}F$", by Corollary 4.22. ⊠

The application of this result is in general limited, because there is no nicer characterization of the filters F the congruence $\ker h$ is compatible with; in the case of quotients, two important ones are the characterization of full g-models of a logic in Proposition 5.94, and the Correspondence Theorem 6.19 for protoalgebraic logics.

Two more classes, one of matrices and one of algebras, arise naturally:

Definition 4.37

Let $\mathcal{L}$ be a logic. The classes of ***reduced models*** *of $\mathcal{L}$ and of its algebraic reducts are denoted as follows:*

$$\begin{aligned} \mathsf{Mod}^*\mathcal{L} &:= \{\langle \boldsymbol{A}, F\rangle \in \mathsf{Mod}\mathcal{L} : \langle \boldsymbol{A}, F\rangle \text{ is reduced}\} \\ \mathsf{Alg}^*\mathcal{L} &:= \{\boldsymbol{A} : \text{ there is some } F \subseteq A \text{ such that } \langle \boldsymbol{A}, F\rangle \in \mathsf{Mod}^*\mathcal{L}\} \\ &= \mathsf{Alg}(\mathsf{Mod}^*\mathcal{L}). \end{aligned}$$

The algebras in the class $\mathsf{Alg}^*\mathcal{L}$ are sometimes called the ***Leibniz-reduced $\mathcal{L}$-algebras***, for obvious reasons. This class, introduced (implicitly) by Wójcicki in 1973 as a generalization of Rasiowa's notion of an "$\mathcal{L}$-algebra" for implicative logics, was believed to be the "canonical" algebraic counterpart of an arbitrary logic $\mathcal{L}$, until the proof in 1991 that $\mathsf{Alg}^*\mathcal{C}\ell_{\wedge\vee}$ is not the class of distributive lattices (see Example 4.47 and Corollary 4.50). However, as Proposition 4.57 and Corollary 4.59 show, for implicative logics and for algebraizable logics it coincides with the classes of algebras introduced in previous chapters as their algebraic counterpart; actually, Wójcicki chose the same notation introduced by Rasiowa (Definition 2.5). The presence of the * in $\mathsf{Alg}^*\mathcal{L}$ does not mean that this class is obtained from another class by some process of "reduction"; it is there just by tradition, and to recall that $\mathsf{Alg}^*\mathcal{L} = \mathsf{Alg}(\mathsf{Mod}^*\mathcal{L})$ (see also the discussion after Definition 5.67).

Proposition 4.38

For any logic $\mathcal{L}$, the classes $\mathsf{Mod}^*\mathcal{L}$ and $\mathsf{Alg}^*\mathcal{L}$ are closed under isomorphisms, and contain all trivial matrices and all trivial algebras, respectively.

Proof: That the class of matrices $\mathsf{Mod}^*\mathcal{L}$ is closed under isomorphisms follows from Propositions 4.32.2 and 4.35.4, and this fact implies that also the class of its algebraic reducts $\mathsf{Alg}^*\mathcal{L}$ is closed under isomorphisms. A trivial matrix has the form $\langle \boldsymbol{A}, \{a\}\rangle$ where $\boldsymbol{A}$ is a trivial algebra with universe $\{a\}$. The total subset is always an $\mathcal{L}$-filter, hence $\langle \boldsymbol{A}, \{a\}\rangle \in \mathsf{Mod}\mathcal{L}$. But $\boldsymbol{\Omega}^{\boldsymbol{A}}\{a\} = \{a\} \times \{a\} = \{\langle a, a\rangle\} = \mathrm{Id}_A$, therefore $\langle \boldsymbol{A}, \{a\}\rangle$ is reduced, therefore $\langle \boldsymbol{A}, \{a\}\rangle \in \mathsf{Mod}^*\mathcal{L}$. As a consequence, trivial algebras belong to $\mathsf{Alg}^*\mathcal{L}$. ⊠

Notice that $\langle \boldsymbol{A}, \varnothing\rangle \in \mathsf{Mod}^*\mathcal{L}$ if and only if $\boldsymbol{A}$ is a trivial algebra (making $\langle \boldsymbol{A}, \varnothing\rangle$ reduced) *and* $\mathcal{L}$ has no theorems (making $\langle \boldsymbol{A}, \varnothing\rangle$ a model of $\mathcal{L}$); this partly explains why matrices like $\langle \boldsymbol{A}, \varnothing\rangle$ are not considered "trivial".

Combining several already proven results, one also gets a subtle point:

Proposition 4.39
For any logic $\mathcal{L}$, $\mathsf{Mod}^*\mathcal{L} = \mathbb{I}\big((\mathsf{Mod}\,\mathcal{L})^*\big)$. That is, a matrix is a reduced model of a logic $\mathcal{L}$ if and only if it is isomorphic to the reduction of some model of $\mathcal{L}$. ⊠

Thus, the reduced models of a logic are *not* obtained by just reducing all its models: one has to add all isomorphic copies of these reductions.

Corollary 4.40
For any logic $\mathcal{L}$ and every algebra $\boldsymbol{A}$, if $F \in \mathcal{F}i_{\mathcal{L}}\boldsymbol{A}$, then $\boldsymbol{A}/\boldsymbol{\Omega}^{\boldsymbol{A}}F \in \mathsf{Alg}^*\mathcal{L}$ and $\boldsymbol{\Omega}^{\boldsymbol{A}}F \in \mathrm{Con}_{\mathsf{Alg}^*\mathcal{L}}\boldsymbol{A}$. ⊠

Thus, the Leibniz operator when restricted to the filters of a logic gives images in the set of congruences of the algebra relative to the class $\mathsf{Alg}^*\mathcal{L}$. But there is more:

Lemma 4.41
For every logic $\mathcal{L}$ and every algebra $\boldsymbol{A}$, the Leibniz operator $\boldsymbol{\Omega}^{\boldsymbol{A}}$ is a surjective function from $\mathcal{F}i_{\mathcal{L}}\boldsymbol{A}$ onto $\mathrm{Con}_{\mathsf{Alg}^*\mathcal{L}}\boldsymbol{A}$.

Proof: By Corollary 4.40, $\boldsymbol{\Omega}^{\boldsymbol{A}} \colon \mathcal{F}i_{\mathcal{L}}\boldsymbol{A} \to \mathrm{Con}_{\mathsf{Alg}^*\mathcal{L}}\boldsymbol{A}$. It remains to see that it is surjective. Let $\theta \in \mathrm{Con}_{\mathsf{Alg}^*\mathcal{L}}\boldsymbol{A}$: Since $\theta \in \mathrm{Con}\boldsymbol{A}$ and $\boldsymbol{A}/\theta \in \mathsf{Alg}^*\mathcal{L}$, there is some $G \in \mathcal{F}i_{\mathcal{L}}(\boldsymbol{A}/\theta)$ such that $\boldsymbol{\Omega}^{\boldsymbol{A}/\theta}G = \mathrm{Id}_{A/\theta}$. Consider the canonical projection $\pi \colon \boldsymbol{A} \to \boldsymbol{A}/\theta$ and put $F := h^{-1}G$. Then, by Proposition 2.24, $F \in \mathcal{F}i_{\mathcal{L}}\boldsymbol{A}$, and we obtain that

$$\theta = \ker\pi = \pi^{-1}\mathrm{Id}_{A/\theta} = \pi^{-1}\boldsymbol{\Omega}^{\boldsymbol{A}/\theta}G = \boldsymbol{\Omega}^{\boldsymbol{A}}\pi^{-1}G = \boldsymbol{\Omega}^{\boldsymbol{A}}F.$$

Notice how in the next-to-last step Proposition 4.25 has been used, thanks to the surjectivity of π. ⊠

Therefore (cf. page 199), the ***Leibniz operator relative to a logic*** $\mathcal{L}$, on an algebra $\boldsymbol{A}$, is the function

$$\begin{aligned} \boldsymbol{\Omega}^{\boldsymbol{A}} : \mathcal{F}i_{\mathcal{L}}\boldsymbol{A} &\longrightarrow \mathrm{Con}_{\mathsf{Alg}^*\mathcal{L}}\boldsymbol{A} \\ F &\longmapsto \boldsymbol{\Omega}^{\boldsymbol{A}}F. \end{aligned}$$

It is the notion of a reduced matrix that makes *the final, truly general version of the Lindenbaum-Tarski process* possible. It involves the following classes:

Definition 4.42
For any logic $\mathcal{L}$, the class of its **Lindenbaum-Tarski models** *is the class*

$$\mathsf{LTMod}^*\mathcal{L} := \big\{\langle \boldsymbol{Fm}, T\rangle^* : T \in \mathcal{T}h\mathcal{L}\big\} = \big\{\langle \boldsymbol{Fm}/\boldsymbol{\Omega}T, T/\boldsymbol{\Omega}T\rangle : T \in \mathcal{T}h\mathcal{L}\big\},$$

and its **Lindenbaum-Tarski algebras** *are the algebraic reducts of the models:*

$$\mathsf{LTAlg}^*\mathcal{L} := \mathsf{Alg}(\mathsf{LTMod}^*\mathcal{L}) = \big\{\boldsymbol{Fm}/\boldsymbol{\Omega}T : T \in \mathcal{T}h\mathcal{L}\big\}.$$

This is a generalization of the definition of $\mathsf{LTAlg}^*\mathcal{L}$ given for algebraizable logics in Section 3.3; as in the case of the notation $\boldsymbol{\Omega}$, Corollary 4.55 shows that there is no conflict in using the same symbols. Now the following fundamental result† is straightforward to prove:

THEOREM 4.43 (COMPLETENESS)
Any logic $\mathcal{L}$ *is complete with respect to any class* M *of matrices satisfying that* $\mathsf{LTMod}^*\mathcal{L} \subseteq \mathsf{M} \subseteq \mathsf{Mod}^*\mathcal{L}$. *In particular, every logic is complete with respect to the class of all its reduced models.*

PROOF: Since reduced models are models, from the definition of model and the inclusions in the assumptions it follows that

$$\vdash_{\mathcal{L}} \subseteq \vdash_{\mathsf{Mod}^*\mathcal{L}} \subseteq \vdash_{\mathsf{M}} \subseteq \vdash_{\mathsf{LTMod}^*\mathcal{L}}$$

and it remains only to prove that $\vdash_{\mathsf{LTMod}^*\mathcal{L}} \subseteq \vdash_{\mathcal{L}}$, which is done by contraposition, by following the Lindenbaum-Tarski process:

(LT1) Assume that $\Gamma \nvdash_{\mathcal{L}} \varphi$ and consider the theory T generated by Γ. Thus $\Gamma \subseteq T$ but $\varphi \notin T$.

(LT2) Consider the relation $\boldsymbol{\Omega} T$ of Definition 4.21.

(LT3) By definition this relation is a congruence of $\boldsymbol{Fm}$. Consider the canonical projection $\pi\colon \boldsymbol{Fm} \to \boldsymbol{Fm}/\boldsymbol{\Omega} T$.

(LT4) $\langle \boldsymbol{Fm}/\boldsymbol{\Omega} T, T/\boldsymbol{\Omega} T\rangle \in \mathsf{LTMod}^*\mathcal{L}$ by Definition 4.42.

(LT5) For any $\alpha \in Fm$, $\alpha \in T$ if and only if $\pi\alpha \in T/\boldsymbol{\Omega} T$, because $\boldsymbol{\Omega} T$ is trivially compatible with T.

The process is finished here. Applying (LT5) to (LT1) we see that $\pi\Gamma \subseteq T/\boldsymbol{\Omega} T$ while $\pi\varphi \notin T/\boldsymbol{\Omega} T$. By (LT4), this establishes that $\Gamma \nvdash_{\mathsf{LTMod}^*\mathcal{L}} \varphi$, and hence completes the proof. ⊠

A comparison with the proof of Theorem 3.27 and with the initial analysis of the process for $\mathcal{Cl}$ in Section 2.1 makes it apparent that, in fact, this completeness theorem is completely [sic] natural. Exercise 4.30 suggests an alternative, less explicit proof. Anyway, we see that every logic has (several) complete reduced matrix semantics: here we have found two natural ones, the class of all its reduced matrix models and the class of its Lindenbaum-Tarski models. The latter are paradigmatic in the following sense:

THEOREM 4.44
For any logic $\mathcal{L}$, *the countably generated matrices in* $\mathsf{Mod}^*\mathcal{L}$ *are the matrices isomorphic to one in* $\mathsf{LTMod}^*\mathcal{L}$.

PROOF: Let $\langle \boldsymbol{A}, F\rangle \in \mathsf{Mod}^*\mathcal{L}$ be countably generated (this means that its algebraic reduct is so). This implies that there exists a surjective $h \in \mathrm{Hom}(\boldsymbol{Fm}, \boldsymbol{A})$. If $\Gamma := h^{-1}F$, then $\Gamma \in \mathcal{Th}\mathcal{L}$ by Corollary 2.20.2, and h becomes a strict surjective homomorphism from $\langle \boldsymbol{Fm}, \Gamma\rangle$ onto $\langle \boldsymbol{A}, F\rangle$. By Proposition 4.32.3 this implies

† The theorem should be credited to Wójcicki, although it is not explicitly stated in [246], where he introduced reduced matrices (he called them "simple"); however, it is clearly implicit in the way he applied his other results to implicative logics.

that $\langle \boldsymbol{A}, F\rangle^* \cong \langle \boldsymbol{Fm}, \Gamma\rangle^*$ but since $\langle \boldsymbol{A}, F\rangle$ is reduced, Corollary 4.33 implies that $\langle \boldsymbol{A}, F\rangle \cong \langle \boldsymbol{Fm}, \Gamma\rangle^*$. The converse implication follows from the fact that the class $\mathsf{Mod}^*\mathcal{L}$ is closed under isomorphisms (Proposition 4.38) and that clearly $\mathsf{LTMod}^*\mathcal{L} \subseteq \mathsf{Mod}^*\mathcal{L}$. Finally, by definition $\langle \boldsymbol{Fm}, \Gamma\rangle^* \in \mathsf{LTMod}^*\mathcal{L}$. ⊠

Corollary 4.45

For any logic $\mathcal{L}$, $\mathsf{LTAlg}^*\mathcal{L} \subseteq \mathsf{Alg}^*\mathcal{L}$, *and the countably generated algebras in* $\mathsf{Alg}^*\mathcal{L}$ *are the algebras isomorphic to one in* $\mathsf{LTAlg}^*\mathcal{L}$. ⊠

Reduced models seem to constitute a better algebra-based semantics than arbitrary models, which can be quite strange. Then the natural question of *how to determine them* for a particular logic $\mathcal{L}$ arises. In general, this depends on having a Hilbert-style presentation of $\mathcal{L}$ and a workable description of $\boldsymbol{\Omega}^{\boldsymbol{A}}F$ for $F \in \mathcal{F}i_{\mathcal{L}}\boldsymbol{A}$; in Section 4.4 this is solved for algebraizable logics in general, and in particular for implicative logics, because in these cases the reduced models are related to the algebras in $\mathsf{Alg}^*\mathcal{L}$ in a one-to-one way, and indeed a natural one. In particular, $\mathsf{Mod}^*\mathcal{C\ell} = \{\langle \boldsymbol{A}, \{1\}\rangle : \boldsymbol{A} \in \mathsf{BA}\}$, and hence $\mathsf{Alg}^*\mathcal{C\ell} = \mathsf{BA}$; and the same holds for $\mathcal{I\ell}$ with HA, for $\mathcal{I\ell}_{\rightarrow}$ with HiA, and so forth. But *there is no algorithm of universal application,* and as a matter of fact, the determination of the reduced models of some less standard logics suggests they might not be so meaningful as it may seem at first glance.

Here two of these non-standard examples, corresponding to logics very simple to define, will be reviewed: the determination of the reduced models of the fragments of classical logic with only conjunction, and with only conjunction and disjunction; the first produces a very simple characterization, and the second a rather complicated one. Although working out (almost) all the details is a useful illustration of how to proceed in non-implicative logics, this in no way showcases a generalizable procedure.

Example 4.46 (The conjunction fragment of $\mathcal{C\ell}$)

Let $\mathcal{C\ell}_{\wedge}$ denote the fragment of classical logic with only conjunction $\wedge$ in the language. By Proposition 4.6, $\mathcal{C\ell}_{\wedge}$ is the logic defined by the two-element matrix $\langle \boldsymbol{2}_{\wedge}, \{1\}\rangle$ where $\boldsymbol{2}_{\wedge}$ denotes the $\langle\wedge\rangle$ reduct of the two-element Boolean algebra $\boldsymbol{2}$ (Example 1.11). Similarly, $\boldsymbol{1}_{\wedge}$ denotes a trivial (one-element) algebra of this type. This logic does not have theorems (by Exercise 1.29; this implies it is not algebraizable) and is finitary (by Exercise 1.28, or by Corollary 4.9).

From the given definition, it is obvious that

$$\begin{aligned} x \wedge y &\vdash_{\mathcal{C\ell}_{\wedge}} x \\ x \wedge y &\vdash_{\mathcal{C\ell}_{\wedge}} y \\ x, y &\vdash_{\mathcal{C\ell}_{\wedge}} x \wedge y \end{aligned} \tag{50}$$

and therefore, in any algebra $\boldsymbol{A} = \langle A, \wedge\rangle$, the $\mathcal{C\ell}_{\wedge}$-filters F satisfy that

$$a \wedge b \in F \iff a, b \in F. \tag{51}$$

Using just this, it is easy to prove (Exercise 4.32) that if $\langle \boldsymbol{A}, F\rangle \in \mathsf{Mod}\mathcal{Cl}_\wedge$, then for all $a,b \in A$, $a \equiv b\ (\boldsymbol{\Omega}^{\boldsymbol{A}}F)$ if and only if $\big[a \in F \iff b \in F\big]$, that is, if and only if either $a,b \in F$ or $a,b \notin F$. Observe that this is to say that the formulas $\delta(x,\vec{z})$ used in the characterization in Theorem 4.23 can be reduced to just one formula: x. As a consequence, the quotient set is $A/\boldsymbol{\Omega}^{\boldsymbol{A}}F = \{F, A \smallsetminus F\}$, which can have one or two elements. It is easy to check, again using just (51), that the quotient algebra is either $\boldsymbol{A}/\boldsymbol{\Omega}^{\boldsymbol{A}}F \cong \mathbf{1}_\wedge$ or $\boldsymbol{A}/\boldsymbol{\Omega}^{\boldsymbol{A}}F \cong \mathbf{2}_\wedge$. Therefore, $\mathsf{Alg}^*\mathcal{Cl}_\wedge = \mathbb{I}\{\mathbf{1}_\wedge, \mathbf{2}_\wedge\}$, a deceptively simple class of algebras! Moreover, the three cases $F = \emptyset$, $F = A$ and F non-empty and proper imply that the quotient matrix is isomorphic either to $\langle \mathbf{1}_\wedge, \emptyset\rangle$ or to $\langle \mathbf{1}_\wedge, \{1\}\rangle$ or to $\langle \mathbf{2}_\wedge, \{1\}\rangle$, respectivley; therefore, $\mathsf{Mod}^*\mathcal{Cl}_\wedge = \mathbb{I}\{\langle \mathbf{1}_\wedge, \emptyset\rangle, \langle \mathbf{1}_\wedge, \{1\}\rangle, \langle \mathbf{2}_\wedge, \{1\}\rangle\}$, also a deceptively simple class of matrices!

On the other hand, the definition of the logic implies that two formulas α and β are interderivable in $\mathcal{Cl}_\wedge$ if and only if the equation $\alpha \approx \beta$ holds in $\mathbf{2}_\wedge$, that is, if and only if $\alpha \approx \beta$ holds in SL, because the class SL of ***semilattices*** can be *defined* as the variety generated by $\mathbf{2}_\wedge$. Thus, in some sense, this variety would also deserve the title of "the (natural) algebraic counterpart of the logic $\mathcal{Cl}_\wedge$"; Exercise 5.53 confirms this. ⊠

Example 4.47 (The conjunction and disjunction fragment of $\mathcal{Cl}$)
The logic $\mathcal{Cl}_{\wedge\vee}$ is the fragment of classical logic for the language containing only the connectives of conjunction $\wedge$ and disjunction $\vee$; the notations $\vdash_{\wedge\vee}$ for $\vdash_{\mathcal{Cl}_{\wedge\vee}}$ and $\dashv\vdash_{\wedge\vee}$ for $\dashv\vdash_{\mathcal{Cl}_{\wedge\vee}}$ will be used. $\mathcal{Cl}_{\wedge\vee}$ is finitary (Exercise 1.28) and satisfies all rules of classical logic involving only $\wedge$ and $\vee$, among them those in (50); moreover, by Proposition 4.6, $\mathcal{Cl}_{\wedge\vee}$ is defined by the matrix $\langle \mathbf{2}_{\wedge\vee}, \{1\}\rangle$, where $\mathbf{2}_{\wedge\vee} := \langle 2, \wedge, \vee\rangle$ is the two-element distributive lattice, with $2 = \{0,1\}$ with $0 < 1$. Using this, it is easy to show that $\alpha \dashv\vdash_{\wedge\vee} \beta$ if and only if $\mathbf{2}_{\wedge\vee} \vDash \alpha \approx \beta$, and if and only if $\mathsf{DL} \vDash \alpha \approx \beta$, because the class DL of ***distributive lattices*** can be *defined* as the variety generated by $\mathbf{2}_{\wedge\vee}$. From this the following imediately follow:

$$(x \wedge y) \vee z \dashv\vdash_{\wedge\vee} (x \vee z) \wedge (y \vee z) \tag{52}$$

$$x \vee (y \vee z) \dashv\vdash_{\wedge\vee} (x \vee y) \vee z \tag{53}$$

$$x \vee y \dashv\vdash_{\wedge\vee} y \vee x \tag{54}$$

The following rules of disjunction will also be used:

$$x \vdash_{\wedge\vee} x \vee y \tag{55}$$

$$x \vee x \vdash_{\wedge\vee} x \tag{56}$$

More importantly, it is also easy to show that, for all $n \geqslant 1$,

$$\begin{aligned} \varphi_1, \dots, \varphi_n \vdash_{\wedge\vee} \varphi &\iff \mathsf{DL} \vDash \varphi_1 \wedge \dots \wedge \varphi_n \preccurlyeq \varphi \\ &\iff \mathsf{DL} \vDash \varphi_1 \wedge \dots \wedge \varphi_n \wedge \varphi \approx \varphi_1 \wedge \dots \wedge \varphi_n \end{aligned} \tag{57}$$

(for these and similar basic properties, see Exercise 4.33). Moreover, the logic does not have theorems (Exercise 1.29; this implies it is not algebraizable) and is finitary. This means that $\mathcal{C}\ell_{\wedge\vee}$ is completely determined by (57), that is, by the equations that hold in DL. Accordingly, one hopes to find some further intrinsic relationship between the class of algebras associated with $\mathcal{C}\ell_{\wedge\vee}$ and DL; actually, in [248] it is stated that $\mathsf{Alg}^*\mathcal{C}\ell_{\wedge\vee} = \mathsf{DL}$, which turns out to be wrong, as discovered in [104, 117], and the truth is more complicated, as the following results show.

PROPOSITION 4.48
Let $\langle \boldsymbol{A}, F\rangle \in \mathsf{Mod}\mathcal{C}\ell_{\wedge\vee}$. *For any* $a, b \in A$, $a \equiv b\ (\boldsymbol{\Omega}^{\boldsymbol{A}}F)$ *if and only if for every* $c \in A$, $\big[a \vee c \in F \Leftrightarrow b \vee c \in F\big]$.

PROOF: This is a simplification of the characterization in Theorem 4.23, with the set of all unary polynomial functions replaced by a single one, defined by the formula $x \vee z$; therefore, the implication from left to right is trivial. To show the converse, it is enough to define, for $a, b \in A$,

$$a \equiv b\ (\theta) \overset{\text{def}}{\Longleftrightarrow} \text{for every } c \in A,\ \big[a \vee c \in F \Leftrightarrow b \vee c \in F\big] \tag{58}$$

and prove that θ is a congruence of $\boldsymbol{A}$ compatible with F. The relation is trivially an equivalence relation. To show that it is compatible with $\wedge$, assume that $a \equiv b\ (\theta)$, take any $d \in A$ and assume that for some $c \in A$, $(a \wedge d) \vee c \in F$. Since F is a filter of the logic, using (52) and (50) we obtain that $(a \vee c) \wedge (d \vee c) \in F$ and $a \vee c \in F$, which by (58) implies that $b \vee c \in F$. Using that also $d \vee c \in F$, a similar process shows that $(b \wedge d) \vee c \in F$. Swapping the roles of a and b we obtain that if $(b \wedge d) \vee c \in F$, then $(a \wedge d) \vee c \in F$. Since c is arbitrary, we conclude that $a \wedge d \equiv b \wedge d\ (\theta)$. A similar argument, using also (54), shows that $d \wedge a \equiv d \wedge b\ (\theta)$. The compatibility with $\vee$ is even easier, thanks to (53): The equivalence $(a \vee d) \vee c \in F$ if and only if $a \vee (d \vee c) \in F$ if and only if $b \vee (d \vee c) \in F$ if and only if $(b \vee d) \vee c \in F$ shows that $a \vee d \equiv b \vee d\ (\theta)$; and a similar one shows that $d \vee a \equiv d \vee b\ (\theta)$. Thus, $\theta \in \mathrm{Con}\boldsymbol{A}$. Finally, θ is compatible with F: if $a \equiv b\ (\theta)$ and $a \in F$, then by (55) also $a \vee b \in F$, and by (58) $b \vee b \in F$ and then by (56) also $b \in F$. ⊠

An interesting feature of this proof is that it does not use a *complete* axiomatization of the logic, but only a few of its rules.

PROPOSITION 4.49
$\langle \boldsymbol{A}, F\rangle \in \mathsf{Mod}^*\mathcal{C}\ell_{\wedge\vee}$ *if and only if either* $\boldsymbol{A} \in \mathsf{DL}$, *with maximum* 1, *satisfying the following condition*:

$$\begin{array}{l}\textit{For all } a, b \in A,\ \textit{if } a < b, \textit{ then there is some } c \in A \\ \qquad\qquad\qquad \textit{such that } a \vee c \neq 1 \textit{ and } b \vee c = 1,\end{array} \tag{59}$$

and $F = \{1\}$, *or* $\boldsymbol{A}$ *is trivial and* $F = \varnothing$.

PROOF: Let $\langle \boldsymbol{A}, F\rangle \in \mathsf{Mod}^*\mathcal{C}\ell_{\wedge\vee}$. To show that $\boldsymbol{A} \in \mathsf{DL}$, since this one is a variety, it is enough to check that $\boldsymbol{A} \vDash \alpha \approx \beta$ for any equation such that $\mathsf{DL} \vDash \alpha \approx \beta$. Take some variable z appearing neither in α nor in β; then $\mathsf{DL} \vDash \alpha \vee z \approx \beta \vee z$, and therefore $\alpha \vee z \dashv\vdash_{\wedge\vee} \beta \vee z$. Since F is a $\mathcal{C}\ell_{\wedge\vee}$-filter, this implies that for any

$h \in \mathrm{Hom}(\boldsymbol{Fm}, \boldsymbol{A})$ and any $c \in A$, $h\alpha \vee c \in F$ if and only if $h\beta \vee c \in F$. By Proposition 4.48 this means that $h\alpha \equiv h\beta\ (\boldsymbol{\Omega}^{\boldsymbol{A}} F)$, but this matrix is reduced, so this actually means that $h\alpha = h\beta$. Thus, $\boldsymbol{A} \vDash \alpha \approx \beta$ and hence $\boldsymbol{A} \in \mathsf{DL}$. Since the logic $\mathcal{C\ell}_{\wedge\vee}$ has no theorems, $F = \emptyset$ is a $\mathcal{C\ell}_{\wedge\vee}$-filter, for any algebra; but in this case this implies that the algebra is trivial because in general $\boldsymbol{\Omega}^{\boldsymbol{A}}\emptyset = A \times A$. So, assume that $F \neq \emptyset$, and first prove that F is a one-element set: If $a, b \in F$, since $x \vdash_{\wedge\vee} x \vee z$, also $a \vee c \in F$ and $b \vee c \in F$ for any $c \in A$. Proposition 4.48 again implies that $a \equiv b\ (\boldsymbol{\Omega}^{\boldsymbol{A}} F)$ and hence that $a = b$. Thus, $F = \{1\}$ for some $1 \in A$. Applying the same rule of $\mathcal{C\ell}_{\wedge\vee}$ now to $\{1\}$, one obtains that $1 \vee c = 1$ for any $c \in A$; this shows that 1 is the maximum of $\boldsymbol{A}$. A matrix is reduced when $a \neq b$ implies $a \not\equiv b\ (\boldsymbol{\Omega}^{\boldsymbol{A}} F)$; in this case, putting $F = \{1\}$ in Proposition 4.48, we directly obtain:

For all $a, b \in A$, if $a \neq b$, then there is some $c \in A$ such that either $a \vee c \neq 1$ and $b \vee c = 1$, or $a \vee c = 1$ and $b \vee c \neq 1$. (60)

It is easy to see (Exercise 4.35) that in a distributive lattice with maximum this condition is equivalent to (59).

In order to show the converse, first observe that for a trivial algebra $\boldsymbol{A}$, $\langle \boldsymbol{A}, \emptyset \rangle \in \mathsf{Mod}^*\mathcal{C\ell}_{\wedge\vee}$. So, assume that $\boldsymbol{A} \in \mathsf{DL}$ with a maximum 1 and satisfies the condition (59). Using (57) it is easy to see that $\langle \boldsymbol{A}, \{1\} \rangle \in \mathsf{Mod}\mathcal{C\ell}_{\wedge\vee}$. Now Proposition 4.48 applies, and condition (59), in its equivalent form (60), says that this matrix is reduced. ☒

COROLLARY 4.50
$\boldsymbol{A} \in \mathsf{Alg}^*\mathcal{C\ell}_{\wedge\vee}$ if and only if $\boldsymbol{A} \in \mathsf{DL}$ with a maximum 1 and satisfies condition (59) *or its equivalent form* (60). ☒

The class $\mathsf{Alg}^*\mathcal{C\ell}_{\wedge\vee}$ happens to be a quite strange class of distributive lattices. It is easy to see that it contains all complemented distributive lattices, that, is all Boolean lattices (the $\langle \wedge, \vee \rangle$ reducts of Boolean algebras). Actually, it is possible to show that its finite members are exactly all finite Boolean lattices, which for instance means that the only finite chain it contains is the two-element one $\mathbf{2}_{\wedge\vee}$. This implies that this class is not closed under subalgebras; therefore (by Theorem 1.77.6) this class is not even a generalized quasivariety.

Lattices satisfying a condition dual to (60) (with $\wedge$ instead of $\vee$ and 0 instead of 1) have appeared briefly in the literature [25, 58, 242]; this dual property is called by Birkhoff and Wallman "disjunctive property", but notice that here the word "disjunctive" does not refer to the lattice operation of disjunction (which in those works was called "addition" and was denoted by "+") but to the metalinguistic disjunction contained in the second line of (60), and also in its dual. ☒

Examples such as the two preceding ones prompt the need for a more refined algebra-based semantics, the *semantics of generalized matrices*, which is dealt with in Chapter 5. It applies to arbitrary logics, as the matrix semantics does, but produces a new notion of *the algebraic counterpart of a logic*; in the cases of $\mathcal{C\ell}_{\wedge}$

and $\mathcal{C\ell}_{\wedge\vee}$, the results are the expected ones, that is, the varieties of semilattices and of distributive lattices, respectively.

The section ends with a different issue: one feature of the notion of $\mathsf{Alg}^*\mathcal{L}$ (which can be viewed as a limitation) is that not every class of algebras is of this form. This can be established by purely algebraic work, due to the following observation: if a given class of algebras is of the form $\mathsf{Alg}^*\mathcal{L}$ for some, unspecified logic $\mathcal{L}$, in particular each algebra in the class supports a reduced matrix, that is, it has at least one subset whose Leibniz congruence is the identity relation. Of course, the subset should be a filter of the logic, but if there is no subset at all with this property, we are done. Here is an example of this strategy:

Example 4.51 (Semigroups)
Let SG denote the variety of ***semigroups***, the algebras $\langle A, \cdot \rangle$ of type $\langle 2 \rangle$ in which the equation $x \cdot (y \cdot z) \approx (x \cdot y) \cdot z$ holds. There is no logic $\mathcal{L}$ such that $\mathsf{Alg}^*\mathcal{L} = \mathsf{SG}$. The same holds for the variety of commutative semigroups.
Proof: Consider the three-element null semigroup $\mathbf{3}$; that is, a three-element set 3 with the binary operation defined by $a \cdot b = 0$ for every $a, b \in 3$, and a fixed $0 \in 3$ as unit (this is trivially a semigroup). It turns out that for every $F \subseteq 3$ the matrix $\langle \mathbf{3}, F \rangle$ is not reduced. If $F = \emptyset$ or $F = 3$, this is clear, for $\boldsymbol{\Omega}^{\mathbf{3}}\emptyset = \boldsymbol{\Omega}^{\mathbf{3}}3 = 3 \times 3 \neq \mathrm{Id}_3$. Then suppose that $\emptyset \subsetneq F \subsetneq 3$. In this case the equivalence relation partitioning 3 into the two (non-empty) classes F and $3 \smallsetminus F$ is a congruence (by the trivial definition of the operation), is compatible with F, and is not the identity relation on 3. This shows that $\boldsymbol{\Omega}^{\mathbf{3}}F \neq \mathrm{Id}_3$, and hence that $\mathbf{3} \notin \mathsf{Alg}^*\mathcal{L}$ for any logic $\mathcal{L}$ whatsoever. Therefore, $\mathsf{Alg}^*\mathcal{L} \neq \mathsf{SG}$ for all $\mathcal{L}$. The example is commutative, and thus it also works as a counterexample for the commutative case. This also shows that $\mathsf{Alg}^*\mathcal{L} \neq \mathsf{K}$ for any logic $\mathcal{L}$ and any class K of algebras of this type larger than the class of commutative semigroups. ⊠

It is *not known* whether there exists a logic $\mathcal{L}$ such that $\mathsf{Alg}^*\mathcal{L} = \mathsf{SL}$, the class of semilattices (which is the subvariety of SG in which the operation is commutative and idempotent); and the same for the class DL of distributive lattices.

The notion of an algebraic counterpart constructed in the semantics of generalized matrices makes it possible to find logics having each of the just mentioned classes as their algebraic counterpart; in fact, in [112] it is shown that *every variety of algebras* (of an arbitrary type) is the algebraic counterpart (in that sense) of some logic.

Exercises for Section 4.3

Exercise 4.25. Consider the matrix $\langle \boldsymbol{A}_3', \{1\} \rangle$, where $\boldsymbol{A}_3'$ is the algebra of Exercise 4.13. Use Theorem 4.34 to show that $\{\boldsymbol{A}_3'\}$ constitutes an algebraic semantics for $\mathcal{C\ell}$ with the usual transformer $\boldsymbol{\tau} x := \{x \approx x \to x\}$, and conclude that this matrix defines classical logic. Thus, $\boldsymbol{A}_3' \in \mathsf{K}(\mathcal{C\ell}, \boldsymbol{\tau})$, but $\boldsymbol{A}_3' \notin \mathsf{Alg}^*\mathcal{C\ell}$. Check that this confirms the claim made on page 121 concerning the existence of small algebraic semantics disjoint from the equivalent one. Try to do the same for the matrix $\langle \boldsymbol{A}_3, \{1/2, 1\} \rangle$, where $\boldsymbol{A}_3$ is the algebra of Exercise 4.12. Does it work? What is the difference?

EXERCISE 4.26. Consider Belnap-Dunn's four-valued logic $\mathcal{B}$ as defined in Example 1.13.

1. Check that if $a \in M_4$, then $t \in a$ if and only if $a \in \{\mathbf{t},\mathbf{b}\}$, and that $f \notin a$ if and only if $a \in \{\mathbf{t},\mathbf{n}\}$, and use these facts to show that the expression (2) on page 20 amounts exactly to saying that $\mathcal{B}$ is the logic defined by the bundle of two matrices $\{\langle \boldsymbol{M}_4, \{\mathbf{t},\mathbf{b}\} \rangle, \langle \boldsymbol{M}_4, \{\mathbf{t},\mathbf{n}\} \rangle\}$, as stated in Example 4.12.
2. Use the last statement in Theorem 4.34 to show that $\mathcal{B}$ is also the logic defined by any of the two matrices $\langle \boldsymbol{M}_4, \{\mathbf{t},\mathbf{b}\} \rangle$ and $\langle \boldsymbol{M}_4, \{\mathbf{t},\mathbf{n}\} \rangle$ alone.

EXERCISE 4.27. Let $h\colon \boldsymbol{A} \to \boldsymbol{B}$ be surjective, and let $F \subseteq A$ be such that $\ker h \subseteq \boldsymbol{\Omega}^{\boldsymbol{A}} F$. Prove that $F \in \mathcal{F}i_{\mathcal{L}} \boldsymbol{A}$ if and only if $hF \in \mathcal{F}i_{\mathcal{L}} \boldsymbol{B}$.

EXERCISE 4.28. Complete the proof of Lemma 4.36 by showing that for every $\theta \in \mathrm{Con} \boldsymbol{A}$, if $\pi\colon \boldsymbol{A} \to \boldsymbol{A}/\theta$ is the canonical projection, then π and π^{-1} establish mutually inverse order isomorphisms between the set $\{F \in \mathcal{F}i_{\mathcal{L}} \boldsymbol{A} : \theta \subseteq \boldsymbol{\Omega}^{\boldsymbol{A}} F\}$ and the set $\mathcal{F}i_{\mathcal{L}}(\boldsymbol{A}/\theta)$.

EXERCISE 4.29. Prove that for any logic $\mathcal{L}$ the class $\mathsf{Alg}^*\mathcal{L}$ includes all simple algebras of the similarity type that have a nonempty, proper $\mathcal{L}$-filter.

EXERCISE 4.30. Prove directly that if a logic is complete with respect to a class M of matrices, then it is also complete with respect to the class M^*. Then, use this fact to obtain an alternative, compact proof of Theorem 4.43.

EXERCISE 4.31. Prove versions of Theorem 4.44 and its Corollary 4.45 for algebras of arbitrary cardinality:

1. $\langle \boldsymbol{A}, F \rangle \in \mathsf{Mod}^*\mathcal{L}$ if and only if there is a cardinal κ and an $\mathcal{L}$-filter Γ on the formula algebra $\boldsymbol{Fm}_\kappa$ with κ variables such that $\langle \boldsymbol{A}, F \rangle \cong \langle \boldsymbol{Fm}_\kappa, \Gamma \rangle^*$.
2. $\boldsymbol{A} \in \mathsf{Alg}^*\mathcal{L}$ if and only if there is a cardinal κ and an $\mathcal{L}$-filter Γ on the formula algebra $\boldsymbol{Fm}_\kappa$ with κ variables such that $\boldsymbol{A} \cong \boldsymbol{Fm}/\boldsymbol{\Omega}^{\boldsymbol{Fm}_\kappa} \Gamma$.

EXERCISE 4.32. Let $\langle \boldsymbol{A}, F \rangle \in \mathsf{Mod}\mathcal{C\ell}_\wedge$ (Example 4.46). Define a binary relation θ on A by $a \equiv b\ (\theta)$ if and only if $[a \in F \Leftrightarrow b \in F]$, that is, if and only if either $a, b \in F$ or $a, b \notin F$. Prove that θ is the largest congruence of $\boldsymbol{A}$ that is compatible with F, and hence that $\theta = \boldsymbol{\Omega}^{\boldsymbol{A}} F$.

HINT. The relation just partitions A into two subsets, F and $A \smallsetminus F$, so that it is obviously the largest binary relation compatible with F; as it is obviously an equivalence relation, you need only prove that it is a congruence.

EXERCISE 4.33. Prove the following facts about the logic $\mathcal{C\ell}_{\wedge\vee}$ considered in Example 4.47.

1. $\varphi \vdash_{\wedge\vee} \psi$ if and only if $\mathbf{2}_{\wedge\vee} \vDash \varphi \preccurlyeq \psi$, and if and only if $\mathbf{2}_{\wedge\vee} \vDash \varphi \wedge \psi \approx \varphi$.
2. $\varphi \dashv\vdash_{\wedge\vee} \psi$ if and only if $\mathbf{2}_{\wedge\vee} \vDash \varphi \approx \psi$, and if and only if $\mathsf{DL} \vDash \varphi \approx \psi$.
3. For all $n \geqslant 1$, $\varphi_1, \dots, \varphi_n \vdash_{\wedge\vee} \varphi$ if and only if $\varphi_1 \wedge \dots \wedge \varphi_n \vdash_{\wedge\vee} \varphi$, and if and only if $\mathsf{DL} \vDash \varphi_1 \wedge \dots \wedge \varphi_n \preccurlyeq \varphi$.

Check also properties (52)–(56) mentioned there.

HINTS. For 1, recall that $\mathbf{2}_{\wedge\vee}$ is a lattice. 2 follows immediately from 1. For 3, take the particular algebraic structure of $\mathbf{2}_{\wedge\vee}$ into account: if $a, b \in 2$, then $a \wedge b = 1$ if and only if $a = 1$ and $b = 1$.

EXERCISE 4.34. Explore the following alternative proof by *reductio ad absurdum* of the implication from right to left in Proposition 4.48: Assume that some $a, b \in A$ satisfy that for every $c \in A$, $\big[a \vee c \in F \Leftrightarrow b \vee c \in F\big]$, but that $a \not\equiv b\ (\boldsymbol{\Omega}^{\boldsymbol{A}}F)$, which, by Theorem 4.23, means that there are $\delta(x,\vec{z}) \in Fm$ and $\vec{c} \in \vec{A}$ such that (without loss of generality) $\delta^{\boldsymbol{A}}(a,\vec{c}) \in F$ while $\delta^{\boldsymbol{A}}(b,\vec{c}) \notin F$. Consider a $\delta(x,\vec{z})$ with least complexity with this property, and show that such a δ must have the form $\delta(x,\vec{z}) = x \vee z$. This clearly contradicts the first assumption on a and b.

HINT. In the process you will have to use rules of $\mathcal{C\ell}_{\wedge\vee}$ similar to those used in the proof of Proposition 4.48, applied to the fact that F is a $\mathcal{C\ell}_{\wedge\vee}$-filter. The initial assumption on a and b has to be used as well.

EXERCISE 4.35. Let $\boldsymbol{A}$ be a distributive lattice with maximum 1. Prove that the following properties are equivalent:

1. For all $a,b \in A$, if $a \neq b$, then there is some $c \in A$ such that either $a \vee c \neq 1$ and $b \vee c = 1$, or $a \vee c = 1$ and $b \vee c \neq 1$.
2. For all $a,b \in A$, if $a < b$, then there is some $c \in A$ such that $a \vee c \neq 1$ and $b \vee c = 1$.

EXERCISE 4.36. Concerning Belnap-Dunn's logic $\mathcal{B}$ of Examples 1.13 and 4.12, consider any $\langle \boldsymbol{A}, F\rangle \in \mathsf{Mod}\mathcal{B}$. Prove that for all $a,b \in A$, $a \equiv b\ (\boldsymbol{\Omega}^{\boldsymbol{A}}F)$ if and only if for every $c \in A$, $[a \vee c \in F \Leftrightarrow b \vee c \in F]$ and $[\neg a \vee c \in F \Leftrightarrow \neg b \vee c \in F]$.

HINTS. Follow the strategy of the proof of Proposition 4.48. Use Exercise 4.6 to show that $\mathcal{B}$ is an expansion of $\mathcal{C\ell}_{\wedge\vee}$, and hence satisfies all rules of the latter, concerning $\wedge$ and $\vee$. As to rules of $\mathcal{B}$ concerning negation, you will have to use the following

$$x \vee z \dashv\vdash_{\mathcal{B}} \neg\neg x \vee z$$
$$\neg(x \vee y) \vee z \dashv\vdash_{\mathcal{B}} (\neg x \wedge \neg y) \vee z$$
$$\neg(x \wedge y) \vee z \dashv\vdash_{\mathcal{B}} (\neg x \vee \neg y) \vee z,$$

and the same without the "$\vee z$". All follow easily from the definition of the logic.

4.4 Applications to algebraizable logics

This section should be regarded both as a sample of the application of the notions of previous sections to the restricted case of algebraizable logics, and as an appetizer of some topics that constitute the *leitmotiv* of the study of the Leibniz hierarchy in Chapter 6: the definability of the Leibniz congruence and of the truth filter in the reduced models of a logic. In algebraizable logics both definabilities hold, as shown here, and considered separately and in several variants they define other kinds of logics, as shown there.

DEFINITION 4.52
Let $\Delta(x,y) \subseteq Fm$ be a set of formulas in at most two variables. For any algebra $\boldsymbol{A}$ and any $F \subseteq A$ the ***analytical relation*** *defined by Δ is the relation defined as follows: $R^{\boldsymbol{A}}(F,\Delta) := \big\{\langle a,b\rangle \in A \times A : \Delta^{\boldsymbol{A}}(a,b) \subseteq F\big\}$. A relation is* ***analytically definable*** *when it is of the form $R^{\boldsymbol{A}}(F,\Delta)$ for some $F \subseteq A$ and $\Delta(x,y) \subseteq Fm$.*

If ρ denotes the structural transformer from equations into formulas defined by the set $\Delta(x,y)$, then the relation $R^{\boldsymbol{A}}(F,\Delta)$ coincides with the evaluated residual of F, denoted by $\rho^{\boldsymbol{A}^{-1}}F$ in Section 3.5; it also coincides with the relation denoted by $\boldsymbol{\Omega}T$ for a theory T in Definition 3.24 (the symbol is used there anticipating

Corollary 4.55). But here transformers are not the initial focus, as the first results do not concern algebraizable logics explicitly.

LEMMA 4.53
Let $\Delta, \boldsymbol{A}, F$ and $R^{\boldsymbol{A}}(F,\Delta)$ be as in Definition 4.52.

1. *If the relation $R^{\boldsymbol{A}}(F,\Delta)$ is reflexive, then $\boldsymbol{\Omega}^{\boldsymbol{A}}F \subseteq R^{\boldsymbol{A}}(F,\Delta)$.*
2. $R^{\boldsymbol{A}}(F,\Delta) \in \mathrm{Con}\langle \boldsymbol{A}, F\rangle$ *if and only if* $\boldsymbol{\Omega}^{\boldsymbol{A}}F = R^{\boldsymbol{A}}(F,\Delta)$.

PROOF: 1. By assumption, $\langle b,b\rangle \in R^{\boldsymbol{A}}(F,\Delta)$, that is, $\delta^{\boldsymbol{A}}(b,b) \in F$ for each $\delta(x,y) \in \Delta(x,y)$. Now, if $a \equiv b\ (\boldsymbol{\Omega}^{\boldsymbol{A}}F)$, then also $b \equiv a\ (\boldsymbol{\Omega}^{\boldsymbol{A}}F)$. By Theorem 4.23 applied to $\delta(x,y)$, taking y as the parameter, from $\delta^{\boldsymbol{A}}(b,b) \in F$ it follows that $\delta^{\boldsymbol{A}}(a,b) \in F$ as well. Thus, $\Delta^{\boldsymbol{A}}(a,b) \subseteq F$, that is, $\langle a,b\rangle \in R^{\boldsymbol{A}}(F,\Delta)$.

2. If $R^{\boldsymbol{A}}(F,\Delta)$ is a strict congruence of $\langle \boldsymbol{A}, F\rangle$, then $R^{\boldsymbol{A}}(F,\Delta) \subseteq \boldsymbol{\Omega}^{\boldsymbol{A}}F$, and since it is reflexive, by 1 the equality follows. The converse is trivial. ⊠

Thus,[†] the Leibniz congruence of a set is the only congruence compatible with the set that is "analytically definable" from it in the sense of Definition 4.52; a similar result [32, Theorem 1.6] holds in general for the weaker notion of definability of Proposition 4.62. While $\boldsymbol{\Omega}^{\boldsymbol{A}}F$ is an algebraic object depending only of the algebraic structure of $\boldsymbol{A}$, its definability is not; it may depend on whether F is an $\mathcal{L}$-filter, for some well-behaved logic $\mathcal{L}$. It is not difficult to characterize syntactically, in reference to a logic $\mathcal{L}$, when this happens to its filters:

THEOREM 4.54
Let $\mathcal{L}$ be a logic, let $\Delta(x,y) \subseteq Fm$ be a set of formulas in at most two variables, and let M be a class of matrices such that $\mathcal{L}$ is complete with respect to M. The following conditions are equivalent:

(i) *The Leibniz congruence is analytically definable in M by $\Delta(x,y)$; that is, for all $\langle \boldsymbol{A}, F\rangle \in \mathsf{M}$, $\boldsymbol{\Omega}^{\boldsymbol{A}}F = R^{\boldsymbol{A}}(F,\Delta)$.*

(ii) *$\mathcal{L}$ satisfies conditions* (R), (Sym), (Trans), (Re) *and* (MP) *with respect to Δ.*

If these conditions hold, then the Leibniz congruence is analytically definable, by the same set of formulas, in all the matrices in $\mathsf{Mod}\mathcal{L}$.

PROOF: By Lemma 4.53, $\boldsymbol{\Omega}^{\boldsymbol{A}}F = R^{\boldsymbol{A}}(F,\Delta)$ holds if and only if $R^{\boldsymbol{A}}(F,\Delta)$ is a compatible congruence of $\langle \boldsymbol{A}, F\rangle$. By the completeness assumption, it is easy to see that having this property for all matrices in M amounts to having $\mathcal{L}$ satisfy the properties in the list; while (R)–(Re) clearly correspond to $R^{\boldsymbol{A}}(F,\Delta)$ being a congruence, (MP) corresponds to it being compatible. Let us check the last fact: (MP) for Δ is $x, \Delta(x,y) \vdash_{\mathcal{L}} y$. By completeness of $\mathcal{L}$ with respect to M, this holds if and only if for all $\langle \boldsymbol{A}, F\rangle \in \mathsf{M}$ and all $a,b \in A$, if $a \in F$ and $\Delta^{\boldsymbol{A}}(a,b) \subseteq F$, then $b \in F$. By Definition 4.52, this is the same as saying that if $a \in F$ and $\langle a,b\rangle \in R^{\boldsymbol{A}}(F,\Delta)$, then $b \in F$. But this is exactly to say that $R^{\boldsymbol{A}}(F,\Delta)$

[†] Lemma 4.53 is often named after the French logician Porte. Apparently, he was the first to call attention to these facts in his interesting, slightly non-standard, and little-known monograph [196]; obviously, he did not use the Leibniz congruence terminology. The first modern version of his remark is Theorem 6 of [49].

is compatible with F. This shows the equivalence between (i) and (ii). Finally, since the syntactic conditions in (ii) do not depend on M, if (ii) holds, then (i) holds for any class of matrices satisfying completeness, and obviously $\mathsf{Mod}\,\mathcal{L}$ is one of them. ⊠

By Theorem 3.21, all algebraizable logics satisfy the conditions in Theorem 4.54; therefore:

COROLLARY 4.55
If $\mathcal{L}$ is an algebraizable logic, with equivalence formulas $\Delta(x,y)$, and $\langle \boldsymbol{A}, F\rangle \in \mathsf{Mod}\,\mathcal{L}$, then $\boldsymbol{\Omega}^{\boldsymbol{A}}F = \rho^{\boldsymbol{A}^{-1}}F = R^{\boldsymbol{A}}(F,\Delta)$. ⊠

This confirms that the earlier usages of the notation $\boldsymbol{\Omega}^{\boldsymbol{A}}F$ in Definitions 2.1 and 2.29 for implicative logics, and in 3.24 for algebraizable logics, coincide with the general one, and in all cases it denotes the largest congruence of $\boldsymbol{A}$ compatible with F. This also implies that the earlier definitions are independent of the particular set Δ used in them, provided it satisfies the conditions in Theorem 4.54. The logics satisfying these conditions are called ***equivalential logics***, and form a larger group; modulo Exercise 3.17, Theorem 3.21 characterizes algebraizable logics as exactly the equivalential logics that in addition satisfy condition (ALG3). They are studied in depth in Chapter 6, where for instance it is proved that conditions (Sym) and (Trans) can be deleted from Theorem 4.54.

Now we arrive at the key results concerning matrix semantics of algebraizable logics. The first one completes Theorems 3.56 and 3.58 by giving a characterization of the isomorphism there established that is totally independent of the transformers.

THEOREM 4.56 (THE ISOMORPHISM THEOREM, THIRD VERSION)
Let $\mathcal{L}$ be a logic and let K be a generalized quasivariety. The following conditions are equivalent:

(i) *The logic $\mathcal{L}$ is algebraizable with equivalent algebraic semantics K.*
(ii) *In any algebra $\boldsymbol{A}$, the Leibniz operator is an isomorphism between the lattices $\mathcal{F}i_{\mathcal{L}}\boldsymbol{A}$ and $\mathrm{Con}_{\mathsf{K}}\boldsymbol{A}$ ($\boldsymbol{\Omega}^{\boldsymbol{A}} \colon \mathcal{F}i_{\mathcal{L}}\boldsymbol{A} \cong \mathrm{Con}_{\mathsf{K}}\boldsymbol{A}$) that commutes with endomorphisms.*
(iii) *In the formula algebra $\boldsymbol{Fm}$, the Leibniz operator is an isomorphism between the lattices $\mathcal{T}h\mathcal{L}$ and $\mathrm{Con}_{\mathsf{K}}\boldsymbol{Fm}$ ($\boldsymbol{\Omega} \colon \mathcal{T}h\mathcal{L} \cong \mathrm{Con}_{\mathsf{K}}\boldsymbol{Fm}$) that commutes with substitutions.*

PROOF: By Proposition 3.13, algebraizability of $\mathcal{L}$ implies it satisfies the conditions in Theorem 4.54 for the set $\Delta(x,y)$ of equations defining the transformer ρ that witnesses algebraizability, and therefore $\boldsymbol{\Omega}^{\boldsymbol{A}}F = R^{\boldsymbol{A}}(F,\Delta)$ in any $\langle \boldsymbol{A}, F\rangle \in \mathsf{Mod}\,\mathcal{L}$. But, as noted before, this equality is the same as $\boldsymbol{\Omega}^{\boldsymbol{A}}F = \rho^{\boldsymbol{A}^{-1}}F$, and by Theorem 3.56, for each algebra $\boldsymbol{A}$ the function $\rho^{\boldsymbol{A}^{-1}}$ is an isomorphim between $\mathcal{F}i_{\mathcal{L}}\boldsymbol{A}$ and $\mathrm{Con}_{\mathsf{K}}\boldsymbol{A}$ that commutes with endomorphisms, therefore the Leibniz operator is. Conversely, by Theorem 3.58 the mere existence of such an isomorphism on the formula algebra commuting with substitutions implies algebraizability with respect to K. ⊠

Several refinements of this theorem are obtained through the study of other order-related properties of the Leibniz operator (Theorems 6.131–6.133). For instance, when $\mathcal{L}$ is finitary and K is a quasivariety, the condition that the Leibniz operator commutes with endomorphisms is superfluous. Anyway, the more useful applications of this result are in the negative, and when it helps in falsifying algebraizability, it is usually the isomorphism part that fails rather than the commutativity with endomorphisms.

Now we can see that the equivalent algebraic semantics K of an algebraizable logic coincides with the algebra reducts of its reduced models; thus, the two general definitions of what the *algebraic counterpart* of a logic is coincide for algebraizable logics.

PROPOSITION 4.57
Let $\mathcal{L}$ be an algebraizable logic, with transformer $\boldsymbol{\tau}$ defined by equations $E(x)$ and equivalent algebraic semantics K. Then

$$\langle \boldsymbol{A}, F\rangle \in \mathsf{Mod}^*\mathcal{L} \iff \boldsymbol{A} \in \mathsf{K} \text{ and } F = Sol_{\boldsymbol{\tau}}^{\boldsymbol{A}} := \{a \in A : \boldsymbol{A} \vDash E(x)\ [\![a]\!]\}.$$

Therefore $\mathsf{Alg}^*\mathcal{L} = \mathsf{K}$, and for each $\boldsymbol{A} \in \mathsf{K}$ the set $Sol_{\boldsymbol{\tau}}^{\boldsymbol{A}}$ just described is the unique $F \in \mathcal{F}i_{\mathcal{L}}\boldsymbol{A}$ such that $\langle \boldsymbol{A}, F\rangle$ is reduced.

PROOF: If $\langle \boldsymbol{A}, F\rangle \in \mathsf{Mod}^*\mathcal{L}$, then $F \in \mathcal{F}i_{\mathcal{L}}\boldsymbol{A}$ and $\boldsymbol{\Omega}^{\boldsymbol{A}}F = \mathrm{Id}_A$, so $\mathrm{Id}_A \in \mathrm{Con}_{\mathsf{K}}\boldsymbol{A}$, and this implies that $\boldsymbol{A} \in \mathsf{K}$. In order to characterize F consider condition (ALG3): $x \dashv\vdash_{\mathcal{L}} \Delta(E(x))$. Since F is an $\mathcal{L}$-filter, for any $a \in A$, $a \in F$ if and only if $\Delta^{\boldsymbol{A}}(E^{\boldsymbol{A}}(a)) \subseteq F$ if and only if $E^{\boldsymbol{A}}(a) \subseteq R^{\boldsymbol{A}}(F, \Delta) = \boldsymbol{\Omega}^{\boldsymbol{A}}F = \mathrm{Id}_A$ if and only if $\boldsymbol{A} \vDash E(x)\ [\![a]\!]$ if and only if $a \in Sol_{\boldsymbol{\tau}}^{\boldsymbol{A}}$. Conversely, let $\boldsymbol{A} \in \mathsf{K}$ and let $Sol_{\boldsymbol{\tau}}^{\boldsymbol{A}}$ be the set described in the statement. Using the ($\Rightarrow$) half of (ALG1), it is easy to see that $Sol_{\boldsymbol{\tau}}^{\boldsymbol{A}} \in \mathcal{F}i_{\mathcal{L}}\boldsymbol{A}$, that is, $\langle \boldsymbol{A}, Sol_{\boldsymbol{\tau}}^{\boldsymbol{A}}\rangle \in \mathsf{Mod}\mathcal{L}$. Moreover by Corollary 4.55 $a \equiv b\ (\boldsymbol{\Omega}^{\boldsymbol{A}} Sol_{\boldsymbol{\tau}}^{\boldsymbol{A}})$ if and only if $\Delta^{\boldsymbol{A}}(a,b) \subseteq Sol_{\boldsymbol{\tau}}^{\boldsymbol{A}}$, which by definition of $Sol_{\boldsymbol{\tau}}^{\boldsymbol{A}}$ is equivalent to $\boldsymbol{A} \vDash E(\Delta(x,y))\ [\![a,b]\!]$. Now, using (ALG4): $E(\Delta(x,y)) \vDash_{\mathsf{K}} x \approx y$, this implies $a = b$. This shows that the matrix $\langle \boldsymbol{A}, Sol_{\boldsymbol{\tau}}^{\boldsymbol{A}}\rangle$ is reduced, and thus belongs to $\mathsf{Mod}^*\mathcal{L}$. Since the set $Sol_{\boldsymbol{\tau}}^{\boldsymbol{A}}$ can be considered in any $\boldsymbol{A} \in \mathsf{K}$, the just proven equivalence implies that $\mathsf{Alg}^*\mathcal{L} = \mathsf{K}$. The last statement (uniqueness) is a consequence of the first part and of the injectivity of $\boldsymbol{\Omega}^{\boldsymbol{A}}$ on the filters of $\mathcal{L}$, guaranteed by algebraizability. ⊠

PROPOSITION 4.58
Let $\mathcal{L}$ be an algebraizable logic with equivalent algebraic semantics K. The logic $\mathcal{L}$ is regularly algebraizable if and only if it is complete with respect to some class of reduced matrices with one-element sets as filters. Then, $\mathsf{Mod}^*\mathcal{L} = \{\langle \boldsymbol{A}, \{\top^{\boldsymbol{A}}\}\rangle : \boldsymbol{A} \in \mathsf{K}\}$ for any theorem $\top$ of $\mathcal{L}$.

PROOF: Let $E(x)$ and $\Delta(x,y)$ be the sets of defining equations and equivalence formulas, respectively, witnessing algebraizability. If $\mathcal{L}$ is regularly algebraizable, by Theorem 3.52 $E(x) := \{x \approx \top\}$, for any theorem $\top$ of $\mathcal{L}$ having at most the variable x, can be taken as set of defining equations. Then the characterization of $\mathsf{Mod}^*\mathcal{L}$ in Proposition 4.57 produces a class of reduced matrices in which the filter is the set $\{\top^{\boldsymbol{A}}\}$; since $\mathcal{L}$ is complete with respect to $\mathsf{Mod}^*\mathcal{L}$, this proves the assertion and the additional statement. Conversely, assume that $\mathcal{L}$ is complete with respect to some class of reduced matrices with a one-element set as the filter. Let $\langle \boldsymbol{A}, F \rangle$ be one of these, and let $a, b \in A$ such that $a, b \in F$. Then $a = b$. Since the matrix is reduced and a model of $\mathcal{L}$, by algebraizability identity in A is defined by $\Delta(x,y)$. Therefore $\Delta^{\boldsymbol{A}}(a,b) \subseteq F$. This shows that $x, y \vdash_{\mathcal{L}} \Delta(x,y)$, which is condition (G): $\mathcal{L}$ is regularly algebraizable. ⊠

This characterization of regularly algebraizable logics inside the algebraizable ones has originated a usage of the term "regularly" in abstract algebraic logic, alternative to its origin as explained next to Theorem 3.63: that of selecting the logics of a certain class that are complete with respect to a class of matrices with a one-element set as the filter.

The result explains why regularly algebraizable logics have been regarded as the "best" algebraized logics: in their reduced models we can actually dispense with the filter, as it can be described by an algebraic constant of the class of algebras. For these logics *their reduced matrix semantics can be identified with their algebraic counterpart*; that is, these logics are characterized by a class of algebras, something that does not happen in general (while every logic is characterized by a class of matrices).

Since implicative logics are regularly algebraizable, this confirms that our initial choice of terminology and notation in Section 2.2 matches the general one of the present section. From Propositions 4.57 and 4.58 we obtain:

COROLLARY 4.59
If $\mathcal{L}$ is an implicative logic, then the class $\mathsf{Alg}^*\mathcal{L}$ of $\mathcal{L}$-algebras of Definition 2.5 coincides with the class $\mathsf{Alg}^*\mathcal{L}$ of Definition 4.37 and with its equivalent algebraic semantics, and $\mathsf{Mod}^*\mathcal{L} = \{\langle \boldsymbol{A}, \{1\} \rangle : \boldsymbol{A} \in \mathsf{Alg}^*\mathcal{L}\}$. ⊠

The results in this section provide the proper perspective to obtain yet another characterization of algebraizable logics, this time in terms of some characteristics of their matrix semantics:

THEOREM 4.60
A logic $\mathcal{L}$ is algebraizable if and only if there exist sets $\Delta(x,y) \subseteq Fm$ and $E(x) \subseteq Eq$ such that for every $\langle \boldsymbol{A}, F \rangle \in \mathsf{Mod}^*\mathcal{L}$ the following conditions hold:

(a) *The identity relation in A is definable from the set F through the formulas in $\Delta(x,y)$ in the sense that $\mathrm{Id}_A = R^{\boldsymbol{A}}(F,\Delta)$.*

(b) *The set F is definable from the identity relation in A through the equations in $E(x)$ in the sense that $F = \{a \in A : E^{\boldsymbol{A}}(a) \subseteq \mathrm{Id}_A\}$.*

PROOF: By Theorem 4.56, if $\mathcal{L}$ is algebraizable, then for any $\langle \boldsymbol{A}, F \rangle \in \mathsf{Mod}\mathcal{L}$, $\boldsymbol{\Omega}^{\boldsymbol{A}}F$ is definable from F through the formulas in $\Delta(x,y)$ as in Definition 4.52. In particular, in the matrices in $\mathsf{Mod}^*\mathcal{L}$ this defines the identity relation. This establishes (a), and Proposition 4.57 establishes (b). Now for the converse: Since $\mathcal{L}$ is complete with respect to $\mathsf{Mod}^*\mathcal{L}$, from (a) and Theorem 4.54 it follows that $\mathcal{L}$ satisfies conditions (R)–(Re) relatively to the set $\Delta(x,y)$. Given this, (b) implies that $\mathcal{L}$ satisfies condition (ALG3): $x \dashv\vdash_{\mathcal{L}} \Delta(E(x))$: To see this, take any $\langle \boldsymbol{A}, F\rangle \in \mathsf{Mod}^*\mathcal{L}$ and any $a \in A$. Then by (b), $a \in F$ if and only if $E^{\boldsymbol{A}}(a) \subseteq \mathrm{Id}_A = \boldsymbol{\Omega}^{\boldsymbol{A}}F$; since by (a) the Leibniz congruence is defined by $\Delta(x,y)$, this is equivalent to $\Delta^{\boldsymbol{A}}(E^{\boldsymbol{A}}(a)) \subseteq F$. By completeness of $\mathcal{L}$ with respect to $\mathsf{Mod}^*\mathcal{L}$, this establishes (ALG3). Now $\mathcal{L}$ satisfies all the conditions in Theorem 3.21 for a logic to be algebraizable. ⊠

We thus see that algebraizability of a logic is equivalent to the *mutual definability* of the identity relation and the truth filter in reduced models of the logic. Considering each of these definability conditions separately leads to several of the classes of logics in the Leibniz hierarchy studied in Chapter 6. See Exercise 4.37 for other versions of this result.

The section ends with a particularly tricky application of the Isomorphism Theorem in its third version (Theorem 4.56). It is interesting because it is a reasoning about filters of a logic, in a particular algebra, without actually knowing anything about the logic (except that it is assumed to be algebraizable); but it takes advantage of the fundamental fact that the Leibniz congruence of a subset is a purely algebraic object, whose computation is independent of any logic.

EXAMPLE 4.61
The variety L of lattices and the variety DL of distributive lattices are not the equivalent algebraic semantics of any algebraizable logic. In particular, they are not varieties of logic in the sense of Definition 3.45.

PROOF: Since DL is a subvariety of L, by Theorem 3.33 it is enough to check the case of DL. Take the four-element chain $\boldsymbol{A}_4 = \langle \{0,a,b,1\}, \wedge, \vee\rangle$, where $0 < a < b < 1$. Since $\boldsymbol{A}_4 \in \mathsf{DL}$, $\mathrm{Con}_{\mathsf{DL}}\boldsymbol{A}_4 = \mathrm{Con}\boldsymbol{A}_4$, and it is not difficult to check† that $\mathrm{Con}\boldsymbol{A}_4$ is an eight-element Boolean algebra. Now, if there is some algebraizable logic $\mathcal{L}$ (in the language with just $\wedge$ and $\vee$) such that DL is its equivalent algebraic semantics, Theorem 4.56 implies that the Leibniz operator $\boldsymbol{\Omega}^{\boldsymbol{A}_4}$ would make the lattice $\mathcal{F}i_{\mathcal{L}}\boldsymbol{A}_4$ isomorphic to $\mathrm{Con}\boldsymbol{A}_4$. Therefore, $\mathcal{F}i_{\mathcal{L}}\boldsymbol{A}_4$ should be an eight-element Boolean algebra of subsets of $\{0,a,b,1\}$. Further, since all algebraizable logics have theorems, the smallest $\mathcal{L}$-filter should be non-empty. Inspection of the power set of $\{0,a,b,1\}$, which has just sixteen elements, shows that both properties are only possible when this smallest $\mathcal{L}$-filter is a one-element subset and the family $\mathcal{F}i_{\mathcal{L}}\boldsymbol{A}_4$ is exactly the family of all subsets containing this one-element subset (see Figure 6 on the next page). The Leibniz operator should be a lattice isomorphism, hence it should send this smallest $\mathcal{L}$-filter to the smallest congruence, the identity relation. There are just four candidates for this role of smallest $\mathcal{L}$-filter. Being a chain, it is easy to compute

† An indirect reasoning goes as follows: Since $\boldsymbol{A}_4$ is a finite distributive lattice, the lattice $\mathrm{Con}\boldsymbol{A}_4$ is isomorphic to the power set of the set of prime filters of $\boldsymbol{A}_4$ [78, Exercise 6.17]. In a chain all proper filters are prime, so $\boldsymbol{A}_4$ has three prime filters, and $\mathrm{Con}\boldsymbol{A}_4$ is (isomorphic to) the power set of a three-element set, i.e., an eight-element Boolean algebra ($2^3 = 8$).

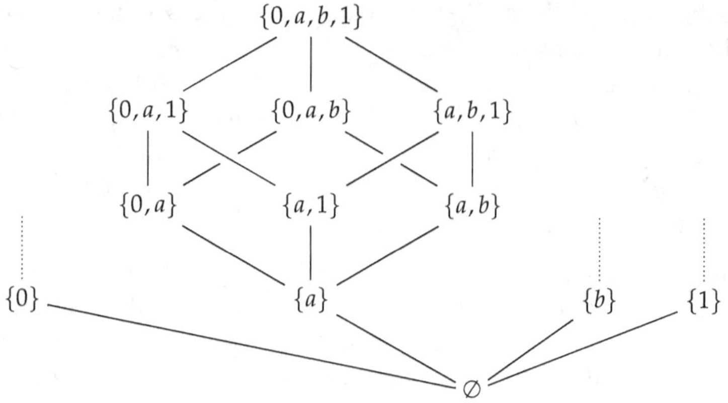

FIGURE 6: One of the possibilities of an eight-element Boolean algebra of subsets of $\{0,a,b,1\}$ with non-empty smallest member (here $\{a\}$). See Example 4.61.

their Leibniz congruences, and realize that neither is the identity relation:

$\boldsymbol{\Omega}^{\boldsymbol{A}_4}\{0\}$ identifies a, b and 1	$\boldsymbol{\Omega}^{\boldsymbol{A}_4}\{a\}$ identifies b and 1
$\boldsymbol{\Omega}^{\boldsymbol{A}_4}\{b\}$ identifies 0 and a	$\boldsymbol{\Omega}^{\boldsymbol{A}_4}\{1\}$ identifies $0, a$ and b

Therefore, the logic $\mathcal{L}$ cannot exist. ⊠

Exercise 4.38 offers an alternative proof of an even stronger result: neither of these two varieties can be the algebraic semantics of any logic!

Exercises for Section 4.4

EXERCISE 4.37. Show that one can replace the class $\mathsf{Mod}^*\mathcal{L}$ in Theorem 4.60 by any class M of reduced matrices such that $\mathcal{L}$ is complete with respect to M, without making any essential change in the proof. Investigate the changes that should be made in the statement (and in the proof) if one wants that the matrices in this class M need not be reduced; treat conditions (a) and (b) separately.

EXERCISE 4.38. Let L be the variety of lattices, formulated in the language $\langle \wedge, \vee \rangle$ of type $\langle 2,2 \rangle$. Prove that $\mathsf{L} \vDash \alpha(x) \approx x$ for any formula $\alpha(x)$ with the single variable x. Use this fact to prove that neither L nor any of its subclasses with at least one non-trivial member can be the algebraic semantics of any non-trivial logic.

4.5 Matrices as relational structures

This section aims to give just a glimpse of a sharply delimited framework, which connects algebraic logic with model theory in the most explicit way. The basic ideas started in Bloom's 1975 paper [39]. The framework is introduced, without giving many details (relying on your model-theoretic background†) and

† Sections V.1 and V.2 of [51] constitute a crash course on model theory which includes almost all the basic material needed here. More extensive and classical reference books on model theory

a few instances of its usage are presented. Here the symbol "$\vDash$" without any subscript denotes the first-order satisfaction and model relations (which include as particular cases the familiar satisfaction and model relations for equations, quasi-equations and "orders").

The goal is to apply ordinary model theory to study the matrices of a logic $\mathcal{L}$ formulated in a language $\boldsymbol{L}$. The key idea is to associate with $\boldsymbol{L}$ a first-order language $\boldsymbol{L}_D$ having $\boldsymbol{L}$ as its set of function symbols and having as logical symbols those of classical predicate logic, which is the logic of ordinary mathematics in which model theory is developed. This is done as follows:

- $\boldsymbol{L}_D$ has $\boldsymbol{L}$ as its set of constant and function symbols (some would say "as algebraic reduct") and has *only one relational symbol D*, which is *unary*.
- The ***individual variables*** of $\boldsymbol{L}_D$ are the sentential variables V.

Therefore:

- The ***terms*** of $\boldsymbol{L}_D$ are the sentential formulas, which are denoted by Fm or $Fm_{\boldsymbol{L}}$, as they have been up to now. Notice that this set constitutes an $\boldsymbol{L}$-algebra, denoted by $\boldsymbol{Fm}_{\boldsymbol{L}}$.
- The ***atomic formulas*** have either the form $D\alpha$, with $\alpha \in Fm_{\boldsymbol{L}}$, or the form $\alpha \approx \beta$, with $\alpha, \beta \in Fm_{\boldsymbol{L}}$.

And finally:

- The ***formulas*** of $\boldsymbol{L}_D$ are constructed from the atomic ones in the standard, classical way, using the *classical* sentential operators and quantification over individual variables. Notice that these sentential operators have nothing to do with the ordinary sentential connectives of the given logic $\mathcal{L}$, which are the operations in $\boldsymbol{L}$. The symbols $\boldsymbol{\wedge}, \boldsymbol{\vee}, \boldsymbol{\rightarrow}$ and $\boldsymbol{\neg}$ stand for these classical conjunction, disjunction, implication and negation of $\boldsymbol{L}_D$; they are used in Section 1.6 to write quasi-equations (which are indeed formulas of $\boldsymbol{L}_D$, of a particular kind), and should not be confused with the sentential connectives $\wedge, \vee, \to$ and $\neg$, which may or may not be present in $\boldsymbol{L}$ (see the examples later on).

Now we examine the ordinary first-order semantics of the language $\boldsymbol{L}_D$:

- A ***structure*** for $\boldsymbol{L}_D$ can be described as a pair $\mathcal{M} = \langle \boldsymbol{A}, F \rangle$ where $\boldsymbol{A}$ is an $\boldsymbol{L}$-algebra and $F \subseteq A$ is the value of D; classically one would write $D^{\mathcal{M}} = F$. Thus, *$\boldsymbol{L}_D$-structures can be identified with $\boldsymbol{L}$-matrices*, with the symbol D evaluated as the designated set of the matrix!
- The ***assignments*** of values to the individual variables are arbitrary functions $h \colon V \to A$. Since they are extended to terms by freeness they are actually the $\boldsymbol{L}$-homomorphisms from $\boldsymbol{Fm}_{\boldsymbol{L}}$ to $\boldsymbol{A}$.
- Since F is the value of D, this means that

$$\langle \boldsymbol{A}, F \rangle \vDash D\alpha \, [\![h]\!] \iff h\alpha \in F, \tag{61}$$

are [22, 56, 144].

for any $\alpha \in Fm_{L}$ and any $h \in \operatorname{Hom}(\boldsymbol{Fm}_{L}, \boldsymbol{A})$.

- This is extended in the standard way to obtain the ***satisfaction relation*** $\vDash$ between a matrix, a formula of $\boldsymbol{L}_{\mathrm{D}}$ and an assignment, and the associated model relation, also denoted by $\vDash$. If Σ is a set of formulas of $\boldsymbol{L}_{\mathrm{D}}$, $\operatorname{Mod}\Sigma$ denotes the class of (matrices that are) ***models*** of the formulas in Σ in this first-order sense.

A first consequence of this setting is a characterization of the Leibniz congruence of a matrix that was used by Blok and Pigozzi to better justify naming it after Leibniz, as announced in the discussion on page 198.

PROPOSITION 4.62
Let $\boldsymbol{A}$ be an algebra, $F \subseteq A$ and $a, b \in A$. Then $a \equiv b \ (\boldsymbol{\Omega}^{\boldsymbol{A}} F)$ if and only if a and b satisfy the same properties of the matrix $\langle \boldsymbol{A}, F \rangle$ definable in the language $\boldsymbol{L}_{\mathrm{D}}$ with parameters and without equality; that is, if and only if for any first-order equality-free formula $\alpha(x, \vec{z})$ of $\boldsymbol{L}_{\mathrm{D}}$ with free variables at most x and those in $\vec{z}$, and for every $\vec{c} \in \vec{A}$, $\langle \boldsymbol{A}, F \rangle \vDash \alpha(x, \vec{z}) \ [\![a, \vec{c}]\!]$ if and only if $\langle \boldsymbol{A}, F \rangle \vDash \alpha(x, \vec{z}) \ [\![b, \vec{c}]\!]$.

PROOF: Observe that (61) can be equivalently written as: $\langle \boldsymbol{A}, F \rangle \vDash D\alpha(x, \vec{z}) \ [\![a, \vec{c}]\!]$ if and only if $\alpha^{\boldsymbol{A}}(a, \vec{c}) \in F$. Using this, Theorem 4.23 establishes that $a \equiv b \ (\boldsymbol{\Omega}^{\boldsymbol{A}} F)$ if and only if for all $\alpha(x, \vec{z}) \in Fm$ and all $\vec{c} \in \vec{A}$, $\langle \boldsymbol{A}, F \rangle \vDash D\alpha(x, \vec{z}) \ [\![a, \vec{c}]\!]$ if and only if $\langle \boldsymbol{A}, F \rangle \vDash D\alpha(x, \vec{z}) \ [\![b, \vec{c}]\!]$, that is, if and only if a and b satisfy the same equality-free *atomic* formulas of $\boldsymbol{L}_{\mathrm{D}}$ with parameters. Then a straightforward inductive argument (Exercise 4.39) shows that this holds if and only if it holds for *arbitrary* equality-free formulas of $\boldsymbol{L}_{\mathrm{D}}$ with parameters. ⊠

The restriction to *equality-free* formulas is unavoidable: If the equality symbol were allowed to appear in the formulas of the proposition, then in particular a and b would both satisfy the (atomic) formula $x \approx z$ for the same values c of z; since for $c := a$ we get $b = b$ which is true, we would obtain $a = b$: we would be defining the real identity relation!

Next we define a "translation" **tr** from sequents of the sentential language $\boldsymbol{L}$ into (equality-free) formulas of $\boldsymbol{L}_{\mathrm{D}}$:

DEFINITION 4.63
If $\Gamma \rhd \varphi$ is a sequent with a finite, non-empty Γ, then

$$\mathbf{tr}(\Gamma \rhd \varphi) := \forall \vec{x} \big(\bigwedge_{\gamma \in \Gamma} D\gamma \rightarrow D\varphi \big),$$

where $\forall \vec{x}$ abbreviates $\forall x_1 \ldots \forall x_n$, with $\vec{x} = \langle x_1, \ldots, x_n \rangle$ being the sequence of all individual variables in the open formula, that is, all sentential variables in the sequent. For the empty antecedent case we take

$$\mathbf{tr}(\varnothing \rhd \varphi) := \forall \vec{x} \, D\varphi .$$

If $\mathcal{L}$ is a logic of type $\boldsymbol{L}$, then we define

$$\mathbf{tr}\mathcal{L} := \big\{ \mathbf{tr}(\Gamma \rhd \varphi) : \Gamma \vdash_{\mathcal{L}} \varphi \,, \Gamma \text{ finite} \big\}.$$

Since Γ is *a set* (rather than a sequence) and $\pmb{\wedge}$ is a *binary* operator, if you are extremely formal, you will recognize that the definition of **tr** requires fixing an order between formulas (in particular, between variables) and an association convention for $\pmb{\wedge}$ in such a way that the finite conjunction $\bigwedge_{\gamma\in\Gamma} D\gamma$ and the string $\forall\vec{x}$ are uniquely defined; it is safe to skip these details.

EXAMPLES.

The *Rule of Adjunction*: $\mathbf{tr}(x, y \rhd x \wedge y) = \forall x \forall y \big(Dx \pmb{\wedge} Dy \pmb{\rightarrow} D(x \wedge y)\big)$.

The *Rule of Modus Ponens*: $\mathbf{tr}(x, x \to y \rhd y) = \forall x \forall y \big(Dx \pmb{\wedge} D(x \to y) \pmb{\rightarrow} Dy\big)$.

Observe the coexistence of two conjunctions, or of two implications; this only happens because we are assuming that $\wedge$ or $\to$ appear in $\boldsymbol{L}$. While $\wedge$ and $\to$ are *logical*, in the sense that they are connectives *of* the sentential logic to be studied, $\pmb{\wedge}$ and $\pmb{\rightarrow}$ are *metalogical*, in the sense that they formalise the conjunction and implication of the (natural) language we use to speak *about* the logic (more precisely, to study its matrix semantics†) and are independent of it.

By its own definition, $\mathbf{tr}\mathcal{L}$ is a set of formulas of $\boldsymbol{L}_D$ that encodes the finitary consequences of $\mathcal{L}$. The presence of the universal quantifier means that validity of the formula $\mathbf{tr}(\Gamma \rhd \varphi)$ involves all assignments of values to the variables occurring in the sequent, or equivalently all $h \in \mathrm{Hom}(\boldsymbol{Fm}, \boldsymbol{A})$. The next basic properties are straightforward:

PROPOSITION 4.64

1. $\Gamma \vdash_{\langle \boldsymbol{A},F\rangle} \varphi$ *if and only if* $\langle \boldsymbol{A}, F\rangle \vDash \mathbf{tr}(\Gamma \rhd \varphi)$, *for any finite* $\Gamma \cup \{\varphi\} \subseteq Fm_{\boldsymbol{L}}$.
2. *If* $\mathcal{L}$ *is a logic and* $\mathcal{L}_{\mathrm{f}}$ *is its finitary companion, then* $\mathsf{Mod}(\mathbf{tr}\mathcal{L}) = \mathsf{Mod}(\mathcal{L}_{\mathrm{f}})$.
3. *If* $\mathcal{L}$ *is a finitary logic, then* $\mathsf{Mod}(\mathbf{tr}\mathcal{L}) = \mathsf{Mod}\,\mathcal{L}$.
4. *If* $\mathcal{L}$ *is a finitary logic, then for all finite* $\Gamma \cup \{\varphi\} \subseteq Fm_{\boldsymbol{L}}$, $\Gamma \vdash_{\mathcal{L}} \varphi$ *if and only if* $\mathbf{tr}\mathcal{L} \vDash \mathbf{tr}(\Gamma \rhd \varphi)$.

PROOF: 1 is just a matter of unravelling the meaning of each side, using (61), and looking at how the connectives of the natural language get translated into those of $\boldsymbol{L}_D$. However, since it is actually the core idea of all this construction, it is best to make this explicit. Assume first that $\Gamma = \{\varphi_1, \dots, \varphi_n\}$ with $n \geqslant 1$ for simplicity. Then

$$\begin{aligned}
&\langle \boldsymbol{A}, F\rangle \vDash \mathbf{tr}(\Gamma \rhd \varphi) \iff \forall h,\ \langle \boldsymbol{A}, F\rangle \vDash (D\varphi_1 \pmb{\wedge} \dots \pmb{\wedge} D\varphi_n) \pmb{\rightarrow} D\varphi\ [\![h]\!] \\
&\iff \forall h,\ \text{if } \langle \boldsymbol{A}, F\rangle \vDash D\varphi_1 \pmb{\wedge} \dots \pmb{\wedge} D\varphi_n\ [\![h]\!],\ \text{then } \langle \boldsymbol{A}, F\rangle \vDash D\varphi\ [\![h]\!] \\
&\iff \forall h,\ \text{if } \langle \boldsymbol{A}, F\rangle \vDash D\varphi_1\ [\![h]\!] \text{ and} \dots\ \langle \boldsymbol{A}, F\rangle \vDash D\varphi_n\ [\![h]\!],\ \text{then } \langle \boldsymbol{A}, F\rangle \vDash D\varphi\ [\![h]\!] \\
&\iff \forall h,\ \text{if } h\varphi_1 \in F \text{ and} \dots\ h\varphi_n \in F,\ \text{then } h\varphi \in F \\
&\iff \{\varphi_1, \dots, \varphi_n\} \vdash_{\langle \boldsymbol{A},F\rangle} \varphi,
\end{aligned}$$

where in each middle step h ranges over $\mathrm{Hom}(\boldsymbol{Fm}, \boldsymbol{A})$.

The case $\Gamma = \varnothing$ is proved similarly.

† The proof of Proposition 4.64.1 illustrates this point very well. See also footnote § on page 163.

2 is a consequence of 1, and 3 is a particular case of 2. The (⇒) half of 4 is a consequence of the definition of $\mathbf{tr}\mathcal{L}$, but the equivalence is shown directly:

$$\begin{aligned} \Gamma \vdash_{\mathcal{L}} \varphi &\iff \Gamma \vdash_{\langle \boldsymbol{A},F\rangle} \varphi \ \text{ for all } \langle \boldsymbol{A},F\rangle \in \mathsf{Mod}\,\mathcal{L} \\ &\iff \langle \boldsymbol{A},F\rangle \vDash \mathbf{tr}(\Gamma \rhd \varphi) \ \text{ for all } \langle \boldsymbol{A},F\rangle \in \mathsf{Mod}(\mathbf{tr}\mathcal{L}) \\ &\iff \mathbf{tr}\mathcal{L} \vDash \mathbf{tr}(\Gamma \rhd \varphi) \end{aligned}$$

The first equivalence holds by Theorem 4.16, the second by points 1 and 3, and the last one is a general model-theoretic property, which holds because all $\mathbf{tr}(\cdot)$ are *sentences*. ⊠

The "Mod" symbols appearing in the left-hand side of the equalities of this result are the "first-order" ones, while those in the right-hand sides are the "sentential" or "matrix" ones; this dual usage of the symbol, and of the notion of a "model" itself, is very convenient and should not confuse anybody.

Now observe that the formulas $\mathbf{tr}(\Gamma \rhd \varphi)$ have a peculiar structure: equality does not appear in them, and they are universally quantified prenex formulas with a fixed structure; in model theory these are known as ***equality-free universal strict basic Horn sentences***† of $\boldsymbol{L}_\mathrm{D}$. Thus:

COROLLARY 4.65
For every finitary logic $\mathcal{L}$, the class $\mathsf{Mod}\,\mathcal{L}$ *is an equality-free universal strict basic Horn class; that is, a class of $\boldsymbol{L}_\mathrm{D}$-structures axiomatized by a set of equality-free universal strict basic Horn sentences. In particular,* $\mathsf{Mod}\,\mathcal{L}$ *is elementary.* ⊠

But there is more: it is easy to see that any sentence of this kind has the form $\mathbf{tr}(\Gamma \rhd \varphi)$ for some sequent $\Gamma \rhd \varphi$ of $\boldsymbol{Fm}$; actually, modulo the formal details mentioned (and skipped) after Definition 4.63, this sequent is unique. Then:

THEOREM 4.66
If $\mathcal{L}$ is a finitary logic of type $\boldsymbol{L}$, then the set $\mathbf{tr}\mathcal{L}$ is an equality-free universal strict basic Horn theory of $\boldsymbol{L}_\mathrm{D}$; that is, a set of equality-free universal strict basic Horn sentences that is closed under (first-order) consequence.

Conversely, for every theory T of this kind in the language $\boldsymbol{L}_\mathrm{D}$ there is a unique finitary logic $\mathcal{L}$ such that $\mathbf{tr}\mathcal{L} = T$.

PROOF: To show that $\mathbf{tr}\mathcal{L}$ is an equality-free universal strict basic Horn theory we have to show that for each sentence α of this kind, if $\mathbf{tr}\mathcal{L} \vDash \alpha$, then $\alpha \in \mathbf{tr}\mathcal{L}$. However, by the previous observation, all sentences of this kind are of the form $\mathbf{tr}(\Gamma \rhd \varphi)$ for some sequent $\Gamma \rhd \varphi$. Therefore what we have to show is that

† Strictly speaking, our formulas are *equivalent to* a sentence of this kind. In general, ***strict basic Horn formulas*** are defined as formulas in prenex form whose non-quantified part is a ***clause***, i.e., a disjunction of literals (the ***literals*** are the atomic formulas and their negations) with the particularity that these literals are all negated atomic formulas except exactly one. Modulo logical equivalence, here it is safe to identify them with prenex formulas whose non-quantified part is either $(\alpha_1 \wedge \ldots \wedge \alpha_n) \to \alpha_{n+1}$ or simply α_1, where the α_i are atomic. When adding ***universal*** and saying ***sentences*** instead of formulas, we mean that all quantifiers are universal and cover all the occurring variables.

if $\mathbf{tr}\mathcal{L} \vDash \mathbf{tr}(\Gamma \rhd \varphi)$, then $\Gamma \vdash_{\mathcal{L}} \varphi$. But this is contained in Proposition 4.64.4 (actually, this property is exactly equivalent to the assertion that $\mathbf{tr}\mathcal{L}$ is a theory, as Exercise 4.40 shows).

For the converse, given an equality-free universal strict basic Horn theory without equality T, define, for every $\Gamma \cup \{\varphi\} \subseteq Fm$,

$$\Gamma \vdash_{\mathcal{L}} \varphi \overset{\text{def}}{\iff} \text{There is some finite } \Gamma_0 \subseteq \Gamma \text{ with } \mathbf{tr}(\Gamma_0 \rhd \varphi) \in T. \tag{62}$$

There is a subtle point here. Since $\forall\vec{x}(\alpha \to \gamma) \vDash \forall\vec{x}(\alpha \wedge \beta \to \gamma)$ holds generally in first-order logic, from (62) it follows that for any finite Γ,

$$\Gamma \vdash_{\mathcal{L}} \varphi \overset{\text{def}}{\iff} \mathbf{tr}(\Gamma \rhd \varphi) \in T. \tag{63}$$

This shows that $\mathbf{tr}\mathcal{L} = T$, and one has to show that $\mathcal{L}$ is a finitary logic, that is, to prove the properties (I), (M), (C), (S) and (F), using in an essential way that T is a (first-order) theory, and hence it is closed under (first-order) consequence; observe that in particular all equality-free universal strict basic Horn sentences that are universally valid belong to T. For instance, since $\forall\vec{x}(D\varphi \to D\varphi)$ is universaly valid, if $\varphi \in \Gamma$, (62) implies that $\Gamma \vdash_{\mathcal{L}} \varphi$, which proves (I). Properties (M) and (F) follow trivially from the definition. Property (C) is proved in a similar, but more elaborated way (Exercise 4.41). As to property (S): We have to show that $\Gamma \vdash_{\mathcal{L}} \varphi$ implies $\sigma\Gamma \vdash_{\mathcal{L}} \sigma\varphi$ for any substitution σ. We can safely assume that Γ is finite and show, by (63), that $\mathbf{tr}(\sigma\Gamma \rhd \sigma\varphi) \in T$ whenever $\mathbf{tr}(\Gamma \rhd \varphi) \in T$. Since T is a theory, it is enough to show that the formula $\mathbf{tr}(\sigma\Gamma \rhd \sigma\varphi)$ is a (first-order) consequence of the formula $\mathbf{tr}(\Gamma \rhd \varphi)$. That is, we have to show that any model $\langle \boldsymbol{A}, F\rangle$ of $\mathbf{tr}(\Gamma \rhd \varphi)$ is a model of $\mathbf{tr}(\sigma\Gamma \rhd \sigma\varphi)$. By Proposition 4.64.1, this amounts to showing that if $\Gamma \vdash_{\langle \boldsymbol{A},F\rangle} \varphi$, then $\sigma\Gamma \vdash_{\langle \boldsymbol{A},F\rangle} \sigma\varphi$, and this is true simply because the relation $\vdash_{\langle \boldsymbol{A},F\rangle}$ defines a logic, hence in particular it is structural. Uniqueness is obvious by the comments before the statement. ⊠

Thus, *from a model-theoretic point of view, the theory of matrix semantics of finitary logics can be considered as a part of the model theory of equality-free languages*. This is a little cultivated discipline, which incorporates intuitions from algebraic logic but develops them with typical model-theoretic techniques, and which can in turn be applied or have some influence in algebraic logic; two members of the Barcelona group, Elgueta [87, 88] and Dellunde [80, 81] significantly developed it, independently, in their doctoral dissertations and further papers; more recent additions to this area are [185, 186, 228].

Model-theoretic characterizations

Let us see a few of the ways in which one can take advantage of this new viewpoint. The operators with (classes of) matrices are the usual model-theoretic ones $\mathbb{I}, \mathbb{S}, \mathbb{P}, \mathbb{P}_{\mathrm{R}}, \mathbb{P}_{\mathrm{U}}$; respectively: isomorphic copies, substructures (which are the *submatrices* of Definition 4.5), products,[†] reduced products and ultraproducts,

[†] Look at the end of the section for some details on products. To be consistent with Section 1.6, the product operator and its derivatives are understood as excluding the empty families. Recall that

plus the less usual ones $\mathbb{H}_s$ and $\mathbb{H}_s^{-1}$ (taking images and inverse images by strict surjective homomorphisms). The operator * of taking reductions, a special case of $\mathbb{H}_s$ (see Definition 4.31), can also be considered.

It is working with equality-free sentences that gives this direction of research its distinctive character. For instance, one can define the notion of a strict homomorphism for structures of an arbitrary similarity type as the (algebraic) homomorphisms that preserve all relations back and forth, extending Definition 4.26; then it is not difficult to see that $\mathbb{H}_s$ and $\mathbb{H}_s^{-1}$ preserve the validity of equality-free sentences.‡ That this preservation actually characterizes this kind of sentences was shown only in 1996 by Dellunde and Jansana in [81]: A sentence of an arbitrary first-order language is logically equivalent to an equality-free sentence if and only if it is preserved by $\mathbb{H}_s$ and $\mathbb{H}_s^{-1}$.

It is not difficult to show:

PROPOSITION 4.67
For every logic $\mathcal{L}$, the class $\mathsf{Mod}\mathcal{L}$ *is closed under the operators* $\mathbb{I}, \mathbb{S}, \mathbb{P}, {}^*, \mathbb{H}_s$ *and* $\mathbb{H}_s^{-1}$. *If $\mathcal{L}$ is finitary, then* $\mathsf{Mod}\mathcal{L}$ *is moreover closed under* $\mathbb{P}_R$ *and* $\mathbb{P}_U$.

PROOF: Proposition 4.35 establishes that $\mathsf{Mod}\mathcal{L}$ is closed under the operators $\mathbb{I}, \mathbb{H}_s, \mathbb{H}_s^{-1}$ and *, and by the result in Exercise 4.3, it is closed under $\mathbb{S}$. Closure under $\mathbb{P}$ is routinely proved by using just the definitions (everything works coordinatewise). In the finitary case, by Corollary 4.65 $\mathsf{Mod}\mathcal{L}$ is the class of models of a set of Horn sentences, and by a general result of model theory (see [56, Proposition 6.2.2], or [144, Theorem 9.4.3]) these classes are closed under reduced products, in particular under ultraproducts. ⊠

We see in Theorem 4.70 that the two last properties characterize finitary logics; the key is the following important fact:

THEOREM 4.68
Let M *be a class of matrices closed under ultraproducts. The logic* $\mathcal{L}_{\mathsf{M}}$ *is finitary.*

PROOF: The proof is modelled after the usual proof of the Compactness Theorem of first-order logic using ultraproducts (see [56, Corollary 4.1.11]), and is done by contraposition. So, let $\Gamma \cup \{\varphi\} \subseteq Fm$ be such that for every finite $\Delta \subseteq \Gamma$, $\Delta \nvdash_{\mathsf{M}} \varphi$. This means that there is some $\mathcal{M}_\Delta = \langle \boldsymbol{A}_\Delta, F_\Delta \rangle \in \mathsf{M}$ and some $h_\Delta \colon \boldsymbol{Fm} \to \boldsymbol{A}_\Delta$ such that $h_\Delta \Delta \subseteq F_\Delta$ but $h_\Delta \varphi \notin F_\Delta$. Put $I := \{\Delta \subseteq \Gamma : \Delta \text{ finite}\}$ and construct the usual ultraproduct of the family $\{\mathcal{M}_\Delta : \Delta \in I\}$ modulo some ultrafilter $\mathcal{U}$ of $\mathscr{P}I$ such that for each $\Delta \in I$, $\Delta^* := \{\Delta' \in I : \Delta' \supseteq \Delta\} \in \mathcal{U}$; such an ultrafilter exists because the family $\{\Delta^* : \Delta \in I\}$ has the finite intersection property (because $\Delta_1 \cup \cdots \cup \Delta_n \in \Delta_1^* \cap \cdots \cap \Delta_n^*$ for any $\Delta_i \in I$). Denote this ultraproduct by $\langle \boldsymbol{A}/\mathcal{U}, F \rangle$, where $\boldsymbol{A} := \prod_{\Delta \in I} \boldsymbol{A}_\Delta$ and, by definition, $\langle a_\Delta : \Delta \in I \rangle / \mathcal{U} \in F$ if and

products can be identified with trivial cases of reduced products, and that ultraproducts are special cases of reduced products.

‡ This is implicit in Proposition 29.8 of Monk's book [175], one of the few that mention (though only briefly) these homomorphisms, calling them "two-way homomorphisms". Homomorphisms in the usual sense of model theory only preserve relations in the "forth" direction.

only if $\{\Delta \in I : a_\Delta \in F_\Delta\} \in \mathcal{U}$. Define $h\colon \boldsymbol{Fm} \to \boldsymbol{A}/\mathcal{F}$ by $h\alpha := \langle h_\Delta \alpha : \Delta \in I \rangle / \mathcal{U}$ for all $\alpha \in Fm$. Then, for each $\delta \in \Gamma$, $\{\delta\} \in I$ and hence $\{\delta\}^* \in \mathcal{U}$. But $\{\delta\}^* \subseteq \{\Delta \in I : h_\Delta \delta \in F_\Delta\}$, because if $\Delta \in \{\delta\}^*$, then $\delta \in \Delta$ and since $h_\Delta \Delta \subseteq F_\Delta$, in particular $h_\Delta \delta \in F_\Delta$. Since $\mathcal{U}$ is a filter, it follows that $\{\Delta \in I : h_\Delta \delta \in F_\Delta\} \in \mathcal{U}$, and hence that $h\delta \in F$. Thus, $h\Gamma \subseteq F$; but $h\varphi \notin F$ because by construction $\{\Delta \in I : h_\Delta \varphi \in F_\Delta\} = \emptyset \notin \mathcal{U}$. Since $\langle \boldsymbol{A}/\mathcal{U}, F \rangle \in \mathsf{M}$ by the assumption, this shows that $\Gamma \nvdash_{\mathsf{M}} \varphi$. ⊠

From this, the following alternative proofs of Theorem 4.4 and of Corollary 4.9 are easily obtained:

Corollary 4.69
If M is a finite set of finite matrices, then the logic $\mathcal{L}_{\mathsf{M}}$ is finitary. In particular, the logic defined by a finite matrix is finitary.

Proof: By Theorem 4.34, $\mathcal{L}_{\mathsf{M}} = \mathcal{L}_{\mathbb{I}\mathsf{M}}$, and it is well known [51, Lemma IV.6.5] that if M is a finite set of finite structures, then $\mathbb{I}\mathsf{M}$ is closed under ultraproducts. Thus, Theorem 4.68 applies. ⊠

The converse of Theorem 4.68 is not true (think of a finitary logic defined by a single matrix), but there is a good approximation to it. By combining several of the previous results with standard facts of model theory, the following characterization theorem can be formulated:

Theorem 4.70
Let $\mathcal{L}$ be a logic of similarity type $\boldsymbol{L}$. The following conditions are equivalent:

(i) *$\mathcal{L}$ is finitary.*
(ii) *$\mathsf{Mod}\mathcal{L}$ is axiomatizable by a set of equality-free universal strict basic Horn sentences of $\boldsymbol{L}_{\mathrm{D}}$.*
(iii) *$\mathsf{Mod}\mathcal{L}$ is axiomatizable in $\boldsymbol{L}_{\mathrm{D}}$. That is, $\mathsf{Mod}\mathcal{L}$ is an elementary class.*
(iv) *$\mathsf{Mod}\mathcal{L}$ is closed under ultraproducts.*
(v) *$\mathcal{L}$ is complete with respect to a class of matrices that is closed under ultraproducts.*

Proof: (i)⇒(ii) is Corollary 4.65, (iii)⇒(iv) is a standard fact of model theory, and (v)⇒(i) is Theorem 4.68. The remaining implications are trivial, taking completeness of $\mathcal{L}$ with respect to $\mathsf{Mod}\mathcal{L}$ into account. ⊠

This result may be slightly surprising: For any logic $\mathcal{L}$, if the class $\mathsf{Mod}\mathcal{L}$ is axiomatizable in some (unknown) way, then it is automatically axiomatizable by formulas of a very specific form, namely by all formulas in $\mathbf{tr}\mathcal{L}$, and moreover $\mathcal{L}$ is finitary. Exercise 4.44 collects other equivalent characterizations that could have been included here as well.

Now consider the following, related issue. Any class M of matrices defines a logic $\mathcal{L}_{\mathsf{M}}$, and then $\mathsf{M} \subseteq \mathsf{Mod}\mathcal{L}_{\mathsf{M}}$. The question now arises as to how large the class $\mathsf{Mod}\mathcal{L}_{\mathsf{M}}$ can be, and how can it be found (or *constructed*) starting from M. These are typical questions in universal algebra or model theory contexts (recall, for instance, Tarski's theorem on varieties). This is solved by a classical result of

Czelakowski. In the following, $\mathcal{M}_t$ denotes a trivial matrix (recall, one with a one-element algebra and its universe as the filter).

LEMMA 4.71
For every class of matrices M, $\mathsf{LTMod}^*\mathcal{L}_\mathsf{M} \subseteq \mathbb{H}_\mathrm{S}\mathbb{SP}(\mathsf{M} \cup \{\mathcal{M}_t\})$.

PROOF: Let $\langle \boldsymbol{Fm}/\boldsymbol{\Omega}\Gamma, \Gamma/\boldsymbol{\Omega}\Gamma\rangle \in \mathsf{LTMod}^*\mathcal{L}_\mathsf{M}$, for some $\Gamma \in \mathcal{T}h\mathcal{L}_\mathsf{M}$. If $\Gamma = Fm$, the inconsistent theory, then $\boldsymbol{\Omega}\Gamma = Fm \times Fm$ and hence $\langle \boldsymbol{Fm}/\boldsymbol{\Omega}\Gamma, \Gamma/\boldsymbol{\Omega}\Gamma\rangle$ is a trivial matrix, isomorphic to $\mathcal{M}_t$, and therefore it belongs to $\mathbb{H}_\mathrm{S}\mathbb{SP}(\mathsf{M} \cup \{\mathcal{M}_t\})$. Now let Γ be a consistent theory of $\mathcal{L}_\mathsf{M}$. Since it can be safely assumed that all matrices in M are proper, Exercise 4.45 implies that there is a non-empty family $\{\langle \boldsymbol{A}_i, F_i\rangle : i \in I\} \subseteq \mathsf{M}$ with corresponding $h_i \in \mathrm{Hom}(\boldsymbol{Fm}, \boldsymbol{A}_i)$ such that $\Gamma = \bigcap\{h_i^{-1}F_i : i \in I\}$. Now consider the product of this family $\prod_{i\in I}\langle \boldsymbol{A}_i, F_i\rangle := \langle \prod \boldsymbol{A}_i, \prod F_i\rangle$, and the product homomorphism $h := \langle h_i : i \in I\rangle$. Then for every formula δ, $\delta \in \Gamma$ if and only if $h_i\delta \in F_i$ for all $i \in I$, if and only if $h\delta \in \prod F_i$. Thus, h is strict from $\langle \boldsymbol{Fm}, \Gamma\rangle$ to $\langle \prod \boldsymbol{A}_i, \prod F_i\rangle$. Now consider the image submatrix, that is, the matrix $\langle h\boldsymbol{Fm}, F\rangle$, where $F := (\prod F_i) \cap hFm$, which makes h become strict and surjective from $\langle \boldsymbol{Fm}, \Gamma\rangle$ to $\langle h\boldsymbol{Fm}, F\rangle$. By Proposition 4.32.3, the reductions of these two matrices are isomorphic: $\langle \boldsymbol{Fm}/\boldsymbol{\Omega}\Gamma, \Gamma/\boldsymbol{\Omega}\Gamma\rangle \cong \langle h\boldsymbol{Fm}, F\rangle^*$. Now, by construction $\langle h\boldsymbol{Fm}, F\rangle \in \mathbb{SP}\mathsf{M}$, reduction is an application of $\mathbb{H}_\mathrm{S}$, and the same is true of isomorphisms. A trivial matrix can be safely added. Therefore, $\langle \boldsymbol{Fm}/\boldsymbol{\Omega}\Gamma, \Gamma/\boldsymbol{\Omega}\Gamma\rangle \in \mathbb{H}_\mathrm{S}\mathbb{SP}(\mathsf{M} \cup \{\mathcal{M}_t\})$. ⊠

This result has an interest by itself; Exercise 4.46 showcases one interesting application. Since every matrix is a strict inverse image of its reduction, and a countably generated reduced model of a logic is isomorphic to one of its Lindenbaum-Tarski models (Theorem 4.44), from Lemma 4.71 we obtain the following, which is what is actually used in the proof of the next theorem.

COROLLARY 4.72
If M *is a class of matrices and* $\mathcal{M} \in \mathsf{Mod}\,\mathcal{L}_\mathsf{M}$ *is a countably generated matrix, then* $\mathcal{M} \in \mathbb{H}_\mathrm{S}^{-1}\mathbb{H}_\mathrm{S}\mathbb{SP}(\mathsf{M} \cup \{\mathcal{M}_t\})$. ⊠

The next property is a typical model-theoretic one, whose proof, I think, can be skipped:

LEMMA 4.73
For any class M *of matrices, the class* $\mathbb{H}_\mathrm{S}^{-1}\mathbb{H}_\mathrm{S}\mathbb{SP}_\mathrm{R}\mathsf{M}$ *is closed under* $\mathbb{S}$ *and* $\mathbb{P}_\mathrm{R}$. ⊠

Actually, a stronger, nicer result holds: For every class M of matrices, the class $\mathbb{H}_\mathrm{S}^{-1}\mathbb{H}_\mathrm{S}\mathbb{SP}_\mathrm{R}(\mathsf{M} \cup \{\mathcal{M}_t\})$ is the smallest class of matrices containing M that is closed under the operators $\mathbb{H}_\mathrm{S}^{-1}$, $\mathbb{H}_\mathrm{S}$, $\mathbb{S}$ and $\mathbb{P}_\mathrm{R}$ and contains a trivial matrix. But we do not need all the strength of this characterization. Now we have all the tools necessary to prove:

THEOREM 4.74
Let M *be a class of matrices, and assume that the logic* $\mathcal{L}_\mathsf{M}$ *defined by* M *is finitary. Then* $\mathsf{Mod}\,\mathcal{L}_\mathsf{M} = \mathbb{H}_\mathrm{S}^{-1}\mathbb{H}_\mathrm{S}\mathbb{SP}_\mathrm{R}(\mathsf{M} \cup \{\mathcal{M}_t\})$.

PROOF: ($\supseteq$) is a consequence of the facts that $\mathsf{M} \subseteq \mathsf{Mod}\mathcal{L}_\mathsf{M}$ and that the latter class is closed under the relevant operations (Proposition 4.67) and contains a trivial matrix (actually all of them).

($\subseteq$) Take any $\mathcal{M} \in \mathsf{Mod}\mathcal{L}_\mathsf{M}$ and let $\{\mathcal{M}_i : i \in I\}$ be the family of all its finitely generated submatrices. These matrices are countably generated, and belong to $\mathsf{Mod}\mathcal{L}_\mathsf{M}$. By Corollary 4.72, $\{\mathcal{M}_i : i \in I\} \subseteq \mathbb{H}_\mathrm{S}^{-1}\mathbb{H}_\mathrm{S}\mathbb{S}\mathbb{P}(\mathsf{M} \cup \{\mathcal{M}_t\})$. By a standard model-theoretic procedure (see [51, Theorem V.2.14] for instance), $\mathcal{M}$ can be embedded into an ultraproduct of the $\mathcal{M}_i$. So, $\mathcal{M} \in \mathbb{I}\mathbb{S}\mathbb{P}_\mathrm{U}\mathbb{H}_\mathrm{S}^{-1}\mathbb{H}_\mathrm{S}\mathbb{S}\mathbb{P}(\mathsf{M} \cup \{\mathcal{M}_t\})$. Now, since (ultra)products are particular cases of reduced products, we also have that $\mathcal{M} \in \mathbb{I}\mathbb{S}\mathbb{P}_\mathrm{R}\mathbb{H}_\mathrm{S}^{-1}\mathbb{H}_\mathrm{S}\mathbb{S}\mathbb{P}_\mathrm{R}(\mathsf{M} \cup \{\mathcal{M}_t\}) = \mathbb{H}_\mathrm{S}^{-1}\mathbb{H}_\mathrm{S}\mathbb{S}\mathbb{P}_\mathrm{R}(\mathsf{M} \cup \{\mathcal{M}_t\})$; the last equality is a consequence of Lemma 4.73 and obvious properties of $\mathbb{I}$. ⊠

This allows us to address another question, similar in spirit, but from a different angle: when is a given class of matrices exactly the class of all models of some (finitary) logic?

THEOREM 4.75
Let M *be a class of matrices. The following conditions are equivalent:*

(i) $\mathsf{M} = \mathsf{Mod}\mathcal{L}$ *for some finitary logic* $\mathcal{L}$.

(ii) M *is closed under the operators* $\mathbb{H}_\mathrm{S}^{-1}$, $\mathbb{H}_\mathrm{S}$, $\mathbb{S}$ *and* $\mathbb{P}_\mathrm{R}$ *and contains a trivial matrix.*

PROOF: (i)⇒(ii) is given by Proposition 4.67.

(ii)⇒(i) The wanted logic is obviously $\mathcal{L}_\mathsf{M}$, the logic defined by the class M of matrices. By assumption this class is closed under $\mathbb{P}_\mathrm{R}$, so in particular it is closed under $\mathbb{P}_\mathrm{U}$, and by Theorem 4.68, $\mathcal{L}_\mathsf{M}$ is finitary. Then we can apply Theorem 4.74, and together with the closure assumptions on M, we conclude that $\mathsf{Mod}\mathcal{L}_\mathsf{M} = \mathsf{M}$. ⊠

Similar results exist without the assumption of finitarity, either for countable languages [70], or in total generality [81]. All this work is very close to the model theory of implicational classes, but since only equality-free formulas are considered, at certain points one has to perform reductions, and this is what makes the operators $\mathbb{H}_\mathrm{S}$ and $\mathbb{H}_\mathrm{S}^{-1}$ appear into the scene. The presence of the operator sequence $\mathbb{H}_\mathrm{S}^{-1}\mathbb{H}_\mathrm{S}$ is characteristic of equality-free model theory; for instance, observe (Exercise 4.47) that if $\mathcal{M}, \mathcal{N}$ are matrices, then $\mathcal{M} \in \mathbb{H}_\mathrm{S}^{-1}\mathbb{H}_\mathrm{S}\mathcal{N}$ if and only if $\mathcal{M}^* \cong \mathcal{N}^*$; this relation has been called *the relativeness relation* in some model-theoretic works.

A natural question at this point is: *Is it possible to have similar results for reduced matrices?* As to Theorem 4.70, the answer is: in general, no. While for a finitary logic $\mathcal{L}$ the class $\mathsf{Mod}\mathcal{L}$ is always elementary, the class $\mathsf{Mod}^*\mathcal{L}$ need not be so; one example is the local logic K^ℓ associated with the normal modal system K, whose class of reduced models was shown by Malinowski [172] not to be closed under ultraproducts. In general, $\mathsf{Mod}^*\mathcal{L}$ need not be closed under $\mathbb{S}$ or $\mathbb{P}$; actually, Corollary 6.74 shows that closure of $\mathsf{Mod}^*\mathcal{L}$ under these operators characterizes equivalential logics. As to Theorem 4.74, the answer is yes, and is contained in the following result of Blok and Pigozzi [34, Theorem 6.2]; its proof largely surpasses the intended length and depth of the present section:

THEOREM 4.76
Let M *be a class of reduced matrices, and denote by* $\mathcal{L} := (\mathcal{L}_{\mathsf{M}})_{\mathrm{f}}$ *the finitary logic defined by* M*. Then* $\mathsf{Mod}^*\mathcal{L} = \mathbb{I}\mathbb{S}^*\mathbb{P}_{\mathrm{U}}^*\mathbb{P}_{\omega}^*(\mathsf{M} \cup \{\mathcal{M}_t\})$*, where* $\mathbb{P}_\omega$ *is the operator of constructing products of finite families, and for any class operator* $\mathbb{O}$*, the operator* $\mathbb{O}^*$ *is defined by* $\mathbb{O}^*\mathsf{M} := \{\mathcal{M}^* : \mathcal{M} \in \mathbb{O}\mathsf{M}\}$ *for any class* M *of matrices.* ⊠

An interesting and elementary result around this, due to Rautenberg [211], involves the following notion: The Leibniz congruence is ***finitizable*** in a class M of matrices when there is a *finite* set $\Phi \subseteq Fm$ such that for every matrix $\langle \boldsymbol{A}, F\rangle \in \mathsf{M}$ and any $a,b \in A$, $a \equiv b\ (\boldsymbol{\Omega}^{\boldsymbol{A}}F)$ if and only if for all $\delta \in \Phi$ and all $\vec{c} \in \vec{A}$, $[\delta^{\boldsymbol{A}}(a,\vec{c}) \in F \Leftrightarrow \delta^{\boldsymbol{A}}(b,\vec{c}) \in F]$; that is, the characterization in Theorem 4.23 can be limited to use formulas in the finite set Φ. In particular, this implies that, for matrices in M, to be reduced is a first-order property; indeed, a matrix in M is reduced if and only if it satisfies the following sentence of $\boldsymbol{L}_D$ (with equality):

$$\forall x \forall y \big(\big(\bigwedge_{\delta \in \Phi} \forall \vec{z} (D\delta(x,\vec{z}) \leftrightarrow D\delta(y,\vec{z})) \big) \to x \approx y \big). \tag{64}$$

THEOREM 4.77
Let $\mathcal{L}$ *be a finitary logic. The following conditions are equivalent:*

(i) *The Leibniz congruence is finitizable in* $\mathsf{Mod}\,\mathcal{L}$.
(ii) *The Leibniz congruence is finitizable in* $\mathsf{Mod}^*\mathcal{L}$.
(iii) $\mathsf{Mod}^*\mathcal{L}$ *is an elementary class.*
(iv) $\mathsf{Mod}^*\mathcal{L}$ *is closed under ultraproducts.*

PROOF SKETCH: That (i) implies (iii) follows from the fact that $\mathsf{Mod}\,\mathcal{L}$ is already an elementary class (by Theorem 4.70) and the assumption of finitizability in $\mathsf{Mod}\,\mathcal{L}$, as discussed above. That (iii) implies (iv) is a general model-theoretic fact. That (iv) implies (ii) is proved by *reductio ad absurdum* and performing an ultraproduct construction, following a strategy similar to that in the proof of Theorem 4.68; here the index set would be the set of all finite sets of formulas, and the assumption would be that none of these defines the Leibniz congruence in $\mathsf{Mod}^*\mathcal{L}$. Finally, the implication from (ii) to (i) is an elementary fact of the theory of matrices and can be safely left to you as Exercise 4.49. ⊠

Notice that in principle equality cannot be eliminated in (64), that is, in (iii), unlike in the case of Theorem 4.70.

To end this section let us see something on the product construction, which has a rather silent role in this chapter, and in general in the book. Construction of the product of a family $\{\mathcal{M}_i := \langle \boldsymbol{A}_i, F_i\rangle : i \in I\}$ of matrices presents no difficulties: one takes the product algebra $\boldsymbol{A} := \prod_{i\in I} \boldsymbol{A}_i$ and the product of the filters $F := \prod_{i \in I} F_i$ and one defines $\prod_{i\in I} \mathcal{M}_i := \langle \boldsymbol{A}, F\rangle := \langle \prod_{i\in I} \boldsymbol{A}_i, \prod_{i\in I} F_i\rangle$.

Everything works coordinatewise, homomorphisms $h \in \mathrm{Hom}(\boldsymbol{Fm}, \boldsymbol{A})$ can be identified with tuples $\langle h_i : i \in I \rangle$ where $h_i \in \mathrm{Hom}(\boldsymbol{Fm}, \boldsymbol{A}_i)$, and the logic $\mathcal{L}_{\prod_{i \in I} \mathcal{M}_i}$ is defined in the usual way: For all $\Gamma \cup \{\varphi\} \subseteq Fm$,

$$\begin{aligned} \Gamma \vdash_{\prod_{i \in I} \mathcal{M}_i} \varphi &\iff \text{for all } h \in \mathrm{Hom}(\boldsymbol{Fm}, \boldsymbol{A}), \text{if } h\Gamma \subseteq F, \text{ then } h\varphi \in F \\ &\iff \text{for all } h \in \mathrm{Hom}(\boldsymbol{Fm}, \boldsymbol{A}), \\ &\qquad \text{if for all } i \in I,\ h_i\Gamma \subseteq F_i, \ \text{ then for all } i \in I,\ h_i\varphi \in F_i. \end{aligned} \tag{65}$$

One then would like that this works as with validity of equations in universal algebra (an equation holds in $\prod_{i \in I} \boldsymbol{A}_i$ if and only if it holds in all the $\boldsymbol{A}_i$). This is certainly what happens with theorems: from (65), it follows that

$$\vdash_{\prod_{i \in I} \mathcal{M}_i} \varphi \iff \vdash_{\mathcal{M}_i} \varphi \ \text{ for all } i \in I.$$

One might then hope that this extends to the consequence relation, that is, that the logic defined by the product of the matrices equals the meet (intersection) of the logics defined by each of them. However, the fact that the logic is related to validity of (infinitary) implicational formulas breaks this in some borderline cases. To see this, consider how this intersection is defined:

$$\begin{aligned} \Gamma \vdash_{\bigwedge_{i \in I} \mathcal{L}_{\mathcal{M}_i}} \varphi &\iff \text{for all } i \in I,\ \Gamma \vdash_{\mathcal{M}_i} \varphi \\ &\iff \text{for all } i \in I, \text{ for all } h_i \in \mathrm{Hom}(\boldsymbol{Fm}, \boldsymbol{A}_i), \\ &\qquad\qquad \text{if } h_i\Gamma \subseteq F_i, \text{ then } h_i\varphi \in F_i. \end{aligned} \tag{66}$$

It is clear that the universal quantifiers, which appear with different scopes in (65) and in (66), cannot be arbitrarily rearranged. The real truth is the following result; it is practical to asume that all matrices are proper, because the addition of some non-proper ones does not modify the product nor the intersection of the family.

PROPOSITION 4.78
Let $\{\mathcal{M}_i : i \in I\}$ be a family of proper matrices.

1. $\bigwedge_{i \in I} \mathcal{L}_{\mathcal{M}_i} \leqslant \mathcal{L}_{\prod_{i \in I} \mathcal{M}_i}$.
2. *If $\Gamma \subseteq Fm$ is $\mathcal{L}_{\mathcal{M}_i}$-inconsistent for some $i \in I$, then Γ is inconsistent in $\mathcal{L}_{\prod_{i \in I} \mathcal{M}_i}$.*
3. *If $\Gamma \subseteq Fm$ is $\mathcal{L}_{\mathcal{M}_i}$-consistent for all $i \in I$, then $\Gamma \vdash_{\prod_{i \in I} \mathcal{M}_i} \varphi$ if and only if $\Gamma \vdash_{\bigwedge_{i \in I} \mathcal{L}_{\mathcal{M}_i}} \varphi$.*
4. *If all the matrices define the same logic, then $\mathcal{L}_{\prod_{i \in I} \mathcal{M}_i} = \bigwedge_{i \in I} \mathcal{L}_{\mathcal{M}_i} = \mathcal{L}_{\mathcal{M}_i}$, for each $i \in I$.*

PROOF: The right-hand side of (66) obviously implies the one of (65); this proves 1. Since all the matrices are proper, Exercise 4.1 applies, and the assumption that Γ is $\mathcal{L}_{\mathcal{M}_i}$-inconsistent means that there is no $h \in \mathrm{Hom}(\boldsymbol{Fm}, \boldsymbol{A}_i)$ such that $h\Gamma \subseteq F_i$. Thus, the antecendent of the implication in the last line of (65) can never hold (as it includes a "for all $i \in I$"), and hence the implication itself is vacuously true, regardless of φ. This proves 2. In order to prove 3, observe that $\Gamma \vdash_{\bigwedge_{i \in I} \mathcal{L}_{\mathcal{M}_i}} \varphi$ implies $\Gamma \vdash_{\prod_{i \in I} \mathcal{M}_i} \varphi$ by 1, and only the converse implication should be proved. So assume that $\Gamma \vdash_{\prod_{i \in I} \mathcal{M}_i} \varphi$, take any $i_0 \in I$ and any $h_{i_0} \in \mathrm{Hom}(\boldsymbol{Fm}, \boldsymbol{A}_{i_0})$

such that $h_{i_0}\Gamma \subseteq F_{i_0}$. By the assumption on Γ and Exercise 4.1 again, for every $i \neq i_0$ there is some $h_i \in \operatorname{Hom}(\boldsymbol{Fm}, \boldsymbol{A}_i)$ such that $h_i\Gamma \subseteq F_i$. Taking these homomorphisms we obtain $h := \langle h_i : i \in I \rangle \in \operatorname{Hom}(\boldsymbol{Fm}, \prod_{i\in I} \boldsymbol{A}_i)$ such that $h\Gamma \subseteq \prod_{i\in I} F_i$. Thefore by the assumption $h\varphi \in \prod_{i\in I} F_i$ and in particular $h_{i_0}\varphi \in F_{i_0}$. This proves that $\Gamma \vdash_{\mathcal{M}_{i_0}} \varphi$ for an arbitrary $i_0 \in I$, that is, $\Gamma \vdash_{\mathcal{M}_i} \varphi$ for all $i \in I$, and hence $\Gamma \vdash_{\mathcal{L}} \varphi$. This completes the proof of 3. To prove 4, by 3 only the case where Γ is $\mathcal{L}_{\mathcal{M}_i}$-inconsistent for some $i \in I$ has to be checked. On the one hand, 2 tells us that Γ is inconsistent in $\mathcal{L}_{\prod_{i\in I}\mathcal{M}_i}$. On the other hand, by the assumption in 4, Γ is inconsistent in all the logics $\mathcal{L}_{\mathcal{M}_i}$, and therefore it is obviously inconsistent in the meet of this family of logics, so its consequences in the logic of the product coincide with those in the meet of the logics. This completes the proof of 4. ⊠

Thus, the equality $\mathcal{L}_{\prod_{i\in I}\mathcal{M}_i} = \bigwedge_{i\in I} \mathcal{L}_{\mathcal{M}_i}$ is guaranteed to hold only in particular cases. One of these, of special interest, is when the family reduces to a single matrix (a particular case of point 4 above): All "powers" of a matrix define the same logic as the original matrix; this has direct applications in simple, finite cases, such as the following one.

EXAMPLE 4.79 (A NON-EQUIVALENT ALGEBRAIC SEMANTICS FOR $\mathcal{Cl}$)
Consider the algebra $\boldsymbol{O}_6$ in the language $\langle \wedge, \vee, \neg, \top, \bot \rangle$ of type $\langle 2,2,1,0,0 \rangle$ defined by the lattice depicted in Figure 7, with the natural values of the constants, and the involutive negation given by the labelling of the elements plus $\neg 0 = 1$. By obvious reasons, this algebra is known as the "benzene ring". Consider the logic defined in this language by the matrix $\langle \boldsymbol{O}_6, \{1\} \rangle$. It is not difficult to check that $\boldsymbol{\Omega}^{\boldsymbol{O}_6}\{1\}$ is the congruence with blocks $\{0\}, \{a,b\}, \{\neg a, \neg b\}, \{1\}$. Thus, this matrix is not reduced and, again, it is easy to check that its reduction is (isomorphic to) the four-element Boolean algebra $\mathbf{2} \times \mathbf{2}$, depicted in Figure 7 as well (here $\mathbf{2}$ represents the two-element Boolean algebra, considered in this language). Thus, there is a strict surjective homomorphism from the matrix $\langle \boldsymbol{O}_6, \{1\} \rangle$ to the product matrix $\langle \mathbf{2} \times \mathbf{2}, \{\langle 1,1 \rangle\} \rangle = \langle \mathbf{2}, \{1\} \rangle \times \langle \mathbf{2}, \{1\} \rangle$. From Theorems 4.34 and 4.78 it follows that the logic defined by the matrix $\langle \boldsymbol{O}_6, \{1\} \rangle$ is the same as that defined by the matrix $\langle \mathbf{2}, \{1\} \rangle$, that is, classical logic $\mathcal{Cl}$.

Since $\top$ is in the similarity type, the algebra $\boldsymbol{O}_6$ constitutes an algebraic semantics for $\mathcal{Cl}$, with the transformer $\boldsymbol{\tau} x := \{x \approx \top\}$. Thus, so is its generated

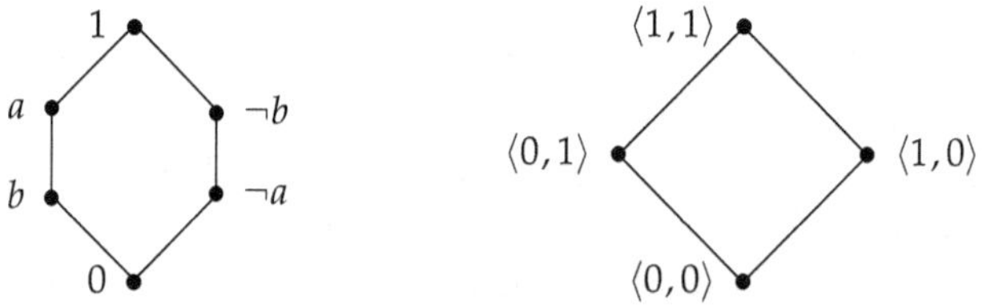

FIGURE 7: The ortholattice $\boldsymbol{O}_6$, commonly called "the benzene ring" (left), alongside with its homomorphic image the four-element Boolean algebra, here represented as $\mathbf{2} \times \mathbf{2}$ (right).

quasivariety $\mathbb{Q}\{O_6\}$. To see that these are *non-equivalent* algebraic semantics for $\mathcal{C}\ell$, it is enough to see that it is larger than $\mathsf{Alg}^*\mathcal{C}\ell = \mathsf{BA}$, the variety of Boolean algebras. Since $\mathbf{2}$ is clearly a subalgebra of O_6, $\mathsf{BA} \subseteq \mathbb{Q}\{O_6\}$ (to see this one needs to use that $\mathsf{BA} = \mathbb{Q}\{\mathbf{2}\}$). But O_6 is clearly not a Boolean algebra, therefore this quasivariety is stricly larger than BA, the largest equivalent algebraic semantics for $\mathcal{C}\ell$. Therefore, it is a non-equivalent one.

The algebra O_6 is one of the best known *non-orthomodular ortholattices*; the class of all these algebras provides natural examples of logics that are weakly algebraizable but not algebraizable (this concerns notions discussed in Section 6.4; see Example 6.122.5 for more details). ⊠

Exercises for Section 4.5

EXERCISE 4.39. Complete the proof of Theorem 4.62 by working out the details of the inductive argument suggested there. Notice that the quantification step works because we are allowing any number of parameters to appear in the formulas under consideration.

EXERCISE 4.40. Show that, for any finitary logic $\mathcal{L}$, the statement that the set of sentences $\mathbf{tr}\mathcal{L}$ is a theory implies the property in Proposition 4.64.4.

EXERCISE 4.41. Complete the proof of Theorem 4.66 by checking that the relation $\vdash_{\mathcal{L}}$ defined in (62) satisfies property (C) of Definition 1.5.

HINT. Check that, for any first-order formulas $\alpha_i, \beta_i, \delta$,

$$\{\forall\vec{x}(\textstyle\bigwedge_{i<k}\beta_i \rightarrow \delta)\} \cup \{\forall\vec{x}(\alpha_i \rightarrow \beta_i) : i<k\} \vDash \forall\vec{x}(\textstyle\bigwedge_{i<k}\alpha_i \rightarrow \delta).$$

EXERCISE 4.42. Provide an alternative, purely first-order proof of property (S) in the second part of the proof of Theorem 4.66, without using Proposition 4.64.1.

HINT. Use the Substitution Lemma (for terms) of first-order semantics [89, Lemma 25B].

EXERCISE 4.43. Use Theorem 4.66 to give an alternative proof (comparing with that in Proposition 4.67) that, for a finitary logic $\mathcal{L}$, the class $\mathsf{Mod}\mathcal{L}$ is closed under $\mathbb{I}, \mathbb{S}$ and $\mathbb{P}$.

EXERCISE 4.44. Show that the following conditions can be added to the list of equivalent ones in Theorem 4.70:

(iii)′ $\mathsf{Mod}\mathcal{L}$ is axiomatizable with Horn sentences of $\boldsymbol{L}_D$.

(iii)″ $\mathsf{Mod}\mathcal{L}$ is axiomatizable with universal sentences of $\boldsymbol{L}_D$.

(iii)‴ $\mathsf{Mod}\mathcal{L}$ is axiomatizable with equality-free sentences of $\boldsymbol{L}_D$.

(iv)′ $\mathsf{Mod}\mathcal{L}$ is closed under reduced products.

(v)′ $\mathcal{L}$ is complete with respect to a class of matrices that is closed under reduced products.

COMMENT. Notice that, by contrast, homologous conditions to the two last ones for closure under products or under submatrices cannot be included in this list, by Proposition 4.67.

EXERCISE 4.45. Let $\mathcal{L}$ be the logic defined by a class M of matrices, and let $\Gamma \in \mathcal{T}h\mathcal{L}$ be consistent. Show that there is a non-empty family $\{\langle \boldsymbol{A}_i, F_i\rangle : i \in I\} \subseteq \mathsf{M}$ and a corresponding family $\{h_i : i \in I\}$ with $h_i \in \mathrm{Hom}(\boldsymbol{Fm}, \boldsymbol{A}_i)$ for each $i \in I$, such that $\Gamma = \bigcap\{h_i^{-1}F_i : i \in I\}$.

HINT. Use Exercise 4.1.

EXERCISE 4.46. Let $\mathcal{L}$ be a logic complete with respect to a class of matrices M containing a trivial matrix. Show that $\mathcal{L}'$ is an extension of $\mathcal{L}$ if and only if $\mathcal{L}'$ is complete with respect to a class of matrices $\mathsf{M}' \subseteq \mathbb{SP}\mathsf{M}$.

HINT. For the ($\Rightarrow$) half, use Lemma 4.71 and consider $\mathsf{M}' := \mathbb{SP}\mathsf{M} \cap \mathsf{Mod}\,\mathcal{L}'$.

EXERCISE 4.47. Let $\mathcal{M}$ and $\mathcal{N}$ be matrices. Prove that $\mathcal{M} \in \mathbb{H}_{\mathrm{s}}^{-1}\mathbb{H}_{\mathrm{s}}\mathcal{N}$ if and only if $\mathcal{N} \in \mathbb{H}_{\mathrm{s}}^{-1}\mathbb{H}_{\mathrm{s}}\mathcal{M}$ if and only if $\mathcal{M}^* \cong \mathcal{N}^*$.

EXERCISE 4.48. Show that if $\mathcal{M}$ is a submatrix of $\mathcal{N}$ and $\mathcal{M}$ is reduced, then $\mathcal{M}$ is isomorphic to a submatrix of $\mathcal{N}^*$.

EXERCISE 4.49. Prove that if there is a strict surjective homomorphism from a matrix $\langle \boldsymbol{A}, F \rangle$ to a matrix $\langle \boldsymbol{B}, G \rangle$, then the Leibniz congruence is finitizable in $\langle \boldsymbol{A}, F \rangle$ if and only if it is finitizable in $\langle \boldsymbol{B}, G \rangle$. Thus, in particular, the Leibniz congruence is finitizable in a matrix if and only if it is finitizable in its reduction, and if and only if it is finitizable in any matrix isomorphic to it. Use these facts to show that for any logic $\mathcal{L}$, if the Leibniz congruence is finitizable in $\mathsf{Mod}^*\mathcal{L}$, then it is finitizable in $\mathsf{Mod}\,\mathcal{L}$ (the missing step in Theorem 4.77).

Chapter 5

The semantics of generalized matrices

The study of generalized matrices is of interest for at least the following reasons:

- The insufficiency of matrices, algebraically speaking: It is true that every logic is complete with respect to several classes of matrices, such as $\mathsf{Mod}\mathcal{L}$ and $\mathsf{Mod}^*\mathcal{L}$ (Theorems 4.16 and 4.43); in this sense, matrix semantics is truly universal. However, there is no evidence that the classes $\mathsf{Mod}^*\mathcal{L}$ and $\mathsf{Alg}^*\mathcal{L}$, which arise naturally in the context of matrix semantics, are *always* the *right* algebra-based counterparts of the logic $\mathcal{L}$; recall the example of $\mathcal{C}\ell_{\wedge\vee}$ and its weird reduced models (Example 4.47).
- The insufficiency of matrices, semantically speaking: Matrix semantics truly makes sense when a notion of *truth in a structure* makes sense for the logic or logical situation to be modelled, and its basic idea is that of *preservation of truth* (see the general discussion on semantics on pages 22–25). The more general idea of a semantics based on the idea of *preservation of degrees of truth* seems to model a larger class of natural situations, and although it is introduced in Example 4.11 by using bundles of matrices, it can be more naturally implemented by using generalized matrices.
- The dual presentation of generalized matrices (as closure systems and as closure operators) allows them to be understood, when used as models of a logic, in a more *semantically neutral* way, more precisely, as *modelling consequence*, without having to ground the semantics on a notion of *truth*. This may be more natural for logics that are not semantically motivated or do not have a natural semantics in any sense, and in any case it is decidedly more general and abstract.

All these issues are discussed further in this chapter; I hope the more specific situations will help to make them clearer.

5.1 Generalized matrices: basic concepts

DEFINITION 5.1
A pair $\mathcal{M} = \langle \boldsymbol{A}, \mathscr{C} \rangle$ *is a* ***generalized matrix*** *(**g-matrix** for short) when* $\boldsymbol{A}$ *is an algebra and* $\mathscr{C} \subseteq \mathscr{P}A$ *is a closure system on* A*. The algebra* $\boldsymbol{A}$ *is called the* ***algebraic reduct*** *of the g-matrix. If* M *is a class of g-matrices, then* $\mathsf{Alg}\mathsf{M}$ *denotes the class of algebraic reducts of the g-matrices in* M.

The terminology of many constructions and properties of g-matrices is inherited from either of their two components, the algebra and the closure system. Thus, for instance, the ***type*** of a g-matrix is that of its algebraic reduct, a g-matrix is ***finite*** when its algebraic reduct is finite, is ***finitary*** when its associated closure operator is finitary, and so forth.

By the duality between closure operators and closure systems described in Theorem 1.27, g-matrices can be presented as pairs $\langle \boldsymbol{A}, C \rangle$ where C is a closure operator on A; many of the terms introduced in Sections 1.3–1.5 for closure operators and closure systems and many of the constructions performed there carry over to g-matrices. Either presentation of g-matrices offers its own advantages, and I interchangeably write "let $\langle \boldsymbol{A}, \mathscr{C} \rangle$ be a g-matrix" and denote the associated closure operator by C, or (less often) write "let $\langle \boldsymbol{A}, C \rangle$ be a g-matrix" and denote the associated closure system by $\mathscr{C}$.[†]

The term "generalized matrix" was introduced by Wójcicki [244], without requiring the family of subsets $\mathscr{C}$ to be a closure system; however, for our purposes the restriction to closure systems is safe and convenient (see the comment after Corollary 5.5.2). G-matrices, presented as pairs $\langle \boldsymbol{A}, C \rangle$ where C is a closure operator on A, were called ***abstract logics*** by Suszko and his co-authors, who did some pioneering work in the 1970s (see [41, 49], among others), and this terminology was followed in many early works of the Barcelona school, up to [107]; however, since the term "logic" may create some confusions, and the term "abstract logic" has also been used in abstract model theory, in later works the term "generalized matrix" has been preferred, along with the less common "second-order matrices" [75, 76]. G-matrices were re-discovered by Dunn and Hardegree [86], who called them "atlases". Abstract logics also appear in Cleave's [65] under the name "logics".

Families of matrices with the same algebraic reduct have been called ***bundles of matrices*** in the literature. It is natural to identify each g-matrix $\langle \boldsymbol{A}, \mathscr{C} \rangle$ with the associated bundle of matrices $\{\langle \boldsymbol{A}, F \rangle : F \in \mathscr{C}\}$; conversely, with each bundle of matrices $\{\langle \boldsymbol{A}, F_i \rangle : i \in I\}$ one can associate the g-matrix $\langle \boldsymbol{A}, \mathscr{C} \rangle$, where $\mathscr{C}$ is the closure system generated by the family of subsets $\{F_i : i \in I\}$ (Definition 1.31). Moreover, with each ordinary matrix $\langle \boldsymbol{A}, F \rangle$ one can associate the g-matrix $\langle \boldsymbol{A}, \{F, A\} \rangle$; thus in some sense a matrix is a particular kind of g-matrix. These

[†] Recall from page 33 that there is still a third equivalent object, the associated *closure relation*, usually denoted by $\vdash_C$; it is seldom used in this book (except in the case of logics, of course).

relations are reflected in many of the notions and constructions that are performed in this chapter.

EXAMPLES 5.2
Let $\mathcal{L}$ be a logic. The following are generalized matrices:

1. The pair $\langle \boldsymbol{Fm}, \mathcal{T}h\mathcal{L}\rangle$, which has the pair $\langle \boldsymbol{Fm}, C_{\mathcal{L}}\rangle$ as its dual presentation.
2. The pair $\langle \boldsymbol{Fm}, (\mathcal{T}h\mathcal{L})^{\Gamma}\rangle$, where $(\mathcal{T}h\mathcal{L})^{\Gamma} := \{\Gamma' \in \mathcal{T}h\mathcal{L} : \Gamma' \supseteq \Gamma\}$, for each $\Gamma \in \mathcal{T}h\mathcal{L}$.
3. The pair $\langle \boldsymbol{A}, \mathcal{F}i_{\mathcal{L}}\boldsymbol{A}\rangle$, for any algebra $\boldsymbol{A}$.
4. The pair $\langle \boldsymbol{A}, (\mathcal{F}i_{\mathcal{L}}\boldsymbol{A})^{F}\rangle$, where $(\mathcal{F}i_{\mathcal{L}}\boldsymbol{A})^{F} := \{G \in \mathcal{F}i_{\mathcal{L}}\boldsymbol{A} : G \supseteq F\}$, for any algebra $\boldsymbol{A}$ and each $F \in \mathcal{F}i_{\mathcal{L}}\boldsymbol{A}$.
5. Pairs of the form $\langle \boldsymbol{A}, \mathcal{F}\rangle$ where $\boldsymbol{A}$ belongs to a specific class of algebras and $\mathcal{F}$ is a closure system of subsets of particular algebraic significance. For instance, $\boldsymbol{A} \in \mathsf{BA}$ and $\mathcal{F}$ is the set of its lattice filters; or $\boldsymbol{A} \in \mathsf{HiA}$ and $\mathcal{F}$ is the set of its implicative filters; or $\boldsymbol{A}$ is a modal algebra and $\mathcal{F}$ is either the set of all its lattice filters or the set of all its open filters. Actually, most of these examples can be viewed as particular cases of 3, but they can also be studied only from the algebraic side, regardless of whether the sets in these families are filters of a certain logic or not.
6. The pair $\langle \boldsymbol{A}, \mathscr{C}^{F}\rangle$, where $\mathscr{C}^{F} := \{G \in \mathscr{C} : G \supseteq F\}$ (Definition 1.34), for any g-matrix $\langle \boldsymbol{A}, \mathscr{C}\rangle$ and any $F \in \mathscr{C}$. The examples in 2 and 4 above are particular cases of this construction, which plays a relevant role in some sections of this and the next chapters. ⊠

It is important to notice that, by Example 5.2.1, *any definition or construction that applies to arbitrary g-matrices applies in particular to logics*; you should bear this in mind. Actually, many concepts originally introduced for logics are found to depend neither on the formula algebra being free, nor on structurality, and can hence be applied to arbitrary g-matrices.

Logics defined by generalized matrices

Recall (Definition 4.2) that, if $\langle \boldsymbol{A}, F\rangle$ is an ordinary matrix, then $\vdash_{\langle \boldsymbol{A},F\rangle}$ denotes the consequence defined by the matrix, that is,

$$\Gamma \vdash_{\langle \boldsymbol{A},F\rangle} \varphi \overset{\text{def}}{\iff} \text{for all } h \in \operatorname{Hom}(\boldsymbol{Fm}, \boldsymbol{A}), \text{ if } h\Gamma \subseteq F, \text{ then } h\varphi \in F, \tag{67}$$

for all $\Gamma \cup \{\varphi\} \subseteq Fm$. Recall also that the consequence defined by a class M of matrices is the intersection of the consequences defined by each of the matrices in the class.

DEFINITION 5.3
Let $\mathcal{M} = \langle \boldsymbol{A}, \mathscr{C}\rangle$ be a g-matrix of type $\boldsymbol{L}$, with associated closure operator C. ***The logic defined by*** *$\mathcal{M}$ is $\mathcal{L}_{\mathcal{M}} := \langle \boldsymbol{L}, \vdash_{\mathcal{M}}\rangle$, where for any $\Gamma \cup \{\varphi\} \subseteq Fm$,*

$$\Gamma \vdash_{\mathcal{M}} \varphi \overset{\text{def}}{\iff} \text{for all } h \in \operatorname{Hom}(\boldsymbol{Fm}, \boldsymbol{A}),\ h\varphi \in Ch\Gamma. \tag{68}$$

If M *is a class of g-matrices of type* $\boldsymbol{L}$, *then* ***the logic defined by*** M *is* $\mathcal{L}_{\mathsf{M}} := \bigwedge_{\mathcal{M}\in\mathsf{M}} \mathcal{L}_{\mathcal{M}}$*; that is,* $\mathcal{L}_{\mathsf{M}} = \langle \boldsymbol{L}, \vdash_{\mathsf{M}} \rangle$ *with* $\vdash_{\mathsf{M}} = \bigcap_{\mathcal{M}\in\mathsf{M}} \vdash_{\mathcal{M}}$.

If one focuses on the identification of a g-matrix with its associated bundle of matrices, it would then be natural to take the logic defined by the bundle of matrices as the logic defined by the g-matrix. The next result confirms it is the same, and moreover justifies the use of the term "logic", because intersections of logics are logics (Section 1.5).

Proposition 5.4
If $\mathcal{M} = \langle \boldsymbol{A}, \mathscr{C} \rangle$ *is a g-matrix, then* $\mathcal{L}_{\mathcal{M}} = \bigwedge_{F\in\mathscr{C}} \mathcal{L}_{\langle \boldsymbol{A},F \rangle}$. *That is, for all* $\Gamma \cup \{\varphi\} \subseteq Fm$,

$$\begin{aligned} \Gamma \vdash_{\mathcal{M}} \varphi \iff & \text{for all } F \in \mathscr{C}, \text{ for all } h \in \mathrm{Hom}(\boldsymbol{Fm}, \boldsymbol{A}), \\ & \text{if } h\Gamma \subseteq F, \text{ then } h\varphi \in F. \end{aligned} \tag{69}$$

Proof: Shuffle the two universal quantifiers in (69) and use that, by (8) on page 36, $C_h\Gamma = \bigcap\{F \in \mathscr{C} : h\Gamma \subseteq F\}$, and you obtain exactly (68). ⊠

Some consequences of the two expressions are straightforward:

Corollary 5.5
1. *If* $\mathcal{M} = \langle \boldsymbol{A}, F \rangle$ *is a matrix, then the logic defined by* $\mathcal{M}$ *coincides with the logic defined by its associated g-matrix* $\langle \boldsymbol{A}, \{F, A\} \rangle$.
2. *If* $\mathcal{M} = \langle \boldsymbol{A}, \mathscr{C} \rangle$ *is a g-matrix and* $\mathscr{B}$ *is a base of* $\mathscr{C}$, *then* $\mathcal{L}_{\mathcal{M}} = \bigwedge_{F\in\mathscr{B}} \mathcal{L}_{\langle \boldsymbol{A},F \rangle}$.
3. *Any logic defined by a finite g-matrix or by a finite family of finite g-matrices is a finitary logic.* ⊠

By Proposition 5.4, a logic defined by a g-matrix or by a class of g-matrices is also defined by a class of matrices; this led Wójcicki to write in [246, p. 36] that

> "[The notion of a generalized matrix] does not provide us with essentially new tools for the semantic analyse of sentential calculi."

Alternatively, in view of Corollary 5.5.1, one can consider that the logic defined by a matrix is a particular case of the logic defined by a g-matrix. Moreover, by Corollary 5.5.2, the logic defined by an arbitrary family of subsets coincides with the logic defined by the g-matrix (that is, by the closure system) it generates; this fact justifies the restriction made on the family $\mathscr{C}$ in the definition of a g-matrix to be a closure system.

Logics defined by a single matrix $\langle \boldsymbol{A}, F \rangle$ as in (67) are sometimes called ***logics defined by preservation of truth***. This comes from the interpretation of the expression (67) when the elements of A are understood as truth-values and the elements of F as representing a notion of truth.

By contrast, logics defined by a single g-matrix $\langle \boldsymbol{A}, \mathscr{C} \rangle$ can be called ***logics defined by preservation of degrees of truth***, when the subsets in $\mathscr{C}$ are understood as representing different "degrees of truth"; here "degree" can be understood in an abstract, non-quantitative sense; then expression (69) may be read as saying that φ follows from Γ if and only if, for every evaluation, every truth degree attained by all formulas in Γ is also attained by φ.

A particular case of this idea is examined in Example 4.11 as one natural instance of a logic defined by a bundle of matrices. Recall that if $\boldsymbol{A}$ is an ordered algebra, then the consequence $\vdash_{\boldsymbol{A}}^{\leqslant}$ is defined there as the one determined by the bundle of matrices $\{\langle \boldsymbol{A}, \uparrow a\rangle : a \in A\}$; the principal up-sets $\uparrow a := \{b \in A : a \leqslant b\}$ can be easily read as meaning "to have a truth degree of at least a". By (69) and Corollary 5.5.2, this is the consequence determined by the g-matrix generated by the family of all principal up-sets; since this family need not be a closure system itself (Exercise 1.38), it may be difficult to identify this g-matrix, but there is a very common case in which this procedure yields a rather natural result.

Recall from Exercise 1.42 that if $\boldsymbol{A}$ is a (meet-)semilattice, $\mathcal{F}ilt_{\wedge}\boldsymbol{A}$ denotes the family of all its filters, that $\mathcal{F}ilt_{\wedge}^{\circ}\boldsymbol{A}$ denotes the closure system generated by this family, and that $\mathcal{F}ilt_{\wedge}^{\circ}\boldsymbol{A} = \mathcal{F}ilt_{\wedge}\boldsymbol{A}$ when $\boldsymbol{A}$ has a maximum, while $\mathcal{F}ilt_{\wedge}^{\circ}\boldsymbol{A} = \mathcal{F}ilt_{\wedge}\boldsymbol{A} \cup \{\varnothing\}$ when it has no maximum. Then:

Proposition 5.6
Let $\boldsymbol{A}$ be a finite (meet-)semilattice,. The consequence $\vdash_{\boldsymbol{A}}^{\leqslant}$ is the consequence determined by the g-matrix $\langle \boldsymbol{A}, \mathcal{F}ilt_{\wedge}^{\circ}\boldsymbol{A}\rangle$.

Proof: When the semilattice has a maximum 1, the family $\mathcal{F}ilt_{\wedge}\boldsymbol{A}$ has a non-empty intersection and constitutes a closure system, with minimum $Fg_{\wedge}^{\boldsymbol{A}}\varnothing = \{1\}$, in which case $\mathcal{F}ilt_{\wedge}^{\circ}\boldsymbol{A} = \mathcal{F}ilt_{\wedge}\boldsymbol{A}$; when it does not have it, the intersection of all filters is empty, and then $\mathcal{F}ilt_{\wedge}^{\circ}\boldsymbol{A} = \mathcal{F}ilt_{\wedge}\boldsymbol{A} \cup \{\varnothing\}$, with minimum $Fg_{\wedge}^{\boldsymbol{A}}\varnothing = \varnothing$. Notice that in a semilattice, $a \leqslant b$ if and only if $b \in \uparrow a = Fg_{\wedge}^{\boldsymbol{A}}\{a\}$. Since any finite semilattice with maximum is complete, $\vdash_{\boldsymbol{A}}^{\leqslant}$ can be defined by (49) on page 188; but when it does not have a maximum, this expression makes sense only for non-empty Γ.

Now assume that $\Gamma \neq \varnothing$. Observe that the set $h\Gamma$ is finite, even if Γ is not, and (this is the key point) the filter generated by a finite, non-empty set is the principal filter generated by its infimum, which always exists. Therefore

$$\bigwedge h\Gamma \leqslant h\varphi \iff h\varphi \in Fg_{\wedge}^{\boldsymbol{A}}\{\bigwedge h\Gamma\} = Fg_{\wedge}^{\boldsymbol{A}} h\Gamma,$$

and (49) becomes here:

$$\Gamma \vdash_{\boldsymbol{A}}^{\leqslant} \varphi \iff \text{for all } h \in \operatorname{Hom}(\boldsymbol{Fm}, \boldsymbol{A}),\ h\varphi \in Fg_{\wedge}^{\boldsymbol{A}} h\Gamma. \tag{70}$$

For $\Gamma = \varnothing$ the general definition (48) on page 188 gives that $\vdash_{\boldsymbol{A}}^{\leqslant} \varphi$ if and only if $a \leqslant h\varphi$ for all $a \in A$ and all $h \in \operatorname{Hom}(\boldsymbol{Fm}, \boldsymbol{A})$. If $\boldsymbol{A}$ has a maximum 1, the condition is equivalent to saying that $h\varphi = 1$ for all $h \in \operatorname{Hom}(\boldsymbol{Fm}, \boldsymbol{A})$, and hence to saying that $h\varphi \in \{1\} = Fg_{\wedge}^{\boldsymbol{A}}\varnothing = Fg_{\wedge}^{\boldsymbol{A}} h\Gamma$. If $\boldsymbol{A}$ has no maximum, the condition never holds, so it is equivalent to saying that $h\varphi \in \varnothing = Fg_{\wedge}^{\boldsymbol{A}}\varnothing = Fg_{\wedge}^{\boldsymbol{A}} h\Gamma$.

Therefore, the equivalence in the expression (70) does hold in all cases. Thus, $\vdash_{\boldsymbol{A}}^{\leqslant}$ is defined by the g-matrix $\langle \boldsymbol{A}, Fg_{\wedge}^{\boldsymbol{A}}\rangle$; described in terms of its closure system, it is as in the statement. ⊠

G-matrices of these and similar forms can be understood as *truth structures* representing *a collective notion of truth* that is not contained in any of the individual matrices $\langle \boldsymbol{A}, F\rangle$ associated with them but resides in the whole structure; in the case of ordered algebras, in the order structure. One popular example is Belnap-Dunn's logic $\mathcal{B}$: in Example 4.12 it is identified as the logic defined by the set of matrices $\{\langle \boldsymbol{M}_4, \{\mathsf{t},\mathsf{n}\}\rangle, \langle \boldsymbol{M}_4, \{\mathsf{t},\mathsf{b}\}\rangle\}$ and it is shown that $\vdash_{\mathcal{B}} = \vdash^{\leqslant}_{\boldsymbol{M}_4}$. By Proposition 5.6, $\mathcal{B}$ is also defined by the g-matrix $\langle \boldsymbol{M}_4, \mathcal{F}ilt_{\wedge} \boldsymbol{M}_4\rangle$. Actually, each of the two mentioned matrices alone defines $\mathcal{B}$ as well (Exercise 4.26); however, neither of these matrices, considered individually, has a clear semantic significance with regard to the initial motivation for the logic, while the g-matrix, as determined by the order structure of the algebra, does indeed have it (reviewing the initial presentation of $\mathcal{B}$ in Example 1.13 may help you confirm this).

Proposals to define logics in ways similar to the ones discussed above abound in the literature, though they have been rather unnoticed until recently. Notice that logics defined as in (48) or (49) from infinite algebras need not be finitary; applied to the standard Łukasiewicz algebra $[0,1]$, the result is an infinitary infinitely-valued logic studied in [95]. For this reason, their finitary companions, called *the logics of order* of a class of ordered algebras, are introduced in Definition 7.26 and studied in Section 7.2; in the case of $[0,1]$, the result is the logic denoted by $Ł^{\leqslant}_{\infty}$. The procedure has been applied to other logics in the family of substructural logics in [46]; see Example 7.47.2.

Generalized matrices as models of logics

Ordinary matrices determine the class $\mathsf{Mod}\mathcal{L}$ of the *models* of a logic $\mathcal{L}$. Analogously, g-matrices originate the notion of a *generalized model* of a logic, but here it is more natural to view them as modelling *the consequence relation* of the logic, without having to introduce a notion of truth. The language of g-sequents is particularly apt to describe these notions, and is useful later on. Recall from Definition 3.65 that a *generalized sequent* (*g-sequent* for short) is a pair $\langle \Gamma, \varphi\rangle$ where $\Gamma \subseteq Fm$ and $\varphi \in Fm$, denoted by $\Gamma \rhd \varphi$, and that a g-sequent $\Gamma \rhd \varphi$ is a (*Hilbert-style*) *g-rule* of a logic $\mathcal{L}$ when $\Gamma \vdash_{\mathcal{L}} \varphi$; in this way, a logic can be identified with the set of all its g-rules. When Γ is finite the "g-" are suppressed.

Definition 5.7

Let $\langle \boldsymbol{A}, \mathscr{C}\rangle$ be a g-matrix with associated closure operator C, and $h \in \mathrm{Hom}(\boldsymbol{Fm}, \boldsymbol{A})$.

1. *h **satisfies a g-sequent** $\Gamma \rhd \varphi$ in $\langle \boldsymbol{A}, \mathscr{C}\rangle$ when $h\varphi \in Ch\Gamma$.*
2. *$\langle \boldsymbol{A}, \mathscr{C}\rangle$ **satisfies** (is a **model**) **of a g-sequent** $\Gamma \rhd \varphi$ when all $h \in \mathrm{Hom}(\boldsymbol{Fm}, \boldsymbol{A})$ satisfy it.*
3. *$\langle \boldsymbol{A}, \mathscr{C}\rangle$ is a **generalized model** (**g-model** for short) **of a logic** $\mathcal{L}$ when for all $\Gamma \cup \{\varphi\} \subseteq Fm$,*

$$\Gamma \vdash_{\mathcal{L}} \varphi \implies \text{for all } h \in \mathrm{Hom}(\boldsymbol{Fm}, \boldsymbol{A}),\ h\varphi \in Ch\Gamma. \tag{71}$$

The class of all the g-models of $\mathcal{L}$ is denoted by $\mathsf{GMod}\mathcal{L}$.

Notice how the term "model" is used when the g-matrix is modelling a g-sequent (this is extended to model a Gentzen system in Definition 5.11), while the term "g-model" is used when the g-matrix is modelling a (sentential) logic, which has matrices as its "models".

We see that a g-matrix is a g-model of a logic $\mathcal{L}$ when it is a model of all the Hilbert-style g-rules of $\mathcal{L}$. Notice that by structurality, a logic $\mathcal{L}$ (viewed as a g-matrix) is itself a model of a g-sequent $\Gamma \rhd \varphi$ if and only if the sequent is one of its g-rules, that is, if and only if $\Gamma \vdash_{\mathcal{L}} \varphi$.

The notion of a g-matrix being a g-model of a logic can be expressed by using the logic defined by the g-matrix itself; comparing (71) with (68) quickly gives:

Proposition 5.8
If $\mathcal{M}$ is a g-matrix, then $\mathcal{M} \in \mathsf{GMod}\,\mathcal{L}$ if and only if $\mathcal{L} \leqslant \mathcal{L}_{\mathcal{M}}$. ⊠

Now, relating the two kinds of "models" of a logic (the *models* and the *g-models*), we obtain:

Proposition 5.9
If $\langle \boldsymbol{A}, \mathscr{C} \rangle$ is a g-matrix, then the following conditions are equivalent:

(i) $\langle \boldsymbol{A}, \mathscr{C} \rangle \in \mathsf{GMod}\,\mathcal{L}$.

(ii) $\langle \boldsymbol{A}, F \rangle \in \mathsf{Mod}\,\mathcal{L}$ *for all* $F \in \mathscr{C}$. *That is,* $\mathscr{C} \subseteq \mathcal{F}i_{\mathcal{L}}\boldsymbol{A}$.

(iii) $h^{-1}\mathscr{C} \subseteq \mathcal{T}h\mathcal{L}$ *for all* $h \in \mathrm{Hom}(\boldsymbol{Fm}, \boldsymbol{A})$.

As a consequence, the g-matrix $\langle \boldsymbol{A}, \mathcal{F}i_{\mathcal{L}}\boldsymbol{A} \rangle$ is the weakest g-model of $\mathcal{L}$ on $\boldsymbol{A}$ (i.e., the one with the largest closure system).

In particular, any logic (viewed as a g-matrix) is a g-model of itself, and is indeed the weakest of all its g-models on $\boldsymbol{Fm}$.

Proof: By Proposition 5.4, $\mathcal{L}_{\langle \boldsymbol{A},\mathscr{C} \rangle} = \bigwedge_{F \in \mathscr{C}} \mathcal{L}_{\langle \boldsymbol{A},F \rangle}$. After comparing Definition 4.13 and Proposition 5.8, we see that this this proves the equivalence between (i) and (ii). That between (ii) and (iii) comes from Lemma 4.15 and Corollary 2.20.2. ⊠

Observe that *any family* of $\mathcal{L}$-filters of the same algebra produces a g-model of $\mathcal{L}$ (formally: after closing the family under intersections, if necessary). This makes these "models" rather unspecific. Section 5.5 introduces a special kind of g-models, the so-called *full g-models*, which seem to have in general a better connection with the logic.

Instantiating Definition 1.16 to the semantics of g-matrices we obtain the obvious notions of soundness, adequateness, completeness, and associated ones.

The following first general result is straightforward:

Theorem 5.10 (Completeness)
Let M be a class of g-matrices such that $\mathcal{L} \in \mathsf{M} \subseteq \mathsf{GMod}\,\mathcal{L}$. The logic $\mathcal{L}$ is complete with respect to M. In particular, every logic is complete with respect to the class $\mathsf{GMod}\,\mathcal{L}$ of all its g-models.

PROOF: Using successively Proposition 5.8, that $\mathsf{M} \subseteq \mathsf{GMod}\mathcal{L}$, and that $\mathcal{L} \in \mathsf{M}$ we obtain the chain of inclusions $\vdash_{\mathcal{L}} \subseteq \vdash_{\mathsf{GMod}\mathcal{L}} \subseteq \vdash_{\mathsf{M}} \subseteq \vdash_{\mathcal{L}}$. Therefore, all these inclusions are equalities, and the theorem is proved. ⊠

As with the case of matrices, "reduced" g-models provide a more interesting completeness result (Proposition 5.66).

Generalized matrices as models of Gentzen-style rules

Generalized matrices are also very natural mathematical objects to model Gentzen-style rules. A ***Gentzen-style g-rule*** is a pair $\langle \Pi, \Gamma \rhd \varphi \rangle$ where Π is an arbitrary set of ordinary sequents, and $\Gamma \rhd \varphi$ is an ordinary sequent. These rules are typically presented in the form of a displayed fraction:

$$\frac{\{\Gamma_i \rhd \varphi_i : i \in I\}}{\Gamma \rhd \varphi} \tag{72}$$

The sequents above the line are the ***premises*** and the one under it is the ***conclusion***. In examples and in some applications the index set I is finite (and then the "g-" disappears), but for many general results this restriction is not necessary. Hilbert-style rules are considered as Gentzen-style rules with an empty set of premises. The natural continuation of the first point of Definition 5.7 gives:

DEFINITION 5.11
*A g-matrix $\langle \boldsymbol{A}, \mathscr{C} \rangle$, with associated closure operator C, **satisfies**, or is a **model** of a Gentzen-style g-rule* (72) *when every evaluation satisfying the premises also satisfies the conclusion; that is, when for all $h \in \mathrm{Hom}(\boldsymbol{Fm}, \boldsymbol{A})$, if $h\varphi_i \in Ch\Gamma_i$ for all $i \in I$, then $h\varphi \in Ch\Gamma$.*

Of course, this notion applies to a logic $\mathcal{L}$, viewed as a g-matrix; it is also said that the g-rule is ***a rule of*** $\mathcal{L}$ when $\mathcal{L}$ satisfies it. Observe that, since the definition involves all evaluations, for g-matrices on the formula algebra this means all substitutions; in particular, if $\mathcal{L}$ is a logic, then

$$\begin{aligned} \mathcal{L} \text{ satisfies a Gentzen-style g-rule (72)} \iff &\text{ for all substitutions } \sigma, \\ &\text{if } \sigma\Gamma_i \vdash_{\mathcal{L}} \sigma\varphi_i \text{ for all } i \in I, \text{ then } \sigma\Gamma \vdash_{\mathcal{L}} \sigma\varphi. \end{aligned} \tag{73}$$

This issue concerning substitutions is not trivial: it would be a mistake to replace the right-hand side of (73) by "if $\Gamma_i \vdash_{\mathcal{L}} \varphi_i$ for all $i \in I$, then $\Gamma \vdash_{\mathcal{L}} \varphi$"; you can check that structurality of $\mathcal{L}$ shows this only when $I = \emptyset$, that is, for Hilbert-style rules. In examples, this issue is often hidden behind the use of *schemata*, in which symbols for arbitrary formulas and for *arbitrary finite sets* of formulas are used; these symbols are *meta-variables*, and their usage produces that in many cases and examples one informally speaks of a single rule, but in fact one deals with a denumerable family of similar rules, one for each cardinality of the sets involved in the rule. Having to handle substitutions, formally the symbols for arbitrary formulas may be replaced by symbols for variables, but it is not possible to do the same with the symbols for arbitrary finite sets of formulas. For instance, the well-known rule of "introduction of implication"

$$\frac{\Gamma, \varphi \rhd \psi}{\Gamma \rhd \varphi \to \psi}$$

is actually a *schema* representing the following (infinite) *set of rules*†

$$\frac{x_1, \dots, x_n, y \rhd z}{x_1, \dots, x_n \rhd y \to z} \qquad \text{for all } n \geqslant 0 .$$

Bearing this in mind may prevent some occasional misunderstandings; moreover, in some cases making the variables explicit as above is almost unavoidable, as in Exercise 5.8.

An immediate consequence of Definition 5.11 is the following property, which may be useful in constructing examples of logics with desired properties:

PROPOSITION 5.12
If a logic $\mathcal{L}$ is complete with respect to a class of g-matrices M, then $\mathcal{L}$ satisfies all the Gentzen-style g-rules that are satisfied by all the g-matrices in M.

PROOF: Let (72) be a g-rule satisfied by all g-matrices in M. In order to show that it is satisfied by $\mathcal{L}$, one has to assume that for some substitution σ, $\sigma\Gamma_i \vdash_{\mathcal{L}} \sigma\varphi_i$ for all $i \in I$, and show that $\sigma\Gamma \vdash_{\mathcal{L}} \sigma\varphi$. To show this, by the completeness assumption it is enough to take any $\langle \boldsymbol{A}, \mathscr{C} \rangle \in \mathsf{M}$ and any $h \in \mathrm{Hom}(\boldsymbol{Fm}, \boldsymbol{A})$ and show that $h\sigma\varphi \in Ch\sigma\Gamma$. But the assumption on σ plus the completeness of $\mathcal{L}$ with respect to M and the fact that $h\sigma \in \mathrm{Hom}(\boldsymbol{Fm}, \boldsymbol{A})$ as well, implies that $h\sigma\varphi_i \in Ch\sigma\Gamma_i$ for all $i \in I$. Now the assumption that $\langle \boldsymbol{A}, \mathscr{C} \rangle$ satisfies the rule implies that $h\sigma\varphi \in Ch\sigma\Gamma$, as is to be shown. ⊠

Exercises for Section 5.1

EXERCISE 5.1. Since a logic $\mathcal{L}$ can be viewed as the g-matrix $\langle \boldsymbol{Fm}, Th\mathcal{L} \rangle$, it makes sense (in principle) to consider the logic $\mathcal{L}_{\langle \boldsymbol{Fm}, Th\mathcal{L} \rangle}$ defined from this g-matrix as in Definition 5.3. Show that $\mathcal{L}_{\langle \boldsymbol{Fm}, Th\mathcal{L} \rangle} = \mathcal{L}$.

EXERCISE 5.2. Prove the points of Corollary 5.5 that you do not find obvious.

EXERCISE 5.3. Let $\mathcal{L}$ be the logic defined by a g-matrix $\mathcal{M} = \langle \boldsymbol{A}, \mathscr{C} \rangle$ of type $\boldsymbol{L}$. Show that for every $\boldsymbol{L}' \subseteq \boldsymbol{L}$, the $\boldsymbol{L}'$ fragment of $\mathcal{L}$ is the logic defined by the g-matrix $\mathcal{M}' = \langle \boldsymbol{A}', \mathscr{C} \rangle$, where $\boldsymbol{A}'$ is the $\boldsymbol{L}'$ reduct of the algebra $\boldsymbol{A}$. Formulate and prove the analogue result for logics defined by classes of g-matrices.

EXERCISE 5.4. Check that the equivalence between points (i) and (iii) of Proposition 5.9 can be proved directly, using only the tools of closure operators and closure systems, that is, the original notion of a g-model of a logic.

† Recall that for $n = 0$ the expression $x_1, \dots, x_n$ represents the empty sequence or the empty set, according to the convention on page 1.

5.2 Basic full generalized models, Tarski-style conditions and transfer theorems

According to Proposition 5.9, the weakest g-model of $\mathcal{L}$ on a given algebra $\boldsymbol{A}$ has the form $\langle \boldsymbol{A}, \mathcal{F}i_{\mathcal{L}}\boldsymbol{A}\rangle$. These g-models receive a special name:

DEFINITION 5.13
*Let $\mathcal{L}$ be a logic. A **basic full g-model** of $\mathcal{L}$ is a g-matrix of the form $\langle \boldsymbol{A}, \mathcal{F}i_{\mathcal{L}}\boldsymbol{A}\rangle$, for some algebra $\boldsymbol{A}$.*

Notice that *there is exactly one basic full g-model on every algebra* of the similarity type of the logic. Since $\mathcal{F}i_{\mathcal{L}}\boldsymbol{Fm} = \mathcal{T}h\mathcal{L}$, the basic full g-model of $\mathcal{L}$ on the formula algebra is the logic itself, viewed as a g-matrix. Hence, as a consequence of Theorem 5.10:

PROPOSITION 5.14 (COMPLETENESS)
Every logic $\mathcal{L}$ is complete with respect to the class of all its basic full g-models, that is, with respect to the class $\{\langle \boldsymbol{A}, \mathcal{F}i_{\mathcal{L}}\boldsymbol{A}\rangle : \boldsymbol{A}$ an algebra$\}$. ☒

The ordered set $\langle \mathcal{F}i_{\mathcal{L}}\boldsymbol{A}, \subseteq\rangle$ is, as any closure system, a complete lattice, with the operations

$$\bigwedge_{k\in I} F_k = \bigcap_{k\in I} F_k \qquad \text{and} \qquad \bigvee_{k\in I} F_k = Fg_{\mathcal{L}}^{\boldsymbol{A}} \bigcup_{k\in I} F_k\,,$$

where $Fg_{\mathcal{L}}^{\boldsymbol{A}}$ denotes the closure operator of $\mathcal{L}$-filter generation associated with $\mathcal{F}i_{\mathcal{L}}\boldsymbol{A}$. Theorem 2.23 proves that if $\mathcal{L}$ is finitary, then $Fg_{\mathcal{L}}^{\boldsymbol{A}}$ is also finitary, and indeed it gives a useful characterization of the $\mathcal{L}$-filter $Fg_{\mathcal{L}}^{\boldsymbol{A}} X$ generated by a given set $X \subseteq A$.

The ideas of ***bridge theorems*** and ***transfer theorems*** are introduced in Section 3.6 in the context of algebraizable logics, but of course they apply to completely general situations. As already discussed there, among the most usual transfer theorems are the ones of the form: "$\langle \boldsymbol{Fm}, \mathcal{T}h\mathcal{L}\rangle$ satisfies property **P** if and only if for all $\boldsymbol{A}$, $\langle \boldsymbol{A}, \mathcal{F}i_{\mathcal{L}}\boldsymbol{A}\rangle$ satisfies **P**", or "[...] if and only if every g-model of $\mathcal{L}$ satisfies **P**". Some may have to add some restrictions like "If $\mathcal{L}$ is of such and such kind, then ...". For this to make sense, the property **P** has to be a property of a g-matrix, so that one can say both that $\langle \boldsymbol{Fm}, \mathcal{T}h\mathcal{L}\rangle$ satisfies **P** and that a g-model satisfies **P**; in this case, one can meaningfully say that "*the property* **P** *transfers from the logic to* [*all g-models/all basic full g-models/all full g-models, etc.*]". A special kind of these properties are the ones that refer to the lattice structure of the closure system of the g-matrix: "The lattice $\langle \mathcal{T}h\mathcal{L}, \wedge, \vee\rangle$ satisfies the (lattice-theoretic) property **P** if and only if for each $\boldsymbol{A}$, the lattice $\langle \mathcal{F}i_{\mathcal{L}}\boldsymbol{A}, \wedge, \vee\rangle$ satisfies **P**"; these properties can be, for instance, completeness, distributivity, or modularity.

The first simple transfer results for g-models are actually established in Proposition 3.66, and can now be rephrased:

PROPOSITION 5.15
The following properties transfer from a logic to all its basic full g-models:

1. *To satisfy (be a model of) a g-sequent.*
2. *To have theorems.*
3. *To have no theorems.*
4. *To be finitary.* ⊠

Point 1 above can be enhanced: by the definition of the notion of a g-model, all Hilbert-style g-rules of a logic transfer to *all its g-models*, and hence so do all metalogical properties that can be expressed by Hilbert-style g-rules (conjunction is one example; see Proposition 5.17). Thus, a way to show that a certain metalogical property cannot be expressed equivalently in terms of Hilbert-style g-rules is to exhibit a logic that has it but which has a g-model that does not have it; this is basically what is done in Theorem 3.75 to show that the Deduction-Detachment Theorem cannot be expressed in terms of Hilbert-style g-rules.

Wójcicki in [249, Chapter 2] coined the expression ***Tarski-style conditions*** to denote the properties of a logic that can be expressed by its closure operator and a single connective; hence, these properties can be formulated for arbitrary g-matrices, and the investigation of their possible transfer from the logic to its basic full g-models is one of the topics specifically addressed in the area of abstract algebraic logic that studies g-matrices. The most important of these conditions are:

DEFINITION 5.16
Let C be a closure operator on a set A, and assume that in each case the mentioned symbol corresponds to a term operation on A.

1. *C satisfies the **Property of Conjunction**, PC, with respect to the binary symbol $\wedge$, when for all $a, b \in A$, $C\{a \wedge b\} = C\{a, b\}$. The operation $\wedge$ is called **a conjunction** for C, and C is called **conjunctive**.*
2. *C satisfies the **Property of Disjunction**, PDI, with respect to the binary symbol $\vee$, when for all $X \cup \{a, b\} \subseteq A$, $C(X, a \vee b) = C(X, a) \cap C(X, b)$. The operation $\vee$ is called **a disjunction** for C, and C is called **disjunctive**.*
3. *C satisfies the **uniterm Deduction-Detachment Theorem**, u-DDT, with respect to the binary symbol $\rightarrow$, when for all $X \cup \{a, b\} \subseteq A$, $a \rightarrow b \in CX$ if and only if $b \in C(X, a)$.*
4. *C satisfies the **Deduction-Detachment Theorem**, DDT, with respect to the set $I(x, y)$ of binary terms, when for all $X \cup \{a, b\} \subseteq A$, $I^{\boldsymbol{A}}(a, b) \subseteq CX$ if and only if $b \in C(X, a)$.*
5. *C satisfies the **Property of Equivalence**, PEQ, with respect to the binary symbol $\leftrightarrow$, when for all $X \cup \{a, b\} \subseteq A$, $a \leftrightarrow b \in CX$ if and only if $C(X, a) = C(X, b)$.*
6. *C satisfies the **Property of Intuitionistic Reductio ad Absurdum**, PIRA, with respect to the unary symbol $\neg$, when for all $X \cup \{a\} \subseteq A$, $\neg a \in CX$ if and only if $X \cup \{a\}$ is inconsistent, that is, $C(X, a) = A$.*

7. *C satisfies the* **Property of Reductio ad Absurdum**, PRA, *with respect to the unary symbol $\neg$, when for all $X \cup \{a\} \subseteq A$, $a \in CX$ if and only if $X \cup \{\neg a\}$ is inconsistent, that is, $C(X, \neg a) = A$.*
8. *C satisfies the* **Property of Introduction of a Modality**, PIM, *with respect to the unary symbol*[†] *$\bigcirc$, when for all $X \cup \{a\} \subseteq A$, $a \in CX$ implies that $\bigcirc a \in C \bigcirc X$, where $\bigcirc X := \{\bigcirc a : a \in X\}$.*

A g-matrix (and in particular a logic) is said to satisfy one of these properties when its associated closure operator satisfies it, with respect to the relevant operation.

Let me insist that the symbols used here correspond to an *arbitrary* unary or binary term, which can be of any shape, and need not be one of the primitive connectives of the language. For instance, it is well known that when classical logic is formulated within the language with just $\vee$ and $\neg$, the term $\neg(\neg x \vee \neg y)$ is a conjunction for it (and, accordingly, it is written as $x \wedge y$).

These properties may be generalized in several ways, in particular by considering a set of terms instead of a single term, as is done in point 4 above for the DDT (it can be called "multiterm" for emphasis; see the comment after Definition 3.76); in the literature this has also been done for the PDI (see Exercise 5.16 and [70, Section 2.5.1], and [63] for a version with parameters), and for the PIRA and the PRA, under the names "Inconsistency Lemma" and "Classical Inconsistency Lemma", respectively (see [205]).

The reason for Wójcicki's terminology is that Tarski gave in [234] an "axiomatic" definition of classical logic as the finitary logic characterized by the u-DDT and the following three conditions on a negation $\neg$:

$$\begin{aligned} &\text{NEG1: There is an } a_0 \in A \text{ such that } C\{a_0\} = A;\\ &\qquad\quad \text{i.e., an inconsistent element.}\\ &\text{NEG2: For all } a \in A,\ \{a, \neg a\} \text{ is inconsistent.}\\ &\text{NEG3: For all } a \in A,\ C\{a\} \cap C\{\neg a\} = C\varnothing. \end{aligned} \tag{74}$$

That these conditions actually characterize classical logic follows from Theorem 3 of [234] together with Theorem 3 of [235], and is commented by Tarski after this one (although he offers no complete proofs). Exercise 5.6 shows that a finitary logic with the DDT (hence, in particular, with the u-DDT) satisfies conditions (74) if and only if it satisfies the PRA.[‡] Together with Tarski's results, this implies that classical logic is also the weakest finitary logic satisfying the u-DDT and the PRA. In the literature there are similar characterizations,[§] using the properties defined above, of intuitionistic logic $\mathcal{IL}$ and all its fragments; see [40, 130, 193, 195].

[†] The generic modal symbol is represented here as $\bigcirc$ because, among the modal unary connectives of usual logics, both $\Box$ and $\Diamond$ satisfy the PIM.

[‡] Thus, the common saying that Tarski proved his characterization with respect to the u-DDT and the PRA seems to be, strictly speaking, inaccurate; but the result is indeed true.

[§] In general, one says that a logic is "characterized" by certain conditions when it is the weakest logic satisfying them.

For finitary g-matrices, most of the Tarski-style conditions considered up to now can be equivalently expressed by *Gentzen-style rules*; while the rules for some conditions are fairly obvious (Exercise 5.7), the ones for negation are slightly less so (Exercise 5.8). Clearly, dealing with Tarski-style conditions is more general, because the expressive power of the Gentzen-style rules is limited by the cardinality of the language; only for finitary g-matrices this is not relevant.

The PC is unique among the conditions of Definition 5.16 in that it is equivalent to a (finite) set of *Hilbert-style rules*; this is established in the next result, together with other easy characterizations, which are left to you as Exercise 5.11.

PROPOSITION 5.17
Let C be a closure operator on the universe of an algebra $\boldsymbol{A}$, and let $\wedge$ be a binary term. The following conditions are equivalent:

(i) *$\wedge$ is a conjunction for C.*
(ii) *The g-matrix $\langle \boldsymbol{A}, C\rangle$ is a model of the sequents $x,y \rhd x\wedge y$, $x\wedge y \rhd x$ and $x \wedge y \rhd y$. That is, for all $a,b \in A$ it holds that $a\wedge b \in C\{a,b\}$, $a \in C\{a\wedge b\}$ and $b \in C\{a\wedge b\}$.*
(iii) *For all $X \subseteq A$, for all $a,b \in A$, $a\wedge b \in CX$ if and only if $a \in CX$ and $b \in CX$.*
(iv) *Every $F \in \mathscr{C}$ is an $\wedge$-**filter**, in the sense that for all $a,b \in A$, $a \wedge b \in F$ if and only if $a \in F$ and $b \in F$.*
(v) *$\mathscr{C}$ has a base made of $\wedge$-filters.* ⊠

You can also check that in (ii) the third sequent can be replaced by $x\wedge y \rhd y\wedge x$, and thus the third condition on C can be replaced by $b\wedge a \in C\{a\wedge b\}$. These Hilbert-style rules can be equivalently formulated as the usual Gentzen-style rules for conjunction (Exercise 5.12). From Propositions 5.15.1 and 5.17(ii) it easily follows that:

COROLLARY 5.18
For each logic $\mathcal{L}$ the following conditions are equivalent:

(i) *$\mathcal{L}$ is conjunctive.*
(ii) *All the g-models of $\mathcal{L}$ are conjunctive.*
(iii) *All the basic full g-models of $\mathcal{L}$ are conjunctive.* ⊠

Thus the PC transfers from a logic to *all* its g-models, and in particular to all its extensions; this is used in a deep way in Section 7.2. Besides, the PC turns out to be a very powerful, yet apparently simple property. As a first glimpse of its strength, observe that if a finitary logic $\mathcal{L}$ satisfies it, then its knowledge can be reduced to that of its set of theorems plus its unitary part (because $\varphi_1,\dots,\varphi_n \vdash_{\mathcal{L}} \varphi$ if and only if $\varphi_1 \wedge\cdots\wedge \varphi_n \vdash_{\mathcal{L}} \varphi$), and that this unitary part can be recovered from its interderivability relation (because $\varphi \vdash_{\mathcal{L}} \psi$ if and only if $\varphi\wedge\psi \dashv\vdash_{\mathcal{L}} \varphi$); thus, conjunctive and finitary logics are determined by their set of theorems and their interderivability relation.

While the transfer of Hilbert-style g-rules is completely solved (they always transfer, and to arbitrary g-models), the transfer of other Tarski-style conditions, and of Gentzen-style rules, seems to be addressed in the literature on a case-by-case strategy. For instance, the case of the DDT is extensively treated already in Section 3.6; there, it is shown that the DDT transfers to all the basic full g-models of a logic, but does not need to transfer to arbitrary g-models, and that it is not equivalent to any set of Hilbert-style rules; moreover, the weakest logic satisfying the u-DDT (essentially, $\mathcal{I\ell}_{\rightarrow}$) is found.

Here is a general result which, together with the ones around Theorem 3.81, illustrates the techniques used in the study of the transfer problem of some Tarski-style conditions. It can be considered as a generalization of the so-called *Praemissentheorem* of Hilbert and Bernays [143]; as an "operation on rules" it was considered, at a proof-theoretic level, by Pogorzelski [192]; see also [194].

THEOREM 5.19
Let $\gamma(x,\vec{y})$ be a fixed formula, and let $\langle x_n : n \geqslant 1\rangle$ be a denumerable sequence of variables, all different from those appearing in γ. Consider the following set of Gentzen-style rules:

$$\frac{x_1,\dots,x_n \rhd x_{n+1}}{\gamma(x_1,\vec{y}),\dots,\gamma(x_n,\vec{y}) \rhd \gamma(x_{n+1},\vec{y})} \qquad \text{for all}^{\dagger}\ n \geqslant 0. \tag{75}$$

For a finitary logic, satisfying this set of rules transfers to all its basic full g-models.

PROOF: Assume that $\mathcal{L}$ satisfies all the rules; to show that $\langle \boldsymbol{A}, \mathcal{F}i_{\mathcal{L}}\boldsymbol{A}\rangle$ is also a model of all the rules we have to show that for all $n \geqslant 0$ and all $\vec{a},\vec{b} \in \vec{A}$, if $a_{n+1} \in Fg_{\mathcal{L}}^{\boldsymbol{A}}\{a_1,\dots,a_n\}$, then $\gamma^{\boldsymbol{A}}(a_{n+1},\vec{b}) \in Fg_{\mathcal{L}}^{\boldsymbol{A}}\{\gamma^{\boldsymbol{A}}(a_1,\vec{b}),\dots,\gamma^{\boldsymbol{A}}(a_n,\vec{b})\}$; the sequence $\vec{b}$ has the same length as $\vec{y}$. The characterization of the operator $Fg_{\mathcal{L}}^{\boldsymbol{A}}$ in Theorem 2.23, in its notation, implies that $Fg_{\mathcal{L}}^{\boldsymbol{A}}\{a_1,\dots,a_n\} = \bigcup_{m\geqslant 0} X_m$ and $Z := Fg_{\mathcal{L}}^{\boldsymbol{A}}\{\gamma^{\boldsymbol{A}}(a_1,\vec{b}),\dots,\gamma^{\boldsymbol{A}}(a_n,\vec{b})\}$. We show by induction on m that

$$\text{for all } a \in A, \text{ if } a \in X_m, \text{ then } \gamma^{\boldsymbol{A}}(a,\vec{b}) \in Z. \tag{76}$$

The elements $a_1,\dots,a_n$ and those in $\vec{b}$ are fixed throughout the rest of the proof.

- $(m = 0)$ $X_0 = \{a_1,\dots,a_n\}$, so $a = a_i$ for some $i \in \{1,\dots,n\}$ and hence $\gamma^{\boldsymbol{A}}(a,\vec{b}) = \gamma^{\boldsymbol{A}}(a_i,\vec{b}) \in Z$ by construction.
- $(m \mapsto m+1)$ Assume that $a \in X_{m+1}$. By the definition of this set, there are some $\varphi_1,\dots,\varphi_{k+1} \in Fm$, for some $k \geqslant 0$, such that

$$\varphi_1,\dots,\varphi_k \vdash_{\mathcal{L}} \varphi_{k+1} \tag{77}$$

and there is some $h \in \operatorname{Hom}(\boldsymbol{Fm},\boldsymbol{A})$ such that $h\varphi_1,\dots,h\varphi_k \in X_m$ and $h\varphi_{k+1} = a$. Let $\vec{z}$ be a sequence of variables not appearing in $\gamma,\varphi_1,\dots,\varphi_{k+1}$, as long as $\vec{y}$; without loss of generality, we can assume that $h\vec{z} = \vec{b}$. By the inductive

† Recall the convention that for $n = 0$ the antecedent of the two sequents is empty (page 1).

assumption for each of the $h\varphi_i \in X_m$, we obtain that $h\gamma(\varphi_i, \vec{z}) = \gamma^{\boldsymbol{A}}(h\varphi_i, \vec{b}) \in Z$. Now, since we are assuming that $\mathcal{L}$ satisfies all the rules (75), by taking the substitution instance with $x_i \mapsto \varphi_i$ and $\vec{y} \mapsto \vec{z}$, we apply the rule for $n := k$ to (77) and obtain that $\gamma(\varphi_1, \vec{z}), \dots, \gamma(\varphi_k, \vec{z}) \vdash_{\mathcal{L}} \gamma(\varphi_{k+1}, \vec{z})$. Since Z is an $\mathcal{L}$-filter, we conclude that $h\gamma(\varphi_{k+1}, \vec{z}) = \gamma^{\boldsymbol{A}}(h\varphi_{k+1}, \vec{b}) = \gamma^{\boldsymbol{A}}(a, \vec{b}) \in Z$.

This completes the proof of (76). Since $a_{n+1} \in Fg_{\mathcal{L}}^{\boldsymbol{A}}\{a_1, \dots, a_n\}$, there is some $m \geqslant 0$ such that $a_{n+1} \in X_m$. By (76) for $a := a_{n+1}$, we obtain that $\gamma^{\boldsymbol{A}}(a_{n+1}, \vec{b}) \in Z = Fg_{\mathcal{L}}^{\boldsymbol{A}}\{\gamma^{\boldsymbol{A}}(a_1, \vec{b}), \dots, \gamma^{\boldsymbol{A}}(a_n, \vec{b})\}$, as desired. ⊠

Actually, the same property holds when the single formula $\gamma(x, \vec{y})$ is replaced by a finite set of formulas (Exercise 5.16). An infinitary extension of this already extended property, for the particular case where $\Gamma(x, \vec{y}) := I(x, y)$ and need not be finite, appears in Proposition 3.80.4, where it is shown that it holds for all logics (unlike Theorem 5.19); this extension is the key to the proof of Theorem 3.81, which in the present terminology can be rephrased as follows.

Theorem 5.20
The DDT *transfers from any logic to all its basic full g-models.* ⊠

The case of the PDI can be handled as a proper application of Theorem 5.19 thanks to the following characterization, which plays for the PDI the role played by Proposition 3.80 (or, more precisely, by Exercise 3.58) for the DDT.

Proposition 5.21
A finitary g-matrix satisfies the PDI *with respect to a binary term* $\vee$ *if and only if it is a model of the Hilbert-style rules* $x \rhd x \vee y$, $x \vee y \rhd y \vee x$, $x \vee x \rhd x$ *and of the Gentzen-style rules* (75) *with* $\gamma(x, y) := x \vee y$.

Proof: Exercise 5.14 gives the necessary clues to show this fact. Notice that the "only if" half does not need finitarity. ⊠

Theorem 5.22
The PDI *transfers from a finitary logic to all its basic full g-models.*

Proof: If $\mathcal{L}$ is finitary, then Proposition 5.21 applies. Since Hilbert-style rules transfer to all g-models, and by Theorem 5.19 the Gentzen-style rules (75) transfer to basic full g-models, it follows that all basic full g-models of $\mathcal{L}$ satisfy them. But all these are finitary, therefore Proposition 5.21 applies again, and therefore basic full g-models of $\mathcal{L}$ have the PDI. ⊠

It seems that the transfer problem of the PDI for arbitrary, non-finitary logics, is *open* [70, Note 2.5.5]; the difficulty resides in the need of finitarity not only in Theorem 5.19 but also in Proposition 5.21; see Exercise 5.15.

Notice that the auxiliary variables $\vec{y}$ need not be present in the formula γ of Theorem 5.19. A case where this happens is that of the PIM: this property, for a finitary g-matrix, is clearly equivalent to the property in Theorem 5.19 for $\gamma(x, \vec{y}) := \bigcirc x$. Therefore:

COROLLARY 5.23
The PIM *transfers from a finitary logic to all its basic full g-models.* ⊠

By contrast, the PIRA and the PRA do not transfer, in general, to basic full g-models. Since by Exercise 5.17 the PRA is equivalent to the PIRA plus a Hilbert-style rule, it is enough to show this for the PIRA:

EXAMPLE 5.24
Let $\mathcal{C\ell}_\neg$ be the fragment of classical logic with just negation. It is easy to see that this logic satisfies the PIRA (actually, the PRA; it can be done directly, using its two-valued definition, but observe that by Exercise 5.10.2, it inherits the property from $\mathcal{C\ell}$) and that it has no theorems. Now consider a trivial algebra $\boldsymbol{A}$, with $A = \{a\}$. Forcefully $\neg a = a$, and $\mathcal{F}i_{\mathcal{C\ell}_\neg}\boldsymbol{A} = \{\varnothing, A\}$ is the basic full g-model of $\mathcal{C\ell}_\neg$ on $\boldsymbol{A}$. But the associated closure operator does not satisfy the PIRA, because $Fg^{\boldsymbol{A}}_{\mathcal{C\ell}_\neg}\{a\} = A$ but $\neg a \notin Fg^{\boldsymbol{A}}_{\mathcal{C\ell}_\neg}\varnothing = \varnothing$. ⊠

A few, miscellaneous transfer results for properties of negation are known. For finitary logics, the PIRA transfers when it is coupled with the u-DDT (this is a particular, easy case of the more advanced result that the PIRA transfers for all finitary protoalgebraic logics, a consequence of Theorem 3.6 of [205]) and when it is coupled with both the PC and the PCONG; the last fact is proved in Theorem 7.22, using more advanced techniques. See also Exercises 5.18 and 5.19.

The just mentioned PCONG is one of the key properties of Chapter 7. Recall from Definition 1.36 that, for a closure operator C, its associated Frege relation is $\Lambda C := \{\langle a,b\rangle \in A \times A : C\{a\} = C\{b\}\}$. The Frege relation of a g-matrix is that of its associated closure operator. For a logic $\mathcal{L}$ its Frege relation is its *interderivability relation* $\dashv\vdash_{\mathcal{L}}$. The Frege relation is always an equivalence relation, but not necessarily a congruence; the fact that it is so deserves to be highlighted:

DEFINITION 5.25
A g-matrix $\langle \boldsymbol{A}, C\rangle$ *has the* ***property of congruence***, PCONG, *when its Frege relation* ΛC *is a congruence of the algebra* $\boldsymbol{A}$, *or, equivalently, when it satisfies the following Gentzen-style rule*

$$\frac{\{x_i \rhd y_i \,,\, y_i \rhd x_i \ : \ i < n\}}{\lambda x_0 \dots x_{n-1} \rhd \lambda y_0 \dots y_{n-1}} \qquad \text{(CONG)}$$

for each of the operations $\lambda \in L$, *with* $n = \operatorname{ar}\lambda \geqslant 1$.

A logic is ***selfextensional*** *when, viewed as a g-matrix, it has the* PCONG.

The equivalence in the statement is obvious.

Classical and intuitionistic logics and their fragments are probably the best-known examples of selfextensional logics (Exercises 5.20 and 5.21).

The following fact is interesting to notice.

PROPOSITION 5.26
Each of the Tarski-style conditions in Definition 5.16 *implies that the Frege relation is a congruence with respect to the relevant operation.*

PROOF: For all the binary connectives, we assume that $a \equiv a' \ (\Lambda C)$ and that $b \equiv b' \ (\Lambda C)$, that is, that $C\{a\} = C\{a'\}$ and that $C\{b\} = C\{b'\}$. In the case of the PC, we have

$$C\{a \wedge b\} = C\{a,b\} = C(C\{a\} \cup C\{b\}) = C(C\{a'\} \cup C\{b'\}) = C\{a' \wedge b'\},$$

which proves that $a \wedge b \equiv a' \wedge b' \ (\Lambda C)$. The case of the PDI is dealt with, under weaker assumptions, in Exercise 5.22. The case of the u-DDT is proposed to you in Exercise 5.23. In the case of PEQ, we have to prove that $C\{a \leftrightarrow b\} = C\{a' \leftrightarrow b'\}$. To show this, we consider any $X \subseteq A$ and show that $a \leftrightarrow b \in CX$ if and only if $a' \leftrightarrow b' \in CX$. By the PEQ, this is the same as showing that $C(X,a) = C(X,b)$ if and only if $C(X,a') = C(X,b')$. But from the assumptions it follows that $C(X,a) = C(X,C\{a\}) = C(X,C\{a'\}) = C(X,a')$, and the same for b and b'. Using these facts the wanted equivalence follows. The case of the PIRA is dealt with in Exercise 5.24, again under weaker assumptions. The case of the PRA holds because it implies the PIRA (Exercise 5.17). Finally, the PIM is again proposed to you in Exercise 5.23. ☒

This result implies that the logics that satisfy the Tarski-style conditions corresponding to all the connectives of their language are directly selfextensional.

Working with g-matrices that satisfy the PCONG turns out to be particularly convenient; this property implies that the largest strict congruence of the g-matrix (a fundamental tool: see next section) is indeed the Frege relation, and therefore has a very simple expression. Selfextensional logics form a significant class of logics, actually the largest among the four classes in the Frege hierarchy, which are all defined by stronger versions of the PCONG. The problem of transfer of the PCONG (and of its stronger versions) is a relevant one in abstract algebraic logic, and has originated some interesting research. It has been solved in the negative in general, but in the positive for restricted classes of logics; for instance, for logics satisfying the PC and for logics satisfying the u-DDT. All these issues are treated in Chapter 7.

Exercises for Section 5.2

EXERCISE 5.5. Let C be a closure operator o a set A that satisfies the PRA with respect to some unary operation $\neg$. Prove:

1. For all $a \in A$, the set $\{a, \neg a\}$ is inconsistent; this is condition NEG2 from page 246.
2. For all $a \in A$, $C\{a\} = C\{\neg\neg a\}$.
3. The ***Property of Contraposition***:

$$\text{For all } X \cup \{a,b\} \subseteq A,\ b \in C(X,a) \text{ if and only if } \neg a \in C(X,\neg b). \qquad \text{(PCO)}$$

4. For all $a \in A$, $C\{a\} \cap C\{\neg a\} = C\varnothing$; this is condition NEG3 from page 246.
5. If moreover C has theorems, then it has an inconsistent element.

Conclude that closure operators with theorems and the PRA satisfy Tarski's conditions (74) for negation (page 246).

EXERCISE 5.6. Let C be a finitary closure operator on a set A that satisfies the DDT and Tarski's conditions (74) for negation (page 246). Prove that C satisfies the PRA, and conclude from this and Exercise 5.5 that a finitary closure operator with the DDT satisfies the PRA if and only if it satisfies the conditions (74).

COMMENT. Condition NEG1 is not used in the first part of the exercise.

EXERCISE 5.7. Prove that a finitary g-matrix satisfies the PDI if and only if it is a model of the usual Gentzen-style rules of Example 3.87:

$$\frac{\Gamma \rhd \varphi}{\Gamma \rhd \varphi \vee \psi} \qquad \frac{\Gamma \rhd \psi}{\Gamma \rhd \varphi \vee \psi} \qquad \frac{\Gamma, \alpha \rhd \varphi \qquad \Gamma, \beta \rhd \varphi}{\Gamma, \alpha \vee \beta \rhd \varphi}$$

Find Gentzen-style rules equivalent, for finitary g-matrices, to the Tarski-style conditions u-DDT, PEQ, and PIM.

EXERCISE 5.8. Prove that a finitary g-matrix satisfies the PIRA if and only if it is a model of the following sets of Gentzen-style rules, for all $n \geqslant 0$:

$$\frac{x_1, \dots, x_n \rhd \neg x_{n+1}}{x_1, \dots, x_n, x_{n+1} \rhd x_{n+2}} \qquad \frac{x_1, \dots, x_n, x_{n+1} \rhd x_{n+2}}{x_1, \dots, x_n \rhd \neg x_{n+1}}$$

Find similar rules for the PRA. Find Gentzen-style rules for condition NEG3, the PCO (Exercise 5.5) and the PWCO (Exercise 5.24), and a Hilbert-style rule for NEG2.

EXERCISE 5.9. Examine whether using g-sequents instead of ordinary sequents in the Gentzen-style rules of Exercise 5.7 would allow you to prove its first part (the one concerning the PDI) for arbitrary (i.e., not necessarily finitary) g-matrices, and explain why this is not successful.

EXERCISE 5.10. Consider the Tarski-style conditions in Definition 5.16. Prove:

1. If a closure operator C satisfies any one of them, then its finitary companion C_{f} (see Definition 1.47) also satisfies it.
2. If a g-matrix satisfies any of these properties, then all its reducts for languages containing the relevant operation also satisfy it (the reduct of a g-matrix is the notion considered in Exercise 5.3)
3. If a g-matrix satisfies any of these properties, save for the PIM, then all its axiomatic extensions also satisfy it.
4. The family of closure operators satisfying one of these conditions is closed under intersections, and its smallest member is finitary.

Recall that these properties apply also to logics.

EXERCISE 5.11. Prove Proposition 5.17 on several equivalent forms of the PC. Prove that the third sequent in its part (ii) can be replaced by $x \wedge y \rhd y \wedge x$; that is, that the third condition on C can be replaced by $b \wedge a \in C\{a \wedge b\}$. Finally, check that Corollary 5.18 follows directly from Propositions 5.15 and 5.17, and that the assertions made in the paragraph after this corollary are obvious consequences of the PC.

EXERCISE 5.12. Prove that a g-matrix $\langle \boldsymbol{A}, C \rangle$ is a model of the sequents $x, y \rhd x \wedge y$, $x \wedge y \rhd x$ and $x \wedge y \rhd y$ if and only if it is a model of the usual Gentzen-style rules for conjunction of Example 3.87:

$$\frac{\Gamma \rhd \varphi \qquad \Gamma \rhd \psi}{\Gamma \rhd \varphi \wedge \psi} \qquad \frac{\Gamma, \alpha, \beta \rhd \varphi}{\Gamma, \alpha \wedge \beta \rhd \varphi}$$

Exercise 5.13. Let $\vee$ be a binary operation in some set A. A subset $X \subseteq A$ is ***$\vee$-prime*** when it satisfies, for all $a, b \in A$, that $a \vee b \in X$ if and only if $a \in X$ or $b \in X$. Prove:

1. If a closure system $\mathscr{C}$ has a base made of $\vee$-prime sets, then its associated closure operator satisfies the PDI with respect to $\vee$.
2. If a closure operator satisfies the PDI with respect to $\vee$, then for any $F \in \mathscr{C}$, F is finitely C-irreducible (Definition 1.59) if and only if F is $\vee$-prime.

Exercise 5.14.

1. Let C be a closure operator that satisfies the PDI with respect to the binary term $\vee$. Prove that it satisfies the following extended form: For all $n \geqslant 1$ and $X \cup \{a_1, \dots, a_n, b\} \subseteq A$, $C(X, a_1 \vee b, \dots, a_n \vee b) = C(X, a_1, \dots, a_n) \cap C(X, b)$.
2. Let $\langle \boldsymbol{A}, C \rangle$ be a finitary g-matrix that is a model of the Hilbert-style rules $x \rhd x \vee y$, $x \vee y \rhd y \vee x$ and $x \vee x \rhd x$ and of the set (75) of Gentzen-style rules of Theorem 5.19 for $\gamma(x,y) := x \vee y$. Prove that for all $X \cup \{a, b, c\} \subseteq A$, if $c \in C(X, a)$, then $c \vee b \in C(X, a \vee b)$.
3. Use points 1 and 2 to prove Proposition 5.21.

Exercise 5.15. Generalize point 2 of Exercise 5.14 to arbitrary g-matrices in the following sense: Let $\langle \boldsymbol{A}, C \rangle$ be a g-matrix satisfying the following conditions for all $X \cup \{a, b\} \subseteq A$: $a \vee b \in C\{a\}$; $a \vee b \in C\{b \vee a\}$; $a \in C\{a \vee a\}$; and if $a \in CX$, then $a \vee b \in C\{c \vee b : c \in X\}$. Prove that C satisfies the PDI. Try to prove the converse and realize you cannot prove the last condition for an arbitrary X. Conclude that Proposition 5.21 does not seem to hold, in a generalized form, for arbitrary g-matrices.

Exercise 5.16. Investigate how should the statement of Theorem 5.19 (and its proof) be modified so that it holds for a finite set of formulas $\Gamma(x, \vec{y})$ in the place of the single formula $\gamma(x, \vec{y})$. Then, apply it to show that the PDI, suitably generalized to use a finite set of terms instead of a single term, also transfers from a finitary logic to all its basic full g-models, thus enhancing Theorem 5.22.

Exercise 5.17. Show that a closure operator satisfies the PRA with respect to a unary operation $\neg$ if and only if it satisfies the PIRA and is a model of the Hilbert-style rule $\neg\neg x \rhd x$. Hence, in particular, the PRA implies the PIRA.

Exercise 5.18. Investigate the transfer of Tarski's conditions (74) for negation (page 246) from a logic to its basic full g-models.

> **Hint.** Show that conditions NEG1 and NEG2 always transfer, and that condition NEG3 does not transfer in general.

Exercise 5.19. Let $\langle \boldsymbol{A}, C \rangle$ be a g-matrix satisfying the u-DDT with respect to a binary operation $\to$. Prove that it satisfies the PIRA with respect to some unary operation $\neg$ if and only if it has an inconsistent element a_0. Show that then $C\{\neg a\} = C\{a \to a_0\}$, and one can always take $\neg a := a \to a_0$. Conclude that when coupled with the u-DDT, the PIRA does transfer from a logic to all its basic full g-models.

Exercise 5.20. Use your knowledge of classical logic and of intuitionistic logic to prove that they are selfextensional.

Exercise 5.21. Show that if a logic is selfextensional, then all its fragments are selfextensional. Formulate and prove the abstract analogue of this property, for arbitrary g-matrices. Conclude that all fragments of classical logic and of intuitionistic logic are selfextensional.

EXERCISE 5.22. Let C be a closure operator on a set A satisfying the following weak form of the PDI, with respect to a binary term $\vee$: For all $a, b \in A$, $C\{a \vee b\} = C\{a\} \cap C\{b\}$. Prove that $\varLambda C$ is a congruence with respect to $\vee$.

EXERCISE 5.23. Prove that if a g-matrix satisfies the u-DDT, then its Frege relation is a congruence with respect to $\rightarrow$; and that if a g-matrix satisfies the PIM, then its Frege relation is a congruence with respect to $\bigcirc$.

EXERCISE 5.24. Let $\boldsymbol{A}$ be an algebra, and let C be a closure operator on A. Consider the ***Property of Weak Contraposition*** (compare with Exercise 5.5.3) with respect to a unary term $\neg$ of the language of $\boldsymbol{A}$:

$$\text{For all } a, b \in A, \text{ if } a \in C\{b\}, \text{ then } \neg b \in C\{\neg a\}. \qquad \text{(PWCO)}$$

Prove:

1. If C satisfies the PIRA with respect to $\neg$, then it satisfies the PWCO.
2. If C satisfies the PWCO, then $\varLambda C$ is a congruence with respect to $\neg$.

5.3 The Tarski operator and the Suszko operator

This section is the g-matrix analogue of Section 4.2. The notions and constructions that are dealt with here apply to arbitrary g-matrices, and are independent of any logic the g-matrix may or may not model. As happened with matrices, g-matrices of an arbitrary nature carry little information on the logic, and the Completeness Theorem 5.10 is not particularly useful. The "real" completeness theorems concern special classes of (g-)models, which are obtained by a process of *reduction*. In Sections 4.2 and 4.3 we saw that for matrices this process is based on the notion of a congruence being compatible with the filter of an ordinary matrix, which defines the *Leibniz congruence* of a matrix. Here we find parallel notions, also based on a notion of *compatibility* of a congruence with a g-matrix, which means just compatibility with all the sets in its closure system.

Congruences in generalized matrices

DEFINITION 5.27
A congruence $\theta \in \operatorname{Con}\boldsymbol{A}$ is a ***congruence of the g-matrix*** *$\langle \boldsymbol{A}, \mathscr{C} \rangle$, or a* ***strict congruence****, when $\theta \subseteq \varLambda\mathscr{C}$, that is, when for all $a, b \in A$, $a \equiv b\ (\theta)$ implies $C\{a\} = C\{b\}$. The set of the congruences of the g-matrix is denoted by $\operatorname{Con}\langle \boldsymbol{A}, \mathscr{C} \rangle$. Strict congruences are also called* ***congruences compatible with*** *the closure system, or with the associated closure operator.*

Thus, a congruence is compatible with a closure operator when it does not identify elements with different closure. The connection with the parallel notion for matrices is quickly established:

LEMMA 5.28
If $\theta \in \operatorname{Con}\boldsymbol{A}$, then $\theta \in \operatorname{Con}\langle \boldsymbol{A}, \mathscr{C} \rangle$ if and only if θ is compatible with all the $F \in \mathscr{C}$. Therefore, $\operatorname{Con}\langle \boldsymbol{A}, \mathscr{C} \rangle = \bigcap_{F \in \mathscr{C}} \operatorname{Con}\langle \boldsymbol{A}, F \rangle$.

PROOF: To say that $C\{a\} = C\{b\}$ is to say that for all $F \in \mathscr{C}$, $a \in F$ if and only if $b \in F$. Therefore, to say that $a \equiv b\ (\theta)$ implies this is to say that θ is compatible with all $F \in \mathscr{C}$. ⊠

Since all the sets $\mathrm{Con}\langle \boldsymbol{A}, F\rangle$ are complete sublattices of $\mathrm{Con}\boldsymbol{A}$ (Theorem 4.20), we quickly get:

PROPOSITION 5.29
The set $\mathrm{Con}\langle \boldsymbol{A}, \mathscr{C}\rangle$ *is a complete sublattice of* $\mathrm{Con}\boldsymbol{A}$ *with least element the identity relation, and it is a down-set of* $\mathrm{Con}\boldsymbol{A}$ *(if* $\theta, \theta' \in \mathrm{Con}\boldsymbol{A}$ *and* $\theta' \subseteq \theta \in \mathrm{Con}\langle \boldsymbol{A}, \mathscr{C}\rangle$, *then* $\theta' \in \mathrm{Con}\langle \boldsymbol{A}, \mathscr{C}\rangle$*) and hence an ideal.* ⊠

Since the sublattice is complete, it has a maximum:

DEFINITION 5.30
The ***Tarski congruence*** *of a g-matrix* $\langle \boldsymbol{A}, \mathscr{C}\rangle$ *is* $\widetilde{\boldsymbol{\Omega}}^{\boldsymbol{A}}\mathscr{C} := \max \mathrm{Con}\langle \boldsymbol{A}, \mathscr{C}\rangle$.

From Definition 5.27 it follows that $\widetilde{\boldsymbol{\Omega}}^{\boldsymbol{A}}\mathscr{C}$ is *the largest congruence below the Frege relation* $\boldsymbol{\Lambda}\mathscr{C}$ of the g-matrix. Following the practice of handling two dual presentations for g-matrices, the notation $\widetilde{\boldsymbol{\Omega}}^{\boldsymbol{A}}C$ may also be used. Moreover, if a notation such as $\mathcal{M} = \langle \boldsymbol{A}, \mathscr{C}\rangle$ is used, then it is natural to use $\mathrm{Con}\mathcal{M}$ and $\widetilde{\boldsymbol{\Omega}}\mathcal{M}$

These notions apply in particular to a logic $\mathcal{L}$, for we consider it as a special case of a g-matrix on the formula algebra, in which case the superscript "$\boldsymbol{Fm}$" is omitted and some simplifications in the notation are adopted; we thus obtain the ***Tarski congruence of the logic***,

$$\widetilde{\boldsymbol{\Omega}}\mathcal{L} := \widetilde{\boldsymbol{\Omega}}\mathcal{T}h\mathcal{L},$$

which is *the largest congruence below the Frege relation* $\boldsymbol{\Lambda}\mathcal{L}$ of the logic.

Naming these notions† after Tarski entails no claim that he did even consider them; it is rather an hommage to his seminal work in studying logics through their closure operators. Actually, in his two papers [234, 235] in which the closure operator approach is prominent, the notion of equivalence he considers is the Frege relation $\boldsymbol{\Lambda}\mathcal{L}$; but since he is studying only classical logic, which is selfextensional, it coincides with $\widetilde{\boldsymbol{\Omega}}\mathcal{L}$ (see Lemma 5.31).

LEMMA 5.31
1. $\mathrm{Con}\langle \boldsymbol{A}, \mathscr{C}\rangle = \{\theta \in \mathrm{Con}\boldsymbol{A} : \theta \subseteq \widetilde{\boldsymbol{\Omega}}^{\boldsymbol{A}}\mathscr{C}\}$.
2. *For any g-matrix* $\langle \boldsymbol{A}, \mathscr{C}\rangle$, $\widetilde{\boldsymbol{\Omega}}^{\boldsymbol{A}}\mathscr{C} = \bigcap_{F \in \mathscr{C}} \boldsymbol{\Omega}^{\boldsymbol{A}}F$.
3. *A g-matrix* $\langle \boldsymbol{A}, \mathscr{C}\rangle$ *has the* PCONG *if and only if* $\widetilde{\boldsymbol{\Omega}}^{\boldsymbol{A}}\mathscr{C} = \boldsymbol{\Lambda}\mathscr{C}$.
4. *A logic* $\mathcal{L}$ *is selfextensional if and only if* $\widetilde{\boldsymbol{\Omega}}\mathcal{L} = \boldsymbol{\Lambda}\mathcal{L} = \dashv\vdash_{\mathcal{L}}$.

PROOF: 1 just says that $\mathrm{Con}\langle \boldsymbol{A}, \mathscr{C}\rangle$ is the principal down-set generated by its largest element, which is $\widetilde{\boldsymbol{\Omega}}^{\boldsymbol{A}}\mathscr{C}$ by definition. From this and Lemma 5.28 it follows that for any $\theta \in \mathrm{Con}\boldsymbol{A}$, $\theta \subseteq \widetilde{\boldsymbol{\Omega}}^{\boldsymbol{A}}\mathscr{C}$ if and only if $\theta \subseteq \boldsymbol{\Omega}^{\boldsymbol{A}}F$ for all $F \in \mathscr{C}$,

† In some papers g-matrices are called "second-order matrices", and accordingly the Tarski congruence is called "the second-order Leibniz congruence".

and this shows 2 (there is an alternative, purely lattice-theoretic proof: in a complete lattice, the intersection of a family of principal ideals is the principal ideal generated by the meet of the generators). Finally, 3 is a consequence of the fact that by definition $\widetilde{\boldsymbol{\Omega}}^{\boldsymbol{A}}\mathscr{C}$ is the largest congruence below $\boldsymbol{\Lambda}\mathscr{C}$, and 4 is a particular case of 3. ⊠

The expression in 5.31.2 is particularly significant; it is often helpful to take it into account. Here, just two of its consequences are recorded: First, that always $\widetilde{\boldsymbol{\Omega}}^{\boldsymbol{A}}\mathscr{C} \subseteq \boldsymbol{\Omega}^{\boldsymbol{A}}F$ for every $F \in \mathscr{C}$, and in particular $\widetilde{\boldsymbol{\Omega}}^{\boldsymbol{A}}\mathscr{C} \subseteq \boldsymbol{\Omega}^{\boldsymbol{A}}C\varnothing$. Second, that applying Theorem 4.23 to it immediately yields:

THEOREM 5.32
Let $\langle \boldsymbol{A}, \mathscr{C}\rangle$ be a g-matrix. For all $a, b \in A$, the following conditions are equivalent:

(i) $a \equiv b \ (\widetilde{\boldsymbol{\Omega}}^{\boldsymbol{A}}\mathscr{C})$.
(ii) *For all unary polynomial functions f on $\boldsymbol{A}$, $C\{fa\} = C\{fb\}$.*
(iii) *For all $\delta(x,\vec{z}) \in Fm$ and all $\vec{c} \in \vec{A}$, $C\{\delta^{\boldsymbol{A}}(a,\vec{c})\} = C\{\delta^{\boldsymbol{A}}(b,\vec{c})\}$.* ⊠

For logics this expression can be simplified, thanks to structurality, as is done in Corollary 4.24:

COROLLARY 5.33
Let $\mathcal{L}$ be a logic. For all $\alpha, \beta \in Fm$,

$$\alpha \equiv \beta \ (\widetilde{\boldsymbol{\Omega}}\mathcal{L}) \iff \text{for all } \delta(x,\vec{z}) \in Fm,\ \delta(\alpha,\vec{z}) \dashv\vdash_{\mathcal{L}} \delta(\beta,\vec{z}).$$ ⊠

This characterization supports an interpretation of $\widetilde{\boldsymbol{\Omega}}\mathcal{L}$ as a formal counterpart of a certain idea of *logical equivalence* of propositional sentences found in some theories in philosophy of logic and philosophy of language. According to this view, two sentences are logically equivalent when they are interchangeable in every context *salva veritate*, that is, without altering the truth-values of the resulting compound sentences. In our situation, one may think that the "context" has both a linguistic aspect, given by the formulas $\delta(x,\vec{z})$ inside which the sentences may be replaced, and a semantic aspect, given by the matrices in which the resulting sentences may be evaluated. Then, applying completeness of $\mathcal{L}$ with respect to the class of models $\mathsf{Mod}\mathcal{L}$ to the right-hand half of the expression in Corollary 5.33, one can read it as saying that two formulas α and β are equivalent modulo $\widetilde{\boldsymbol{\Omega}}\mathcal{L}$ if and only if for every linguistic context $\delta(x,\vec{z}) \in Fm$, the formulas $\delta(\alpha,\vec{z})$ and $\delta(\beta,\vec{z})$ have the same truth value (they are either both designated or both undesignated) in all semantic contexts, that is, under all evaluations in all models of the logic. Since only models of $\mathcal{L}$ are contemplated, this would mean that $\widetilde{\boldsymbol{\Omega}}\mathcal{L}$ corresponds to the mentioned idea of logical equivalence, but relative to the logic $\mathcal{L}$.

The main idea (and the phrase *salva veritate*) can be traced back to Leibniz's definition of identity as indiscernibility, as discussed on page 198, although he was thinking of linguistic expressions describing objects rather than meaningful sentences. The application of a similar idea to all linguistic expressions, including sentences, starts with Frege; see the discussion and references on page 416 for the application of his *principle of compositionality* in this formal context.

The following characterization of the Tarski congruence, which extends Porte's Theorem for matrices (Lemma 4.53), is useful in a number of examples of well-behaved logics:

PROPOSITION 5.34
If a strict congruence $\theta \in \mathrm{Con}\langle \boldsymbol{A}, \mathscr{C} \rangle$ *is analytically definable in terms of* $C\emptyset$, *in the sense that there is some* $\Delta(x,y) \subseteq Fm$ *such that for all* $a,b \in A$, $a \equiv b\ (\theta)$ *if and only if* $\Delta^{\boldsymbol{A}}(a,b) \subseteq C\emptyset$, *then* $\theta = \boldsymbol{\Omega}^{\boldsymbol{A}} C\emptyset = \widetilde{\boldsymbol{\Omega}}^{\boldsymbol{A}} \mathscr{C}$.

PROOF: Since $C\emptyset \in \mathscr{C}$, if $\theta \in \mathrm{Con}\langle \boldsymbol{A}, \mathscr{C} \rangle$, then by Lemma 5.28, $\theta \in \mathrm{Con}\langle \boldsymbol{A}, C\emptyset \rangle$, and then Lemma 4.53 implies that $\theta = \boldsymbol{\Omega}^{\boldsymbol{A}} C\emptyset$. From the expression in Lemma 5.31.2 it follows that $\widetilde{\boldsymbol{\Omega}}^{\boldsymbol{A}} \mathscr{C} \subseteq \theta$, and from the assumption that $\theta \in \mathrm{Con}\langle \boldsymbol{A}, \mathscr{C} \rangle$ and Lemma 5.31.1 it follows that $\theta \subseteq \widetilde{\boldsymbol{\Omega}}^{\boldsymbol{A}} \mathscr{C}$. Therefore, $\theta = \widetilde{\boldsymbol{\Omega}}^{\boldsymbol{A}} \mathscr{C}$. ⊠

Notice the equality $\boldsymbol{\Omega}^{\boldsymbol{A}} C\emptyset = \widetilde{\boldsymbol{\Omega}}^{\boldsymbol{A}} \mathscr{C}$, which does not hold in general; actually, requiring it for all g-models of a logic characterizes protoalgebraicity (Proposition 6.14).

On any algebra $\boldsymbol{A}$, the ***Tarski operator*** is the function

$$\widetilde{\boldsymbol{\Omega}}^{\boldsymbol{A}} : \mathscr{C} \longmapsto \widetilde{\boldsymbol{\Omega}}^{\boldsymbol{A}} \mathscr{C}.$$

This function is in principle defined over the set of all closure systems on A (or, dually, over the set of all closure operators on A). However, our main interest is to restrict it to the set of all g-models of $\mathcal{L}$ on $\boldsymbol{A}$, and even further to the set of all *full* g-models of $\mathcal{L}$ on $\boldsymbol{A}$, a notion introduced in Section 5.5.

A first property is that, unlike the Leibniz operator, the Tarski operator is *(anti-)monotone*:

PROPOSITION 5.35
Let $\boldsymbol{A}$ *be an algebra and let* $\mathscr{C}_1, \mathscr{C}_2 \in \mathrm{CS}(A)$. *If* $\mathscr{C}_1 \subseteq \mathscr{C}_2$, *then* $\widetilde{\boldsymbol{\Omega}}^{\boldsymbol{A}} \mathscr{C}_2 \subseteq \widetilde{\boldsymbol{\Omega}}^{\boldsymbol{A}} \mathscr{C}_1$. *Dually, if* $C_1, C_2 \in \mathrm{CO}(A)$ *and* $C_2 \leqslant C_1$, *then* $\widetilde{\boldsymbol{\Omega}}^{\boldsymbol{A}} C_2 \subseteq \widetilde{\boldsymbol{\Omega}}^{\boldsymbol{A}} C_1$.

PROOF: If $\mathscr{C}_1 \subseteq \mathscr{C}_2$, then by Lemma 1.38.2, $\boldsymbol{\Lambda} \mathscr{C}_2 \subseteq \boldsymbol{\Lambda} \mathscr{C}_1$. Since by definition $\widetilde{\boldsymbol{\Omega}}^{\boldsymbol{A}} \mathscr{C}_2 \subseteq \boldsymbol{\Lambda} \mathscr{C}_2$, it follows that $\widetilde{\boldsymbol{\Omega}}^{\boldsymbol{A}} \mathscr{C}_2 \subseteq \boldsymbol{\Lambda} \mathscr{C}_1$, that is, $\widetilde{\boldsymbol{\Omega}}^{\boldsymbol{A}} \mathscr{C}_2$ is a congruence of the g-matrix $\langle \boldsymbol{A}, \mathscr{C}_1 \rangle$, and since $\widetilde{\boldsymbol{\Omega}}^{\boldsymbol{A}} \mathscr{C}_1$ is the largest one, it follows that $\widetilde{\boldsymbol{\Omega}}^{\boldsymbol{A}} \mathscr{C}_2 \subseteq \widetilde{\boldsymbol{\Omega}}^{\boldsymbol{A}} \mathscr{C}_1$. The dual statement follows because $C_2 \leqslant C_1$ if and only if $\mathscr{C}_1 \subseteq \mathscr{C}_2$ (Theorem 1.33). In the closure operator case, an alternative proof is suggested in Exercise 5.25. ⊠

There is still a third special congruence, which somehow mixes the Tarksi and Leibniz ideas:

DEFINITION 5.36
Let $\langle \boldsymbol{A}, \mathscr{C} \rangle$ *be a g-matrix, and* $F \in \mathscr{C}$. *The* ***Suszko congruence of*** F ***relative to*** $\mathscr{C}$ *is*

$$\widetilde{\boldsymbol{\Omega}}^{\boldsymbol{A}}_{\mathscr{C}} F := \widetilde{\boldsymbol{\Omega}}^{\boldsymbol{A}} \mathscr{C}^{F} = \bigcap \{ \boldsymbol{\Omega}^{\boldsymbol{A}} G : G \in \mathscr{C}, G \supseteq F \}.$$

The ***Suszko operator*** *(**relative to*** $\mathscr{C}$*) is the function* $\widetilde{\boldsymbol{\Omega}}^{\boldsymbol{A}}_{\mathscr{C}} : \mathscr{C} \longmapsto \widetilde{\boldsymbol{\Omega}}^{\boldsymbol{A}}_{\mathscr{C}} F$.

Recall that $\mathscr{C}^F := \{G \in \mathscr{C} : G \supseteq F\}$, which explains the right-most equality above. Notice also that $\mathscr{C}^F \subseteq \mathscr{C}$ and that $\mathscr{C}^{C\emptyset} = \mathscr{C}$. The straightforward proof of the following properties is left to you as Exercise 5.26:

LEMMA 5.37
1. *For all* $F \in \mathscr{C}$, $\widetilde{\Omega}^{A}\mathscr{C} \subseteq \widetilde{\Omega}^{A}_{\mathscr{C}}F \subseteq \Omega^{A}F$, *and* $\widetilde{\Omega}^{A}_{\mathscr{C}}C\emptyset = \widetilde{\Omega}^{A}\mathscr{C}$.
2. $\widetilde{\Omega}^{A}_{\mathscr{C}}F \subseteq \Lambda_{\mathscr{C}}F$, *and* $\widetilde{\Omega}^{A}_{\mathscr{C}}F$ *is the largest congruence of* A *below* $\Lambda_{\mathscr{C}}F$.
3. $\widetilde{\Omega}^{A}_{\mathscr{C}}$ *is monotone: If* $F, G \in \mathscr{C}$ *with* $F \subseteq G$, *then* $\widetilde{\Omega}^{A}_{\mathscr{C}}F \subseteq \widetilde{\Omega}^{A}_{\mathscr{C}}G$. ⊠

The Suszko operator was formally introduced by Czelakowski in [71],[†] in the special case where $\mathscr{C} = \mathcal{F}i_{\mathcal{L}}A$; in this case, it is in fact the logic $\mathcal{L}$ that determines the relativization, and it thus makes sense to use the notational shortcuts

$$\widetilde{\Omega}^{A}_{\mathcal{L}}F := \widetilde{\Omega}^{A}_{\mathcal{F}i_{\mathcal{L}}A}F$$
$$\Lambda^{A}_{\mathcal{L}}F := \Lambda_{\mathcal{F}i_{\mathcal{L}}A}F$$

for $F \in \mathcal{F}i_{\mathcal{L}}A$. The operator $\widetilde{\Omega}^{A}_{\mathcal{L}}$ is then called the ***Suszko operator relative to*** $\mathcal{L}$ (on the algebra A). As usual, in the formula algebra the superscript is omitted, and we get the ***Suszko operator of the logic,*** $\widetilde{\Omega}_{\mathcal{L}}$, defined by

$$\widetilde{\Omega}_{\mathcal{L}}\Gamma := \widetilde{\Omega}_{Th\mathcal{L}}\Gamma = \widetilde{\Omega}(Th\mathcal{L})^{\Gamma}.$$

The importance of the Suszko operator for abstract algebraic logic is highlighted by Czelakowski when he writes [70, p. 9]:

> "An algebraic instrument which would make all logics amenable to its methods is available—it is the Suszko operator. For protoalgebraic logics, the Suszko and the Leibniz operators coincide."

However, a general classification of logics based on the properties of the Suszko operator (perhaps supersedding the Leibniz hierarchy of Chapter 6, based on the properties of the Leibniz operator) has still to be developed. Anyway, the Suszko operator plays some role in the treatment of protoalgebraic logics (Section 6.2), of truth-equational logics (Section 6.4) and of Fregean logics (Section 7.3). One aspect of this study is to compare the three[‡] operators Ω^{A}, $\widetilde{\Omega}^{A}_{\mathcal{L}}$ and $\Lambda^{A}_{\mathcal{L}}$, all defined over $\mathcal{F}i_{\mathcal{L}}A$. As in the previous lemma, for all $F \in \mathcal{F}i_{\mathcal{L}}A$,

$$\widetilde{\Omega}^{A}\mathcal{F}i_{\mathcal{L}}A \subseteq \widetilde{\Omega}^{A}_{\mathcal{L}}F \subseteq \Omega^{A}F, \text{ and } \widetilde{\Omega}^{A}_{\mathcal{L}}F \subseteq \Lambda^{A}_{\mathcal{L}}F,$$

and for all $\Gamma \in Th\mathcal{L}$,

$$\widetilde{\Omega}\mathcal{L} \subseteq \widetilde{\Omega}_{\mathcal{L}}\Gamma \subseteq \Omega\Gamma, \text{ and } \widetilde{\Omega}_{\mathcal{L}}\Gamma \subseteq \Lambda_{\mathcal{L}}\Gamma.$$

The coincidence of $\widetilde{\Omega}^{A}_{\mathcal{L}}$ and Ω^{A} on $\mathcal{F}i_{\mathcal{L}}A$ (or just of $\widetilde{\Omega}_{\mathcal{L}}$ and Ω on $Th\mathcal{L}$) characterizes the class of protoalgebraic logics (Proposition 6.14), and that of $\widetilde{\Omega}_{\mathcal{L}}$ and $\Lambda_{\mathcal{L}}$ on $Th\mathcal{L}$ defines the class of Fregean logics (Definition 7.3), while

[†] He attributes its invention and first characterization to Suszko, in unpublished lectures.

[‡] Since the output of the Frege operator is in general not a congruence, its role in this kind of study is mostly limited to Chapter 7.

that of $\widetilde{\Omega}^{A}_{\mathcal{L}}$ and $\Lambda^{A}_{\mathcal{L}}$ on $\mathcal{F}i_{\mathcal{L}}\boldsymbol{A}$ characterizes the class of fully Fregean logics (Proposition 7.4).

In Definition 5.36 the Suszko operator is obtained from the Tarski operator, but since it is definable directly from the Leibniz operator, it is also possible to view the Tarski operator as defined from the Suszko operator through the equality $\widetilde{\Omega}^{A}\mathscr{C} = \widetilde{\Omega}^{A}_{\mathscr{C}} C\emptyset$. Thus, in some sense they are interdefinable.† Two important differences are:

- $\widetilde{\Omega}^{A}$ is *absolute* (i.e., it depends only on the algebraic structure), while $\widetilde{\Omega}^{A}_{\mathscr{C}}$ is *relative* to the closure system $\mathscr{C}$, and $\widetilde{\Omega}^{A}_{\mathcal{L}}$ is *relative* to the logic $\mathcal{L}$.
- $\widetilde{\Omega}^{A}$ is an operator defined on closure systems (on A), while $\widetilde{\Omega}^{A}_{\mathscr{C}}$ and $\widetilde{\Omega}^{A}_{\mathcal{L}}$ are operators defined on subsets (of A).

The second point makes it easier to integrate the Suszko operator into investigations in which the Leibniz operator (which is also an operator defined on subsets) is the central tool, and makes it an alternative to the Tarski operator in approaches to abstract algebraic logic that do not want to use g-matrices; however, one should not forget that these are actually hidden in the definition of the Suszko operator and its relative character.

The interpretative discussions on the Leibniz and the Tarski congruences (concerning synonymy, indiscernibility, logical equivalence, etc.) on pages 198 and 256 apply to the Suszko congruence as well, *mutatis mutandis*.

Strict homomorphisms

NOTATION.
Recall (page 1) that each function $h\colon A_1 \to A_2$ induces a power set extension $h\colon \mathscr{P}A_1 \to \mathscr{P}A_2$ and this one has a residual $h^{-1}\colon \mathscr{P}A_2 \to \mathscr{P}A_1$. In turn, these two functions on the power sets can be extended to families of subsets, still keeping the same symbol: if $\mathscr{C}_1 \subseteq \mathscr{P}A_1$ and $\mathscr{C}_2 \subseteq \mathscr{P}A_2$, one defines

$$h\mathscr{C}_1 := \{hF : F \in \mathscr{C}_1\} \qquad \text{and} \qquad h^{-1}\mathscr{C}_2 := \{h^{-1}G : G \in \mathscr{C}_2\}.$$

In spite of the notation, the extended h^{-1} is neither the residual nor the inverse of the twice extended h appearing at its left; it is rather the power set extension of the residual h^{-1} mentioned before. Using the same symbol h for what are really three different functions (and h^{-1} for two) need not lead to confusion if you pay attention to the arguments of the functions each time they are used, and in particular if you check whether they are used as a *point function* (the original h) or as set functions (the rest). ⊠

Observe that a closure operator is always a power set function, and never a point function.

The following observations are straightforward:

† Actualy, it is best to regard both as deriving from the Leibniz operator; this view is explored, at a more abstract level, in [5].

- If $X \subseteq A_1$, then $X \subseteq h^{-1}hX$, and the equality holds for all $X \subseteq A_1$ if and only if h is injective (one-to-one).
- If $Y \subseteq A_2$, then $hh^{-1}Y \subseteq Y$, and the equality holds for all $Y \subseteq A_2$ if and only if h is surjective.

DEFINITION 5.38
An $h \in \mathrm{Hom}(\boldsymbol{A}_1, \boldsymbol{A}_2)$ is an ***homomorphism from*** *the g-matrix $\mathcal{M}_1 = \langle \boldsymbol{A}_1, \mathscr{C}_1 \rangle$* ***to*** *the g-matrix $\mathcal{M}_2 = \langle \boldsymbol{A}_2, \mathscr{C}_2 \rangle$ when for all $X \subseteq A_1$ and all $a \in A_1$, if $a \in C_1X$, then $ha \in C_2hX$; that is, when $hC_1X \subseteq C_2hX$ for all $X \subseteq A_1$; using the relation $\leqslant$ defined on page 2, this fact is denoted by $hC_1 \leqslant C_2h$. The set of homomorphisms from $\mathcal{M}_1$ to $\mathcal{M}_2$ is denoted by $\mathrm{Hom}(\mathcal{M}_1, \mathcal{M}_2)$. An* ***endomorphism*** *of a g-matrix is a homomorphism from the g-matrix to itself.*

PROPOSITION 5.39
The following conditions are equivalent:

(i) $h \in \mathrm{Hom}(\mathcal{M}_1, \mathcal{M}_2)$.

(ii) $C_1 \leqslant h^{-1}C_2h$. *That is,* $C_1X \subseteq h^{-1}C_2hX$ *for all* $X \subseteq A_1$

(iii) $C_1h^{-1} \leqslant h^{-1}C_2$. *That is,* $C_1h^{-1}Y \subseteq h^{-1}C_2Y$ *for all* $Y \subseteq A_2$

(iv) $h^{-1}\mathscr{C}_2 \subseteq \mathscr{C}_1$. *That is,* $h^{-1}G \in \mathscr{C}_1$ *for all* $G \in \mathscr{C}_2$.

PROOF: Clearly (ii) is equivalent to the definition of (i), by residuation.

(ii)⇒(iii) Put $h^{-1}Y$ for X in (ii) and use that $h^{-1}hY \subseteq Y$ and the monotonicity of C_2.

(iii)⇒(iv) If $G \in \mathscr{C}_2$, then $G = C_2G$. By (iii) $C_1h^{-1}G \subseteq h^{-1}C_2G = h^{-1}G$ therefore $C_1h^{-1}G = h^{-1}G$, that is, $h^{-1}G \in \mathscr{C}_1$.

(iv)⇒(ii) It is always true that $C_2hX \in \mathscr{C}_2$, and by (iv) $h^{-1}C_2hX \in \mathscr{C}_1$. Since $X \subseteq h^{-1}hX \subseteq h^{-1}C_2hX$, it follows that $C_1X \subseteq h^{-1}C_2hX$. ⊠

Property (iv) is the most convenient alternative definition of the notion of a homomorphism between g-matrices in terms of the corresponding closure systems. I gave the details of the proof so that you can appreciate the typical mechanisms of this kind of work. Other similar properties are left as exercises.

EXAMPLES 5.40

1. Any algebraic homomorphism between two Boolean (Heyting, etc.) algebras is a homomorphism between the corresponding g-matrices given by the closure systems of all their filters.
2. A continuous function between two topological spaces is also a "homomorphism" between the closure systems of their closed sets (no algebra here, so it would be improper to speak of "g-matrices").
3. $\langle \boldsymbol{A}, \mathscr{C} \rangle \in \mathsf{GMod}\mathcal{L}$ if and only if $\mathrm{Hom}(\boldsymbol{Fm}, \boldsymbol{A}) = \mathrm{Hom}(\langle \boldsymbol{Fm}, \mathcal{T}h\mathcal{L} \rangle, \langle \boldsymbol{A}, \mathscr{C} \rangle)$.
4. A closure operator C on the universe of an algebra $\boldsymbol{A}$ is $\mathrm{End}\boldsymbol{A}$-structural if and only if the set of endomorphisms of the g-matrix $\langle \boldsymbol{A}, C \rangle$ is $\mathrm{End}\boldsymbol{A}$. ⊠

However, this notion is too weak and has a limited role in the general theory. The important one is the following.

DEFINITION 5.41
An $h \in \mathrm{Hom}(\boldsymbol{A}_1, \boldsymbol{A}_2)$ is a **strict homomorphism from** *the g-matrix $\mathcal{M}_1 = \langle \boldsymbol{A}_1, \mathscr{C}_1 \rangle$* **to** *the g-matrix $\mathcal{M}_2 = \langle \boldsymbol{A}_2, \mathscr{C}_2 \rangle$ when $C_1 = h^{-1} C_2 h$. That is, when*

$$a \in C_1 X \iff ha \in C_2 hX \qquad \textit{for all } X \cup \{a\} \subseteq A_1.$$

An **isomorphism** *between g-matrices is a strict and bijective homomorphism from one of them to the other, and is denoted by the symbol $\cong$.*

The g-matrix $\mathcal{M}_2$ is an **image** *of $\mathcal{M}_1$, and $\mathcal{M}_1$ is an* **inverse image** *of $\mathcal{M}_2$, when there is a strict and surjective homomorphism from $\mathcal{M}_1$ to $\mathcal{M}_2$; the notations $\mathcal{M}_2 = h\mathcal{M}_1$ and $\mathcal{M}_1 = h^{-1}\mathcal{M}_2$ may be occasionally used. The operator of taking all g-matrices that are images (resp., inverse images) of a given one (or of those in a class) by strict surjective homomorphisms is denoted by $\mathbb{H}_s$ (resp., by $\mathbb{H}_s^{-1}$); the operator $\mathbb{I}$ forms all isomorphic copies of the g-matrices in a class.*

This definition expresses the idea that strict homomorphisms preserve both the algebraic structure (algebraic homomorphisms) and the "logical" structure (in both directions); this idea is enhanced for strict and surjective homomorphisms in Theorem 5.48. Observe that a strict homomorphism is a homomorphism, and hence satisfies the properties of Proposition 5.39.

The image construction, defined above only for strict surjective homomorphisms, is also possible when starting from an arbitrary strict homomorphism, provided that certain (natural) notion of a sub-g-matrix is used; see Exercises 5.29 and 5.30.

The next characterization of strict homomorphisms is equally useful, and expresses the preservation character from the side of the closure systems. It would equally well make sense to take it as the initial definition of the notion.

PROPOSITION 5.42
An $h \in \mathrm{Hom}(\boldsymbol{A}_1, \boldsymbol{A}_2)$ is a strict homomorphism from $\langle \boldsymbol{A}_1, \mathscr{C}_1 \rangle$ to $\langle \boldsymbol{A}_2, \mathscr{C}_2 \rangle$ if and only if $h^{-1}\mathscr{C}_2 = \mathscr{C}_1$.

PROOF: ($\Rightarrow$) Since every strict homomorphism is in particular a homomorphism, by Proposition 5.39 $h^{-1}\mathscr{C}_2 \subseteq \mathscr{C}_1$. To show the reverse inclusion, take any $F \in \mathscr{C}_1$; then $F = C_1 F = h^{-1} C_2 h F$, but $C_2 h F \in \mathscr{C}_2$ and hence $F \in h^{-1}\mathscr{C}_2$.

($\Leftarrow$) In particular $h^{-1}\mathscr{C}_2 \subseteq \mathscr{C}_1$, so that h is a homomorphism and $C_1 \leqslant h^{-1} C_2 h$. To show the reverse order take any $a \in h^{-1} C_2 h X$ for some $X \subseteq A_1$. We have that $ha \in C_2 h X$. In order to see that $a \in C_1 X$ consider any $F \in \mathscr{C}_1$ with $X \subseteq F$. By assumption $F = h^{-1} G$ for some $G \in \mathscr{C}_2$. Thus $X \subseteq h^{-1} G$ and so $hX \subseteq hh^{-1} G \subseteq G \in \mathscr{C}_2$, which implies that $ha \in G$, that is, that $a \in h^{-1} G = F$. Thus, $a \in F$ for all $F \in \mathscr{C}_1$ with $X \subseteq F$, which means that $a \in C_1 X$. ⊠

Given $h \in \mathrm{Hom}(\boldsymbol{A}_1, \boldsymbol{A}_2)$ and a g-matrix $\langle \boldsymbol{A}_2, \mathscr{C}_2 \rangle$, the family $h^{-1}\mathscr{C}_2$ is a closure system on A_1, therefore by Proposition 5.42 there is always a g-matrix on $\boldsymbol{A}_1$ such that the function h is a strict homomorphism between them, and it is moreover unique: it is the g-matrix $\langle \boldsymbol{A}_1, h^{-1}\mathscr{C}_2 \rangle$. The reverse process is, however, more delicate: given a closure system $\mathscr{C}_1$ on A_1, $h\mathscr{C}_1$ need not be a closure system on A_2 (see Exercise 5.31), and it is not always possible to find a closure system $\mathscr{C}_2$ on A_2 such that h becomes a strict homomorphism between the resulting g-matrices; an additional condition on h is needed, and even then, the wanted closure system is not the family $h\mathscr{C}_1$, unless h is surjective.

The following lemma, parallel to Proposition 4.27.1, is simple but very useful.

LEMMA 5.43
Let $h \in \mathrm{Hom}(\boldsymbol{A}_1, \boldsymbol{A}_2)$ and let $\langle \boldsymbol{A}_1, \mathscr{C}_1 \rangle$ be a g-matrix. The relation $\ker h$ *is a strict congruence of $\langle \boldsymbol{A}_1, \mathscr{C}_1 \rangle$ if and only if for all $F \in \mathscr{C}_1,$, $h^{-1}hF = F$.*

PROOF: Left to you as Exercise 5.32. ⊠

Recall that $\ker h$ is a congruence of $\langle \boldsymbol{A}_1, \mathscr{C}_1 \rangle$ if and only if $\ker h \subseteq \boldsymbol{\Lambda}\mathscr{C}_1$, or equivalently $\ker h \subseteq \widetilde{\Omega}^{\boldsymbol{A}_1}\mathscr{C}_1$; the first fact amounts to saying that if $ha = hb$, then $C_1\{a\} = C_1\{b\}$. As to the equivalent property in the lemma, observe that the equality would hold for all subsets of A_1 if and only if h is one-to-one. Thus, the fact that it holds for some selected subsets (the closed sets of C_1) is a significant property; in fact, it is a necessary property for the existence of a closure system $\mathscr{C}_2$ on A_2 such that h becomes a homomorphism between the resulting g-matrices:

LEMMA 5.44
If h is a strict homomorphism from $\langle \boldsymbol{A}_1, \mathscr{C}_1 \rangle$ to $\langle \boldsymbol{A}_2, \mathscr{C}_2 \rangle$, then $h^{-1}hF = F$ for all $F \in \mathscr{C}_1$. As a consequence, $\ker h \in \mathrm{Con}\langle \boldsymbol{A}_1, \mathscr{C}_1 \rangle$.

PROOF: Let $F \in \mathscr{C}_1$. That $F \subseteq h^{-1}hF$ always holds. If h is strict, by Proposition 5.42, $F \in h^{-1}\mathscr{C}_2$, that is, $F = h^{-1}G$ for some $G \in \mathscr{C}_2$. Then $hF = hh^{-1}G \subseteq G$, which implies $h^{-1}hF \subseteq h^{-1}G = F$. This shows that $F = h^{-1}hF$. The final conclusion follows by Lemma 5.43. ⊠

Now the above existence problem can be solved:

PROPOSITION 5.45
Let $h \in \mathrm{Hom}(\boldsymbol{A}_1, \boldsymbol{A}_2)$, let $\langle \boldsymbol{A}_1, \mathscr{C}_1 \rangle$ be a g-matrix, and define the following family $\mathscr{C}_2 := \{G \subseteq A_2 : h^{-1}G \in \mathscr{C}_1\}$. This family is a closure system, and:

1. *h is a homomorphism from $\langle \boldsymbol{A}_1, \mathscr{C}_1 \rangle$ to $\langle \boldsymbol{A}_2, \mathscr{C}_2 \rangle$, and $\langle \boldsymbol{A}_2, \mathscr{C}_2 \rangle$ is the weakest g-matrix on $\boldsymbol{A}_2$ with this property.*
2. *h is strict from $\langle \boldsymbol{A}_1, \mathscr{C}_1 \rangle$ to $\langle \boldsymbol{A}_2, \mathscr{C}_2 \rangle$ if and only if* $\ker h \in \mathrm{Con}\langle \boldsymbol{A}_1, \mathscr{C}_1 \rangle$.
3. *If h is strict from $\langle \boldsymbol{A}_1, \mathscr{C}_1 \rangle$ to $\langle \boldsymbol{A}_2, \mathscr{C}_2 \rangle$ and it is surjective, then $\mathscr{C}_2 = h\mathscr{C}_1$.*

PROOF: 1. It is easy to check that $\mathscr{C}_2$ is a closure system. By construction $h^{-1}\mathscr{C}_2 \subseteq \mathscr{C}_1$, therefore h is a homomorphism, and, precisely by its definition, if $h^{-1}\mathscr{C} \subseteq \mathscr{C}_1$, then $\mathscr{C} \subseteq \mathscr{C}_2$. Therefore, $\mathscr{C}_2$ is the weakest closure system with the said property.

2. Using Lemma 5.43, we need just show that h is strict if and only if for all $F \in \mathscr{C}_1$, $h^{-1}hF = F$.

($\Rightarrow$) This holds by Lemma 5.44, regardless of how $\mathscr{C}_2$ is defined

($\Leftarrow$) By definition of $\mathscr{C}_2$, $h^{-1}\mathscr{C}_2 \subseteq \mathscr{C}_1$, and the condition implies that $hF \in \mathscr{C}_2$ for all $F \in \mathscr{C}_1$; hence $F \in h^{-1}\mathscr{C}_2$, and thus $\mathscr{C}_1 = h^{-1}\mathscr{C}_2$, which shows that h is strict.

3. The proof of 2 showed that $h\mathscr{C}_1 \subseteq \mathscr{C}_2$. If moreover h is surjective, then for all $G \in \mathscr{C}_2$, $G = hh^{-1}G$, but $h^{-1}G \in \mathscr{C}_1$ by definition of $\mathscr{C}_2$. Therefore $G \in h\mathscr{C}_1$. Thus $\mathscr{C}_2 = h\mathscr{C}_1$. ⊠

In [49], the closure system $\mathscr{C}_2$ defined in this proposition is called the closure system *inductively defined* from $\mathscr{C}_1$; it is a particular case of a universal construction on closure systems studied there.

As a consequence of this proposition, if $h \in \mathrm{Hom}(\boldsymbol{A}_1, \boldsymbol{A}_2)$ is surjective and satisfies $\ker h \subseteq \boldsymbol{\Lambda}\mathscr{C}_1$, then instead of the definition of $\mathscr{C}_2$ given there one can just take $\mathscr{C}_2 := h\mathscr{C}_1$, and one obtains the weakest strict image of the original g-matrix, which moreover satisfies $\mathscr{C}_1 = h^{-1}\mathscr{C}_2$. Strict and surjective homomorphisms have many different but equivalent characterizations; what is important is to realize all the properties they exhibit rather than the fact that certain groups are equivalent (although this is technically useful at certain points):

PROPOSITION 5.46
Let $\langle \boldsymbol{A}_1, \mathscr{C}_1 \rangle$ and $\langle \boldsymbol{A}_2, \mathscr{C}_2 \rangle$ be two g-matrices, and let $h \in \mathrm{Hom}(\boldsymbol{A}_1, \boldsymbol{A}_2)$ be surjective and such that $\ker h \subseteq \boldsymbol{\Lambda}\mathscr{C}_1$. The following conditions are equivalent:

(i) *h is a strict homomorphism from $\langle \boldsymbol{A}_1, \mathscr{C}_1 \rangle$ to $\langle \boldsymbol{A}_2, \mathscr{C}_2 \rangle$.*
(ii) $C_2 = hC_1h^{-1}$.
(iii) $hC_1 = C_2h$.
(iv) $h^{-1}C_2 = C_1h^{-1}$.
(v) $\mathscr{C}_2 = h\mathscr{C}_1$.

PROOF: At this point, proving this is a relatively routine exercise, but doing it is interesting in order to master the typical techniques of working with closure operators and g-matrices. ⊠

Putting the same facts in a different way:

COROLLARY 5.47
Let $h \in \mathrm{Hom}(\boldsymbol{A}_1, \boldsymbol{A}_2)$ be surjective, and let $\langle \boldsymbol{A}_1, \mathscr{C}_1 \rangle$ be a g-matrix. There is a closure system $\mathscr{C}_2$ on A_2 such that h is a strict homomorphism from $\langle \boldsymbol{A}_1, \mathscr{C}_1 \rangle$ to $\langle \boldsymbol{A}_2, \mathscr{C}_2 \rangle$ if and only if $\ker h \subseteq \boldsymbol{\Lambda}\mathscr{C}_1$. In this case, the closure operator is unique, defined by $\mathscr{C}_2 := h\mathscr{C}_1$, and satisfies that $\mathscr{C}_1 = h^{-1}\mathscr{C}_2$. ⊠

Probably the strongest and the most useful characterization of strict surjective homomorphisms is the following result:

THEOREM 5.48
A surjective $h \in \mathrm{Hom}(\boldsymbol{A}_1, \boldsymbol{A}_2)$ is a strict homomorphism from $\langle \boldsymbol{A}_1, \mathscr{C}_1 \rangle$ to $\langle \boldsymbol{A}_2, \mathscr{C}_2 \rangle$ if and only if the extended function $h \colon \mathscr{P}A_1 \to \mathscr{P}A_2$ restricts to an order isomorphism from the set $\mathscr{C}_1$ to the set $\mathscr{C}_2$, both ordered under set inclusion (and hence to a lattice isomorphism between them) and its inverse is the (restricted) residual function h^{-1}.

PROOF: Assume that h is strict. Then, by Lemma 5.43 and Proposition 5.45, $h^{-1}hF = F$ for all $F \in \mathscr{C}_1$; and since it is surjective, in particular $hh^{-1}G = G$ for all $G \in \mathscr{C}_2$. Moreover, by Propositions 5.42 and 5.46, $h\mathscr{C}_1 = \mathscr{C}_2$ and $h^{-1}\mathscr{C}_2 = \mathscr{C}_1$. Thus, h and h^{-1} are mutually inverse bijections between $\mathscr{C}_1$ and $\mathscr{C}_2$, and by definition they are both order-preserving. Therefore the extended h restricts to an order isomorphism from $\mathscr{C}_1$ to $\mathscr{C}_2$, and its residual h^{-1} restricts to the inverse of the extended h.

For the converse, notice that the assumption implies that the residual h^{-1} of h coincides with the inverse of the function extended to the power sets, and as a consequence, $h^{-1}hF = F$ for all $F \in \mathscr{C}_1$. By Lemma 5.43 this implies that $\ker h \subseteq \boldsymbol{\Lambda}\mathscr{C}_1$. Since $\mathscr{C}_2 = h\mathscr{C}_1$, Proposition 5.46 closes the proof. ⊠

Notice that, since the closure systems are complete lattices, the order isomorphisms are, in fact, complete lattice isomorphisms. It is thus clear, as announced after Definition 5.41, that *strict and surjective homomorphisms*† are the functions that best preserve the structure of g-matrices at the "logical" level.

In order to study the effects of strict homomorphisms on relations, it is helpful to first review some notations and elementary properties. Given any function $h\colon A_1 \to A_2$, consider its extension $h\colon A_1 \times A_1 \to A_2 \times A_2$ to the cartesian products, defined by $h\langle a,b\rangle := \langle ha, hb\rangle$. The power set extension of this function is given by $hR := \{\langle ha,hb\rangle : \langle a,b\rangle \in R\}$ for $R \subseteq A_1 \times A_1$, and in turn this one has a residual, given by $h^{-1}S := \{\langle a,b\rangle \in A_1 \times A_1 : \langle ha,hb\rangle \in S\}$ for $S \subseteq A_2 \times A_2$. Then the following facts are easy to show, for all $R \subseteq A_1 \times A_1$, and all $S \subseteq A_2 \times A_2$; in the last two h is assumed to be a homomorphism from an algebra $\boldsymbol{A}_1$ to an algebra $\boldsymbol{A}_2$.

- $R \subseteq h^{-1}S$ if and only if $hR \subseteq S$.
- $R \subseteq h^{-1}hR$, and $hh^{-1}S \subseteq S$.
- h is surjective if and only if $hh^{-1}S = S$ for all $S \subseteq A_2 \times A_2$.
- If $S \in \operatorname{Con}\boldsymbol{A}_2$, then $h^{-1}S \in \operatorname{Con}\boldsymbol{A}_1$.
- If $R \in \operatorname{Con}\boldsymbol{A}_1$, h is surjective and $\ker h \subseteq R$, then $hR \in \operatorname{Con}\boldsymbol{A}_2$.

If you are surprised by the last property, then you should do Exercise 5.34.

PROPOSITION 5.49
Let h be a strict surjective homomorphism from $\mathcal{M}_1 = \langle \boldsymbol{A}_1, \mathscr{C}_1\rangle$ to $\mathcal{M}_2 = \langle \boldsymbol{A}_2, \mathscr{C}_2\rangle$.

1. $\boldsymbol{\Lambda}\mathscr{C}_1 = h^{-1}\boldsymbol{\Lambda}\mathscr{C}_2$ *and* $\boldsymbol{\Lambda}\mathscr{C}_2 = h\boldsymbol{\Lambda}\mathscr{C}_1$.
2. *If* $\theta_2 \in \operatorname{Con}\boldsymbol{A}_2$, *then* $\theta_2 \in \operatorname{Con}\mathcal{M}_2$ *if and only if* $h^{-1}\theta_2 \in \operatorname{Con}\mathcal{M}_1$. *Moreover, then* $\theta_2 = hh^{-1}\theta_2$.
3. *If* $\theta_1 \in \operatorname{Con}\boldsymbol{A}_1$, *then* $\ker h \subseteq \theta_1$ *if and only if* $\theta_1 = h^{-1}h\theta_1$.
4. *If* $\theta_1 \in \operatorname{Con}\boldsymbol{A}_1$ *and* $\ker h \subseteq \theta_1$, *then* $\theta_1 \in \operatorname{Con}\mathcal{M}_1$ *if and only if* $h\theta_1 \in \operatorname{Con}\mathcal{M}_2$.
5. $h\colon [\ker h, \widetilde{\boldsymbol{\Omega}}\mathcal{M}_1] \cong \operatorname{Con}\mathcal{M}_2$, *with inverse isomorphism the residual function* h^{-1}.

† They are called *bilogical morphisms* in the works where g-matrices are called abstract logics, such as [41] or [107].

6. $\widetilde{\Omega}^{A_2}\mathscr{C}_2 = h\widetilde{\Omega}^{A_1}\mathscr{C}_1$ *and* $\widetilde{\Omega}^{A_1}\mathscr{C}_1 = h^{-1}\widetilde{\Omega}^{A_2}\mathscr{C}_2$*, or, in the compact notation,* $\widetilde{\Omega}\mathcal{M}_2 = h\widetilde{\Omega}\mathcal{M}_1$ *and* $\widetilde{\Omega}\mathcal{M}_1 = h^{-1}\widetilde{\Omega}\mathcal{M}_2$.

PROOF: Points 1 to 4 are left to you as Exercise 5.35.

5. Notice that since h is strict, $\ker h \subseteq \widetilde{\Omega}\mathcal{M}_1$, therefore considering the segment $[\ker h, \widetilde{\Omega}\mathcal{M}_1]$ in $\mathrm{Con}\boldsymbol{A}_1$ makes sense. If $\ker h \subseteq \theta_1 \subseteq \widetilde{\Omega}\mathcal{M}_1$, in particular $\theta_1 \in \mathrm{Con}\mathcal{M}_1$, and by 3 and 4, $h\theta_1 \in \mathrm{Con}\mathcal{M}_2$ and $\theta_1 = h^{-1}h\theta_1$. Conversely, if $\theta_2 \in \mathrm{Con}\mathcal{M}_2$, then by 2, $h^{-1}\theta_2 \in \mathrm{Con}\mathcal{M}_1$, hence $h^{-1}\theta_2 \subseteq \widetilde{\Omega}\mathcal{M}_1$, and $\theta_2 = hh^{-1}\theta_2$, which implies $h^{-1}\theta_2 = h^{-1}hh^{-1}\theta_2$. Using 3 for $h^{-1}\theta_2$ we obtain $\ker h \subseteq h^{-1}\theta_2$, that is, $h^{-1}\theta_2 \in [\ker h, \widetilde{\Omega}\mathcal{M}_1]$. And we have seen that the two compositions hh^{-1} and $h^{-1}h$, when restricted to the said sets of congruences, become the identity functions; hence, h and h^{-1} are isomorphisms between these sets and are inverse to one another.

6. This is a consequence of 5: since h is an isomorphism, it must put the maximum elements of the two sets into correspondence; these elements are $\widetilde{\Omega}\mathcal{M}_1$ and $\widetilde{\Omega}\mathcal{M}_2$, respectively. ⊠

Exercise 5.36 asks you to supply an alternative, direct proof of the property in point 6 above.

Quotients

A particular case where the construction in Proposition 5.45 applies by default is the quotient under some $\theta \in \mathrm{Con}\langle \boldsymbol{A}, \mathscr{C}\rangle$, because by definition there is a surjective $\pi\colon \boldsymbol{A} \to \boldsymbol{A}/\theta$ such that $\ker \pi = \theta \subseteq \boldsymbol{\Lambda}\mathscr{C}$. According to the comment after Proposition 5.45, the following makes sense:

DEFINITION 5.50
Let θ *be a strict congruence of a g-matrix* $\mathcal{M} = \langle \boldsymbol{A}, \mathscr{C}\rangle$. *The* ***quotient g-matrix*** $\mathcal{M}/\theta := \langle \boldsymbol{A}/\theta, \mathscr{C}/\theta\rangle$ *is given by the closure system* $\mathscr{C}/\theta := \{F/\theta : F \in \mathscr{C}\} = \pi\mathscr{C}$.

By Proposition 5.45, it also holds that $\mathscr{C}/\theta = \{G \subseteq A/\theta : \pi^{-1}G \in \mathscr{C}\}$, and the canonical projection $\pi\colon \boldsymbol{A} \to \boldsymbol{A}/\theta$ is a strict surjective homomorphism from $\langle \boldsymbol{A}, \mathscr{C}\rangle$ to $\langle \boldsymbol{A}/\theta, \mathscr{C}/\theta\rangle$. Therefore, as a particular case of Proposition 5.49, we immediately obtain:

COROLLARY 5.51
Let $\mathcal{M} = \langle \boldsymbol{A}, C\rangle$ *be a g-matrix, and let* $\theta \in \mathrm{Con}\mathcal{M}$.

1. $\boldsymbol{\Lambda}(\mathscr{C}/\theta) = (\boldsymbol{\Lambda}\mathscr{C})/\theta$.
2. $\widetilde{\Omega}(\mathcal{M}/\theta) = (\widetilde{\Omega}\mathcal{M})/\theta$.
3. $\mathrm{Con}(\mathcal{M}/\theta) = \{\theta'/\theta : \theta' \in \mathrm{Con}\mathcal{M}, \theta' \supseteq \theta\}$*, and the function* $\theta' \longmapsto \theta'/\theta$ *defines a lattice isomorphism* $[\theta, \widetilde{\Omega}\mathcal{M}] \cong \mathrm{Con}(\mathcal{M}/\theta)$. ⊠

Point 3 is a universal-algebraic-style *Correspondence Theorem*, parallel to point 3 of Theorem 4.30 for matrices; its other two points also have g-matrix parallels, which could be called *First* and *Second Isomorphism Theorems*:

THEOREM 5.52

Let h be a strict surjective homomorphism from the g-matrix $\mathcal{M}_1$ to the g-matrix $\mathcal{M}_2$. Then $\mathcal{M}_1/\ker h \cong \mathcal{M}_2$ through a unique isomorphism g such that $h = g \circ \pi$:

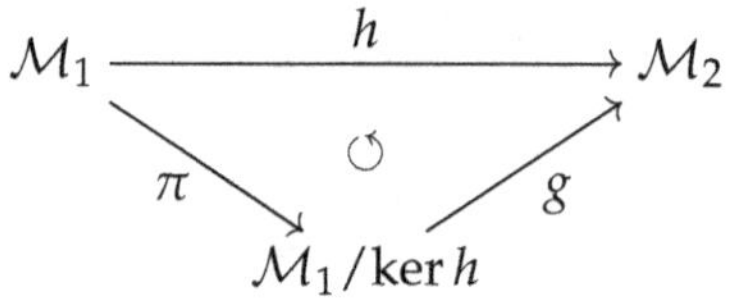

⊠

THEOREM 5.53

If $\theta, \theta' \in \operatorname{Con}\mathcal{M}$ are such that $\theta \subseteq \theta'$, then $\theta'/\theta \in \operatorname{Con}(\mathcal{M}/\theta)$ and $(\mathcal{M}/\theta)/(\theta'/\theta) \cong \mathcal{M}/\theta'$ in the obvious way. ⊠

As happened with Theorem 4.30, the needed constructions are the well-known algebraic ones, and only the facts concerning the closure systems or the closure operators have to be checked (Exercise 5.37). For an application or generalization of Theorem 5.52, see Exercise 5.38.

The process of reduction

Among the quotients by strict congruences of a g-matrix there is a special one, namely the quotient by the Tarski congruence of the g-matrix. The process of forming this quotient is called *reduction*.

DEFINITION 5.54

*Let $\mathcal{M} = \langle \boldsymbol{A}, \mathscr{C} \rangle$ be a g-matrix. The **reduction of** $\mathcal{M}$ is defined as the g-matrix $\mathcal{M}^* := \mathcal{M}/\widetilde{\boldsymbol{\Omega}}\mathcal{M} = \langle \boldsymbol{A}^*, \mathscr{C}^* \rangle$ where $\boldsymbol{A}^* := \boldsymbol{A}/\widetilde{\boldsymbol{\Omega}}\mathcal{M}$ and $\mathscr{C}^* := \mathscr{C}/\widetilde{\boldsymbol{\Omega}}\mathcal{M}$.*

If M is a class of g-matrices, then $\mathsf{M}^* := \{\mathcal{M}^* : \mathcal{M} \in \mathsf{M}\}$.

*A g-matrix $\mathcal{M} = \langle \boldsymbol{A}, \mathscr{C} \rangle$ is **reduced** when $\widetilde{\boldsymbol{\Omega}}\mathcal{M} = \mathrm{Id}_A$, or, equivalently, when $\operatorname{Con}\mathcal{M} = \{\mathrm{Id}_A\}$.*

Strictly speaking, the notations $\boldsymbol{A}^*$ and $\mathscr{C}^*$ are ambiguous, because $\boldsymbol{A}$ denotes only the algebraic structure and does not presuppose any particular closure system, and dually $\mathscr{C}$ does not presuppose any algebraic structure; they are used only when the missing component is clear from context. Observe that $\mathsf{M}^* \subseteq \mathbb{H}_s\mathsf{M}$.

The following trivial fact tells us why it is so nice to work with g-matrices with the PCONG:

PROPOSITION 5.55

A g-matrix $\langle \boldsymbol{A}, \mathscr{C} \rangle$ has the PCONG *and is reduced if and only if $\boldsymbol{\Lambda}\mathscr{C} = \mathrm{Id}_A$, that is, if and only if for all $a, b \in A$, if $C\{a\} = C\{b\}$, then $a = b$.* ⊠

The property $\boldsymbol{\Lambda}\mathscr{C} = \mathrm{Id}_A$, viewed from its closure system side, corresponds to the topological separation property commonly called "T_0": If $a \neq b$, then there is some $F \in \mathscr{C}$ such that $a \in F$ and $b \notin F$ or that $a \notin F$ and $b \in F$; this property is described by saying that $\mathscr{C}$ *separates points* (see page 42). From previous results it easily follows:

PROPOSITION 5.56
Let $\mathcal{M}, \mathcal{M}_1, \mathcal{M}_2$ be g-matrices.

1. $\mathcal{M}^*$ *is reduced.*
2. *If $\mathcal{M}_1 \cong \mathcal{M}_2$, then $\mathcal{M}_1$ is reduced if and only if $\mathcal{M}_2$ is reduced.*
3. *$\mathcal{M}$ is reduced if and only if $\mathcal{M} \cong \mathcal{M}^*$.*
4. *If $\theta \in \mathrm{Con}\mathcal{M}$, then $(\mathcal{M}/\theta)^* \cong \mathcal{M}^*$.*
5. $(\mathcal{M}^*)^* \cong \mathcal{M}^*$.
6. *If $\mathcal{M}_1 \cong \mathcal{M}_2$, then $\mathcal{M}_1^* \cong \mathcal{M}_2^*$.*

PROOF: 1. By Corollary 5.51.2, $\widetilde{\Omega}\mathcal{M}^* = \widetilde{\Omega}(\mathcal{M}/\widetilde{\Omega}\mathcal{M}) = (\widetilde{\Omega}\mathcal{M})/\widetilde{\Omega}\mathcal{M} = \mathrm{Id}_{A^*}$.

2. Obvious, by Proposition 5.49.6.

3. If $\widetilde{\Omega}\mathcal{M} = \mathrm{Id}_A$, then trivially $\mathcal{M} \cong \mathcal{M}/\widetilde{\Omega}\mathcal{M}$. The converse is a consequence of 1 and 2.

4. Using Theorem 5.53, and again Corollary 5.51.2, we have that if $\theta \subseteq \widetilde{\Omega}\mathcal{M}$, then $(\mathcal{M}/\theta)^* = (\mathcal{M}/\theta)/\widetilde{\Omega}(\mathcal{M}/\theta) = (\mathcal{M}/\theta)/(\widetilde{\Omega}\mathcal{M}/\theta) \cong \mathcal{M}/\widetilde{\Omega}\mathcal{M} = \mathcal{M}^*$.

5. $\widetilde{\Omega}\mathcal{M} \in \mathrm{Con}\mathcal{M}$, so this is a particular case of 4.

6. Notice that the composition $\pi_2 h$ of the isomorphism h with the canonical projection $\pi_2 \colon \mathcal{M}_2 \to \mathcal{M}_2^*$ is a strict surjective homomorphism from $\mathcal{M}_1$ onto $\mathcal{M}_2^*$ (Exercise 5.41), and that $\ker(\pi_2 h) = h^{-1}\ker \pi_2 = h^{-1}\widetilde{\Omega}\mathcal{M}_2 = \widetilde{\Omega}\mathcal{M}_1$; the last equality is a consequence of Proposition 5.49.6. Then by Theorem 5.52 applied to $\pi_2 h$, there is an isomorphism between $\mathcal{M}_1^*$ and $\mathcal{M}_2^*$. ⊠

There is a reading of points 4 and 5 that somehow justifies calling "reduction" the process of factoring out a g-matrix by its Tarski congruence: they say that this factorization produces the "smallest" or "most reduced" of the quotients by strict congruences. The second part of the next result also expresses this. Exercise 5.39 proposes an alternative, but basically equivalent formulation of this property.

PROPOSITION 5.57
If there is a strict surjective homomorphism between $\mathcal{M}_1$ and $\mathcal{M}_2$, then $\mathcal{M}_1^* \cong \mathcal{M}_2^*$. If moreover $\mathcal{M}_1$ is reduced, then the homomorphism is actually an isomorphism, and $\mathcal{M}_2$ is reduced as well.

PROOF: Let $h \colon \mathcal{M}_1 \to \mathcal{M}_2$ be a strict surjective homomorphism. By Lemma 5.44, $\ker h \in \mathrm{Con}\mathcal{M}_1$, and by Theorem 5.52, $\mathcal{M}_1/\ker h \cong \mathcal{M}_2$. Now, using several points of Proposition 5.56 we conclude that $\mathcal{M}_1^* \cong (\mathcal{M}_1/\ker h)^* \cong \mathcal{M}_2^*$. If moreover $\mathcal{M}_1$ is reduced, since $\ker h \subseteq \widetilde{\Omega}\mathcal{M}_1$, it follows that $\ker h$ is the identity relation and h is an isomorphism establishing that $\mathcal{M}_1 \cong \mathcal{M}_2$. Then by 5.56.2, $\mathcal{M}_2$ is reduced as well. ⊠

The converse of the first implication cannot be true in general (for instance, just for cardinality reasons); the best approximation is:

PROPOSITION 5.58
$\mathcal{M}_1^* \cong \mathcal{M}_2^*$ if and only if there is a g-matrix $\mathcal{M}_3$ such that there are strict surjective homomorphisms from $\mathcal{M}_1$ onto $\mathcal{M}_3$ and from $\mathcal{M}_2$ onto $\mathcal{M}_3$.

PROOF: Put $\mathcal{M}_3 := \mathcal{M}_1^*$, and take the corresponding canonical projection as the first required strict surjective homomorphism; the other one is obtained as the composition of the other canonical projection with the isomorphism. To prove the converse, applying Proposition 5.57 to the assumption, you obtain that $\mathcal{M}_1^* \cong \mathcal{M}_3^* \cong \mathcal{M}_2^*$. ⊠

If you noticed that Proposition 5.57 follows, as a particular case, from Proposition 5.58, you should also notice that the given proof of the latter actually uses the former; in Exercise 5.40 you are asked to supply alternative proofs of the two properties that reverse the situation.

In [110] two g-matrices are called *bilogically morphic* when they have isomorphic reductions. It is not difficult to see that this is the symmetric and transitive closure of the relation of existence of a strict surjective homomorphism from one onto the other (actually, the relation is presented in [110] in this way), but it is easy to see, suing Proposition 5.58, that this property can be simplified (Exercise 5.41). As a consequence, one can show that many of the properties given here for g-matrices with a strict surjective homomorphism between them in fact do hold for bilogically morphic pairs of g-matrices.

Exercises for Section 5.3

EXERCISE 5.25. Use the characterization of the Tarski congruence in Theorem 5.32 to prove directly that the Tarski operator, when considered on closure operators, is monotone (Proposition 5.35).

EXERCISE 5.26. Prove the properties in the three points of Lemma 5.37.

EXERCISE 5.27. Let $\mathcal{L}$ be a logic having an algebraic semantics, with $E(x)$ as set of defining equations. Use the result in Exercise 3.6 to prove that for each $\varepsilon \approx \delta \in E(x)$ and each $\langle \boldsymbol{A}, F\rangle \in \mathsf{Mod}\mathcal{L}$, if $a \in F$, then $\varepsilon^{\boldsymbol{A}}(a) \equiv \delta^{\boldsymbol{A}}(a)\ (\widetilde{\Omega}_{\mathcal{L}}^{\boldsymbol{A}}F)$. Deduce from this that for each $\varepsilon \approx \delta \in E(x)$, $\varepsilon \equiv \delta\ (\widetilde{\Omega}_{\mathcal{L}} C_{\mathcal{L}}\{x\})$. Compare with Exercise 4.16.

EXERCISE 5.28. Check all the non-obvious statements in Example 5.40.

EXERCISE 5.29. Let $\mathcal{M} = \langle \boldsymbol{A}, \mathscr{C}\rangle$ be a g-matrix, with associated closure operator C, and let $\boldsymbol{B}$ be a subalgebra of $\boldsymbol{A}$. Define $\mathscr{C}\restriction B := \{F \cap B : F \in \mathscr{C}\}$ and $(C\restriction B)X := (CX) \cap B$ for all $X \subseteq B$. Prove that $\mathcal{M}\restriction B := \langle \boldsymbol{B}, \mathscr{C}\restriction B\rangle$ is a g-matrix with associated closure operator $C\restriction B$, and that the inclusion from $\boldsymbol{B}$ into $\boldsymbol{A}$ is a strict homomorphism from $\mathcal{M}\restriction B$ into $\mathcal{M}$.
COMMENT. This construction provides the natural notion of a ***sub-g-matrix***. It does not have a prominent role in this book, but may appear in some exercises and is used in Lemmas 6.35 and 6.36; it is studied in depth in [111, Section 2].

EXERCISE 5.30. Let $h\colon \langle \boldsymbol{A}_1, \mathscr{C}_1\rangle \to \langle \boldsymbol{A}_2, \mathscr{C}_2\rangle$ be a strict homomorphism. Prove that h is a strict surjective homomorphism from $\langle \boldsymbol{A}_1, \mathscr{C}_1\rangle$ to $\langle h\boldsymbol{A}_1, \mathscr{C}_2\restriction hA_1\rangle$ and that this one is a sub-g-matrix of $\langle \boldsymbol{A}_2, \mathscr{C}_2\rangle$ in the sense of Exercise 5.29.

COMMENT. This would justify defining the g-matrix $h\mathcal{M}_1 := \langle h\boldsymbol{A}_1, \mathscr{C}_2 \upharpoonright hA_1 \rangle$ as the ***image*** of $\mathcal{M}_1 = \langle \boldsymbol{A}_1, \mathscr{C}_1 \rangle$ by the strict homomorphism h. See Exercise 5.38.

EXERCISE 5.31. Provide an example of two algebras $\boldsymbol{A}_1$ and $\boldsymbol{A}_2$, a homomorphism $h\colon \boldsymbol{A}_1 \to \boldsymbol{A}_2$ and a closure system $\mathscr{C}_1$ on A_1 such that $h\mathscr{C}_1$ is not a closure system on A_2; and of a closure system $\mathscr{C}_1'$ (perhaps the same) such that there is no closure system $\mathscr{C}_2$ on A_2 making h a strict homomorphism from $\langle \boldsymbol{A}_1, \mathscr{C}_1' \rangle$ to $\langle \boldsymbol{A}_2, \mathscr{C}_2 \rangle$.

EXERCISE 5.32. Prove Lemma 5.43 by using Proposition 4.27.1 and Lemma 5.28. Conversely, show that Proposition 4.27.1 can be considered as a particular case of Lemma 5.43.

EXERCISE 5.33. Prove the equivalence of the five conditions in Proposition 5.46.

EXERCISE 5.34. Let A_1, A_2 be sets, $h\colon A_1 \to A_2$ a surjective function, and $R \subseteq A_1 \times A_1$ an equivalence relation on A_1 such that $\ker h \subseteq R$. Prove that $hR \subseteq A_2 \times A_2$ is an equivalence relation on A_2. Find counterexamples showing the necessity of including the assumptions that h is surjective and that $\ker h \subseteq R$.

EXERCISE 5.35. Prove the assertions of points 1 to 4 of Proposition 5.49.

EXERCISE 5.36. Supply a proof of point 6 of Proposition 5.49 that does not use the other points of this proposition.

EXERCISE 5.37. Prove Theorems 5.52 and 5.53, relying on the parallel universal algebraic results (see the hints for Exercises 4.17 and 4.18).

EXERCISE 5.38. Prove that every strict homomorphism $h\colon \mathcal{M}_1 \to \mathcal{M}_2$ decomposes into the strict surjective homomorphism $\mathcal{M}_1 \to \mathcal{M}_1/\ker h$ followed by the isomorphism $\mathcal{M}_1/\ker h \cong h\mathcal{M}_1$ and the inclusion of $h\mathcal{M}_1$ into $\mathcal{M}_2$ as a sub-g-matrix, as in the diagram:

$$\begin{array}{ccc}
\mathcal{M}_1 & \xrightarrow{\quad h \quad} & \mathcal{M}_2 \\
{\scriptstyle \pi}\big\downarrow & \circlearrowleft & \big\uparrow \\
\mathcal{M}_1/\ker h & \xrightarrow[\cong]{\qquad} & h\mathcal{M}_1
\end{array}$$

Here the notation $h\mathcal{M}_1$ refers to the construction in Exercise 5.30. Notice that the mentioned inclusion is strict by Exercise 5.29. Theorem 5.52 can be considered a particular case of this property, but can also be used to prove it.

EXERCISE 5.39. Prove that if $\mathcal{M}_1, \mathcal{M}_2$ are two g-matrices such that $\mathcal{M}_2 \in \mathbb{H}_s \mathcal{M}_1$, then $\mathcal{M}_2 \in \mathbb{H}_s^{-1} \mathcal{M}_1^*$. That is, show that if there is a strict surjective homomorphism from $\mathcal{M}_1$ onto $\mathcal{M}_2$, then there is another one from $\mathcal{M}_2$ onto $\mathcal{M}_1^*$.

EXERCISE 5.40. Find a direct proof of Proposition 5.58 that does not use Proposition 5.57, and then obtain the latter as a particular case of the former.

EXERCISE 5.41. Check that the composition of two strict homomorphisms is again a strict homomorphism. Use this to show that the symmetric and transitive closure of the relation of existence of a strict surjective homomorphism between g-matrices (in either direction) is the relation of having isomorphic reductions.

5.4 The algebraic counterpart of a logic

This section deals with the interplay between the (purely algebraic) constructions of the preceding section and the role of g-matrices as g-models of logics. It is convenient to recall here a few basic properties from previous chapters (Lemma 1.2 and Propositions 2.24 and 2.25), which are used often without notice:

PROPOSITION 5.59
Let $\mathcal{L}$ be a logic, $\boldsymbol{A}_1,, \boldsymbol{A}_2$ algebras, and $h \in \operatorname{Hom}(\boldsymbol{A}_1, \boldsymbol{A}_2)$.

1. *If h is surjective, then for each $f \in \operatorname{Hom}(\boldsymbol{Fm}, \boldsymbol{A}_2)$ there is a $g \in \operatorname{Hom}(\boldsymbol{Fm}, \boldsymbol{A}_1)$ such that $hg = f$.*
2. *If $G \in \mathcal{F}i_{\mathcal{L}}\boldsymbol{A}_2$, then $h^{-1}G \in \mathcal{F}i_{\mathcal{L}}\boldsymbol{A}_1$. That is, $h^{-1}\mathcal{F}i_{\mathcal{L}}\boldsymbol{A}_2 \subseteq \mathcal{F}i_{\mathcal{L}}\boldsymbol{A}_1$. Thus, every algebraic homomorphism is a g-matrix homomorphism between the corresponding basic full g-models of a given logic.*
3. *If h is surjective and $G \subseteq A_2$ is such that $h^{-1}G \in \mathcal{F}i_{\mathcal{L}}\boldsymbol{A}_1$, then $G \in \mathcal{F}i_{\mathcal{L}}\boldsymbol{A}_2$.*
4. *If h is surjective and $F \in \mathcal{F}i_{\mathcal{L}}\boldsymbol{A}_1$ is such that $\ker h$ is compatible with F (that is, $\ker h \subseteq \boldsymbol{\Omega}^{\boldsymbol{A}}F$), then $hF \in \mathcal{F}i_{\mathcal{L}}\boldsymbol{A}_2$.* ⊠

PROPOSITION 5.60
The class of basic full g-models of a logic is closed under images by strict surjective homomorphisms and under isomorphisms.

PROOF: If $h\colon \langle \boldsymbol{A}_1, \mathcal{F}i_{\mathcal{L}}\boldsymbol{A}_1\rangle \to \langle \boldsymbol{A}_2, \mathscr{C}_2\rangle$ is a strict surjective homomorphism, then $h^{-1}\mathscr{C}_2 = \mathcal{F}i_{\mathcal{L}}\boldsymbol{A}_1$, so by 5.59.3, if $G \in \mathscr{C}_2$, then $G \in \mathcal{F}i_{\mathcal{L}}\boldsymbol{A}_2$. Conversely, if $G \in \mathcal{F}i_{\mathcal{L}}\boldsymbol{A}_2$ we know that $h^{-1}G \in \mathcal{F}i_{\mathcal{L}}\boldsymbol{A}_1$, and by Theorem 5.48, $G \in \mathscr{C}_2$. That is, we have shown that $\mathscr{C}_2 = \mathcal{F}i_{\mathcal{L}}\boldsymbol{A}_2$, which shows closure under images by strict surjective homomorphisms. The fact that every isomorphism is in particular a strict surjective homomorphism in either direction shows the second statement.⊠

This class is not closed under *inverse* images by strict surjective homomorphisms. However, since many properties are preserved in both directions, it is practical to say that a certain property is ***preserved by strict surjective homomorphisms*** (resp., that a certain class is ***closed under strict surjective homomorphisms***), when the property is preserved by (resp., the class is closed under) the process of taking images *and* inverse images by strict surjective homomorphisms.

THEOREM 5.61
If there is a strict surjective homomorphism between two g-matrices $\mathcal{M}_1$ and $\mathcal{M}_2$, then:

1. *$\mathcal{M}_1$ and $\mathcal{M}_2$ are models of the same g-sequents, that is, they satisfy the same Hilbert-style g-rules.*
2. *$\mathcal{M}_1$ and $\mathcal{M}_2$ define the same logic: $\mathcal{L}_{\mathcal{M}_1} = \mathcal{L}_{\mathcal{M}_2}$.*
3. *$\mathcal{M}_1$ and $\mathcal{M}_2$ are models of the same Gentzen-style g-rules.*

In particular all these properties hold when $\mathcal{M}_1$ and $\mathcal{M}_2$ are isomorphic.

PROOF: 1. Assume that $\mathcal{M}_2$ is a model of $\Gamma \rhd \varphi$ and let $g \in \operatorname{Hom}(\boldsymbol{Fm}, \boldsymbol{A}_1)$. Then $hg \in \operatorname{Hom}(\boldsymbol{Fm}, \boldsymbol{A}_2)$ and $g\varphi \in C_1 g\Gamma = h^{-1}C_2 hg\Gamma$ if and only if

$hg\varphi \in C_2 hg\Gamma$, which is true. Conversely, if $\mathcal{M}_1$ is a model of $\Gamma \rhd \varphi$ and $f \in \mathrm{Hom}(\boldsymbol{Fm}, \boldsymbol{A}_2)$, then by 5.59.1 $f = hg$ for some $g \in \mathrm{Hom}(\boldsymbol{Fm}, \boldsymbol{A}_1)$, and a similar reasoning as before proves that f satisfies $\Gamma \rhd \varphi$ in $\mathcal{M}_2$.

2 follows from 1 because the logic $\mathcal{L}_{\mathcal{M}}$ is defined by the g-sequents satisfied by $\mathcal{M}$, and 3 is proved in a way similar to 1 (in fact, 1 is a particular case of 3). ⊠

COROLLARY 5.62
For any logic $\mathcal{L}$ the class $\mathsf{GMod}\,\mathcal{L}$ *is closed under strict surjective homomorphisms, and hence under isomorphisms.* ⊠

It would be nice to be able to add to the preceding list of Theorem 5.61 the property that $\mathcal{M}_1$ and $\mathcal{M}_2$ satisfy the same Tarski-style conditions; but the general notion of a "Tarski-style condition" is too vague to admit such a treatment, and these general statements are void. However, we can see that all the particular conditions we have dealt with so far are preserved by strict surjective homomorphisms:

THEOREM 5.63
The following properties of a g-matrix are preserved by strict surjective homomorphisms:

1. *To be finitary.*
2. *To have theorems.*
3. *To have no theorems.*
4. *Any property expressed by a set of Gentzen-style g-rules.*
5. *All the Tarski-style conditions previously considered:* PC, u-DDT, PDI, PIRA, PRA, PIM, PEQ, NEG1, NEG2, NEG3, PCONG.
6. *Any property of a g-matrix that is expressible as a lattice-theoretic property (i.e., a property preserved by lattice isomorphisms) of its closure system.*

PROOF: To see that a property is preserved by both images and inverse images by strict surjective homomorphisms, assume that h is a strict surjective homomorphism from $\mathcal{M}_1 = \langle \boldsymbol{A}_1, \mathscr{C}_1 \rangle$ to $\mathcal{M}_2 = \langle \boldsymbol{A}_2, \mathscr{C}_2 \rangle$, and show that $\mathcal{M}_1$ has the property if and only if $\mathcal{M}_2$ has the property. Here only a few of the properties are checked by way of example; the remaining ones follow a very similar pattern and are left to you as Exercise 5.44. Going from $\mathcal{M}_1$ to $\mathcal{M}_2$ is often slightly trickier than the reverse direction.

1. First assume that C_1 is finitary and that $X \cup \{a\} \subseteq A_2$ with $a \in C_2 X$; since $C_2 X = hC_1 h^{-1} X$, $a = hb$ for some $b \in C_1 h^{-1} X$. By finitarity, $b \in C_1 Z$ for some finite $Z \subseteq h^{-1} X$. If we put $Y := hZ$, then Y is finite, and by residuation $Y = hZ \subseteq X$ and also $Z \subseteq h^{-1} Y$. Then $b \in C_1 h^{-1} Y$, and therefore $a = hb \in hC_1 h^{-1} Y = C_2 Y$, and $Y \subseteq X$ is finite. For the converse, assume that C_2 is finitary and that $X \cup \{a\} \subseteq A_1$ with $a \in C_1 X = h^{-1} C_2 hX$. Then $ha \in C_2 hX$, which implies $ha \in C_2 Z$ for a finite $Z \subseteq hX$; therefore, $ha \in C_2 hY$ for a finite $Y \subseteq X$. Thus, $a \in h^{-1} C_2 hY = C_1 Y$.

5. Only the case of PIRA is checked here: First assume that $\mathcal{M}_1$ satisfies it and

let $X \cup \{a\} \subseteq A_2$. Consider any $b \in A_1$ such that $hb = a$. Using first Lemma 5.44 and then two of the main properties of strict surjective homomorphisms in Proposition 5.46,

$$C_1\{b\} = h^{-1}hC_1\{b\} = h^{-1}C_2\{hb\} = h^{-1}C_2\{a\} = C_1h^{-1}\{a\}.$$

Now the following chain of equivalences (where the PIRA is used) should be obvious:

$$\begin{aligned} \neg a \in C_2X \iff h\neg b = \neg hb = \neg a \in C_2X \iff \neg b \in h^{-1}C_2X = C_1h^{-1}X \\ \iff C_1(h^{-1}X, b) = A_1 \iff C_1h^{-1}(X,a) = C_1(h^{-1}X, h^{-1}\{a\}) = A_1 \\ \iff C_2(X,a) = hC_1h^{-1}(X,a) = hA_1 = A_2. \end{aligned}$$

For the converse, assume that $\mathcal{M}_2$ satisfies the PIRA and let $X \cup \{a\} \subseteq A_1$. Then

$$\begin{aligned} \neg a \in C_1X = h^{-1}C_2hX &\iff \neg ha = h\neg a \in C_2hX \\ &\iff hC_1(X,a) = C_2h(X,a) = C_2(hX, ha) = A_2 \\ &\iff C_1(X,a) = h^{-1}hC_1(X,a) = h^{-1}A_2 = A_1. \end{aligned}$$

6. This is an obvious consequence of the lattice isomorphism established in Theorem 5.48. ⊠

We know that for finitary g-matrices, satisfying some of these Tarski-style conditions is equivalent to satisfying a set of Gentzen-style rules (Exercises 5.7 and 5.8) and hence their preservation between finitary g-matrices would follow from point 4. However, you can see that they are preserved in general, and this has to be proved directly in all cases. The PCONG would be an exception,[†] by the equivalence in Definition 5.25; moreover, in this case there is an alternative, more interesting proof: By points 1 and 6 of Proposition 5.49, it is clear that $\widetilde{\Omega}^{A_1}\mathscr{C}_1 = \Lambda\mathscr{C}_1$ if and only if $\widetilde{\Omega}^{A_2}\mathscr{C}_2 = \Lambda\mathscr{C}_2$; by Lemma 5.31.3, this just says that $\mathcal{M}_1$ has the PCONG if and only if $\mathcal{M}_2$ has it. Finally, notice that finitarity of the g-matrix (that is, of the closure operator) is equivalent to inductiveness of the closure system, but this is *not* a lattice-theoretic property, as it involves set-theoretic union (recall the discussion at the end of Section 1.5); hence preservation of finitarity cannot be proved by using the general property of point 6.

All the properties given for strict surjective homomorphisms do hold in particular for reductions. It is convenient to highlight some of them:

PROPOSITION 5.64

1. *For every algebra* $\boldsymbol{A}$, $(\mathcal{F}i_{\mathcal{L}}\boldsymbol{A})^* = \mathcal{F}i_{\mathcal{L}}(\boldsymbol{A}^*)$, *where* $\boldsymbol{A}^* = \boldsymbol{A}/\widetilde{\Omega}^{\boldsymbol{A}}\mathcal{F}i_{\mathcal{L}}\boldsymbol{A}$.
2. *A g-matrix* $\mathcal{M}$ *and its reduction define the same logic:* $\mathcal{L}_{\mathcal{M}} = \mathcal{L}_{\mathcal{M}^*}$.
3. $\mathcal{M} \in \mathsf{GMod}\mathcal{L}$ *if and only if* $\mathcal{M}^* \in \mathsf{GMod}\mathcal{L}$, *for any g-matrix* $\mathcal{M}$.
4. *If* $\mathcal{L}$ *is complete with respect to a certain class* M *of g-matrices, then* $\mathcal{L}$ *is also complete with respect to the class* M^* *of their reductions.* ⊠

[†] The PC and NEG2 are also exceptions, as they are equivalent to Hilbert-style rules.

PROOF: 1 is a consequence of the fact that the class of basic full g-models of $\mathcal{L}$ is closed under images by strict surjective homomorphisms (Proposition 5.60), applied to the particular case of the reduction function: since $\langle \boldsymbol{A}, \mathcal{F}i_{\mathcal{L}}\boldsymbol{A}\rangle$ is always a basic full g-model of $\mathcal{L}$, $\langle \boldsymbol{A}^*, (\mathcal{F}i_{\mathcal{L}}\boldsymbol{A})^*\rangle$ must also be a basic full g-model, and hence its closure system has to contain all the $\mathcal{L}$-filters of the reduced algebra. 2 is a particular case of Theorem 5.61.2. Finally, 3 and 4 are consequences of 2. ⊠

Notice that in general the class M^* of the reductions of the g-matrices in a class M need not be closed under isomorphisms. However, there are situations in which this closure is desirable. If, in parallel to Definition 4.37, we define, for each logic $\mathcal{L}$, the class

$$\mathsf{GMod}^*\mathcal{L} := \text{the class of all reduced g-models of } \mathcal{L},$$

then we obtain:

PROPOSITION 5.65
$\mathsf{GMod}^*\mathcal{L} := \mathbb{I}\big((\mathsf{GMod}\,\mathcal{L})^*\big)$. *That is, a g-matrix is a reduced g-model of a logic if and only if it is isomorphic to the reduction of some g-model.*

PROOF: If $\mathcal{M} \in \mathsf{GMod}\,\mathcal{L}$ and is reduced, then by Proposition 5.56.3 $\mathcal{M} \cong \mathcal{M}^*$, and hence $\mathcal{M} \in \mathbb{I}\big((\mathsf{GMod}\,\mathcal{L})^*\big)$. Conversely, if $\mathcal{M} \cong \mathcal{N}^*$ for some $\mathcal{N} \in \mathsf{GMod}\,\mathcal{L}$, then $\mathcal{N}^*$ is reduced (by Proposition 5.56.1) and a g-model of $\mathcal{L}$ (by Proposition 5.64.3); since the two properties are preserved by $\mathbb{I}$, $\mathcal{M} \in \mathsf{GMod}^*\mathcal{L}$. ⊠

From Theorem 5.10 and Proposition 5.64.4 it readily follows:

PROPOSITION 5.66 (COMPLETENESS)
Every logic $\mathcal{L}$ is complete with respect to any class M of g-matrices such that $\mathcal{L}^* \in \mathsf{M}$ and $\mathsf{M} \subseteq \mathsf{GMod}^*\mathcal{L}$, where $\mathcal{L}^* = \langle \boldsymbol{Fm}^*, (\mathcal{T}h\mathcal{L})^*\rangle = \langle \boldsymbol{Fm}/\widetilde{\Omega}\mathcal{L}, (\mathcal{T}h\mathcal{L})/\widetilde{\Omega}\mathcal{L}\rangle$. In particular, $\mathcal{L}$ is complete with respect to the class $\mathsf{GMod}^*\mathcal{L}$. ⊠

The $\mathcal{L}$-algebras and the intrinsic variety of a logic

Recall from Definition 4.37 that in the framework of matrix semantics the class of algebras naturally associated with a logic $\mathcal{L}$ is

$$\mathsf{Alg}^*\mathcal{L} := \mathsf{Alg}\big(\mathsf{Mod}^*\mathcal{L}\big) = \big\{\boldsymbol{A} : \boldsymbol{\Omega}^{\boldsymbol{A}}F = \mathrm{Id}_A \text{ for some } F \in \mathcal{F}i_{\mathcal{L}}\boldsymbol{A}\big\}.$$

The semantics of g-matrices suggests at least two more classes of algebras that are naturally associated with a logic:

DEFINITION 5.67
*Let $\mathcal{L}$ be a logic. The class of **$\mathcal{L}$-algebras** is the class of the algebraic reducts of the reduced g-models of $\mathcal{L}$:*

$$\mathsf{Alg}\,\mathcal{L} := \mathsf{Alg}\big(\mathsf{GMod}^*\mathcal{L}\big)$$

*The **intrinsic variety** of $\mathcal{L}$ is the variety generated by the algebra $\boldsymbol{Fm}/\widetilde{\Omega}\mathcal{L}$:*

$$\mathbb{V}\mathcal{L} := \mathbb{V}\{\boldsymbol{Fm}/\widetilde{\Omega}\mathcal{L}\}$$

The coexistence of the notations $\mathsf{Alg}\mathcal{L}$ and $\mathsf{Alg}^*\mathcal{L}$ *does not imply* that the class $\mathsf{Alg}^*\mathcal{L}$ is obtained from the class $\mathsf{Alg}\mathcal{L}$ by some "reduction" process.† The name "$\mathcal{L}$-algebras" and the notation $\mathsf{Alg}^*\mathcal{L}$ had been extensively used in the literature much before $\mathsf{Alg}\mathcal{L}$ was introduced [107, 1996], although, before the 1980s, they were used only for implicative logics. The original notation now just recalls that $\mathsf{Alg}^*\mathcal{L} = \mathsf{Alg}(\mathsf{Mod}^*\mathcal{L})$. Anyway, there is indeed a very close relation between the two classes, which is found soon (Theorem 5.70). Corollary 5.73 supports the general usage of the name and the notation, because for algebraizable logics (thus, for implicative logics) the two classes of algebras coincide; in fact, they coincide for the even larger class of protoalgebraic logics (Corollary 6.15).

According to the definition, $\boldsymbol{A} \in \mathsf{Alg}\mathcal{L}$ if and only if there is some closure system $\mathscr{C} \subseteq \mathcal{F}i_{\mathcal{L}}\boldsymbol{A}$ such that the g-matrix $\langle \boldsymbol{A}, \mathscr{C} \rangle$ is reduced. On a given algebra there can be several closure systems with these properties; however, if there is one, then there is a special, uniquely determined one:

PROPOSITION 5.68
$\boldsymbol{A} \in \mathsf{Alg}\mathcal{L}$ *if and only if* $\langle \boldsymbol{A}, \mathcal{F}i_{\mathcal{L}}\boldsymbol{A} \rangle$ *is reduced. That is,* $\mathsf{Alg}\mathcal{L}$ *is the class of the algebraic reducts of the reduced basic full g-models of* $\mathcal{L}$.

PROOF: If $\mathscr{C} \subseteq \mathcal{F}i_{\mathcal{L}}\boldsymbol{A}$, then $\widetilde{\boldsymbol{\Omega}}^{\boldsymbol{A}}\mathcal{F}i_{\mathcal{L}}\boldsymbol{A} \subseteq \widetilde{\boldsymbol{\Omega}}^{\boldsymbol{A}}\mathscr{C}$, therefore if $\boldsymbol{A} \in \mathsf{Alg}\mathcal{L}$ with $\langle \boldsymbol{A}, \mathscr{C} \rangle$ reduced, then also $\langle \boldsymbol{A}, \mathcal{F}i_{\mathcal{L}}\boldsymbol{A} \rangle$ is reduced. There is nothing to prove for the converse implication. ⊠

Notice that, since $(\mathcal{F}i_{\mathcal{L}}\boldsymbol{A})^* = \mathcal{F}i_{\mathcal{L}}(\boldsymbol{A}^*)$, the class of g-matrices involved in this result is, up to isomorphism, the one obtained by reduction from the class of all basic full g-models of $\mathcal{L}$. Hence, from Proposition 5.64.3 the following follows.

COROLLARY 5.69 (COMPLETENESS)
Every logic $\mathcal{L}$ *is complete with respect to the class* $\{\langle \boldsymbol{A}, \mathcal{F}i_{\mathcal{L}}\boldsymbol{A} \rangle : \boldsymbol{A} \in \mathsf{Alg}\mathcal{L}\}$. ⊠

As we shall see, this class of g-matrices, which are the basic full g-models of algebras in $\mathsf{Alg}\mathcal{L}$, and the class of algebras $\mathsf{Alg}\mathcal{L}$ itself, are very significant in the semantics of logics.

Now, the main relation between the two classes of algebras so far associated with a logic can be established:

THEOREM 5.70
For every logic $\mathcal{L}$*, the* $\mathcal{L}$*-algebras are the algebras isomorphic to a subdirect product of algebras in* $\mathsf{Alg}^*\mathcal{L}$:

$$\mathsf{Alg}\mathcal{L} = \mathbb{I}\mathbb{P}_{\mathrm{SD}}\mathsf{Alg}^*\mathcal{L}$$

As a consequence, $\mathsf{Alg}^*\mathcal{L} \subseteq \mathsf{Alg}\mathcal{L}$.

PROOF: ($\subseteq$) If $\boldsymbol{A} \in \mathsf{Alg}\mathcal{L}$, $\mathrm{Id}_A = \widetilde{\boldsymbol{\Omega}}^{\boldsymbol{A}}\mathcal{F}i_{\mathcal{L}}\boldsymbol{A} = \bigcap\{\boldsymbol{\Omega}^{\boldsymbol{A}}F : F \in \mathcal{F}i_{\mathcal{L}}\boldsymbol{A}\}$. It is well known (see [51, Lemma II.8.2], for instance) that this implies that $\boldsymbol{A}$ is isomorphic to a subdirect product of the family $\{\boldsymbol{A}/\boldsymbol{\Omega}^{\boldsymbol{A}}F : F \in \mathcal{F}i_{\mathcal{L}}\boldsymbol{A}\}$. Since for each $F \in \mathcal{F}i_{\mathcal{L}}\boldsymbol{A}$, $\boldsymbol{A}/\boldsymbol{\Omega}^{\boldsymbol{A}}F \in \mathsf{Alg}^*\mathcal{L}$, it follows that $\boldsymbol{A} \in \mathbb{I}\mathbb{P}_{\mathrm{SD}}\mathsf{Alg}^*\mathcal{L}$.

† In contrast to the relation between $\mathsf{Mod}\mathcal{L}$ and $\mathsf{Mod}^*\mathcal{L}$, between $\mathsf{GMod}\mathcal{L}$ and $\mathsf{GMod}^*\mathcal{L}$, etc.

($\supseteq$) Let $\boldsymbol{A}$ be isomorphic to a subdirect product of a family $\{\boldsymbol{A}_i : i \in I\} \subseteq \mathsf{Alg}^*\mathcal{L}$. This means that for each $i \in I$ there is some $F_i \in \mathcal{F}i_{\mathcal{L}}\boldsymbol{A}_i$ such that $\boldsymbol{\Omega}^{\boldsymbol{A}_i} F_i = \mathrm{Id}_{A_i}$, and there is a surjective $p_i \colon \boldsymbol{A} \to \boldsymbol{A}_i$, namely the composition of the subdirect embedding with the projection onto the i-th component. Then, as a matrix homomorphism, $p_i \colon \langle \boldsymbol{A}, p_i^{-1}F_i \rangle \to \langle \boldsymbol{A}_i, F_i \rangle$ is strict and surjective, therefore by Proposition 4.25 $\boldsymbol{\Omega}^{\boldsymbol{A}} p_i^{-1} F_i = p_i^{-1} \boldsymbol{\Omega}^{\boldsymbol{A}_i} F_i = p_i^{-1} \mathrm{Id}_{A_i} = \ker p_i$. But $p_i^{-1}F_i \in \mathcal{F}i_{\mathcal{L}}\boldsymbol{A}$ for each $i \in I$, therefore $\widetilde{\boldsymbol{\Omega}}^{\boldsymbol{A}} \mathcal{F}i_{\mathcal{L}}\boldsymbol{A} \subseteq \bigcap_{i \in I} \boldsymbol{\Omega}^{\boldsymbol{A}} p_i^{-1} F_i = \bigcap_{i \in I} \ker p_i = \mathrm{Id}_A$. This shows that $\widetilde{\boldsymbol{\Omega}}^{\boldsymbol{A}} \mathcal{F}i_{\mathcal{L}}\boldsymbol{A} = \mathrm{Id}_A$, that is, that the g-matrix $\langle \boldsymbol{A}, \mathcal{F}i_{\mathcal{L}}\boldsymbol{A} \rangle$ is reduced. Thus, $\boldsymbol{A} \in \mathsf{Alg}\mathcal{L}$. ⊠

COROLLARY 5.71

1. $\mathsf{Alg}\mathcal{L}$ *is always closed under* $\mathbb{P}_{\mathrm{SD}}, \mathbb{P}, \mathbb{I}$.
2. $\mathsf{Alg}^*\mathcal{L} = \mathsf{Alg}\mathcal{L}$ *if and only if* $\mathsf{Alg}^*\mathcal{L}$ *is closed under* $\mathbb{P}_{\mathrm{SD}}$.
3. *The classes* $\mathsf{Alg}^*\mathcal{L}$ *and* $\mathsf{Alg}\mathcal{L}$ *generate the same variety and the same quasivariety.*
4. *If* $\mathsf{Alg}^*\mathcal{L}$ *is a quasivariety (resp., a variety), then* $\mathsf{Alg}^*\mathcal{L} = \mathsf{Alg}\mathcal{L}$*, and therefore* $\mathsf{Alg}\mathcal{L}$ *is also a quasivariety (resp., a variety).*
5. $\mathsf{Alg}\mathcal{L}$ *contains all trivial algebras.*

PROOF: 1 and 2 are direct consequences of Theorem 5.70 and the facts that the operator $\mathbb{IP}_{\mathrm{SD}}$ is idempotent and that $\mathbb{P}$ is a particular case of it. The proof of 3 needs some universal algebra. Since $\mathbb{IP}_{\mathrm{SD}} \leqslant \mathbb{Q}$,[†] from Theorem 5.70 it follows that $\mathbb{Q}\mathsf{Alg}\mathcal{L} = \mathbb{QIP}_{\mathrm{SD}}\mathsf{Alg}^*\mathcal{L} \subseteq \mathbb{QQ}\mathsf{Alg}^*\mathcal{L} = \mathbb{Q}\mathsf{Alg}^*\mathcal{L}$. On the other hand, from $\mathsf{Alg}^*\mathcal{L} \subseteq \mathsf{Alg}\mathcal{L}$ it follows that $\mathbb{Q}\mathsf{Alg}^*\mathcal{L} \subseteq \mathbb{Q}\mathsf{Alg}\mathcal{L}$. All this proves that $\mathbb{Q}\mathsf{Alg}\mathcal{L} = \mathbb{Q}\mathsf{Alg}^*\mathcal{L}$. From this, since $\mathbb{V} = \mathbb{HQ}$, it follows that $\mathbb{V}\mathsf{Alg}\mathcal{L} = \mathbb{V}\mathsf{Alg}^*\mathcal{L}$. 4 is a direct consequence of 3. Finally, 5 follows from the fact that if $\boldsymbol{A}_t$ is a trivial algebra, then $\mathcal{F}i_{\mathcal{L}}\boldsymbol{A}_t$ is either $\{A_t\}$, when $\mathcal{L}$ has theorems, or $\{\varnothing, A_t\}$, when it does not; since $\widetilde{\boldsymbol{\Omega}}^{\boldsymbol{A}_t}\{A_t\} = \widetilde{\boldsymbol{\Omega}}^{\boldsymbol{A}_t}\{\varnothing, A_t\} = A_t \times A_t = \mathrm{Id}_{A_t}$, the g-matrix $\langle \boldsymbol{A}_t, \mathcal{F}i_{\mathcal{L}}\boldsymbol{A}_t \rangle$ is reduced, therefore $\boldsymbol{A}_t \in \mathsf{Alg}\mathcal{L}$. ⊠

Point 4 gives a first explanation of why the two classes of algebras coincide so often in examples: the class $\mathsf{Alg}^*\mathcal{L}$ is a quasivariety for a very large number of logics. This happens, for instance, when $\mathcal{L}$ is BP-algebraizable, which includes all finitary implicative logics. But this coincidence also holds in more cases. To give a rather generic example, recall that if $F \subseteq A$ and $\Delta(x,y) \subseteq Fm$, Definition 4.52 sets $R^{\boldsymbol{A}}(F, \Delta) := \{\langle a,b \rangle \in A \times A : \Delta^{\boldsymbol{A}}(a,b) \subseteq F\}$; then:

LEMMA 5.72

Let $\langle \boldsymbol{A}, \mathscr{C} \rangle$ *be a g-matrix such that the Leibniz congruences of all the* $F \in \mathscr{C}$ *are uniformly analytically definable, in the sense that there is* $\Delta(x,y) \subseteq Fm$ *such that for each* $F \in \mathscr{C}$, $\boldsymbol{\Omega}^{\boldsymbol{A}} F = R^{\boldsymbol{A}}(F, \Delta)$. *Then* $\widetilde{\boldsymbol{\Omega}}^{\boldsymbol{A}} \mathscr{C} = \boldsymbol{\Omega}^{\boldsymbol{A}} C\varnothing$. *If the assumption holds for all basic full g-models of a logic* $\mathcal{L}$, *then* $\mathsf{Alg}\mathcal{L} = \mathsf{Alg}^*\mathcal{L}$.

PROOF: Left to you as Exercise 5.47. ⊠

[†] This means that $\mathbb{IP}_{\mathrm{SD}}\mathsf{K} \subseteq \mathbb{Q}\mathsf{K}$ for any class of algebras K.

One large class of logics in which the assumption of this lemma holds for all basic full g-models, and indeed with the same set Δ in all of them, is that of algebraizable logics (Corollary 4.55). Adding Proposition 4.57 to round up the result, we get:

COROLLARY 5.73
If $\mathcal{L}$ *is an algebraizable logic, with equivalent algebraic semantics the class of algebras* K*, then* $\mathsf{K} = \mathsf{Alg}\mathcal{L} = \mathsf{Alg}^*\mathcal{L}$. ⊠

The coincidence of the classes $\mathsf{Alg}\mathcal{L}$ and $\mathsf{Alg}^*\mathcal{L}$ does *not* characterize algebraizable logics: as a matter of fact, in Corollary 6.15 we find a much larger class of logics, the protoalgebraic ones, for which the two classes also coincide (and still this class is not characterized by this coincidence). The converse to the implication in 5.71.4 does not hold either, and not even among algebraizable logics, because there are algebraizable logics with an equivalent equational consequence that is not finitary; we have already met two of them: the infinitary infinitely-valued logic of Łukasiewicz $\mathcal{Ł}_\infty$, which is finitely algebraizable (Example 3.41), and Herrmann's $\mathcal{LJ}$, which is finitary but not finitely algebraizable (Example 3.42). See also Exercise 5.48

Now let us examine the intrinsic variety of the logic, $\mathbb{V}\mathcal{L} = \mathbb{V}\{\boldsymbol{Fm}/\widetilde{\Omega}\mathcal{L}\}$. Its significance as an algebra-based semantics for $\mathcal{L}$ is in general weak, because no general theory asserts that the algebraic counterpart of a logic should be a variety. However, *if one insists on having a variety associated with a logic*, then $\mathbb{V}\mathcal{L}$ is the natural choice; this becomes clearer after the next results. The characterization in Corollary 5.33 immediately implies:

LEMMA 5.74
$\widetilde{\Omega}\mathcal{L}$ *is a fully invariant congruence; that is, if* $\sigma \in \mathrm{End}\boldsymbol{Fm}$ *and* $\alpha \equiv \beta \ (\widetilde{\Omega}\mathcal{L})$*, then* $\sigma\alpha \equiv \sigma\beta \ (\widetilde{\Omega}\mathcal{L})$. ⊠

PROPOSITION 5.75
1. $\mathbb{V}\mathcal{L} \vDash \alpha \approx \beta$ *if and only if* $\alpha \equiv \beta \ (\widetilde{\Omega}\mathcal{L})$.
2. $\boldsymbol{Fm}/\widetilde{\Omega}\mathcal{L}$ *is the free algebra in the variety* $\mathbb{V}\mathcal{L}$ *with denumerably many generators; one set of generators is* $\{x/\widetilde{\Omega}\mathcal{L} : x \in V\}$*, the set of equivalence classes of the variables.*
3. $\mathcal{L}$ *is selfextensional if and only if for all* $\alpha, \beta \in Fm$, $\mathbb{V}\mathcal{L} \vDash \alpha \approx \beta$ *if and only if* $\alpha \dashv\vdash_{\mathcal{L}} \beta$.
4. $\mathcal{L}$ *is selfextensional if and only it there is a variety* V *such that for all* $\alpha, \beta \in Fm$, $\alpha \dashv\vdash_{\mathcal{L}} \beta$ *if and only if* $\mathsf{V} \vDash \alpha \approx \beta$*. Then exactly* $\mathsf{V} = \mathbb{V}\mathcal{L}$.

PROOF: 1 and 2 follow from the definition of $\mathbb{V}\mathcal{L}$ and Lemma 1.68, given the result in Lemma 5.74. If $\mathcal{L}$ is selfextensional, then by Lemma 5.31.4, $\widetilde{\Omega}\mathcal{L} = \dashv\vdash_{\mathcal{L}}$, therefore the direct implication in 3 follows from 1. To show the backwards implication, assume the stated equivalence; then clearly $\dashv\vdash_{\mathcal{L}}$ is a congruence of $\boldsymbol{Fm}$, therefore $\mathcal{L}$ is selfextensional. This proves 3. The proof of 4 is similar, and left to you as Exercise 5.50. ⊠

If $\widetilde{\Omega}\mathcal{L}$ is understood as the natural relation of "logical equivalence" relative to $\mathcal{L}$, as discussed on page 256, then point 1 reinforces the importance of the intrinsic variety for the logic $\mathcal{L}$, as it is characterized by the equations that identify exactly the pairs of formulas that are equivalent in this sense.

Point 4 is a useful technique to determine the class $\mathbb{V}\mathcal{L}$ for selfextensional logics (and to prove at the same time that they are so); Exercise 5.50 proposes you some applications. Corollary 5.77 and Proposition 5.79 contain other practical criteria.

The variety $\mathbb{V}\mathcal{L}$ is even more intrinsic to $\mathcal{L}$ than its definition seems to indicate: as the next result tells us, it also arises from the two main classes of algebras intrinsically associated with $\mathcal{L}$.

THEOREM 5.76
For any logic $\mathcal{L}$, $\mathbb{V}\mathsf{Alg}^*\mathcal{L} = \mathbb{V}\mathsf{Alg}\mathcal{L} = \mathbb{V}\mathcal{L}$.

PROOF: By Corollary 5.71.3, $\mathbb{V}\mathsf{Alg}^*\mathcal{L} = \mathbb{V}\mathsf{Alg}\mathcal{L}$. Since $\boldsymbol{Fm}/\widetilde{\Omega}\mathcal{L} \in \mathsf{Alg}\mathcal{L}$, $\mathbb{V}\mathcal{L} \subseteq \mathbb{V}\mathsf{Alg}\mathcal{L}$. So we need only prove that $\mathsf{Alg}\mathcal{L} \subseteq \mathbb{V}\mathcal{L}$. Assume that $\alpha \approx \beta$ holds in $\mathbb{V}\mathcal{L}$. By Proposition 5.75.1 and Corollary 5.33, this means that for all $\delta(x,\vec{z}) \in \boldsymbol{Fm}$, $\delta(\alpha,\vec{z}) \dashv\vdash_{\mathcal{L}} \delta(\beta,\vec{z})$. Since the g-matrix $\langle \boldsymbol{A}, \mathcal{F}i_{\mathcal{L}}\boldsymbol{A}\rangle$ is always a g-model of $\mathcal{L}$, we know that for all $\boldsymbol{A}$, all $h \in \mathrm{Hom}(\boldsymbol{Fm}, \boldsymbol{A})$ and all $\delta(x,\vec{z}) \in \boldsymbol{Fm}$, $Fg^{\boldsymbol{A}}_{\mathcal{L}}\{h\delta(\alpha,\vec{z})\} = Fg^{\boldsymbol{A}}_{\mathcal{L}}\{h\delta(\beta,\vec{z})\}$, that is, $Fg^{\boldsymbol{A}}_{\mathcal{L}}\{\delta^{\boldsymbol{A}}(h\alpha,h\vec{z})\} = Fg^{\boldsymbol{A}}_{\mathcal{L}}\{\delta^{\boldsymbol{A}}(h\beta,h\vec{z})\}$. Since we are taking *all* the homomorphisms (and if necessary, by structurality we can replace the parameters $\vec{z}$ by completely new ones), this is the same as to say that $Fg^{\boldsymbol{A}}_{\mathcal{L}}\{\delta^{\boldsymbol{A}}(h\alpha,\vec{c})\} = Fg^{\boldsymbol{A}}_{\mathcal{L}}\{\delta^{\boldsymbol{A}}(h\beta,\vec{c})\}$ for all $\vec{c} \in \vec{A}$ and all $\delta(x,\vec{z}) \in \boldsymbol{Fm}$; but by Theorem 5.32 this means that $h\alpha \equiv h\beta \ (\widetilde{\Omega}^{\boldsymbol{A}}\mathcal{F}i_{\mathcal{L}}\boldsymbol{A})$. Now, if $\boldsymbol{A} \in \mathsf{Alg}\mathcal{L}$, $\widetilde{\Omega}^{\boldsymbol{A}}\mathcal{F}i_{\mathcal{L}}\boldsymbol{A} = \mathrm{Id}_A$, and thus $h\alpha = h\beta$. Since the evaluation h is arbitrary, this shows that $\alpha \approx \beta$ holds in $\boldsymbol{A}$, and since this is done for all $\boldsymbol{A} \in \mathsf{Alg}\mathcal{L}$, the equation holds in $\mathsf{Alg}\mathcal{L}$. This proves that $\mathbb{V}\mathsf{Alg}\mathcal{L} \subseteq \mathbb{V}\mathcal{L}$ and thus completes the proof of the theorem. ⊠

As a consequence, the relative positions of the three classes of algebras are determined, with some more facts:

COROLLARY 5.77
1. $\mathsf{Alg}^*\mathcal{L} \subseteq \mathsf{Alg}\mathcal{L} \subseteq \mathbb{V}\mathcal{L}$.
2. $\boldsymbol{Fm}/\widetilde{\Omega}\mathcal{L}$ *is the free algebra in the classes* $\mathsf{Alg}\mathcal{L}$ *and* $\mathsf{Alg}^*\mathcal{L}$ *with denumerably many generators.*
3. *The class* $\mathsf{Alg}\mathcal{L}$ *is a variety if and only if* $\mathsf{Alg}\mathcal{L} = \mathbb{V}\mathcal{L}$.
4. *The class* $\mathsf{Alg}^*\mathcal{L}$ *is a variety if and only if* $\mathsf{Alg}^*\mathcal{L} = \mathsf{Alg}\mathcal{L} = \mathbb{V}\mathcal{L}$. ⊠

This solves the determination of the intrinsic variety of logics for which either $\mathsf{Alg}^*\mathcal{L}$ or $\mathsf{Alg}\mathcal{L}$ has been determined and found to be a variety: then this variety is itself the intrinsic variety of the logic. This applies to a large number of the logics reviewed in the book so far, among them all the implicative logics $\mathcal{L}$ such that $\mathsf{Alg}^*\mathcal{L}$ is a variety ($\mathcal{C}\ell, \mathcal{I}\ell, \mathcal{I}\ell_{\rightarrow}$, S^g for any normal modal system S, $Ł_{\infty f}$, etc.).

Theorem 5.76 may also be useful in cases where neither of the two classes is a variety, provided the variety they generate can be determined.

Another consequence of the characterization of $\mathbb{V}\mathcal{L}$ in Proposition 5.75 is the following technical fact, which is useful in a number of situations.

LEMMA 5.78

Let $\mathcal{L}$ be the logic defined from (or complete with respect to) a class M of matrices or of g-matrices, and put $\mathsf{K} := \mathsf{Alg}\,\mathsf{M}$.

1. *For all $\alpha, \beta \in Fm$, if $\mathsf{K} \vDash \alpha \approx \beta$, then $\alpha \dashv\vdash_{\mathcal{L}} \beta$.*
2. *For all $\alpha, \beta \in Fm$, if $\mathsf{K} \vDash \alpha \approx \beta$, then $\alpha \equiv \beta\ (\widetilde{\boldsymbol{\Omega}}\mathcal{L})$.*
3. *$\mathbb{V}\mathcal{L} \subseteq \mathbb{V}\mathsf{K}$, and hence also $\mathsf{Alg}^*\mathcal{L} \subseteq \mathbb{V}\mathsf{K}$ and $\mathsf{Alg}\,\mathcal{L} \subseteq \mathbb{V}\mathsf{K}$.*

PROOF: 1. To deal with the matrix case, let $\langle \boldsymbol{A}, F \rangle \in \mathsf{M}$. Then $\boldsymbol{A} \in \mathsf{K}$, and by assumption $h\alpha = h\beta$ for all $h \in \mathrm{Hom}(\boldsymbol{Fm}, \boldsymbol{A})$, therefore $h\alpha \in F$ if and only if $h\beta \in F$. Since $\mathcal{L}$ is complete with respect to M, this implies that $\alpha \dashv\vdash_{\mathcal{L}} \beta$. The g-matrix case is similar. 2 follows from 1, because $\widetilde{\boldsymbol{\Omega}}\mathcal{L}$ is the largest congruence of $\boldsymbol{Fm}$ contained in $\boldsymbol{\Lambda}\mathcal{L}$, and the relation $\{\langle \alpha, \beta \rangle \in Fm \times Fm : \mathsf{K} \vDash \alpha \approx \beta\}$, which is certainly a congruence of $\boldsymbol{Fm}$, is asserted in 1 to be contained in $\boldsymbol{\Lambda}\mathcal{L}$. Finally, 3 follows from 2, Proposition 5.75.1 and Corollary 5.77. ⊠

In practice this means that if a logic is defined from a class of (g-)matrices, then the equations that hold in the algebraic reducts of these (g-)matrices hold in the (three) algebraic counterparts of the logic. Exercise 5.52 asks you to prove the following strengthening, which further confirms the *intrinsic* character of $\mathbb{V}\mathcal{L}$ for the logic $\mathcal{L}$.

PROPOSITION 5.79

If $\mathcal{L}$ is the logic defined by (or complete with respect to) a class M of reduced matrices or of reduced g-matrices, and $\mathsf{K} := \mathsf{Alg}\,\mathsf{M}$, then $\mathbb{V}\mathcal{L} = \mathbb{V}\mathsf{K}$. ⊠

This quickly solves the determination of $\mathbb{V}\mathcal{L}$ when $\mathcal{L}$ is defined by a single reduced matrix, or by a small family of reduced matrices, which is a common situation in real examples. Another consequence of Theorem 5.70 gives the following procedure, also of very practical application.

PROPOSITION 5.80

If all subdirectly irreducible algebras in $\mathbb{V}\mathcal{L}$ belong to $\mathsf{Alg}^*\mathcal{L}$, then $\mathsf{Alg}\,\mathcal{L} = \mathbb{V}\mathcal{L}$.

PROOF: It is well known that every algebra in a variety is isomorphic to a subdirect product of subdirectly irreducible algebras in the variety [51, Theorem II.9.6]. The assumption implies that $\mathbb{V}\mathcal{L} \subseteq \mathbb{I}\mathbb{P}_{\mathrm{SD}}\mathsf{Alg}^*\mathcal{L} = \mathsf{Alg}\,\mathcal{L}$, and since $\mathsf{Alg}\,\mathcal{L} \subseteq \mathbb{V}\mathcal{L}$ always holds, this means that $\mathsf{Alg}\,\mathcal{L} = \mathbb{V}\mathcal{L}$. ⊠

This is used in Example 5.83 below, and in Exercise 5.53.

EXAMPLES.

There are examples of all the possibilities for the relative positions of the three classes, compatible with the inclusions in Corollary 5.77.1:

- $\mathsf{Alg}^*\mathcal{L} = \mathsf{Alg}\mathcal{L} = \mathbb{V}\mathcal{L}$: By Theorem 5.76, this happens whenever $\mathsf{Alg}^*\mathcal{L}$ is a variety, which is the case of, for instance, $\mathcal{C}\ell$ (Boolean algebras), $\mathcal{I}\ell$ (Heyting algebras), $\mathcal{I}\ell_{\rightarrow}$ (Hilbert algebras), and many more (modal logics, many-valued logics, etc). As a matter of fact, a very large number (one is tempted to say: almost all) of the logics investigated in the literature have varieties as their algebraic counterparts. Most fall under this case, and some (less) under the next one.
- $\mathsf{Alg}^*\mathcal{L} \subsetneq \mathsf{Alg}\mathcal{L} = \mathbb{V}\mathcal{L}$: This is the case of Belnap-Dunn's logic $\mathcal{B}$ (Example 5.82), of $\mathcal{C}\ell_{\wedge\vee}$ (Example 5.93), and of their expansions with constants (Exercises 5.63 and 5.64).
- $\mathsf{Alg}^*\mathcal{L} = \mathsf{Alg}\mathcal{L} \subsetneq \mathbb{V}\mathcal{L}$: By Proposition 4.57, this happens for every algebraizable logic $\mathcal{L}$ such that $\mathsf{Alg}^*\mathcal{L}$ is *not* a variety; examples are the logic $\mathcal{BCK}$, where $\mathsf{Alg}^*\mathcal{BCK}$ is the class of BCK algebras, a well-known proper quasivariety, and the weakest implicative logic $\mathcal{I}mp$, where $\mathsf{Alg}^*\mathcal{I}mp = \mathsf{IA}$, the class of implicative algebras, which is a proper quasivariety (Exercise 2.14).
- $\mathsf{Alg}^*\mathcal{L} \subsetneq \mathsf{Alg}\mathcal{L} \subsetneq \mathbb{V}\mathcal{L}$: The examples of this situation are rare.† Bou [45] found a couple of examples in the field of subintuitionistic logics (Example 6.16.8). Another, highly ad hoc example was independently constructed by Babyonyshev (Example 7.6.4). Finally, a more natural one is the paraconsistent weak Kleene logic $\mathcal{PWK}$, a three-valued logic (Example 6.122.12).

Discussion.

I contend that the definition of $\mathsf{Alg}\mathcal{L}$ provides *the right general notion of* ***the true algebraic counterpart*** *of a logic* $\mathcal{L}$, *and certainly a better one than* $\mathsf{Alg}^*\mathcal{L}$. This is a kind of "Church-Turing Thesis" for abstract algebraic logic: as it is not a technical result, it cannot be *proved*, but only *supported*, both by empirical evidence and by the mathematical results connected with it. Here is some of the evidence put forward:

- The class $\mathsf{Alg}\mathcal{L}$ coincides with $\mathsf{Alg}^*\mathcal{L}$ in cases in which there is an alternative and strong general theory establishing $\mathsf{Alg}^*\mathcal{L}$ as the right algebraic counterpart of the logic $\mathcal{L}$. The situation in which this can be asserted with most conviction is when $\mathcal{L}$ is implicative. However, they also coincide in the larger classes of algebraizable and of protoalgebraic logics, in which the connections between filters and congruences are good enough to produce a nice mathematical theory, as shown in Section 6.2.
- The class $\mathsf{Alg}\mathcal{L}$ has been originally obtained in the approach through g-models and the Tarski congruence, but it is also the class of algebras obtained by the alternative approach that uses ordinary matrices and the Suszko congruence, as is proved in Proposition 6.97. Although this approach is not fully developed in this book, it is introduced in Section 5.3, and a few further results appear in Section 6.4.

† Their existence was even posed as an open problem in the first edition of [107].

- In some of the classes of logics of the Frege hierarchy (Chapter 7), there are general results determining $\mathsf{Alg}\mathcal{L}$, proving that it is always a variety (and therefore coincides with $\mathbb{V}\mathcal{L}$), and showing a smooth integration with an approach to these logics through Gentzen calculi; by contrast, $\mathsf{Alg}^*\mathcal{L}$ seems to play no special role there.
- In many particular non-protoalgebraic logics, the determination of $\mathsf{Alg}\mathcal{L}$ yields better results than that of $\mathsf{Alg}^*\mathcal{L}$ regarding other, rather empirical associations between $\mathcal{L}$ and a class of algebras, mainly those coming from the very definition or motivation of the logic. For instance, in Example 5.93 we see that $\mathsf{Alg}\mathcal{C}\ell_{\wedge\vee}$ is the class of distributive lattices, while we know that $\mathsf{Alg}^*\mathcal{C}\ell_{\wedge\vee}$ is a rather strange class. Something similar happens with Belnap-Dunn's logic $\mathcal{B}$, where $\mathsf{Alg}\mathcal{B}$ is the variety of De Morgan lattices while $\mathsf{Alg}^*\mathcal{B}$ is again a weird subclass (Example 5.82 below); or with the implication-less fragment of $\mathcal{I}\ell$ and the variety of pseudo-complemented distributive lattices (Example 6.16.6).
- The semantics of g-matrices, which is the appropriate context for the definition of $\mathsf{Alg}\mathcal{L}$, has produced a rich, deep and interesting general theory, with new and relevant notions and results, as witnessed by the material in this chapter and in the literature.

WARNING.

However important the determination of $\mathsf{Alg}\mathcal{L}$ (or of $\mathsf{Alg}^*\mathcal{L}$, or of $\mathbb{V}\mathcal{L}$) may seem for the algebraic study of a logic $\mathcal{L}$, it is also important to realize that even in many reasonably well-behaved cases (for instance, the equivalential but non-algebraizable logics studied in Chapter 6) the class of algebras alone may not contain enough information to characterize the logic. Bearing witness to this are the examples of the pairs of logics with the same class of algebras studied in Section 7.2, such as the local and the global logics $\mathcal{S}^{\ell}$ and $\mathcal{S}^{g}$ associated with each normal modal system S, or the semilattice-based logics with an algebraizable assertional companion: they have the same $\mathsf{Alg}^*\mathcal{L}$, and hence the same $\mathsf{Alg}\mathcal{L}$ and the same $\mathbb{V}\mathcal{L}$, but different $\mathsf{Mod}^*\mathcal{L}$ and different $\mathsf{GMod}^*\mathcal{L}$.

In examples, the following obvious facts are often used (without notice):

PROPOSITION 5.81

If $\mathcal{L}_1, \mathcal{L}_2$ are logics and $\mathcal{L}_1 \leqslant \mathcal{L}_2$, then $\mathsf{Mod}\mathcal{L}_2 \subseteq \mathsf{Mod}\mathcal{L}_1$, $\mathsf{GMod}\mathcal{L}_2 \subseteq \mathsf{GMod}\mathcal{L}_1$ and $\mathsf{Alg}\mathcal{L}_2 \subseteq \mathsf{Alg}\mathcal{L}_1$. The same inclusions hold with a $*$ in the appropriate places. ⊠

EXAMPLES.

Several of the preceding results in conjunction solve the determination of the three classes of algebras associated with a logic for a number of cases. For algebraizable logics whose equivalent algebraic semantics is a variety, Corollary 5.77.4 tells us that this class coincides with the intrinsic variety. For instance, this shows that $\mathsf{Alg}\mathcal{I}\ell_{\rightarrow} = \mathbb{V}\mathcal{I}\ell_{\rightarrow} = \mathsf{HiA}$, because in Exercise 2.28 it is shown that $\mathsf{Alg}^*\mathcal{I}\ell_{\rightarrow} = \mathsf{HiA}$ and this class is a variety. For other logics, the issue can be more complicated, particularly when $\mathsf{Alg}^*\mathcal{L}$ is unknown or is known to be a strange class; the same result solves the question whenever $\mathsf{Alg}^*\mathcal{L}$ has been determined and is a variety, while Theorem 5.76 does it when it is not, provided that the variety it generates can be determined. We have seen several tools to determine $\mathbb{V}\mathcal{L}$ in well-behaved cases, but often $\mathsf{Alg}\mathcal{L}$ can be determined only by finding the (basic) full g-models. The case of $\mathcal{C}\ell_{\wedge\vee}$ is dealt with in Example 5.93, together with other points; but a direct, more algebraic way of finding $\mathsf{Alg}\mathcal{C}\ell_{\wedge\vee}$ and $\mathbb{V}\mathcal{C}\ell_{\wedge\vee}$ (indeed, of proving that $\mathsf{Alg}\mathcal{C}\ell_{\wedge\vee} = \mathbb{V}\mathcal{C}\ell_{\wedge\vee} = \mathsf{DL}$) is indicated in Exercises 5.50 and 5.53. Another example where $\mathsf{Alg}^*\mathcal{L}$ has been found to be a weird class is now analysed in detail.

EXAMPLE 5.82 (BELNAP-DUNN'S FOUR-VALUED LOGIC)
The logic $\mathcal{B}$ is introduced in Example 4.12 as defined (in several ways) from the algebra $\boldsymbol{M}_4$ described in Figure 5 on page 189, and is shown on page 240 to be the logic defined by the g-matrix $\langle \boldsymbol{M}_4, \mathcal{F}ilt_\wedge \boldsymbol{M}_4\rangle$. Here I just want to show how the reduced full basic g-models, and the reduced models, of $\mathcal{B}$ can be determined by direct work on the logic, starting from its definition, and as a by-product how the classes $\mathsf{Alg}^*\mathcal{B}$, $\mathsf{Alg}\mathcal{B}$ and $\mathbb{V}\mathcal{B}$ are obtained, without using some of the general results, either in this section or in Chapter 7 (which, in any case, cannot help to the determine $\mathsf{Alg}^*\mathcal{B}$). A detailed, systematic study of this logic with the tools of abstract algebraic logic is contained in [93]; there you can find the missing details.

A ***De Morgan lattice*** is an algebra $\boldsymbol{A} = \langle A, \wedge, \vee, \neg\rangle$ of type $\langle 2,2,1\rangle$ such that the reduct $\langle A, \wedge, \vee\rangle$ is a distributive lattice and the equations $x \approx \neg\neg x$ and $\neg(x \vee y) \approx \neg x \wedge \neg y$, $\neg(x \wedge y) \approx \neg x \vee \neg y$ hold in $\boldsymbol{A}$ (these are the *Law of Double Negation* and the *De Morgan Laws*; of course, given the former, it is enough to have one of the latter). Bounded De Morgan lattices, called ***De Morgan algebras***, are more popular† in the literature; the bounds are denoted by 0 and 1. The variety DM of De Morgan lattices is generated by the four-element $\boldsymbol{M}_4$. Notice that any chain (totally ordered set) with an order-reversing involution‡ is a De Morgan lattice, and it is a De Morgan algebra if and only if it is bounded. The subdirectly irreducible members of DM are, up to isomorphism, $\boldsymbol{M}_4$ itself and its subalgebras $\boldsymbol{M}_2$ and $\boldsymbol{M}_3$, with universes $\{\mathbf{f},\mathbf{t}\}$ and $\{\mathbf{f},\mathbf{n},\mathbf{t}\}$ respectively. Similar properties hold for De Morgan algebras, but one has to add at least one of the two constants to the type (for one of the bounds); see [19].

(B1) The logic $\mathcal{B}$ is originally defined using the order relation of $\boldsymbol{M}_4$; since it is actually defined by a finite generalized matrix, it is a finitary logic. As a consequence it is determined by points (B2) and (B3) below:

(B2) $\varphi_1,\dots,\varphi_n \vdash_{\mathcal{B}} \varphi \Leftrightarrow \boldsymbol{M}_4 \vDash \varphi_1 \wedge \cdots \wedge \varphi_n \preccurlyeq \varphi \Leftrightarrow \mathsf{DM} \vDash \varphi_1 \wedge \cdots \wedge \varphi_n \preccurlyeq \varphi$. Recall the notation $\preccurlyeq$ from page 7, and the observation made there on its equivalence with an equation, which justifies the last equivalence.

(B3) $\mathcal{B}$ has no theorems (because the maximum of $\boldsymbol{M}_4$ is not an algebraic constant). Hence, $\emptyset \in \mathcal{F}i_{\mathcal{B}}\boldsymbol{A}$ for any $\boldsymbol{A}$.

† De Morgan lattices were first studied by Moisil [174]. Naming the two classes after De Morgan is due to Monteiro [177]; the names "quasi-Boolean algebras" and "i-lattices" have also been used.
‡ A unary operation $\neg$ such that $a \leqslant b$ implies $\neg b \leqslant \neg a$, and $\neg\neg a = a$, for all a, b.

(B4) $\mathcal{B}$ is complete with respect to the class $\{\langle \boldsymbol{A}, \mathcal{F}ilt^{\circ}_{\wedge}\boldsymbol{A}\rangle : \boldsymbol{A} \in \mathsf{DM}\}$, where $\mathcal{F}ilt^{\circ}_{\wedge}\boldsymbol{A}$ is the closure system generated by the family $\mathcal{F}ilt_{\wedge}\boldsymbol{A}$ of all the lattice filters of $\boldsymbol{A}$; recall that $\mathcal{F}ilt^{\circ}_{\wedge}\boldsymbol{A}$ equals $\mathcal{F}ilt_{\wedge}\boldsymbol{A}$ when $\boldsymbol{A}$ has a maximum, and $\mathcal{F}ilt_{\wedge}\boldsymbol{A} \cup \{\varnothing\}$ otherwise (Exercise 1.42). The completeness for consequences from non-empty assumptions follows from (B2), due to the characterization of the lattice filter $Fg^{\boldsymbol{A}}_{\wedge}X$ generated by a (non-empty) set X as $Fg^{\boldsymbol{A}}_{\wedge}X = \{a \in A : a \geqslant a_1 \wedge \cdots \wedge a_n \text{ for some } a_i \in X, n \geqslant 1\}$ (see Exercise 1.53). For the empty case, it follows from (B3) and the fact that there are De Morgan lattices without a maximum (for instance, any chain with order type $\omega^* \oplus \omega$, such as the set of integers).

(B5) $\alpha \dashv\vdash_{\mathcal{B}} \beta \Leftrightarrow \boldsymbol{M}_4 \vDash \alpha \approx \beta \Leftrightarrow \mathsf{DM} \vDash \alpha \approx \beta$. This follows directly from (B2).

(B6) $\mathcal{B}$ is selfextensional and $\mathbb{V}\mathcal{B} = \mathsf{DM}$. The two facts follow from (B5), by Corollary 5.75.4.

(B7) $\mathsf{Alg}^*\mathcal{B} \subseteq \mathsf{Alg}\mathcal{B} \subseteq \mathsf{DM}$. This is the instantiation of Corollary 5.77.1 to (B6).

(B8) If $\boldsymbol{A} \in \mathsf{DM}$, then $\mathcal{F}i_{\mathcal{B}}\boldsymbol{A} = \mathcal{F}ilt_{\wedge}\boldsymbol{A} \cup \{\varnothing\}$. The inclusion ($\supseteq$) is a consequence of (B3) and (B4). In order to show the inclusion ($\subseteq$) we first check, using (B2), that $x \wedge y \vdash_{\mathcal{B}} x$, that $x \wedge y \vdash_{\mathcal{B}} y$, and that $x, y \vdash_{\mathcal{B}} x \wedge y$. The logic $\mathcal{B}$ has, thus, PC. Then Proposition 5.17 implies that every non-empty $\mathcal{B}$-filter of a De Morgan lattice is a lattice filter.

Usually filters of a logic are determined by a Hilbert-style presentation, and this is probably the only way if we want to find them on arbitrary algebras; but note how here we do not need it to determine the $\mathcal{B}$-filters of De Morgan lattices.

(B9) If $\boldsymbol{A} \in \mathsf{DM}$, the g-matrix $\langle \boldsymbol{A}, \mathcal{F}i_{\mathcal{B}}\boldsymbol{A}\rangle$ has the PCONG and is reduced. This follows from (B8), because filters separate points in every lattice (Exercise 1.42) and, clearly, the addition of the empty set to a closure system does not alter its interderivability relation; therefore $\boldsymbol{\Lambda}\mathcal{F}i_{\mathcal{B}}\boldsymbol{A} = \mathrm{Id}_A$. By Proposition 5.55, this implies that the g-matrix $\langle \boldsymbol{A}, \mathcal{F}i_{\mathcal{B}}\boldsymbol{A}\rangle$ has the PCONG and is reduced.

(B10) $\mathsf{Alg}\mathcal{B} = \mathsf{DM}$. That $\mathsf{Alg}\mathcal{B} \subseteq \mathsf{DM}$ is shown in (B7), and that $\mathsf{Alg}\mathcal{B} \supseteq \mathsf{DM}$ is a consequence of (B9).

(B11) The reduced basic full g-models of $\mathcal{B}$ are all the g-matrices of the form $\langle \boldsymbol{A}, \mathcal{F}ilt_{\wedge}\boldsymbol{A} \cup \{\varnothing\}\rangle$ with $\boldsymbol{A} \in \mathsf{DM}$. This is a consequence of (B10) and (B8).

(B12) If $\langle \boldsymbol{A}, F\rangle \in \mathsf{Mod}\mathcal{B}$ and $a, b \in A$, then $a \equiv b \ (\boldsymbol{\Omega}^{\boldsymbol{A}}F)$ if and only if for all $c \in A$, $[a \vee c \in F \Leftrightarrow b \vee c \in F]$ and $[\neg a \vee c \in F \Leftrightarrow \neg b \vee c \in F]$. You have proved this in Exercise 4.36.

(B13) The non-trivial reduced models of $\mathcal{B}$ are characterized in the following way: $\langle \boldsymbol{A}, F\rangle \in \mathsf{Mod}^*\mathcal{B}$ with $F \neq \varnothing$ if and only if $\boldsymbol{A} \in \mathsf{DM}$ and F is a lattice filter of A such that for all $a, b \in A$, if $a < b$, then there is $c \in A$ with $\big[a \vee c \notin F \text{ and } b \vee c \in D\big]$ or $\big[\neg b \vee c \notin F \text{ and } \neg a \vee c \in F\big]$. This is left to you as Exercise 5.54.

(B14) $\mathsf{Alg}^*\mathcal{B} = \{ \boldsymbol{A} \in \mathsf{DM} :$ there is a lattice filter F of $\boldsymbol{A}$ satisfying the condition in (B13)$\} \cup \{\boldsymbol{A}_t\}$.

Thus, the algebraic counterpart of $\mathcal{B}$, according to the paradigm of g-matrices, is $\mathsf{Alg}\mathcal{B} = \mathsf{DM}$, while according to the matrix semantics paradigm it is a rather strange class $\mathsf{Alg}^*\mathcal{B}$. Just to mention a few facts about this one:

- $\boldsymbol{M}_2, \boldsymbol{M}_3, \boldsymbol{M}_4 \in \mathsf{Alg}^*\mathcal{B}$.
- There are three different reduced matrices with the same algebraic reduct $\boldsymbol{M}_4$, namely $\langle \boldsymbol{M}_4, \{\mathbf{t}\}\rangle$, $\langle \boldsymbol{M}_4, \{\mathbf{t},\mathbf{n}\}\rangle$ and $\langle \boldsymbol{M}_4, \{\mathbf{t},\mathbf{b}\}\rangle$. In particular, observe that $\mathcal{B}$ has non-trivial reduced models in which the filter is not a one-element set (this is relevant for Section 7.3).
- $\mathsf{Alg}^*\mathcal{B} \subsetneq \mathsf{Alg}\mathcal{B} = \mathsf{DM}$: Not every De Morgan lattice (even bounded) belongs to $\mathsf{Alg}^*\mathcal{B}$; for instance the six-element algebra $\boldsymbol{M}_6$ depicted in Figure 8 on page 299 does not support any reduced model of $\mathcal{B}$.
- $\mathsf{Alg}^*\mathcal{B}$ is neither a variety nor a quasivariety, by Corollary 5.71.4 and the previous fact. Another reason for $\mathsf{Alg}^*\mathcal{B}$ not being a variety is that it contains $\boldsymbol{M}_4$, the algebra that generates the whole class DM. ⊠

If one is not interested in finding the class $\mathsf{Alg}^*\mathcal{B}$, then one can prove that $\mathsf{Alg}\mathcal{B} = \mathbb{V}\mathcal{B} = \mathsf{DM}$ by the method of Proposition 5.80; see Exercises 5.50 and 5.53. A similar application, requiring more specialized algebraic knowledge, is:

Example 5.83 (Łukasiewicz's many-valued logics)
For each $n > 2$, $Ł_n$ is the logic defined by the matrix $\langle [\mathbf{0},\mathbf{1}]_n, \{1\}\rangle$. By Proposition 5.79, this implies that $\mathbb{V}Ł_n = \mathbb{V}\{[\mathbf{0},\mathbf{1}]_n\}$, the subvariety of MV generated by the single MV-algebra $[\mathbf{0},\mathbf{1}]_n$. On the other hand, the logic is algebraizable (Example 3.34.6), therefore $\mathsf{Alg}Ł_n = \mathsf{Alg}^*Ł_n$. Now we can see that $\mathsf{Alg}Ł_n = \mathbb{V}Ł_n = \mathbb{V}\{[\mathbf{0},\mathbf{1}]_n\}$. The key fact† is that the subdirectly irreducible algebras in the variety $\mathbb{V}\{[\mathbf{0},\mathbf{1}]_n\}$ are (up to isomorphism) just the subalgebras of $[\mathbf{0},\mathbf{1}]_n$. Let $\boldsymbol{A}$ be one‡ of these subalgebras. Since it is a subalgebra of $[\mathbf{0},\mathbf{1}]$, the matrix $\langle \boldsymbol{A}, \{1\}\rangle$ is a submatrix of $\langle [\mathbf{0},\mathbf{1}]_n, \{1\}\rangle$, hence by Propositions 4.6.1 and 5.81, $\mathsf{Alg}^*\mathcal{L}_{\langle \boldsymbol{A},\{1\}\rangle} \subseteq \mathsf{Alg}^*Ł_n$. Since the matrix $\langle \boldsymbol{A}, \{1\}\rangle$ is reduced and defines the logic, clearly $\boldsymbol{A} \in \mathsf{Alg}^*\mathcal{L}_{\langle \boldsymbol{A},\{1\}\rangle}$, therefore $\boldsymbol{A} \in \mathsf{Alg}^*Ł_n$. Now, applying Proposition 5.80, we obtain that $\mathsf{Alg}Ł_n = \mathbb{V}Ł_n = \mathbb{V}\{[\mathbf{0},\mathbf{1}]_n\}$ as claimed.

The logics $Ł_\infty$ and $Ł_{\infty f}$ associated with the infinite matrix $\langle [\mathbf{0},\mathbf{1}], \{1\}\rangle$ are also algebraizable (Example 3.34.6), and from Proposition 5.79 and Exercise 5.49 we obtain that $\mathbb{V}Ł_\infty = \mathbb{V}Ł_{\infty f} = \mathbb{V}\{[\mathbf{0},\mathbf{1}]\} = \mathsf{MV}$, the variety of MV-algebras. However, the above strategy of using these facts plus Proposition 5.80 to find $\mathsf{Alg}Ł_\infty$ or $\mathsf{Alg}Ł_{\infty f}$ does not work, because there are many more subdirectly irreducible MV-algebras. For $Ł_{\infty f}$, one can check directly (using its syntactic presentation,

† This is a consequence of general results of universal algebra.
‡ By Exercise 3.7, these are the algebras $[\mathbf{0},\mathbf{1}]_m$ with $m-1$ a divisor of $n-1$, and the logic $\mathcal{L}_{\langle \boldsymbol{A},\{1\}\rangle}$ is actually $Ł_m$; but the argument here does not need this extra knowledge.

as is done in [218]) that MV is its equivalent algebraic semantics, and therefore $\mathsf{Alg}^*\text{Ł}_{\infty f} = \mathsf{Alg}\text{Ł}_{\infty f} = \mathsf{MV}$. As to Ł_∞, we already know (Example 3.41) that $\mathsf{Alg}^*\text{Ł}_\infty = \mathsf{Alg}\text{Ł}_\infty = \mathbb{GQ}\{[0,1]\}$, a proper subclass of MV which is not a quasivariety. ⊠

Exercises for Section 5.4

EXERCISE 5.42. Prove that if a g-matrix is a model of a Gentzen-style g-rule, then all its sub-g-matrices are also models of that rule. Use this to show that for any logic $\mathcal{L}$, the class of g-matrices $\mathsf{GMod}\mathcal{L}$ is closed under the construction of sub-g-matrices (Exercise 5.29).

EXERCISE 5.43. Let $\mathcal{M}$ be a finitary g-matrix. Show that a Gentzen-style rule is valid in $\mathcal{M}$ if and only if for every finitary, finitely generated sub-g-matrix $\mathcal{N}$ of $\mathcal{M}$ there is a finitary, countably generated g-matrix $\mathcal{N}'$ such that $\mathcal{N} \leqslant \mathcal{N}' \leqslant \mathcal{M}$ and the rule is valid in $\mathcal{N}'$. (Here $\leqslant$ means "is a sub-g-matrix of".)

EXERCISE 5.44. Complete the proof of Theorem 5.63. In the case of the PCONG, do it withotu using point 4 in the theorem.

EXERCISE 5.45. Prove that if $\mathcal{L}$ is a logic and $\boldsymbol{A}, \boldsymbol{B}$ are two isomorphic algebras, then the corresponding basic full g-models of $\mathcal{L}$, that is, the g-matrices $\langle \boldsymbol{A}, \mathcal{F}i_{\mathcal{L}}\boldsymbol{A}\rangle$ and $\langle \boldsymbol{B}, \mathcal{F}i_{\mathcal{L}}\boldsymbol{B}\rangle$, are isomorphic as well.

EXERCISE 5.46. Theorem 5.70 establishes that for every logic $\mathcal{L}$, $\mathsf{Alg}^*\mathcal{L} \subseteq \mathsf{Alg}\mathcal{L}$, as a consequence of the equality $\mathsf{Alg}\mathcal{L} = \mathbb{P}_{\mathrm{SD}}\mathsf{Alg}^*\mathcal{L}$. Prove the inclusion directly from the definitions of the two classes of algebras.

EXERCISE 5.47. Use Proposition 5.34 to prove Lemma 5.72. Observe that in the second part it is not necessary to demand that the set Δ is the same in all algebras in order to obtain the final conclusion (however, it is natural to do so, and this is indeed what happens, for instance, in algebraizable logics).

EXERCISE 5.48. Look in later chapters for a joint counterexample for the converses of the implications in Corollaries 5.71.4 and 5.73.

HINT. Look for a protoalgebraic but non-algebraizable logic $\mathcal{L}$ such that the class $\mathsf{Alg}\mathcal{L}(= \mathsf{Alg}^*\mathcal{L})$ is not a quasivariety.

EXERCISE 5.49. Let $\mathcal{L}$ be a logic, and $\mathcal{L}_{\mathrm{f}}$ its finitary companion. Use Proposition 1.49 to prove that $\mathbb{V}\mathcal{L} = \mathbb{V}\mathcal{L}_{\mathrm{f}}$. Look into examples in this section to see that, by contrast, the classes $\mathsf{Alg}\mathcal{L}$ and $\mathsf{Alg}\mathcal{L}_{\mathrm{f}}$ need not coincide; observe that this implies that the classes $\mathsf{Alg}^*\mathcal{L}$ and $\mathsf{Alg}^*\mathcal{L}_{\mathrm{f}}$ do not coincide either.

EXERCISE 5.50. Prove Proposition 5.75.4. Apply it to show, using just the definitions of the logics, that $\mathbb{V}\mathcal{C}\ell = \mathsf{BA}$, $\mathbb{V}\mathcal{C}\ell_{\wedge} = \mathsf{SL}$, $\mathbb{V}\mathcal{C}\ell_{\wedge\vee} = \mathsf{DL}$ and $\mathbb{V}\mathcal{B} = \mathsf{DM}$. You can also apply it to show that $\mathbb{V}\mathcal{I}\ell = \mathsf{HA}$ and $\mathbb{V}\mathcal{I}\ell_{\to} = \mathsf{HiA}$, but you will need to use the Deduction Theorem, the Algebraic Completeness Theorem, and the condition (LALG2) of the algebras of implicative logics (Theorem 5.76 offers an alternative way to determine these classes).

EXERCISE 5.51. Check that the proof of Theorem 5.76 actually contains a proof of the following interesting property: For any $\alpha, \beta \in Fm$, $\alpha \equiv \beta\ (\widetilde{\Omega}\mathcal{L})$ if and only if for all algebras $\boldsymbol{A}$ and all $h \in \mathrm{Hom}(\boldsymbol{Fm}, \boldsymbol{A})$, $h\alpha \equiv h\beta\ \big(\widetilde{\Omega}^{\boldsymbol{A}}\mathcal{F}i_{\mathcal{L}}\boldsymbol{A}\big)$. That is, for any logic $\mathcal{L}$, $\widetilde{\Omega}\mathcal{L} = \bigcap\big\{h^{-1}\widetilde{\Omega}^{\boldsymbol{A}}\mathcal{F}i_{\mathcal{L}}\boldsymbol{A} : \boldsymbol{A} \text{ an algebra}, h \in \mathrm{Hom}(\boldsymbol{Fm}, \boldsymbol{A})\big\}$. Check that Lemma 5.74 is in fact a consequence of this property.

EXERCISE 5.52. Let $\mathcal{L}$ be the logic defined by a class M of reduced matrices or of reduced g-matrices, and put $\mathsf{K} := \mathsf{Alg}\,\mathsf{M}$. Prove that $\mathbb{V}\mathcal{L} = \mathbb{V}\mathsf{K}$. Show that Theorem 5.76 is a particular case of this result.

HINT. Review the proof of Theorem 5.76, and use Lemma 5.78.

EXERCISE 5.53. Show that $\mathsf{Alg}\mathcal{C}\ell_{\wedge\vee} = \mathbb{V}\mathcal{C}\ell_{\wedge\vee} = \mathsf{DL}$, that $\mathsf{Alg}\mathcal{C}\ell_{\wedge} = \mathbb{V}\mathcal{C}\ell_{\wedge} = \mathsf{SL}$, and that $\mathsf{Alg}\mathcal{B} = \mathbb{V}\mathcal{B} = \mathsf{DM}$, by the combined use of Propositions 5.79 and 5.80.

HINT. The subdirectly irreducible De Morgan lattices are mentioned in Example 5.82.

EXERCISE 5.54. Prove the characterization of the non-trivial reduced models of $\mathcal{B}$ stated in point (B13) of Example 5.82.

HINT. Look at the proof of Proposition 4.49, and use points (B7), (B8) and (B12).

5.5 Full generalized models

The main concept

We finally arrive at one of the central notions of this chapter:

DEFINITION 5.84

A ***full g-model of*** *a logic $\mathcal{L}$ is an inverse image of a basic full g-model of $\mathcal{L}$. That is, $\langle \boldsymbol{A}, \mathscr{C} \rangle$ is a full g-model of $\mathcal{L}$ if and only if $\mathscr{C} = h^{-1}\mathcal{F}i_{\mathcal{L}}\boldsymbol{B}$ for some algebra $\boldsymbol{B}$ and some surjective $h \colon \boldsymbol{A} \to \boldsymbol{B}$.*

The class of the full g-models of $\mathcal{L}$ is denoted by $\mathsf{FGMod}\mathcal{L}$, and the class of the reduced ones by $\mathsf{FGMod}^*\mathcal{L}$.

Recall from Definition 5.41 that the inverse image construction is considered only for surjective homomorphisms, and that with this definition h becomes strict from $\langle \boldsymbol{A}, \mathscr{C} \rangle \in \mathsf{FGMod}\mathcal{L}$ to $\langle \boldsymbol{B}, \mathcal{F}i_{\mathcal{L}}\boldsymbol{B} \rangle$. Notice that, by Theorem 5.61, a full g-model of $\mathcal{L}$ is a g-model of $\mathcal{L}$, as the name suggests; thus we are dealing with a special kind of g-models. Moreover, since identity is a (trivial) surjective homomorphism, all basic full g-models are indeed full g-models (hence their name), and in particular, extending Proposition 5.9, we see that *the logic itself is its weakest full g-model on the formula algebra*. As a consequence, any interesting characterization of full g-models of a given logic as "the g-matrices that satisfy such and such properties" produces an equally interesting characterization of that logic as "the weakest logic (actually: the weakest g-matrix) on the formula algebra satisfying such and such properties".

The conclusion of the general results of this and the next section, and the analysis of examples, is that full g-models are g-models that behave particularly well, show interesting general properties, and produce interesting descriptions of the algebra-based models of the logic, and even, as commented in the previous paragraph, of the logic itself.

The unspecified basic full g-model appearing in Definition 5.84 can be chosen in a particular, significant way:

PROPOSITION 5.85
A g-matrix $\mathcal{M}$ is a full g-model of $\mathcal{L}$ if and only if its reduction $\mathcal{M}^*$ is a basic full g-model of $\mathcal{L}$. That is, $\mathcal{M} = \langle \boldsymbol{A}, \mathscr{C} \rangle \in \mathsf{FGMod}\mathcal{L}$ if and only if $\mathscr{C}^* = \mathcal{F}i_{\mathcal{L}}(\boldsymbol{A}^*)$.

PROOF: Assume that $\mathcal{M} = h^{-1}\mathcal{N}$ for some basic full g-model $\mathcal{N} = \langle \boldsymbol{B}, \mathcal{F}i_{\mathcal{L}}\boldsymbol{B} \rangle$ and some strict surjective $h \colon \boldsymbol{A} \to \boldsymbol{B}$. By Propositions 5.57 and 5.64.1, $\mathcal{M}^* \cong \mathcal{N}^* = \langle \boldsymbol{B}^*, (\mathcal{F}i_{\mathcal{L}}\boldsymbol{B})^* \rangle = \langle \boldsymbol{B}^*, \mathcal{F}i_{\mathcal{L}}(\boldsymbol{B}^*) \rangle$. Now by Proposition 5.60 the isomorphism implies that $\mathscr{C}^* = \mathcal{F}i_{\mathcal{L}}(\boldsymbol{A}^*)$, and $\mathcal{M}^*$ is a basic full g-model of $\mathcal{L}$. The converse holds because the reduction function is a strict surjective homomorphism. ⊠

PROPOSITION 5.86
The class $\mathsf{FGMod}\mathcal{L}$ is closed under images and inverse images by strict surjective homomorphisms, and it is the smallest class of g-matrices containing the basic full g-models of $\mathcal{L}$ and being closed under these two operators.

PROOF: The very definition tells us that $\mathsf{FGMod}\mathcal{L}$ is closed under inverse images by strict surjective homomorphisms, because the composition of two of these is again one of them (Exercise 5.41). If $\mathcal{M}, \mathcal{N}$ are two g-models of $\mathcal{L}$ with $\mathcal{N} = h\mathcal{M}$ and h is strict and surjective, then by Proposition 5.57, $\mathcal{N}^* \cong \mathcal{M}^*$. If $\mathcal{M}$ is full, by Proposition 5.85 $\mathcal{M}^*$ is basic full, and then Proposition 5.60 implies that $\mathcal{N}^*$ is basic full as well, and thus $\mathcal{N}$ is full. Finally, I already remarked that $\mathsf{FGMod}\mathcal{L}$ contains all basic full g-models of $\mathcal{L}$, and if another class contains all these and is closed under the two operators, in particular closure under inverse images implies by Definition 5.84 that it contains all full g-models of $\mathcal{L}$. ⊠

COROLLARY 5.87
1. *$\mathcal{M} \in \mathsf{FGMod}\mathcal{L}$ if and only if $\mathcal{M}^* \in \mathsf{FGMod}^*\mathcal{L}$.*
2. *$\mathsf{FGMod}^*\mathcal{L} = \mathbb{I}((\mathsf{FGMod}\mathcal{L})^*)$. That is, $\mathcal{M}$ is a reduced full g-model of $\mathcal{L}$ if and only if $\mathcal{M} \cong \mathcal{N}^*$ for a full g-model $\mathcal{N}$ of $\mathcal{L}$.* ⊠

To complete this panorama, from the previous properties and Proposition 5.68 we obtain:

COROLLARY 5.88
Let $\mathcal{L}$ be any logic.
1. *Every reduced full g-model of $\mathcal{L}$ is basic.*
2. *The following four classes of g-matrices coincide:*
   - *The reduced full g-models of $\mathcal{L}$.*
   - *The reduced basic full g-models of $\mathcal{L}$.*
   - *The reduced g-matrices of the form $\langle \boldsymbol{A}, \mathcal{F}i_{\mathcal{L}}\boldsymbol{A} \rangle$, for some algebra $\boldsymbol{A}$.*
   - *The g-matrices of the form $\langle \boldsymbol{A}, \mathcal{F}i_{\mathcal{L}}\boldsymbol{A} \rangle$, for some algebra $\boldsymbol{A} \in \mathsf{Alg}\mathcal{L}$.*

   Therefore, $\mathsf{FGMod}^*\mathcal{L} = \{\langle \boldsymbol{A}, \mathcal{F}i_{\mathcal{L}}\boldsymbol{A} \rangle : \boldsymbol{A} \in \mathsf{Alg}\mathcal{L}\}$.
3. *$\mathsf{Alg}\mathcal{L} = \mathsf{Alg}\,\mathsf{FGMod}^*\mathcal{L}$. That is, $\mathsf{Alg}\mathcal{L}$ is the class of algebraic reducts of the reduced full g-models of $\mathcal{L}$.*

4. *The full g-models of $\mathcal{L}$ are the inverse images of g-matrices of the form $\langle \boldsymbol{A}, \mathcal{F}i_{\mathcal{L}}\boldsymbol{A}\rangle$ with $\boldsymbol{A} \in \mathsf{Alg}\mathcal{L}$.* ⊠

Point 3 constitutes one more argument for considering $\mathsf{Alg}\mathcal{L}$ as the right general notion of "the algebraic counterpart" of a logic.

Now the basic "logical" properties of full g-models are straightforward:

PROPOSITION 5.89 (COMPLETENESS)
Any logic $\mathcal{L}$ is complete with respect to the class $\mathsf{FGMod}\mathcal{L}$ and with respect to the class $\mathsf{FGMod}^*\mathcal{L} = \{\langle \boldsymbol{A}, \mathcal{F}i_{\mathcal{L}}\boldsymbol{A}\rangle : \boldsymbol{A} \in \mathsf{Alg}\mathcal{L}\}$.

PROOF: Every logic $\mathcal{L}$ is complete with respect to the class of its basic full g-models (Proposition 5.14). This class is contained in $\mathsf{FGMod}\mathcal{L}$, which is a class of g-models. Thus, $\mathcal{L}$ is complete with respect to $\mathsf{FGMod}\mathcal{L}$. By Theorem 5.61, completeness is preserved by reductions and by isomorphisms. ⊠

This completeness result can have an algebraic interest, in the cases in which there are good algebraic characterizations of reduced full g-models. You are advised to check the examples below, and others by yourself.

As the first transfer results for full g-models, by Proposition 5.86 we have:

PROPOSITION 5.90
The full g-models of $\mathcal{L}$ have all properties of $\mathcal{L}$ that are transferred to its basic full g-models and are preserved by strict surjective homomorphisms.

In particular, full g-models of a logic inherit the following properties from the logic:

1. *To be finitary.*
2. *To have theorems.*
3. *To have no theorems.*
4. *All properties expressible as a set of Hilbert-style g-rules, for instance the* PC.
5. *All Gentzen-style rules that transfer to the basic full g-models of $\mathcal{L}$.*
6. *The* u-DDT.

If $\mathcal{L}$ is finitary, also:

7. *All properties expressible as a set of Gentzen-style rules of the form given in Theorem 5.19, for instance the* PDI *and the* PIM. ⊠

By Example 5.24 there is no hope that either the PIRA or the PRA transfer to full g-models. As to the other conditions on negation, see Exercise 5.55. In Chapter 7 it is shown that the PCONG does not transfer in general.

❧ ❧ ❧

It is clear that *determination of the full g-models of a logic is an important step in its algebraic analysis*, and this becomes clearer after the general results in the second part of this section. Here "determination" means characterizing them in a more interesting way than just the bare definition. By Corollary 5.88.4, full g-models reflect the abstract properties that the $\mathcal{L}$-filters of $\mathcal{L}$-algebras have. The study of these properties and of their preservation by strict surjective homomorphisms is one of the typical tasks in this branch of abstract algebraic logic; in most cases, this produces some really interesting characterizations of the full g-models of the logic (and, as a by product, of the logics themselves, as already observed).

Moreover, by Corollary 5.88.3, if we have an interesting characterization of full g-models of $\mathcal{L}$ by conditions of some logical significance, then by adding the condition of being reduced we find a characterization of $\mathcal{L}$-algebras in terms of properties of a closure operator with a logical reading. Particular results obtained in this way can be called "logical characterizations" of certain (classes of) algebras. We can say that, besides *the algebraization of logics*, we are also dealing with *the logification of algebras*; an example of an early paper written in this spirit is [239].

Three case studies

There are no general recipes on how to determine the full g-models of a logic, but one can find some clues arising from experience with particular examples. Here I am going to take three paradigmatic examples, one of a syntactic inspiration, and two of a semantic origin, and explain in detail the process of finding and characterizing their full g-models in different ways. As in the case of Example 5.82, some points might be obtained more quickly by using results of future chapters, but here what is interesting is not the final result, but the detailed process to reach it, and in particular the striking difference in treatment for logics of the two kinds. By Theorem 6.117, for implicative and algebraizable logics this task can be greatly simplified: their full g-models correspond to families of filters containing a given one, but still, as the next example shows, this does not cover other, potentially more interesting characterizations.

Example 5.91 (The full g-models of $\mathcal{I\ell}_{\to}$)
The implication fragment $\mathcal{I\ell}_{\to}$ of intuitionistic logic is introduced on page 85. We already know that this logic is implicative, hence algebraizable, with equivalence formulas the set $\{x \to y, y \to x\}$ and defining equation $x \approx x \to x$. It is well known that $\mathcal{I\ell}_{\to}$ is axiomatized by the two axioms

$$x \to (y \to x) \quad \text{and} \quad (x \to (y \to z)) \to ((x \to y) \to (x \to z))$$

together with the rule of *Modus Ponens* $(x, x \to y \rhd y)$. Usage of this axiomatization enabled you to prove, in Exercise 2.28, that $\mathsf{Alg}^*\mathcal{I\ell}_{\to}$ is the class HiA of Hilbert algebras; since this class is a variety, Corollary 5.77.4 implies that $\mathsf{Alg}\mathcal{I\ell}_{\to} = \mathbb{V}\mathcal{I\ell}_{\to} = \mathsf{HiA}$ as well. Here its full g-models are identified, by following what can be viewed as *an exercise in the algebraic analysis of the Deduction Theorem*. This is just a detailed skeleton of the process; Exercise 5.56 asks you to check all the successive steps.

Unless stated otherwise, in what follows $\mathcal{M} = \langle \boldsymbol{A}, \mathscr{C} \rangle$ is an arbitrary g-matrix where $\boldsymbol{A} = \langle A, \to \rangle$ is an algebra of type $\langle 2 \rangle$.

(I1) If $\mathcal{M} \in \mathsf{GMod}\mathcal{I}\ell_{\to}$, then for all $a, b \in A$, $a \equiv b\ (\widetilde{\Omega}\mathcal{M})$ if and only if $a \to b, b \to a \in C\emptyset$. This is a consequence of Corollary 4.55 and Lemma 5.72, but it can be proved by just using the axiomatization of $\mathcal{I}\ell_{\to}$.

(I2) If $\mathcal{M} \in \mathsf{GMod}\mathcal{I}\ell_{\to}$ and satisfies the WDT ($a \to b \in C\emptyset$ if and only if $b \in C\{a\}$), then it satisfies the PCONG.

(I3) If $\mathcal{M}$ satisfies the u-DDT then:

1. For all $a \in A$, $a \to a \in C\emptyset$.
2. $\mathcal{M} \in \mathsf{GMod}\mathcal{I}\ell_{\to}$. That is, $\mathscr{C} \subseteq \{F \in \mathcal{F}i_{\mathcal{I}\ell_{\to}}\boldsymbol{A} : F \supseteq C\emptyset\}$.
3. $\mathcal{M}$ satisfies the PCONG.

(I4) If $\mathcal{M}$ is finitary and satisfies the u-DDT, then $\mathscr{C} = \{F \in \mathcal{F}i_{\mathcal{I}\ell_{\to}}\boldsymbol{A} : F \supseteq C\emptyset\}$.

Recall that in Definition 2.27 the notion of an *implicative filter* is introduced for implicative algebras: it is a set $F \subseteq A$ such that $1 \in F$ and F is closed under *Modus Ponens*. The family of all implicative filters of a Hilbert algebra $\boldsymbol{A}$ is denoted by $\mathcal{F}ilt_{\to}\boldsymbol{A}$. Then:

(I5) Let $\boldsymbol{A} \in \mathsf{HiA}$.

1. $\{1\}$ is the smallest implicative filter of $\boldsymbol{A}$.
2. The g-matrix $\langle \boldsymbol{A}, \mathcal{F}ilt_{\to}\boldsymbol{A}\rangle$ is finitary, satisfies the u-DDT and the PCONG, and is reduced.
3. $\mathcal{F}i_{\mathcal{I}\ell_{\to}}\boldsymbol{A} = \mathcal{F}ilt_{\to}\boldsymbol{A}$. Actually, after what we already know, this is just Proposition 2.28.
4. $\langle \boldsymbol{A}, \mathcal{F}ilt_{\to}\boldsymbol{A}\rangle$ is a reduced full g-model of $\mathcal{I}\ell_{\to}$.

(I6) The following conditions are equivalent:

(i) $\mathcal{M}$ is finitary and satisfies the u-DDT.
(ii) $\boldsymbol{A}^* \in \mathsf{HiA}$ and $\mathscr{C}^* = \mathcal{F}ilt_{\to}(\boldsymbol{A}^*)$.
(iii) There is a strict surjective homomorphism from $\mathcal{M}$ onto a g-matrix of the form $\langle \boldsymbol{B}, \mathcal{F}ilt_{\to}\boldsymbol{B}\rangle$ for some $\boldsymbol{B} \in \mathsf{HiA}$.

After this and by Corollary 5.88.4 and (I5)(3), the condition (I6)(iii) can be rewritten as "$\mathcal{M}$ is a full g-model of $\mathcal{I}\ell_{\to}$". Therefore in particular we obtain:

(I7) $\mathcal{M} \in \mathsf{FGMod}\mathcal{I}\ell_{\to}$ if and only if $\mathcal{M}$ is finitary and satisfies the u-DDT.

(I8) $\mathcal{I}\ell_{\to}$ is the weakest finitary logic that satisfies the u-DDT.

(I9) $\mathcal{M} \in \mathsf{FGMod}\mathcal{I}\ell_{\to}$ if and only if there is some $F_0 \in \mathcal{F}i_{\mathcal{I}\ell_{\to}}\boldsymbol{A}$ such that $\mathscr{C} = \{F \in \mathcal{F}i_{\mathcal{I}\ell_{\to}}\boldsymbol{A} : F \supseteq F_0\}$.

(I10) $\boldsymbol{A} \in \mathsf{HiA}$ if and only if there is an inductive closure system $\mathscr{C} \subseteq \mathscr{P}A$ such that the g-matrix $\langle \boldsymbol{A}, \mathscr{C}\rangle$ satisfies the u-DDT and $\mathscr{C}$ separates points; that is, if $a \neq b$, then there is some $F \in \mathscr{C}$ such that either $a \in F$ and $b \notin F$, or $a \notin F$ and $b \in F$.

The characterization in (I7) is perhaps the most interesting result here: it offers another view of the importance of the u-DDT for $\mathcal{I}\ell_{\to}$: it characterizes not only the logic, but its full g-models. Then (I8) follows from this and the fact that any logic is its weakest full g-model on the formula algebra; this characterization (obtained directly in Corollary 3.73) can be further strengthened (Exercise 5.57). Property (I9) contains an important point in the description of full g-models: their closure system consists of *all* the filters of the logic containing a given one; as Theorem 6.117 shows, this property is exclusive of weakly algebraizable logics (which include the algebraizable ones). Finally, (I10) is an example of the announced characterizations of a class of algebras in terms of "logical" properties. The separation property amounts to saying that "if $C\{a\} = C\{b\}$, then $a = b$"; thus, all the properties that appear in this characterization can be formulated in terms of the closure operator of the g-matrix. ⊠

The characterizations of full g-models obtained for $\mathcal{IL}_{\rightarrow}$, being based on the u-DDT, have all a syntactic flavour; the next example shows that in some cases characterizations with a semantic flavour are also possible.

EXAMPLE 5.92 (THE FULL G-MODELS OF $\mathcal{B}$)
This continues Example 5.82 and its numbering of the properties that appear. The first two are straightforward, while the other require more technical work than is intended here. I hope the sketches of proofs given are understandable enough; if you want to see all details, you should read [93].

(B15) $\mathcal{M} \in \mathsf{FGMod}^*\mathcal{B}$ if and only if $\mathcal{M} = \langle \boldsymbol{A}, \mathcal{F}ilt_{\wedge}\boldsymbol{A} \cup \{\emptyset\}\rangle$ with $\boldsymbol{A} \in \mathsf{DM}$. This is just property (B11) expressed using the present notions, taking property (B8) and Corollary 5.88.2 into account.

(B16) The following conditions are equivalent:

(i) $\mathcal{M} \in \mathsf{FGMod}\mathcal{B}$.

(ii) $\boldsymbol{A}^* \in \mathsf{DM}$ and $\mathscr{C}^* = \mathcal{F}ilt_{\wedge}(\boldsymbol{A}^*) \cup \{\emptyset\}$.

(iii) There is a strict surjective homomorphism from $\mathcal{M}$ onto a g-matrix of the form $\langle \boldsymbol{A}, \mathcal{F}ilt_{\wedge}\boldsymbol{A} \cup \{\emptyset\}\rangle$, for some $\boldsymbol{A} \in \mathsf{DM}$.

Observe that the closure system $\mathcal{F}ilt_{\wedge}\boldsymbol{A} \cup \{\emptyset\}$ is intrinsically associated with the algebraic structure of a De Morgan lattice, and in principle no "logical" elements remain in this characterization; it is purely algebraic.

(B17) $\mathcal{M} = \langle \boldsymbol{A}, \mathscr{C}\rangle \in \mathsf{FGMod}\mathcal{B}$ if and only if C is finitary and there is a base $\mathscr{E}$ of $\mathscr{C}$ such that $\emptyset \in \mathscr{E}$ and for every non-empty and proper $F \in \mathscr{E}$, the following hold:

(a) F is an $\wedge$-filter ($a \wedge b \in F$ if and only if $a \in F$ and $b \in F$).

(b) F is $\vee$-prime ($a \vee b \in F$ if and only if $a \in F$ or $b \in F$).

(c) $\Phi F \in \mathscr{E}$ and $\Phi^2 F = F$.

The function Φ is defined,[†] for each $X \subseteq A$, by $\Phi X := \{a \in A : \neg a \notin X\}$.

[†] In some purely algebraic works the function Φ is defined by $\Phi X = A \smallsetminus \neg X$, where obviously $\neg X := \{\neg a : a \in X\}$. This is equivalent to the given one in De Morgan lattices (see Exercise 5.59), but does not work correctly in arbitrary algebras, while the one used above does.

Observe that $\Phi\varnothing = A$ and $\Phi A = \varnothing$.

SKETCH OF THE PROOF: ($\Rightarrow$) First one shows that all the intended properties are preserved by strict surjective homomorphisms; thus, by (B16) it is enough to prove them for g-matrices of the form $\langle \boldsymbol{A}, \mathcal{F}ilt_{\wedge}\boldsymbol{A} \cup \{\varnothing\} \rangle$ with $\boldsymbol{A} \in \mathsf{DM}$. Since $\boldsymbol{A}$ is a distributive lattice, one takes as $\mathscr{E}$ the family of all prime filters of $\boldsymbol{A}$ plus $\varnothing$ [19, Theorem III.6.2]. This family obviously satisfies (a) and (b), and (c) is a well-known algebraic property of De Morgan lattices related to their representation theory [19, Section XI.2]. Finitarity is also well-known.

($\Leftarrow$) One begins by showing that $\mathcal{M}$ has the PCONG, that is, that $\boldsymbol{\Lambda}\mathscr{C}$ is a congruence; this follows easily from the three properties (a), (b) and (c) of the base of $\mathscr{C}$. Now, it is easy to show that $\boldsymbol{A}^* := \boldsymbol{A}/\boldsymbol{\Lambda}\mathscr{C} \in \mathsf{DM}$, and that $\mathscr{C}^* := \mathscr{C}/\boldsymbol{\Lambda}\mathscr{C} \subseteq \mathcal{F}ilt_{\wedge}\boldsymbol{A} \cup \{\varnothing\}$. Finally, finitarity shows that the reverse inclusion is also true, that is, that $\mathscr{C}^* = \mathcal{F}ilt_{\wedge}\boldsymbol{A} \cup \{\varnothing\}$. According to (B16), this entails that $\mathcal{M} \in \mathsf{FGMod}\mathcal{B}$.

Methodologically, it is interesting to highlight that in order to check that $\boldsymbol{A}/\boldsymbol{\Lambda}\mathscr{C}$ satisfies the equations that define the variety DM, it is enough to check them in $\boldsymbol{A}$ but "modulo C"; it is best to describe this by an example: that $\boldsymbol{A}/\boldsymbol{\Lambda}\mathscr{C} \vDash x \approx \neg\neg x$ is proved by showing that $C\{a\} = C\{\neg\neg a\}$ for all $a \in A$. This is the typical situation in which the PCONG makes life particularly easy. Moreover, in this case, $\mathscr{E}$ is a base of $\mathscr{C}$, therefore $C\{a\} = C\{b\}$ if and only if for all $F \in \mathscr{E}$, $[a \in F \Leftrightarrow b \in F]$. ⊠

The crucial role of finitarity is condensed (and generalized) in Exercise 5.60.

This characterization helps in showing that every full g-model can be obtained from the four-element De Morgan lattice $\boldsymbol{M}_4$ in a way that is similar to the definition of the logic $\mathcal{B}$ itself; this can be considered as a "semantical" characterization of the full g-models of $\mathcal{B}$:

(B18) $\mathcal{M} \in \mathsf{FGMod}\mathcal{B}$ if and only if $\mathcal{M}$ is finitary and is generated from the g-matrix $\langle \boldsymbol{M}_4, \mathcal{F}ilt_{\wedge}\boldsymbol{M}_4 \cup \{\varnothing\} \rangle$ through some family $\mathscr{H} \subseteq \mathrm{Hom}(\boldsymbol{A}, \boldsymbol{M}_4)$ such that $h_{\mathbf{b}} \in \mathscr{H}$, where $h_{\mathbf{b}}$ is the constant function equal to $\mathbf{b}$, in the following way: For all $X \cup \{a\} \subseteq A$,

$$a \in CX \iff \begin{array}{l}\text{for all } h \in \mathscr{H} \text{ and all } F \in \mathcal{F}ilt_{\wedge}\boldsymbol{M}_4 \cup \{\varnothing\}, \\ \text{if } hX \subseteq F, \text{ then } ha \in F.\end{array} \tag{78}$$

The PROOF is too complicated to detail here, but it is interesting to see one of the constructions it uses. If $\mathcal{M} \in \mathsf{FGMod}\mathcal{B}$, from (B17) it follows that for each $F \in \mathscr{E}$, the family $\{\varnothing, F, \Phi F, A\}$ satisfies properties (a)–(c) as well, and it is the base of the closure system $\mathscr{C}_F := \{\varnothing, F \cap \Phi F, F, \Phi F, A\}$; therefore, this one also produces a full g-model of $\mathcal{B}$. Observe that, in contrast to point (I9) of Example 5.91, the closure system is not constituted by all filters containing a given one. Then, by (B16), factoring out $\boldsymbol{A}$ by $\boldsymbol{\Lambda}\mathscr{C}_F$ produces a De Morgan lattice; the interesting point is to see that, depending on the relative positions of F and ΦF, the following quotients are obtained:

- If $F = \Phi F$, the quotient is (isomorphic to) $\boldsymbol{M}_2$.
- If $F \subsetneq \Phi F$ or $\Phi F \subsetneq F$, the quotient is (isomorphic to) $\boldsymbol{M}_3$.
- If F and ΦF are not comparable, the quotient is (isomorphic to) $\boldsymbol{M}_4$.

Since $\boldsymbol{M}_2$ and $\boldsymbol{M}_3$ are subalgebras of $\boldsymbol{M}_4$, each $F \in \mathscr{E}$ is the inverse image of some set of the base $\{\varnothing, \{\mathbf{t},\mathbf{n}\}, \{\mathbf{t},\mathbf{b}\}, M_4\}$ of $\mathcal{F}ilt_\wedge \boldsymbol{M}_4 \cup \{\varnothing\}$ by some homomorphism into $\boldsymbol{M}_4$. This is basically how an $\mathscr{H} \subseteq \mathrm{Hom}(\boldsymbol{A}, \boldsymbol{M}_4)$ with the intended property is obtained.

Observe that the way in which the full g-models of $\mathcal{B}$ are "generated" from the g-matrix $\langle \boldsymbol{M}_4, \mathcal{F}ilt_\wedge \boldsymbol{M}_4 \cup \{\varnothing\}\rangle$, given by (78), is a kind of image of the way this g-matrix generates the logic itself. That is why this characterization is said to have a "semantical" flavour. As a further application of (B16), another characterization can be obtained, this time with a syntactic flavour, in terms of Tarski-style conditions:

(B19) $\mathcal{M} = \langle \boldsymbol{A}, \mathscr{C}\rangle \in \mathsf{FGMod}\mathcal{B}$ if and only if $\mathcal{M}$ is finitary and satisfies, for all $X \cup \{a,b\} \subseteq A$:

E	$C\varnothing = \varnothing$	Empty set
PC	$C\{a \wedge b\} = C\{a,b\}$	Conjunction
PDI	$C(X, a \vee b) = C(X,a) \cap C(X,b)$	Disjunction
PDN	$C\{a\} = C\{\neg\neg a\}$	Double Negation
PWCO	$b \in C\{a\} \Rightarrow \neg a \in C\{\neg b\}$	Weak Contraposition

For ($\Rightarrow$) one shows that the five conditions in the statement hold in the reduced case, using (B16), and that they are preserved by strict surjective homomorphisms. The proof of ($\Leftarrow$) is similar to that of the corresponding point of (B17): one shows that $\boldsymbol{\Lambda}\mathscr{C} \in \mathrm{Con}\boldsymbol{A}$, that $\boldsymbol{A}/\boldsymbol{\Lambda}\mathscr{C} \in \mathsf{DM}$, and so forth, finally arriving at (B16)(ii).

For finitary g-matrices, almost all these properties can be formulated as Gentzen-style rules: Those for the PC are in fact Hilbert-style rules, and have already appeared (Proposition 5.17), and so have the ones for the PDI (Exercise 5.7); as to the PDN and the PWCO, it is clear that they correspond to the following ones:[†]

$$\frac{\Gamma, \varphi \rhd \psi}{\Gamma, \neg\neg\varphi \rhd \psi} \qquad \frac{\Gamma \rhd \varphi}{\Gamma \rhd \neg\neg\varphi} \qquad \frac{\varphi \rhd \psi}{\neg\psi \rhd \neg\varphi} \tag{79}$$

Therefore:

(B20) $\mathcal{M} \in \mathsf{FGMod}\mathcal{B}$ if and only if $\mathcal{M}$ is a model of the Gentzen-style rules (79) together with those corresponding to the PC and the PDI, is finitary and has no theorems.

Of course, not every g-model of $\mathcal{B}$ is full; for instance, in $\boldsymbol{M}_4$ the closure systems $\{\{\mathbf{n},\mathbf{t}\}, M_4\}$ and $\{\{\mathbf{t}\}, M_4\}$ do not define full g-models of $\mathcal{B}$. What is missing

[†] The rules for the PDN are clearly equivalent to the Hilbert-style rules $\neg\neg x \rhd x$ and $x \rhd \neg\neg x$.

here? Can the full g-models be characterized among the arbitrary g-models? Yes, and by a key property:

(B21) Let $\mathcal{M}$ be a g-model of $\mathcal{B}$ that is finitary and has no theorems. The g-matrix $\mathcal{M}$ is a full g-model of $\mathcal{B}$ if and only if it satisfies the PCONG.
SKETCH OF THE PROOF: For ($\Rightarrow$) use (B19) plus Proposition 5.26 for $\wedge$ and $\vee$, and the PWCO for $\neg$. The key point for ($\Leftarrow$) is the following reasoning: If $\mathsf{DM} \vDash \alpha \approx \beta$, then $\alpha \dashv\vdash_{\mathcal{B}} \beta$, and therefore, in any g-model of $\mathcal{B}$ and for any h, $C\{h\alpha\} = C\{h\beta\}$. Since the g-matrix satisfies the PCONG by assumption, this implies that $h\alpha \equiv h\beta \ (\widetilde{\Omega}^{A}\mathscr{C})$. Using this, one shows that $A^* \vDash \alpha \approx \beta$. Thus, $A^* \in \mathsf{DM}$. Then, one can use (B16) to show that $\mathcal{M}$ is in fact full, by showing that $\mathscr{C}^* = \mathcal{F}ilt_{\wedge}(A^*) \cup \{\emptyset\}$. It is in this point where finitarity again plays as crucial role. ⊠

This shows the importance of the PCONG.

Each of the characterizations of the full g-models of $\mathcal{B}$ yields a "logical" characterization of the variety of De Morgan lattices:

(B22) $A \in \mathsf{DM}$ if and only if there is an inductive closure system $\mathscr{C} \subseteq \mathscr{P}A$ that separates points and satisfies the conditions of any of the previous characterizations.

Finally,

(B23) $\mathcal{B}$ is the weakest logic satisfying the properties of any of the previous characterizations.

This last property is especially interesting regarding the characterizations in (B19) and in (B20): they witness that other logics, besides the usual ones (such as $\mathcal{C\ell}$, $\mathcal{I\ell}$, their fragments, or some modal expansions) can be characterized by Tarski-style conditions or by Gentzen-style rules, as in [249, Chapter 2]. ⊠

EXAMPLE 5.93 (THE FULL G-MODELS OF $\mathcal{C\ell}_{\wedge\vee}$)
The conjunction and disjunction fragment $\mathcal{C\ell}_{\wedge\vee}$ of classical logic is analysed in Example 4.47, where the class $\mathsf{Alg}^*\mathcal{C\ell}_{\wedge\vee}$ is determined. Almost everything done for $\mathcal{B}$ in Examples 5.82 and 5.92 can also be done, in a simpler form, to study the full g-models of $\mathcal{C\ell}_{\wedge\vee}$: one just deletes all facts concerning the negation operation and the function Φ (Exercise 5.62). The logic is defined by the matrix $\langle \mathbf{2}_{\wedge\vee}, \{1\}\rangle$, and the characterization similar to that in (B18) uses only this matrix. The other characterizations that can be obtained in this way can be summarized as follows: For an arbitrary g-matrix $\mathcal{M} = \langle A, \mathscr{C}\rangle$, the following conditions are equivalent:

(i) $\mathcal{M} \in \mathsf{FGMod}\mathcal{C\ell}_{\wedge\vee}$.

(ii) There is a strict surjective homomorphism from $\mathcal{M}$ onto $\langle A, \mathcal{F}ilt_{\wedge} A \cup \{\emptyset\}\rangle$ for some $A \in \mathsf{DL}$.

(iii) $A^* \in \mathsf{DL}$ and $\mathscr{C}^* = \mathcal{F}ilt_{\wedge} A \cup \{\emptyset\}$.

(iv) The closure system $\mathscr{C}$ is inductive and has a base $\mathscr{E}$ made of $\wedge$-filters that are $\vee$-prime, with $\emptyset \in \mathscr{E}$.

(v) $\mathcal{M}$ is finitary, satisfies the PC and the PDI and has no theorems.

(vi) $\mathcal{M}$ is a model of the Gentzen-style rules corresponding to the PC and the PDI, is finitary and has no theorems.

(vii) $\mathcal{M}$ is a g-model of $\mathcal{Cl}_{\wedge\vee}$ that satisfies the PCONG, is finitary and has no theorems.

The equivalence between points (v) and (vi) appears in Exercises 5.7 and 5.12.

From (iii) it follows that $\mathsf{Alg}\mathcal{Cl}_{\wedge\vee} = \mathsf{DL}$; since this is a variety, it also follows that $\mathbb{V}\mathcal{Cl}_{\wedge\vee} = \mathsf{DL}$. Exercise 5.53 suggested a direct, purely algebraic way to show these two facts, without going through the full g-models. Finally, by Proposition 5.26, the properties in (v) imply that the logic is selfextensional; this is also established in Exercise 5.21, and also follows from the fact that $\alpha \dashv\vdash_{\wedge\vee} \beta \iff \mathsf{DL} \vDash \alpha \approx \beta$, thanks to Exercise 5.50, which also implies that $\mathbb{V}\mathcal{Cl}_{\wedge\vee} = \mathsf{DL}$. ⊠

In the two preceding examples, the condition that the g-matrix has no theorems is ubiquitous. It appears because the logics themselves do not have theorems, and this property is transferred to all full g-models (Proposition 5.90). If you prefer logics with theorems (or suspect that it is precisely the lack of theorems that leads to some of the non-standard results we have found), then you may be interested in the logic $\mathcal{Cl}_{\wedge\vee\top\bot}$, the $\langle\wedge,\vee,\top,\bot\rangle$ fragment of $\mathcal{Cl}$, whose algebraic counterpart is the variety of bounded distributive lattices, or in the expansion of $\mathcal{B}$ whose algebraic counterpart is the variety of De Morgan algebras (bounded De Morgan lattices). For these logics very similar results are obtained, as suggested in Exercises 5.63 and 5.64.

The full g-models of $\mathcal{Il}$ are determined in Exercise 5.65, and those of $\mathcal{Cl}$ in Exercise 5.66.

As we see in property (I9) of Example 5.91 and in the just mentioned exercises, the full g-models of well-behaved logics are determined by filters of the logic; several variations of this property are proved for protoalgebraic logics in Theorems 6.39 and 6.41, and for the weakly algebraizable ones in Theorem 6.117. In these cases, the goal of their study is to find characterizations providing further specific information. However, there are no general rules on how to obtain Tarski-style, Gentzen-style and semantic-style characterizations of full g-models; this is each time a different, often challenging enterprise.

The Isomorphism Theorem

The rest of this section is devoted to the study of the set of full g-models of a logic on a fixed algebra $\boldsymbol{A}$; this set is denoted by $\mathcal{FGMod}_{\mathcal{L}}\boldsymbol{A}$. Since $\boldsymbol{A}$ is fixed, it is convenient to identify this set with the set of the corresponding closure systems, denoted† by $\mathrm{FCS}_{\mathcal{L}}\boldsymbol{A} := \{\mathscr{C} \subseteq \mathscr{P}A : \langle \boldsymbol{A}, \mathscr{C}\rangle \in \mathsf{FGMod}\,\mathcal{L}\}$, and to apply the term ***full*** to the closure systems themselves; if necessary the set $\mathrm{FCO}_{\mathcal{L}}\boldsymbol{A}$ of the corresponding closure operators can be considered as well. These sets,

† "FCS" stands for Full Closure Systems, and "FCO" stands for Full Closure Operators.

which are subsets of the sets $\mathrm{CS}(A)$ and $\mathrm{CO}(A)$ considered in Section 1.5, are ordered under the set inclusion relation $\subseteq$ and are dually order isomorphic by the restriction of the dual order isomorphism of Theorem 1.33.

It is useful to start with yet another characterization of full g-models, which gives new meanings to the term "full".

PROPOSITION 5.94
$\langle \boldsymbol{A}, \mathscr{C} \rangle \in \mathsf{FGMod}\mathcal{L}$ if and only if $\mathscr{C} = \{F \in \mathcal{F}i_{\mathcal{L}}\boldsymbol{A} : \widetilde{\Omega}^{\boldsymbol{A}}\mathscr{C} \subseteq \Omega^{\boldsymbol{A}}F\}$.

PROOF: A g-matrix $\langle \boldsymbol{A}, \mathscr{C} \rangle$ is a full g-model of $\mathcal{L}$ if and only if $\mathscr{C} = \pi^{-1}\mathcal{F}i_{\mathcal{L}}(\boldsymbol{A}^*)$, where $\boldsymbol{A}^* = \boldsymbol{A}/\widetilde{\Omega}^{\boldsymbol{A}}\mathscr{C}$ and $\pi \colon \boldsymbol{A} \to \boldsymbol{A}^*$ is the reduction function. By Lemma 4.36 for $\theta := \widetilde{\Omega}^{\boldsymbol{A}}\mathscr{C}$, we obtain that $\pi^{-1}\mathcal{F}i_{\mathcal{L}}(\boldsymbol{A}^*) = \{F \in \mathcal{F}i_{\mathcal{L}}\boldsymbol{A} : \widetilde{\Omega}^{\boldsymbol{A}}\mathscr{C} \subseteq \Omega^{\boldsymbol{A}}F\}$, and this produces the expression in the statement. ⊠

This simple property offers two interesting readings of what the "full" in "full g-model" refers to:

- It says that a g-matrix is a full g-model of $\mathcal{L}$ if and only if its closure system contains *all* the $\mathcal{L}$-filters which the Tarski congruence of the closure system is compatible with.
- Notice that $\mathscr{C} \subseteq \{F \in \mathcal{F}i_{\mathcal{L}}\boldsymbol{A} : \widetilde{\Omega}^{\boldsymbol{A}}\mathscr{C} \subseteq \Omega^{\boldsymbol{A}}F\}$ for any g-model $\langle \boldsymbol{A}, \mathscr{C} \rangle$ of $\mathcal{L}$; thus, 5.94 says that a g-model of $\mathcal{L}$ is full if and only if it is *the largest* of all the g-models with the same Tarski congruence.

A direct consequence of this property is that full g-models are uniquely determined by their own Tarski congruence, and hence for every algebra $\boldsymbol{A}$, the operator $\widetilde{\Omega}^{\boldsymbol{A}}$ is one-to-one over $\mathrm{FCS}_{\mathcal{L}}\boldsymbol{A}$. This is one of the key points in the proof of the following important result [107]:

THEOREM 5.95 (THE ISOMORPHISM THEOREM FOR FULL G-MODELS)
For each algebra $\boldsymbol{A}$, the Tarski operator $\widetilde{\Omega}^{\boldsymbol{A}}$ is a dual order isomorphism between $\mathrm{FCS}_{\mathcal{L}}\boldsymbol{A}$ and $\mathrm{Con}_{\mathsf{Alg}\mathcal{L}}\boldsymbol{A}$ as ordered sets (order is set inclusion):

$$\widetilde{\Omega}^{\boldsymbol{A}} \colon \mathrm{FCS}_{\mathcal{L}}\boldsymbol{A} \overset{\mathrm{D}}{\cong} \mathrm{Con}_{\mathsf{Alg}\mathcal{L}}\boldsymbol{A}$$

PROOF: First, the function goes as stated: If $\langle \boldsymbol{A}, \mathscr{C} \rangle$ is full, then $\langle \boldsymbol{A}^*, \mathscr{C}^* \rangle$ is reduced and full, hence $\boldsymbol{A}^* \in \mathsf{Alg}\mathcal{L}$, which proves that $\widetilde{\Omega}^{\boldsymbol{A}}\mathscr{C} \in \mathrm{Con}_{\mathsf{Alg}}\mathcal{L}$. Second, the function $\widetilde{\Omega}^{\boldsymbol{A}}$ is order-reversing by Proposition 5.35. We now calculate what will be its inverse. Take any $\theta \in \mathrm{Con}_{\mathsf{Alg}\mathcal{L}}\boldsymbol{A}$. Since $\boldsymbol{A}/\theta \in \mathsf{Alg}\mathcal{L}$, by Corollary 5.88, $\langle \boldsymbol{A}/\theta, \mathcal{F}i_{\mathcal{L}}(\boldsymbol{A}/\theta) \rangle$ is reduced, that is, $\widetilde{\Omega}^{\boldsymbol{A}/\theta}\mathcal{F}i_{\mathcal{L}}(\boldsymbol{A}/\theta) = \mathrm{Id}_{A/\theta}$. Now let us put $\mathscr{C}_\theta := \pi_\theta^{-1}\mathcal{F}i_{\mathcal{L}}(\boldsymbol{A}/\theta)$, where $\pi_\theta \colon \boldsymbol{A} \to \boldsymbol{A}/\theta$ is the canonical projection. By its own definition, $\langle \boldsymbol{A}, \mathscr{C}_\theta \rangle$ is full, that is, $\mathscr{C}_\theta \in \mathrm{FCS}_{\mathcal{L}}\boldsymbol{A}$. Moreover, π_θ becomes then a strict surjective homomorphism from $\langle \boldsymbol{A}, \mathscr{C}_\theta \rangle$ to $\langle \boldsymbol{A}/\theta, \mathcal{F}i_{\mathcal{L}}(\boldsymbol{A}/\theta) \rangle$, and by Proposition 5.49.6 this implies that

$$\begin{aligned}\widetilde{\Omega}^{\boldsymbol{A}}\mathscr{C}_\theta &= \widetilde{\Omega}^{\boldsymbol{A}}\pi_\theta^{-1}\mathcal{F}i_{\mathcal{L}}(\boldsymbol{A}/\theta) \\ &= \pi_\theta^{-1}\widetilde{\Omega}^{\boldsymbol{A}/\theta}\mathcal{F}i_{\mathcal{L}}(\boldsymbol{A}/\theta) = \pi_\theta^{-1}\mathrm{Id}_{A/\theta} = \ker \pi_\theta = \theta.\end{aligned}$$

Thus, the composition of the function $\theta \mapsto \mathscr{C}_\theta$ with $\widetilde{\Omega}^{\boldsymbol{A}}$ is the identity relation on $\mathrm{Con}_{\mathsf{Alg}\mathcal{L}}\boldsymbol{A}$. The composition in the reverse order is the identity relation on $\mathrm{FCS}_{\mathcal{L}}\boldsymbol{A}$: If $\mathscr{C} \in \mathrm{FCS}_{\mathcal{L}}\boldsymbol{A}$ then its reduction function π is a strict surjective homomorphism from $\langle \boldsymbol{A}, \mathscr{C} \rangle$ onto $\langle \boldsymbol{A}^*, \mathcal{F}i_{\mathcal{L}}(\boldsymbol{A}^*) \rangle$, and therefore $\mathscr{C} = \pi^{-1}\mathcal{F}i_{\mathcal{L}}(\boldsymbol{A}^*) = \pi^{-1}\mathcal{F}i_{\mathcal{L}}(\boldsymbol{A}/\widetilde{\Omega}^{\boldsymbol{A}}\mathscr{C})$ which by definition is $\mathscr{C}_{\widetilde{\Omega}^{\boldsymbol{A}}\mathscr{C}}$. This establishes at the same time that both functions are bijective, and inverse to one another, between $\mathrm{FCS}_{\mathcal{L}}\boldsymbol{A}$ and $\mathrm{Con}_{\mathsf{Alg}\mathcal{L}}\boldsymbol{A}$. It only remains to show that the function $\theta \mapsto \mathscr{C}_\theta$ is order-reversing, as $\widetilde{\Omega}^{\boldsymbol{A}}$ is, and the dual order isomorphism will be established. Take $\theta_1, \theta_2 \in \mathrm{Con}_{\mathsf{Alg}\mathcal{L}}\boldsymbol{A}$ with $\theta_1 \subseteq \theta_2$. By the usual universal-algebraic construction, $(\boldsymbol{A}/\theta_1)/(\theta_2/\theta_1) \cong \boldsymbol{A}/\theta_2$, therefore there is some (algebraic) surjective homomorphism j such that the diagram

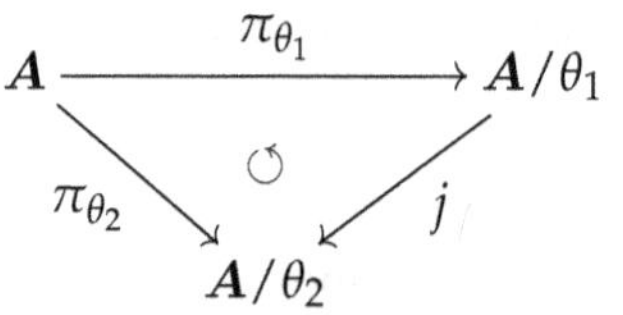

commutes. Now, if $F \in \mathscr{C}_{\theta_2} = \pi_{\theta_2}^{-1}\mathcal{F}i_{\mathcal{L}}(\boldsymbol{A}/\theta_2)$, then there is some $G \in \mathcal{F}i_{\mathcal{L}}(\boldsymbol{A}/\theta_2)$ such that $F = \pi_{\theta_2}^{-1}G$. But $j^{-1}G \in \mathcal{F}i_{\mathcal{L}}(\boldsymbol{A}/\theta_1)$ by Proposition 2.24, and by construction $\pi_{\theta_2} = j \circ \pi_{\theta_1}$; therefore $F = \pi_{\theta_1}^{-1}j^{-1}G \in \mathscr{C}_{\theta_1}$. This shows that $\mathscr{C}_{\theta_2} \subseteq \mathscr{C}_{\theta_1}$ and the proof is finished. ⊠

From the proof it is clear that the inverse process to that of constructing the Tarski congruence of a full g-model of $\mathcal{L}$ is the process of "lifting up" the closure system of all $\mathcal{L}$-filters of a quotient algebra. The two ordered sets are complete lattices, and this fact can indeed be proved independently for each of them:

Proposition 5.96

For each algebra $\boldsymbol{A}$, the set $\mathrm{FCS}_{\mathcal{L}}\boldsymbol{A}$ is closed under intersections of non-empty families and has a maximum. Thus it is a complete lattice and its infimum operation is intersection. Its maximum is $\mathcal{F}i_{\mathcal{L}}\boldsymbol{A}$. Its minimum is $\{A\}$ when $\mathcal{L}$ has theorems and it is $\{\varnothing, A\}$ when $\mathcal{L}$ has no theorems.

Proof: Let $\{\mathscr{C}_i : i \in I\} \subseteq \mathrm{FCS}_{\mathcal{L}}\boldsymbol{A}$ be a non-empty family, and put $\mathscr{C} := \bigcap \mathscr{C}_i$. This is a closure system, and since in general $\widetilde{\Omega}^{\boldsymbol{A}}\mathscr{C} = \bigcap_{F \in \mathscr{C}} \Omega^{\boldsymbol{A}}F$, one always has that $\mathscr{C} \subseteq \{F \in \mathcal{F}i_{\mathcal{L}}\boldsymbol{A} : \widetilde{\Omega}^{\boldsymbol{A}}\mathscr{C} \subseteq \Omega^{\boldsymbol{A}}F\}$. Now take some $F \in \mathcal{F}i_{\mathcal{L}}\boldsymbol{A}$ such that $\widetilde{\Omega}^{\boldsymbol{A}}\mathscr{C} \subseteq \Omega^{\boldsymbol{A}}F$. Since $\mathscr{C} \subseteq \mathscr{C}_i$, $\widetilde{\Omega}^{\boldsymbol{A}}\mathscr{C} \supseteq \widetilde{\Omega}^{\boldsymbol{A}}\mathscr{C}_i$ and so $\widetilde{\Omega}^{\boldsymbol{A}}\mathscr{C}_i \subseteq \Omega^{\boldsymbol{A}}F$. Since $\langle \boldsymbol{A}, \mathscr{C}_i \rangle$ is full, by Proposition 5.94, $F \in \mathscr{C}_i$. This works for an arbitrary $i \in I$, therefore $F \in \mathscr{C}$. As a conclusion, $\mathscr{C} = \{F \in \mathcal{F}i_{\mathcal{L}}\boldsymbol{A} : \widetilde{\Omega}^{\boldsymbol{A}}\mathscr{C} \subseteq \Omega^{\boldsymbol{A}}F\}$ which, by the same proposition, says that $\langle \boldsymbol{A}, \mathscr{C} \rangle$ is full, that is, $\mathscr{C} \in \mathrm{FCS}_{\mathcal{L}}\boldsymbol{A}$. This shows that $\mathrm{FCS}_{\mathcal{L}}\boldsymbol{A}$ is closed under intersections of non-empty families. Clearly, $\mathcal{F}i_{\mathcal{L}}\boldsymbol{A}$ is the maximum. Since $\widetilde{\Omega}^{\boldsymbol{A}}\{A\} = \widetilde{\Omega}^{\boldsymbol{A}}\{\varnothing, A\} = A \times A$, factoring out $\boldsymbol{A}$ by any of these two g-matrices produces the trivial (one-element) algebra $\boldsymbol{A}_t$, and then $\mathcal{F}i_{\mathcal{L}}\boldsymbol{A}_t = \{A_t\}$ when $\mathcal{L}$ has theorems, while $\mathcal{F}i_{\mathcal{L}}\boldsymbol{A}_t = \{\varnothing, A_t\}$ when it has no theorems. In the first case, $\langle \boldsymbol{A}, \{A\} \rangle$ is full, and $\{A\}$ is clearly contained in all full closure systems, therefore it is the minimum one. In the second case, it is $\langle \boldsymbol{A}, \{\emptyset, A\}\rangle$ the one that is full, and since in this case full g-models have no theorems (Proposition 5.90), $\{\emptyset, A\}$ is contained in the closure system of every full g-model of $\mathcal{L}$ on $\boldsymbol{A}$, and in this case the minimum in $\mathrm{FCS}_{\mathcal{L}}\boldsymbol{A}$ is $\{\emptyset, A\}$. ⊠

Then, Theorem 5.95 readily implies that $\mathrm{Con}_{\mathsf{Alg}\mathcal{L}}\boldsymbol{A}$ is also a complete lattice, dually isomorphic to $\mathrm{FCS}_{\mathcal{L}}\boldsymbol{A}$; however, that the infimum is also intersection requires an explicit proof:

PROPOSITION 5.97
For each algebra $\boldsymbol{A}$, the set $\mathrm{Con}_{\mathsf{Alg}\mathcal{L}}\boldsymbol{A}$ is closed under intersections of non-empty families and has a maximum. Thus, it is a complete lattice and its infimum operation is intersection. Its maximum is $A \times A$ and its minimum is $\widetilde{\boldsymbol{\Omega}}^{\boldsymbol{A}} \mathcal{F}i_{\mathcal{L}}\boldsymbol{A}$. Therefore the Tarski operator $\widetilde{\boldsymbol{\Omega}}^{\boldsymbol{A}}$ is a dual lattice isomorphism between the complete lattices $\mathrm{FCS}_{\mathcal{L}}\boldsymbol{A}$ and $\mathrm{Con}_{\mathsf{Alg}\mathcal{L}}\boldsymbol{A}$.

PROOF: That $\mathrm{Con}_{\mathsf{Alg}\mathcal{L}}\boldsymbol{A}$ is closed under intersections of non-empty families follows, by Proposition 1.75, from the fact that the class $\mathsf{Alg}\mathcal{L}$ is closed under subdirect products and isomorphisms (Corollary 5.71.1); Exercise 5.67 suggests a direct proof. Clearly $A \times A$ is its maximum, for it is the largest congruence in $\boldsymbol{A}$ and its quotient, the trivial algebra, is in $\mathsf{Alg}\mathcal{L}$ (Corollary 5.71.5). That its minimum is $\widetilde{\boldsymbol{\Omega}}^{\boldsymbol{A}} \mathcal{F}i_{\mathcal{L}}\boldsymbol{A}$ is best shown by using the dual order isomorphism of Theorem 5.95 between $\mathrm{FCS}_{\mathcal{L}}\boldsymbol{A}$ and $\mathrm{Con}_{\mathsf{Alg}\mathcal{L}}\boldsymbol{A}$, because $\mathcal{F}i_{\mathcal{L}}\boldsymbol{A}$ is the maximum of $\mathrm{FCS}_{\mathcal{L}}\boldsymbol{A}$ (Proposition 5.96). ⊠

Thus, we can say that on a fixed algebra $\boldsymbol{A}$, the set of full g-models of a given logic $\mathcal{L}$ has a complete lattice structure, which can be identified with the complete lattice of their closure systems $\langle \mathrm{FCS}_{\mathcal{L}}\boldsymbol{A}, \bigcap, \bigvee^{\mathrm{f}}\rangle$ (the superscript "f" is for "full"). Thanks to the dual isomorphism, an expression for its join operation can be obtained: If $\{\mathscr{C}_i : i \in I\} \subseteq \mathrm{FCS}_{\mathcal{L}}\boldsymbol{A}$, then $\bigvee^{\mathrm{f}} \mathscr{C}_i = \pi^{-1}\mathcal{F}i_{\mathcal{L}}(\boldsymbol{A}/\bigcap \widetilde{\boldsymbol{\Omega}}^{\boldsymbol{A}}\mathscr{C}_i)$. As far as I know, there is no better, more direct description, along the lines of that obtained for the join operation in the lattice of all closure systems on a set in Proposition 1.58. The Galois adjunction studied on pages 300ff. below presents this dual isomorphism from another perspective.

The fact that we are obtaining *dual* order isomorphisms should not upset anyone: after using closure operators instead of closure systems, that is, replacing $\mathrm{FCS}_{\mathcal{L}}\boldsymbol{A}$ by $\mathrm{FCO}_{\mathcal{L}}\boldsymbol{A}$, the Tarski operator becomes order-preserving and hence an (ordinary) order isomorphism.

Going back to Theorem 5.95: I hope you agree that it is an important result. One of its most remarkable aspects is that *it holds for every logic*, as it has no restrictive assumptions (not even finitarity); this contrasts, for instance, with the Isomorphism Theorem for the Leibniz operator $\boldsymbol{\Omega}^{\boldsymbol{A}}$ over $\mathcal{F}i_{\mathcal{L}}\boldsymbol{A}$, which only holds for algebraizable logics (Theorem 4.56). The theorem has applications to particular logics $\mathcal{L}$ as well, provided there are good descriptions of the full g-models of $\mathcal{L}$ and of the class $\mathsf{Alg}\mathcal{L}$. In some cases, one also obtains results of independent, purely algebraic interest, or alternative views of already known facts. Things behave particularly well when $\mathsf{Alg}\mathcal{L}$ is a variety; see Exercise 5.68. Let us review the customary examples.

EXAMPLE 5.98 (MORE ON THE FULL G-MODELS OF $\mathcal{IL}_{\rightarrow}$)
This is an implicative logic, and we know that $\mathsf{Alg}\mathcal{IL}_{\rightarrow} = \mathsf{Alg}^*\mathcal{IL}_{\rightarrow} = \mathsf{HiA}$ (Example 5.91). Therefore, by Theorem 2.32, for every $\boldsymbol{A}$, $\boldsymbol{\Omega}^{\boldsymbol{A}} \colon \mathcal{F}i_{\mathcal{IL}_{\rightarrow}} \cong \mathrm{Con}_{\mathsf{HiA}}\boldsymbol{A}$. By Theorem 5.95, $\widetilde{\boldsymbol{\Omega}}^{\boldsymbol{A}} \colon \mathrm{FCS}_{\mathcal{IL}_{\rightarrow}} \overset{\mathrm{D}}{\cong} \mathrm{Con}_{\mathsf{HiA}}\boldsymbol{A}$. As a consequence, we obtain a dual isomorphism $\mathcal{F}i_{\mathcal{IL}_{\rightarrow}}\boldsymbol{A} \overset{\mathrm{D}}{\cong} \mathrm{FCS}_{\mathcal{IL}_{\rightarrow}}\boldsymbol{A}$ between filters of $\mathcal{IL}_{\rightarrow}$ and its full g-models on $\boldsymbol{A}$. In fact, it is not difficult to show (Exercise 5.69) that this isomorphism can be described with the result in point (I9) in Example 5.91. By Theorem 6.117, this situation is common to all (weakly) algebraizable logics, and any of the three isomorphisms can be derived from the other two.

When $\boldsymbol{A} \in \mathsf{HiA}$, by point (I5) of Example 5.91, $\mathcal{F}i_{\mathcal{IL}_{\rightarrow}}\boldsymbol{A} = \mathcal{F}ilt_{\rightarrow}\boldsymbol{A}$, and since HiA is a variety, $\mathrm{Con}_{\mathsf{Hi}}\boldsymbol{A} = \mathrm{Con}\boldsymbol{A}$. Therefore, $\boldsymbol{\Omega}^{\boldsymbol{A}} \colon \mathcal{F}ilt_{\rightarrow}\boldsymbol{A} \cong \mathrm{Con}\boldsymbol{A}$; this offers another view of the well-known isomorphism between congruences and implicative filters in Hilbert algebras, which is usually described in purely algebraic terms. ⊠

EXAMPLE 5.99 (MORE ON THE FULL G-MODELS OF $\mathcal{CL}_{\wedge\vee}$)
The full g-models of $\mathcal{CL}_{\wedge\vee}$ are characterized in several ways in Example 5.93. By its part (v) and Proposition 5.26, they satisfy the PCONG, thus for them $\widetilde{\boldsymbol{\Omega}}^{\boldsymbol{A}}\mathscr{C} = \boldsymbol{\Lambda}\mathscr{C}$; in this case Theorem 5.95 turns out to say that on every algebra $\boldsymbol{A}$, $\boldsymbol{\Lambda} \colon \mathrm{FCS}_{\mathcal{CL}_{\wedge\vee}}\boldsymbol{A} \overset{\mathrm{D}}{\cong} \mathrm{Con}_{\mathsf{DL}}\boldsymbol{A}$. This is interesting because the relation $\boldsymbol{\Lambda}\mathscr{C}$ is easier to understand, even on purely algebraic grounds, than $\widetilde{\boldsymbol{\Omega}}^{\boldsymbol{A}}\mathscr{C}$.

In particular, since DL is a variety, if $\boldsymbol{A} \in \mathsf{DL}$, then $\mathrm{Con}_{\mathsf{DL}}\boldsymbol{A} = \mathrm{Con}\boldsymbol{A}$, and hence $\boldsymbol{\Lambda} \colon \mathrm{FCS}_{\mathcal{CL}_{\wedge\vee}}\boldsymbol{A} \overset{\mathrm{D}}{\cong} \mathrm{Con}\boldsymbol{A}$. As a side result, this yields a *representation theorem of the congruences of a distributive lattice* as closure systems made of lattice filters plus the empty set, of the precise kind determined in point (iv) of Example 5.93. This is important, because in distributive lattices there is no isomorphism between congruences and a family of filters of certain kind, as there is in other well-known classes of algebras, such as Hilbert algebras (Example 5.98), Boolean algebras, Heyting algebras, or closure algebras.

More in general, by Proposition 5.97 and Exercise 5.68, for any $\boldsymbol{A}$ we know that $\mathrm{Con}_{\mathsf{DL}}\boldsymbol{A} \cong \mathrm{Con}(\boldsymbol{A}/\theta_0)$, where $\theta_0 := \widetilde{\boldsymbol{\Omega}}^{\boldsymbol{A}}\mathcal{F}i_{\mathcal{CL}_{\wedge\vee}}\boldsymbol{A}$. Since $\boldsymbol{A}/\theta_0 \in \mathsf{Alg}\mathcal{CL}_{\wedge\vee} = \mathsf{DL}$, it results that $\boldsymbol{A}/\theta_0$ is a distributive lattice. This already gives lots of information on the lattice $\mathrm{Con}_{\mathsf{DL}}\boldsymbol{A}$ (that it is algebraic and distributive, that its compacts elements form a relatively complemented lattice, and so forth). The dually isomorphic lattice $\mathrm{FCS}_{\mathcal{CL}_{\wedge\vee}}\boldsymbol{A}$ inherits the duals of these properties.

Moreover, it is possible to prove that for an arbitrary algebra $\boldsymbol{A}$ the correspondence $\mathscr{C} \mapsto \mathscr{E}$ found in characterization 5.93(iv) is one-to-one. As a consequence, one obtains a representation of the relative congruences $\mathrm{Con}_{\mathsf{DL}}\boldsymbol{A}$ as the closed sets of certain topology over the family of $\vee$-prime $\wedge$-filters of $\boldsymbol{A}$. This and related results, contained in [117, 118], generalize several well-known representation results in the theory of distributive lattices. ⊠

Example 5.100 (More on the full g-models of $\mathcal{B}$)
The main consequences of Theorem 5.95 for the logic $\mathcal{B}$ are similar to the ones for $\mathcal{Cl}_{\wedge\vee}$. Example 5.82 shows that $\mathsf{Alg}\mathcal{B} = \mathsf{DM}$, the class of De Morgan lattices, which is a variety. So, if $\boldsymbol{A} \in \mathsf{DM}$, then $\mathrm{FCS}_{\mathcal{B}}\boldsymbol{A} \overset{\mathrm{D}}{\cong} \mathrm{Con}_{\mathsf{DM}}\boldsymbol{A} = \mathrm{Con}\boldsymbol{A}$, and hence the congruences of a De Morgan lattice are determined by the full g-models of $\mathcal{B}$ on $\boldsymbol{A}$. By point (B17) of Example 5.92, the full g-models of $\mathcal{B}$ on $\boldsymbol{A}$ are characterized through the families of prime filters of $\boldsymbol{A}$ (in a lattice the $\vee$-prime $\wedge$-filters are the prime filters) that are closed under Φ, to which one has to add $\emptyset$ and A, and, when the algebra is not finite, the condition that the generated closure system is inductive; as in the case of $\mathcal{Cl}_{\wedge\vee}$, here there is no isomorphism between filters and congruences.

Let us consider the less trivial De Morgan lattice $\boldsymbol{M}_6$ obtained by adding two supplementary bounds to $\boldsymbol{M}_4$; it has the lattice structure and the negation depicted in Figure 8.

It is not difficult to check, simply by counting, that $\boldsymbol{M}_6$ has four congruences: $M_6 \times M_6, \mathrm{Id}_{M_6}$ and the two congruences θ_1 and θ_2 described in Figure 8 by their blocks. It is illustrative to see how they are also determined by the full g-models; in practice this means by the families of prime filters that are closed under Φ (the empty set and the total set are always included, and the generated closure system is always inductive because the algebra is finite). Now, there are four prime filters in $\boldsymbol{M}_6$, and the function Φ on them is described by the table in Figure 8. There are thus four eligible families for $\mathscr{E}$:

$$\{\{1\},\{1,c,a,b,d\}\} \qquad\qquad \emptyset$$
$$\{\{1,c,a\},\{1,c,b\}\} \qquad\qquad \{\text{all the prime filters}\}$$

The congruences can be easily obtained from each of these $\mathscr{E}$, because the fact that $\mathscr{E} \cup \{\emptyset\}$ is a base of $\mathscr{C}$ implies that $a \equiv b \ (\theta_{\mathscr{E}})$ if and only if for all $F \in \mathscr{E}$, $[a \in F \iff b \in F]$, where $\theta_{\mathscr{E}}$ denotes here the Tarski congruence of the closure system generated by $\mathscr{E}$. It is straightforward to check that the resulting congruences are indeed $\theta_1, \theta_2, M_6 \times M_6$ and Id_{M_6}, respectively.

Here we see that one cannot associate a single $\mathcal{B}$-filter with each congruence; it is not difficult to check that $\boldsymbol{\Omega}^{\boldsymbol{M}_6}F \neq \mathrm{Id}_{M_6}$ for all $F \in \mathcal{F}i_{\mathcal{B}}\boldsymbol{M}_6$, thus $\boldsymbol{M}_6 \notin \mathsf{Alg}^*\mathcal{B}$.

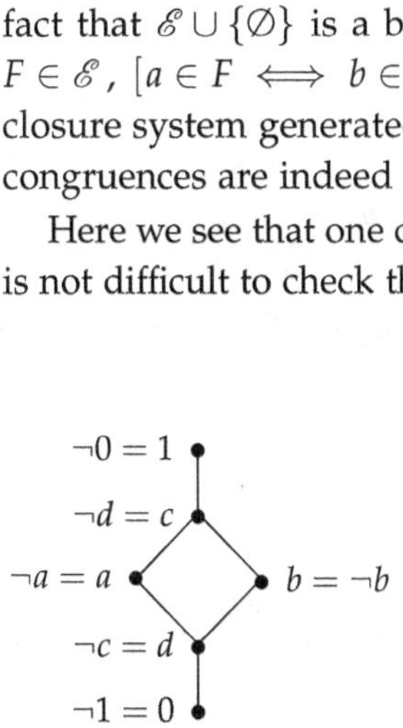

θ_1 has blocks:
$\{0\}$, $\{a,b,c,d\}$, $\{1\}$

θ_2 has blocks:
$\{0,d\}$, $\{a\}$, $\{b\}$, $\{c,1\}$

F	ΦF
$\{1\}$	$\{1,c,a,b,d\}$
$\{1,c,a\}$	$\{1,c,b\}$
$\{1,c,b\}$	$\{1,c,a\}$
$\{1,c,a,b,d\}$	$\{1\}$

Figure 8: The De Morgan lattice $\boldsymbol{M}_6$, two of its congruences, and the function Φ on its prime filters.

What does hold is that $M_6 \in \mathsf{Alg}\mathcal{B} = \mathsf{DM}$, a consequence of the fact that $\mathrm{Id}_{M_6} = \widetilde{\Omega}^{M_6}(\mathcal{F}ilt_\wedge M_6 \cup \{\emptyset\}) = \widetilde{\Omega}^{M_6}(\mathcal{F}i_{\mathcal{B}} M_6)$. However, the three other congruences of M_6 do correspond to $\mathcal{B}$-filters: Trivially, $\Omega^{M_6}\emptyset = \Omega^{M_6} M_6 = M_6 \times M_6$, and it is not difficult to check that $\Omega^{M_6}\{1\} = \Omega^{M_6}\{1,a,b,c,d\} = \theta_1$ and that $\Omega^{M_6}\{1,c\} = \Omega^{M_6}\{1,c,a\} = \Omega^{M_6}\{1,c,b\} = \theta_2$. ⊠

The Galois adjunction of the compatibility relation

It is well known that any binary relation between two sets generates two functions between their power sets with specific order-theoretic properties related to the subset relation; these functions are said to form an *adjoint pair* (also called a *Galois adjunction* or *connection*, or a *residuated pair*). Here for any algebra $\boldsymbol{A}$ we consider the relation COMP between $\mathrm{Con}\boldsymbol{A}$ and $\mathcal{F}i_{\mathcal{L}}\boldsymbol{A}$; the relation "$\theta$ COMP F" holds when θ is compatible with F in the usual sense of the theory of matrices, that is, when $\theta \subseteq \Omega^{\boldsymbol{A}}F$. By the standard procedure we obtain two functions $\mathrm{Cf}\colon \mathscr{P}\mathrm{Con}\boldsymbol{A} \to \mathscr{P}\mathcal{F}i_{\mathcal{L}}\boldsymbol{A}$ and $\mathrm{Cc} : \mathscr{P}\mathcal{F}i_{\mathcal{L}}\boldsymbol{A} \to \mathscr{P}\mathrm{Con}\boldsymbol{A}$ given, for $X \subseteq \mathrm{Con}\boldsymbol{A}$ and $\mathscr{Y} \subseteq \mathcal{F}i_{\mathcal{L}}\boldsymbol{A}$, by:

$$\begin{aligned} \mathrm{Cf}X &:= \{F \in \mathcal{F}i_{\mathcal{L}}\boldsymbol{A} : \theta \text{ COMP } F \text{ for all } \theta \in X\} \\ \mathrm{Cc}\mathscr{Y} &:= \{\theta \in \mathrm{Con}\boldsymbol{A} : \theta \text{ COMP } F \text{ for all } F \in \mathscr{Y}\} \end{aligned} \tag{80}$$

The next result contains the instantiations for our case of a bunch of properties that hold in general for Galois adjunctions; general references for this topic[†] are [25, pp. 124–126], [78, Chapter 7] and [129, Section 3.13].

LEMMA 5.101

1. *$\mathscr{Y} \subseteq \mathrm{Cf}X$ if and only if $\mathrm{Cc}\mathscr{Y} \subseteq X$, for all $X \subseteq \mathrm{Con}\boldsymbol{A}$ and for all $\mathscr{Y} \subseteq \mathcal{F}i_{\mathcal{L}}\boldsymbol{A}$.*
2. *$\mathrm{Cf}X = \bigcap_{\theta\in X}\mathrm{Cf}\{\theta\}$ and $\mathrm{Cc}\mathscr{Y} = \bigcap_{F\in\mathscr{Y}}\mathrm{Cc}\{F\}$, for all $X \subseteq \mathrm{Con}\boldsymbol{A}$ and for all $\mathscr{Y} \subseteq \mathcal{F}i_{\mathcal{L}}\boldsymbol{A}$.*
3. *The functions Cc and Cf are order-reversing, with respect to the subset relation.*
4. *$\mathrm{Cc}\circ\mathrm{Cf}\circ\mathrm{Cc} = \mathrm{Cc}$ and $\mathrm{Cf}\circ\mathrm{Cc}\circ\mathrm{Cf} = \mathrm{Cf}$.*
5. *The compositions $\mathrm{Cc}\circ\mathrm{Cf}$ and $\mathrm{Cf}\circ\mathrm{Cc}$ are closure operators on the sets $\mathrm{Con}\boldsymbol{A}$ and $\mathcal{F}i_{\mathcal{L}}\boldsymbol{A}$, respectively.*
6. *The fixed points of these compositions are the sets in the images of Cc and Cf, respectively. More precisely, if $X \subseteq \mathrm{Con}\boldsymbol{A}$, then $X = \mathrm{Cc}\,\mathrm{Cf}X$ if and only if $X = \mathrm{Cc}\mathscr{Y}$ for some $\mathscr{Y} \subseteq \mathcal{F}i_{\mathcal{L}}\boldsymbol{A}$, the set $\mathrm{Cf}X$ being the largest such $\mathscr{Y}$. Dually, if $\mathscr{Y} \subseteq \mathcal{F}i_{\mathcal{L}}\boldsymbol{A}$, then $\mathscr{Y} = \mathrm{Cf}\,\mathrm{Cc}\mathscr{Y}$ if and only if $\mathscr{Y} = \mathrm{Cf}X$ for some $X \subseteq \mathrm{Con}\boldsymbol{A}$, the set $\mathrm{Cc}\mathscr{Y}$ being the largest such X.*
7. *The sets of fixed points are closed under intersections of non-empty families and each has a largest element; thus, they are complete lattices, and they are dually isomorphic through the functions Cc and Cf.* ⊠

[†] Notice that in some expositions of Galois adjunctions one of the orders (here, the subset relations) is reversed. The symbols "Cf" and "Cc" stand for "Compatible filters" and "Compatible congruences", respectively.

Using the specific definition of the relation COMP and the associated machinery, it is easy to show:

LEMMA 5.102
1. $\mathrm{Cc}\{F\} = \{\theta \in \mathrm{Con}\boldsymbol{A} : \theta \subseteq \boldsymbol{\Omega}^{\boldsymbol{A}}F\}$ *for all* $F \in \mathcal{F}i_{\mathcal{L}}\boldsymbol{A}$.
2. $\mathrm{Cf}\{\theta\} = \{F \in \mathcal{F}i_{\mathcal{L}}\boldsymbol{A} : \theta \subseteq \boldsymbol{\Omega}^{\boldsymbol{A}}F\}$ *for all* $\theta \in \mathrm{Con}\boldsymbol{A}$.
3. $\mathrm{Cc}\mathscr{Y} = \{\theta \in \mathrm{Con}\boldsymbol{A} : \theta \subseteq \bigcap_{F\in\mathscr{Y}} \boldsymbol{\Omega}^{\boldsymbol{A}}F\}$ *for all* $\mathscr{Y} \subseteq \mathcal{F}i_{\mathcal{L}}\boldsymbol{A}$.
4. $\mathrm{Cc}\mathscr{C} = \{\theta \in \mathrm{Con}\boldsymbol{A} : \theta \subseteq \widetilde{\boldsymbol{\Omega}}^{\boldsymbol{A}}\mathscr{C}\}$ *for all closure systems* $\mathscr{C} \subseteq \mathcal{F}i_{\mathcal{L}}\boldsymbol{A}$.
5. $\mathrm{Cf}X = \mathrm{Cf}\{\bigvee X\} = \mathrm{Cf}\{\theta \in \mathrm{Con}\boldsymbol{A} : \theta \subseteq \bigvee X\}$, *for all* $X \subseteq \mathrm{Con}\boldsymbol{A}$, *where* $\bigvee$ *is the supremum operation in* $\mathrm{Con}\boldsymbol{A}$.

PROOF: The first two points are just particular cases of (80), and the proof of the next two is straightforward. The last point requires some thought and is left to you as Exercise 5.70. ⊠

Observe that by point 3 all sets in the image of Cc are principal down-sets in the lattice $\mathrm{Con}\boldsymbol{A}$, and that by point 5 all sets in the image of Cf are actually images of a one-element set, and at the same time they are images of a principal down-set of $\mathrm{Con}\boldsymbol{A}$. Thus, as far as the functions Cc and Cf are concerned, instead of sets of congruences one can safely restrict attention to single congruences (identifying each one with the principal down-set it determines). The result in the second part of Lemma 4.36 can now be rewritten in the following way:

PROPOSITION 5.103
Let $\theta \in \mathrm{Con}\boldsymbol{A}$ *be arbitrary. Then* $\mathrm{Cf}\{\theta\} = \pi^{-1}\mathcal{F}i_{\mathcal{L}}(\boldsymbol{A}/\theta)$, *where* $\pi\colon \boldsymbol{A} \to \boldsymbol{A}/\theta$ *is the canonical projection.* ⊠

THEOREM 5.104
The set of fixed points of $\mathrm{Cf} \circ \mathrm{Cc}$ *is* $\mathrm{FCS}_{\mathcal{L}}\boldsymbol{A}$. *More precisely,* $\mathscr{C} \in \mathrm{FCS}_{\mathcal{L}}\boldsymbol{A}$ *if and only if* $\mathscr{C} = \mathrm{Cf}\{\widetilde{\boldsymbol{\Omega}}^{\boldsymbol{A}}\mathscr{C}\}$. *The fixed points of* $\mathrm{Cc} \circ \mathrm{Cf}$ *are the sets of congruences having the form* $\{\theta \in \mathrm{Con}\boldsymbol{A} : \theta \subseteq \widetilde{\boldsymbol{\Omega}}^{\boldsymbol{A}}\mathscr{C}\}$ *for some* $\mathscr{C} \in \mathrm{FCS}_{\mathcal{L}}\boldsymbol{A}$.

PROOF: By Lemma 5.102.2, the first equivalence is actually Proposition 5.94, reformulated in the present terms. Notice that the set of fixed points of $\mathrm{Cf} \circ \mathrm{Cc}$ is the image of the function Cf. The equivalence just shown implies that if $\mathscr{C} \in \mathrm{FCS}_{\mathcal{L}}\boldsymbol{A}$, then $\mathscr{C}$ belongs to the image of Cf. Conversely, if $X \subseteq \mathrm{Con}\boldsymbol{A}$, then by 5.102.5 and 5.103 it is clear that $\mathrm{Cf}X \in \mathrm{FCS}_{\mathcal{L}}\boldsymbol{A}$. The fixed points of the function $\mathrm{Cc} \circ \mathrm{Cf}$ are the images by Cc of the fixed points of $\mathrm{Cf} \circ \mathrm{Cc}$; by the previous characterization and 5.102.4, the wanted result is obtained. ⊠

This approach offers alternative proofs to several facts obtained before; to detail just one:

PROPOSITION 5.105
For every algebra $\boldsymbol{A}$, *the set* $\mathrm{FCS}_{\mathcal{L}}\boldsymbol{A}$ *is a complete lattice. The infimum operation of this lattice is intersection, and the supremum operation of this lattice is given by* $\bigvee^{\mathrm{f}} \mathscr{C}_i = \mathrm{Cf}\{\bigcap \widetilde{\boldsymbol{\Omega}}^{\boldsymbol{A}}\mathscr{C}_i\} = \pi^{-1}\mathcal{F}i_{\mathcal{L}}(\boldsymbol{A}/\bigcap \widetilde{\boldsymbol{\Omega}}^{\boldsymbol{A}}\mathscr{C}_i)$.

PROOF: The first assertion comes from Lemma 5.101 and Theorem 5.104. Then using that $\bigvee^{\mathrm{f}}$ is the supremum in the set of fixed points of $\mathrm{Cf} \circ \mathrm{Cc}$, and points 4 and 5 of Lemma 5.102, we obtain

$$\begin{aligned}\textstyle\bigvee^{\mathrm{f}} \mathscr{C}_i &= (\mathrm{Cf} \circ \mathrm{Cc}) \textstyle\bigvee^{\mathrm{f}} \mathscr{C}_i = \mathrm{Cf} \textstyle\bigcap \mathrm{Cc}\mathscr{C}_i \\ &= \mathrm{Cf}\{\theta \in \mathrm{Con}\boldsymbol{A} : \theta \subseteq \textstyle\bigcap \widetilde{\boldsymbol{\Omega}}^{\boldsymbol{A}}\mathscr{C}_i\} = \mathrm{Cf}\{\textstyle\bigcap \widetilde{\boldsymbol{\Omega}}^{\boldsymbol{A}}\mathscr{C}_i\},\end{aligned}$$

which gives the first equality; the second one results from Proposition 5.103. ⊠

Following this approach one would eventually find an alternative, and in a sense more direct proof of the Isomorphism Theorem 5.95. By the last statement in Theorem 5.104, the set of fixed points of $\mathrm{Cc} \circ \mathrm{Cf}$ can be identified with the set of congruences of the form $\widetilde{\boldsymbol{\Omega}}^{\boldsymbol{A}}\mathscr{C}$ for some $\mathscr{C} \in \mathrm{FCS}_{\mathcal{L}}\boldsymbol{A}$. Then one would prove that these are exactly the congruences of $\mathrm{Con}_{\mathsf{Alg}\mathcal{L}}\boldsymbol{A}$. Since by 5.102.4, $\widetilde{\boldsymbol{\Omega}}^{\boldsymbol{A}}\mathscr{C} = \max \mathrm{Cc}\mathscr{C}$, one sees that $\widetilde{\boldsymbol{\Omega}}^{\boldsymbol{A}} \colon \mathrm{FCS}_{\mathcal{L}}\boldsymbol{A} \to \mathrm{Con}_{\mathsf{Alg}\mathcal{L}}\boldsymbol{A}$, and one has to check that it is really a dual isomorphism, but this time one works directly with the lattice operations. For instance, one can prove that

$$\widetilde{\boldsymbol{\Omega}} \textstyle\bigvee^{\mathrm{f}} \mathscr{C}_i = \max \mathrm{Cc} \textstyle\bigvee^{\mathrm{f}} \mathscr{C}_i = \max \textstyle\bigcap \mathrm{Cc}\mathscr{C}_i = \textstyle\bigcap \widetilde{\boldsymbol{\Omega}}^{\boldsymbol{A}}\mathscr{C}_i .$$

Exercise 5.71 proposes you to fill in all the details. This approach was introduced in [111], where it is further exploited; in this paper everything is assumed to be finitary, but the points discussed here do not need this restriction.

Exercises for Section 5.5

EXERCISE 5.55. Examine which of the Tarski-style conditions (74) for negation (page 246) are preserved by strict surjective homomorphisms, and which ones transfer to full g-models of a logic.

EXERCISE 5.56. Work out all details of Example 5.91 on the algebraic study of the u-DDT and the determination of the full g-models of $\mathcal{I}\ell_{\rightarrow}$.

HINTS. Everything is more or less straightforward except point 2 of property (I5), the key step in this example: To prove this you have to use some properties of implicative filters in Hilbert algebras, which you can find for instance in [208, Section II.3]. The other tricky step is the crucial role of finitarity in (I4) and in (i)⇒(ii) of (I6); you may find some inspiration in the proof of Theorem 7.53.5.

EXERCISE 5.57. Prove the following strengthening of the characterization of point (I8) of Example 5.91: $\mathcal{I}\ell_{\rightarrow}$, viewed as the g-matrix $\langle \boldsymbol{Fm}, \mathcal{T}h\mathcal{I}\ell_{\rightarrow} \rangle$, is the weakest of all g-matrices on the formula algebra that satisfy the u-DDT.

HINT. Use Exercise 5.10.

COMMENT. Recall that *not all* g-matrices on the formula algebra are logics; see the warning on page 40.

EXERCISE 5.58. Prove that in every Hilbert algebra the following equations and quasi-equations hold:

$$\top \to y \approx \top \rightarrow y \approx \top \quad \text{(that is, } \{1\} \text{ is closed under } \textit{Modus Ponens}\text{)}$$

$$x \to \top \approx \top$$

$$x \to y \approx \top \wedge y \to z \approx \top \rightarrow x \to z \approx \top$$

Conclude that every Hilbert algebra is an implicative algebra.

EXERCISE 5.59. Let $\boldsymbol{A} = \langle A, \neg \rangle$ be an algebra of type $\langle 2 \rangle$ such that $a = \neg\neg a$ for all $a \in A$. Prove that $\{a \in A : \neg a \notin X\} = A \smallsetminus \neg X$, where $\neg X := \{\neg a : a \in X\}$, for all $X \subseteq A$.

EXERCISE 5.60. Let $\boldsymbol{A}$ be a (semi)lattice and $\mathscr{C} \subseteq \mathcal{F}ilt_{\wedge}\boldsymbol{A} \cup \{\varnothing\}$ an inductive closure system. Assume that in addition $\boldsymbol{\Lambda}\mathscr{C} = \mathrm{Id}_A$. Prove that $\mathscr{C} \smallsetminus \{\varnothing\} = \mathcal{F}ilt_{\wedge}\boldsymbol{A}$.

HINT. Observe that the associated closure operator C satisfies the PC, and use the following trick: under this assumption, $a \in C\{b\}$ if and only if $C\{a \wedge b\} = C\{b\}$. The strategy needed here is similar to that in the last part of the proof of Proposition 7.31.

EXERCISE 5.61. Reconstruct as much as possible the sketched proofs of points (B17)–(B21) of Example 5.92.

EXERCISE 5.62. Strip off all considerations concerning negation from the process in Example 5.92, as indicated in Example 5.93, and prove that the reduced full g-models of $\mathcal{C}\ell_{\wedge\vee}$ are the g-matrices of the form $\langle \boldsymbol{A}, \mathcal{F}ilt_{\wedge}\boldsymbol{A} \cup \{\varnothing\}\rangle$ for some $\boldsymbol{A} \in \mathsf{DL}$, and as many of the characterizations of the full g-models of $\mathcal{C}\ell_{\wedge\vee}$ listed in Example 5.93 as possible.

EXERCISE 5.63. Rework Examples 4.47 and 5.93, but this time for the logic $\mathcal{C}\ell_{\wedge\vee\top\bot}$, the $\langle \wedge, \vee, \top, \bot \rangle$ fragment of $\mathcal{C}\ell$. In particular, show that this logic is selfextensional; that $\mathsf{Alg}\mathcal{C}\ell_{\wedge\vee\top\bot} = \mathbb{V}\mathcal{C}\ell_{\wedge\vee\top\bot} = \mathsf{DL}_{01}$, the variety of bounded distributive lattices; and find characterizations of its full g-models. Show directly, using Theorem 5.95, that this logic is not algebraizable.

HINT. Notice that the addition of the constants to the similarity type does not alter the congruences, but it does alter the homomorphisms and the subalgebras; in particular, the sets $\{1\}$ and $\{0\}$ are no longer subuniverses of $\mathbf{2}_{\wedge\vee\top\bot}$, and the constant functions are no longer homomorphisms from another algebra to $\mathbf{2}_{\wedge\vee\top\bot}$. The algebras in the expanded similarity type have two constant elements, conventionally denoted by 1 and 0; in the full g-models, they have to be forced to be a theorem and an inconsistent element, respectively.

EXERCISE 5.64. If you finished Exercise 5.63 and want more of the same style, rework Examples 5.82 and 5.92, this time for a logic in the language $\langle \wedge, \vee, \neg, \top \rangle$ defined by the g-matrix $\langle \boldsymbol{M}_4^{\top}, \mathcal{F}ilt_{\wedge}\boldsymbol{M}_4^{\top} \rangle$ similar to the Belnap-Dunn's four-valued logic $\mathcal{B}$ ($\boldsymbol{M}_4^{\top}$ denotes the algebra $\boldsymbol{M}_4$ expanded with its maximum element); show that it is still selfextensional and non-protoalgebraic; that its algebraic counterpart is the variety of De Morgan algebras; and find characterizations of its full g-models. (Notice that, in contrast to the case of $\mathcal{C}\ell_{\wedge\vee\top\bot}$ treated in Exercise 5.63, only one constant is necessary here, due to the presence of negation satisfying the PWCO.)

EXERCISE 5.65. Let $\langle \boldsymbol{A}, \mathscr{C} \rangle$ be an arbitrary g-matrix with $\boldsymbol{A} = \langle A, \wedge, \vee, \to, \neg \rangle$ of type $\langle 2, 2, 2, 1 \rangle$. Using that $\mathsf{Alg}\mathcal{I}\ell = \mathsf{HA}$ (Exercise 2.31) and getting some inspiration from Example 5.91, prove that the following conditions are equivalent:

(i) $\langle \boldsymbol{A}, \mathscr{C} \rangle$ is a full g-model of $\mathcal{I}\ell$.

(ii) $\boldsymbol{A}^* \in \mathsf{HA}$ and $\mathscr{C}^* = \mathcal{F}ilt_{\to}\boldsymbol{A}$.

(iii) There is a strict surjective homomorphism from $\langle \boldsymbol{A}, \mathscr{C} \rangle$ to a g-matrix of the form $\langle \boldsymbol{B}, \mathcal{F}ilt_{\to}\boldsymbol{B} \rangle$ for some $\boldsymbol{B} \in \mathsf{HA}$.

(iv) There is some $F_0 \in \mathcal{F}i_{\mathcal{I}\ell}\boldsymbol{A}$ such that $\mathscr{C} = \{F \in \mathcal{F}i_{\mathcal{I}\ell}\boldsymbol{A} : F \supseteq F_0\}$.

(v) C is finitary and satisfies the properties PC, PDI, u-DDT and PIRA with respect to the operations $\wedge, \vee, \rightarrow$ and $\neg$, respectively.

Conclude that $\mathcal{I}\ell$ is the weakest logic that satisfies the Tarski-style conditions appearing in point (v).

COMMENT. It is possible to see that everything can be formulated in the language $\langle \wedge, \vee, \rightarrow, \bot \rangle$ of type $\langle 2,2,2,0 \rangle$; the PIRA has to be replaced by the condition that $0 := \bot^{\boldsymbol{A}}$ is C-inconsistent. For a presentation of Heyting algebras in this similarity type, see [19, Chapter IX].

EXERCISE 5.66. Formulate and prove characterizations of the full g-models of $\mathcal{C}\ell$ in terms of the class of Boolean algebras parallel to the ones for $\mathcal{I}\ell$ in Exercise 5.65.

HINT. In the same language as that of $\mathcal{I}\ell$, you need only replace the PIRA by the PRA. However, it is possible to have similar results in more restricted languages, because in Boolean algebras some operations are definable in terms of others.

EXERCISE 5.67. Let $\mathcal{L}$ be a logic and $\boldsymbol{A}$ an algebra, and let $\{\theta_i : i \in I\} \subseteq \mathrm{Con}_{\mathsf{Alg}\mathcal{L}}\boldsymbol{A}$ for some $I \neq \emptyset$. Put $\theta := \bigcap_{i \in I} \theta_i$. Prove directly, without using that $\mathsf{Alg}\mathcal{L}$ is closed under subdirect products, that the g-matrix $\langle \boldsymbol{A}/\theta, \mathcal{F}i_{\mathcal{L}}(\boldsymbol{A}/\theta) \rangle$ is reduced. Conclude that the set $\mathrm{Con}_{\mathsf{Alg}\mathcal{L}}\boldsymbol{A}$ is closed under intersections of arbitrary non-empty families.

EXERCISE 5.68. Assume that $\mathsf{Alg}\mathcal{L}$ is a variety and let $\boldsymbol{A}$ be any algebra. Prove that if $\theta \in \mathrm{Con}_{\mathsf{Alg}\mathcal{L}}\boldsymbol{A}$ and $\theta' \in \mathrm{Con}\boldsymbol{A}$ is such that $\theta \subseteq \theta'$, then $\theta' \in \mathrm{Con}_{\mathsf{Alg}\mathcal{L}}\boldsymbol{A}$. Conclude that in this case $\mathrm{Con}_{\mathsf{Alg}\mathcal{L}}\boldsymbol{A} = [\theta_0, , A \times A]$, where $\theta_0 := \widetilde{\Omega}^{\boldsymbol{A}} \mathcal{F}i_{\mathcal{L}}\boldsymbol{A}$ (thus, a filter in the lattice $\mathrm{Con}\boldsymbol{A}$, and hence a complete sublattice). Finally, show that $\mathrm{Con}_{\mathsf{Alg}\mathcal{L}}\boldsymbol{A} \cong \mathrm{Con}(\boldsymbol{A}/\theta_0)$.

EXERCISE 5.69. Show that the result in point (I9) of Example 5.91 defines for each algebra $\boldsymbol{A}$ a dual order isomorphism between the set of all $\mathcal{I}\ell_{\rightarrow}$-filters of $\boldsymbol{A}$ and the set of all full g-models of $\mathcal{I}\ell_{\rightarrow}$ on $\boldsymbol{A}$. Do the same for condition (iv) of Exercise 5.65, relatively to $\mathcal{I}\ell$.

COMMENT. A general statement of this property, for all weakly algebraizable logics, appears in Theorem 6.117.

EXERCISE 5.70. Prove that $\mathrm{Cf}X = \mathrm{Cf}\{\bigvee X\} = \mathrm{Cf}\{\theta \in \mathrm{Con}\boldsymbol{A} : \theta \subseteq \bigvee X\}$ for $X \subseteq \mathrm{Con}\boldsymbol{A}$, with $\bigvee$ denoting the supremum operation in the lattice $\mathrm{Con}\boldsymbol{A}$.

EXERCISE 5.71. Fill in all the missing details in the hint given at the end of the section for an alternative proof of Theorem 5.95 using exclusively the approach of the Galois adjunction of the compatibility relation.

5.6 Generalized matrices as models of Gentzen systems

This section is an introduction to one of the directions of research on one of the most natural application of g-matrices, with particular emphasis on the contribution of the notion of a full g-model to establish a link between sentential logics and Gentzen systems. The section just includes the basic definitions and a few of the results obtained so far. For more information you may read [107, 109, 111] and Raftery's [200]; this one treats the more comprehensive notion of a "Gentzen relation", which is based on two-sided sequents of the form $\Gamma \rhd \Delta$ in which Γ and Δ are finite sequences of formulas.

Recall from page 8 that a *sequent* is a pair $\Gamma \rhd \varphi$ with Γ a *finite set* of formulas and φ a formula, and that two sets of sequents are particularly highlighted: the set *Seq* of all sequents, and the set Seq° of all sequents with $\Gamma \neq \emptyset$. It turns out that it makes little sense to consider sequents with an empty antecedent when dealing with a logic without theorems; hence the distinction.

Recall (Definition 1.18) that a *Gentzen system* is a pair $\langle \boldsymbol{L}, \vdash_{\mathfrak{G}} \rangle$ where $\boldsymbol{L}$ is an algebraic language, and $\vdash_{\mathfrak{G}}$ is a substitution-invariant closure relation on either the set *Seq* or the set Seq° of sequents built over the set of formulas of type $\boldsymbol{L}$; it is practical to denote the set by $Seq\,\mathfrak{G}$. The theorems of $\vdash_{\mathfrak{G}}$ are called the *derivable sequents* of $\mathfrak{G}$. A Gentzen-style (g-)rule

$$\frac{\{\Gamma_i \rhd \varphi_i : i \in I\}}{\Gamma \rhd \varphi} \tag{81}$$

is ***a (derivable) (g-)rule of*** $\mathfrak{G}$ when $\{\Gamma_i \rhd \varphi_i : i \in I\} \vdash_{\mathfrak{G}} \Gamma \rhd \varphi$. By contrast, (81) is an ***admissible (g-)rule of*** $\mathfrak{G}$ when for any substitution σ, if for all $i \in I$ the sequent $\sigma\Gamma_i \rhd \sigma\varphi_i$ is a derivable sequent of $\mathfrak{G}$, then the sequent $\sigma\Gamma \rhd \sigma\varphi$ is also a derivable sequent of $\mathfrak{G}$. As with Hilbert-style rules, the "g-" is suppressed when I is finite.

If a Gentzen system is *finitary* then, as happens with sentential logics, it can be defined by a ***Gentzen calculus***, which is just an axiomatic or Hilbert-style calculus with sequents instead of formulas; its axioms are usually called ***initial sequents***. Because of the (crucial) fact that sequents contain finite sets of formulas *of arbitrary size*, no axiomatization can be *finite* in the strict sense of the term. However, we can also consider *schematizations*, which are axiomatizations by means of rules written with the help of *metavariables* that represent arbitrary finite sets of formulas (this issue is discussed after Definition 5.11). Most of the Gentzen calculi in the literature are, strictly speaking, finitely schematized rather than finitely axiomatized.

The notion of a g-matrix satisfying or being a model of a Gentzen-style g-rule is introduced in Definition 5.11. Clearly, *every class of g-matrices defines a Gentzen system* in the obvious way: the g-rules of this system are the g-rules that are satisfied by all the g-matrices in the class. That is, if M is a class of g-matrices, then the Gentzen system $\mathfrak{G}\mathsf{M}$ it defines is given by:

$$\{\Gamma_i \rhd \varphi_i : i \in I\} \vdash_{\mathfrak{G}\mathsf{M}} \Gamma \rhd \varphi \iff \text{for all } h \in \mathrm{Hom}(\boldsymbol{Fm}, \boldsymbol{A}) \text{ and all } \langle \boldsymbol{A}, \mathscr{C} \rangle \in \mathsf{M}, \\ \text{if } h\varphi_i \in C h\Gamma_i \text{ for all } i \in I, \text{ then } h\varphi \in C h\Gamma.$$

Now consider the so-called ***structural rules***. Since Exchange and Contraction are built-in into the notation, because the Γ are sets rather than sequences, here we have to deal only with Axiom (AX), Weakening (W) and Cut (CUT). Due to the just mentioned fact and because we consider them together, they can be formulated in a simple way:

$$\varphi \rhd \varphi \quad (\text{AX}) \qquad \frac{\Gamma \rhd \varphi}{\Gamma, \psi \rhd \varphi} \quad (\text{W}) \qquad \frac{\Gamma \rhd \psi \qquad \Gamma, \psi \rhd \varphi}{\Gamma \rhd \varphi} \quad (\text{CUT})$$

It is easy to see that any g-matrix satisfies all the structural rules, therefore these are all derivable rules of any Gentzen system defined by a class of g-matrices.

Thus, g-matrices can only be useful in the study of this kind of Gentzen systems, which from now on are called ***structural Gentzen systems*.**[†] Using g-matrices as models of Gentzen systems has an algebraic and perhaps a metalogical interest rather than a proof-theoretic one; this notion is defined by extending Definition 5.11 in a natural way:

DEFINITION 5.106
A g-matrix is a **model of a (structural) Gentzen system** $\mathfrak{G}$ *when it is a model of all the g-rules of* $\mathfrak{G}$*. The class of all models of* $\mathfrak{G}$ *is denoted by* $\mathsf{Mod}\mathfrak{G}$.

Thus g-matrices can be at the same time *generalized models* of some (sentential) logics, and *models* of some Gentzen systems; *this dual role is one of the points that give g-matrices their special interest.* The relation of g-matrices to Gentzen systems is parallel to that of matrices to sentential logics. Thus, many of the next definitions and results look familiar and are proved in a standard way.

PROPOSITION 5.107
For any (structural) Gentzen system $\mathfrak{G}$*, the class* $\mathsf{Mod}\mathfrak{G}$ *is closed under the formation of sub-g-matrices and under strict surjective homomorphisms. In particular, it is closed under isomorphisms and under reductions.*

PROOF: The first part is a straightforward consequence of Exercise 5.42, and the second part follows immediately from Theorem 5.61.3. ⊠

The following lemma implies that attention can always be restricted to finitary g-matrices when studying the models of a Gentzen system; this is due to the already mentioned crucial facts that the antecendents of sequents are finite sets of formulas, and is independent of whether the Gentzen system itself is finitary or not. Recall the notion of the finitary companion C_{f} of a closure operator C from Definition 1.47.

LEMMA 5.108
A g-matrix $\langle \boldsymbol{A}, \mathscr{C} \rangle$ *and its finitary companion* $\langle \boldsymbol{A}, \mathscr{C}_{\mathrm{f}} \rangle$ *are models of the same Gentzen-style rules. If a (structural) Gentzen system is complete with respect to some class* M *of g-matrices, then it is also complete with respect to* $\mathsf{M}_{\mathrm{f}} := \{ \langle \boldsymbol{A}, \mathscr{C}_{\mathrm{f}} \rangle : \langle \boldsymbol{A}, \mathscr{C} \rangle \in \mathsf{M} \}$.

PROOF: The first statement is a consequence of the fact (Theorem 1.48) that if $X \subseteq A$ is finite, then $C_{\mathrm{f}}X = CX$, because when evaluating sequents only finite sets of formulas appear. The second statement follows as a straightforward consequence of the first. ⊠

The models of a (structural) Gentzen system are g-matrices, and as such they are subject to all the constructions of the general sections of this chapter. By analogy with what can be done with models of logics, the following natural notions can be introduced, and the following facts can be routinely proved:

[†] Here "structural" is used as is customary in proof theory, and not as referring to the property of structurality (substitution-invariance) of sentential logics.

DEFINITION 5.109
*$\mathsf{Mod}^*\mathfrak{G}$ is the class of all reduced models of $\mathfrak{G}$, and $\mathsf{Alg}\mathfrak{G} := \mathsf{AlgMod}^*\mathfrak{G}$. The algebras in this class are called **$\mathfrak{G}$-algebras**. Also, we put $\mathsf{Mod_f}\mathfrak{G} := \{\mathcal{M} \in \mathsf{Mod}\mathfrak{G} : \mathcal{M}$ is finitary$\}$ and $\mathsf{Mod}^*_\mathsf{f}\mathfrak{G} := \{\mathcal{M} \in \mathsf{Mod}^*\mathfrak{G} : \mathcal{M}$ is finitary$\}$.*

COROLLARY 5.110
$\mathsf{Mod}^*\mathfrak{G} = \mathbb{I}\big((\mathsf{Mod}\mathfrak{G})^*\big)$. ⊠

PROPOSITION 5.111
Every structural Gentzen system $\mathfrak{G}$ is complete with respect to the class $\mathsf{Mod}\mathfrak{G}$ of all its models, with respect to the class $\mathsf{Mod}^*\mathfrak{G}$ of all its reduced models, and with respect of their finitary subclasses $\mathsf{Mod_f}\mathfrak{G}$ and $\mathsf{Mod}^*_\mathsf{f}\mathfrak{G}$. ⊠

In particular, Lemma 5.108 implies that a (sentential) logic $\mathcal{L}$ and its finitary companion $\mathcal{L}_\mathrm{f}$ satisfy the same Gentzen-style rules. Thus, in what follows we limit our attention to the relations between Gentzen systems and *finitary* logics. The following is but one of the main relations between a Gentzen system and a (sentential) logic.

DEFINITION 5.112
*Let $\mathfrak{G}$ be a structural Gentzen system, and $\mathcal{L}$ a finitary logic. $\mathfrak{G}$ is **adequate for** $\mathcal{L}$ when for every finite $\Gamma \cup \{\varphi\} \subseteq Fm$,*

$$\Gamma \vdash_{\mathcal{L}} \varphi \iff \text{the sequent } \Gamma \rhd \varphi \text{ is a derivable sequent of } \mathfrak{G}, \tag{82}$$

and either $Seq\,\mathfrak{G} = Seq$ (if $\mathcal{L}$ has theorems) or $Seq\,\mathfrak{G} = Seq^\circ$ (if $\mathcal{L}$ has not theorems).

Notice that if $Seq\,\mathfrak{G} = Seq$ and $\mathfrak{G}$ satisfies (82) for a logic $\mathcal{L}$ without theorems, then its restriction to Seq° is also a Gentzen system, and also satisfies (82). Thus, there is no point in handling sequents of the form $\emptyset \rhd \varphi$ in the study a logic without theorems; it is more convenient to eliminate them. This justifies the distinction made in the definition.

PROPOSITION 5.113
Let $\mathfrak{G}$ be a structural Gentzen system that is adequate for a finitary logic $\mathcal{L}$.

1. *$\mathcal{L}$ satisfies all the derivable g-rules of $\mathfrak{G}$.*
2. $\mathsf{Mod}\mathfrak{G} \subseteq \mathsf{GMod}\mathcal{L}$.
3. $\mathsf{Alg}\mathfrak{G} \subseteq \mathsf{Alg}\mathcal{L}$.

PROOF: 1. Let (81) be a derivable g-rule of $\mathfrak{G}$, and suppose that σ is a substitution such that $\sigma\Gamma_i \vdash_{\mathcal{L}} \sigma\varphi_i$ for all $i \in I$. By the assumption this means that the sequents $\sigma\Gamma_i \rhd \sigma\varphi_i$ are derivable in $\mathfrak{G}$. But the set of derivable sequents of $\mathfrak{G}$ is obviously closed under (the substitution instances of) its own derivable g-rules, therefore the sequent $\sigma\Gamma \rhd \sigma\varphi$ is also derivable in $\mathfrak{G}$, that is, $\sigma\Gamma \vdash_{\mathcal{L}} \sigma\varphi$. Thus, $\mathcal{L}$ satisfies the rule (81).

2. Assume that $\Gamma \vdash_{\mathcal{L}} \varphi$ holds, and let $\langle \boldsymbol{A}, C\rangle$ be any model of $\mathfrak{G}$. By finitarity of $\mathcal{L}$, we can assume that Γ is finite. Then by assumption the sequent $\Gamma \rhd \varphi$ is derivable in $\mathfrak{G}$, that is, the rule $\langle \emptyset, \Gamma \rhd \varphi\rangle$ is one of the rules of $\mathfrak{G}$. Thus, $\langle \boldsymbol{A}, C\rangle$

is a model of this rule, but this amounts to satisfying the sequent $\Gamma \rhd \varphi$, which means that the g-matrix is a g-model of $\mathcal{L}$.

3 follows from 2 by taking algebraic reducts. ⊠

Every finitary logic has at least one adequate structural Gentzen system:

DEFINITION 5.114

Let $\mathcal{L}$ be a finitary logic. The Gentzen system $\mathfrak{G}_{\mathcal{L}}$ is defined by the Gentzen calculus determined as follows:

(a) *$Seq\,\mathfrak{G}_{\mathcal{L}} = Seq$ if $\mathcal{L}$ has theorems, and $Seq\,\mathfrak{G}_{\mathcal{L}} = Seq^{\circ}$ if $\mathcal{L}$ has no theorems.*

(b) *The initial sequents of $\mathfrak{G}_{\mathcal{L}}$ are all sequents $\Gamma \rhd \varphi \in Seq\,\mathfrak{G}_{\mathcal{L}}$ such that $\Gamma \vdash_{\mathcal{L}} \varphi$.*

(c) *The rules of $\mathfrak{G}_{\mathcal{L}}$ are only the structural rules.*

By definition $\mathfrak{G}_{\mathcal{L}}$ is structural. Then:

PROPOSITION 5.115

For every finitary logic $\mathcal{L}$, the Gentzen system $\mathfrak{G}_{\mathcal{L}}$ is adequate for $\mathcal{L}$, is the weakest of all structural Gentzen systems that are adequate for $\mathcal{L}$, and is the only structural Gentzen system $\mathfrak{G}$ such that $\mathsf{Mod}\,\mathfrak{G} = \mathsf{GMod}\,\mathcal{L}$.

PROOF: Since the relation $\vdash_{\mathcal{L}}$ satisfies the structural rules, obviously the derivable sequents of $\mathfrak{G}_{\mathcal{L}}$ are actually its initial sequents, and (82) is satisfied: $\mathfrak{G}_{\mathcal{L}}$ is adequate for $\mathcal{L}$. It is also clear that any Gentzen system adequate for $\mathcal{L}$ must have the initial sequents of $\mathfrak{G}_{\mathcal{L}}$ as derivable ones, and that, being structural, must be closed under the structural rules. Therefore, $\mathfrak{G}_{\mathcal{L}}$ is the weakest of such systems. Since any g-matrix is a model of the structural rules, Definition 5.114 readily implies that $\mathsf{Mod}\,\mathfrak{G}_{\mathcal{L}} = \mathsf{GMod}\,\mathcal{L}$. Now, since by Theorem 5.111 any Gentzen system is determined by the class of its models, the equation $\mathsf{Mod}\,\mathfrak{G} = \mathsf{GMod}\,\mathcal{L}$ implies $\mathfrak{G} = \mathfrak{G}_{\mathcal{L}}$, giving the uniqueness property. ⊠

It is thus clear that the "equation" $\mathsf{Mod}\,\mathfrak{G} = \mathsf{GMod}\,\mathcal{L}$ does not have a particular interest (when restricted to structural $\mathfrak{G}$ and finitary $\mathcal{L}$), because a solution always exists, and is unique: Given a finitary logic $\mathcal{L}$, by Proposition 5.115 the system $\mathfrak{G}_{\mathcal{L}}$ is its unique solution; and given a structural $\mathfrak{G}$, the finitary logic $\mathcal{L}$ obtained by taking (82) as a definition is forcefully its unique solution.

The notion of full adequacy

The equation $\mathsf{Mod}\,\mathfrak{G} = \mathsf{FGMod}\,\mathcal{L}$ may be more interesting. However, when $\mathcal{L}$ is finitary we know that all its full g-models are also finitary, while we do not impose finitarity in general to models of Gentzen systems, so it might be reasonable to restrict the equation to the class $\mathsf{Mod}_{\mathsf{f}}\,\mathfrak{G}$. Unfortunately, even in this case it need not have a solution:

PROPOSITION 5.116

If $\mathcal{L}$ is finitary and has no theorems, then there is no structural Gentzen system $\mathfrak{G}$ such that $\mathsf{Mod}_{\mathsf{f}}\,\mathfrak{G} = \mathsf{FGMod}\,\mathcal{L}$.

PROOF: If $\mathcal{L}$ has no theorems, then neither do the g-matrices in $\mathsf{FGMod}\,\mathcal{L}$ have them. However, for any $\mathfrak{G}$ and any algebra $\boldsymbol{A}$, the inconsistent g-matrix $\langle \boldsymbol{A}, \{A\} \rangle$ is trivially a finitary model of $\mathfrak{G}$, and it does have theorems. ⊠

However, a second slight reformulation of the equation originates a notion that has revealed to be an interesting one:

DEFINITION 5.117
*A structural Gentzen system $\mathfrak{G}$ is **fully adequate**† for a finitary logic $\mathcal{L}$ when one of the following conditions holds:*

(a) *$\mathcal{L}$ has theorems, $Seq\,\mathfrak{G} = Seq$, and the full g-models of $\mathcal{L}$ are exactly the finitary models of $\mathfrak{G}$, i.e., $\mathsf{FGMod}\,\mathcal{L} = \mathsf{Mod_f}\mathfrak{G}$.*

(b) *$\mathcal{L}$ has no theorems, $Seq\,\mathfrak{G} = Seq^{\circ}$, and the full g-models of $\mathcal{L}$ are the finitary models of $\mathfrak{G}$ without theorems, i.e., $\mathsf{FGMod}\,\mathcal{L} = \{\mathcal{M} \in \mathsf{Mod_f}\mathfrak{G} : \mathcal{M}$ has no theorems$\}$.*

PROPOSITION 5.118
Every fully adequate Gentzen system is adequate.

PROOF: Let $\mathfrak{G}$ be a Gentzen system fully adequate for $\mathcal{L}$. Assume first that $\mathcal{L}$ has theorems. By Proposition 5.89, $\Gamma \vdash_{\mathcal{L}} \varphi$ holds if and only if the sequent $\Gamma \rhd \varphi$ holds in all full g-models of $\mathcal{L}$; by full adequacy, this amounts to holding in all finitary models of $\mathfrak{G}$, and by Proposition 5.111 this happens if and only if the sequent is derivable in $\mathfrak{G}$. In the case without theorems, it is easy to make a few small adjustments to basically the same reasoning in order to obtain the same conclusion. ⊠

Since any Gentzen system $\mathfrak{G}$ is complete with respect to the class $\mathsf{Mod_f}\mathfrak{G}$, and every logic $\mathcal{L}$ determines the class $\mathsf{FGMod}\,\mathcal{L}$, if a logic $\mathcal{L}$ has a fully adequate Gentzen system $\mathfrak{G}$, then this one is clearly *unique*. In essence, the central idea of the notion of full adequacy is that the g-rules of the Gentzen system completely characterize the full g-models of the logic; or equivalently, its basic full g-models, that is, the closure systems made of all filters of the logic, on arbitrary algebras. This implies the following interesting properties:

PROPOSITION 5.119
Let $\mathfrak{G}$ be a Gentzen system fully adequate for a finitary logic $\mathcal{L}$. Then:

1. *$\mathcal{L}$ is the weakest logic that satisfies the g-rules of $\mathfrak{G}$.*
2. *The g-rules of $\mathfrak{G}$ are the Gentzen-style g-rules that are satisfied in one (equivalently: in each) of the following classes of g-matrices:*
   (a) *All the basic full g-models of $\mathcal{L}$; that is, all the g-matrices of the form $\langle \boldsymbol{A}, \mathcal{F}i_{\mathcal{L}}\boldsymbol{A} \rangle$, for some algebra $\boldsymbol{A}$.*
   (b) *All the full g-models of $\mathcal{L}$.*

PROOF: 1 is a consequence of the fact that $\mathcal{L}$ is the weakest of all its own full g-models on the formula algebra. 2(b) follows from the definition of the notion

† This notion was introduced in the first edition of [107] under the name "strongly adequate".

of full adequacy and the completeness of every structural Gentzen system with respect to the class of all its finitary models (Proposition 5.111). 2(a) follows from (b) and Theorem 5.63.4. ⊠

While a fully adequate Gentzen system is unique, provided it exists, as observed before, its existence cannot be guaranteed in general, as shown by Example 5.123 below. This also tells us that the converse of Proposition 5.119 does not hold (because the Gentzen system described in it always exists). It is easy to see that, if $\mathfrak{G}$ is defined by a Gentzen calculus, then the expression "the g-rules of $\mathfrak{G}$" in point 1 of the proposition can be replaced by "the rules of the calculus". Thus, the interest of finding a fully adequate Gentzen system for a given logic depends on how nicely can it be presented, that is, on whether one can exhibit a calculus for it and on how nice this calculus is.

EXAMPLES 5.120

From the characterizations of the full g-models of several logics in terms of Gentzen-style rules found on pages 288ff. and elsewhere, it is clear that in each case the relevant rules constitute a finite presentation of a Gentzen system (to be formally defined on the convenient set of sequents) that is fully adequate for the corresponding logic:

1. By point (I7) of Example 5.91, the Gentzen system having as proper rules those corresponding to the u-DDT:

$$\frac{\Gamma, \varphi \rhd \psi}{\Gamma \rhd \varphi \to \psi} \ (\textsc{dt}) \qquad\qquad \frac{\Gamma \rhd \varphi \qquad \Gamma, \psi \rhd \xi}{\Gamma, \varphi \to \psi \rhd \xi} \ (\textsc{mp})$$

   is the Gentzen system fully adequate for $\mathcal{IL}_{\to}$.
2. The characterizations of the full g-models of $\mathcal{IL}$ and of $\mathcal{CL}$ in terms of Tarski-style conditions contained in Exercises 5.65 and 5.66, respectively, provide the definitions of the Gentzen systems fully adequate for these logics, modulo Exercises 5.7 and 5.8.
3. By Example 5.93(vi), the Gentzen system defined by the usual rules for conjunction and disjunction of Example 3.87 (see also Exercises 5.7 and 5.12) is fully adequate for $\mathcal{CL}_{\wedge\vee}$. Hence, by Proposition 5.119, this logic is the weakest logic satisfying these rules; this fact is mentioned on page 177.
4. By point (B20) of Example 5.92, the Gentzen system expanding the previous one with the rules for negation (79) displayed there is fully adequate for $\mathcal{B}$.

In each case, this means not just that the logic is characterized by the corresponding Gentzen-style rules, but also that its "algebraic images", the g-matrices made of all filters of any algebra, are characterized by the same rules as well. ⊠

A more elaborate and technically useful characterization is the following:

PROPOSITION 5.121

Assume that $\mathcal{L}$ is a finitary logic and that $\mathfrak{G}$ is a structural Gentzen system such that $Seq\,\mathfrak{G} = Seq$ in case $\mathcal{L}$ has theorems, and $Seq\,\mathfrak{G} = Seq^{\circ}$ in case $\mathcal{L}$ has no theorems. The *system $\mathfrak{G}$ is fully adequate for $\mathcal{L}$ if and only if* $\mathsf{Alg}\mathfrak{G} = \mathsf{Alg}\mathcal{L}$ *and for each* $\boldsymbol{A} \in \mathsf{Alg}\mathcal{L}$ *the g-matrix* $\langle \boldsymbol{A}, \mathcal{F}i_{\mathcal{L}}\boldsymbol{A} \rangle$ *is the only reduced finitary model of* $\mathfrak{G}$ *on* $\boldsymbol{A}$ *(without theorems, if* $\mathcal{L}$ *has no theorems).*

PROOF: ($\Rightarrow$) By Proposition 5.118 $\mathfrak{G}$ is adequate for $\mathcal{L}$, and by Proposition 5.113.3 this implies that $\mathsf{Alg}\mathfrak{G} \subseteq \mathsf{Alg}\mathcal{L}$. If $\boldsymbol{A} \in \mathsf{Alg}\mathcal{L}$, then $\langle \boldsymbol{A}, \mathcal{F}i_{\mathcal{L}}\boldsymbol{A} \rangle$ is a reduced full basic g-model of $\mathcal{L}$, hence it is a reduced model of $\mathfrak{G}$, which implies that $\boldsymbol{A} \in \mathsf{Alg}\mathfrak{G}$. Now, if $\boldsymbol{A} \in \mathsf{Alg}\mathcal{L}$, then $\langle \boldsymbol{A}, \mathcal{F}i_{\mathcal{L}}\boldsymbol{A} \rangle$ is a reduced full g-model of $\mathcal{L}$, and by Theorem 5.95 it is the only such one. Moreover, by Proposition 5.15 it is finitary (because $\mathcal{L}$ is) and has theorems if and only if $\mathcal{L}$ has them, therefore by full adequacy it is the only reduced finitary model of $\mathfrak{G}$ on $\boldsymbol{A}$ (without theorems, if $\mathcal{L}$ has no theorems).

($\Leftarrow$) In the case with theorems, let $\langle \boldsymbol{A}, \mathscr{C} \rangle$ be a g-matrix. The following chain of equivalences should be obvious, given the assumptions:

$$\begin{aligned} \langle \boldsymbol{A}, \mathscr{C} \rangle \in \mathsf{FGMod}\mathcal{L} &\iff \langle \boldsymbol{A}^*, \mathscr{C}^* \rangle \text{ is a reduced full g-model of } \mathcal{L} \\ &\iff \boldsymbol{A}^* \in \mathsf{Alg}\mathcal{L} \text{ and } \mathscr{C}^* = \mathcal{F}i_{\mathcal{L}}(\boldsymbol{A}^*) \\ &\iff \boldsymbol{A}^* \in \mathsf{Alg}\mathfrak{G} \text{ and } \mathscr{C}^* \text{ is a reduced finitary model of } \mathfrak{G} \\ &\iff \langle \boldsymbol{A}, \mathscr{C} \rangle \text{ is a finitary model of } \mathfrak{G}. \end{aligned}$$

In the case without theorems, just add "without theorems" where needed; recall that in this case $\mathcal{F}i_{\mathcal{L}}(\boldsymbol{A}^*)$ has no theorems. ⊠

The investigation of the issue of when a given logic has a fully adequate Gentzen system, either for particular cases, or looking for necessary or sufficient conditions (or both), has been a source of interesting results and has developed in several directions. To close this chapter, a sketchy description of three of them, mentioning some of the key results, follows.

- An important direction of research is to exploit **the presentation of finitary g-matrices as relational structures** and the re-writing of Gentzen-style rules as equality-free universal strict basic Horn sentences in a first-order language as outlined below; this is similar to what is done for matrices and Hilbert-style rules in Section 4.5.

The starting idea is that every finitary g-matrix $\langle \boldsymbol{A}, \mathscr{C} \rangle$ can be equivalently presented as a *finite closure relation,* which is a structure $\langle \boldsymbol{A}, R \rangle$ where $R \subseteq A^{<\omega}$ satisfies the natural conditions expressing the properties that define a closure operator; the connection is that, for all $n \geqslant 1$,

$$\langle a_1, \dots, a_{n-1}, a_n \rangle \in R \iff a_n \in C\{a_1, \dots, a_{n-1}\}.$$

Then one "splits" the relation R into its fixed-arity components $R_n = R \cap A^n$ for all $n \geqslant 1$, obtaining an ordinary relational structure $\mathscr{A} = \langle \boldsymbol{A}, R_1, R_2, \dots \rangle$; its similarity type has algebraic reduct $\boldsymbol{L}$ and exactly one relational symbol, say D_n, of arity n for each $n \geqslant 1$. Of course, one sets $D_n^{\mathscr{A}} = R_n$. Observe that only $R_1 = \{a \in A : a \in C\varnothing\}$ may be empty, by condition (I).

To establish the connection between sentential satisfaction and first-order satisfaction, the sequents are translated in a way different from that of Section 4.5:

$$\mathbf{tr}(\alpha_1, \ldots, \alpha_{n-1} \rhd \alpha_n) := D_n \alpha_1 \ldots \alpha_{n-1} \alpha_n$$

Thus, sequents are translated into the atomic formulas of the new, first-order language. Notice that only equality-free formulas are obtained. Then the translation is extended to Gentzen-style rules in the expected way, producing universal strict basic Horn sentences.

Since $D_n^{\mathscr{A}} = R_n$, it is straightforward that

$$\mathscr{A} \vDash \mathbf{tr}(\alpha_1 \ldots \alpha_{n-1} \rhd \alpha_n) \, [\![h]\!] \iff h\alpha_n \in C\{h\alpha_1, \ldots, h\alpha_{n-1}\}.$$

This makes a first-order model-theoretic treatment of the notions of a finitary g-matrix being a g-model of a (sentential) logic and being a model of a Gentzen-style rule possible.

Finitary g-matrices were first presented as finite closure relations in [109], and the transition to the first-order presentation was done and fully exploited in [111]. The next results are not difficult to show ($\mathcal{L}$ is always a finitary logic, and $\mathfrak{G}$ a structural Gentzen system):

- The class $\mathsf{Mod_f}\mathfrak{G}$ is an equality-free universal strict Horn class.
- If $\mathcal{L}$ has theorems, then $\mathcal{L}$ has a fully adequate Gentzen system if and only if $\mathsf{FGMod}\,\mathcal{L}$ is an equality-free universal strict Horn class.
- If $\mathcal{L}$ has no theorems, then $\mathcal{L}$ has a fully adequate Gentzen system if and only if $\mathsf{FGMod}\,\mathcal{L}$ is an equality-free universal Horn class that is not strict.†

In these results $\mathsf{FGMod}\,\mathcal{L}$ denotes the class of relational structures corresponding to the finitary g-matrices of the full g-models of $\mathcal{L}$.

Then, one can apply the techniques and results on the model theory of equality-free languages developed in [81] inspired in abstract algebraic logic developments. The key result for us is that a class of relational structures (for an arbitrary language) is an equality-free universal (strict) Horn class if and only if it is closed under $\mathbb{H}_\mathrm{S}$, $\mathbb{H}_\mathrm{S}^{-1}$, $\mathbb{S}$ and $\mathbb{P}_\mathrm{R}$ and contains a trivial structure; these operators are the ones corresponding in the first-order framework to the analogue operators for g-matrices. Since we know that the class of full g-models of any logic is closed under $\mathbb{H}_\mathrm{S}$ and $\mathbb{H}_\mathrm{S}^{-1}$ (Proposition 5.86), it is not difficult to obtain the final result:

THEOREM 5.122
A finitary logic $\mathcal{L}$ *has a fully adequate Gentzen system if and only if the class (of relational structures corresponding to)* $\mathsf{FGMod}\,\mathcal{L}$ *is closed under* $\mathbb{S}$ *and* $\mathbb{P}_\mathrm{R}$. ⊠

This result makes it possible to show that not every logic has a fully adequate Gentzen system:

† In *strict* Horn formulas (see footnote † on page 224) the literals are all negated except exactly one, while in general (i.e., the not necessarily strict ones) there may be *at most* one non-negated literal.

EXAMPLE 5.123
Consider the $\Box$ fragment of the local logic K^{ℓ} associated with the weakest normal modal system K. In [152] it is proved that a g-matrix $\langle \boldsymbol{A}, \mathscr{C} \rangle$ is a full g-model of this logic if and only if it satisfies:

(a) For all $X \subseteq A$, $\Box C X \subseteq C \Box X$.

(b) $C\varnothing = \varnothing$.

(c) For all $a, b \in A$, $a \in C\{b\}$ if and only if $b \in C\{a\}$.

(d) C is unitary. That is, if $a \in CX$, then there is some $b \in X$ such that $a \in C\{b\}$.

Condition (d) does not look as being expressible by a Gentzen-style rule, and it is actually possible to prove rigorously this fact by seeing that this condition is not preserved by products. Thus, in particular the class of full g-models of this logic is not closed under $\mathbb{P}_{\mathrm{R}}$, and hence this logic cannot have a fully adequate Gentzen system. ⊠

Other relevant results obtained with similar techniques are:

- $\mathcal{L}$ has fully adequate Gentzen system if and only if for every algebra $\boldsymbol{A}$, the lattice $\mathrm{FCS}_{\mathcal{L}}\boldsymbol{A}$ is a complete sublattice of the lattice of all inductive closure systems on A (which amounts to saying that their supremum operations coincide, because the infimum is intersection in the two cases).
- $\mathcal{L}$ has fully adequate Gentzen system if and only if for every algebra $\boldsymbol{A}$, and for every family $\{\langle \boldsymbol{A}, \mathscr{C}_i \rangle : i \in I\} \subseteq \mathsf{FGMod}\,\mathcal{L}$, if we define $CX := \bigcap_{i\in I} C_i X$ for all $X \subseteq A$, then $\langle \boldsymbol{A}, \mathscr{C} \rangle \in \mathsf{FGMod}\,\mathcal{L}$.
- If $\mathcal{L}$ is weakly algebraizable (Definition 6.115) and has a fully adequate Gentzen system, then $\mathcal{L}$ is filter-distributive (which means that for every algebra $\boldsymbol{A}$, the lattice $\mathcal{F}i_{\mathcal{L}}\boldsymbol{A}$ is distributive).

With this methodology one can also treat the relation between the existence of a fully adequate Gentzen system and the multiterm DDT, but the proofs are not constructive, unlike in the case of Theorem 5.124.

- One further line of research tries to exploit the special form of **full g-models for protoalgebraic logics**; the essentials of this work appear in Section 6.2 (pages 341ff.). As we shall see in Theorem 6.41, they are determined by a special kind of $\mathcal{L}$-filters, the so-called *Leibniz filters*. Among other results, which are more general but much more technical as well, let me highlight one from [109], which does not mention these filters but is obtained with their help:

THEOREM 5.124
A finitary and weakly algebraizable logic has a fully adequate Gentzen system if and only if it satisfies the multiterm DDT. ⊠

This result settles the problem of existence of a fully adequate Gentzen system for some non-selfextensional logics, not covered by the result in Section 7.2 (see Theorem 5.125 below); for instance, it implies that $S4^{g}$ and $S5^{g}$ do have a fully adequate Gentzen system.

• Finally, another fruitful line of research is to connect this issue with the **theory of algebraization of Gentzen systems** briefly discussed in Section 3.7. Almost no work has been done at a completely general level, but in [107, 155, 156] this issue is analysed for two very large classes of selfextensional logics. The main results rely on typical techniques of selfextensional logics and are proved in Section 7.2; they are summarized in the following:

THEOREM 5.125
Let $\mathcal{L}$ be a finitary and selfextensional logic with either the PC *or the* u-DDT.

1. *$\mathcal{L}$ has a fully adequate Gentzen system $\mathfrak{G}$.*
2. *This $\mathfrak{G}$ is algebraizable.*
3. $\mathsf{Alg}\mathfrak{G} = \mathsf{Alg}\mathcal{L} = \mathbb{V}\mathcal{L}$. *Thus, in particular,* $\mathsf{Alg}\mathcal{L}$ *is a variety.* ⊠

This general result corresponds to facts empirically observed throughout the years in many particular cases. It clearly suggests the notion of a ***G-algebraizable logic***, which would be one having a fully adequate Gentzen system which is itself algebraizable in the sense of this extended theory. This class of logics would not contain, nor would be contained in the class of algebraizable logics:

- By Theorem 5.124, an algebraizable logic without a multiterm DDT cannot be G-algebraizable; there are many examples of this kind, among them the global modal logics K^g and T^g, $Ł_\infty$, $\mathcal{BCK}$ logic, and relevance logic $\mathcal{R}$.
- According to Theorem 5.125, all logics satisfying its assumptions would be G-algebraizable; again, there are many of them (in the PC case) that are not even protoalgebraic (hence, not algebraizable); $\mathcal{C}\ell_{\wedge\vee}$ and $\mathcal{B}$ are but two simple examples.

Thus, both notions are independent. Perhaps, in order to avoid confusions, a different name should be chosen for this new notion, which is still waiting to be investigated in depth.

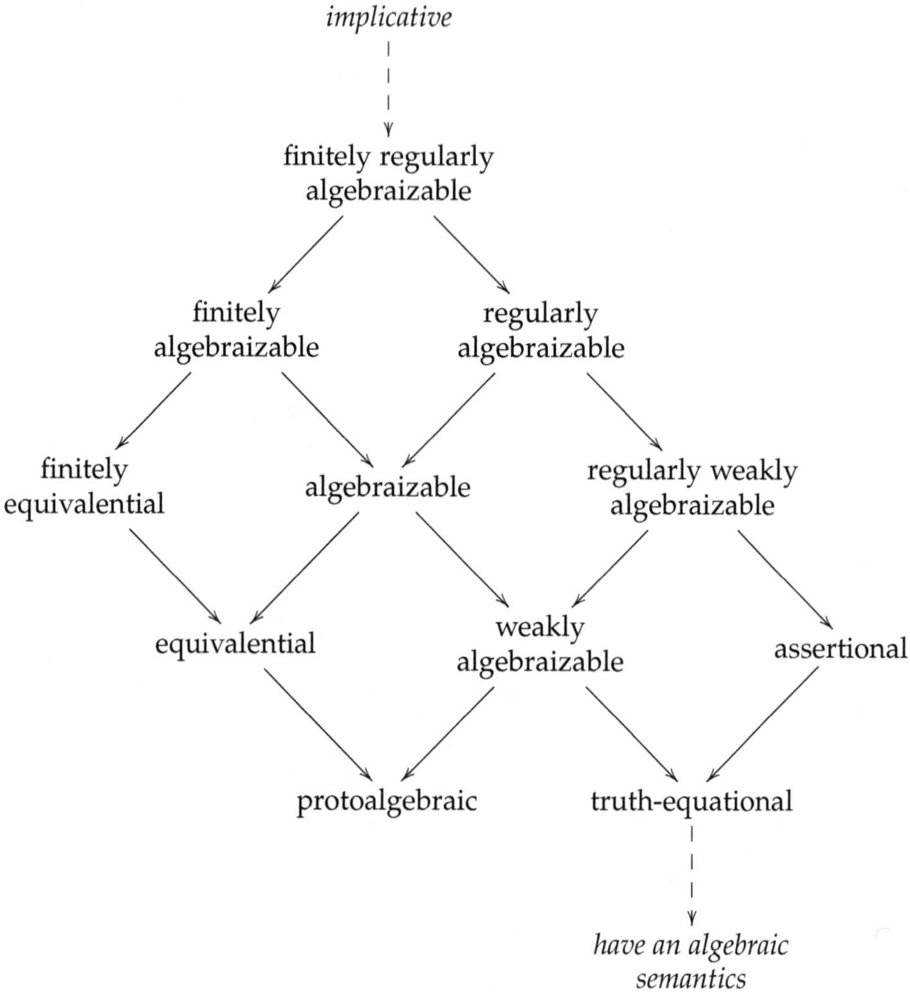

FIGURE 9: A diagram of the *poset* of the eleven classes in the Leibniz hierarchy, showing also two classes not belonging to it (in *italics* and separated by dashed arrows). Arrows mean class inclusion (or implication of properties). Therefore, the classes near the top are smaller and have stronger properties, while the ones in lower positions are larger and have weaker properties. Notice however that this is *not a lattice* diagram: while joins in the graph correspond to intersections of classes, meets do not correspond to unions.

Chapter 6

Introduction to the Leibniz hierarchy

6.1 Overview

This chapter is devoted to the study of several classes of logics, which have been found to enjoy a good behaviour with regard to their matrix semantics and their lattices of filters, and which can be characterized, among other methods, by the behaviour of the Leibniz operator on their theories and on their filters of arbitrary algebras; this justifies the name "Leibniz hierarchy" given by Czelakowski to the resulting classsification and organization of these classes. They are shown in Figure 9 on the facing page.

❧ ❧ ❧

Some of these classes have already appeared in Chapter 3 (compare with Figure 4 on page 143), and here they are just revisited, especially as far as their relationship to the new classes is concerned. To gain some perspective, and to facilitate the reference to any of them even before they are formally studied, it is useful to have a summary of several of their characterizations. One of the most interesting facts of the Leibniz hierarchy is that almost all its classes can be defined, or equivalently characterized, from a variety of quite different perspectives; depending on the approach, any of these may be taken as the official definition, and then the others are proved equivalent, but this sublety is not important just now.

Order-theoretic characterizations. These are the properties that give the core of the hierarchy its distinctive *abstract character*, and concern properties of the Leibniz operator $\boldsymbol{\Omega}^{\boldsymbol{A}}$ as a function, mostly relating the order structure of its domain $\mathcal{F}i_{\mathcal{L}}\boldsymbol{A}$ with that of its codomain $\mathrm{Con}_{\mathsf{Alg}^*\mathcal{L}}\boldsymbol{A}$; these sets are ordered under set inclusion, and recall (Lemma 4.41) that for any algebra $\boldsymbol{A}$, the function $\boldsymbol{\Omega}^{\boldsymbol{A}}$ on $\mathcal{F}i_{\mathcal{L}}\boldsymbol{A}$ is always onto $\mathrm{Con}_{\mathsf{Alg}^*\mathcal{L}}\boldsymbol{A}$.

$\mathcal{L}$ is ...	if and only if over every $\boldsymbol{A}$, $\boldsymbol{\Omega}^{\boldsymbol{A}}$ is ... (equivalently: over $\boldsymbol{Fm}$, $\boldsymbol{\Omega}$ is ...)
Protoalgebraic	monotone
Equivalential	monotone and commutes with endomorphisms
Finitely equivalential	continuous
Truth-equational	completely order-reflecting
Weakly algebraizable	monotone and injective (i.e., an isomorphism)
Algebraizable	an isomorphism that commutes with endomorphisms
Finitely algebraizable	a continuous isomorphism

The "equivalently" in the table's headline means that each group of properties, as given in the table, holds for all algebras if and only if it holds in the formula algebra (then, the endomorphisms are the substitutions); that is, there are *transfer theorems* for all these groups of properties. The properties in the table with a not-so-obvious meaning are defined as follows:

- $\boldsymbol{\Omega}^{\boldsymbol{A}}$ is ***continuous*** when for any family $\{F_i : i \in I\} \subseteq \mathcal{F}i_{\mathcal{L}}\boldsymbol{A}$ that is upwards-directed and such that $\bigcup_{i\in I} F_i \in \mathcal{F}i_{\mathcal{L}}\boldsymbol{A}$, it holds that $\boldsymbol{\Omega}^{\boldsymbol{A}} \bigcup_{i\in I} F_i = \bigcup_{i\in I} \boldsymbol{\Omega}^{\boldsymbol{A}} F_i$. Notice that when $\mathcal{L}$ is finitary, the condition that the union of the family is a filter is automatically satisfied. Clearly, continuity implies monotonicity.
- $\boldsymbol{\Omega}^{\boldsymbol{A}}$ is ***completely order-reflecting*** when for any $\{F_i : i \in I\} \cup \{G\} \subseteq \mathcal{F}i_{\mathcal{L}}\boldsymbol{A}$, if $\bigcap_{i\in I} \boldsymbol{\Omega}^{\boldsymbol{A}} F_i \subseteq \boldsymbol{\Omega}^{\boldsymbol{A}} G$, then $\bigcap_{i\in I} F_i \subseteq G$.

The four classes in the hierarchy not present in the table are characterized by adding another property of the Leibniz operator, of a different character. By Theorem 6.112, a logic $\mathcal{L}$ is assertional if and only if it has theorems and

$$x \equiv y \ (\boldsymbol{\Omega}\Gamma) \text{ for every } \Gamma \in \mathcal{T}h\mathcal{L} \text{ such that } x, y \in \Gamma. \tag{83}$$

By adding to this condition those for being protoalgebraic, equivalential or finitely equivalential, we obtain characterizations, all in terms of the Leibniz operator, for the three "regularly" classes; moreover, when protoalgebracity is assumed, monotonicity of $\boldsymbol{\Omega}$ makes it possible to simplify condition (83) and write just $x \equiv y \ (\boldsymbol{\Omega} C_{\mathcal{L}}\{x, y\})$.

Definability characterizations. There are two main kinds of definabilities, that of the truth filter and that of the Leibniz congruence. Moreover, they can be required to hold in models or in reduced models of the logic. Here I give only some versions of the characterizations concerning these properties.

$\mathcal{L}$ is ...	if and only if ...
Protoalgebraic	$\boldsymbol{\Omega}^{\boldsymbol{A}}F$ is definable from $F \in \mathcal{F}i_{\mathcal{L}}\boldsymbol{A}$ by some $\Delta(x,y,\vec{z})$ with parameters, for every $\boldsymbol{A}$
Equivalential	$\boldsymbol{\Omega}^{\boldsymbol{A}}F$ is definable from $F \in \mathcal{F}i_{\mathcal{L}}\boldsymbol{A}$ by some $\Delta(x,y)$ without parameters, for every $\boldsymbol{A}$
Truth-equational	The truth filter is definable in $\mathsf{Mod}^*\mathcal{L}$ by some $E(x) \subseteq Eq$
Assertional	The truth filter is definable in $\mathsf{Mod}^*\mathcal{L}$ by $x \approx \top$, where $\top$ is an algebraic constant of $\mathsf{Alg}^*\mathcal{L}$

By putting several of the conditions together we obtain characterizations of four more classes: the (regularly) (weakly) algebraizable. By requiring the non-parametrised set $\Delta(x,y)$ to be finite we obtain the three remaining classes, the "finitely" ones.

It should be understood that in each case the definition is common to all involved algebras or matrices (that is, this is about *uniform* definability).

The definability of the truth filter has already been found, for algebraizable logics, in Theorem 4.60, while that of $\boldsymbol{\Omega}^{\boldsymbol{A}}F$ corresponds to the condition called "to be analytically definable" in Theorem 4.54, and is naturally generalized to the *parametrised* case of the first row as follows:

- If $\Delta(x,y,\vec{z}) \subseteq Fm$ is a set of formulas in two variables x,y and possibly other variables $\vec{z}$, called parameters, then $\boldsymbol{\Omega}^{\boldsymbol{A}}F$ is ***definable*** from $F \in \mathcal{F}i_{\mathcal{L}}\boldsymbol{A}$ by $\Delta(x,y,\vec{z})$ when for any $a,b \in A$,

$$a \equiv b \ (\boldsymbol{\Omega}^{\boldsymbol{A}}F) \iff \bigcup\{\Delta^{\boldsymbol{A}}(a,b,\vec{c}) : \vec{c} \in \vec{A}\} \subseteq F.$$

The condition given here for assertional logics is equivalent to saying that the truth filter in $\mathsf{Mod}^*\mathcal{L}$ is a one-element set; but as a "definability" property this formulation looks weaker.

❧ ❧ ❧

Syntactic characterizations. This refers to characterizations of the classes by putting conditions exclusively on the consequence relation $\vdash_{\mathcal{L}}$ of the logic $\mathcal{L}$, relative to certain sets of formulas or of equations.

$\mathcal{L}$ is ...	if and only if there is some $\varDelta(x,y) \subseteq Fm$ such that $\vdash_{\mathcal{L}}$ satisfies ...
Protoalgebraic	(R) + (MP)
Equivalential	(R) + (MP) + (Re)
Algebraizable	(R) + (MP) + (Re) + (ALG3) for some $E(x) \subseteq Eq$
Regularly algebraizable	(R) + (MP) + (Re) + (G)

and

$\mathcal{L}$ is ...	if and only if there is a finite $\varDelta(x,y) \subseteq Fm$ such that $\vdash_{\mathcal{L}}$ satisfies ...
Finitely equivalential	(R) + (MP) + (Re)
Finitely algebraizable	(R) + (MP) + (Re) + (ALG3) for some $E(x) \subseteq Eq$
Finitely regularly algebraizable	(R) + (MP) + (Re) + (G)

These conditions are those already found in Chapter 3. Notice that the characterizations of the two algebraizable classes obtained there included the conditions (Sym) and (Trans); but these two conditions will be shown to be superfluous in this chapter. The conditions (R), (MP), (Re) and (G) concern only the set $\varDelta(x,y)$, while (ALG3) concerns both a set $\varDelta(x,y)$ of formulas and a set $E(x)$ of equations.

In this chapter we meet *parametrised versions* of some of the above conditions, concerning a set $\varDelta(x,y,\vec{z})$; they are indicated by the corresonding acronym with a "P" subscript. These parametrised conditions provide syntactic characterizations of the classes of protoalgebraic, weakly algebraizable and regularly weakly algebraizable logics. By contrast, there are no known purely syntactic characterizations of the classes of assertional and of truth-equational logics.

Model-theoretic characterizations. These use properties of closure of the class of matrices $\mathsf{Mod}^*\mathcal{L}$ or of the class of algebras $\mathsf{Alg}^*\mathcal{L}$ under well-known model-theoretic or algebraic operators.

$\mathcal{L}$ is ...	if and only if ...
Protoalgebraic	$\mathsf{Mod}^*\mathcal{L}$ is closed under $\mathbb{P}_{SD}$
Equivalential	$\mathsf{Mod}^*\mathcal{L}$ is closed under $\mathbb{S}$ and $\mathbb{P}$
Finitely equivalential and finitary	$\mathsf{Mod}^*\mathcal{L}$ is closed under $\mathbb{S}$, $\mathbb{P}$ and $\mathbb{P}_U$, i.e., it is a "quasivariety" of matrices

and

$\mathcal{L}$ is ...	if and only if $\mathcal{L}$ is truth-equational and ...
Weakly algebraizable	$\mathsf{Alg}^*\mathcal{L}$ is closed under $\mathbb{P}_{SD}$
Algebraizable	$\mathsf{Alg}^*\mathcal{L}$ is closed under $\mathbb{S}$ and $\mathbb{P}$
Finitely algebraizable	$\mathsf{Alg}^*\mathcal{L}$ is closed under $\mathbb{S}$, $\mathbb{P}$ and $\mathbb{P}_U$, i.e., it is a quasivariety

Concerning the third rows of each of these two tables, recall that the classes $\mathsf{Mod}^*\mathcal{L}$ and $\mathsf{Alg}^*\mathcal{L}$ are always closed under $\mathbb{I}$, and always contain all trivial matrices and all trivial algebras, respectively (Proposition 4.38). One noticeable difference betweeen these two rows is finitarity, which appears in the first but not in the second.

❧ ❧ ❧

Two further classes, not belonging to the Leibniz hierarchy, are displayed in Figure 9 on page 316, in order to show their relative position with respect to the classes in the hierarchy. At the top we see the class of *implicative logics*. Its definition by conditions (IL1)–(IL5) in Definition 2.3 certainly fits into the group of syntactic characterizations, with these conditions concerning a single formula $x \to y$; but this class of logics has none of the characterizations of other kinds enjoyed by the classes in the Leibniz hierarchy, and normally is not considered as belonging to it.

The other is the class of *logics having an algebraic semantics*, which we met in Chapter 3. In some sense, these logics are characterized in ways that resemble some of the above kinds: A logic $\mathcal{L}$ has an algebraic semantics if and only if it is complete with respect to some class of matrices in which the filters are equationally definable (a definability characterization); if and only if it satisfies condition (ALG1) with respect to some $E(x) \subseteq Eq$ and some class of algebras (a syntactic characterization); and if and only if there is an order isomorphism (which need not be the Leibniz operator) between the lattice of theories of the logic and a join-complete sub-semilattice of $\mathrm{Con}_{\mathsf{K}}\boldsymbol{Fm}$ that commutes with substitutions (an order-theoretic characterization; see Exercise 3.46). However, these characterizations do not concern the Leibniz operator, and the classes of matrices and of algebras present in the definition need not be intrinsically associated with the logic by the use of constructions related to the Leibniz operator (like as the reduction process). Therefore, this class of logics is not considered as belonging to the Leibniz hierarchy; in fact, there is no known characterization of the concept explicitly in terms of a property of the Leibniz operator.

Along the chapter, the necessary examples and counterexamples are mentioned to establish that the diagram in Figure 9 on page 316 is really as depicted, that is, that there are no other inclusions between classes than those shown there (explicitly, or derived from them by transitivity). Many logics lying at the top of the hierarchy are mentioned in Examples 3.50, while (fewer) logics located in one of the hierarchy's two bottoms appear in Examples 6.122.8–11.

6.2 Protoalgebraic logics

Basic concepts

Among the classes in the Leibniz hierarchy, the class of protoalgebraic logics is probably the one with the largest number of essentially different characterizations. My choice is to approach them by comparing several equivalence relations (some of which are congruences) that in previous chapters are associated with matrices and g-matrices, and in particular with a logic $\mathcal{L}$ and its theories. These are the Frege relation $\boldsymbol{\Lambda}\mathcal{L}$ and the Tarski congruence $\widetilde{\boldsymbol{\Omega}}\mathcal{L}$ of the logic itself; and, for each theory Γ of $\mathcal{L}$, its Frege relation and its Suszko congruence, both relative to $\mathcal{L}$, which are denoted by $\boldsymbol{\Lambda}_{\mathcal{L}}\Gamma$ and $\widetilde{\boldsymbol{\Omega}}_{\mathcal{L}}\Gamma$ respectively, and of course its Leibniz congruence $\boldsymbol{\Omega}\Gamma$. In this way, several "operators" on the set of theories of the logic are defined, and analogously they have correlates in the set of $\mathcal{L}$-filters in an arbitrary algebra $\boldsymbol{A}$; here we deal with the Suszko operator relative to the logic $\mathcal{L}$, denoted by $\widetilde{\boldsymbol{\Omega}}^{\boldsymbol{A}}_{\mathcal{L}}$. It may be helpful for you to review now their definitions and most fundamental properties, mainly from Sections 1.3, 4.2 and 5.3, and in particular Lemma 5.37.

In order to give Blok and Pigozzi's original definition (1986) of protoalgebraic logics in their same wording, given a theory Γ of a logic, two formulas α, β are called ***Γ-indiscernible*** when $\alpha \equiv \beta\ (\boldsymbol{\Omega}\Gamma)$, and are called ***$\Gamma$-interderivable*** when $\Gamma, \alpha \dashv\vdash_{\mathcal{L}} \Gamma, \beta$, that is, when $\alpha \equiv \beta\ (\boldsymbol{\Lambda}_{\mathcal{L}}\Gamma)$; while the justification of the latter term is obvious, that of the former is grounded on the discussion on page 198.

DEFINITION 6.1
*A logic $\mathcal{L}$ is **protoalgebraic** when for every $\Gamma \in \mathcal{T}h\mathcal{L}$, every two formulas that are Γ-indiscernible are Γ-interderivable; that is, when $\boldsymbol{\Omega}\Gamma \subseteq \boldsymbol{\Lambda}_{\mathcal{L}}\Gamma$ for every $\Gamma \in \mathcal{T}h\mathcal{L}$.*

Alternative definitions arise from the following straightforward fact:

LEMMA 6.2
For an arbitrary logic $\mathcal{L}$ and every $\Gamma \in \mathcal{T}h\mathcal{L}$, the following conditions are equivalent:

(i) $\boldsymbol{\Omega}\Gamma \subseteq \boldsymbol{\Lambda}_{\mathcal{L}}\Gamma$,

(ii) $\boldsymbol{\Omega}\Gamma \subseteq \widetilde{\boldsymbol{\Omega}}_{\mathcal{L}}\Gamma$.

(iii) $\boldsymbol{\Omega}\Gamma = \widetilde{\boldsymbol{\Omega}}_{\mathcal{L}}\Gamma$.

PROOF: (i) is equivalent to (ii) because $\widetilde{\boldsymbol{\Omega}}_{\mathcal{L}}\Gamma$ is the largest congruence below the equivalence relation $\boldsymbol{\Lambda}_{\mathcal{L}}\Gamma$, and (ii) is equivalent to (iii) because the reverse inclusion $\widetilde{\boldsymbol{\Omega}}_{\mathcal{L}}\Gamma \subseteq \boldsymbol{\Omega}\Gamma$ holds in general. ⊠

COROLLARY 6.3
A logic $\mathcal{L}$ is protoalgebraic if and only if the Leibniz operator and the Suszko operator coincide on the theories of $\mathcal{L}$. ⊠

This view of protoalgebraicity as related to the Suszko operator produces a quick proof of the basic characterization of this notion:

THEOREM 6.4
A logic $\mathcal{L}$ is protoalgebraic if and only if the Leibniz operator $\boldsymbol{\Omega}$ is ***monotone*** *over $\mathcal{T}h\mathcal{L}$; that is, if and only if for all $\Gamma, \Gamma' \in \mathcal{T}h\mathcal{L}$, if $\Gamma \subseteq \Gamma'$, then $\boldsymbol{\Omega}\Gamma \subseteq \boldsymbol{\Omega}\Gamma'$.*

PROOF: ($\Rightarrow$) By Corollary 6.3, because the Suszko operator is always monotone on theories (Lemma 5.37.3).

($\Leftarrow$) If $\boldsymbol{\Omega}$ is monotone, then clearly $\boldsymbol{\Omega}\Gamma = \bigcap\{\boldsymbol{\Omega}\Gamma' : \Gamma' \in \mathcal{T}h\mathcal{L},\ \Gamma \subseteq \Gamma'\} = \widetilde{\boldsymbol{\Omega}}_{\mathcal{L}}\Gamma$. By Corollary 6.3 again, this amounts to protoalgebraicity. ⊠

Exercise 6.1 shows that the original definition of protoalgebraicity and the monotonicity characterization are in fact the same property.

Blok and Pigozzi [34, p. 22] suggest a reading of this property in terms of "states of knowledge":

> "[...] think of $\mathcal{L}$-theories T and S [with $T \subseteq S$] as representing two possible states of knowledge of the world. If $\mathcal{L}$ is protoalgebraic, then Leibniz equality is monotone in the sense that the additional knowledge we gain about the world in passing from T to S can only result in the denotations of more terms being identified; it can never be used to distinguish two terms that were formerly identified."

See also their interpretation of $\boldsymbol{\Omega}T$ with respect to T in [32, p. 15]. The real meaning of these interpretations is not clear to me, but they may perhaps be useful to you, especially when dealing with logics related to applications.

The fundamental set and the syntactic characterization

NOTATION.
Recall from Section 1.1 that the notation $\varphi(x,y)$ for a formula φ indicates that its variables are among x and y; given this, the notation $\varphi(\alpha,\beta)$ is shorthand for the formula $\sigma\varphi$, where σ is any substitution such that $\sigma x = \alpha$ and $\sigma y = \beta$. In

this section substitutions replacing variables by other variables play a prominent role, and in particular this produces notations such as $\varphi(x,x)$ and $\varphi(y,x)$, which might be confusing. You should bear in mind that these notations do not have a meaning by themselves, but make sense only when they denote the results of performing the corresponding substitutions in a formula denoted initially by $\varphi(x,y)$. The same shorthand notations are used in other situations, and might also produce similar odd-looking results; for instance, when the formulas may have other variables and are hence denoted by $\varphi(x,y,\vec{z})$; or for sets of formulas, where $\Delta(x,y,\vec{z})$ might produce $\Delta(x,x,\vec{z})$ and even $\Delta(x,x,\vec{x})$; here $\vec{x}$ denotes the sequence made up entirely of xs (the result of a substitution), while $\vec{z}$ initially denotes an arbitrary sequence of variables (distinct from x and y). By contrast, for evaluated formulas, a notation such as $\varphi^{\boldsymbol{A}}(a,a)$ just denotes the value of the formula $\varphi(x,y)$ under some h such that $hx = hy = a$.

Two variables $x,y \in V$ are chosen and fixed. When writing $\varphi(x,y,\vec{z})$ and similar expressions, in general it is understood that the variables in $\vec{z}$ (if any) are all different from x and y. Soon it will be clear that the choice of these two variables is irrelevant. The following technical notion was introduced by Blok and Pigozzi, but it was Herrmann who fully exploited its crucial role in the theory of protoalgebraic logics.

DEFINITION 6.5
*The **fundamental set** of a logic $\mathcal{L}$ is the set $\Sigma_{\mathcal{L}} := \sigma^{-1} C_{\mathcal{L}} \emptyset$, where σ is the substitution defined by $\sigma y := x$ and $\sigma u := u$ for all variables $u \neq y$ (thus, in particular $\sigma x = x$). Equivalently, $\Sigma_{\mathcal{L}} := \{\varphi \in Fm : \vdash_{\mathcal{L}} \sigma\varphi\} = \{\varphi(x,y,\vec{z}) \in Fm : \vdash_{\mathcal{L}} \varphi(x,x,\vec{z})\}$.*

Thus, $\Sigma_{\mathcal{L}}$ is the set of *all* formulas that become a theorem of $\mathcal{L}$ after just replacing the variable y by x. The formulas in this set may, it is obvious, have other variables in them, which are then called ***parameters***. Handling formulas with parameters is one of the crucial techniques in the study of protoalgebraic logics; at certain points parameters may be eliminated (usually, by structurality), but in others parameters are essential. To make their presence explicit, the set $\Sigma_{\mathcal{L}}$ can be also denoted by $\Sigma_{\mathcal{L}}(x,y,\vec{z})$. Observe that $\vdash_{\mathcal{L}} \Sigma_{\mathcal{L}}(x,x,\vec{z})$, simply by definition of $\Sigma_{\mathcal{L}}$; other basic properties of this set are:

PROPOSITION 6.6

1. *$\Sigma_{\mathcal{L}} \in Th\mathcal{L}$. That is, the fundamental set is a theory of the logic.*
2. *$\delta(x,\vec{z}) \in \Sigma_{\mathcal{L}}$ if and only if $\delta(y,\vec{z}) \in \Sigma_{\mathcal{L}}$, for every formula $\delta(u,\vec{z})$, every variable u and every $\vec{z} \in \vec{V}$.*
3. *$x \equiv y \ (\boldsymbol{\Omega}\Sigma_{\mathcal{L}})$.*
4. *$\Sigma_{\mathcal{L}}$ is the smallest theory Γ of $\mathcal{L}$ such that $x \equiv y \ (\boldsymbol{\Omega}\Gamma)$.*
5. *$\varphi(x,y,\vec{z}) \in \Sigma_{\mathcal{L}}$ if and only if $\varphi(y,x,\vec{z}) \in \Sigma_{\mathcal{L}}$ if and only if $\varphi(x,x,\vec{z}) \in \Sigma_{\mathcal{L}}$ if and only if $\varphi(y,y,\vec{z}) \in \Sigma_{\mathcal{L}}$, for every formula $\varphi(u,v,\vec{z})$.*
6. *$\varphi(x,y,\vec{z}) \in \Sigma_{\mathcal{L}}$ if and only if $\big[\varphi(x,y,\vec{\gamma}) \in \Sigma_{\mathcal{L}}$ for every $\vec{\gamma} \in \vec{Fm}\big]$.*

PROOF: It is left to you as Exercise 6.2; this is a good exercise to check understanding of structurality, substitutions, and the notations. ⊠

Some of these properties are rather technical and are used at a few points in this or the next section. The most important ones for now are 1, 3 and 4.

THEOREM 6.7
For every logic $\mathcal{L}$ the following conditions are equivalent:

(i) *$\mathcal{L}$ is protoalgebraic.*

(ii) $x, \Sigma_{\mathcal{L}} \vdash_{\mathcal{L}} y$.

(iii) *There is a set $\Delta(x,y) \subseteq Fm$ such that* $\vdash_{\mathcal{L}} \Delta(x,x)$ (R)

and $x, \Delta(x,y) \vdash_{\mathcal{L}} y$. (MP)

PROOF: (i)⇒(ii) By Proposition 6.6, $x \equiv y \ (\boldsymbol{\Omega}\Sigma_{\mathcal{L}})$, but by protoalgebraicity $\boldsymbol{\Omega}\Sigma_{\mathcal{L}} \subseteq \boldsymbol{\Lambda}_{\mathcal{L}}\Sigma_{\mathcal{L}}$, which in particular implies that $x, \Sigma_{\mathcal{L}} \vdash_{\mathcal{L}} y$.
(ii)⇒(iii) Take $\Delta(x,y) := \Sigma_{\mathcal{L}}(x,y,\vec{x})$. Then (R) is $\vdash_{\mathcal{L}} \Sigma_{\mathcal{L}}(x,x,\vec{x})$, which follows from $\vdash_{\mathcal{L}} \Sigma_{\mathcal{L}}(x,x,\vec{z})$ by structurality, and (MP) is a consequence of (ii), again by structurality.
(iii)⇒(i) Assume that $\alpha \equiv \beta \ (\boldsymbol{\Omega}\Gamma)$ for some $\Gamma \in \mathcal{T}h\mathcal{L}$. In order to show that $\alpha \equiv \beta \ (\boldsymbol{\Lambda}_{\mathcal{L}}\Gamma)$, let $\Gamma' \in \mathcal{T}h\mathcal{L}$ be such that $\Gamma \cup \{\alpha\} \subseteq \Gamma'$. Now consider the analytical relation $R^{\boldsymbol{Fm}}(\Gamma, \Delta) := \{\langle \varepsilon, \delta \rangle \in Fm \times Fm : \Delta(\varepsilon, \delta) \subseteq \Gamma\}$ from Definition 4.52. Since $\Gamma \in \mathcal{T}h\mathcal{L}$, condition (R) tells us that this relation is reflexive. We can thus apply Lemma 4.53.1, and obtain that $\boldsymbol{\Omega}\Gamma \subseteq R^{\boldsymbol{Fm}}(\Gamma, \Delta)$. This implies that $\Delta(\alpha, \beta) \subseteq \Gamma$, and hence $\Delta(\alpha, \beta) \subseteq \Gamma'$. By structurality, (MP) yields $\alpha, \Delta(\alpha, \beta) \vdash_{\mathcal{L}} \beta$, and we conclude that $\beta \in \Gamma'$. Symmetrically, if $\Gamma \cup \{\beta\} \subseteq \Gamma'$, then $\alpha \in \Gamma'$. Therefore $\Gamma, \alpha \dashv\vdash_{\mathcal{L}} \Gamma, \beta$, that is, $\alpha \equiv \beta \ (\boldsymbol{\Lambda}_{\mathcal{L}}\Gamma)$, as desired. ⊠

Exercise 6.5 provides a simpler definition of the set Δ. Moreover, for finitary logics the previous characterization can be enhanced (Exercise 6.6):

PROPOSITION 6.8
A finitary logic is protoalgebraic if and only if there is a finite set $\Delta \subseteq Fm$ satisfying (R) *and* (MP). ⊠

Property (iii) of Theorem 6.7 is a *syntactic characterization*: a logic is protoalgebraic if and only if it has a set $\Delta(x,y)$ of "implication formulas" in the very weak sense that the set just satisfies (R) and (MP); such a set is sometimes called ***a protoimplication set***[†] or a set ***witnessing protoalgebraicity of*** $\mathcal{L}$ (particularly when this characterization is taken as the definition).

It is now clear that the choice of the two variables x, y is irrelevant: by structurality, a similar set on any two desired variables can be obtained that satisfies the same properties, and the proof of Theorem 6.7 would also work.

This characterization facilitates the classification of a large number of logics as protoalgebraic. First, recalling the syntactic properties of algebraizability in Proposition 3.13, one immediately gets:

[†] The term "protoequivalence set" has also been used in the literature, with less success.

COROLLARY 6.9
All algebraizable logics are protoalgebraic, with the set of equivalence formulas witnessing protoalgebraicity. ⊠

In particular, all implicative logics in the sense of Rasiowa are protoalgebraic, and in this case the set $\{x \to y\}$ witnesses protoalgebraicity. You already know that this includes $\mathcal{Cl}$, $\mathcal{Il}$ and all their extensions and expansions, and all their fragments with implication as well, and also many-valued logics, the global logics associated with normal modal systems, and many substructural logics.

PROPOSITION 6.10
All logics that satisfy the (multiterm) DDT *are protoalgebraic.*

PROOF: It is easy to check (Exercise 6.7) that any DD set for $\mathcal{L}$ satisfies (R) and (MP) and thus witnesses protoalgebraicity through Theorem 6.7. ⊠

Thus, the local logics associated with normal modal systems, shown in Example 3.61 to be non-algebraizable, are nevertheless protoalgebraic. More in general, all logics satisfying (R) and (MP) for some set of formulas $\Delta(x,y)$ are protoalgebraic. Examples of this, not falling in previous groups, are several weak logics of the relevance world such as $\mathcal{T}$ (ticket entailment) or $\mathcal{E}$ (entailment).

Logics not having a "natural" implication connective (for instance, positive modal logics, or implication-less fragments of other logics) have a chance of not being protoalgebraic; but often negative results are achieved with the help of other, semantic characterizations of protoalgebraicity.

Still, Theorem 6.7 has some direct, interesting applications:

PROPOSITION 6.11

1. *Every expansion and every extension of a protoalgebraic logic is protoalgebraic, with the same protoimplication set.*
2. *Every fragment of a protoalgebraic logic whose language contains all the connectives appearing in its protoimplication set of formulas is protoalgebraic, with the same protoimplication set.*
3. *For every language, the inconsistent logic is protoalgebraic, and any set $\Delta(x,y) \subseteq Fm$ is a protoimplication set for it, including the empty set.*
4. *For every language, the almost inconsistent logic is protoalgebraic, and its only protoimplication set is the empty set.*
5. *The almost inconsistent logics are the only protoalgebraic logics without theorems, and the inconsistent and the almost inconsitent logics are the only protoalgebraic logics with an empty protoimplication set.*
6. *A protoalgebraic logic with theorems must have theorems with at least one variable.*
7. *If a language has no connective of arity 2 or greater, then the only protoalgebraic logics in this language are the inconsistent and the almost inconsistent ones.*

PROOF: Left to you as Exercise 6.8. Notice that Theorem 6.7 does not require the set Δ appearing in it to be non-empty. ⊠

In view of the first two points of this proposition, a reasonable question to ask is about the weakest protoalgebraic logic, but it has been proved in [101] that this logic does not exist (in any language). However, it does make sense to consider the *simplest* conceivable protoalgebraic logic, which should be one with the simplest language and the simplest protoimplication set: It is the logic $\mathcal{I}$, which is formulated in a language with just a binary connective $\rightarrow$ and is axiomatized by just the axiom (R) $x \rightarrow x$ and the rule (MP) $x, x \rightarrow y \rhd y$; this logic is studied in [99, 100] and provides a natural counterexample to a variety of issues. Example 6.78.1 contains more information about it.

The protoimplication set is extremely useful in a variety of situations. As an example, consider the following characterization of filter-generation for finitary protoalgebraic logics, and compare it with the general one of Theorem 2.23.

PROPOSITION 6.12
Let $\mathcal{L}$ be a finitary and protoalgebraic logic and $\boldsymbol{A}$ any algebra. For any $X \cup \{a\} \subseteq A$, $a \in Fg_{\mathcal{L}}^{\boldsymbol{A}} X$ if and only if there is a finite $\Gamma \cup \{\varphi\} \subseteq Fm$ such that $\Gamma \vdash_{\mathcal{L}} \varphi$ and there is $h \in \mathrm{Hom}(\boldsymbol{Fm}, \boldsymbol{A})$ such that $h\Gamma \subseteq X \cup Fg_{\mathcal{L}}^{\boldsymbol{A}}\varnothing$ and $h\varphi = a$.

PROOF: Denote by Z the set of points $a \in A$ satisfying the stated condition for a given X; we have to show that $Z = Fg_{\mathcal{L}}^{\boldsymbol{A}} X$. It is easy to see that $X \subseteq Z \subseteq Fg_{\mathcal{L}}^{\boldsymbol{A}} X$, so the only point to be shown is that Z is actually an $\mathcal{L}$-filter. By finitarity, start by assuming that $\varphi_1, \dots, \varphi_n \vdash_{\mathcal{L}} \varphi$ and that for some $h \in \mathrm{Hom}(\boldsymbol{Fm}, \boldsymbol{A})$, $h\varphi_i \in Z$ for each $i = 1, \dots, n$. By definition, there are finite $\Gamma_i \cup \{\psi_i\} \subseteq Fm$ with $\Gamma_i \vdash_{\mathcal{L}} \psi_i$, and $h_i \in \mathrm{Hom}(\boldsymbol{Fm}, \boldsymbol{A})$ such that $h_i\Gamma_i \subseteq X \cup Fg_{\mathcal{L}}^{\boldsymbol{A}}\varnothing$ and $h_i\psi_i = h\varphi_i$. By structurality, the formulas in $\Gamma_i \cup \{\psi_i\}$ can be chosen so that the variables occurring in each case (for i) are different, and are disjoint from those occurring in $\{\varphi_1, \dots, \varphi_n, \varphi\}$, modifying the h_i accordingly to obtain the wanted effect. And h can also be modified on the variables not occurring in $\{\varphi_1, \dots, \varphi_n, \varphi\}$, so that all the h_i actually coincide with h on the relevant formulas; that is, one can assume that $h\psi_i = h_i\psi_i = h\varphi_i$ and that $h\Gamma_i = h_i\Gamma_i \subseteq X \cup Fg_{\mathcal{L}}^{\boldsymbol{A}}\varnothing$ for each i. Now, let Δ be a protoimplication set for $\mathcal{L}$. By (MP), $\{\psi_i\} \cup \Delta(\psi_i, \varphi_i) \vdash_{\mathcal{L}} \varphi_i$; therefore, if $\Gamma := (\bigcup \Gamma_i) \cup (\bigcup \Delta(\psi_i, \varphi_i))$, then $\Gamma \vdash_{\mathcal{L}} \varphi$. But by (R) $h\Delta(\psi_i, \varphi_i) = \Delta(h\psi_i, h\varphi_i) \subseteq Fg_{\mathcal{L}}^{\boldsymbol{A}}\varnothing$, therefore $h\Gamma \subseteq X \cup Fg_{\mathcal{L}}^{\boldsymbol{A}}\varnothing$. The shown properties ensure that $h\varphi \in Z$, and this completes the proof that $Z = Fg_{\mathcal{L}}^{\boldsymbol{A}} X$. ⊠

Despite its greater simplicity, application of this characterization rests on having one for $Fg_{\mathcal{L}}^{\boldsymbol{A}}\varnothing$, the smallest $\mathcal{L}$-filter of the algebra. Exercise 6.9 shows that protoalgebraicity is essential in this result.

Monotonicity, and its applications

The first important application of the protoimplication set of protoalgebraic logics (Theorem 6.7) is actually a *transfer theorem* for an order-theoretic property of the Leibniz operator: For every logic, the Leibniz operator is monotone on its theories if and only if it is monotone on the set of its filters of an arbitrary algebra.

THEOREM 6.13
A logic $\mathcal{L}$ is protoalgebraic if and only if for every algebra $\boldsymbol{A}$, the Leibniz operator $\boldsymbol{\Omega}^{\boldsymbol{A}}$ is ***monotone*** *over the set $\mathcal{F}i_{\mathcal{L}}\boldsymbol{A}$ of all $\mathcal{L}$-filters of $\boldsymbol{A}$; that is, if and only if for every $\boldsymbol{A}$ and every $F, G \in \mathcal{F}i_{\mathcal{L}}\boldsymbol{A}$, if $F \subseteq G$, then $\boldsymbol{\Omega}^{\boldsymbol{A}}F \subseteq \boldsymbol{\Omega}^{\boldsymbol{B}}G$.*

PROOF: If $\mathcal{L}$ is protoalgebraic, then there is a set Δ satisfying (R) and (MP). A proof similar to that in the step (iii)⇒(i) of Theorem 6.7, but performed on an algebra $\boldsymbol{A}$ and with filters instead of theories, easily shows monotonicity of $\boldsymbol{\Omega}^{\boldsymbol{A}}$ over $\mathcal{F}i_{\mathcal{L}}\boldsymbol{A}$; the details are left to you as Exercise 6.10. The converse is trivial. ⊠

This is more useful than the preceding characterizations to show that a given logic $\mathcal{L}$ is *not* protoalgebraic: it is enough to find a small algebra $\boldsymbol{A}$ in which it is not difficult to compute the congruences, and hence the Leibniz operator, and where one can find two $\mathcal{L}$-filters showing the failure of monotonicity.

Using Theorem 6.13 (and recalling that $\widetilde{\boldsymbol{\Omega}}^{\boldsymbol{A}}_{\mathcal{L}}F \subseteq \boldsymbol{\Omega}^{\boldsymbol{A}}F$), one can readily show:

PROPOSITION 6.14
For every logic $\mathcal{L}$ the following conditions are equivalent:

(i) *$\mathcal{L}$ is protoalgebraic.*

(ii) *For every algebra $\boldsymbol{A}$, the operator $\boldsymbol{\Omega}^{\boldsymbol{A}}$ preserves intersections of arbitrary families of $\mathcal{L}$-filters; that is, for any family $\{F_i : i \in I\} \subseteq \mathcal{F}i_{\mathcal{L}}\boldsymbol{A}$, $\boldsymbol{\Omega}^{\boldsymbol{A}}(\bigcap F_i) = \bigcap \boldsymbol{\Omega}^{\boldsymbol{A}}F_i$.*

(iii) *For every algebra $\boldsymbol{A}$ and every $F \in \mathcal{F}i_{\mathcal{L}}\boldsymbol{A}$, $\boldsymbol{\Omega}^{\boldsymbol{A}}F \subseteq \widetilde{\boldsymbol{\Omega}}^{\boldsymbol{A}}_{\mathcal{L}}F$, or, equivalently, $\boldsymbol{\Omega}^{\boldsymbol{A}}F = \widetilde{\boldsymbol{\Omega}}^{\boldsymbol{A}}_{\mathcal{L}}F$.*

(iv) *For every algebra $\boldsymbol{A}$ and every $\langle \boldsymbol{A}, \mathscr{C} \rangle \in \mathsf{GMod}\mathcal{L}$, $\widetilde{\boldsymbol{\Omega}}^{\boldsymbol{A}}\mathscr{C} = \boldsymbol{\Omega}^{\boldsymbol{A}}C\varnothing$.* ⊠

This is again a *transfer theorem*: The equivalence between (i) and (iii) says that the result of Corollary 6.3 transfers to arbitrary algebras: For every logic $\mathcal{L}$, the Leibniz operator and the Suszko operator relative to $\mathcal{L}$ coincide on the theories of $\mathcal{L}$ if and only if they coincide on the $\mathcal{L}$-filters of an arbitrary algebra. The equivalence between (i) and (ii) is also worth remarking: in this case, monotonicity (of the Leibniz operator) is equivalent to preservation of arbitrary meets (and, by Exercise 6.11, to preservation of finite meets); this is not true in general for functions between ordered sets.

COROLLARY 6.15
If $\mathcal{L}$ is protoalgebraic, then $\mathsf{Alg}^*\mathcal{L} = \mathsf{Alg}\mathcal{L}$.

PROOF: The inclusion $\subseteq$ always holds (Proposition 5.70), and $\supseteq$ follows from condition (iv) in Proposition 6.14, because it implies that a g-model $\langle \boldsymbol{A}, \mathscr{C} \rangle$ of $\mathcal{L}$ is reduced if and only if the model $\langle \boldsymbol{A}, C\varnothing \rangle$ is reduced. ⊠

The converse implication is not true, as shown in Example 6.16.7 below.

EXAMPLES 6.16 (NON-PROTOALGEBRAIC LOGICS)
The creators of protoalgebraic logics, Blok and Pigozzi, wrote in [30, p. 355] that

> "all the logics we are aware of in the literature,
> except for pathological cases, are protoalgebraic."

Newer investigations have revealed that this was a rather optimistic (or pessimistic?) statement. Here are some of the best-known non-protoalgebraic logics encountered so far in the literature.

1. $\mathcal{C\ell}_{\wedge\vee}$ is not protoalgebraic: We already know that $\mathsf{Alg}\mathcal{C\ell}_{\wedge\vee} = \mathsf{DL} \supsetneq \mathsf{Alg}^*\mathcal{C\ell}_{\wedge\vee}$. Corollary 6.15 implies that this logic cannot be protoalgebraic. Another reason for it not being protoalgebraic is that $\mathcal{C\ell}_{\wedge\vee}$ does not have theorems but is not almost inconsistent. One example showcasing both the difference of the classes of algebras and the failure of monotonicity over the $\mathcal{C\ell}_{\wedge\vee}$-filters is the three-element chain $\boldsymbol{D}_3$, with universe $D_3 = \{0, 1/2, 1\}$ and $0 < 1/2 < 1$. $\boldsymbol{D}_3 \in \mathsf{DL} = \mathsf{Alg}\mathcal{C\ell}_{\wedge\vee}$ but $\boldsymbol{D}_3 \notin \mathsf{Alg}^*\mathcal{C\ell}_{\wedge\vee}$, because it is not difficult to see that if F is one of its three lattice filters $\{1\}, \{1, 1/2\}$ or $\{1, 1/2, 0\}$, then $\boldsymbol{\Omega}^{\boldsymbol{D}_3}F \neq \mathrm{Id}_{D_3}$. Moreover, $\{1\} \subseteq \{1, 1/2\}$ but $\boldsymbol{\Omega}^{\boldsymbol{D}_3}\{1\} \nsubseteq \boldsymbol{\Omega}^{\boldsymbol{D}_3}\{1, 1/2\}$, because the former congruence identifies only the points 0 and 1/2 while the latter congruence identifies only 1 and 1/2.

   All this applies to $\mathcal{C\ell}_{\wedge}$ as well; just delete everything concerning disjunction.
2. The "logic of lattices" $\mathcal{L}_{\mathsf{L}}^{\leqslant}$ is the logic in the same language as $\mathcal{C\ell}_{\wedge\vee}$, defined as in Definition 5.112 from the Gentzen calculus that has the structural rules, the usual rules for conjunction (Proposition 5.17 or Exercise 5.12) and the rules for disjunction of Exercise 5.7 but with the rightmost rule weakened to

$$\frac{\alpha \rhd \varphi \qquad \beta \rhd \varphi}{\alpha \vee \beta \rhd \varphi}.$$

   Therefore, this logic is weaker than $\mathcal{C\ell}_{\wedge\vee}$. In Section 7.2, where an alternative definition justifying the notation $\mathcal{L}_{\mathsf{L}}^{\leqslant}$ is given (Example 7.34.5), it is shown that its algebraic counterpart $\mathsf{Alg}\mathcal{L}_{\mathsf{L}}^{\leqslant}$ is the class L of all lattices. By Proposition 6.11 and the previous point, it cannot be protoalgebraic. Other properties are found in Example 7.60.7.
3. Belnap-Dunn's logic $\mathcal{B}$ is not protoalgebraic: It is shown in Example 5.82 that $\mathsf{Alg}\mathcal{B} = \mathsf{DM} \supsetneq \mathsf{Alg}^*\mathcal{B}$. By Corollary 6.15, this logic cannot be protoalgebraic. Indeed, the algebra $\boldsymbol{M}_6$ described in Figure 8 on page 299 witnesses this: $\boldsymbol{M}_6 \in \mathsf{DM}$ but $\boldsymbol{M}_6 \notin \mathsf{Alg}^*\mathcal{B}$ because no $\mathcal{B}$-filter of it is reduced. Moreover, in [93, p. 428] there are examples of (small) algebras $\boldsymbol{A}$ such that $\boldsymbol{\Omega}^{\boldsymbol{A}}$ is not monotone over the $\mathcal{B}$-filters of $\boldsymbol{A}$. A more elementary reason is that this logic does not have theorems and is not almost inconsistent.
4. One of the more direct reasons for the non-protoalgebraicity of $\mathcal{C\ell}_{\wedge\vee}$ and of $\mathcal{B}$ is that these logics do not have theorems but are not almost inconsistent. This is, however, not a "definitive" reason, in the sense that the expansions of these logics with constants, considered in Exercises 5.63 and 5.64 respectively, are still non-protoalgebraic; you can easily see this by adapting the examples in points 1 and 3.
5. Positive modal logic $\mathcal{PML}$ is the $\langle \wedge, \vee, \Box, \Diamond, \top, \bot \rangle$ fragment of the local logic K^{ℓ} associated with the normal modal system K. It was characterized by a

Gentzen calculus and studied in depth with abstract algebraic logic techniques in [153]. It does have theorems, but it is non-protoalgebraic: In the cited paper it is shown that $\mathsf{Alg}^*\mathcal{PML} \subsetneq \mathsf{Alg}\mathcal{PML}$; the latter is the class of *positive modal algebras*, which are bounded distributive lattices with two modal-like unary operators with suitable properties. On the other hand, expanding the algebra $\boldsymbol{D}_3$ of point 1 with operators $\Box$ and $\Diamond$ evaluated as the identity function, one obtains an algebra $\boldsymbol{D}_3' := \langle \boldsymbol{D}_3, \Box, \Diamond \rangle$, with the same filters except $\varnothing$, and similar arguments show non-monotonicity of the Leibniz operator.

6. The implication-less fragment† of intuitionistic logic, denoted by $\mathcal{I}\ell^*$, is non-protoalgebraic, as shown in [32, Theorem 5.13] by using the monotonicity characterization. This logic was chronologically the first non-trivial one to be shown non-protoalgebraic, and is more thoroughly studied in [212, 213]; for a summary see [107, Section 5.1.4]. In [212] it is shown that $\mathsf{Alg}^*\mathcal{I}\ell^* \subsetneq \mathsf{Alg}\mathcal{I}\ell^*$; the latter is the variety PCDL of *pseudo-complemented distributive lattices*. Also, a denumerable chain of logics extending $\mathcal{I}\ell^*$ is considered, whose algebraic counterparts are all the subvarieties of PCDL, and it is also proved that all these logics are non-protoalgebraic.
7. The weak relevance logic $\mathcal{WR}$ is studied in [116]. It is defined, following Wójcicki's insightful discussions in [248, pp. 52–73] and in [249, pp. 163–170], as a weak, purely inferential version of the logic $\mathcal{R}$ of Example 3.34.7. More precisely, for all $\Gamma \cup \{\varphi\} \subseteq Fm$, $\Gamma \vdash_{\mathcal{WR}} \varphi$ is defined to hold if and only if there are $\varphi_1, \ldots, \varphi_n \in \Gamma$ (for some $n \geqslant 1$) such that $\vdash_{\mathcal{R}} \varphi_1 \wedge \cdots \wedge \varphi_n \to \varphi$. According to this definition, $\mathcal{WR}$ has no theorems. In [116] it is shown that $\mathcal{WR}$ is not almost inconsistent, hence it is not protoalgebraic, but that $\mathsf{Alg}\mathcal{WR} = \mathsf{Alg}^*\mathcal{WR}$, in contrast to the preceding examples. This is the variety of the so-called *R-algebras*, and coincides with $\mathsf{Alg}\mathcal{R}$, the equivalent algebraic semantics of $\mathcal{R}$, identified and studied in [114]. This example is interesting because, unlike others, its language contains a natural implication connective satisfying (MP); what it lacks is (R), and it turns out that the logic $\mathcal{R}$ is exactly the extension of $\mathcal{WR}$ with (R) as additional axiom. Further details may be found in [107, Section 5.4.1] and in the cited papers.
8. Subintuitionistic logics [215, 243] are logics defined in the intuitionistic language by several classes of Kripke-style frames, with the same forcing conditions as those for intuitionistic logic, but without requiring the accessibility relation to be reflexive and transitive, and without requiring valuations to be persistent. Each class of frames defines both a local and a global logic, similarly to the modal case. Intuitionistic logic $\mathcal{I}\ell$ corresponds to the class of frames with all these properties, where the global and the local coincide, and is hence the strongest in the family. Almost as strong is $\mathcal{BPL}$, the global (equivalently, the local) logic corresponding to the class of frames (equivalently, of transitive

† While in [32] the $\langle \wedge, \vee, \neg, \top, \bot \rangle$ fragment is considered, $\top$ and $\bot$ are definable, therefore it is enough to consider the $\langle \wedge, \vee, \neg \rangle$ fragment, as in the other papers and here (see page 384).

frames) with persistent valuations; this logic, introduced by Visser in [240], is often called "basic propositional logic" in the literature, and has arisen some interest. $\mathcal{BPL}$ was proved to be non-protoalgebraic in [232, Theorem 14]; as a consequence, four other logics in the group, weaker than $\mathcal{BPL}$, are not protoalgebraic either, namely the local and the global consequences of the classes of all frames and of transitive frames. This group of ten different logics has been studied with abstract algebraic logic techniques in [4, 6, 45, 54], and is particularly rich in examples and counterexamples; for instance, of its five protoalgebraic members, three are finitely regularly algebraizable (including $\mathcal{I}\ell$) while two are finitely equivalential but not weakly algebraizable, and two of the non-protoalgebraic ones are among the few known $\mathcal{L}$ such that $\mathsf{Alg}^*\mathcal{L} \subsetneq \mathsf{Alg}\,\mathcal{L} \subsetneq \mathbb{V}\mathcal{L}$ (see page 279).

9. The (finitary) infinitely-valued logic $Ł_\infty^{\leqslant}$ that preserves degrees of truth (in the sense of Example 4.11) with respect to Łukasiewicz's algebra $[\mathbf{0,1}]$ on the real unit interval was shown in [103] to be non-protoalgebraic. Generalizing this approach, the logics that preserve degrees of truth with respect to varieties of commutative integral residuated lattices are analysed in [46]. Several algebraic criteria are given to separate the protoalgebraic logics in this family from the non-protoalgebraic ones, and as a result one finds that many natural logics in this family are non-protoalgebraic. In particular, the logics preserving degrees of truth from the following varieties are non-protoalgebraic: all commutative integral residuated lattices, MV-algebras, product algebras, BL-algebras, MTL-algebras, $\mathrm{FL_{ew}}$-algebras, and any variety generated by a family of continuous t-norms over the real unit interval that is not the variety of Gödel algebras; general references for these classes of algebras are [59, 123, 132]. See also Example 7.47.2. ⊠

A model-theoretic characterization

Although the full force of monotonicity lies in the soon-to-be-proved Correspondence Theorem, already in its bare form (or in its equivalent form of $\boldsymbol{\Omega}^{\boldsymbol{A}}$ commuting with intersections) it allows us to obtain significant results.

A matrix $\langle \boldsymbol{A}, F\rangle$ is a ***subdirect product*** of a family $\{\langle \boldsymbol{A}_i, F_i\rangle : i \in I\}$ when it is a submatrix of the product matrix $\langle \prod_i \boldsymbol{A}_i, \prod_i F_i\rangle$ such that $\pi_i A = A_i$ for all $i \in I$, where π_i denotes the i-th projection from the product set onto its i-th component; that is, the algebra $\boldsymbol{A}$ is a subdirect product of the $\boldsymbol{A}_i$, and $F = (\prod_i F_i) \cap A$. If M is a class of matrices, then $\mathbb{P}_{\mathrm{SD}}\mathsf{M}$ denotes the class of matrices that are a subdirect product of a non-empty family of matrices in M.

THEOREM 6.17
A logic $\mathcal{L}$ is protoalgebraic if and only if the class $\mathsf{Mod}^*\mathcal{L}$ is closed under $\mathbb{P}_{\mathrm{SD}}$.

PROOF: ($\Rightarrow$) Let $\langle \boldsymbol{A}, F\rangle$ be a subdirect product of $\{\langle \boldsymbol{A}_i, F_i\rangle : i \in I\} \subseteq \mathsf{Mod}^*\mathcal{L}$. By Proposition 4.67, $\mathsf{Mod}\,\mathcal{L}$ is closed under $\mathbb{S}$ and $\mathbb{P}$, so in particular under $\mathbb{P}_{\mathrm{SD}}$, hence $\langle \boldsymbol{A}, F\rangle \in \mathsf{Mod}\mathcal{L}$. Protoalgebraicity of $\mathcal{L}$ implies that it is reduced: By assumption the π_i are surjective homomorphisms, and it is easy to see that $F = (\prod F_i) \cap A = \bigcap \pi_i^{-1} F_i$. Now, using that by protoalgebraicity $\boldsymbol{\Omega}^{\boldsymbol{A}}$ commutes with intersections of $\mathcal{L}$-filters (Proposition 6.14), Proposition 4.25, and that the matrices are reduced, one obtains

$$\begin{aligned}\boldsymbol{\Omega}^{\boldsymbol{A}} F = \boldsymbol{\Omega}^{\boldsymbol{A}} \bigcap_{i \in I} \pi_i^{-1} F_i &= \bigcap_{i \in I} \boldsymbol{\Omega}^{\boldsymbol{A}} \pi_i^{-1} F_i \\ &= \bigcap_{i \in I} \pi_i^{-1} \boldsymbol{\Omega}^{\boldsymbol{A}_i} F_i = \bigcap_{i \in I} \pi_i^{-1} \mathrm{Id}_{A_i} = \mathrm{Id}_A .\end{aligned}$$

This proves that $\langle \boldsymbol{A}, F\rangle \in \mathsf{Mod}^*\mathcal{L}$.

($\Leftarrow$) Let us see that for each $\boldsymbol{A}$, $\boldsymbol{\Omega}^{\boldsymbol{A}}$ is monotone over $\mathcal{F}i_{\mathcal{L}}\boldsymbol{A}$: Let $F, G \in \mathcal{F}i_{\mathcal{L}}\boldsymbol{A}$ with $F \subseteq G$. Consider $\theta := \boldsymbol{\Omega}^{\boldsymbol{A}} F \cap \boldsymbol{\Omega}^{\boldsymbol{A}} G \in \mathrm{Con}\boldsymbol{A}$ and apply the construction in Exercise 4.20 taking into account that here $F \cap G = F$. As a result, the matrix $\langle \boldsymbol{A}/\theta, F/\theta\rangle$ is isomorphic to a subdirect product of the matrices $\langle \boldsymbol{A}/\boldsymbol{\Omega}^{\boldsymbol{A}} F, F/\boldsymbol{\Omega}^{\boldsymbol{A}} F\rangle$ and $\langle \boldsymbol{A}/\boldsymbol{\Omega}^{\boldsymbol{A}} G, G/\boldsymbol{\Omega}^{\boldsymbol{A}} G\rangle$. Since these are reduced models of $\mathcal{L}$, and the class $\mathsf{Mod}^*\mathcal{L}$ is closed under isomorphisms, by the assumption the first matrix is also a reduced model of $\mathcal{L}$, so in particular $\mathrm{Id}_{A/\theta} = \boldsymbol{\Omega}^{\boldsymbol{A}/\theta}(F/\theta) = (\boldsymbol{\Omega}^{\boldsymbol{A}} F)/\theta$; but by construction $\theta \subseteq \boldsymbol{\Omega}^{\boldsymbol{A}} F$, therefore $\boldsymbol{\Omega}^{\boldsymbol{A}} F = \theta$, which amounts to $\boldsymbol{\Omega}^{\boldsymbol{A}} F \subseteq \boldsymbol{\Omega}^{\boldsymbol{A}} G$. ⊠

COROLLARY 6.18

If $\mathcal{L}$ is protoalgebraic, then $\mathsf{Alg}^*\mathcal{L}$ is closed under $\mathbb{P}_{\mathrm{SD}}$. ⊠

In fact, this is already known to us, because, by Corollary 6.15, if the logic $\mathcal{L}$ is protoalgebraic, then $\mathsf{Alg}^*\mathcal{L} = \mathsf{Alg}\mathcal{L}$, and by Corollary 5.71, the latter is always closed under subdirect products. The property is particularly interesting when formulated as in Corollary 6.18, because $\mathsf{Alg}^*\mathcal{L}$ is in general not closed under subdirect products; thus this can be useful to *disprove* that a particular logic $\mathcal{L}$ is protoalgebraic, assuming that $\mathsf{Alg}^*\mathcal{L}$ is known. Moreover, notice that, in contrast to Theorem 6.17, the converse of Corollary 6.18 is not true: the weak relevance logic $\mathcal{WR}$ of Example 6.16.7 is not protoalgebraic but $\mathsf{Alg}^*\mathcal{WR}$ is a variety.

The Correspondence Theorem

This is another of the most useful and significant characterizations of protoalgebraicity. It tells us that for these logics the lattices of filters behave, in a sense, as well as the lattices of congruences of ordinary algebras, and that strict surjective homomorphisms between matrices (and in particular reductions) show a good behaviour as far as the filters containing those in the matrices are concerned. Roughly speaking, this means that some methods of universal algebra may be adapted for matrices and produce interesting results when dealing with models of protoalgebraic logics.

Theorem 6.19 (The Correspondence Theorem)
For any logic $\mathcal{L}$ the following conditions are equivalent:

(i) *$\mathcal{L}$ is protoalgebraic.*

(ii) *For every strict surjective homomorphism $h\colon \langle \boldsymbol{A}, F\rangle \to \langle \boldsymbol{B}, G\rangle$ from a model of $\mathcal{L}$ to another, if $F' \in \mathcal{F}i_{\mathcal{L}}\boldsymbol{A}$ is such that $F' \supseteq F$, then $h^{-1}hF' = F'$; that is, h is strict from $\langle \boldsymbol{A}, F'\rangle$ to $\langle \boldsymbol{B}, hF'\rangle$*

(iii) *Every strict surjective homomorphism $h\colon \langle \boldsymbol{A}, F\rangle \to \langle \boldsymbol{B}, G\rangle$ from a model of $\mathcal{L}$ to another induces a strict surjective homomorphism from the g-matrix $\langle \boldsymbol{A}, (\mathcal{F}i_{\mathcal{L}}\boldsymbol{A})^{F}\rangle$ to the g-matrix $\langle \boldsymbol{B}, (\mathcal{F}i_{\mathcal{L}}\boldsymbol{B})^{G}\rangle$, which are g-models of $\mathcal{L}$.*

Proof: (i)⇒(ii) Assume that $\mathcal{L}$ is protoalgebraic and let $h\colon \langle \boldsymbol{A}, F\rangle \to \langle \boldsymbol{B}, G\rangle$ and F' be as in the statement. Since h is strict, by Proposition 4.27.1 $\ker h \subseteq \boldsymbol{\Omega}^{\boldsymbol{A}}F$, and by protoalgebraicity $\boldsymbol{\Omega}^{\boldsymbol{A}}F \subseteq \boldsymbol{\Omega}^{\boldsymbol{A}}F'$, therefore $\ker h \subseteq \boldsymbol{\Omega}^{\boldsymbol{A}}F'$, which implies that $h^{-1}hF' = F'$ by the same proposition.

(ii)⇒(iii) By Lemma 4.36, h induces in general an order isomorphism between $\{F' \in \mathcal{F}i_{\mathcal{L}}\boldsymbol{A} : \ker h \subseteq \boldsymbol{\Omega}^{\boldsymbol{A}}F'\}$ and $\mathcal{F}i_{\mathcal{L}}\boldsymbol{B}$, with inverse the residual h^{-1}. Since $F = h^{-1}G$ and $G = hF$, this isomorphism restricts to one between $\{F' \in \mathcal{F}i_{\mathcal{L}}\boldsymbol{A} : \ker h \subseteq \boldsymbol{\Omega}^{\boldsymbol{A}}F' \text{ and } F \subseteq F'\}$ and $(\mathcal{F}i_{\mathcal{L}}\boldsymbol{B})^{G}$. By Theorem 5.48, in order to establish the wanted strict homomorphism we need only check that the first set, obviously included in $(\mathcal{F}i_{\mathcal{L}}\boldsymbol{A})^{F}$, actually equals it. But by (ii), if $F' \in (\mathcal{F}i_{\mathcal{L}}\boldsymbol{A})^{F}$, then $h^{-1}hF' = F'$, which is the same as $\ker h \subseteq \boldsymbol{\Omega}^{\boldsymbol{A}}F'$.

(iii)⇒(i) To show monotonicity, take any $F, F' \in \mathcal{F}i_{\mathcal{L}}\boldsymbol{A}$ with $F \subseteq F'$ and consider the canonical projection $\pi\colon \langle \boldsymbol{A}, F\rangle \to \langle \boldsymbol{A}/\boldsymbol{\Omega}^{\boldsymbol{A}}F, F/\boldsymbol{\Omega}^{\boldsymbol{A}}F\rangle$. It is a strict surjective homomorphism between two models of $\mathcal{L}$, therefore by (iii) it establishes a strict surjective homomorphism from $(\mathcal{F}i_{\mathcal{L}}\boldsymbol{A})^{F}$ to $\big(\mathcal{F}i_{\mathcal{L}}(\boldsymbol{A}/\boldsymbol{\Omega}^{\boldsymbol{A}}F)\big)^{F/\boldsymbol{\Omega}^{\boldsymbol{A}}F}$. Since $F' \in (\mathcal{F}i_{\mathcal{L}}\boldsymbol{A})^{F}$, this implies that $F' = \pi^{-1}\pi F'$, which amounts to saying that $\boldsymbol{\Omega}^{\boldsymbol{A}}F$ is compatible with F', or that $\boldsymbol{\Omega}^{\boldsymbol{A}}F \subseteq \boldsymbol{\Omega}^{\boldsymbol{A}}F'$. ⊠

You may have noticed that the property (ii) is actually part of the property in (iii), by the very definition of a strict surjective homomorphism between g-matrices; thus, the proof of (iii)⇒(i) is actually a proof of (ii)⇒(i). Condition (ii) is included here to emphasize this aspect of the correspondence property: that each strict surjective homomorphism between two models of the logic is also strict between all the pairs of models obtained by taking the sets of filters that contain the given ones.

However, the formulation of (iii) in terms of g-models is often more useful, since strict surjective homomorphisms between g-matrices enjoy a good number of nice properties; for instance, recall that, by Theorem 5.48, in the described situation $h\colon (\mathcal{F}i_{\mathcal{L}}\boldsymbol{A})^{F} \cong (\mathcal{F}i_{\mathcal{L}}\boldsymbol{B})^{G}$. In particular, Theorem 6.19 holds for quotients and for reductions, and there are other applications for particular situations, which may also be useful to record here:

COROLLARY 6.20

Let $\mathcal{L}$ be protoalgebraic and let $h\colon \boldsymbol{A} \to \boldsymbol{B}$ be a surjective homomorphism.

1. *For each $G \in \mathcal{F}i_{\mathcal{L}}\boldsymbol{B}$ there is an $F \in \mathcal{F}i_{\mathcal{L}}\boldsymbol{A}$ such that $(\mathcal{F}i_{\mathcal{L}}\boldsymbol{A})^F \cong (\mathcal{F}i_{\mathcal{L}}\boldsymbol{B})^G$. In particular, there is an $F_0 \in \mathcal{F}i_{\mathcal{L}}\boldsymbol{A}$ such that $(\mathcal{F}i_{\mathcal{L}}\boldsymbol{A})^{F_0} \cong \mathcal{F}i_{\mathcal{L}}\boldsymbol{B}$.*
2. *If $G \in \mathcal{F}i_{\mathcal{L}}\boldsymbol{B}$ and $a \in A$, then $Fg_{\mathcal{L}}^{\boldsymbol{A}}(h^{-1}G,a) = h^{-1}Fg_{\mathcal{L}}^{\boldsymbol{B}}(G,ha)$.*
3. *If $\bigcap \mathcal{F}i_{\mathcal{L}}\boldsymbol{B} \subseteq Y \subseteq B$, then $Fg_{\mathcal{L}}^{\boldsymbol{B}}Y = hFg_{\mathcal{L}}^{\boldsymbol{A}}h^{-1}Y$.*
4. *If $\bigcap \mathcal{F}i_{\mathcal{L}}\boldsymbol{B} \subseteq Y \subseteq B$ and $Y = hX$ for some $X \subseteq A$, then $Fg_{\mathcal{L}}^{\boldsymbol{B}}Y = hFg_{\mathcal{L}}^{\boldsymbol{A}}X$.*

PROOF: Left to you as Exercise 6.12. ⊠

One can imagine other variations of these properties, with added subleties (for instance, if the first assumption on Y in the last two points is deleted, then $h^{-1}\bigcap \mathcal{F}i_{\mathcal{L}}\boldsymbol{B}$ must be added somewhere in the conclusion); all are consequences of the surjective homomorphism being strict from $\langle \boldsymbol{A}, (\mathcal{F}i_{\mathcal{L}}\boldsymbol{A})^{h^{-1}G}\rangle$ to $\langle \boldsymbol{B}, (\mathcal{F}i_{\mathcal{L}}\boldsymbol{B})^{G}\rangle$. The phrase "by the Correspondence Theorem" actually means any of its consequences in Corollary 6.20. To be historically accurate, the term "correspondence property" in this context was coined by Blok and Pigozzi in their 1986 seminal paper [30] to refer to one of these variations, recorded in Exercise 6.13; in his monograph [70] on protoalgebraic logics, Czelakowski uses the same name to refer to the property in Exercise 6.14, which is property 6.19(ii) but for arbitrary strict homomorphisms, and is again equivalent to protoalgebraicity.

Protoalgebraic logics and the Deduction Theorem

Protoalgebraic logics happen to be characterized by a very general form of the Deduction Theorem:

DEFINITION 6.21

Let $\Phi \subseteq \mathscr{P}Fm$ be a family of sets of formulas, each of them of the form $I(x,y,\vec{z})$, and let $\mathcal{L}$ be a logic. Φ is a ***family of Deduction-Detachment (DD) sets with parameters for $\mathcal{L}$*** *when for all $\Gamma \cup \{\varphi,\psi\} \subseteq Fm$,*

$$\Gamma,\varphi \vdash_{\mathcal{L}} \psi \iff \text{there is some } I \in \Phi \text{ and some } \vec{\delta} \in \vec{Fm} \text{ such that } \Gamma \vdash_{\mathcal{L}} I(\varphi,\psi,\vec{\delta}). \qquad \text{PLDDT}$$

A logic $\mathcal{L}$ satisfies the ***Parametrised Local Deduction-Detachment Theorem****, PLDDT, when it has a family of DD sets with parameters.*

In practice all the sets in the family Φ are assumed to be non-empty; it is easy to see that if $\varnothing \in \Phi$, then this Φ can be a family of DD sets only for the inconsistent and the almost inconsistent logics.

THEOREM 6.22

A logic is protoalgebraic if and only if it satisfies the PLDDT.

PROOF: ($\Rightarrow$) Take $\Phi := \{T \in \mathcal{T}h\mathcal{L} : x, T \vdash_{\mathcal{L}} y\}$. Notice that $C_{\mathcal{L}}\{y\} \in \Phi$ and that protoalgebraicity implies that $\Sigma_{\mathcal{L}} \in \Phi$ as well, by Theorem 6.7; so $\Phi \neq \varnothing$. We are going to see that this Φ is a family of DD sets with parameters for $\mathcal{L}$.

Assume that $\Gamma, \varphi \vdash_{\mathcal{L}} \psi$. Take any surjective substitution $\sigma\colon \boldsymbol{Fm} \to \boldsymbol{Fm}$ such that $\sigma x := \varphi$ and $\sigma y := \psi$, and put $I := \sigma^{-1} C_{\mathcal{L}} \Gamma$. The surjectivity of σ implies that $\sigma I = \sigma\sigma^{-1} C_{\mathcal{L}} \Gamma = C_{\mathcal{L}} \Gamma$, that is, $\Gamma \vdash_{\mathcal{L}} \sigma I = I(\varphi, \psi, \vec{\delta})$, where $\vec{\delta} := \sigma \vec{z}$ for the variables $\vec{z}$ in I different from x, y. It remains only to show that $I \in \Phi$, that is, that $I \in \mathcal{T}h\mathcal{L}$ and $x, I \vdash_{\mathcal{L}} y$. The definition of I implies that $I \in \mathcal{T}h\mathcal{L}$. We have to show that $y \in C_{\mathcal{L}}(I, x)$. But $C_{\mathcal{L}}(I, x) \in \mathcal{T}h\mathcal{L}$, and $I \subseteq C_{\mathcal{L}}(I, x)$, and σ is a strict surjective matrix homomorphism from $\langle \boldsymbol{Fm}, I \rangle$ onto $\langle \boldsymbol{Fm}, C_{\mathcal{L}} \Gamma \rangle$. Since $\mathcal{L}$ is protoalgebraic, the Correspondence Theorem implies that $\sigma C_{\mathcal{L}}(I, x) \in \mathcal{T}h\mathcal{L}$ and $\Gamma \subseteq C_{\mathcal{L}} \Gamma \subseteq \sigma C_{\mathcal{L}}(I, x)$. By structurality, $x \in C_{\mathcal{L}}(I, x)$ implies $\varphi = \sigma x \in \sigma C_{\mathcal{L}}(I, x)$. Then, since $\Gamma, \varphi \vdash_{\mathcal{L}} \psi$, it follows that $\sigma y = \psi \in \sigma C_{\mathcal{L}}(I, x)$, that is, $y \in \sigma^{-1} \sigma C_{\mathcal{L}}(I, x) = C_{\mathcal{L}}(I, x)$, by the Correspondence Theorem again.

Conversely, assume that $\Gamma \vdash_{\mathcal{L}} I(\varphi, \psi, \vec{\delta})$ for some $I(x, y, \vec{z}) \in \Phi$ and some $\vec{\delta} \in \vec{Fm}$. Since $I \in \Phi$, $I(x, y\vec{z}), x \vdash_{\mathcal{L}} y$, and by structurality $I(\varphi, \psi, \vec{\delta}), \varphi \vdash_{\mathcal{L}} \psi$, which implies $\Gamma, \varphi \vdash_{\mathcal{L}} \psi$.

($\Leftarrow$) Applying the PLDDT to $x \vdash_{\mathcal{L}} x$, there is some $I \in \Phi$ and some $\vec{\delta} \in \vec{Fm}$ such that $\emptyset \vdash_{\mathcal{L}} I(x, x, \vec{\delta})$. Now apply the PLDDT (in the reverse sense) to $I(x, y, \vec{\delta}) \vdash_{\mathcal{L}} I(x, y, \vec{\delta})$, which gives $I(x, y, \vec{\delta}), x \vdash_{\mathcal{L}} y$. Since the formulas $\vec{\delta}$ may contain the variables x, y as well as other variables $\vec{z}$, the same set can be described as $I'(x, y, \vec{z}) := I\big(x, y, \vec{\delta}(x, y, \vec{z})\big)$. Now $I'(x, y, \vec{z}), x \vdash_{\mathcal{L}} y$, and from $\vdash_{\mathcal{L}} I(x, x, \vec{\delta})$, after applying the substitution $y \mapsto x$, by structurality it follows that $\vdash_{\mathcal{L}} I\big(x, x, \vec{\delta}(x, x, \vec{z})\big)$, that is, $\vdash_{\mathcal{L}} I'(x, x, \vec{z})$. Finally, by taking $\Delta(x, y) := I'(x, y, \vec{x})$, where here $\vec{x}$ denotes a sequence with all its members equal to x, structurality implies that this set satisfies the conditions (R) and (MP) of Theorem 6.7(iii), and this shows that $\mathcal{L}$ is protoalgebraic. ⊠

Exercise 6.15 proposes you to find, hidden in the previous proof, another, purely syntactic characterization of protoalgebraicity. Moreover, it is not difficult to prove (Exercise 6.16) the following generalization of a classical result (Theorem 3.72) concerning the DDT:

Proposition 6.23

Let $\mathcal{L}$ be a finitary protoalgebraic logic.

1. *$\mathcal{L}$ satisfies the* PLDDT *with respect to a family Φ such that each $I \in \Phi$ is finite.*
2. *For each family Φ as in 1, $\mathcal{L}$ can be axiomatized by the Hilbert-style calculus that has the formulas in $C_{\mathcal{L}}\emptyset$ as axioms and the rules* (MP_I), $I, x \rhd y$, *for each $I \in \Phi$, as inference rules.* ⊠

The PLDDT is probably the weakest among the many forms of the Deduction Theorem that have been considered in the literature. In Section 3.6 you already got acquainted with its strongest and most classical form, the ***Deduction-Detachment Theorem***, DDT, which can be described as a PLDDT for a family Φ consisting of a single set I and without parameters; this makes the PLDDT adopt the very simple form

$$\Gamma, \varphi \vdash_{\mathcal{L}} \psi \iff \Gamma \vdash_{\mathcal{L}} I(\varphi, \psi). \qquad \text{DDT}$$

where the set $I(x,y)$ is the same for all Γ,φ,ψ; this is why this is a "non local" form of the theorem.

In between the PLDDT and the DDT there are several kinds of Deduction Theorems:

- The ***Local Deduction-Detachment Theorem***, LDDT, where the sets $I \in \Phi$ do not have parameters, that is, are of the form $I(x,y)$, and the PLDDT takes the form

$$\Gamma,\varphi \vdash_{\mathcal{L}} \psi \iff \text{there is some } I \in \Phi \text{ such that } \Gamma \vdash_{\mathcal{L}} I(\varphi,\psi). \qquad \text{LDDT}$$

  The term *local* refers to the fact that the set I (and the parameters $\vec{\delta}$ in the case of the PLDDT) are not fixed but depend on the formulas Γ,φ,ψ.

- The ***Parametrised Deduction-Detachment Theorem***, PDDT, where Φ consists of a single set I but possibly with parameters, and the PLDDT takes the form

$$\Gamma,\varphi \vdash_{\mathcal{L}} \psi \iff \text{there is some } \vec{\delta} \in \vec{Fm} \text{ such that } \Gamma \vdash_{\mathcal{L}} I(\varphi,\psi,\vec{\delta}). \qquad \text{PDDT}$$

There is also the Contextual Deduction Theorem, a more recent (and more complicated to state) version, which has been briefly reviewed on page 174.

The variety of *Deduction Theorems* lies at the heart of the creation of abstract algebraic logic as a modern discipline, due to their equivalence with purely algebraic properties. Actually, Theorem 6.22 states the equivalence between the PLDDT and protoalgebraicity, which can be considered either as a syntactic property or as an algebraic one, due to its multi-faceted characterizations. Just for the record, let me state the most popular *bridge theorem* for protoalgebraic logics related to the Local Deduction Theorem:

THEOREM 6.24
Let $\mathcal{L}$ be a finitary protoalgebraic logic. The following conditions are equivalent:

(i) *$\mathcal{L}$ satisfies the* LDDT.

(ii) *The class* $\mathsf{Mod}\,\mathcal{L}$ *satisfies the* **Filter Extension Property**, FEP: *If $\langle \boldsymbol{A},F\rangle \in \mathsf{Mod}\,\mathcal{L}$ and $\langle \boldsymbol{B},G\rangle$ is a submatrix of $\langle \boldsymbol{A},F\rangle$, then for each $G' \in \mathcal{F}i_{\mathcal{L}}\boldsymbol{B}$ with $G \subseteq G'$ there is an $F' \in \mathcal{F}i_{\mathcal{L}}\boldsymbol{A}$ with $F \subseteq F'$ and $F' \cap B = G'$.*

(iii) *The class* $\mathsf{Mod}^*\mathcal{L}$ *satisfies the* FEP *(in the same sense as above).* ☒

Without the assumption of finitarity, the equivalence holds for the *principal* $\mathcal{L}$-filters only. More information on these issues can be found in [70, Section 2.3].

EXAMPLES 6.25

1. Of the global logics associated with normal modal systems, the ones associated with $S4$ and with $S5$ satisfy the DDT for $I(x,y) = \{\Box x \to y\}$, while the ones associated with K and T satisfy the LDDT for the family $\Phi = \{\{(x \wedge \Box x \wedge \Box^2 x \wedge \cdots \wedge \Box^n x) \to y\} : n \geqslant 0\}$. The finitary infinitely-valued logic of Łukasiewicz, $Ł_{\infty f}$, satisfies the LDDT for the family $\{\{x \to^n y\} : n \geqslant 0\}$, with the terms $x \to^n y$ defined on page 165. The logic $\mathcal{BCK}$ satisfies the same form of the LDDT. All these cases correspond to algebraizable logics; one can use

the bridge theorem connecting the DDT with EDPRC (Corollary 3.86) to show that the mentioned logics, except the first two, do *not* satisfy the DDT for any set of binary formulas.

2. The first natural example to be found in the literature of a real PLDDT where parameters cannot be eliminated is the relevance logic $\mathcal{R}$ of Example 3.34.7: it satisfies the PLDDT for the family

$$\Phi = \Big\{ \{ \big(\big(\bigwedge_{1 \leqslant i \leqslant n} (z_i \to z_i) \big) \wedge x \big) \to y \} : n \geqslant 1 \Big\}.$$

The discovery of this result is attributed in [70, p. 126] to Dziobiak (as unpublished). It is remarkable that if $\top$ is added to the language (with some appropriate axioms), then parameters can be dispensed with. In general, however, they cannot; this is a consequence of Theorem 6.24 plus the fact, proved in [33, Application 5.4.3], that the class of reduced models of $\mathcal{R}$ does not satisfy the FEP.

3. The logics associated with varieties of residuated lattices are algebraizable (Example 3.34.2); a fortiori, they are protoalgebraic, and by Theorem 6.22 they satisfy some form of the PLDDT. Galatos *et al.* have found in [123, Section 2.4] the form of the families of DD sets with parameters for these logics, which is in general very complicated. They also show that parameters can be eliminated for logics that are extensions of $\mathcal{FL}_e$ (Full Lambek with exchange), so that these logics satisfy the LDDT and the DD sets have a simpler and more understandable form.

4. The equivalence fragment of intuitionistic logic, $\mathcal{I}\ell_{\leftrightarrow}$, is algebraizable and hence protoalgebraic; the form of the family of DD sets involved in its PLDDT is too complicated to reproduce it here; you can find it in [70, Theorem 6.6.7].

5. Let Ri denote the class of commutative rings with unit, presented as algebras $\langle A, +, -, \cdot, 0, 1 \rangle$ of type $\langle 2, 2, 2, 0, 0 \rangle$, and consider the logic $\mathcal{R}i$ defined by the class of matrices $\{\langle \boldsymbol{A}, \{0\} \rangle : \boldsymbol{A} \in \mathsf{Ri}\}$. This class is clearly closed under ultraproducts, therefore the logic is finitary. It is easy to see that if we take $\Delta(x, y) := \{x - y\}$, then this set satisfies (R) and (MP), thus showing that this logic is protoalgebraic. Hence it satisfies the PLDDT, but locality can be eliminated: in [70, Note 2.4.2] it is shown that $\mathcal{R}i$ satisfies the PDDT with the (one-parametrised) single formula $\{y - x \cdot z\}$ as DD set. It is also stated there (without proof) that its lattice of theories is not distributive; by Corollary 6.30 below, this implies that the logic cannot satisfy any form of DDT, that is, that the parameter in the PDDT cannot be eliminated.

6. The first explicit example in the literature of a protoalgebraic logic *not* satisfying the PDDT, as far as I know, is the "simplest protoalgebraic logic" $\mathcal{I}$ of Example 6.78.1; that it does not satisfy the PDDT is shown in [100] as a negative application of the bridge theorem for the PDDT established by Czelakowski [70, Theorem 2.4.1]. ⊠

Now we head towards the main bridge theorem concerning the DDT in the protoalgebraic environment. It all starts from the following observation, which is actually a particular case of the more general property contained in Proposition 1.64.

PROPOSITION 6.26
For any logic $\mathcal{L}$, the family of finitely generated $\mathcal{L}$-filters of an algebra $\boldsymbol{A}$,

$$\mathcal{F}i^{\omega}_{\mathcal{L}}\boldsymbol{A} := \{F \in \mathcal{F}i_{\mathcal{L}}\boldsymbol{A} : F = Fg^{\boldsymbol{A}}_{\mathcal{L}}X,\ X \subseteq A, X \textit{ finite}\},$$

ordered by inclusion, is a join-semilattice, and a join-sub-semilattice of the filter lattice $\mathcal{F}i_{\mathcal{L}}\boldsymbol{A}$, and is generated by the principal $\mathcal{L}$-filters of $\boldsymbol{A}$. This applies in particular to the family $\mathcal{T}h^{\omega}\mathcal{L}$ of the finitely generated theories of $\mathcal{L}$. ⊠

A join-semilattice $\boldsymbol{A} = \langle A, \vee\rangle$ is ***dually residuated***† when for all $a, b \in A$ there exists $a \star b = \min\{c \in A : b \leqslant a \vee c\}$. This defines (not necessarily as a term) a binary operation $\star$, called the ***dual residual*** of $\vee$, which is equivalently characterized by the property

$$a \star b \leqslant c \iff b \leqslant a \vee c.$$

Only the following two (easy) properties of dually residuated join-semilattices are needed. They are left to you as Exercise 6.17:

PROPOSITION 6.27
Let $\boldsymbol{A} = \langle A, \vee\rangle$ be a join-semilattice.

1. *If $\boldsymbol{A}$ is generated by a set $X \subseteq A$, then $\boldsymbol{A}$ is dually residuated if and only if the dual residual $a \star b$ exists in A for all pairs of generators $a, b \in X$.*
2. *If $\boldsymbol{A}$ is dually residuated, then its lattice of ideals is distributive.* ⊠

The next result, due to Czelakowski [68], is probably the oldest published *bridge theorem* in abstract algebraic logic, though not the best known one. It holds for protoalgebraic logics that are finitary; this assumption is not necessary for Proposition 6.26, on which it is based, but is needed in a crucial step in the proof.

THEOREM 6.28
Let $\mathcal{L}$ be a finitary and protoalgebraic logic. The following conditions are equivalent:

(i) *$\mathcal{L}$ satisfies the* DDT.

(ii) *For all $\boldsymbol{A}$, the join-semilattice $\mathcal{F}i^{\omega}_{\mathcal{L}}\boldsymbol{A}$ is dually residuated.*

(iii) *The join-semilattice $\mathcal{T}h^{\omega}\mathcal{L}$ is dually residuated.*

PROOF: (i)⇒(ii) By Proposition 5.90, the DDT transfers from $\mathcal{L}$ to its basic full g-models. Therefore, in particular, for all $a, b \in A$ and all $F \in \mathcal{F}i^{\omega}_{\mathcal{L}}\boldsymbol{A}$, we have that $b \in Fg^{\boldsymbol{A}}_{\mathcal{L}}(F, a) \Leftrightarrow I^{\boldsymbol{A}}(a, b) \subseteq F$, where I is the DD set for $\mathcal{L}$. Expressed in terms of the lattice of filters, this is to say that $Fg^{\boldsymbol{A}}_{\mathcal{L}}\{b\} \subseteq F \vee Fg^{\boldsymbol{A}}_{\mathcal{L}}\{a\} \Leftrightarrow Fg^{\boldsymbol{A}}_{\mathcal{L}} I^{\boldsymbol{A}}(a, b) \subseteq F$.

† The terms "dually Brouwerian" for the structure and "dual relative pseudo-complement" for the operation have also been used in the literature.

Since by finitarity of $\mathcal{L}$ the set I can be assumed to be finite, $Fg^{\boldsymbol{A}}_{\mathcal{L}} I^{\boldsymbol{A}}(a,b) \in \mathcal{F}i^{\omega}_{\mathcal{L}}\boldsymbol{A}$, and hence the last equivalence says that $Fg^{\boldsymbol{A}}_{\mathcal{L}} I^{\boldsymbol{A}}(a,b) = Fg^{\boldsymbol{A}}_{\mathcal{L}}\{a\} \star Fg^{\boldsymbol{A}}_{\mathcal{L}}\{b\}$. Proposition 6.26 tells us that the join operation of $\mathcal{F}i^{\omega}_{\mathcal{L}}\boldsymbol{A}$ is that of $\mathcal{F}i_{\mathcal{L}}\boldsymbol{A}$, and that, as a join-semilattice, $\mathcal{F}i^{\omega}_{\mathcal{L}}\boldsymbol{A}$ is generated by the principal $\mathcal{L}$-filters of $\boldsymbol{A}$. Therefore, by Proposition 6.27.1, what we have proved is enough to show that $\mathcal{F}i^{\omega}_{\mathcal{L}}\boldsymbol{A}$ is dually residuated.

(ii)⇒(iii) because (iii) is a particular case of (ii).

(iii)⇒(i) Let $x, y \in V$. The theories $C_{\mathcal{L}}\{x\}$ and $C_{\mathcal{L}}\{y\}$ are finitely generated, therefore by assumption there is a finite $\Sigma \subseteq Fm$ such that $C_{\mathcal{L}}\Sigma = C_{\mathcal{L}}\{x\} \star C_{\mathcal{L}}\{y\}$. This means that for all finite $\Gamma \subseteq Fm$,

$$y \in C_{\mathcal{L}}(\Gamma, x) \iff C_{\mathcal{L}}\{y\} \subseteq C_{\mathcal{L}}\Gamma \vee C_{\mathcal{L}}\{x\} \iff C_{\mathcal{L}}\Sigma \subseteq C_{\mathcal{L}}\Gamma \iff \Sigma \subseteq C_{\mathcal{L}}\Gamma.$$

Using finitarity of $\mathcal{L}$ and finiteness of Σ, one can see (Exercise 6.18) that the same holds for an arbitrary $\Gamma \subseteq Fm$, not necessarily finite, that is:

$$y \in C_{\mathcal{L}}(\Gamma, x) \iff \Sigma \subseteq C_{\mathcal{L}}\Gamma. \tag{84}$$

It seems as if we almost have it: we need only extract from Σ a subset with only two variables and apply structurality to prove the same thing for all formulas φ, ψ and all sets Γ, but this turns out to be slightly tricky. If Σ is $\Sigma(x, y, \vec{z})$, we take $I(x,y) := \Sigma(x, y, \vec{x})$, where here $\vec{x}$ denotes a sequence entirely made of xs, and show that this is a DD set for $\mathcal{L}$. Take any $\varphi, \psi \in Fm$ and any $\Gamma \subseteq Fm$. Consider a surjective substitution $\sigma\colon \boldsymbol{Fm} \to \boldsymbol{Fm}$ such that $\sigma x := \varphi$, $\sigma y := \psi$ and $\sigma z := \varphi$ for all $z \in \operatorname{Var}\Sigma, z \neq x, y$ (finiteness of Σ guarantees the existence of such a surjective substitution). Then $I(\varphi, \psi) = \Sigma(\varphi, \psi, \vec{\varphi}) = \sigma\Sigma$. Now, σ is a strict surjective homomorphism from the matrix $\langle \boldsymbol{Fm}, \sigma^{-1}C_{\mathcal{L}}\Gamma \rangle$ to the matrix $\langle \boldsymbol{Fm}, C_{\mathcal{L}}\Gamma \rangle$, and both are models of $\mathcal{L}$. Since $\mathcal{L}$ is protoalgebraic and $C_{\mathcal{L}}\Gamma \subseteq C_{\mathcal{L}}(\Gamma, \varphi)$, the Correspondence Theorem can be applied, in the form of its Corollary 6.20.2, and it follows that

$$\sigma^{-1}C_{\mathcal{L}}(\Gamma, \varphi) = C_{\mathcal{L}}(\sigma^{-1}C_{\mathcal{L}}\Gamma, x).$$

Therefore, $\sigma y = \psi \in C_{\mathcal{L}}(\Gamma, \varphi) \Leftrightarrow y \in C_{\mathcal{L}}(\sigma^{-1}C_{\mathcal{L}}\Gamma, x)$. Now by (84) applied to the set $\sigma^{-1}C_{\mathcal{L}}\Gamma$, this is equivalent to $\Sigma \subseteq C_{\mathcal{L}}\sigma^{-1}C_{\mathcal{L}}\Gamma = \sigma^{-1}C_{\mathcal{L}}\Gamma \Leftrightarrow \sigma\Sigma \subseteq C_{\mathcal{L}}\Gamma \Leftrightarrow I(\varphi, \psi) \subseteq C_{\mathcal{L}}\Gamma$. This shows that $\mathcal{L}$ satisfies the DDT. ⊠

This proof exemplifies one of the typical techniques of abstract algebraic logic, namely using the free algebra of terms as a support for matrices to which the general results are applied.

By this bridge theorem for protoalgebraic logics, the DDT turns out to be equivalent to a purely algebraic property of its lattice of filters. In Section 6.5 it is shown how the well-known bridge theorem for BP-algebraizable logics, which asserts the equivalence between DDT and EDPRC (Corollary 3.86), can be derived from this one. Notice that the DDT implies protoalgebraicity of $\mathcal{L}$, hence this assumption can be removed from Theorem 6.28, provided it is added to its points (ii) and (iii); however, the given form highlights that it contains both a *bridge theorem*, the equivalence between (i) and (ii), and a *transfer theorem*, the implication from (iii) to (ii).

EXAMPLE 6.29
It is mentioned earlier (Example 6.25.1) that the finitary infinitely-valued logic of Łukasiewicz $Ł_{\infty f}$ satisfies the LDDT. Using the preceding result it is possible to show that it does *not* satisfy any form of the DDT. Assume it does. Since the logic is finitary and conjunctive, it is easy to see, using first Proposition 3.79 and then Lemma 3.78, that $Ł_{\infty f}$ would satisfy it with respect to a one-element DD set, that is, with respect to a single term, which is denoted here by $x \rightsquigarrow y$. The algebra $[\mathbf{0,1}]$ is an MV-algebra (or Wajsberg algebra), and it is well known that the $Ł_{\infty f}$-filters of $[\mathbf{0,1}]$ are its implicative filters, and that for any $a \in A$, $Fg^{[\mathbf{0,1}]}_{Ł_{\infty f}}\{a\} = \{b \in [0,1] : n(1-a)+b \geqslant 1 \text{ for some } n \in \omega\}$. By Theorem 6.28, the join-semilattice of finitely generated implicative filters should be dually residuated, and from its proof it is clear that the dual residual should be determined by the same term $x \rightsquigarrow y$. In particular, take any $a, b \in [0,1]$ and consider the three principal implicative filters generated by a, b and 1: we should have that $Fg^{[\mathbf{0,1}]}_{Ł_{\infty f}}\{b\} \subseteq Fg^{[\mathbf{0,1}]}_{Ł_{\infty f}}\{1\} \vee Fg^{[\mathbf{0,1}]}_{Ł_{\infty f}}\{a\}$ if and only if

$$Fg^{[\mathbf{0,1}]}_{Ł_{\infty f}}\{a\} \star Fg^{[\mathbf{0,1}]}_{Ł_{\infty f}}\{b\} = Fg^{[\mathbf{0,1}]}_{Ł_{\infty f}}\{a \rightsquigarrow b\} \subseteq Fg^{[\mathbf{0,1}]}_{Ł_{\infty f}}\{1\}.$$

Now, using that $Fg^{[\mathbf{0,1}]}_{Ł_{\infty f}}\{1\} = \{1\}$ and the above mentioned characterization of the principal filters, this would simply entail that, for all $a, b \in [0,1]$,

$$a \rightsquigarrow b = 1 \iff \text{there is some } m \in \omega \text{ such that } m(1-a)+b \geqslant 1. \tag{85}$$

Using this for $a = 1$, it follows that for all $b \in [0,1]$,

$$1 \rightsquigarrow b = 1 \iff b = 1. \tag{86}$$

On the other hand, if $a < 1$ it is possible to take $m \geqslant \frac{1}{1-a}$ and use (85), which implies that $a \rightsquigarrow b = 1$ for all $b \in [0,1]$. Now consider any sequence $\langle a_n : n \in \omega \rangle$ of $[0,1]$ with $a_n < 1$ for all $n \in \omega$ and such that $\lim_{n\to\infty} a_n = 1$. By the preceding properties, $a_n \rightsquigarrow b = 1$ for all $b \in [0,1]$. Now it is time to use the fact (crucial in the algebraic theory of this logic) that all term operations are continuous functions on the unit interval, and in particular the function evaluating $\rightsquigarrow$ should be continuous. Therefore

$$1 \rightsquigarrow b = (\lim_{n\to\infty} a_n) \rightsquigarrow b = \lim_{n\to\infty} (a_n \rightsquigarrow b) = 1, \text{ for all } b \in [0,1],$$

which by (86) implies that $b = 1$ for all $b \in [0,1]$, which is absurd. ⊠

The following important property of protoalgebraic logics satisfying the DDT is a by-product of the characterization of the DDT in Theorem 6.28.

COROLLARY 6.30
If a finitary and protoalgebraic logic $\mathcal{L}$ satisfies the DDT, *then for every alegbra $\boldsymbol{A}$ the lattice $\mathcal{F}i_{\mathcal{L}}\boldsymbol{A}$ is distributive.*

PROOF: Finitarity of $\mathcal{L}$ implies that for any $\boldsymbol{A}$, the closure system $\mathcal{F}i_{\mathcal{L}}\boldsymbol{A}$ is inductive, and hence that as a lattice it is algebraic (by Proposition 1.67). It is a well known property of lattice theory (see [129, Theorem 42] for instance) that an algebraic lattice is isomorphic to the lattice of ideals of the join-semilattice of its compact elements. But by Proposition 1.65, the compact elements of the lattice of closed sets of a finitary closure operator are the finitely generated closed sets. Thus, $\mathcal{F}i_{\mathcal{L}}\boldsymbol{A}$ is isomorphic to the lattice of ideals of the join-semilattice $\mathcal{F}i^{\omega}_{\mathcal{L}}\boldsymbol{A}$, which by Theorem 6.28 is dually residuated. As a consequence of this and of Proposition 6.27.2, this lattice of ideals is distributive. Since distributivity is a property preserved by isomorphisms between lattices, it follows that $\mathcal{F}i_{\mathcal{L}}\boldsymbol{A}$ is distributive. ⊠

Logics such that for every $\boldsymbol{A}$, the lattice $\mathcal{F}i_{\mathcal{L}}\boldsymbol{A}$ is distributive are called ***filter-distributive***. It is possible to see that a finitary protoalgebraic logic is filter-distributive if and only if the lattice of theories $\mathcal{T}h\mathcal{L}$ is distributive (and this is another *transfer theorem*); this can be proved directly, or as an example of the application of Theorem 6.32 below.

Filter-distributive logics enjoy some really nice features from the abstract algebraic logic point of view, resulting from generalization to filter lattices of some universal algebraic properties related to the distributivity of congruence lattices, such as Jónsson's Lemma or Baker's Theorem. More in general, by protoalgebraicity many lattice-theoretic features of congruences and their consequences can be reproduced for matrices. Just as a sample of what can be achieved in this line of research, here is one:

THEOREM 6.31
Let $\mathcal{L}$ be a finitary logic, in a finite language, that is weakly complete with respect to a finite matrix. If $\mathcal{L}$ is protoalgebraic and filter-distributive, then $\mathcal{L}$ is finitely axiomatizable, and it is so with a finite number of axioms and a single proper inference rule. ⊠

More information can be found in [70, Chapter 2].

Full generalized models of protoalgebraic logics and Leibniz filters

Theorem 6.13 states that for every logic $\mathcal{L}$, the monotonicity of $\boldsymbol{\Omega}$ transfers from $\mathcal{T}h\mathcal{L}$ to $\mathcal{F}i_{\mathcal{L}}\boldsymbol{A}$. Hence, it transfers to every g-model $\mathscr{C}$ of $\mathcal{L}$, because $\mathscr{C} \subseteq \mathcal{F}i_{\mathcal{L}}\boldsymbol{A}$, and in particular to every full g-model.

Other properties transfer only to full g-models, and are proved with the help of the Correspondence Theorem or some of its corollaries. Among these properties are the ones that are (or can be expressed as) lattice-theoretic properties of the relevant closure systems. The following important result is one of the few general existing results about the transfer problem in abstract algebraic logic, and deserves to be recorded here:

THEOREM 6.32
If $\mathcal{L}$ is a finitary protoalgebraic logic, then every universal sentence of the first-order language of lattices that holds in $\mathcal{T}h\mathcal{L}$ also holds in the lattice $\mathcal{F}i_{\mathcal{L}}\boldsymbol{A}$ for all $\boldsymbol{A}$, and hence also in the lattices $\mathscr{C}$ of closed sets of all full g-models of $\mathcal{L}$. ⊠

The proof, which can be found in [70, pp. 107ff.], consists in constructing for every $\boldsymbol{A}$ an embedding of the lattice $\mathcal{F}i_{\mathcal{L}}\boldsymbol{A}$ into an ultrapower of the lattice of theories $\mathcal{T}h\mathcal{L}$; this construction is based in that of *natural extensions* of a logic; all this goes beyond the limits of this book. The last part of the statement, not mentioned in [70], follows easily from the first part because every full g-model is the inverse image by a strict surjective homomorphism of a basic full g-model, which has the form $\langle \boldsymbol{A}, \mathcal{F}i_{\mathcal{L}}\boldsymbol{A}\rangle$; by Theorem 5.48, this defines a lattice isomorphism between the closure systems, and all the mentioned properties are preserved by lattice isomorphisms.

Another application of the Correspondence Theorem allows us to prove a transfer theorem for Gentzen-style rules.

DEFINITION 6.33
A set G of Gentzen-style rules is **accumulative** *when for every rule of the form*

$$\frac{\{\Gamma_i \rhd \varphi_i \ : \ i < n\}}{\Gamma \rhd \varphi}$$

in G, and every finite $\Delta \subseteq Fm$, the "accumulated" rule $\dfrac{\{\Delta, \Gamma_i \rhd \varphi_i \ : \ i < n\}}{\Delta, \Gamma \rhd \varphi}$ *also belongs to G.*

Given a rule (or set of rules) G, the **accumulative form** *of G is the smallest accumulative set of rules that contains it, that is, its "closure" under addition of arbitrary finite sets to the antecedents of all the sequents appearing in G.*

Of course, if G is an accumulative set of Gentzen-style rules, then it coincides with its accumulative form.

EXAMPLES.
You have shown (Exercise 5.7) that the DDT for a finitary logic is equivalent to satisfying the following two Gentzen-style rules:

$$\frac{\Gamma \rhd \varphi \qquad \Gamma, \psi \rhd \xi}{\Gamma, \varphi \to \psi \rhd \xi} \qquad \text{and} \qquad \frac{\Gamma, \varphi \rhd \psi}{\Gamma \rhd \varphi \to \psi}$$

Since the Γ that appears in the antecedent of all sequents represents an arbitrary finite set, this is actually an accumulative set of Gentzen-style rules. Other examples are the rules for disjunction corresponding to the PDI (Exercise 5.7 again), and those characterizing the SPCONG for finitary g-matrices (Corollary 7.24). By contrast, the PCONG can be expressed by a single (non-accumulative) Gentzen-style rule for each connective (see Definition 5.25), and it is easy to see that its accumulative form defines the stronger SPCONG. These last properties and their transfer problems are treated in Chapter 7. ⊠

The following lemma, which is easy to show, is left to you as Exercise 6.19.

LEMMA 6.34
Let $\langle \boldsymbol{A}, \mathscr{C} \rangle$ be a finitary g-matrix and let G be a Gentzen-style rule. The following conditions are equivalent:

(i) *$\langle \boldsymbol{A}, \mathscr{C} \rangle$ is a model of the accumulative form of G.*

(ii) *For every $F \in \mathscr{C}$, the g-matrix $\langle \boldsymbol{A}, \mathscr{C}^F \rangle$ is a model of G.*

(iii) *For every finitely generated $F \in \mathscr{C}$, the g-matrix $\langle \boldsymbol{A}, \mathscr{C}^F \rangle$ is a model of G.* ⊠

In the following, for an infinite cardinal κ, we let $\boldsymbol{Fm}_\kappa$ denote the formula algebra with a set V_κ of variables of cardinality κ, assuming that $V \subseteq V_\kappa$. Recall the definition of a sub-g-matrix given in Exercise 5.29.

LEMMA 6.35
If κ is an infinite cardinal and $X \subseteq V_\kappa$ is a denumerable subset and $\boldsymbol{Fm}_\kappa(X)$ is the subalgebra of $\boldsymbol{Fm}_\kappa$ generated by X, then the sub-g-matrix of $\langle \boldsymbol{Fm}_\kappa, \mathcal{F}i_{\mathcal{L}}\boldsymbol{Fm}_\kappa \rangle$ determined by $\boldsymbol{Fm}_\kappa(X)$ is $\langle \boldsymbol{Fm}_\kappa(X), \mathcal{F}i_{\mathcal{L}}\boldsymbol{Fm}_\kappa(X) \rangle$ and is isomorphic to $\langle \boldsymbol{Fm}, \mathcal{T}h\mathcal{L} \rangle$.

PROOF: Since X is denumerable, $\boldsymbol{Fm}$ is clearly isomorphic to $\boldsymbol{Fm}_\kappa(X)$. Therefore, by Exercise 5.45, $\langle \boldsymbol{Fm}, \mathcal{T}h\mathcal{L} \rangle = \langle \boldsymbol{Fm}, \mathcal{F}i_{\mathcal{L}}\boldsymbol{Fm} \rangle \cong \langle \boldsymbol{Fm}_\kappa(X), \mathcal{F}i_{\mathcal{L}}\boldsymbol{Fm}_\kappa(X) \rangle$. Thus, it is enough to show that the latter is the sub-g-matrix described in the statement, which means to show that $\mathcal{F}i_{\mathcal{L}}\boldsymbol{Fm}_\kappa(X) = \{F \cap Fm_\kappa(X) : F \in \mathcal{F}i_{\mathcal{L}}\boldsymbol{Fm}_\kappa\}$. The inclusion ($\supseteq$) is general and holds by Exercise 4.3. To show ($\subseteq$), consider any retraction $h \colon \boldsymbol{Fm}_\kappa \to \boldsymbol{Fm}_\kappa(X)$, that is, any surjective homomorphism such that its restriction to $\boldsymbol{Fm}_\kappa(X)$ is the identity function (its existence is a straightforward consequence of the free character of $\boldsymbol{Fm}_\kappa$). Then, if $G \in \mathcal{F}i_{\mathcal{L}}\boldsymbol{Fm}_\kappa(X)$, $h^{-1}G \in \mathcal{F}i_{\mathcal{L}}\boldsymbol{Fm}_\kappa$ and $G \subseteq h^{-1}G$, hence $G = h^{-1}G \cap Fm_\kappa(X)$. ⊠

LEMMA 6.36
Let $\mathcal{L}$ be a finitary logic and G a Gentzen-style rule. The following conditions are equivalent:

(i) *$\mathcal{L}$ satisfies G.*

(ii) *The g-matrix $\langle \boldsymbol{Fm}_\kappa, \mathcal{F}i_{\mathcal{L}}\boldsymbol{Fm}_\kappa \rangle$ is a model of G, for every infinite cardinal κ.*

(iii) *The g-matrix $\langle \boldsymbol{Fm}_\kappa, \mathcal{F}i_{\mathcal{L}}\boldsymbol{Fm}_\kappa \rangle$ is a model of G, for some infinite cardinal κ.*

PROOF: (i)⇒(ii) Let G be a rule valid in $\mathcal{L}$ and κ an infinite cardinal. Since $\mathcal{L}$ is finitary, the g-matrix $\mathcal{M}_\kappa := \langle \boldsymbol{Fm}_\kappa, \mathcal{F}i_{\mathcal{L}}\boldsymbol{Fm}_\kappa \rangle$ is finitary, therefore Exercise 5.43 can be used in order to show that G holds in $\mathcal{M}_\kappa$. Thus, let $\mathcal{M} = \langle \boldsymbol{A}, \mathscr{C} \rangle$ be any finitary, finitely-generated sub-g-matrix of $\mathcal{M}_\kappa$, and let X be any denumerable set of variables that includes those appearing in the generators of $\boldsymbol{A}$ (a subalgebra of $\boldsymbol{Fm}_\kappa$). Thus, $\mathscr{C} = \{F \cap A : F \in \mathcal{F}i_{\mathcal{L}}\boldsymbol{Fm}_\kappa\}$, but since $\boldsymbol{A}$ is also a subalgebra of $\boldsymbol{Fm}_\kappa(X)$, this can also be written as $\mathscr{C} = \{F \cap A \cap Fm_\kappa(X) : F \in \mathcal{F}i_{\mathcal{L}}\boldsymbol{Fm}_\kappa\}$, which by Lemma 6.35 implies that $\mathscr{C} = \{G \cap A : G \in \mathcal{F}i_{\mathcal{L}}\boldsymbol{Fm}_\kappa(X)\}$. Therefore $\mathcal{M}$ is a sub-g-matrix of $\langle \boldsymbol{Fm}_\kappa(X), \mathcal{F}i_{\mathcal{L}}\boldsymbol{Fm}_\kappa(X) \rangle$, and by Lemma 6.35 this one is a sub-g-matrix of $\mathcal{M}_\kappa$. But by the isomorphy proved in the same lemma, G holds in $\langle \boldsymbol{Fm}_\kappa(X), \mathcal{F}i_{\mathcal{L}}\boldsymbol{Fm}_\kappa(X) \rangle$. Since this matrix is countably generated, the result in Exercise 5.43 implies that G holds in $\mathcal{M}_\kappa$.

(ii)⇒(iii) is trivial.

(iii)⇒(i) If G holds in $\langle \boldsymbol{Fm}_\kappa, \mathcal{F}i_{\mathcal{L}}\boldsymbol{Fm}_\kappa\rangle$, then by Exercise 5.42 it also holds in all its sub-g-matrices, and by Lemma 6.35 the g-matrix $\langle \boldsymbol{Fm}, \mathcal{T}h\mathcal{L}\rangle$ is isomorphic to many of them, so it also satisfies G. ⊠

THEOREM 6.37
If $\mathcal{L}$ is a finitary protoalgebraic logic, then every property of $\mathcal{L}$ expressible as an accumulative set of Gentzen-style rules transfers to all full g-models of $\mathcal{L}$.

PROOF: By Theorem 5.61, the property of being a model of a Gentzen-style rule is preserved by strict surjective homomorphisms between g-matrices; it is thus enough to prove the transfer to basic full g-models $\langle \boldsymbol{A}, \mathcal{F}i_{\mathcal{L}}\boldsymbol{A}\rangle$. Let κ be an infinite cardinal such that $\kappa \geqslant |A|$, and hence that there is some surjective $h\colon \boldsymbol{Fm}_\kappa \to \boldsymbol{A}$. Putting $F_0 := \min \mathcal{F}i_{\mathcal{L}}\boldsymbol{A} = \bigcap \mathcal{F}i_{\mathcal{L}}\boldsymbol{A}$, now h is strict from the matrix $\langle \boldsymbol{Fm}_\kappa, h^{-1}F_0\rangle$ to the matrix $\langle \boldsymbol{A}, F_0\rangle$. Since $\mathcal{L}$ is protoalgebraic, by the Correspondence Theorem h is a strict surjective homomorphism from the g-matrix $\langle \boldsymbol{Fm}_\kappa, (\mathcal{F}i_{\mathcal{L}}\boldsymbol{Fm}_\kappa)^{h^{-1}F_0}\rangle$ to the g-matrix $\langle \boldsymbol{A}, \mathcal{F}i_{\mathcal{L}}\boldsymbol{A}\rangle$. By assumption $\mathcal{L}$ satisfies all rules in the assumed accumulative set G, hence by Lemma 6.36 $\langle \boldsymbol{Fm}_\kappa, \mathcal{F}i_{\mathcal{L}}\boldsymbol{Fm}_\kappa\rangle$ satisfies them as well. Since G is closed under taking accumulative forms, Lemma 6.34 implies that $\langle \boldsymbol{Fm}_\kappa, (\mathcal{F}i_{\mathcal{L}}\boldsymbol{Fm}_\kappa)^{h^{-1}F_0}\rangle$ is also a model of all rules in G. By Theorem 5.61 again, $\langle \boldsymbol{A}, \mathcal{F}i_{\mathcal{L}}\boldsymbol{A}\rangle$ is a model of the rules in G, and hence has the property. ⊠

An application of this result, concerning the property of being Fregean, is given in Theorem 7.71.

Concerning the full g-models of protoalgebraic logics, one may ask: How are they? Do they have some peculiarity? Can they be characterized in some way? We will see that the answer is *yes*. First, a simple observation:

LEMMA 6.38
The full g-models of a protoalgebraic logic are determined by their theorems.

PROOF: Assume that $\langle \boldsymbol{A}, \mathscr{C}\rangle, \langle \boldsymbol{A}, \mathscr{C}'\rangle \in \mathsf{FGMod}\mathcal{L}$ with $C\varnothing = C'\varnothing$. Since in particular they are g-models and $\mathcal{L}$ is protoalgebraic, by Proposition 6.14(iv) $\widetilde{\boldsymbol{\Omega}}^{\boldsymbol{A}}\mathscr{C} = \boldsymbol{\Omega}^{\boldsymbol{A}}C\varnothing = \boldsymbol{\Omega}^{\boldsymbol{A}}C'\varnothing = \widetilde{\boldsymbol{\Omega}}^{\boldsymbol{A}}\mathscr{C}'$. Since both g-models are full, the Isomorphism Theorem 5.95 implies that $\mathscr{C} = \mathscr{C}'$. ⊠

We now see that the full g-models of protoalgebraic logics have a precise form, and moreover that having this form is exclusive of these logics; thus, we have one more characterization of protoalgebraicity.

THEOREM 6.39
A logic $\mathcal{L}$ is protoalgebraic if and only if every full g-model of $\mathcal{L}$ has the form $\langle \boldsymbol{A}, (\mathcal{F}i_{\mathcal{L}}\boldsymbol{A})^G\rangle$ for some $G \in \mathcal{F}i_{\mathcal{L}}\boldsymbol{A}$.

PROOF: (⇒) Let $\langle \boldsymbol{A}, \mathscr{C}\rangle \in \mathsf{FGMod}\mathcal{L}$. By Propositions 5.94 and 6.14(iv), we have that $\mathscr{C} = \{F \in \mathcal{F}i_{\mathcal{L}}\boldsymbol{A} : \widetilde{\boldsymbol{\Omega}}^{\boldsymbol{A}}\mathscr{C} \subseteq \boldsymbol{\Omega}^{\boldsymbol{A}}F\} = \{F \in \mathcal{F}i_{\mathcal{L}}\boldsymbol{A} : \boldsymbol{\Omega}^{\boldsymbol{A}}C\varnothing \subseteq \boldsymbol{\Omega}^{\boldsymbol{A}}F\}$. Since $\mathscr{C} \subseteq \mathcal{F}i_{\mathcal{L}}\boldsymbol{A}$, we also have that $\mathscr{C} \subseteq \{F \in \mathcal{F}i_{\mathcal{L}}\boldsymbol{A} : C\varnothing \subseteq F\}$, and by monotonicity $\{F \in \mathcal{F}i_{\mathcal{L}}\boldsymbol{A} : C\emptyset \subseteq F\} \subseteq \{F \in \mathcal{F}i_{\mathcal{L}}\boldsymbol{A} : \boldsymbol{\Omega}^{\boldsymbol{A}}C\emptyset \subseteq \boldsymbol{\Omega}^{\boldsymbol{A}}F\}$. Therefore, $\mathscr{C} = \{F \in \mathcal{F}i_{\mathcal{L}}\boldsymbol{A} : C\emptyset \subseteq F\} = (\mathcal{F}i_{\mathcal{L}}\boldsymbol{A})^{C\emptyset}$, which shows that $G := C\emptyset \in \mathcal{F}i_{\mathcal{L}}\boldsymbol{A}$ is the required $\mathcal{L}$-filter.

($\Leftarrow$) Protoalgebraicity of $\mathcal{L}$ is proved through monotonicity of $\boldsymbol{\Omega}^{\boldsymbol{A}}$. Let $F, F' \in \mathcal{F}i_{\mathcal{L}}\boldsymbol{A}$ with $F \subseteq F'$. Since $\mathsf{Alg}^*\mathcal{L} \subseteq \mathsf{Alg}\mathcal{L}$, $\boldsymbol{\Omega}^{\boldsymbol{A}}F \in \mathrm{Con}_{\mathsf{Alg}\mathcal{L}}\boldsymbol{A}$ and by Theorem 5.95 there is a full g-model $\langle \boldsymbol{A}, \mathscr{C} \rangle$ of $\mathcal{L}$ such that $\widetilde{\boldsymbol{\Omega}}^{\boldsymbol{A}}\mathscr{C} = \boldsymbol{\Omega}^{\boldsymbol{A}}F$. By Proposition 5.94, $\mathscr{C} = \{H \in \mathcal{F}i_{\mathcal{L}}\boldsymbol{A} : \widetilde{\boldsymbol{\Omega}}^{\boldsymbol{A}}\mathscr{C} \subseteq \boldsymbol{\Omega}^{\boldsymbol{A}}H\}$, therefore $F \in \mathscr{C}$. Now, the assumption tells us that there is some $G \in \mathcal{F}i_{\mathcal{L}}\boldsymbol{A}$ such that $\mathscr{C} = (\mathcal{F}i_{\mathcal{L}}\boldsymbol{A})^{G}$. Thus, $F \supseteq G$, and hence $F' \supseteq G$, which implies $F' \in \mathscr{C}$ and then $\widetilde{\boldsymbol{\Omega}}^{\boldsymbol{A}}\mathscr{C} \subseteq \boldsymbol{\Omega}^{\boldsymbol{A}}F'$; this implies that $\boldsymbol{\Omega}^{\boldsymbol{A}}F \subseteq \boldsymbol{\Omega}^{\boldsymbol{A}}F'$, thus proving monotonicity of $\boldsymbol{\Omega}^{\boldsymbol{A}}$. ⊠

For a stronger form of this result, see Exercise 6.20. We see that full g-models of protoalgebraic logics are determined by their filters. However, not every $\mathcal{L}$-filter determines a full g-model as in Theorem 6.39; the ones that do should be the set of theorems of a full g-model, and not all are, as we will see. An important fact is that these filters can be characterized in a way that does not involve the notion of a full g-model, and hence they are of an independent interest. The theory of Leibniz filters was introduced in [106] and has been further studied in [72, 108, 154], and generalized using the Suszko operator in [4, 5]; the starting ideas (but not the name) already appeared in the first edition of [107].

DEFINITION 6.40
If $\mathcal{L}$ is a protoalgebraic logic, then an $\mathcal{L}$-filter F on an algebra $\boldsymbol{A}$ is ***Leibniz*** *when it is the least one among all those (on the same algebra) having the same Leibniz congruence; that is, when for every $G \in \mathcal{F}i_{\mathcal{L}}\boldsymbol{A}$, if $\boldsymbol{\Omega}^{\boldsymbol{A}}F = \boldsymbol{\Omega}^{\boldsymbol{A}}G$, then $F \subseteq G$, or, in other words, when $F = \min\{G \in \mathcal{F}i_{\mathcal{L}}\boldsymbol{A} : \boldsymbol{\Omega}^{\boldsymbol{A}}G = \boldsymbol{\Omega}^{\boldsymbol{A}}F\}$.*

NOTATIONS: $\mathcal{F}i^{+}_{\mathcal{L}}\boldsymbol{A} := \{F \in \mathcal{F}i_{\mathcal{L}}\boldsymbol{A} : F \text{ is Leibniz}\}$
$\mathcal{T}h^{+}\mathcal{L} := \{\Gamma \in \mathcal{T}h\mathcal{L} : \Gamma \text{ is Leibniz}\}$.

Notice that "being Leibniz" is a property of a filter (or of a theory) relative to the logic $\mathcal{L}$, and it would be more accurate to say "$\mathcal{L}$-Leibniz", but normally the context makes this clear. In principle, the definition does not depend on $\mathcal{L}$ being protoalgebraic; Leibniz filters can be considered for arbitrary logics (see Theorem 7.38 for an example).

Natural questions are: Do Leibniz filters exist? Are they relevant for the theory of the full g-models of protoalgebraic logics? To the first question there is a trivial, affirmative answer: For every logic $\mathcal{L}$ and every algebra $\boldsymbol{A}$, the least $\mathcal{L}$-filter $\bigcap \mathcal{F}i_{\mathcal{L}}\boldsymbol{A}$ is always Leibniz. So, on every algebra $\boldsymbol{A}$ there is always at least one Leibniz filter: $\mathcal{F}i^{+}_{\mathcal{L}}\boldsymbol{A} \neq \emptyset$. In the protoalgebraic case there is a trivial situation: if the Leibniz operator $\boldsymbol{\Omega}^{\boldsymbol{A}}$ is injective on every $\boldsymbol{A}$ (something that, for instance, we know happens when $\mathcal{L}$ is algebraizable), then every filter is Leibniz, and the notion loses its interest (see also Exercises 6.21 and 6.22). This also explains why there are no examples of Leibniz vs. non-Leibniz filters for the best-known logics, such as the algebraizable ones. The second question has an interesting answer:

THEOREM 6.41
Let $\mathcal{L}$ be protoalgebraic, and let $F \in \mathcal{F}i_{\mathcal{L}}\boldsymbol{A}$. The following conditions are equivalent:

(i) *F is Leibniz.*

(ii) *The g-matrix $\langle \boldsymbol{A}, (\mathcal{F}i_{\mathcal{L}}\boldsymbol{A})^F \rangle$ is a full g-model of $\mathcal{L}$.*

(iii) *$F/\boldsymbol{\Omega}^{\boldsymbol{A}}F$ is the smallest $\mathcal{L}$-filter of $\boldsymbol{A}/\boldsymbol{\Omega}^{\boldsymbol{A}}F$.*

PROOF: (i)⇒(ii) Since by protoalgebraicity $\widetilde{\boldsymbol{\Omega}}^{\boldsymbol{A}}(\mathcal{F}i_{\mathcal{L}}\boldsymbol{A})^F = \boldsymbol{\Omega}^{\boldsymbol{A}}F$, by Proposition 5.94 it is enough to show that $(\mathcal{F}i_{\mathcal{L}}\boldsymbol{A})^F = \{G \in \mathcal{F}i_{\mathcal{L}}\boldsymbol{A} : \boldsymbol{\Omega}^{\boldsymbol{A}}F \subseteq \boldsymbol{\Omega}^{\boldsymbol{A}}G\}$. The ($\subseteq$) part is a consequence of monotonicity, so only the ($\supseteq$) part needs a proof: If $\boldsymbol{\Omega}^{\boldsymbol{A}}F \subseteq \boldsymbol{\Omega}^{\boldsymbol{A}}G$, then by protoalgebraicity $\boldsymbol{\Omega}^{\boldsymbol{A}}(F \cap G) = \boldsymbol{\Omega}^{\boldsymbol{A}}F \cap \boldsymbol{\Omega}^{\boldsymbol{A}}G = \boldsymbol{\Omega}^{\boldsymbol{A}}F$; now, the assumption that F is Leibniz implies that $F \subseteq F \cap G$ and hence $F \subseteq G$.

(ii)⇒(iii) The assumption implies that the reduction of the g-matrix $\langle \boldsymbol{A}, (\mathcal{F}i_{\mathcal{L}}\boldsymbol{A})^F \rangle$ is a basic full g-model. But to reduce is to factor out by $\widetilde{\boldsymbol{\Omega}}^{\boldsymbol{A}}(\mathcal{F}i_{\mathcal{L}}\boldsymbol{A})^F = \boldsymbol{\Omega}^{\boldsymbol{A}}F$. Thus, the isomorphism induced by the canonical projection is $(\mathcal{F}i_{\mathcal{L}}\boldsymbol{A})^F \cong (\mathcal{F}i_{\mathcal{L}}\boldsymbol{A})^F/\boldsymbol{\Omega}^{\boldsymbol{A}}F = \mathcal{F}i_{\mathcal{L}}(\boldsymbol{A}/\boldsymbol{\Omega}^{\boldsymbol{A}}F)$, and in particular the least members of each of these lattices are put into correspondence by this isomorphism $G \mapsto G/\boldsymbol{\Omega}^{\boldsymbol{A}}F$. Since F is the least member of $(\mathcal{F}i_{\mathcal{L}}\boldsymbol{A})^F$, $F/\boldsymbol{\Omega}^{\boldsymbol{A}}F$ must be the least member of $\mathcal{F}i_{\mathcal{L}}(\boldsymbol{A}/\boldsymbol{\Omega}^{\boldsymbol{A}}F)$.

(iii)⇒(i) Take any $G \in \mathcal{F}i_{\mathcal{L}}\boldsymbol{A}$ with $\boldsymbol{\Omega}^{\boldsymbol{A}}F = \boldsymbol{\Omega}^{\boldsymbol{A}}G$. Then

$$G/\boldsymbol{\Omega}^{\boldsymbol{A}}F = G/\boldsymbol{\Omega}^{\boldsymbol{A}}G \in \mathcal{F}i_{\mathcal{L}}(\boldsymbol{A}/\boldsymbol{\Omega}^{\boldsymbol{A}}G) = \mathcal{F}i_{\mathcal{L}}(\boldsymbol{A}/\boldsymbol{\Omega}^{\boldsymbol{A}}F),$$

therefore by the assumption $G/\boldsymbol{\Omega}^{\boldsymbol{A}}F \supseteq F/\boldsymbol{\Omega}^{\boldsymbol{A}}F$. Then the Correspondence Theorem can be applied to the homomorphism $\langle \boldsymbol{A}, F \rangle \to \langle \boldsymbol{A}/\boldsymbol{\Omega}^{\boldsymbol{A}}F, F/\boldsymbol{\Omega}^{\boldsymbol{A}}F \rangle$, and we obtain that $G \supseteq F$. ⊠

You should realize the necessity of using the Correspondence Theorem in the last step of the proof (because, without protoalgebraicity, $G/\boldsymbol{\Omega}^{\boldsymbol{A}}F \supseteq F/\boldsymbol{\Omega}^{\boldsymbol{A}}F$ does not imply $G \supseteq F$).

COROLLARY 6.42
Let $\mathcal{L}$ be a protoalgebraic logic, and let $\langle \boldsymbol{A}, F \rangle \in \mathsf{Mod}^*\mathcal{L}$. The $\mathcal{L}$-filter F is Leibniz if and only if $F = \min \mathcal{F}i_{\mathcal{L}}\boldsymbol{A}$. ⊠

So the answer to the second question is *yes*, these are the filters that characterize the full g-models of protoalgebraic logics. They are also those that when reduced become the smallest filter in the quotient.

The main property of the set of Leibniz filters is given in the next results.

THEOREM 6.43
If $\mathcal{L}$ is protoalgebraic, then for every $\boldsymbol{A}$ the set of Leibniz filters $\mathcal{F}i^+_{\mathcal{L}}\boldsymbol{A}$, ordered under set inclusion, is a complete lattice and it is dually isomorphic to the lattice $\mathrm{FCS}_{\mathcal{L}}\boldsymbol{A}$ through the function $F \mapsto (\mathcal{F}i_{\mathcal{L}}\boldsymbol{A})^F$.

PROOF: We have already seen that if $F \in \mathcal{F}i^+_{\mathcal{L}}\boldsymbol{A}$, then $\langle \boldsymbol{A}, (\mathcal{F}i_{\mathcal{L}}\boldsymbol{A})^F \rangle \in \mathsf{FGMod}\,\mathcal{L}$, that is, that $(\mathcal{F}i_{\mathcal{L}}\boldsymbol{A})^F \in \mathrm{FCS}_{\mathcal{L}}\boldsymbol{A}$, and that the function $F \mapsto (\mathcal{F}i_{\mathcal{L}}\boldsymbol{A})^F$ is onto $\mathrm{FCS}_{\mathcal{L}}\boldsymbol{A}$. But $F = \bigcap(\mathcal{F}i_{\mathcal{L}}\boldsymbol{A})^F$, so it is also one-to-one. Finally, $F \subseteq G$ if and only if $(\mathcal{F}i_{\mathcal{L}}\boldsymbol{A})^F \supseteq (\mathcal{F}i_{\mathcal{L}}\boldsymbol{A})^G$. Thus, the function is a dual order isomorphism. Since $\mathrm{FCS}_{\mathcal{L}}\boldsymbol{A}$ is a complete lattice by Proposition 5.96, $\mathcal{F}i^{+}_{\mathcal{L}}\boldsymbol{A}$ is also one. ☒

COROLLARY 6.44
If $\mathcal{L}$ is protoalgebraic, then for every $\boldsymbol{A}$, the Leibniz operator $\boldsymbol{\Omega}^{\boldsymbol{A}}$ is an isomorphism between the complete lattices $\mathcal{F}i^{+}_{\mathcal{L}}\boldsymbol{A}$ and $\mathrm{Con}_{\mathsf{Alg}\mathcal{L}}\boldsymbol{A}$.

PROOF: The composition of two dual order isomorphisms, the previous one and that given by the Isomorphism Theorem 5.95, is obviously an order isomorphism. Then, using protoalgebraicity, the resulting function can be determined: it is $F \mapsto (\mathcal{F}i_{\mathcal{L}}\boldsymbol{A})^F \mapsto \widetilde{\boldsymbol{\Omega}}^{\boldsymbol{A}}(\mathcal{F}i_{\mathcal{L}}\boldsymbol{A})^F = \boldsymbol{\Omega}^{\boldsymbol{A}}F$, the Leibniz operator. ☒

Thus, for a protoalgebraic logic $\mathcal{L}$ there is an isomorphic representation of the congruences of any algebra "relative to the logic $\mathcal{L}$", that is, relative to the class $\mathsf{Alg}\mathcal{L}$, in terms of (a certain family of) $\mathcal{L}$-filters. We have already observed that if $\mathcal{L}$ is algebraizable, then every $\mathcal{L}$-filter is Leibniz; thus, this result specialises to the Isomorphism Theorem for algebraizable logics. The present proof is particularly simple thanks to the use of full g-models and of Theorem 5.95. However, it is possible to prove it directly, using just matrices, as was originally done in [106].

It is possible to associate a unique Leibniz filter with each arbitrary filter:

PROPOSITION 6.45
If $\mathcal{L}$ is protoalgebraic, then for each $F \in \mathcal{F}i_{\mathcal{L}}\boldsymbol{A}$ there is a unique $F^+ \in \mathcal{F}i^{+}_{\mathcal{L}}\boldsymbol{A}$ such that $\boldsymbol{\Omega}^{\boldsymbol{A}}F = \boldsymbol{\Omega}^{\boldsymbol{A}}F^+$. Moreover, $F^+ \subseteq F$, and F is Leibniz if and only if $F = F^+$.

PROOF: For each $F \in \mathcal{F}i_{\mathcal{L}}\boldsymbol{A}$ define $F^+ := \bigcap\{G \in \mathcal{F}i_{\mathcal{L}}\boldsymbol{A} : \boldsymbol{\Omega}^{\boldsymbol{A}}G = \boldsymbol{\Omega}^{\boldsymbol{A}}F\}$. Since $F^+ \in \mathcal{F}i_{\mathcal{L}}\boldsymbol{A}$ and by Theorem 6.14(ii) $\boldsymbol{\Omega}^{\boldsymbol{A}}F^+ = \boldsymbol{\Omega}^{\boldsymbol{A}}F$, it follows that $F^+ = \bigcap\{G \in \mathcal{F}i_{\mathcal{L}}\boldsymbol{A} : \boldsymbol{\Omega}^{\boldsymbol{A}}G = \boldsymbol{\Omega}^{\boldsymbol{A}}F^+\}$. This shows that F^+ is Leibniz. Clearly, the minimality in the definition implies that in general there is at most one Leibniz filter associated with a given filter, so in this case F^+ is unique. By construction $F^+ \subseteq F$, and if F is itself Leibniz, then $F \subseteq F^+$, therefore $F = F^+$. ☒

It is clear that this filter F^+ depends exclusively on the congruence $\boldsymbol{\Omega}^{\boldsymbol{A}}F$; in the next section F^+ is given an explicit characterization in terms of $\boldsymbol{\Omega}^{\boldsymbol{A}}F$ (Lemma 6.58 and Corollary 6.59).

The function $F \mapsto F^+$ has some additional properties, and helps further clarifying the structure of the set of Leibniz filters. To start with, from the previous observation on the dependency of F^+ on $\boldsymbol{\Omega}^{\boldsymbol{A}}F$ it immediately follows:

COROLLARY 6.46
If $\mathcal{L}$ is protoalgebraic and $\langle \boldsymbol{A}, F\rangle, \langle \boldsymbol{A}, G\rangle \in \mathsf{Mod}^*\mathcal{L}$, then $F^+ = G^+$. ☒

PROPOSITION 6.47
If $\mathcal{L}$ is protoalgebraic and $F, G \in \mathcal{F}i_{\mathcal{L}}\boldsymbol{A}$ with $G \subseteq F$, then $G^+ \subseteq F^+$.

PROOF: By protoalgebraicity and the "+" construction, we have that if $G \subseteq F$, then $\boldsymbol{\Omega}^{\boldsymbol{A}}G^+ = \boldsymbol{\Omega}^{\boldsymbol{A}}G \subseteq \boldsymbol{\Omega}^{\boldsymbol{A}}F = \boldsymbol{\Omega}^{\boldsymbol{A}}F^+$. Since both G^+ and F^+ are Leibniz, Corollary 6.44 implies that $G^+ \subseteq F^+$. ☒

Using this, it immediately follows:

Corollary 6.48
Let $\mathcal{L}$ be a protoalgebraic logic, and let $F \in \mathcal{F}i_{\mathcal{L}}\boldsymbol{A}$. The $\mathcal{L}$-filter F^+ is the largest Leibniz $\mathcal{L}$-filter of $\boldsymbol{A}$ contained in F. ⊠

Corollary 6.49
For every protoalgebraic logic $\mathcal{L}$ and every $\boldsymbol{A}$, the ordered set $\mathcal{F}i^+_{\mathcal{L}}\boldsymbol{A}$ is a join-complete sub-semilattice of $\mathcal{F}i_{\mathcal{L}}\boldsymbol{A}$.

Proof: Let $\{F_i : i \in I\} \subseteq \mathcal{F}i^+_{\mathcal{L}}\boldsymbol{A}$, and consider the supremum $\bigvee F_i$ in the lattice $\mathcal{F}i_{\mathcal{L}}\boldsymbol{A}$. Since $F_i \subseteq \bigvee F_i$ and the F_i are Leibniz, by Proposition 6.47 $F_i \subseteq (\bigvee F_i)^+$. Therefore $\bigvee F_i \subseteq (\bigvee F_i)^+$, that is, $\bigvee F_i = (\bigvee F_i)^+$, which is to say that $\bigvee F_i$ is Leibniz. This implies that the supremum in the lattice $\mathcal{F}i^+_{\mathcal{L}}\boldsymbol{A}$ coincides with that in $\mathcal{F}i_{\mathcal{L}}\boldsymbol{A}$. ⊠

Thus, the set of Leibniz filters is closed under the supremum operation of $\mathcal{F}i_{\mathcal{L}}\boldsymbol{A}$. Curiously enough, closure under its infimum, which is intersection, cannot be guaranteed; see the comment after Proposition 6.52 for more information on this point.

The reduced models of protoalgebraic logics can be characterized using the notion of a Leibniz filter.

Theorem 6.50
If $\mathcal{L}$ is a protoalgebraic logic, then $\langle \boldsymbol{A}, F\rangle \in \mathsf{Mod}^*\mathcal{L}$ if and only if $\boldsymbol{A} \in \mathsf{Alg}^*\mathcal{L}$, $F \in \mathcal{F}i_{\mathcal{L}}\boldsymbol{A}$ and F^+ is the only Leibniz $\mathcal{L}$-filter of $\boldsymbol{A}$ contained in F.

Proof: If $\langle \boldsymbol{A}, F\rangle \in \mathsf{Mod}^*\mathcal{L}$, then, by definition, $\boldsymbol{A} \in \mathsf{Alg}^*\mathcal{L}$ and $F \in \mathcal{F}i_{\mathcal{L}}\boldsymbol{A}$. Let $G \in \mathcal{F}i^+_{\mathcal{L}}\boldsymbol{A}$ with $G \subseteq F$. Then, by Proposition 6.47, $G = G^+ \subseteq F^+$ and by protoalgebraicity $\boldsymbol{\Omega}^{\boldsymbol{A}}G \subseteq \boldsymbol{\Omega}^{\boldsymbol{A}}F^+ = \boldsymbol{\Omega}^{\boldsymbol{A}}F = \mathrm{Id}_A$. Therefore, we have that $\boldsymbol{\Omega}^{\boldsymbol{A}}G = \mathrm{Id}_A = \boldsymbol{\Omega}^{\boldsymbol{A}}F^+$, which by Corollary 6.44 implies $G = F^+$: this shows that F^+ is the only Leibniz $\mathcal{L}$-filter of $\boldsymbol{A}$ included in F.

Conversely, assume that $\boldsymbol{A} \in \mathsf{Alg}^*\mathcal{L}$, $F \in \mathcal{F}i_{\mathcal{L}}\boldsymbol{A}$, and F^+ is the only Leibniz $\mathcal{L}$-filter of $\boldsymbol{A}$ included in F. Since $\boldsymbol{A} \in \mathsf{Alg}^*\mathcal{L}$, there is some $G \in \mathcal{F}i_{\mathcal{L}}\boldsymbol{A}$ with $\langle \boldsymbol{A}, G\rangle \in \mathsf{Mod}^*\mathcal{L}$, that is, such that $\boldsymbol{\Omega}^{\boldsymbol{A}}G = \mathrm{Id}_A$. Now consider $G^+ \in \mathcal{F}i^+_{\mathcal{L}}\boldsymbol{A}$: $\boldsymbol{\Omega}^{\boldsymbol{A}}G^+ = \boldsymbol{\Omega}^{\boldsymbol{A}}G = \mathrm{Id}_A \subseteq \boldsymbol{\Omega}^{\boldsymbol{A}}F^+$, and, by Corollary 6.44, $G^+ \subseteq F^+ \subseteq F$. The assumption on F^+ implies that $F^+ = G^+$, and therefore $\boldsymbol{\Omega}^{\boldsymbol{A}}F = \boldsymbol{\Omega}^{\boldsymbol{A}}F^+ = \boldsymbol{\Omega}^{\boldsymbol{A}}G^+ = \mathrm{Id}_A$, which means that $\langle \boldsymbol{A}, F\rangle$ is a reduced model for $\mathcal{L}$. ⊠

For particular logics this may give a workable description of these reduced models; this is done, for instance, in Example 6.76.

A more theoretical application of the notion of a Leibniz filter is the following enhancement of the Correspondence Theorem.

Proposition 6.51
Let $\mathcal{L}$ be protoalgebraic, and $h\colon \langle \boldsymbol{A}, F\rangle \to \langle \boldsymbol{B}, G\rangle$ be a strict surjective homomorphism from one $\mathcal{L}$-model to another.

1. *$F^+ = h^{-1}G^+$. That is, h is also strict from $\langle \boldsymbol{A}, F^+\rangle$ to $\langle \boldsymbol{B}, G^+\rangle$.*

2. $h\colon (\mathcal{F}i_{\mathcal{L}}\boldsymbol{A})^{F^+} \cong (\mathcal{F}i_{\mathcal{L}}\boldsymbol{B})^{G^+}$.

3. *F is Leibniz if and only if G is Leibniz.*

In particular, an $\mathcal{L}$-filter is Leibniz if and only if its reduction is Leibniz.

PROOF: 1. $h^{-1}G^+ \in \mathcal{F}i_{\mathcal{L}}\boldsymbol{A}$ and $\boldsymbol{\Omega}^{\boldsymbol{A}}h^{-1}G^+ = h^{-1}\boldsymbol{\Omega}^{\boldsymbol{B}}G^+ = h^{-1}\boldsymbol{\Omega}^{\boldsymbol{B}}G = \boldsymbol{\Omega}^{\boldsymbol{A}}h^{-1}G = \boldsymbol{\Omega}^{\boldsymbol{A}}F$, therefore $F^+ \subseteq h^{-1}G^+$. To prove the reverse inclusion let us first show that $h^{-1}hF^+ = F^+$. The $\supseteq$ inclusion always holds. Now take $a \in h^{-1}hF^+$, that is, $ha = hb$ for some $b \in F^+$. Then $ha \equiv hb\ (\boldsymbol{\Omega}^{\boldsymbol{B}}G)$, that is, $a \equiv b\ (h^{-1}\boldsymbol{\Omega}^{\boldsymbol{B}}G)$, and $h^{-1}\boldsymbol{\Omega}^{\boldsymbol{B}}G = \boldsymbol{\Omega}^{\boldsymbol{A}}F = \boldsymbol{\Omega}^{\boldsymbol{A}}F^+$, therefore by compatibility $a \in F^+$. This shows that $h^{-1}hF^+ = F^+$. Then $hF^+ \in \mathcal{F}i_{\mathcal{L}}\boldsymbol{B}$, and now $h^{-1}\boldsymbol{\Omega}^{\boldsymbol{B}}hF^+ = \boldsymbol{\Omega}^{\boldsymbol{A}}h^{-1}hF^+ = \boldsymbol{\Omega}^{\boldsymbol{A}}F^+ = \boldsymbol{\Omega}^{\boldsymbol{A}}F = h^{-1}\boldsymbol{\Omega}^{\boldsymbol{B}}G$. Since h is surjective, this implies that $\boldsymbol{\Omega}^{\boldsymbol{B}}hF^+ = \boldsymbol{\Omega}^{\boldsymbol{B}}G$, and from this it follows that $G^+ \subseteq hF^+$, that is, $h^{-1}G^+ \subseteq F^+$. This completes the proof of 1.

2 is a consequence of 1, taking Theorem 6.19 into account, and 3 follows from previous facts and that F is Leibniz if and only if $F = F^+$. ☒

As one of the applications of this notion, let us analyse the logic defined by the Leibniz filters of a given protoalgebraic logic $\mathcal{L}$, or, more precisely, by the class of matrices

$$\mathsf{Mod}^+\mathcal{L} := \big\{\langle \boldsymbol{A}, F\rangle \in \mathsf{Mod}\,\mathcal{L} : F \text{ is Leibniz}\big\}.$$

This logic is called ***the strong version of*** $\mathcal{L}$, and is denoted by $\mathcal{L}^+$. Some elementary properties of these notions are:

PROPOSITION 6.52

Let $\mathcal{L}$ be a protoalgebraic logic, and let $\mathcal{L}^+$ be its strong version.

1. *$\mathcal{L}^+$ is an extension of $\mathcal{L}$, and hence it is protoalgebraic.*
2. *$\mathsf{Mod}^+\mathcal{L} \subseteq \mathsf{Mod}(\mathcal{L}^+) \subseteq \mathsf{Mod}\,\mathcal{L}$, and for every $\boldsymbol{A}$, $\mathcal{F}i^+_{\mathcal{L}}\boldsymbol{A} \subseteq \mathcal{F}i_{\mathcal{L}^+}\boldsymbol{A} \subseteq \mathcal{F}i_{\mathcal{L}}\boldsymbol{A}$.*
3. *$\mathsf{Alg}\,\mathcal{L} = \mathsf{Alg}(\mathcal{L}^+)$.*
4. *For every $\boldsymbol{A}$, $\min \mathcal{F}i^+_{\mathcal{L}}\boldsymbol{A} = \min \mathcal{F}i_{\mathcal{L}^+}\boldsymbol{A} = \min \mathcal{F}i_{\mathcal{L}}\boldsymbol{A}$.*

In particular $\mathcal{L}$ and $\mathcal{L}^+$ have the same theorems.

PROOF: 1. That $\mathcal{L}^+$ an extension of $\mathcal{L}$ is a consequence of its definition by a subclass of the class of models of $\mathcal{L}$.

2. By the previous point, $\mathsf{Mod}(\mathcal{L}^+) \subseteq \mathsf{Mod}\,\mathcal{L}$. And since $\mathcal{L}^+$ is defined by $\mathsf{Mod}^+\mathcal{L}$, all these matrices are among its models, therefore $\mathsf{Mod}^+\mathcal{L} \subseteq \mathsf{Mod}(\mathcal{L}^+)$. By restricting the models under consideration to a fixed algebra, the second statement appears.

3. Since $\mathcal{L}^+$ is an extension of $\mathcal{L}$, $\mathsf{Alg}(\mathcal{L}^+) \subseteq \mathsf{Alg}\,\mathcal{L}$. As to the other inclusion, let $\boldsymbol{A} \in \mathsf{Alg}\,\mathcal{L} = \mathsf{Alg}^*\mathcal{L}$, with some F such that $\langle \boldsymbol{A}, F\rangle \in \mathsf{Mod}^*\mathcal{L}$. Then $\langle \boldsymbol{A}, F^+\rangle$ is also a reduced model of $\mathcal{L}$, but F^+ is Leibniz, hence it is actually a reduced model of $\mathcal{L}^+$. This shows that $\boldsymbol{A} \in \mathsf{Alg}^*(\mathcal{L}^+) = \mathsf{Alg}(\mathcal{L}^+)$.

4. We have already showed that $\min \mathcal{F}i_{\mathcal{L}}\boldsymbol{A} \in \mathcal{F}i^+_{\mathcal{L}}\boldsymbol{A}$, therefore it belongs to the three sets and is hence the minimum of each of them. ☒

It would be nice to find that the $\mathcal{L}^+$-filters are exactly the Leibniz filters of $\mathcal{L}$, i.e., that $\mathsf{Mod}^+\mathcal{L} = \mathsf{Mod}(\mathcal{L}^+)$, but this is not always the case: in general $\mathcal{L}^+$ may have other filters. Using more complicated tools, in [106, Theorem 19] it is shown that this holds if and only if for every $\boldsymbol{A}$, the family $\mathcal{F}i^+_{\mathcal{L}}\boldsymbol{A}$ is closed under intersections (compare with Corollary 6.49 and a comment after it), and also if and only if $\mathcal{L}^+$ is weakly algebraizable.

The interest of this study is that often the Leibniz filters can be characterized as the $\mathcal{L}$-filters that are moreover closed under an additional inference rule, so that $\mathcal{L}^+$ is axiomatized by adding that rule to an axiomatization of $\mathcal{L}$. In this aspect it generalizes a situation found in many particular logics; here are a few of them (see [106] for details).

EXAMPLES 6.53

1. The global logic S^g associated with a normal modal system S is the strong version (in the present sense) of the local logic S^ℓ, and the filters of S^g are exactly the Leibniz filters of S^ℓ; they are the filters of S^ℓ that are closed under the Necessitation Rule $x \rhd \Box x$.
2. Each n-valued Łukasiewicz logic $Ł_n$ (for $n > 2$) is the strong version of the logic $Ł_n^{\leqslant}$ defined by preservation of degrees of truth from $[\mathbf{0,1}]_n$ (see more details in Example 6.122.6); in this case the rule that characterizes the Leibniz filters among those of the original logic is *Modus Ponens* for $\rightarrow$.
3. Another example is found in the field of quantum logic, in which two logics have been associated in the literature to the class of orthomodular lattices; it happens that one of them (described in Example 3.64) is the strong version of the other. ⊠

To some extent, the theory of Leibniz filters and the notion of the strong version of a protoalgebraic logic allow a certain general analysis of the common situation of pairs of logics with the same class of algebras found in the literature; this issue has been more thoroughly studied in [5, 108]. Leibniz filters also appear, somehow surprisingly, in the proof of the existence of a fully adequate Gentzen system for logics with the DDT in [109].

Finally, observe that, while clearly Leibniz filters may be *defined* for an arbitrary logic, they need not *exist* in all cases, besides the smallest filter of each algebra. It is thus surprising that Proposition 6.45 may also hold for some non-protoalgebraic logics; a class of these logics is studied in Section 7.2.

Exercises for Section 6.2

EXERCISE 6.1. Let $\mathcal{L}$ be an arbitrary logic, and let $\Gamma \in \mathcal{T}h\mathcal{L}$. Using only the definitions of $\boldsymbol{\Omega}$ and $\boldsymbol{\Lambda}_{\mathcal{L}}$, show that $\boldsymbol{\Omega}\Gamma \subseteq \boldsymbol{\Lambda}_{\mathcal{L}}\Gamma$ if and only if for all $\Gamma' \in \mathcal{T}h\mathcal{L}$ with $\Gamma \subseteq \Gamma'$, $\boldsymbol{\Omega}\Gamma \subseteq \boldsymbol{\Omega}\Gamma'$. Use this fact to obtain an alternative proof of Theorem 6.4, without using the Suszko congruence, nor any fact concerning g-matrices.

EXERCISE 6.2. Prove the six properties appearing in Proposition 6.6.

Exercise 6.3. Prove that a logic $\mathcal{L}$ is protoalgebraic if and only if for every $\Gamma \in \mathcal{T}h\mathcal{L}$, $x \equiv y\ (\boldsymbol{\Omega}\Gamma)$ if and only if $\Sigma_{\mathcal{L}} \subseteq \Gamma$.

Exercise 6.4. Prove that a logic $\mathcal{L}$ is protoalgebraic if and only if $\delta(x,\vec{z}), \Sigma_{\mathcal{L}} \vdash_{\mathcal{L}} \delta(y,\vec{z})$ for every $\delta(x,\vec{z}) \in Fm$.

Exercise 6.5. Check that the set Δ appearing in the proof of Theorem 6.7 can be defined simply as $\Delta(x,y) = \{\varphi(x,y) \in Fm : \vdash_{\mathcal{L}} \varphi(x,x)\}$. Prove directly, with this definition, the equivalence between points (i) and (iii) of that theorem.

Exercise 6.6. By applying finitarity to (MP), show that for a finitary protoalgebraic logic, the sets Δ in Theorem 6.7 and in Exercise 6.5 can be taken finite.

Exercise 6.7. Let $\mathcal{L}$ be a logic satisfying the DDT with respect to some DD set $I(x,y)$. Show that this set satisfies the properties (R) and (MP), and hence that the logic is protoalgebraic.

Exercise 6.8. Use Theorem 6.7 to prove the points of Proposition 6.11 that you do not find obvious. Use Theorem 6.4 to give alternative proofs that every algebraizable logic is protoalgebraic, and that all extensions of a protoalgebraic logic are protoalgebraic as well.

Exercise 6.9. Let $\mathcal{L}$ be the logic, on the language that has only two unary operations λ, μ, defined by the axiomatic system with no axioms and the only rule $\mu x \rhd \lambda\mu x$. Observe that this logic is not protoalgebraic. Let $\boldsymbol{A}$ be the algebra with universe the set ω of natural numbers and where $\lambda^{\boldsymbol{A}}$ is the successor function and $\mu^{\boldsymbol{A}}$ is the identity function. Let $X := \{0\}$. Show that $Fg_{\mathcal{L}}^{\boldsymbol{A}} X = \omega$ while the set Z constructed in the proof of Proposition 6.12 is $\{0,1\}$. Conclude that protoalgebraicity is essential in this result.

Exercise 6.10. Complete the proofs of Theorem 6.13 and of Proposition 6.14.

Exercise 6.11. Prove that a logic is protoalgebraic if and only if for any $\boldsymbol{A}$ and any $F, G \in \mathcal{F}i_{\mathcal{L}}\boldsymbol{A}$ it holds that $\boldsymbol{\Omega}^{\boldsymbol{A}}(F \cap G) = \boldsymbol{\Omega}^{\boldsymbol{A}}F \cap \boldsymbol{\Omega}^{\boldsymbol{A}}G$.

Exercise 6.12. Prove Corollary 6.20.

Hint. You may need to review Section 5.3 in order to find the needed properties of strict surjective homomorphisms between g-matrices.

Exercise 6.13. Prove that a logic $\mathcal{L}$ is protoalgebraic if and only if for every surjective (algebraic) homomorphism $h\colon \boldsymbol{A} \to \boldsymbol{B}$, every $F \in \mathcal{F}i_{\mathcal{L}}\boldsymbol{A}$ and every $G \in \mathcal{F}i_{\mathcal{L}}\boldsymbol{B}$, $h^{-1}Fg_{\mathcal{L}}^{\boldsymbol{B}}\big((Fg_{\mathcal{L}}^{\boldsymbol{B}} hF) \cup G\big) = Fg_{\mathcal{L}}^{\boldsymbol{A}}(F \cup h^{-1}G)$.

Exercise 6.14. Strengthen Theorem 6.19 by showing that a logic is protoalgebraic if and only if the property stated in its point (ii) holds for arbitrary strict homomorphisms (not just for the surjective ones).

Exercise 6.15. By a careful examination of the proof of Theorem 6.22, show that a logic $\mathcal{L}$ is protoalgebraic if and only if it satisfies the following property: For all $\Gamma \cup \{\alpha, \beta\} \subseteq Fm$, $\Gamma, \alpha \vdash_{\mathcal{L}} \beta$ if and only if there is $\Gamma' \in \mathcal{T}h\mathcal{L}$ satisfying $\Gamma', x \vdash_{\mathcal{L}} y$ and $\Gamma \vdash_{\mathcal{L}} \sigma\Gamma'$ for a substitution σ such that $\sigma x = \alpha, \sigma y = \beta$.

Comment. The implication from right to left is trivial and holds for all logics.

Exercise 6.16. Prove the two points of Proposition 6.23.

Exercise 6.17. Prove the two points of Proposition 6.27.

Hint. Recall (Exercise 1.42) that an ideal of a join-semilattice $\langle A, \vee \rangle$ is a non-empty set $I \subseteq A$ such that for all $a, b \in A$, $a \vee b \in I$ if and only if $a, b \in I$.

EXERCISE 6.18. Complete the step skipped in the proof of Theorem 6.28.

EXERCISE 6.19. Prove the equivalence of the three conditions in Lemma 6.34.

EXERCISE 6.20. Prove the following enhanced form of Theorem 6.39: A logic $\mathcal{L}$ is protoalgebraic if and only if every full g-model of $\mathcal{L}$ on $\boldsymbol{Fm}$ has the form $\langle \boldsymbol{Fm}, (\mathcal{T}h\mathcal{L})^{\Gamma} \rangle$ for some $\Gamma \in \mathcal{T}h\mathcal{L}$.

EXERCISE 6.21. Let $\mathcal{L}$ be a protoalgebraic logic. Prove that an $\mathcal{L}$-filter F on an algebra $\boldsymbol{A}$ is Leibniz if and only if for every $G \in \mathcal{F}i_{\mathcal{L}}\boldsymbol{A}$, if $\boldsymbol{\Omega}^{\boldsymbol{A}}F \subseteq \boldsymbol{\Omega}^{\boldsymbol{A}}G$, then $F \subseteq G$; in other words, when $F = \min\{G \in \mathcal{F}i_{\mathcal{L}}\boldsymbol{A} : \boldsymbol{\Omega}^{\boldsymbol{A}}F \subseteq \boldsymbol{\Omega}^{\boldsymbol{A}}G\}$.

EXERCISE 6.22. Let $\mathcal{L}$ be a protoalgebraic logic. Prove that the following conditions are equivalent:

(i) All $\mathcal{L}$-filters are Leibniz.

(ii) The Leibniz operator $\boldsymbol{\Omega}^{\boldsymbol{A}}$ is injective over $\mathcal{F}i_{\mathcal{L}}\boldsymbol{A}$, for every $\boldsymbol{A}$.

(iii) For every $\boldsymbol{A}$ and every $F \in \mathcal{F}i_{\mathcal{L}}\boldsymbol{A}$, $F/\boldsymbol{\Omega}^{\boldsymbol{A}}F = \min \mathcal{F}i_{\mathcal{L}}(\boldsymbol{A}/\boldsymbol{\Omega}^{\boldsymbol{A}}F)$.

EXERCISE 6.23. You might be tempted to rephrase Corollary 6.42 in the following way: If $\mathcal{L}$ is protoalgebraic and $\boldsymbol{A} \in \mathsf{Alg}^*\mathcal{L}$, then there is only one Leibniz $\mathcal{L}$-filter of $\boldsymbol{A}$, namely $\min \mathcal{F}i_{\mathcal{L}}\boldsymbol{A}$. Show that this is wrong.

6.3 Definability of equivalence (protoalgebraic and equivalential logics)

This section is devoted to the analysis of the property that the Leibniz congruence of a filter is definable, in terms of the filter, by a set of formulas. This happens in all models of a protoalgebraic logic, but parameters must be allowed to appear in the set of formulas. When no parameters are needed, this property defines the class of equivalential logics, which plays a important role in the Leibniz hierarchy.

Definability of the Leibniz congruence, with or without parameters

In Section 6.2 we see that certain sets of formulas that are relevant for the study of protoalgebraic logics may have *parameters*; for instance, this happens in the fundamental set $\Sigma_{\mathcal{L}}$ and in the family of DD sets involved in the PLDDT. The role of these parameters, as in ordinary mathematics, is to vary over all possible values, either in the formula algebra or when evaluated in a particular algebra. In some cases, such as in the PLDDT, only particular values of the parameters play a role in each instance of the property at hand; in other cases, all their values have to be considered at the same time, and a convenient notation for this is established. Please pay attention, as the chosen notation may not be typographically too distinctive!

NOTATION.
If $\Delta(x,y,\vec{z}) \subseteq Fm$ is a set of formulas in the variables x,y and possibly parameters $\vec{z}$, then for each $\alpha,\beta \in Fm$ the following set of formulas is defined:

$$\Delta\langle\alpha,\beta\rangle := \bigcup\{\Delta(\alpha,\beta,\vec{\delta}) : \vec{\delta} \in \vec{Fm}\}$$

This is the set of formulas resulting from the set $\Delta(\alpha,\beta,\vec{z})$ after replacing the parameters $\vec{z}$ by *all* possible formulas.

Similarly, if $\boldsymbol{A}$ is an algebra and $a,b \in A$, then:

$$\Delta^{\boldsymbol{A}}\langle a,b\rangle := \bigcup\{\Delta^{\boldsymbol{A}}(a,b,\vec{c}) : \vec{c} \in \vec{A}\},$$

where the parameters now range over *all* the elements of the algebra. Exercises 6.24 and 6.29 show how careful one must be when dealing with these notations. For instance, $\Delta\langle\alpha,\beta\rangle$ is *not* the result of substituting α for x and β for y in $\Delta\langle x,y\rangle$, and $\Delta^{\boldsymbol{A}}\langle a,b\rangle$ is *not* the result of evaluating the set $\Delta\langle x,y\rangle$ by giving the value a to x and the value b to y. Actually, the latter assertion makes no sense: the set $\Delta\langle x,y\rangle$ always contains all the variables and not just x and y; notice by contrast that the $\vec{z}$ in $\Delta(x,y,\vec{z})$ is a generic notation, and in each particular case the set may contain a different number of parameters, either none, or a finite number, or an infinite one. Finally, notice that even in the last case, each particular formula in $\Delta(x,y,\vec{z})$ will contain only a finite number of the $\vec{z}$.

A set without parameters is here considered as a limiting case of a set with parameters. Therefore, all the definitions and constructions done for a set with parameters apply also when the set has no parameters at all; if $\Delta(x,y)$ is a set of formulas in two variables without parameters, then $\Delta\langle\alpha,\beta\rangle$ is $\Delta(\alpha,\beta)$, and $\Delta^{\boldsymbol{A}}\langle a,b\rangle$ is $\Delta^{\boldsymbol{A}}(a,b)$. ⊠

DEFINITION 6.54
*A set $\Delta(x,y,\vec{z}) \subseteq Fm$ **defines the Leibniz congruence** in a matrix $\langle \boldsymbol{A},F\rangle$ when for all $a,b \in A$,*

$$a \equiv b\ (\Omega^{\boldsymbol{A}}F) \iff \Delta^{\boldsymbol{A}}\langle a,b\rangle \subseteq F. \tag{87}$$

*A **set of congruence formulas with parameters for a logic** $\mathcal{L}$ is a set that defines the Leibniz congruence in all models of $\mathcal{L}$. When the set has no parameters it is called simply a **set of congruence formulas for** $\mathcal{L}$.*

The following fact is very useful.

LEMMA 6.55
Let $h\colon \langle \boldsymbol{A},F\rangle \to \langle \boldsymbol{B},G\rangle$ be a strict surjective homomorphism. A set defines the Leibniz congruence in $\langle \boldsymbol{A},F\rangle$ if and only if it defines the Leibniz congruence in $\langle \boldsymbol{B},G\rangle$.

PROOF: Left to you as Exercise 6.25. Surjectivity is essential here. ⊠

In particular this holds for reductions. But the Leibniz congruence in a reduced matrix is the identity relation. Therefore:

COROLLARY 6.56

1. *A set Δ defines the Leibniz congruence in $\langle \boldsymbol{A}, F\rangle$ if and only if it defines the identity relation in $\langle \boldsymbol{A}/\boldsymbol{\Omega}^{\boldsymbol{A}}F, F/\boldsymbol{\Omega}^{\boldsymbol{A}}F\rangle$; that is, if and only if for all $a,b \in A/\boldsymbol{\Omega}^{\boldsymbol{A}}F$,*

$$\text{if } \Delta^{\boldsymbol{A}/\boldsymbol{\Omega}^{\boldsymbol{A}}F}\langle a,b\rangle \subseteq F/\boldsymbol{\Omega}^{\boldsymbol{A}}F, \text{ then } a = b.$$

2. *A set Δ is a set of congruence formulas with parameters for a logic $\mathcal{L}$ when it defines the identity relation in the class $\mathsf{Mod}^*\mathcal{L}$ of reduced models of $\mathcal{L}$.* ⊠

The first important result here is:

THEOREM 6.57

Let $\mathcal{L}$ be a logic. $\mathcal{L}$ is protoalgebraic if and only if $\mathcal{L}$ has a set of congruence formulas with parameters.

PROOF: ($\Rightarrow$) The wanted set of congruence formulas with parameters will be the fundamental set $\Sigma_{\mathcal{L}}(x,y,\vec{z})$. Let $\langle \boldsymbol{A}, F\rangle \in \mathsf{Mod}\mathcal{L}$ and assume that $a \equiv b \ (\boldsymbol{\Omega}^{\boldsymbol{A}}F)$. Then by congruence also $\varphi^{\boldsymbol{A}}(a,a,\vec{c}) \equiv \varphi^{\boldsymbol{A}}(a,b,\vec{c}) \ (\boldsymbol{\Omega}^{\boldsymbol{A}}F)$ for all $\vec{c} \in \vec{A}$ and all $\varphi(x,y,\vec{z}) \in \Sigma_{\mathcal{L}}$. Since $\vdash_{\mathcal{L}} \varphi(x,x,\vec{z})$, it follows that $\varphi^{\boldsymbol{A}}(a,a,\vec{c}) \in F$ and by compatibility $\varphi^{\boldsymbol{A}}(a,b,\vec{c}) \in F$. Thus, $\Sigma_{\mathcal{L}}^{\boldsymbol{A}}(a,b,\vec{c}) \subseteq F$ for all $\vec{c} \in \vec{A}$, that is, $\Sigma_{\mathcal{L}}^{\boldsymbol{A}}\langle a,b\rangle \subseteq F$. Conversely, assume now that $\Sigma_{\mathcal{L}}^{\boldsymbol{A}}\langle a,b\rangle \subseteq F$. By Proposition 6.6.5, also $\Sigma_{\mathcal{L}}^{\boldsymbol{A}}\langle b,a\rangle \subseteq F$. Now, given an arbitrary $\delta(x,\vec{z}) \in Fm$, using the result in Exercise 6.4 and the fact that F is an $\mathcal{L}$-filter, it follows that $\delta^{\boldsymbol{A}}(a,\vec{c}) \in F$ if and only if $\delta^{\boldsymbol{A}}(b,\vec{c}) \in F$ for all $\vec{c} \in \vec{A}$. That is, $a \equiv b \ (\boldsymbol{\Omega}^{\boldsymbol{A}}F)$.

($\Leftarrow$) The monotonicity of $\boldsymbol{\Omega}^{\boldsymbol{A}}$ over $\mathcal{F}i_{\mathcal{L}}\boldsymbol{A}$ is a straightforward consequence of its definability. ⊠

Notice that when $\mathcal{L}$ is finitary, the set $\Delta(x,y,\vec{z})$ cannot be guaranteed to be finite, unlike what happens with the protoimplication set Δ (without parameters) of Proposition 6.8. This is so because the property in Exercise 6.4, used in the proof, concerns only $\Sigma_{\mathcal{L}}$, and need not hold for an arbitrary protoimplication set. Exercises 6.27–6.31 contain additional properties concerning congruence sets *with parameters*, which are interesting but less central here (but are used in a particular point in Section 6.4). Exercise 6.31 shows another standard way of obtaining a congruence set with parameters for a protoalgebraic logic.

Having a syntactic characterization of the Leibniz congruence facilitates many things. First, and in a rather lateral issue here, it facilitates the characterization of the Leibniz filter associated with a given filter of a protoalgebraic logic in Proposition 6.45:

LEMMA 6.58

Let $\mathcal{L}$ be protoalgebraic, with $\Delta(x,y,\vec{z})$ as set of congruence formulas with parameters. For each $\theta \in \mathrm{Con}_{\mathsf{Alg}^*\mathcal{L}}\boldsymbol{A}$, the set $Fg_{\mathcal{L}}^{\boldsymbol{A}} \bigcup\{\Delta^{\boldsymbol{A}}\langle a,b\rangle : a \equiv b \ (\theta)\}$ is the least $\mathcal{L}$-filter F such that $\boldsymbol{\Omega}^{\boldsymbol{A}}F = \theta$.

PROOF: Put $F_\theta := Fg_{\mathcal{L}}^{\boldsymbol{A}} \bigcup\{\Delta^{\boldsymbol{A}}\langle a,b\rangle : a \equiv b \ (\theta)\}$ for simplicity. Clearly $F_\theta \in \mathcal{F}i_{\mathcal{L}}\boldsymbol{A}$. Now take any $F \in \mathcal{F}i_{\mathcal{L}}\boldsymbol{A}$ such that $\boldsymbol{\Omega}^{\boldsymbol{A}}F = \theta$; such an F exists, because $\boldsymbol{\Omega}^{\boldsymbol{A}}$ is always onto $\mathrm{Con}_{\mathsf{Alg}^*\mathcal{L}}\boldsymbol{A}$ by Lemma 4.41. If $a \equiv b \ (\theta)$, then $a \equiv b \ (\boldsymbol{\Omega}^{\boldsymbol{A}}F)$

and hence $\Delta^{\boldsymbol{A}}\langle a,b\rangle \subseteq F$. Thus, $F_\theta \subseteq F$. By protoalgebraicity, this implies that $\boldsymbol{\Omega}^{\boldsymbol{A}}F_\theta \subseteq \boldsymbol{\Omega}^{\boldsymbol{A}}F = \theta$. Conversely, if $a \equiv b\ (\theta)$, then by the construction of F_θ, $\Delta^{\boldsymbol{A}}\langle a,b\rangle \subseteq F_\theta$, and therefore $a \equiv b\ (\boldsymbol{\Omega}^{\boldsymbol{A}}F_\theta)$. This shows that $\theta = \boldsymbol{\Omega}^{\boldsymbol{A}}F_\theta$, and thus completes the proof. ⊠

Since for every $F \in \mathcal{F}i_{\mathcal{L}}\boldsymbol{A}$ it is always true that $\boldsymbol{\Omega}^{\boldsymbol{A}}F \in \mathrm{Con}_{\mathsf{Alg}^*\mathcal{L}}\boldsymbol{A}$, this result directly implies:

COROLLARY 6.59
Let $\mathcal{L}$ be protoalgebraic, with $\Delta(x,y,\vec{z})$ as set of congruence formulas with parameters, and $F \in \mathcal{F}i_{\mathcal{L}}\boldsymbol{A}$.

1. $F^+ = Fg^{\boldsymbol{A}}_{\mathcal{L}} \bigcup \{\Delta^{\boldsymbol{A}}\langle a,b\rangle : a \equiv b\ (\boldsymbol{\Omega}^{\boldsymbol{A}}F)\}$.
2. *F is Leibniz if and only if* $F = Fg^{\boldsymbol{A}}_{\mathcal{L}} \bigcup \{\Delta^{\boldsymbol{A}}\langle a,b\rangle : a \equiv b\ (\boldsymbol{\Omega}^{\boldsymbol{A}}F)\}$. ⊠

For another, miscellaneous example of the use of the sets of congruence formulas, see Exercise 6.32.

Congruence is typically characterized by four properties: reflexivity, symmetry, transitivity and replacement (that is, compatibility with the operations). However, the Leibniz congruence of a set is a rather special congruence, because it satisfies a condition of compatibility with the set, which is reflected in a form of *Modus Ponens*; with its help one can show that just two of the first mentioned properties suffice (this implies an enhancement of Theorem 4.54). The proof is detailed here for the parameter-less case, and Exercise 6.28 suggests you to prove the corresponding property for sets with parameters.

THEOREM 6.60
A set $\Delta(x,y) \subseteq Fm$ is a set of congruence formulas for a logic $\mathcal{L}$ if and only if the following conditions are satisfied:

(R) $\vdash_{\mathcal{L}} \Delta(x,x)$

(MP) $x, \Delta(x,y) \vdash_{\mathcal{L}} y$

(Re) $\bigcup_{i=1}^{n} \Delta(x_i,y_i) \vdash_{\mathcal{L}} \Delta(\lambda x_1 \dots x_n, \lambda y_1 \dots y_n)$ *for each* $\lambda \in L$, $n = \mathrm{ar}\lambda \geqslant 1$

PROOF: ($\Rightarrow$) Apply (87) from page 353 to matrices of the form $\langle \boldsymbol{Fm}, \Gamma\rangle$ with $\Gamma \in \mathcal{T}h\mathcal{L}$. First, for $\Gamma := C_{\mathcal{L}}\varnothing$, since $x \equiv x\ (\boldsymbol{\Omega} C_{\mathcal{L}}\varnothing)$, we have $\Delta(x,x) \subseteq C_{\mathcal{L}}\varnothing$, which is (R). Now consider an arbitrary $\Gamma \in \mathcal{T}h\mathcal{L}$. If $x \in \Gamma$ and $\Delta(x,y) \subseteq \Gamma$, then $x \equiv y\ (\boldsymbol{\Omega}\Gamma)$, which implies that $y \in \Gamma$ by compatibility. This shows (MP). Finally, if $\Delta(x_i,y_i) \subseteq \Gamma$ for all $i = 1,\dots,n$, then $x_i \equiv y_i\ (\boldsymbol{\Omega}\Gamma)$, which by congruence implies that $\lambda x_1 \dots x_n \equiv \lambda y_1 \dots y_n\ (\boldsymbol{\Omega}\Gamma)$, that is, $\Delta(\lambda x_1 \dots x_n, \lambda y_1 \dots y_n) \subseteq \Gamma$; this shows (Re).

($\Leftarrow$) Take any $F \in \mathcal{F}i_{\mathcal{L}}\boldsymbol{A}$ and consider the binary relation denoted by $R^{\boldsymbol{A}}(F,\Delta)$ in Definition 4.52, and call it here R; that is, define $a R b \overset{\text{def}}{\Longleftrightarrow} \Delta^{\boldsymbol{A}}(a,b) \subseteq F$. To prove (87) is to prove that $\boldsymbol{\Omega}^{\boldsymbol{A}}F = R$, which is done in several steps.

(a) That $\boldsymbol{\Omega}^{\boldsymbol{A}}F \subseteq R$ is proved in Lemma 4.53; notice that this only requires property (R), which actually means that R is reflexive.

(b) If $a R b$ and $a \in F$, then $b \in F$: This is actually equivalent to (MP), given the definition of R.

(c) If $a R b$, then $\varphi^{\boldsymbol{A}}(a,\vec{c}) R \varphi^{\boldsymbol{A}}(b,\vec{c})$ for every $\varphi(x,\vec{z}) \in Fm$ and every $\vec{c} \in \vec{A}$: This is shown by an inductive argument on the complexity of φ, using (Re) and (R); this is left to you as Exercise 6.26.

(d) R is symmetric: Assume that $a R b$, that is, that $\Delta^{\boldsymbol{A}}(a,b) \subseteq F$. By applying (c) to each $\delta(x,y) \in \Delta$, using y as the parameter and giving it the value a, one obtains that $\delta^{\boldsymbol{A}}(a,a) R \delta^{\boldsymbol{A}}(b,a)$. But since $\delta^{\boldsymbol{A}}(a,a) \in F$ by (R), using (b) one obtains that $\delta^{\boldsymbol{A}}(b,a) \in F$. Hence, $\Delta^{\boldsymbol{A}}(b,a) \subseteq F$, that is, $b R a$.

(e) If $a R b$, then by (d) also $b R a$, and using (c) for both relations and then (b), it follows that for every $\varphi(x,\vec{z}) \in Fm$ and every $\vec{c} \in \vec{A}$, $\varphi^{\boldsymbol{A}}(a,\vec{c}) \in F$ if and only if $\varphi^{\boldsymbol{A}}(b,\vec{c}) \in F$, which is the characterization of $\boldsymbol{\Omega}^{\boldsymbol{A}} F$ in Theorem 4.23. Thus, $R \subseteq \boldsymbol{\Omega}^{\boldsymbol{A}} F$.

Together with (a), this completes the proof of the Theorem. ⊠

Notice the proof strategy: it does not show directly that R is a congruence, therefore we cannot obtain that $R \subseteq \boldsymbol{\Omega}^{\boldsymbol{A}} F$ directly from (b). Nevertheless, the final result is that, being equal to $\boldsymbol{\Omega}^{\boldsymbol{A}} F$, R *is* a congruence. Hence:

COROLLARY 6.61
If a set $\Delta(x,y) \subseteq Fm$ satisfies conditions (R), (MP) *and* (Re) *with respect to some logic $\mathcal{L}$, then it also satisfies the conditions*

(Sym) $\Delta(x,y) \vdash_{\mathcal{L}} \Delta(y,x)$

(Trans) $\Delta(x,y) \cup \Delta(y,z) \vdash_{\mathcal{L}} \Delta(x,z)$. ⊠

Observe that there is no assumption about protoalgebraicity in the last two results; they hold for every logic, because it is not necessary to assume that such sets of formulas exist. The following technical result uses the fundamental set $\Sigma_{\mathcal{L}}$ introduced in Definition 6.5, and its properties established in Proposition 6.6, and now, yes, protoalgebraicity is needed.

LEMMA 6.62
Let $\mathcal{L}$ be protoalgebraic and let $\Delta(x,y) \subseteq Fm$. The following conditions are equivalent:

(i) *$\Delta(x,y)$ is a set of congruence formulas for $\mathcal{L}$.*

(ii) *$\Delta(x,y) \subseteq \Sigma_{\mathcal{L}}$ and $x \equiv y \; (\boldsymbol{\Omega} C_{\mathcal{L}} \Delta(x,y))$.*

(iii) *$C_{\mathcal{L}} \Delta(x,y) = \Sigma_{\mathcal{L}}$.*

PROOF: (i)⇒(ii) The assumption implies that $\vdash_{\mathcal{L}} \Delta(x,x)$, therefore $\Delta(x,y) \subseteq \Sigma_{\mathcal{L}}$. Moreover, $\Delta(x,y) \subseteq C_{\mathcal{L}} \Delta(x,y)$, so by the assumption $x \equiv y \; (\boldsymbol{\Omega} C_{\mathcal{L}} \Delta(x,y))$.

(ii)⇒(iii) Since $\Sigma_{\mathcal{L}}$ is a theory, the first property implies that $C_{\mathcal{L}} \Delta(x,y) \subseteq \Sigma_{\mathcal{L}}$. By Proposition 6.6.4, $\Sigma_{\mathcal{L}}$ is the smallest theory T of $\mathcal{L}$ such that $x \equiv y \; (\boldsymbol{\Omega} T)$, therefore the second property implies that $\Sigma_{\mathcal{L}} \subseteq C_{\mathcal{L}} \Delta(x,y)$. This shows that $\Sigma_{\mathcal{L}} = C_{\mathcal{L}} \Delta(x,y)$.

(iii)⇒(i) The proof of Theorem 6.57, which requires $\mathcal{L}$ to be protoalgebraic, showed that $\Sigma_{\mathcal{L}}$ is a set of congruence formulas for $\mathcal{L}$ (with parameters). By Exercise 6.27, the assumption implies that $\Delta(x,y)$ also defines the Leibniz congruence in the models of $\mathcal{L}$. ⊠

These two characterizations of being a set of congruence formulas are of primarily technical interest, and are used in several proofs in this section.

Equivalential logics: Definition and general properties

DEFINITION 6.63
A logic is ***equivalential*** *when it has a set of congruence formulas (without parameters). A logic is* ***finitely equivalential*** *when it has a finite set of congruence formulas (without parameters).*

The notion of an equivalential logic is due to Prucnal and Wroński. Their 1974 original definition [199] uses the five conditions (R), (Sim), (Trans), (Re) and (MP); the simplification of the conditions and the definition involving the Leibniz operator appear in Czelakowski's fundamental 1981 paper [67]. In these early papers the term "equivalence formulas" is used for what are now called congruence formulas; hence, it might seem more natural that these logics be now called "congruential" rather than equivalential, but the term has been also used in the literature with other meanings (see the discussion on page 416).

From Theorems 6.7 and 6.60 (or from the definition and Theorem 6.57) it immediately follows:

PROPOSITION 6.64
Every equivalential logic is protoalgebraic. ⊠

The following facts are also straightforward:

PROPOSITION 6.65
1. *The only logics that are equivalential with the empty set of congruence formulas are the inconsistent and the almost inconsistent logics (in each language); they are actually finitely equivalential.*
2. *Any extension of an equivalential logic is also equivalential. The same holds for finitely equivalential logics.*
3. *An expansion of an equivalential logic is also equivalential if and only if it satisfies* (Re) *for the new connectives. The same holds for finitely equivalential logics.*
4. *Any fragment of an equivalential logic such that the set of congruence formulas belongs to the fragment language is also equivalential. The same holds for finitely equivalential logics.*
5. *Let $\mathcal{L}$ be equivalential, with Δ as a set of congruence formulas. A set $\Delta'(x,y) \subseteq Fm$ is a set of congruence formulas for $\mathcal{L}$ if and only if $\Delta(x,y) \dashv\vdash_{\mathcal{L}} \Delta'(x,y)$.*
6. *Let $\mathcal{L}$ be a finitary and finitely equivalential logic. Given an arbitrary set Δ of congruence formulas for $\mathcal{L}$, there is a finite subset $\Delta' \subseteq \Delta$ such that Δ' is also a set of congruence formulas for $\mathcal{L}$.* ⊠

Points 5 and 6 are crucial when establishing that a certain equivalential logic is not finitely equivalential, and in general when dealing with the cardinality of the sets of congruence formulas.

Also, from Theorem 6.60 and Proposition 3.13 it immediately follows:

PROPOSITION 6.66
Every algebraizable logic is equivalential, and every finitely algebraizable logic is finitely equivalential. A logic is finitely algebraizable if and only if it is both algebraizable and finitely equivalential. ⊠

The congruence formulas coincide with the so-called "equivalence formulas" in the theory of algebraizable logics; there, the term evoked another role of these formulas, namely that of defining one of the transformers that establish the "deductive equivalence" between the consequence of the logic and the relative equational consequence of its algebraic counterpart; both denominations may be used indistinctly.

Notice that a finitary equivalential logic need not be finitely equivalential. Witness to this are the finitary algebraizable logics that are not finitely algebraizable, such as Herrmann's $\mathcal{LJ}$ (Example 3.42), or Dellunde's $\mathcal{L}_D$ (Example 3.53). Below, some non-algebraizable examples are given.

EXAMPLES 6.67

1. It is clear that all implicative logics are finitely equivalential, with set of congruence formulas $\Delta(x,y) := \{x \to y, y \to x\}$.

2. Exercise 6.33 describes a different group of finitely equivalential logics, which are not necessarily implicative for the same implication connective (they are algebraizable, though). This covers the implication fragments of relevance logic $\mathcal{R}$ and of entailment logic $\mathcal{E}$.

3. The local logic K^ℓ associated with the weakest normal modal system K is equivalential, with set of congruence formulas $\Delta(x,y) := \{\Box^n(x \leftrightarrow y) : n \geqslant 0\}$, an infinite one. Hence all its extensions are also equivalential. However, K^ℓ is not finitely equivalential: using Kripke models it is not difficult to see that $\{\Box^n(x \leftrightarrow y) : 0 \leqslant n \leqslant m\} \nvdash^\ell_K \Box^{m+1}(x \leftrightarrow y)$, which would contradict Proposition 6.65.6. The same holds for the modal system T. These logics are shown in Example 3.61 not to be algebraizable.

4. By contrast, the local logic $K4^\ell$ associated with the system $K4$ and its extensions are finitely equivalential, because the axiom $\Box\Box x \to \Box x$ implies that set of congruence formulas can be reduced to just $\{x \leftrightarrow y, \Box(x \leftrightarrow y)\}$. In the extensions of $S4^\ell$, the axiom $\Box x \to x$ further reduces it to $\{\Box(x \leftrightarrow y)\}$. The logics $K4^\ell$ and $S4^\ell$ are not algebraizable either (this is shown in the same example as the previous point).

5. A very curious fact is that, among the logics that preserve degrees of truth with respect to a variety K of commutative integral residuated lattices mentioned

in Example 6.16.9, all those that are protoalgebraic are automatically finitely equivalential as well; this is proved in [46, Corollary 4.11]. These logics are not algebraizable except when K is a variety of generalized Heyting algebras.

6. Another interesting example of an equivalential logic is *propositional dynamic logic* $\mathcal{PDL}$. This is a multi-modal logic with a family of necessity-like operators $\{\boxed{a} : a \in Prog\}$, where *Prog* is an algebra representing programs or program instructions, which has its own operations (thus, this logic is built on a two-sorted language, formulas and programs being constructed by simultaneous recursion) denoted by $;$, $\cup$ and $*$, and which can be understood as follows: if $a, b \in Prog$ then:
   - $a\,;b$ means "perform a and then perform b".
   - $a \cup b$ means "perform a or b, non-deterministically".
   - a^* means "perform a some finite number of times".

   and there are also two operations mixing formulas and programs:
   - For each formula φ, the expression $\varphi?$, called a "test", is the instruction "continue if φ holds, stop if φ does not hold".
   - For each program a, the expression $\boxed{a}\varphi$ is a formula meaning "after every terminating execution of a, φ holds".

   Using these operations, some familiar constructs of programming languages can be defined; for instance, "if-then-else" instructions are represented as expressions of the form $(\varphi?\,;a) \cup ((\neg\varphi)?\,;b)$.

   First-order dynamic logic was introduced as related to the logic of program verification by Pratt in 1974. Later on its propositional fragment was isolated and axiomatized as a multi-modal logic. Some recent, more liberal settings prefer to speak of "actions" instead of programs or instructions, and this has brought other applications to dynamic logic. See [136] for more information.

   The logics in this family (one for each choice of *Prog*) can be syntactically axiomatized, have their own relational (Kripke-style) semantics, and also an algebra-based semantics made of the so-called *dynamic algebras*, which are a specific kind of Boolean algebras with operators. The "local" logics of this family, as studied in [70, Section 3.6], are finitary and equivalential, with set of congruence formulas $\Delta(x,y) := \{x \leftrightarrow y\} \cup \{\boxed{a}(x \leftrightarrow y) : a \in Prog\}$. These logics turn out to be finitely equivalential if and only if the algebra *Prog* is finitely generated, and they are not algebraizable. The reduced models of these logics are described in [70, Theorem 3.6.1]. ⊠

The results and examples given up to now establish the structure of one good portion of the diagram of the Leibniz hierarchy in Figure 9 on page 316, namely that concerning the classes of (finitely) algebraizable and (finitely) equivalential logics: All shown inclusions are proper, and the point (of these four) represented graphically as a join is really an intersection.

Equivalentiality and properties of the Leibniz operator

The classes of (finitely) equivalential logics are intermediate between the class of (finitely) algebraizable logics and that of protoalgebraic logics. Their definitions directly involve a definability property of the Leibniz congruence. There are also characterizations in terms of properties of the Leibniz operator as a function from filters to congruences. Recall that the equalities below hold for arbitrary logics only when the homomorphism (or the substitution) is surjective (Proposition 4.25).

THEOREM 6.68
Let $\mathcal{L}$ be a protoalgebraic logic. The following conditions are equivalent:

(i) *$\mathcal{L}$ is equivalential.*

(ii) *For all $\boldsymbol{A}, \boldsymbol{B}$, all $h \in \operatorname{Hom}(\boldsymbol{A}, \boldsymbol{B})$ and all $F \in \mathcal{F}i_{\mathcal{L}}\boldsymbol{B}$, $\boldsymbol{\Omega}^{\boldsymbol{A}} h^{-1} F = h^{-1} \boldsymbol{\Omega}^{\boldsymbol{B}} F$.*

(iii) *For all $\boldsymbol{A}$, all $h \in \operatorname{End}\boldsymbol{A}$ and all $F \in \mathcal{F}i_{\mathcal{L}}\boldsymbol{A}$, $\boldsymbol{\Omega}^{\boldsymbol{A}} h^{-1} F = h^{-1} \boldsymbol{\Omega}^{\boldsymbol{A}} F$.*

(iv) *For all $\Gamma \in Th\mathcal{L}$ and all substitutions σ, $\boldsymbol{\Omega}\sigma^{-1}\Gamma = \sigma^{-1}\boldsymbol{\Omega}\Gamma$.*

PROOF: (i)⇒(ii) This is routine, using the definability of $\boldsymbol{\Omega}^{\boldsymbol{A}}$ by a set of congruence formulas. You will observe that it is the lack of parameters that makes this hold for an arbitrary homomorphism.

(iii) is a particular case of (ii), and (iv) is a particular case of (iii).

(iv)⇒(i) Consider the fundamental set $\Sigma_{\mathcal{L}}$ and take $\Delta(x,y) := \sigma\Sigma_{\mathcal{L}}$, where σ is the substitution that leaves x, y fixed and sends all other variables to x; that is, $\Delta(x,y) := \Sigma_{\mathcal{L}}(x,y,\vec{x})$. By Proposition 6.6.6 $\Delta(x,y) \subseteq \Sigma_{\mathcal{L}}$. Let us show that $x \equiv y \ (\boldsymbol{\Omega} C_{\mathcal{L}}\Delta(x,y))$. We have that $\Sigma_{\mathcal{L}} \subseteq \sigma^{-1}\sigma\Sigma_{\mathcal{L}} = \sigma^{-1}\Delta(x,y) \subseteq \sigma^{-1}C_{\mathcal{L}}\Delta(x,y)$. Since always $x \equiv y \ (\boldsymbol{\Omega}\Sigma_{\mathcal{L}})$, by protoalgebraicity $x \equiv y \ (\boldsymbol{\Omega}\sigma^{-1}C_{\mathcal{L}}\Delta(x,y))$. But the assumption implies that $\boldsymbol{\Omega}\sigma^{-1}C_{\mathcal{L}}\Delta(x,y) = \sigma^{-1}\boldsymbol{\Omega}C_{\mathcal{L}}\Delta(x,y)$, and by construction $\sigma x = x$ and $\sigma y = y$; therefore, $x \equiv y \ (\boldsymbol{\Omega}C_{\mathcal{L}}\Delta(x,y))$. This set Δ thus satisfies the two conditions in Lemma 6.62(ii), which implies that Δ is a set of congruence formulas for $\mathcal{L}$, and hence that $\mathcal{L}$ is equivalential. ⊠

The above properties (ii)–(iv) are described by saying that the Leibniz operator ***commutes with homomorphisms*** (respectively, ***with endomorphisms***, or ***with substitutions***); this terminology is chosen to be consistent with the one used in the algebraizable case, which is justified after Theorem 3.58.

Taking Theorems 6.4 and 6.13 and Proposition 6.64 into account, the previous result gives:

COROLLARY 6.69
For any logic $\mathcal{L}$ the following conditions are equivalent:

(i) *$\mathcal{L}$ is equivalential.*

(ii) *The Leibniz operator $\boldsymbol{\Omega}$, over the theories of $\mathcal{L}$, is monotone and commutes with substitutions.*

(iii) *For every algebra $\boldsymbol{A}$, the Leibniz operator $\boldsymbol{\Omega}^{\boldsymbol{A}}$, over the $\mathcal{L}$-filters of $\boldsymbol{A}$, is monotone and commutes with endomorphisms.* ⊠

Corollary 6.69 establishes the *transfer theorem* of the property of commuting with endomorphisms for protoalgebraic logics. It is *open* whether the transfer holds for arbitrary logics; that is, whether its points (ii) and iii remain equivalent after deleting the assumption of monotonicity they contain. In the case of finitely equivalential logics there is another lattice-theoretic property of the Leibniz operator, namely continuity, that plays a role in locating them in the Leibniz hierarchy. It may be useful to define it with more generality:

DEFINITION 6.70
*Let $\boldsymbol{A}$ be an algebra and $\mathscr{C}$ a closure system on A. A function $\Phi\colon \mathscr{C} \to \mathrm{Con}\boldsymbol{A}$ is **continuous** over $\mathscr{C}$ when $\Phi \bigcup_{i\in I} F_i = \bigcup_{i\in I} \Phi F_i$ for any upwards directed family $\{F_i : i \in I\} \subseteq \mathscr{C}$ such that $\bigcup_{i\in I} F_i \in \mathscr{C}$.*

One is tempted to simply say "Φ commutes with unions of upwards directed families", but one has to limit the property to families whose union belongs to the closure system; if the associated closure operator is finitary, then this condition is automatically satisfied (Theorem 1.42), but not in general.

Observe that continuity implies monotonicity, because if $F \subseteq G$, then $\{F, G\}$ is an upwards directed family and $F \cup G = G$.

THEOREM 6.71
For every logic $\mathcal{L}$ the following conditions are equivalent:

(i) *$\mathcal{L}$ is finitely equivalential.*

(ii) *For every $\boldsymbol{A}$, $\boldsymbol{\Omega}^{\boldsymbol{A}}$ is continuous over $\mathcal{F}i_{\mathcal{L}}\boldsymbol{A}$.*

(iii) *$\boldsymbol{\Omega}$ is continuous over $Th\mathcal{L}$.*

PROOF: (i)⇒(ii) This is routinely checked using the finiteness of the set of congruence formulas.

(ii)⇒(iii) because (iii) is a particular case of (ii).

(iii)⇒(i) Since continuity implies monotonicity, $\mathcal{L}$ is protoalgebraic. Now, show that $\Sigma_{\mathcal{L}} = C_{\mathcal{L}}\Delta(x,y)$ for a finite and parameter-less $\Delta(x,y) \subseteq Fm$. Since $\Sigma_{\mathcal{L}}$ is a theory, by Proposition 1.29 it can be expressed as

$$\Sigma_{\mathcal{L}} = \bigcup\{C_{\mathcal{L}}\Gamma : \Gamma \subseteq \Sigma_{\mathcal{L}}, \Gamma \text{ finite}\}. \tag{88}$$

Finiteness of the Γ and monotonicity of $C_{\mathcal{L}}$ ensure that this is an upwards directed family of theories, and since its union is a theory, the assumption implies that $\boldsymbol{\Omega}\Sigma_{\mathcal{L}} = \bigcup\{\boldsymbol{\Omega}C_{\mathcal{L}}\Gamma : \Gamma \subseteq \Sigma_{\mathcal{L}}, \Gamma \text{ finite}\}$. Since $x \equiv y\ (\boldsymbol{\Omega}\Sigma_{\mathcal{L}})$, there is some finite $\Gamma \subseteq \Sigma_{\mathcal{L}}$ such that $x \equiv y\ (\boldsymbol{\Omega}C_{\mathcal{L}}\Gamma)$. From the result in Proposition 6.6.4 it follows that $\Sigma_{\mathcal{L}} = C_{\mathcal{L}}\Gamma$. The parameters are eliminated as usual: If $\Gamma(x,y,\vec{z})$, then take $\Delta(x,y) := \Gamma(x,y,\vec{x})$. By Proposition 6.6.6, $\Delta(x,y) \subseteq \Sigma_{\mathcal{L}}$ and so $C_{\mathcal{L}}\Delta(x,y) \subseteq \Sigma_{\mathcal{L}}$. In order to show equality, take $\delta(x,y,v_1,\dots,v_m) \in \Sigma_{\mathcal{L}}$ and choose new variables[†] $w_1,\dots,w_m$ not appearing in formulas of Γ. Then $\delta(x,y,w_1,\dots,w_m) \in \Sigma_{\mathcal{L}}$, and

[†] We need to do this: a potentially simpler proof using just structurality does not work, because some of the v_i may coincide with some of the z_j. Notice that the stated choice of the w_k is possible only because of the finiteness of Γ.

so $\Gamma(x,y,\vec{z}) \vdash_{\mathcal{L}} \delta(x,y,\vec{w})$. By structurality, $\Gamma(x,y,\vec{x}) \vdash_{\mathcal{L}} \delta(x,y,\vec{w})$, but the new variables are chosen so that those in $\vec{w}$ can be replaced by those in $\vec{v}$, in order to obtain that $\Gamma(x,y,\vec{x}) \vdash_{\mathcal{L}} \delta(x,y,\vec{v})$. This shows that $\Sigma_{\mathcal{L}} \subseteq C_{\mathcal{L}}\Delta(x,y)$, and hence that $\Sigma_{\mathcal{L}} = C_{\mathcal{L}}\Delta(x,y)$. By Lemma 6.62(iii), $\Delta(x,y)$ is a set of congruence formulas for $\mathcal{L}$; since it is finite, this shows that $\mathcal{L}$ is finitely equivalential. ⊠

Recall that the decomposition (88), in spite of its aspect, is not related to finitarity, but always holds for a theory. Exercise 6.34, also related to this decomposition, contains an interesting, though marginal result.

Model-theoretic characterizations

Recall that $\langle \boldsymbol{B}, G\rangle$ is a submatrix of $\langle \boldsymbol{A}, F\rangle$ when $\boldsymbol{B}$ is a subalgebra of $\boldsymbol{A}$ and $G = F \cap B$, and that for a class M of matrices, the class of all the submatrices of the matrices in M is denoted by $\mathbb{S}\mathsf{M}$. By Proposition 4.67, in general the class $\mathsf{Mod}\mathcal{L}$ is closed under $\mathbb{S}$: if $F \in \mathcal{F}i_{\mathcal{L}}\boldsymbol{A}$ and $\boldsymbol{B}$ is a subalgebra of $\boldsymbol{A}$, then $G = F \cap B \in \mathcal{F}i_{\mathcal{L}}\boldsymbol{B}$. The following is a general property concerning closure of $\mathsf{Mod}^*\mathcal{L}$ under this operator:

LEMMA 6.72
For any logic $\mathcal{L}$, the class $\mathsf{Mod}^*\mathcal{L}$ is closed under $\mathbb{S}$ if and only if for every matrix $\langle \boldsymbol{A}, F\rangle \in \mathsf{Mod}\mathcal{L}$ and every subalgebra $\boldsymbol{B}$ of $\boldsymbol{A}$, $\boldsymbol{\Omega}^{\boldsymbol{B}}(F \cap B) = \boldsymbol{\Omega}^{\boldsymbol{A}}F \cap B^2$.

PROOF: ($\Rightarrow$) Put $\theta := \boldsymbol{\Omega}^{\boldsymbol{A}}F \cap B^2$ and $G := F \cap B$. Then $\theta \in \mathrm{Con}\boldsymbol{B}$ and it is clearly compatible with G, that is, $\theta \subseteq \boldsymbol{\Omega}^{\boldsymbol{B}}G$. To prove the equality, consider the reduction $\pi\colon \langle \boldsymbol{A}, F\rangle \to \langle \boldsymbol{A}/\boldsymbol{\Omega}^{\boldsymbol{A}}F, F/\boldsymbol{\Omega}^{\boldsymbol{A}}F\rangle \in \mathsf{Mod}^*\mathcal{L}$, restrict it to $B \subseteq A$, and take its image: this produces a strict surjective homomorphism $\pi{\restriction}B\colon \langle \boldsymbol{B}, G\rangle \to \langle \boldsymbol{B}/\boldsymbol{\Omega}^{\boldsymbol{A}}F, (F/\boldsymbol{\Omega}^{\boldsymbol{A}}F) \cap (B/\boldsymbol{\Omega}^{\boldsymbol{A}}F)\rangle$. Since the latter is a submatrix of $\langle \boldsymbol{A}/\boldsymbol{\Omega}^{\boldsymbol{A}}F, F/\boldsymbol{\Omega}^{\boldsymbol{A}}F\rangle$, by assumption it is reduced. Since $\ker\pi = \boldsymbol{\Omega}^{\boldsymbol{A}}F$, $\ker(\pi{\restriction}B) = \theta$, and thus $\langle \boldsymbol{B}/\theta, G/\theta\rangle \cong \langle \boldsymbol{B}/\boldsymbol{\Omega}^{\boldsymbol{A}}F, (F/\boldsymbol{\Omega}^{\boldsymbol{A}}F) \cap (B/\boldsymbol{\Omega}^{\boldsymbol{A}}F)\rangle$, which implies that $\langle \boldsymbol{B}/\theta, G/\theta\rangle$ is also reduced. But, since $\theta \subseteq \boldsymbol{\Omega}^{\boldsymbol{B}}G$, $(\boldsymbol{\Omega}^{\boldsymbol{B}}G)/\theta = \boldsymbol{\Omega}^{\boldsymbol{B}/\theta}(G/\theta) = \mathrm{Id}_{B/\theta}$, which implies that $\theta = \boldsymbol{\Omega}^{\boldsymbol{B}}G$.

($\Leftarrow$) If $\boldsymbol{\Omega}^{\boldsymbol{A}}F = \mathrm{Id}_A$, then the assumption implies that $\boldsymbol{\Omega}^{\boldsymbol{B}}(F \cap B) = \mathrm{Id}_B$. ⊠

As a further application of Lemma 6.62, another characterization of equivalentiality is obtained. Its proof introduces the technique of working with a formula algebra generated by a subset of variables (here two), which is a subalgebra of the usual formula algebra. This provides a new way of eliminating the parameters, different from that of applying a substitution, used in several proofs up to now.

THEOREM 6.73
A protoalgebraic logic $\mathcal{L}$ is equivalential if and only if $\mathsf{Mod}^*\mathcal{L}$ is closed under $\mathbb{S}$.

PROOF: ($\Rightarrow$) If $\boldsymbol{\Omega}^{\boldsymbol{A}}F$ is defined by the condition $\Delta^{\boldsymbol{A}}(a,b) \subseteq F$, then clearly the condition in Lemma 6.72 is satisfied, because $\Delta^{\boldsymbol{A}}(a,b) = \Delta^{\boldsymbol{B}}(a,b)$ when $a,b \in B$; thus, $\mathsf{Mod}^*\mathcal{L}$ is closed under $\mathbb{S}$.

($\Leftarrow$) Let $\boldsymbol{Fm}(x,y)$ be the formula algebra generated by just the variables x

and y, with universe $Fm(x,y)$. This is a subalgebra of $\boldsymbol{Fm}$. Let the Leibniz congruence in either algebra be denoted here by $\boldsymbol{\Omega}$. Put $\Delta(x,y) := \Sigma_{\mathcal{L}} \cap Fm(x,y)$ and $\Gamma := C_{\mathcal{L}}\Delta(x,y) \subseteq Fm$. Let us see that $x \equiv y\ (\boldsymbol{\Omega}\Gamma)$. Since $\Sigma_{\mathcal{L}}$ is a theory, $\Delta(x,y) \subseteq \Gamma \subseteq \Sigma_{\mathcal{L}}$, and then $\Delta(x,y) = \Gamma \cap Fm(x,y) = \Sigma_{\mathcal{L}} \cap Fm(x,y)$. Now the assumption is used in the equivalent form given by Lemma 6.72: Since $x \equiv y\ (\boldsymbol{\Omega}\Sigma_{\mathcal{L}})$, also $x \equiv y\ (\boldsymbol{\Omega}\Sigma_{\mathcal{L}} \cap Fm(x,y)^2)$; but

$$\boldsymbol{\Omega}\Sigma_{\mathcal{L}} \cap Fm(x,y)^2 = \boldsymbol{\Omega}(\Sigma_{\mathcal{L}} \cap Fm(x,y)) = \boldsymbol{\Omega}(\Gamma \cap Fm(x,y)) = \boldsymbol{\Omega}\Gamma \cap Fm(x,y)^2,$$

therefore $x \equiv y\ (\boldsymbol{\Omega}\Gamma)$, that is, $x \equiv y\ (\boldsymbol{\Omega}C_{\mathcal{L}}\Delta(x,y))$. Since by construction $\Delta(x,y) \subseteq \Sigma_{\mathcal{L}}$, the two conditions in Lemma 6.62(ii) are satisfied, and hence $\Delta(x,y)$ is a set of congruence formulas for $\mathcal{L}$. Therefore, $\mathcal{L}$ is equivalential. ⊠

By Theorem 6.17, $\mathcal{L}$ is protoalgebraic if and only if $\mathsf{Mod}^*\mathcal{L}$ is closed under $\mathbb{P}_{SD}$; hence in particular it is closed under $\mathbb{P}$. As a consequence:

Corollary 6.74
A logic $\mathcal{L}$ is equivalential if and only if $\mathsf{Mod}^*\mathcal{L}$ is closed under $\mathbb{S}$ and $\mathbb{P}$. ⊠

Corollary 6.75
If a logic $\mathcal{L}$ is equivalential, then $\mathsf{Alg}\mathcal{L} = \mathsf{Alg}^*\mathcal{L}$ is closed under $\mathbb{S}$ and $\mathbb{P}$. ⊠

Therefore, the algebraic counterpart of an arbitrary equivalential logic is an "S-P-class" (see the discussion on page 66). At this point a natural conjecture might be that it is a generalized quasivariety, but this conjecture has been refuted by Moraschini [personal communication], who has defined an equivalential logic whose algebraic counterpart is not closed under $\mathbb{U}$ (in view of Theorem 1.77.6, this was the only way to obtain a counterexample to the conjecture). By contrast, in Corollary 6.80 we see that the algebraic counterpart of a finitary and finitely equivalential logic is indeed a quasivariety; after it, we see why the converse of the implication in Corollary 6.75 does not hold.

The characterizations of this kind are suitable to be falsified (usually in small models), so they can be useful to show that certain logics are *not* equivalential.

Example 6.76 (Classical systems of modal logic)
Let E denote the least classical *system* of modal logic: this is the set of theorems of the logic in the language $\langle \wedge, \vee, \rightarrow, \neg, \Box \rangle$ axiomatized by all non-modal classical tautologies, *Modus Ponens* and the *Extensionality Rule* (ER): $x \leftrightarrow y \rhd \Box x \leftrightarrow \Box y$. The system E is the weakest of the so-called *classical* systems of modal logic, which may be weaker than the normal systems of Example 1.14, but are still treatable with relational semantics (the so-called "neighbourhood models") and with algebraic tools; for an introduction to this family in a more general modal context, see [57, Chapter 8].

Let E^{MP} denote the logic (in our sense) axiomatized by taking all the formulas in E as axioms and *Modus Ponens* as the only rule. Observe that (ER) is not a rule of E^{MP}, that is, not all its theories are closed under all its instances; only its set of theorems (which is the set of formulas E) is guaranteed to be closed under them. Then one can see [70, Section 3.4] that:

1. E^{MP} is protoalgebraic, as the ordinary $\to$ satisfies (R) and (MP).
2. If $\boldsymbol{A}$ is a Boolean algebra expanded with an arbitrary unary operation $\Box$ and F is a lattice filter, then $\langle \boldsymbol{A}, F\rangle \in \mathsf{Mod}E^{\mathrm{MP}}$.
3. $\langle \boldsymbol{A}, F\rangle \in \mathsf{Mod}^*E^{\mathrm{MP}}$ if and only if $\boldsymbol{A}$ is a Boolean algebra with an arbitrary unary operation $\Box$ and F is a filter such that $\{1\}$ is the only filter closed under (ER) included in F.
4. It is easy to see that, as a consequence, $\mathsf{Alg}E^{\mathrm{MP}} = \mathsf{Alg}^*E^{\mathrm{MP}}$ is the class of Boolean algebras expanded with an arbitrary unary operator $\Box$.
5. $\mathsf{Mod}^*E^{\mathrm{MP}}$ is not closed under $\mathbb{S}$. To see this, take $\boldsymbol{A}$ to be the eight-element Boolean algebra with universe $A = \{0, a, b, c, \neg a, \neg b, \neg c, 1\}$, with a, b, c as atoms, expanded with the unary operator $\Box$ defined by: $\Box a = \Box b = \Box c = \Box 0 = 0$, $\Box 1 = \Box\neg a = \Box\neg c = 1$ and $\Box\neg b = b$. Then $F := \{1, \neg a\}$ is a Boolean filter but it is not closed under (ER), because $\neg b \leftrightarrow c = \neg a \in F$ while $\Box\neg b \leftrightarrow \Box c = \neg b \notin F$. Thus, $\{1\}$ is the only filter closed under (ER) contained in F, which shows that $\langle \boldsymbol{A}, F\rangle \in \mathsf{Mod}^*E^{\mathrm{MP}}$. Now consider the subalgebra $\boldsymbol{B}$ of $\boldsymbol{A}$ with universe $\{0, a, \neg a, 1\}$. The submatrix $\langle \boldsymbol{B}, F\rangle$ of $\langle \boldsymbol{A}, F\rangle$ is not reduced, because the set F is closed under (ER) in $\boldsymbol{B}$.
6. It follows from Theorem 6.73 that E^{MP} is not equivalential. For more properties of this logic see Examples 6.122.8 and 7.45.

Applying the above procedure starting from the modal system K instead of E one finds a logic, which here would be denoted by K^{MP}, that coincides with the one we denote by K^{ℓ} (Example 1.14); it is equivalential (Example 6.67.3). ⊠

EXAMPLE 6.77 (DA COSTA'S FAMILY OF PARACONSISTENT LOGICS $\mathcal{C}_n$)
This is a denumerable family of logics belonging to the wide domain of *paraconsistent logics*, recently also called *logics of formal inconsistency*. The unifying feature of all paraconsistent logics is their refusal of the classical dictum *"ex contradictione quodlibet"*, represented by the inference $\varphi, \neg\varphi \vdash \psi$. These are logics "tolerating inconsistency" in the sense that for a certain formula φ, both φ and $\neg\varphi$ may hold in a certain theory but still this theory need not be inconsistent in the usual sense of closure operators (that is, it need not be the total set of formulas).† Clearly, it is the behaviour of their negation that characterizes them. Johansson's *minimal logic*, for instance, has been considered as a paraconsistent counterpart of intuitionistic logic. For more information on this topic, see [52, 53, 198].

The logics $\mathcal{C}_n$, proposed by Da Costa in 1963, were formally introduced in [77]. Their common starting point is a Hilbert-style axiomatics with the axioms of minimal logic for $\wedge, \vee, \to$ (page 84) plus two laws of classical negation (namely,

† In the paraconsistent logic literature this property is often called "to be explosive", while "to be inconsistent" or "contradictory" means that both a formula and its negation hold. Accordingly, paraconsistent logics may be described as those in which being inconsistent (or contradictory) does not imply being explosive.

$x \vee \neg x$ and $\neg\neg x \to x$), and *Modus Ponens* as the sole rule. This defines the weakest logic $\mathcal{C}_\omega$. The others add a few axioms concerning a derived unary connective x° with the intended meaning of "x is consistent". More precisely, the following formula is postulated

$$y^\circ \to \big((x \to y) \to ((x \to \neg y)) \to \neg x\big),$$

with the intended meaning that *Reductio ad Absurdum* holds when the absurdum concerns a consistent statement. Axioms expressing that the set of consistent sentences is closed under all connectives are also added. Several logics are obtained by varying the definition of x°: if one defines $x^{(1)} := \neg(x \wedge \neg x)$ and, recursively, $x^{(n+1)} := x^{(n)} \wedge \neg\big(x^{(n)} \wedge \neg(x^{(n)})\big)$, then for each $n \geqslant 1$ the logic $\mathcal{C}_n$ is obtained by taking $x^\circ := x^{(n)}$. One obtains a denumerable sequence of logics

$$\mathcal{C}_\omega < \cdots < \mathcal{C}_{n+1} < \mathcal{C}_n < \cdots < \mathcal{C}_1 < \mathcal{C}\ell .$$

They are all protoalgebraic, as $\mathcal{C}_\omega$ clearly is. In [163] it is proved that $\mathcal{C}_1$ is not algebraizable, by exhibiting a five-element algebra with three different $\mathcal{C}_1$-filters whose Leibniz congruence is the identity relation. It is easy to extract, from the same example, a reduced model of $\mathcal{C}_1$ with a non-reduced submodel. Thus, $\mathsf{Mod}^*\mathcal{C}_1$ is not closed under $\mathbb{S}$, therefore, by Theorem 6.73, $\mathcal{C}_1$ not equivalential; by Proposition 6.65, none of the other logics is equivalential either. ⊠

Examples 6.78 (Two very weak logics of implication)
The following are examples of rather weak protoalgebraic but non-equivalential logics in the language $\langle \to \rangle$ of type $\langle 2 \rangle$.

1. The logic $\mathcal{I}$, defined by the Hilbert-style system with axiom (R) $x \to x$ and rule (MP) $x\,,x \to y \rhd y$, is described on page 327 as the simplest conceivable protoalgebraic logic. This logic, which is protoalgebraic by definition, is certainly very simple to manipulate, as the elementary properties in Exercise 6.35 witness. This logic has no algebraic semantics, and has other unusual features; for instance in many cases from a set of assumptions you can only deduce the assumptions themselves and (obviously) the theorems, which are just the formulas of the form $\alpha \to \alpha$ for any α. Its Lindenbaum-Tarski algebra $\boldsymbol{Fm}/\widetilde{\Omega}\mathcal{I}$ is (isomorphic to) the formula algebra itself, and hence its intrinsic variety $\mathbb{V}\mathcal{I}$ is the class of all algebras (of this similarity type). By contrast, in [100] it is proved that $\mathsf{Alg}^*\mathcal{I}$ is a smaller class, hence it cannot be a variety. That this logic is not equivalential is shown indirectly, by considering the next example. This logic appears also in Exercise 7.5.
2. The logic $\mathcal{G}_1$ is defined, as in Definition 5.112, by the Gentzen system whose non-structural rules are (MP) of Example 5.120.1 plus the two rules

$$\frac{\varphi \rhd \psi}{\rhd\, \varphi \to \psi}\ (\text{WDT}) \qquad\qquad \frac{\xi\,,\varphi \rhd \psi}{\xi \rhd \varphi \to \psi}\ (\text{DT1}).$$

   Trivially, this logic has all formulas $\alpha \to \alpha$ as theorems, therefore it satisfies (R) and (MP); this implies that $\mathcal{G}_1$ is an extension of $\mathcal{I}$ and is protoalgebraic.

By Exercise 6.35, the logic $\mathcal{I}$ does not satisfy (WDT). Thus, $\mathcal{I} < \mathcal{G}_1$. The logic $\mathcal{G}_1$ is motivated by the issue (discussed on page 175) of Deduction Theorems with side assumptions of limited cardinality, and has been extensively studied in [47]. In this paper a reduced model of $\mathcal{G}_1$ is exhibited that has a non-reduced submatrix, hence by Theorem 6.73 it cannot be equivalential. Therefore, $\mathcal{I}$ cannot be equivalential either.

The above rules also appear in the Gentzen calculi for some subintuitionistic logics (Example 6.16.8) obtained in [54]. ⊠

Now, using typical model-theoretic techniques, one can obtain the following related result, which is given only in sketch; if you have some basic model-theoretic experience, you will have no difficulty in filling the gaps in. The context is that of Section 4.5.

THEOREM 6.79
A logic $\mathcal{L}$ is finitary and finitely equivalential if and only if $\mathsf{Mod}^*\mathcal{L}$ *is closed under* $\mathbb{S}, \mathbb{P}$ *and* $\mathbb{P}_U$.

SKETCH OF THE PROOF: ($\Rightarrow$) By Corollary 6.74 it only remains to prove that $\mathsf{Mod}^*\mathcal{L}$ is closed under ultraproducts. Since $\mathcal{L}$ is finitary, an ultraproduct of a family of reduced models of $\mathcal{L}$ is again a model of $\mathcal{L}$ (Proposition 4.67), and it only remains to prove that it is reduced. For a matrix, to be reduced is to satisfy the implication "$\forall a, b \in A,\ a \equiv b\ (\boldsymbol{\Omega}^{\boldsymbol{A}} F) \Rightarrow a = b$". When $\mathcal{L}$ is finitely equivalential and the matrix is a model of $\mathcal{L}$, the expression "$a \equiv b\ (\boldsymbol{\Omega}^{\boldsymbol{A}} F)$" can be replaced by the conjunction of a finite number of conditions of the form "$\delta^{\boldsymbol{A}}(a,b) \in F$", with δ ranging over the (finite) set of congruence formulas. Therefore, the whole property can be expressed as satisfaction of a first-order formula of the language $\boldsymbol{L}_D$, and hence it is preserved by ultraproducts.

($\Leftarrow$) Since $\mathcal{L}$ is complete with respect to $\mathsf{Mod}^*\mathcal{L}$, that this class is closed under ultraproducts implies that $\mathcal{L}$ is finitary (Theorem 4.68). By Corollary 6.74, that it is closed under $\mathbb{S}$ and $\mathbb{P}$ implies that it is equivalential. It remains to show that it actually has a finite set of congruence formulas. Suppose that it has not one. This means that, if Δ is a set of congruence formulas for $\mathcal{L}$ and I denotes the family of all its finite subsets, then in particular no $i \in I$ is a set of congruence formulas for $\mathcal{L}$. Thus, for each $i \in I$ there is $\langle \boldsymbol{A}_i, F_i \rangle \in \mathsf{Mod}^*\mathcal{L}$ and there are $a_i, b_i \in A_i$ with $a_i \neq b_i$ but such that $i^{\boldsymbol{A}_i}(a_i, b_i) \subseteq F_i$. Then a convenient ultrafilter $\mathcal{U}$ over I is considered; the ultraproduct $\langle \boldsymbol{A}, F \rangle := \left(\prod_{i \in I} \langle \boldsymbol{A}_i, F_i \rangle\right)/\mathcal{U}$ is such that $\Delta^{\boldsymbol{A}}(a,b) \subseteq F$ for $a := \langle a_i : i \in I \rangle/\mathcal{U}$ and $b := \langle b_i : i \in I \rangle/\mathcal{U}$, while $a \neq b$, but this contradicts that $\langle \boldsymbol{A}, F \rangle \in \mathsf{Mod}^*\mathcal{L}$, which is a consequence of the assumptions. ⊠

Observe that in particular this tells us that the class $\mathsf{Mod}^*\mathcal{L}$ is elementary whenever $\mathcal{L}$ is finitary and finitely equivalential; this is the first class of logics for which this property can be guaranteed. A fortiori, we also obtain:

COROLLARY 6.80
If $\mathcal{L}$ is finitary and finitely equivalential, then $\mathsf{Alg}^*\mathcal{L}$ *is a quasivariety.* ⊠

Thus, the algebraic counterpart of a finitary, finitely equivalential logic is always a quasivariety; by contrast, that of an arbitrary equivalential logic need not even be a generalized quasivariety (see the comment after Corollary 6.75). The weak relevance logic $\mathcal{WR}$ of Example 6.16.7 is a counterexample to the converse implications in the two results, even with a stronger assumption: it is a finitary and non-protoalgebraic logic, hence a fortiori non-equivalential, and the class $\mathsf{Alg}^*\mathcal{WR}$ is a variety.

Theorem 6.79 allows us to obtain an answer to a natural question posed on page 229. While Theorem 4.74 shows that if a finitary logic $\mathcal{L}$ is complete with respect to a class of matrices M (which for simplicity is assumed here to contain a trivial matrix), then $\mathsf{Mod}\mathcal{L} = \mathbb{H}_{\mathrm{S}}^{-1}\mathbb{H}_{\mathrm{S}}\mathbb{S}\mathbb{P}_{\mathrm{R}}\mathsf{M} = \mathbb{H}_{\mathrm{S}}^{-1}\mathbb{H}_{\mathrm{S}}\mathbb{S}\mathbb{P}\mathbb{P}_{\mathrm{U}}\mathsf{M}$, there it is asked about similar results characterizing the class of reduced models of a logic out of a class of matrices that generates the logic, and it is said that there are no general results. At this point we can obtain one that holds for a restricted class of equivalential logics.

THEOREM 6.81
Let $\mathcal{L}$ be a finitary and finitely equivalential logic which is complete with respect to a class M of reduced matrices that contains a trivial one. Then $\mathsf{Mod}^*\mathcal{L} = \mathbb{I}\mathbb{S}\mathbb{P}\mathbb{P}_{\mathrm{U}}\mathsf{M}$, and $\mathsf{Alg}^*\mathcal{L} = \mathsf{Alg}\mathcal{L} = \mathbb{Q}\mathsf{Alg}\mathsf{M}$.

PROOF: We have only to show the first equality; the second follows by taking algebraic reducts and recalling Corollary 6.15.

($\supseteq$) Obviously, if $\mathcal{L}$ is complete with respect to M and all these matrices are reduced, $\mathsf{M} \subseteq \mathsf{Mod}^*\mathcal{L}$. By Theorem 6.79, this implies that $\mathbb{I}\mathbb{S}\mathbb{P}\mathbb{P}_{\mathrm{U}}\mathsf{M} \subseteq \mathsf{Mod}^*\mathcal{L}$.

($\subseteq$) If $\mathcal{M} \in \mathsf{Mod}^*\mathcal{L}$, then $\mathcal{M} \in \mathsf{Mod}\mathcal{L}$, so by Theorem 4.74 $\mathcal{M} \in \mathbb{H}_{\mathrm{S}}^{-1}\mathbb{H}_{\mathrm{S}}\mathbb{S}\mathbb{P}\mathbb{P}_{\mathrm{U}}\mathsf{M}$, and thus $\mathcal{M} \in \mathbb{H}_{\mathrm{S}}^{-1}\mathbb{H}_{\mathrm{S}}\mathcal{N}$ for some $\mathcal{N} \in \mathbb{S}\mathbb{P}\mathbb{P}_{\mathrm{U}}\mathsf{M}$. The first point, by Proposition 4.32.3, implies that $\mathcal{M}^* \cong \mathcal{N}^*$. But $\mathbb{S}\mathbb{P}\mathbb{P}_{\mathrm{U}}\mathsf{M} \subseteq \mathsf{Mod}^*\mathcal{L}$ by the part ($\supseteq$) of the proof. In particular, $\mathcal{N}$ is reduced, and since $\mathcal{M}$ is reduced as well, it follows that $\mathcal{M} \cong \mathcal{N}$, and hence that $\mathcal{M} \in \mathbb{I}\mathbb{S}\mathbb{P}\mathbb{P}_{\mathrm{U}}\mathsf{M}$. ⊠

Recall that, by Proposition 5.79, under the same assumptions, $\mathbb{V}\mathcal{L} = \mathbb{V}\mathsf{Alg}\mathsf{M}$.

Relation with relative equational consequences

As in the case of algebraizable logics, the presence of a set $\Delta(x, y)$ in at most two variables prompts for the consideration of the associated *structural transformer* $\rho\colon \mathscr{P}Eq \to \mathscr{P}Fm$ (recall that $Eq := Fm \times Fm$ is the set of equations of the language $\boldsymbol{L}$), defined as follows.

$$\rho(\alpha \approx \beta) := \Delta(\alpha, \beta) \qquad \text{for every } \alpha \approx \beta \in Eq$$

$$\rho\Theta := \bigcup_{\alpha \approx \beta \in \Theta} \rho(\alpha \approx \beta) \qquad \text{for every } \Theta \subseteq Eq$$

The full force of these transformers as establishing relations between the logic $\mathcal{L}$ and the relative equational consequence $\vDash_{\mathsf{K}}$ of a certain class of algebras K is attained at the level of algebraizable logics; however, already at lower levels of the hierarchy there are some interesting properties. Recall from Definition 3.4 (applied to the reverse transformer) that a structural transformer ρ, or a set Δ, is said to be an ***interpretation*** of the consequence $\vDash_{\mathsf{K}}$ ***into*** the consequence $\vdash_{\mathcal{L}}$ when, for all $\Theta \cup \{\varepsilon \approx \delta\} \subseteq Eq$,

$$\Theta \vDash_{\mathsf{K}} \varepsilon \approx \delta \iff \rho\Theta \vdash_{\mathcal{L}} \rho(\varepsilon \approx \delta). \tag{ALG2}$$

LEMMA 6.82
Let $\mathcal{L}$ be a logic, and let $\Delta(x,y) \subseteq Fm$ satisfy (MP) *for $\mathcal{L}$. The set Δ satisfies* (R) *and* (Re) *for $\mathcal{L}$ if and only if there is a class K of algebras such that Δ interprets $\vDash_{\mathsf{K}}$ into $\vdash_{\mathcal{L}}$. When it exists, the class K can be chosen to be $\mathsf{Alg}^*\mathcal{L}$.*

PROOF: ($\Rightarrow$) By Theorem 6.60, Δ is a set of congruence formulas for it, and by Corollary 6.56, it defines the identity relation in reduced models of $\mathcal{L}$. Then, the completeness of $\mathcal{L}$ with respect to the class $\mathsf{Mod}^*\mathcal{L}$ proves that Δ defines the wanted interpretation, for $\mathsf{K} := \mathsf{Alg}^*\mathcal{L}$:

$$\begin{aligned}
\Theta \vDash_{\mathsf{Alg}^*\mathcal{L}} \varepsilon \approx \delta \iff & \text{For all } \boldsymbol{A} \in \mathsf{Alg}^*\mathcal{L} \text{ and all } h \in \mathrm{Hom}(\boldsymbol{Fm}, \boldsymbol{A}), \\
& h\alpha = h\beta \ \forall \alpha \approx \beta \in \Theta \text{ implies } h\varepsilon = h\delta \\
\iff & \text{For all } \langle \boldsymbol{A}, F\rangle \in \mathsf{Mod}^*\mathcal{L} \text{ and all } h \in \mathrm{Hom}(\boldsymbol{Fm}, \boldsymbol{A}), \\
& h\alpha \equiv h\beta \ (\boldsymbol{\Omega}^{\boldsymbol{A}}F) \ \forall \alpha \approx \beta \in \Theta \text{ implies } h\varepsilon \equiv h\delta \ (\boldsymbol{\Omega}^{\boldsymbol{A}}F) \\
\iff & \text{For all } \langle \boldsymbol{A}, F\rangle \in \mathsf{Mod}^*\mathcal{L} \text{ and all } h \in \mathrm{Hom}(\boldsymbol{Fm}, \boldsymbol{A}), \\
& \Delta^{\boldsymbol{A}}(h\alpha, h\beta) \subseteq F \ \forall \alpha \approx \beta \in \Theta \text{ implies } \Delta^{\boldsymbol{A}}(h\varepsilon, h\delta) \subseteq F \\
\iff & \text{For all } \langle \boldsymbol{A}, F\rangle \in \mathsf{Mod}^*\mathcal{L} \text{ and all } h \in \mathrm{Hom}(\boldsymbol{Fm}, \boldsymbol{A}), \\
& h \bigcup_{\alpha \approx \beta \in \Theta} \Delta(\alpha, \beta) \subseteq F \text{ implies } h\Delta(\varepsilon, \delta) \subseteq F \\
\iff & \text{For all } \langle \boldsymbol{A}, F\rangle \in \mathsf{Mod}^*\mathcal{L} \text{ and all } h \in \mathrm{Hom}(\boldsymbol{Fm}, \boldsymbol{A}), \\
& h\rho\Theta \subseteq F \text{ implies } h\rho(\varepsilon \approx \delta) \subseteq F \\
\iff & \rho\Theta \vdash_{\mathcal{L}} \rho(\varepsilon \approx \delta).
\end{aligned}$$

The mentioned completeness is applied in the last step.

($\Leftarrow$) As in the proof of Proposition 3.13, observe that the conditions (R) and (Re) for Δ are the ρ-translations of the properties (Reflexivity) and (Congruence) on page 61, which hold in every relative equational consequence $\vDash_{\mathsf{K}}$. By (ALG2), the translated conditions hold for $\mathcal{L}$. ⊠

COROLLARY 6.83

1. *A set $\Delta(x,y) \subseteq Fm$ is a set of congruence formulas for a logic $\mathcal{L}$ if and only if $\mathcal{L}$ satisfies* (MP) *for Δ and Δ interprets $\vDash_{\mathsf{K}}$ into $\vdash_{\mathcal{L}}$ for some class K of algebras.*
2. *A logic $\mathcal{L}$ is equivalential if and only if there are a set $\Delta(x,y) \subseteq Fm$ and a class K of algebras such that $\mathcal{L}$ satisfies* (MP) *for Δ and Δ interprets $\vDash_{\mathsf{K}}$ into $\vdash_{\mathcal{L}}$.*

In both cases the class K can be chosen to be $\mathsf{Alg}^*\mathcal{L}$.

In fact, these two results can be strengthened: only one of the implications in (ALG2) is required for them to hold (Exercise 6.37).

There are versions of Lemma 6.82 and of Corollary 6.83.1 with parameters everywhere, just as Exercise 6.28 establishes a parametrised version of Theorem 6.60. The transformer defined by Δ is $\varphi \approx \psi \mapsto \Delta\langle\varphi, \psi\rangle$; of course this is never finite, which makes it lose most of its interest as a syntactic transformation between consequences. In view of Theorem 6.57, as a corollary one would obtain *another characterization of protoalgebraicity*, parallel to Corollary 6.83.2, that is, in terms of parametrised versions of (MP) and a notion of a *parametrised interpretation* of $\vDash_{\mathsf{Alg}^*\mathcal{L}}$ (or of some $\vDash_{\mathsf{K}}$) into $\vdash_{\mathcal{L}}$. See [72] and Theorem 6.121.

Exercises for Section 6.3

Exercise 6.24. By working out simple counterexamples of your own, show that the set $\Delta\langle\alpha, \beta\rangle$ is *not* the result of substituting α for x and β for y in the set $\Delta\langle x, y\rangle$; this set is defined by $\Delta\langle x, y\rangle := \bigcup\{\Delta(x, y, \vec{\delta}) : \vec{\delta} \in \vec{Fm}\}$. See also Exercise 6.29.

Exercise 6.25. Prove the property in Lemma 6.55.

Exercise 6.26. Complete the inductive argument in step (c) of the proof of Theorem 6.60.

Exercise 6.27. Show that if Δ and Δ' are sets of congruence formulas with parameters for the same logic $\mathcal{L}$, then they are equivalent modulo $\mathcal{L}$ in the sense that $\Delta\langle x, y\rangle \dashv\vdash_{\mathcal{L}} \Delta'\langle x, y\rangle$. Show that, conversely, if Δ and Δ' are two sets of formulas satisfying this condition, then one of them is a set of congruence formulas with parameters for $\mathcal{L}$ if and only if the other one is.

Exercise 6.28. Take the proof of Theorem 6.60 as a model to find a proof of the following Theorem: A set $\Delta(x, y, \vec{z})$ is a set of congruence formulas with parameters for a logic $\mathcal{L}$ if and only if the following conditions are satisfied (obviously, the subscript P means "parametrised version"):

$(\mathrm{R_P})$ $\quad \vdash_{\mathcal{L}} \Delta\langle x, x\rangle$

$(\mathrm{MP_P})$ $\quad x\,,\ \Delta\langle x, y\rangle \vdash_{\mathcal{L}} y$

$(\mathrm{Re_P})$ $\quad \bigcup_{i=1}^{n} \Delta\langle x_i, y_i\rangle \vdash_{\mathcal{L}} \Delta\langle \lambda x_1 \dots x_n, \lambda y_1 \dots y_n\rangle \quad$ for each $\lambda \in L$, $n = \mathrm{ar}\lambda \geqslant 1$

As a consequence, the following conditions hold as well:

$(\mathrm{Sym_P})$ $\quad \Delta\langle x, y\rangle \vdash_{\mathcal{L}} \Delta\langle y, x\rangle$

$(\mathrm{Trans_P})$ $\quad \Delta\langle x, y\rangle \cup \Delta\langle y, z\rangle \vdash_{\mathcal{L}} \Delta\langle x, z\rangle$

> Comment. Notice that, by structurality, condition $(\mathrm{R_P})$ may be replaced by the weaker $\vdash_{\mathcal{L}} \Delta(x, x, \vec{z})$, and condition $(\mathrm{MP_P})$ by $x,\ \Delta(x, y, \vec{z}) \vdash_{\mathcal{L}} y$, but, by contrast, a similar simplification is not available for $(\mathrm{Re_P})$.

Exercise 6.29. Let $\Delta(x, y, \vec{z}) \subseteq Fm$ be a set satisfying the conditions $(\mathrm{R_P})$, $(\mathrm{MP_P})$ and $(\mathrm{Re_P})$ of Exercise 6.28. Prove that it also satisfies the same conditions for arbitrary formulas instead of variables, that is:

$(\mathrm{R_P})'$ $\quad \vdash_{\mathcal{L}} \Delta\langle \alpha, \alpha\rangle$

$(\mathrm{MP_P})'$ $\quad \alpha\,,\ \Delta\langle \alpha, \beta\rangle \vdash_{\mathcal{L}} \beta$

$(\mathrm{Re_P})'$ $\quad \bigcup_{i=1}^{n} \Delta\langle \alpha_i, \beta_i\rangle \vdash_{\mathcal{L}} \Delta\langle \lambda \alpha_1 \dots \alpha_n, \lambda \beta_1 \dots \beta_n\rangle \quad$ for each $\lambda \in L$, $n = \mathrm{ar}\lambda \geqslant 1$

COMMENT. This is not as straightforward as it may seem. As discussed in Exercise 6.24, the set $\Delta\langle\alpha,\beta\rangle$ is *not* the result of replacing x by α and y by β in $\Delta\langle x,y\rangle$. Therefore, that $\vdash_{\mathcal{L}} \Delta\langle\alpha,\alpha\rangle$ is not the straightforward result of applying structurality to $\vdash_{\mathcal{L}} \Delta\langle x,x\rangle$. And so on.

EXERCISE 6.30. Prove that Δ is a set of congruence formulas with parameters for a logic $\mathcal{L}$ if and only if for any theory Γ of $\mathcal{L}$ and any $\alpha,\beta \in Fm$, $\alpha \equiv \beta\ (\boldsymbol{\Omega}\Gamma)$ if and only if $\Delta\langle\alpha,\beta\rangle \subseteq \Gamma$. When coupled with Definition 6.54, this establishes the *transfer theorem* of the property that Leibniz congruences are definable (with or without parameters, because the result still holds if the set Δ has no parameters).

HINT. The direct implication is a particular case of Definition 6.54. The converse implication follows easily thanks to Exercise 6.28.

EXERCISE 6.31. Let $\mathcal{L}$ be a protoalgebraic logic with protoimplication set $\Delta(x,y)$. Consider the sets

$$\Delta'(x,y,\vec{z}) := \bigcup\{\Delta(\varphi(x,\vec{z}),\varphi(y,\vec{z})) : \varphi(x,\vec{z}) \in Fm\}$$

$$\Delta''(x,y,\vec{z}) := \bigcup\{\Delta(\varphi(x,\vec{z}),\varphi(y,\vec{z})) \cup \Delta(\varphi(y,\vec{z}),\varphi(x,\vec{z})) : \varphi(x,\vec{z}) \in Fm\}.$$

Prove that the sets Δ' and Δ'' are sets of congruence formulas with parameters for $\mathcal{L}$.

HINT. Observe that $\Delta \subseteq \Delta' \subseteq \Delta'' \subseteq \Sigma_{\mathcal{L}}$. Show first that Δ'' is a set of congruence formulas with parameters for $\mathcal{L}$, using Exercise 6.30. Then show that $\Delta' \dashv\vdash_{\mathcal{L}} \Delta''$ (this is slightly tricky) and use Exercise 6.27.

EXERCISE 6.32. Let $\mathcal{L}$ be a protoalgebraic logic, $\Gamma \subseteq Fm$ and σ a substitution. Use a set of congruence formulas with parameters for $\mathcal{L}$ to show that, if σ is surjective or Γ is finite, then $\sigma\boldsymbol{\Omega}C_{\mathcal{L}}\Gamma \subseteq \boldsymbol{\Omega}C_{\mathcal{L}}\sigma\Gamma$. Show that if $\mathcal{L}$ is equivalential, then it satisfies this property for all substitutions and all Γ.

COMMENT. In some cases (as in Exercise 6.35) this property can be used to disprove that a particular protoalgebraic logic is equivalential. Exercises 6.50 and 7.34 contain generalizations of these properties.

EXERCISE 6.33. Let $\mathcal{L}$ be a logic in the language $\boldsymbol{L} = \langle\to\rangle$ of type $\langle 2\rangle$ that satisfies (MP) for $\to$ and has the following formulas as theorems:

$$x \to x$$

$$(x \to y) \to ((y \to z) \to (x \to z))$$

$$(x \to ((y \to z) \to t)) \to ((y \to z) \to (x \to t))$$

Show that $\mathcal{L}$ is finitely equivalential.

EXERCISE 6.34. Looking at the proof of Theorem 6.71, you realize that its key point to obtain a finite set of congruence formulas is the decomposition (88) of the set $\Sigma_{\mathcal{L}}$, justified in Proposition 1.29. Now you look at this proposition and you see that the following similar decomposition is also true: $\Sigma_{\mathcal{L}} = \bigcup\{C_{\mathcal{L}}\{\gamma\} : \gamma \in \Sigma_{\mathcal{L}}\}$. Check that using this one instead, and making some indispensable alterations to the proof of Theorem 6.71, the following is obtained:

THEOREM. For every logic $\mathcal{L}$ the following conditions are equivalent:

(i) $\mathcal{L}$ is finitely equivalential and has a unitary set of equivalence formulas.

(ii) For every $\boldsymbol{A}$, $\boldsymbol{\Omega}^{\boldsymbol{A}}$ commutes with unions of arbitrary families over $\mathcal{F}i_{\mathcal{L}}\boldsymbol{A}$ in the sense that if $\{F_i : i \in I\} \subseteq \mathcal{F}i_{\mathcal{L}}\boldsymbol{A}$ and $\bigcup_{i\in I} F_i \in \mathcal{F}i_{\mathcal{L}}\boldsymbol{A}$, then $\boldsymbol{\Omega}^{\boldsymbol{A}} \bigcup_{i\in I} F_i = \bigcup_{i\in I} \boldsymbol{\Omega}^{\boldsymbol{A}} F_i$.

(iii) $\boldsymbol{\Omega}$ commutes with unions of arbitrary families over $Th\mathcal{L}$ in the analogous sense.

EXERCISE 6.35. Consider the logic $\mathcal{I}$ defined in Example 6.78.1.

1. Prove that its theorems are exactly all formulas of the form $\alpha \to \alpha$ for any $\alpha \in Fm$.
2. Deduce that $\mathcal{I}$ does not satisfy the rule (WDT) of Example 6.78.2, and therefore $\mathcal{I} < \mathcal{G}_1$.
3. Find two of its sets of congruence formulas with parameters: its fundamental set $\Sigma_{\mathcal{I}}$ and the simpler set Δ' of Exercise 6.31. Realize that all variables occur in these sets.
4. Show that if $\Gamma \vdash_{\mathcal{I}} \beta$ and β is not a theorem, then β must be a subformula of some formula in Γ, and hence each variable of β must occur in some formula in Γ.
5. Show that if Γ is a set of formulas where not all variables occur, then $\boldsymbol{\Omega} C_{\mathcal{I}} \Gamma$ is the identity relation. Deduce that $\boldsymbol{Fm} \in \mathsf{Alg}^*\mathcal{I}$ and that $\mathbb{V}\mathcal{I}$ is the class of all algebras of the type.
6. Use Exercise 6.32 (and some previous points) to show directly that $\mathcal{I}$ is not equivalential.
7. Use Exercise 4.16 (and several of the previous points) to show that $\mathcal{I}$ does not have an algebraic semantics.

EXERCISE 6.36. Check all details in the proof of Theorem 6.79.

EXERCISE 6.37. By careful inspection of the proof of Lemma 6.82, show that it is enough to require, in it and in Corollary 6.83, that the set Δ satisfies the implication ($\Rightarrow$) of the condition (ALG2) on page 368.

COMMENT. A transformer satisfying this implication has received a variety of names in the literature: "semi-interpretation", "representation", "translation" (when the equivalence is called "conservative translation"), and even "interpretation" (when the equivalence is called "faithful interpretation").

EXERCISE 6.38. Given a set $\Delta(x,y) \subseteq Fm$ and a matrix $\langle \boldsymbol{A}, F \rangle$, consider the relation $R^{\boldsymbol{A}}(F,\Delta)$ from Definition 4.52.

1. Prove that, for a logic $\mathcal{L}$, this relation is an order relation in every $\langle \boldsymbol{A}, F \rangle \in \mathsf{Mod}^*\mathcal{L}$ if and only if $\mathcal{L}$ satisfies conditions (R) and (Trans) and
$$\Delta(x,y) \cup \Delta(y,x) \cup \{\delta(x,\vec{z})\} \vdash_{\mathcal{L}} \delta(y,\vec{z})$$
for every $\delta(w,\vec{z}) \in Fm$ where x and y do not appear.
2. Assume that the conditions in point 1 hold. Prove that for every $\langle \boldsymbol{A}, F \rangle \in \mathsf{Mod}\mathcal{L}$, the Leibniz congruence of the matrix is the symmetrization of the relation $R^{\boldsymbol{A}}(F,\Delta)$, that is, that for any $a,b \in A$, $a \equiv b$ $(\boldsymbol{\Omega}^{\boldsymbol{A}} F)$ if and only if $\Delta^{\boldsymbol{A}}(a,b) \cup \Delta^{\boldsymbol{A}}(b,a) \subseteq F$, and deduce that $\mathcal{L}$ is equivalential.

6.4 Definability of truth (truth-equational, assertional and weakly algebraizable logics)

In a matrix $\langle \boldsymbol{A}, F \rangle$, the designated set F is often referred to as the "truth filter" of the matrix, for obvious reasons (recall the discussion on algebra-based semantics for logics preserving truth on pages 22–25). In the best situation for an algebra-based semantics, the truth filters of the reduced models of a logic should be determined by its algebraic reduct alone. An extreme example of this situation is found in the regularly algebraizable logics (which include the implicative logics): by Proposition 4.58, the truth filter of their reduced models has the form $\{1\}$

where 1 is an algebraic constant of the relevant class of algebras. In general, algebraizable logics show a less impressive, but still very strong behaviour: By Proposition 4.57, the truth filter in their reduced models is the solution set of some equations in one variable, the defining equations.

More in general, one can ask about the bare fact that the algebraic structure determines the truth filter and whether this has some significance in abstract algebraic logic. These considerations, strictly speaking, originate five classes of logics: two already known ones, namely the logics with an algebraic semantics (Definition 3.4) and the assertional logics (Definition 3.5), and three new ones, the truth-equational, the weakly algebraizable, and the regularly weakly algebraizable. In this section the new ones are introduced, the main ones (the truth-equational and the weakly algebraizable) are studied in some depth, and the other three just in passing.

Notice that *protoalgebraicity is not assumed* in the first half of this section, but only from page 387 onwards.

NOTATION.
Given a set $E(x) \subseteq Eq$ of equations in at most one variable x, the structural transformer it induces is in general denoted by τ; that is, $\tau\varphi := E(\varphi)$ for all $\varphi \in Fm$. It is convenient to set the notation

$$\begin{aligned} Sol^{\boldsymbol{A}}_{\tau} &:= \{a \in A : \boldsymbol{A} \vDash E(x)\,[\![a]\!]\} \\ &= \{a \in A : \varepsilon^{\boldsymbol{A}}(a) = \delta^{\boldsymbol{A}}(a) \ \text{ for all } \varepsilon \approx \delta \in E(x)\} \subseteq A \end{aligned}$$

for the set of *solutions* of the equations defining τ in the algebra $\boldsymbol{A}$.† Moreover, recall the following notations set up on page 146: for each $a \in A$ and each $F \subseteq A$ we put

$$\begin{aligned} \tau^{\boldsymbol{A}} a &:= \{\langle \varepsilon^{\boldsymbol{A}}(a), \delta^{\boldsymbol{A}}(a)\rangle : \varepsilon \approx \delta \in E(x)\} \subseteq A \times A \\ \tau^{\boldsymbol{A}} F &:= \bigcup_{a \in F} \tau^{\boldsymbol{A}} a = \{\langle \varepsilon^{\boldsymbol{A}}(a), \delta^{\boldsymbol{A}}(a)\rangle : \varepsilon \approx \delta \in E(x), a \in F\} \subseteq A \times A. \end{aligned}$$

It is easy to see that, using the last notations, $Sol^{\boldsymbol{A}}_{\tau} = \{a \in A : \tau^{\boldsymbol{A}} a \subseteq \mathrm{Id}_A\}$. Notice that, in particular, $\tau^{\boldsymbol{A}}\, Sol^{\boldsymbol{A}}_{\tau} \subseteq \mathrm{Id}_A$.

The following easy observation (Exercise 6.39) is also helpful:

LEMMA 6.84
Given a structural transformer τ and a class of algebras K, consider the class of matrices $\mathsf{M} := \{\langle \boldsymbol{A}, Sol^{\boldsymbol{A}}_{\tau}\rangle : \boldsymbol{A} \in \mathsf{K}\}$. Then

$$\Gamma \vdash_{\mathsf{M}} \varphi \iff \tau\Gamma \vDash_{\mathsf{K}} \tau\varphi,$$

for all $\Gamma \cup \{\varphi\} \subseteq Fm$. ⊠

† In the notation of Section 3.5, this would be written as $\tau^{\boldsymbol{A}^{-1}}\mathrm{Id}_A$. There is no uniform notation for this set in the literature; it appears as $\tau^{\boldsymbol{A}}/\mathrm{Id}_A$ in [37], as $Eq^{\boldsymbol{A}}\tau$ in [201], and as $F^{\varepsilon\approx\delta}_{\boldsymbol{A}}$ in [32] for the simpler case where $E(x) = \{\varepsilon \approx \delta\}$.

Implicit and explicit (equational) definitions of truth

DEFINITION 6.85
Let M *be a class of matrices.*

1. ***Truth is implicitly definable*** *in* M *when the matrices in* M *are uniquely determined by their algebraic reducts; that is, when the assumption that* $\langle \boldsymbol{A}, F\rangle, \langle \boldsymbol{A}, G\rangle \in \mathsf{M}$ *implies that* $F = G$.
2. ***Truth is equationally definable*** *in* M *by* $E(x)$ *(or by the associated* $\boldsymbol{\tau}$*) when for every* $\langle \boldsymbol{A}, F\rangle \in \mathsf{M}$, $F = Sol_{\boldsymbol{\tau}}^{\boldsymbol{A}}$. *In this case, the equations in* $E(x)$ *are referred to as the* ***defining equations****.*

An alternative, equivalent terminology is to speak of the ***implicit/equational definability of filters****.*

Equational definability is a restricted form of the usual first-order notion of *explicit definability* present in model theory (applicable here if the matrices are viewed as relational structures as in Section 4.5). Obviously, equational definability implies implicit definability. Very recently, several new kinds of definability of truth, intermediate between these two or close to them, have been considered by Moraschini in [180].

When speaking about a logic, these issues make sense only if predicated of some class of matrices related to the logic. It is not difficult to see that truth is implicitly definable in the whole of $\mathsf{Mod}\mathcal{L}$ only when $\mathcal{L}$ is inconsistent (Exercise 6.40). Thus, $\mathsf{Mod}^*\mathcal{L}$ seems to be the most natural choice to look at. Trivially, in this case the property can be equivalently formulated as follows:

COROLLARY 6.86
Truth is implicitly definable in $\mathsf{Mod}^*\mathcal{L}$ *if and only if for every* $\boldsymbol{A} \in \mathsf{Alg}^*\mathcal{L}$ *there is exactly one* $F \in \mathcal{F}i_{\mathcal{L}}\boldsymbol{A}$ *such that* $\langle \boldsymbol{A}, F\rangle \in \mathsf{Mod}^*\mathcal{L}$. ⊠

It is also interesting to observe:

PROPOSITION 6.87
If truth is implicitly definable in $\mathsf{Mod}^*\mathcal{L}$*, then* $\mathcal{L}$ *has theorems.*

PROOF: If $\mathcal{L}$ has no theorems, then on the trivial algebra, which belongs to $\mathsf{Alg}^*\mathcal{L}$, there are two distinct $\mathcal{L}$-filters making the resulting matrix a reduced model of $\mathcal{L}$: the empty set and the total subset. This contradicts Corollary 6.86. ⊠

It turns out that implicit definability of truth in this class is equivalent to a nice property of the Leibniz operator:

PROPOSITION 6.88
Truth is implicitly definable in $\mathsf{Mod}^*\mathcal{L}$ *if and only if for every algebra* $\boldsymbol{A}$*, the Leibniz operator* $\boldsymbol{\Omega}^{\boldsymbol{A}}$ *is injective over* $\mathcal{F}i_{\mathcal{L}}\boldsymbol{A}$.

PROOF: Assume that truth is implicitly definable in $\mathsf{Mod}^*\mathcal{L}$, and let $F, G \in \mathcal{F}i_{\mathcal{L}}\boldsymbol{A}$ with $\boldsymbol{\Omega}^{\boldsymbol{A}}F = \boldsymbol{\Omega}^{\boldsymbol{A}}G$, for an arbitrary $\boldsymbol{A}$. Then the reductions of $\langle \boldsymbol{A}, F\rangle$ and $\langle \boldsymbol{A}, G\rangle$ are reduced models of $\mathcal{L}$ on the same algebra $\boldsymbol{A}/\boldsymbol{\Omega}^{\boldsymbol{A}}F = \boldsymbol{A}/\boldsymbol{\Omega}^{\boldsymbol{A}}G$, hence by

Corollary 6.86 $F/\boldsymbol{\Omega}^{\boldsymbol{A}}F = G/\boldsymbol{\Omega}^{\boldsymbol{A}}G$. Since each filter is the saturation of its own reduction and the congruence is the same, this implies that $F = G$ (this may be viewed as an application of Exercise 4.14), that is, $\boldsymbol{\Omega}^{\boldsymbol{A}}$ is injective. The converse implication is obvious. ⊠

In a similar way one can prove (Exercise 6.41) that for an arbitrary logic $\mathcal{L}$, truth is implicitly definable in the class of its Lindenbaum-Tarski models $\mathsf{LTMod}^*\mathcal{L}$ if and only if the Leibniz operator $\boldsymbol{\Omega}$ is injective over $\mathcal{T}h\mathcal{L}$, the theories of the logic. Moraschini has recently proved in [180] that this property implies the implicit definability of truth in $\mathsf{Mod}^*\mathcal{L}$ (that is, that injectivity of the Leibniz operator *transfers* from the theories of the logic to its filters of arbitrary algebras), but only for logics on countable languages, while the transfer may fail in general. As a matter of fact, it does not seem that injectivity alone plays a particular role in the Leibniz hierarchy; however, it does when also monotonicity holds, as in Corollary 6.114 and Theorem 6.120.

Now let us leave implicit definability and let us look at equational definability. In its full generality, this property turns out to be related to an already known notion. Notice that truth is equationally definable by $\boldsymbol{\tau}$ in a class M of matrices if and only if M is of the form $\{\langle \boldsymbol{A}, Sol^{\boldsymbol{A}}_{\boldsymbol{\tau}}\rangle : \boldsymbol{A} \in \mathsf{K}\}$ for some class K of algebras, namely $\mathsf{K} := \mathsf{Alg}\mathsf{M}$. Then, a straightforward consequence of Lemma 6.84 (see Exercise 6.39) is:

COROLLARY 6.89
A logic has an algebraic semantics if and only if it is complete with respect to a class of matrices in which truth is equationally definable. ⊠

As shown in Section 3.1, algebraic semantics for a logic $\mathcal{L}$, even the largest one $\mathsf{K}(\mathcal{L},\boldsymbol{\tau})$ for a given interpretation $\boldsymbol{\tau}$, can be quite weird. The same can be said about the resulting class of matrices. It is thus natural to single out the logics that have the property for a natural class of matrices.

Truth-equational logics

DEFINITION 6.90
A logic $\mathcal{L}$ is **truth-equational** *when truth is equationally definable in* $\mathsf{Mod}^*\mathcal{L}$.

The notion of a truth-equational logic is due to Raftery, who introduced† and studied it in [201]. He proved that these logics need not be protoalgebraic, but can be characterized by properties of the Leibniz operator (Theorem 6.101), and hence deserve a natural place in the Leibniz hierarchy.‡

Simply reformulating the definition we obtain the following useful characterization, and its obvious consequence:

† The original definition is that a logic is truth-equational when truth is equationally definable in its Lindenbaum-Tarski models $\mathsf{LTMod}^*\mathcal{L}$; Theorem 6.96 shows that this amounts to the more intuitive definition given here.

‡ This is why the term "Leibniz hierarchy" is preferred to "protoalgebraic hierarchy", used in some early papers.

LEMMA 6.91
Let $\mathcal{L}$ be a logic, and $\boldsymbol{\tau}$ a structural transformer defined by a set $E(x) \subseteq Eq$.

1. *$\mathcal{L}$ is truth-equational with defining equations $E(x)$ if and only if* $\mathsf{Mod}^*\mathcal{L} = \{\langle \boldsymbol{A}, Sol^{\boldsymbol{A}}_{\boldsymbol{\tau}}\rangle : \boldsymbol{A} \in \mathsf{Alg}^*\mathcal{L}\}$.
2. *If $\mathcal{L}$ is truth-equational with defining equations $E(x)$, then for every algebra $\boldsymbol{A}$ it holds that $\boldsymbol{A} \in \mathsf{Alg}^*\mathcal{L}$ if and only if $\langle \boldsymbol{A}, Sol^{\boldsymbol{A}}_{\boldsymbol{\tau}}\rangle \in \mathsf{Mod}^*\mathcal{L}$.*

All algebraizable logics are truth-equational; this is a consequence of Proposition 4.57, and thus provides a first large class of examples (Theorem 6.109 provides a second, also large class). It follows directly from the definition that any extension of a truth-equational logic is also truth-equational (because if $\mathcal{L} \leqslant \mathcal{L}'$, then $\mathsf{Mod}^*\mathcal{L}' \subseteq \mathsf{Mod}^*\mathcal{L}$). Moreover, from Proposition 6.87 and Corollary 6.89 (plus equational definability implying implicit definability) it follows at once:

COROLLARY 6.92
If $\mathcal{L}$ is a truth-equational logic, then it has theorems, it has an algebraic semantics, namely the class $\mathsf{Alg}^*\mathcal{L}$, and for every $\boldsymbol{A}$, the Leibniz operator is injective on $\mathcal{F}i_{\mathcal{L}}\boldsymbol{A}$. ⊠

This may help to quickly classify some logics as *non*-truth-equational. Logics without theorems are automatically excluded, and failure of injectivity of the Leibniz operator in some small model of a logic may also witness its non-truth-equationality. Notice, however, that the converses of neither of the three implications contained in Corollary 6.92, nor of that in Corollary 6.91.2, hold:

EXAMPLES 6.93
1. There are plenty of logics with theorems that are not truth-equational. For instance, all equivalential logics† that are not algebraizable, such as those in Example 6.67; by Theorem 6.127 they are not truth-equational. Non-protoalgebraic examples of non-truth-equational logics with theorems are positive modal logic $\mathcal{PML}$ (Example 6.16.5) and two of the local, non-protoalgebraic subintuitionistic logics (Example 6.16.8); they are shown to be non-truth-equational in [4, 6].
2. There are logics that have an algebraic semantics but are not truth-equational, and indeed some have the class $\mathsf{Alg}^*\mathcal{L}$ as an algebraic semantics; that is, even if a logic is complete with respect to a class of matrices of the form $\{\langle \boldsymbol{A}, Sol^{\boldsymbol{A}}_{\boldsymbol{\tau}}\rangle : \boldsymbol{A} \in \mathsf{Alg}^*\mathcal{L}\}$ for some transformer $\boldsymbol{\tau}$, this class need not be equal to $\mathsf{Mod}^*\mathcal{L}$. One example is the logic introduced in Example 3.9. There we saw that it has the class of all algebras of the type as an algebraic semantics. In [200, p. 117] it is shown that it also has the classes $\mathsf{Alg}\mathcal{L}$ and $\mathsf{Alg}^*\mathcal{L}$ as algebraic semantics. Since this logic has no theorems, by Corollary 6.92 it cannot be truth-equational. The lack of theorems implies that the logic is not protoalgebraic either (another reason is that it has no connectives of arity 2 or greater; see Proposition 6.11).

† Excluding the almost inconsitent logic.

3. As witnessed by Example 6.102, injectivity is not enough to make a logic truth-equational. Theorem 6.101 shows that the property of the Leibniz operator that places truth-equational logics in the hierarchy is a stronger one: to be completely order-reflecting. ⊠

The logic in Example 6.102 does have an algebraic semantics and an injective Leibniz operator, while it fails to be truth-equational for less trivial reasons. However, under restricted conditions, having an algebraic semantics may imply truth-equationality; Exercises 6.42 and 6.43 show some of them.

Other examples of properly truth-equational logics are best considered in the context of weakly algebraizable logics; see Examples 6.122.11–12.

The notion of a truth-equational logic is initially defined by a condition on the reduced models of the logic, but there is a nice equivalent way in terms of all its models, which turns out to be a very useful tool:

PROPOSITION 6.94
Let $\mathcal{L}$ be a logic, and let $\langle \boldsymbol{A}, F \rangle$ be any matrix.

1. *A structural transformer $\boldsymbol{\tau}$ defines truth in the matrix $\langle \boldsymbol{A}/\boldsymbol{\Omega}^{\boldsymbol{A}}F, F/\boldsymbol{\Omega}^{\boldsymbol{A}}F \rangle$ if and only if $F = \{a \in A : \boldsymbol{\tau}^{\boldsymbol{A}}a \subseteq \boldsymbol{\Omega}^{\boldsymbol{A}}F\}$.*
2. *A structural transformer $\boldsymbol{\tau}$ defines truth in $\mathsf{Mod}^*\mathcal{L}$ if and only if for every $\boldsymbol{A}$ and every $F \in \mathcal{F}i_{\mathcal{L}}\boldsymbol{A}$ it holds that $F = \{a \in A : \boldsymbol{\tau}^{\boldsymbol{A}}a \subseteq \boldsymbol{\Omega}^{\boldsymbol{A}}F\}$.*

Thus, $\mathcal{L}$ is truth-equational if and only if there is a structural transformer $\boldsymbol{\tau}$ such that for every $\langle \boldsymbol{A}, F \rangle \in \mathsf{Mod}\mathcal{L}$, $F = \{a \in A : \boldsymbol{\tau}^{\boldsymbol{A}}a \subseteq \boldsymbol{\Omega}^{\boldsymbol{A}}F\}$.

PROOF: 1 is left to you as Exercise 6.44, while 2 is an obvious consequence of 1, and the final statement is a reformulation of the definition using 2. ⊠

Thus, *a logic is truth-equational if and only if each of its filters is definable from its Leibniz congruence through a set of equations.* This property, viewed as the equivalence "$a \in F \iff \boldsymbol{\tau}^{\boldsymbol{A}}a \subseteq \boldsymbol{\Omega}^{\boldsymbol{A}}F$" and compared with the equivalence "$a \equiv b\ (\boldsymbol{\Omega}^{\boldsymbol{A}}F) \iff \Delta^{\boldsymbol{A}}(a,b) \subseteq F$" of (87) on page 353, makes truth-equational logics appear as a sort of *dual notion* to that of equivalential logics.

In Lemma 6.91 we see that if $\mathcal{L}$ is truth-equational, then the set $Sol_{\boldsymbol{\tau}}^{\boldsymbol{A}}$ is an $\mathcal{L}$-filter in algebras of $\mathsf{Alg}^*\mathcal{L}$. This fact, however, does not characterize this class of algebras, but the largest $\boldsymbol{\tau}$-algebraic semantics of $\mathcal{L}$; see Exercise 6.45. Anyway, as a first application of Proposition 6.94, we see that in reduced models of a truth-equational logic, the set $Sol_{\boldsymbol{\tau}}^{\boldsymbol{A}}$ is not just the unique $\mathcal{L}$-filter making the model reduced: it is the smallest $\mathcal{L}$-filter of the algebra.

PROPOSITION 6.95
If $\mathcal{L}$ is a truth-equational logic, then for all $\boldsymbol{A} \in \mathsf{Alg}^*\mathcal{L}$, $Sol_{\boldsymbol{\tau}}^{\boldsymbol{A}} = \min \mathcal{F}i_{\mathcal{L}}\boldsymbol{A}$.

PROOF: Let $a \in Sol_{\boldsymbol{\tau}}^{\boldsymbol{A}}$, that is, $\boldsymbol{\tau}^{\boldsymbol{A}}a \subseteq \mathrm{Id}_A$. Since $\mathrm{Id}_A \subseteq \boldsymbol{\Omega}^{\boldsymbol{A}}F$ for every $F \in \mathcal{F}i_{\mathcal{L}}\boldsymbol{A}$, using Proposition 6.94 we see that $a \in \bigcap \mathcal{F}i_{\mathcal{L}}\boldsymbol{A} = \min \mathcal{F}i_{\mathcal{L}}\boldsymbol{A}$. Conversely, if $a \in \min \mathcal{F}i_{\mathcal{L}}\boldsymbol{A}$, in particular $a \in Sol_{\boldsymbol{\tau}}^{\boldsymbol{A}}$, which by truth-equationality is one of the filters of the logic. ⊠

Observe that the inclusion $Sol_{\boldsymbol{\tau}}^{\boldsymbol{A}} \subseteq \min \mathcal{F}i_{\mathcal{L}}\boldsymbol{A}$ holds in arbitrary algebras, as it is proved without using that $\boldsymbol{A} \in \mathsf{Alg}^*\mathcal{L}$. Exercise 6.46 asks you to prove a refinement of this property, and the extension of the property in the proposition to all algebras in $\mathsf{Alg}\,\mathcal{L}$.

The second application uses a typical technique of abstract algebraic logic, already introduced in Section 4.3: the work with Lindenbaum-Tarski models (Definition 4.42). The inescapable fact that by definition they are countably generated makes its proof slightly complicated.

Theorem 6.96
A logic $\mathcal{L}$ is truth-equational if and only if truth is equationally definable in the class of matrices $\mathsf{LTMod}^*\mathcal{L}$.

Proof: In one direction there is nothing to prove, because $\mathsf{LTMod}^*\mathcal{L} \subseteq \mathsf{Mod}^*\mathcal{L}$. So, let $\boldsymbol{\tau}$ be a structural transformer that defines truth in $\mathsf{LTMod}^*\mathcal{L}$. This property is clearly preserved by isomorphisms, therefore by Theorem 4.44, $\boldsymbol{\tau}$ defines truth in the class of all countably generated reduced models of $\mathcal{L}$. In order to show that $\boldsymbol{\tau}$ defines truth in $\mathsf{Mod}^*\mathcal{L}$, we take any algebra $\boldsymbol{A}$ and any $F \in \mathcal{F}i_{\mathcal{L}}\boldsymbol{A}$, and prove the property in Proposition 6.94. Assume first that $\boldsymbol{\tau}^{\boldsymbol{A}}a \subseteq \boldsymbol{\Omega}^{\boldsymbol{A}}F$, consider the subalgebra $\boldsymbol{B}$ of $\boldsymbol{A}$ generated by a, and $G := F \cap B$. Clearly $G \in \mathcal{F}i_{\mathcal{L}}\boldsymbol{B}$ and $\boldsymbol{\tau}^{\boldsymbol{B}}a = \boldsymbol{\tau}^{\boldsymbol{A}}a \subseteq (\boldsymbol{\Omega}^{\boldsymbol{A}}F) \cap B^2 \subseteq \boldsymbol{\Omega}^{\boldsymbol{B}}G$; the last inclusion holds because $(\boldsymbol{\Omega}^{\boldsymbol{A}}F) \cap B^2$ is a congruence of $\boldsymbol{B}$ that is clearly compatible with G. Since by assumption $\boldsymbol{\tau}$ defines truth in the matrix $\langle \boldsymbol{B}/\boldsymbol{\Omega}^{\boldsymbol{B}}G, G/\boldsymbol{\Omega}^{\boldsymbol{B}}G \rangle$, which is reduced and countably generated, from Proposition 6.94.1 it follows that $a \in G$ and hence that $a \in F$. Now assume that $a \in F$, and take any $\alpha \approx \beta \in E(x)$, the set of equations that define $\boldsymbol{\tau}$. We show that $\alpha^{\boldsymbol{A}}(a) \equiv \beta^{\boldsymbol{A}}(a)\ (\boldsymbol{\Omega}^{\boldsymbol{A}}F)$ by using Theorem 4.23, so take any $\delta(x,\vec{z}) \in Fm$ and any $\vec{c} \in \vec{A}$ and assume that $\delta^{\boldsymbol{A}}(\alpha^{\boldsymbol{A}}(a),\vec{c}) \in F$. Let $\boldsymbol{B}$ be the subalgebra of $\boldsymbol{A}$ generated by the set $\{a, c_1, \dots, c_k\}$, assuming that the $\vec{z}$ in $\delta(x,\vec{z})$ has length k. Now consider $G := F \cap B$; then $a \in G$. Using the assumption and Proposition 6.94.1 as before, it follows that $\boldsymbol{\tau}^{\boldsymbol{B}}a \subseteq \boldsymbol{\Omega}^{\boldsymbol{B}}G$. This implies that $\alpha^{\boldsymbol{B}}(a) \equiv \beta^{\boldsymbol{B}}(a)\ (\boldsymbol{\Omega}^{\boldsymbol{B}}G)$, and since $\delta^{\boldsymbol{B}}(\alpha^{\boldsymbol{B}}(a),\vec{c}) = \delta^{\boldsymbol{A}}(\alpha^{\boldsymbol{A}}(a),\vec{c}) \in G$, by compatibility it follows that $\delta^{\boldsymbol{B}}(\beta^{\boldsymbol{B}}(a),\vec{c}) \in G$, that is, $\delta^{\boldsymbol{A}}(\beta^{\boldsymbol{A}}(a),\vec{c}) \in F$, as is to be proved. ⊠

To go on we need a minimal knowledge of the Suszko operator. Recall from Definition 5.36 and the ensuing discussion that on any algebra $\boldsymbol{A}$ *the Suszko operator relative to a logic* $\mathcal{L}$ is the function $\widetilde{\boldsymbol{\Omega}}{}_{\mathcal{L}}^{\boldsymbol{A}} \colon \mathcal{F}i_{\mathcal{L}}\boldsymbol{A} \to \mathrm{Con}\boldsymbol{A}$ that sends every $F \in \mathcal{F}i_{\mathcal{L}}\boldsymbol{A}$ to the congruence $\widetilde{\boldsymbol{\Omega}}{}_{\mathcal{L}}^{\boldsymbol{A}}F := \bigcap\{\boldsymbol{\Omega}^{\boldsymbol{A}}G : G \in \mathcal{F}i_{\mathcal{L}}\boldsymbol{A}, F \subseteq G\}$. Naturally, a model† $\langle \boldsymbol{A}, F \rangle$ of $\mathcal{L}$ is ***Suszko-reduced*** when its Suszko congruence $\widetilde{\boldsymbol{\Omega}}{}_{\mathcal{L}}^{\boldsymbol{A}}F$ is the identity relation.

Here are some further properties; the first four follow immediately from the corresponding definitions.

† Observe that an absolute notion of an arbitrary matrix being "Suszko-reduced" would make little sense, as the operator is relative to a logic $\mathcal{L}$; therefore the notion is given only for models of $\mathcal{L}$, that is, for matrices in which the filter is an $\mathcal{L}$-filter.

PROPOSITION 6.97
For any logic $\mathcal{L}$, any algebra $\boldsymbol{A}$ and any $F \in \mathcal{F}i_{\mathcal{L}}\boldsymbol{A}$:

1. $\widetilde{\Omega}_{\mathcal{L}}^{\boldsymbol{A}} F = \widetilde{\Omega}^{\boldsymbol{A}}(\mathcal{F}i_{\mathcal{L}}\boldsymbol{A})^F$.
2. *The function $\widetilde{\Omega}_{\mathcal{L}}^{\boldsymbol{A}}$ is monotone over $\mathcal{F}i_{\mathcal{L}}\boldsymbol{A}$.*
3. *For any $a, b \in A$ the following conditions are equivalent:*
   (i) $a \equiv b\ (\widetilde{\Omega}_{\mathcal{L}}^{\boldsymbol{A}} F)$.
   (ii) *For all $\delta(x, \vec{z}) \in Fm$ and all $\vec{c} \in \vec{A}$, $Fg_{\mathcal{L}}^{\boldsymbol{A}}(F, \delta^{\boldsymbol{A}}(a, \vec{c})) = Fg_{\mathcal{L}}^{\boldsymbol{A}}(F, \delta^{\boldsymbol{A}}(b, \vec{c}))$.*
   (iii) *For all $\delta(x, \vec{z}) \in Fm$, all $\vec{c} \in \vec{A}$ and all $G \in \mathcal{F}i_{\mathcal{L}}\boldsymbol{A}$ with $F \subseteq G$, $\left[\delta^{\boldsymbol{A}}(a, \vec{c}) \in G\right.$ if and only if $\left.\delta^{\boldsymbol{A}}(b, \vec{c}) \in G\right]$.*
4. *The Suszko congruence $\widetilde{\Omega}_{\mathcal{L}}^{\boldsymbol{A}} F$ is compatible with all $G \in \mathcal{F}i_{\mathcal{L}}\boldsymbol{A}$ such that $F \subseteq G$. Hence for each of these the canonical projection $\langle \boldsymbol{A}, G\rangle \longrightarrow \langle \boldsymbol{A}/\widetilde{\Omega}_{\mathcal{L}}^{\boldsymbol{A}} F, G/\widetilde{\Omega}_{\mathcal{L}}^{\boldsymbol{A}} F\rangle$ is a strict surjective homomorphism.*
5. *The matrix $\langle \boldsymbol{A}/\widetilde{\Omega}_{\mathcal{L}}^{\boldsymbol{A}} F, F/\widetilde{\Omega}_{\mathcal{L}}^{\boldsymbol{A}} F\rangle$ is Suszko-reduced.* (See Exercise 6.48)
6. $\mathsf{Alg}\mathcal{L}$ *is the class of algebraic reducts of the Suszko-reduced models of $\mathcal{L}$.* (See Exercise 6.49) ☒

All these facts make good support for the statement that the Suszko operator in arbitrary logics has a similar role in some respects to that of the Leibniz operator in protoalgebraic logics. Recall that a logic is protoalgebraic if and only if the Leibniz and the Suszko operators coincide on its models; however, it is still not determined how much of the theory of protoalgebraic logics might admit a more general development in terms of the Suszko operator instead of the Leibniz operator. The property in point 6 is one of the main reasons why basing the algebraic study of an arbitrary logic on the Suszko operator may be regarded as a natural alternative to doing it via generalized models, as developed in Chapter 5. In particular, from points 5 and 6 it follows that $\mathsf{Alg}\mathcal{L}$ contains all quotients $\boldsymbol{Fm}/\widetilde{\Omega}_{\mathcal{L}}\Gamma$ for $\Gamma \in \mathcal{T}h\mathcal{L}$; these algebras could also be legitimately considered as the "Lindenbaum-Tarski algebras" of the logic $\mathcal{L}$. Anyway, the first point already makes it clear that g-matrices and the Tarski operator are actually *hidden* but *present* when working with the Suszko operator.

At this point the only really new property of the Suszko operator we need is the following:

LEMMA 6.98
Let $\mathcal{L}$ be any logic. For any substitution σ, $\sigma\widetilde{\Omega}_{\mathcal{L}}C_{\mathcal{L}}\{x\} \subseteq \widetilde{\Omega}_{\mathcal{L}}C_{\mathcal{L}}\{\sigma x\}$.

PROOF: Let $\alpha \equiv \beta\ (\widetilde{\Omega}_{\mathcal{L}}C_{\mathcal{L}}\{x\})$. By the first characterization in Proposition 6.97.3, in the particular case of the formula algebra, we have that for any $\psi(x, \vec{z}) \in Fm$, $x, \psi(\alpha, \vec{z}) \vdash_{\mathcal{L}} \psi(\beta, \vec{z})$ and $x, \psi(\beta, \vec{z}) \vdash_{\mathcal{L}} \psi(\alpha, \vec{z})$. By structurality this implies that for any substitution σ',

$$\sigma' x, \psi(\sigma'\alpha, \sigma'\vec{z}) \vdash_{\mathcal{L}} \psi(\sigma'\beta, \sigma'\vec{z}) \quad \text{and} \quad \sigma' x, \psi(\sigma'\beta, \sigma'\vec{z}) \vdash_{\mathcal{L}} \psi(\sigma'\alpha, \sigma'\vec{z}). \tag{89}$$

In order to show that $\sigma\alpha \equiv \sigma\beta \ (\widetilde{\Omega}_{\mathcal{L}} C_{\mathcal{L}}\{\sigma x\})$, by the same characterization, we need to show that for any $\varphi(x, \vec{u}) \in Fm$,

$$\sigma x, \varphi(\sigma\alpha, \vec{u}) \vdash_{\mathcal{L}} \varphi(\sigma\beta, \vec{u}) \quad \text{and} \quad \sigma x, \varphi(\sigma\beta, \vec{u}) \vdash_{\mathcal{L}} \varphi(\sigma\alpha, \vec{u}). \tag{90}$$

It is left to you as Exercise 6.50 to check that, given σ and φ, it is possible to find σ' and ψ so that (90) can be obtained from (89). This concludes the proof. ☒

Next the order-theoretic property for the Leibniz operator associated with truth-equational logics is introduced:

Definition 6.99

Let $\boldsymbol{A}$ *be an algebra. The Leibniz operator* $\boldsymbol{\Omega}^{\boldsymbol{A}}$ *is* **order-reflecting** *over* $\mathcal{F}i_{\mathcal{L}}\boldsymbol{A}$ *when for any* $F, G \in \mathcal{F}i_{\mathcal{L}}\boldsymbol{A}$, *if* $\boldsymbol{\Omega}^{\boldsymbol{A}}F \subseteq \boldsymbol{\Omega}^{\boldsymbol{A}}G$, *then* $F \subseteq G$. *The operator is* **completely order-reflecting** *over* $\mathcal{F}i_{\mathcal{L}}\boldsymbol{A}$ *when for all* $\mathcal{F} \cup \{G\} \subseteq \mathcal{F}i_{\mathcal{L}}\boldsymbol{A}$, *if* $\bigcap_{F\in\mathcal{F}} \boldsymbol{\Omega}^{\boldsymbol{A}}F \subseteq \boldsymbol{\Omega}^{\boldsymbol{A}}G$, *then* $\bigcap \mathcal{F} \subseteq G$.

These are particular cases of obvious general lattice-theoretic notions, and hence apply also to the Suszko operator. Notice that a completely order-reflecting operator is also order-reflecting, and in turn this property implies it is *injective* over its domain. The alternative expression in the following lemma is more technical than intuitive, but helpful; its proof is left to you as Exercise 6.51.

Lemma 6.100

For an algebra $\boldsymbol{A}$, *the Leibniz operator* $\boldsymbol{\Omega}^{\boldsymbol{A}}$ *is completely order-reflecting over* $\mathcal{F}i_{\mathcal{L}}\boldsymbol{A}$ *if and only if for all* $F, G \in \mathcal{F}i_{\mathcal{L}}\boldsymbol{A}$, *if* $\widetilde{\boldsymbol{\Omega}}^{\boldsymbol{A}}_{\mathcal{L}}F \subseteq \boldsymbol{\Omega}^{\boldsymbol{A}}G$, *then* $F \subseteq G$. ☒

Theorem 6.101

For any logic $\mathcal{L}$ *the following conditions are equivalent:*

(i) *$\mathcal{L}$ is truth-equational.*

(ii) *The Leibniz operator is completely order-reflecting over its filters of arbitrary algebras.*

(iii) *The Leibniz operator is completely order-reflecting over its theories.*

Proof: (i)⇒(ii) Assume that $\mathcal{F} \cup \{G\} \subseteq \mathcal{F}i_{\mathcal{L}}\boldsymbol{A}$ with $\bigcap_{F\in\mathcal{F}} \boldsymbol{\Omega}^{\boldsymbol{A}}F \subseteq \boldsymbol{\Omega}^{\boldsymbol{A}}G$, and let $a \in \bigcap \mathcal{F}$. Then, from the assumption and Proposition 6.94 it follows that $\boldsymbol{\tau}^{\boldsymbol{A}}a \subseteq \boldsymbol{\Omega}^{\boldsymbol{A}}F$ for each $F \in \mathcal{F}$, therefore $\boldsymbol{\tau}^{\boldsymbol{A}}a \subseteq \boldsymbol{\Omega}^{\boldsymbol{A}}G$. Again by the same proposition, this amounts to $a \in G$.

(ii)⇒(iii) as a particular case.

(iii)⇒(i) Let σ_x be the substitution that sends every variable to x, and consider the structural transformer $\boldsymbol{\tau}$ defined by the set $E(x) := \sigma_x \widetilde{\boldsymbol{\Omega}}_{\mathcal{L}} C_{\mathcal{L}}\{x\}$; this is a set of equations (pairs of formulas) in at most the variable x, therefore $\boldsymbol{\tau}x = \sigma_x \widetilde{\boldsymbol{\Omega}}_{\mathcal{L}} C_{\mathcal{L}}\{x\}$. By Theorem 6.96, it is enough to show that $\boldsymbol{\tau}$ defines truth in $\mathsf{LTMod}^*\mathcal{L} = \{\langle \boldsymbol{Fm}/\boldsymbol{\Omega}\Gamma, \Gamma/\boldsymbol{\Omega}\Gamma\rangle : \Gamma \in Th\mathcal{L}\}$; by Proposition 6.94.1, this amounts to showing that $\Gamma = \{\varphi \in Fm : \boldsymbol{\tau}\varphi \subseteq \boldsymbol{\Omega}\Gamma\}$ for any theory Γ of $\mathcal{L}$. Take any $\varphi \in \Gamma$ and consider any substitution σ_0 such that $\sigma_0 x := \varphi$; observe that also

$\varphi = \sigma_0\sigma_x x$. Now we can apply Lemma 6.98 to the substitution $\sigma_0\sigma_x$ and obtain

$$\boldsymbol{\tau}\varphi = \boldsymbol{\tau}\sigma_0 x = \sigma_0 \boldsymbol{\tau} x = \sigma_0\sigma_x \widetilde{\boldsymbol{\Omega}}_{\mathcal{L}} C_{\mathcal{L}}\{x\} \subseteq \widetilde{\boldsymbol{\Omega}}_{\mathcal{L}} C_{\mathcal{L}} \sigma_0\sigma_x\{x\} = \\ = \widetilde{\boldsymbol{\Omega}}_{\mathcal{L}} C_{\mathcal{L}}\{\varphi\} \subseteq \widetilde{\boldsymbol{\Omega}}_{\mathcal{L}} \Gamma \subseteq \boldsymbol{\Omega}\Gamma.$$

For the converse, let $\varphi \in Fm$ be such that $\boldsymbol{\tau}\varphi \subseteq \boldsymbol{\Omega}\Gamma$, and consider the substitution σ_0 again. Since $\sigma_0\sigma_x \widetilde{\boldsymbol{\Omega}}_{\mathcal{L}} C_{\mathcal{L}}\{x\} = \sigma_0 \boldsymbol{\tau} x = \boldsymbol{\tau}\sigma_0 x = \boldsymbol{\tau}\varphi \subseteq \boldsymbol{\Omega}\Gamma$, we have that $\widetilde{\boldsymbol{\Omega}}_{\mathcal{L}} C_{\mathcal{L}}\{x\} \subseteq (\sigma_0\sigma_x)^{-1}\boldsymbol{\Omega}\Gamma \subseteq \boldsymbol{\Omega}(\sigma_0\sigma_x)^{-1}\Gamma$; the last inclusion follows from Proposition 4.25. Since $(\sigma_0\sigma_x)^{-1}\Gamma$ is a theory, the assumption that $\boldsymbol{\Omega}$ is completely order-reflecting on the theories of $\mathcal{L}$, in the form of Lemma 6.100, implies that $C_{\mathcal{L}}\{x\} \subseteq (\sigma_0\sigma_x)^{-1}\Gamma$, which amounts to $\varphi = \sigma_0\sigma_x x \in \Gamma$, the desired result. ⊠

This confirms that the class of truth-equational logics belongs naturally to the Leibniz hierarchy. Its location as shown in Figure 9 on page 316 is justified by several examples and results in this section.

Theorem 6.101 (containing both a bridge and a transfer theorem) is the kind of result that allows one to *disprove* that a given logic is truth-equational, similar to the results on monotonicity for protoalgebraic logics, and on continuity for finitely equivalential logics, to mention but a few. Besides, it makes possible to show that injectivity of the Leibniz operator on the filters of a logic does not imply its truth-equationality:

EXAMPLE 6.102
Consider the language $\langle \Box, \Diamond, \top \rangle$ of type $\langle 1,1,0 \rangle$, and the logic $\mathcal{L}$ on this language defined by the axioms $\top$, $\Box\Box x$, $\Diamond\Diamond x$, and the three rules $\Box\top \rhd x$, $\Box x \rhd \Diamond x$ and $\Diamond x \rhd \Box x$. The following points are proved in [201, Example 2].

1. $\mathcal{L}$ has an algebraic semantics, with $\Box x \approx \Diamond x$ as defining equation. By contrast, the fragments of $\mathcal{L}$ with $\Box$ and with $\Box$ and $\top$ do not:[†] it is easy to check that the $\Box$ fragment does not satisfy the condition in Example 3.9, while the case of the $\langle \Box, \top \rangle$ fragment is trickier.
2. Filters are first-order definable in $\mathsf{Mod}^*\mathcal{L}$ by a formula of $\boldsymbol{L_D}$, the disjunction $(x \approx \Box\Box x) \vee (x \approx \top)$. This follows from the fact that for all $\langle \boldsymbol{A}, F \rangle \in \mathsf{Mod}\mathcal{L}$ and all $a \in A$, $a \in F \iff \big[a \equiv \Box\Box a \ (\boldsymbol{\Omega}^{\boldsymbol{A}}F)$ or $a \equiv \top^{\boldsymbol{A}} \ (\boldsymbol{\Omega}^{\boldsymbol{A}}F)\big]$. This implies that truth is implicitly definable in $\mathsf{Mod}^*\mathcal{L}$, and hence, by Proposition 6.88, that $\boldsymbol{\Omega}^{\boldsymbol{A}}$ is injective over its filters of any $\boldsymbol{A}$.
3. $\mathcal{L}$ is not truth-equational, that is, truth is not equationally definable in $\mathsf{Mod}^*\mathcal{L}$. The proof consists in showing that $\boldsymbol{\Omega}$ is not completely order-reflecting over the filters of $\mathcal{L}$ on the algebra $\boldsymbol{Fm}(x)$, by considering some suitable theories, and using Theorem 6.101.
4. The class $\mathsf{Mod}^*\mathcal{L}$ is an elementary class. This is because for all $\langle \boldsymbol{A}, F \rangle \in \mathsf{Mod}\mathcal{L}$ and all $a, b \in A$, $a \equiv b \ (\boldsymbol{\Omega}^{\boldsymbol{A}}F) \iff \big[a \in F \Leftrightarrow b \in F\big]$ and $\big[\Box a \in F \Leftrightarrow \Box b \in F\big]$. Since $\mathcal{L}$ is finitary, by Theorem 4.70 the class $\mathsf{Mod}\mathcal{L}$ is elementary,

[†] Hence, the $\Diamond$ operator is not dispensable in $\mathcal{L}$, in spite of the fact that $\Box x \dashv\vdash_{\mathcal{L}} \Diamond x$.

and adding the sentence $\forall x \forall y ((Dx \leftrightarrow Dy) \wedge (D\Box x \leftrightarrow D\Box y) \rightarrow x \approx y)$ one obtains $\mathsf{Mod}^*\mathcal{L}$. This fact is relevant in the digression after Theorem 6.113.

Most of the proofs are of a combinatorial character, taking advantage of the particularly simple structure of the terms of this language: a sequence of operators $\Box$ and $\Diamond$ followed by a single variable or by $\top$. For instance, it is easy to show that the theorems of $\mathcal{L}$ are $\top$ and all the formulas of "modal depth" at least 2 (the formulas that begin with at least two modal operators). ⊠

The following technical but important general property helps in the proof of two further characterizations of truth-equationality, one in terms of the full g-models of the logic and the other in terms of the Suszko operator.

PROPOSITION 6.103
Let $\mathcal{L}$ be any logic and $\boldsymbol{A}$ any algebra. For each $F \in \mathcal{F}i_{\mathcal{L}}\boldsymbol{A}$, the following conditions are equivalent:

(i) *For all $G \in \mathcal{F}i_{\mathcal{L}}\boldsymbol{A}$, if $\widetilde{\Omega}^{\boldsymbol{A}}_{\mathcal{L}}F \subseteq \Omega^{\boldsymbol{A}}G$, then $F \subseteq G$.*

(ii) *The g-matrix $\langle \boldsymbol{A}, (\mathcal{F}i_{\mathcal{L}}\boldsymbol{A})^F \rangle$ is a full g-model of $\mathcal{L}$.*

(iii) *$F/\widetilde{\Omega}^{\boldsymbol{A}}_{\mathcal{L}}F$ is the least $\mathcal{L}$-filter of the quotient algebra $\boldsymbol{A}/\widetilde{\Omega}^{\boldsymbol{A}}_{\mathcal{L}}F$.*

PROOF: (i)⇒(ii) By the characterization of full g-models in Proposition 5.94, and taking into account that $\widetilde{\Omega}^{\boldsymbol{A}}(\mathcal{F}i_{\mathcal{L}}\boldsymbol{A})^F = \widetilde{\Omega}^{\boldsymbol{A}}_{\mathcal{L}}F$, what we have to show amounts to $(\mathcal{F}i_{\mathcal{L}}\boldsymbol{A})^F = \{G \in \mathcal{F}i_{\mathcal{L}}\boldsymbol{A} : \widetilde{\Omega}^{\boldsymbol{A}}_{\mathcal{L}}F \subseteq \Omega^{\boldsymbol{A}}G\}$. The direct inclusion holds by the definition of $\widetilde{\Omega}^{\boldsymbol{A}}_{\mathcal{L}}F$, and the reverse inclusion is exactly the assumption (i).

(ii)⇒(iii) If the g-matrix $\langle \boldsymbol{A}, (\mathcal{F}i_{\mathcal{L}}\boldsymbol{A})^F \rangle$ is full, then its reduction is the g-matrix $\langle \boldsymbol{A}/\widetilde{\Omega}^{\boldsymbol{A}}_{\mathcal{L}}F, \mathcal{F}i_{\mathcal{L}}(\boldsymbol{A}/\widetilde{\Omega}^{\boldsymbol{A}}_{\mathcal{L}}F) \rangle$, and the canonical projection is a strict surjective homomorphism. By Theorem 5.48, it induces an order isomorphism between the associated closure systems $(\mathcal{F}i_{\mathcal{L}}\boldsymbol{A})^F$ and $\mathcal{F}i_{\mathcal{L}}(\boldsymbol{A}/\widetilde{\Omega}^{\boldsymbol{A}}_{\mathcal{L}}F)$; in particular it puts their least elements into correspondence, that is, $F/\widetilde{\Omega}^{\boldsymbol{A}}_{\mathcal{L}}F = \min \mathcal{F}i_{\mathcal{L}}(\boldsymbol{A}/\widetilde{\Omega}^{\boldsymbol{A}}_{\mathcal{L}}F)$ as is to be proved.

(iii)⇒(i) If $G \in \mathcal{F}i_{\mathcal{L}}\boldsymbol{A}$ is such that $\widetilde{\Omega}^{\boldsymbol{A}}_{\mathcal{L}}F \subseteq \Omega^{\boldsymbol{A}}G$, this means that $\widetilde{\Omega}^{\boldsymbol{A}}_{\mathcal{L}}F$ is compatible with G and hence that $G/\widetilde{\Omega}^{\boldsymbol{A}}_{\mathcal{L}}F$ is an $\mathcal{L}$-filter of $\boldsymbol{A}/\widetilde{\Omega}^{\boldsymbol{A}}_{\mathcal{L}}F$. From the assumption it follows that $F/\widetilde{\Omega}^{\boldsymbol{A}}_{\mathcal{L}}F \subseteq G/\widetilde{\Omega}^{\boldsymbol{A}}_{\mathcal{L}}F$. Using again that $\widetilde{\Omega}^{\boldsymbol{A}}_{\mathcal{L}}F$ is compatible with G, this implies that $F \subseteq G$ (Exercise 4.14). ⊠

Now the first announced characterization immediately follows. Notice that it contains a kind of *transfer theorem*.

THEOREM 6.104
For any logic $\mathcal{L}$ the following conditions are equivalent:

(i) *$\mathcal{L}$ is truth-equational.*

(ii) *For every algebra $\boldsymbol{A}$ and every $F \in \mathcal{F}i_{\mathcal{L}}\boldsymbol{A}$, the g-matrix $\langle \boldsymbol{A}, (\mathcal{F}i_{\mathcal{L}}\boldsymbol{A})^F \rangle$ is a full g-model of $\mathcal{L}$.*

(iii) *For every $\Gamma \in \mathcal{T}h\mathcal{L}$, the g-matrix $\langle \boldsymbol{Fm}, (\mathcal{T}h\mathcal{L})^\Gamma \rangle$ is a full g-model of $\mathcal{L}$.*

PROOF: Observe that Lemma 6.100 actually says that for an algebra $\boldsymbol{A}$, the operator $\boldsymbol{\Omega}^{\boldsymbol{A}}$ is completely order-reflecting over $\mathcal{F}i_{\mathcal{L}}\boldsymbol{A}$ if and only if every $\mathcal{L}$-filter F of $\boldsymbol{A}$ satisfies condition (i) in Proposition 6.103. Now, the equivalence between (i) and (ii) follows from this and Theorem 6.101. Checking that the equivalence between (i) and (iii) would follow from the same results in the same way, by just working on the formula algebra, is left as Exercise 6.52 to you. ⊠

In some sense this is *dual* to Theorem 6.39, which asserts that $\mathcal{L}$ is protoalgebraic if and only if the converse property follows: that every full g-model of $\mathcal{L}$ is of the form $\langle \boldsymbol{A}, (\mathcal{F}i_{\mathcal{L}}\boldsymbol{A})^F \rangle$ for some $F \in \mathcal{F}i_{\mathcal{L}}\boldsymbol{A}$ (the parallel property just on the formula algebra also holds, and appears in Exercise 6.20). The following extension of the previous characterization is used in Chapter 7:

PROPOSITION 6.105
A logic $\mathcal{L}$ is truth-equational if and only if for every full g-model $\langle \boldsymbol{A}, \mathscr{C} \rangle$ of $\mathcal{L}$ and every $F \in \mathscr{C}$, the g-matrix $\langle \boldsymbol{A}, \mathscr{C}^F \rangle$ is a full g-model of $\mathcal{L}$.

PROOF: Start by assuming that $\mathcal{L}$ is truth-equational, and let $\langle \boldsymbol{A}, \mathscr{C} \rangle$ be any full g-model of $\mathcal{L}$ and $F \in \mathscr{C}$. We know that the reduction of $\langle \boldsymbol{A}, \mathscr{C} \rangle$ is the basic full g-model $\langle \boldsymbol{A}/\widetilde{\boldsymbol{\Omega}}^{\boldsymbol{A}}\mathscr{C}, \mathcal{F}i_{\mathcal{L}}(\boldsymbol{A}/\widetilde{\boldsymbol{\Omega}}^{\boldsymbol{A}}\mathscr{C}) \rangle$ and that the reduction function is a strict surjective homomorphism between the two g-matrices. As a consequence of Theorem 5.48, it establishes a lattice isomorphism between the closure systems $\mathscr{C}^F$ and $\big(\mathcal{F}i_{\mathcal{L}}(\boldsymbol{A}/\widetilde{\boldsymbol{\Omega}}^{\boldsymbol{A}}\mathscr{C})\big)^{F/\widetilde{\boldsymbol{\Omega}}^{\boldsymbol{A}}\mathscr{C}}$, and hence a strict surjective homomorphism between the corresponding g-matrices. Since $F/\widetilde{\boldsymbol{\Omega}}^{\boldsymbol{A}}\mathscr{C} \in \mathcal{F}i_{\mathcal{L}}(\boldsymbol{A}/\widetilde{\boldsymbol{\Omega}}^{\boldsymbol{A}}\mathscr{C})$, we can apply Theorem 6.104 and conclude that $\langle \boldsymbol{A}/\widetilde{\boldsymbol{\Omega}}^{\boldsymbol{A}}\mathscr{C}, \big(\mathcal{F}i_{\mathcal{L}}(\boldsymbol{A}/\widetilde{\boldsymbol{\Omega}}^{\boldsymbol{A}}\mathscr{C})\big)^{F/\widetilde{\boldsymbol{\Omega}}^{\boldsymbol{A}}\mathscr{C}} \rangle$ is a full g-model of $\mathcal{L}$, hence $\langle \boldsymbol{A}, \mathscr{C}^F \rangle$ is one as well. The converse is trivial because the g-matrix $\langle \boldsymbol{A}, \mathcal{F}i_{\mathcal{L}}\boldsymbol{A} \rangle$ is always full, and so the condition in Theorem 6.104(ii) follows as a particular case from the present one. ⊠

The parallel property limited to full g-models on the formula algebra also holds (Exercise 6.53). Besides its contribution to the characterization of truth-equationality, it is interesting to keep in mind this property: in truth-equational logics, any axiomatic extension of a full g-model is again full; this holds a fortiori in weakly algebraizable and algebraizable logics. The second announced characterization is:

THEOREM 6.106
A logic is truth-equational if and only if the Suszko operator is injective over the set of its filters, for any algebra.

PROOF: Left to you, with detailed hints, as Exercise 6.54. Actually, all the necessary technical work is already done, and you need only connect the properties appearing in Proposition 6.103 with injectivity of the Suszko operator. ⊠

The comparison of this result with Proposition 6.88 is interesting: the two results, taken in conjunction, can be rephrased as saying that truth is equationally (resp., implicitly) definable in $\mathsf{Mod}^*\mathcal{L}$ if and only if for every algebra $\boldsymbol{A}$, the Suszko operator $\widetilde{\boldsymbol{\Omega}}^{\boldsymbol{A}}_{\mathcal{L}}$ (resp., the Leibniz operator $\boldsymbol{\Omega}^{\boldsymbol{A}}$) is injective over $\mathcal{F}i_{\mathcal{L}}\boldsymbol{A}$.

Assertional logics

As said before, having an algebraic semantics does not imply being truth-equational, not even when the algebraic semantics is the class $\mathsf{Alg}^*\mathcal{L}$; however, when the truth definition has a particularly simple form, then this is true. Exploring this issue leads us to an already known class of logics.

Recall from Definition 3.5 the notion of *the assertional logic* $\mathcal{L}_\mathsf{K}^\top$ *of a pointed class* K *of algebras*, with $\top$ denoting a term with at most the variable x that is constant over K: for all $\Gamma \cup \{\varphi\} \subseteq Fm$,

$$\Gamma \vdash_\mathsf{K}^\top \varphi \overset{\text{def}}{\Longleftrightarrow} \{\gamma \approx \top : \gamma \in \Gamma\} \vDash_\mathsf{K} \varphi \approx \top .$$

In other words, the assertional logic of K is the logic that has K as algebraic semantics, with defining equation $x \approx \top$, where $\top$ is the constant term witnessing pointedness of K; an ***assertional logic*** is one of the form $\mathcal{L}_\mathsf{K}^\top$ for some pointed class K of algebras, with constant term $\top$.

The notation $\top^{\boldsymbol{A}}$ denotes the unique value of the term in any $\boldsymbol{A} \in \mathsf{K}$; notice, however, that this notation makes no sense in arbitrary algebras, where $\top$ may not be constant. Since the possibility that x effectively occcurs in $\top$ is not excluded, for some purposes it may be more convenient to write $\top(x)$. The property that this term is constant over K is then expressed by an equation, namely that $\mathsf{K} \vDash \top(x) \approx \top(y)$. Therefore, if a class K is pointed, the variety $\mathbb{V}\mathsf{K}$ it generates is also pointed, and hence so are all its subclasses. This is relevant because of the following first fact about assertional logics:

LEMMA 6.107
If $\mathcal{L}$ is the assertional logic of a pointed class of algebras K, then $\mathbb{V}\mathcal{L} \subseteq \mathbb{V}\mathsf{K}$, and as a consequence the classes $\mathsf{Alg}^*\mathcal{L}$ and $\mathsf{Alg}\,\mathcal{L}$ are also included in $\mathbb{V}\mathsf{K}$; in particular, they are pointed, with the same constant term as K.

PROOF: Observe that $\mathcal{L}$ is the assertional logic of K if and only if it is complete with respect to the class of matrices $\{\langle \boldsymbol{A}, \{\top^{\boldsymbol{A}}\}\rangle : \boldsymbol{A} \in \mathsf{K}\}$. In this situation, Lemma 5.78 tells us directly that $\mathbb{V}\mathcal{L} \subseteq \mathbb{V}\mathsf{K}$. By the previous comment, $\mathbb{V}\mathsf{K}$ is pointed, and since $\mathsf{Alg}^*\mathcal{L} \subseteq \mathsf{Alg}\,\mathcal{L} \subseteq \mathbb{V}\mathcal{L}$, these classes are also included in $\mathbb{V}\mathsf{K}$ and hence they are pointed as well. ⊠

Since a matrix and its reduction define the same logic (Theorem 4.34), the assertional logic $\mathcal{L}$ is also complete with respect to the class of reduced matrices

$$\{\langle \boldsymbol{A}/\boldsymbol{\Omega}^{\boldsymbol{A}}\{\top^{\boldsymbol{A}}\}, \{\top^{\boldsymbol{A}}\}/\boldsymbol{\Omega}^{\boldsymbol{A}}\{\top^{\boldsymbol{A}}\}\rangle : \boldsymbol{A} \in \mathsf{K}\}.$$

Then, from Proposition 5.79 it follows:

COROLLARY 6.108
If $\mathcal{L}$ is the assertional logic of a pointed class of algebras K, then

$$\mathbb{V}\mathcal{L} = \mathbb{V}\{\boldsymbol{A}/\boldsymbol{\Omega}^{\boldsymbol{A}}\{\top^{\boldsymbol{A}}\} : \boldsymbol{A} \in \mathsf{K}\}. \quad ⊠$$

Observe that, since the classes $\mathsf{Alg}^*\mathcal{L}$ and $\mathsf{Alg}\mathcal{L}$ generate the same variety (Corollary 5.71.3), a term is constant in one of them if and only if it is constant in the other.

Now we see, as announced, that assertional logics are the truth-equational logics with a defining equation of the form $x \approx \top$; you have already seen (Exercise 3.3) that the term $\top$ must be a theorem of $\mathcal{L}$, but this property is too weak to obtain the desired result.

THEOREM 6.109
For any logic $\mathcal{L}$ the following conditions are equivalent:

(i) *$\mathcal{L}$ is an assertional logic.*

(ii) *$\mathcal{L}$ is truth-equational, with a truth definition of the form $x \approx \top$, where $\top$ is a constant term of $\mathsf{Alg}^*\mathcal{L}$.*

(iii) *$\mathcal{L}$ has $\mathsf{Alg}^*\mathcal{L}$ as an algebraic semantics with $x \approx \top$ as defining equation, where $\top$ is a constant term of $\mathsf{Alg}^*\mathcal{L}$.*

In particular, if $\mathcal{L}$ is an assertional logic, then it is the assertional logic of the class $\mathsf{Alg}^*\mathcal{L}$, which is pointed.

PROOF: (i)⇒(ii) Let $\langle \boldsymbol{A}, F\rangle \in \mathsf{Mod}\mathcal{L}$. We can apply the result in Exercise 4.16, which tells us that if $a \in F$, then $a \equiv \top^{\boldsymbol{A}}(a)\ (\boldsymbol{\Omega}^{\boldsymbol{A}}F)$; denoting by $\boldsymbol{\tau}$ the transformer defined by the equation $x \approx \top$, the last equivalence amounts to saying that $\boldsymbol{\tau}^{\boldsymbol{A}}a \subseteq \boldsymbol{\Omega}^{\boldsymbol{A}}F$. Conversely, assume that $\boldsymbol{\tau}^{\boldsymbol{A}}a \subseteq \boldsymbol{\Omega}^{\boldsymbol{A}}F$, that is, that $a \equiv \top^{\boldsymbol{A}}(a)\ (\boldsymbol{\Omega}^{\boldsymbol{A}}F)$. Since $\top$ is a theorem of $\mathcal{L}$ (Exercise 3.3), $\top^{\boldsymbol{A}}(a) \in F$, and by compatibility this implies that $a \in F$. Thus, $a \in F$ if and only if $\boldsymbol{\tau}^{\boldsymbol{A}}a \subseteq \boldsymbol{\Omega}^{\boldsymbol{A}}F$. By Proposition 6.94, $\mathcal{L}$ is truth-equational, with the transformer given by $x \approx \top$ as truth definition. By Lemma 6.107, $\top$ is a constant term of $\mathsf{Alg}^*\mathcal{L}$.

(ii)⇒(iii) appears in Corollary 6.92.

(iii)⇒(i) is a direct consequence of the involved definitions.

The final statement follows from (iii). ⊠

Thus, in particular, *all assertional logics are truth-equational*. This brings back into the Leibniz hierarchy many non-protoalgebraic logics that had for some time seemed excluded from it; for instance, the logic $\mathcal{C}\ell_{\wedge\vee\top\bot}$, which is the assertional logic of the variety DL_{01} of bounded distributive lattices (Exercise 5.63), the implication-less fragment $\mathcal{I}\ell^*$ of intuitionistic logic and its denumerable axiomatic extensions (Example 6.16.6), or Visser's basic logic $\mathcal{BPL}$ (Example 6.16.8). Incidentally, in $\mathcal{I}\ell^*$ and its extensions the constant term is less standard than average, namely $\top := \neg(x \wedge \neg x)$.

Some truth-equational logics are assertional, for instance all regularly algebraizable logics (Theorem 3.52); but there are many truth-equational logics that are not assertional, for instance all algebraizable logics that are not regularly algebraizable. All these logics are protoalgebraic; Example 6.122.11 describes a truth-equational logic that is neither protoalgebraic nor assertional. This example also witnesses that the condition that $\top$ be constant over $\mathsf{Alg}^*\mathcal{L}$ cannot be weakened to that of being a theorem of the logic.

A first consequence of assertional logics being truth-equational is the following characterization of its reduced models; it has an interesting application in Exercise 6.55.

COROLLARY 6.110
If $\mathcal{L}$ is assertional, then $\mathsf{Mod}^*\mathcal{L} = \{\langle \boldsymbol{A}, \{\top^{\boldsymbol{A}}\}\rangle : \boldsymbol{A} \in \mathsf{Alg}^*\mathcal{L}\}$.

PROOF: This is just a specialisation of the result in Lemma 6.91.1 for the transformer $\tau x := \{x \approx \top\}$. Since by Theorem 6.109 the term $\top$ is constant over $\mathsf{Alg}^*\mathcal{L}$, in this case $Sol_{\tau}^{\boldsymbol{A}} = \{a \in A : a = \top^{\boldsymbol{A}}(a)\} = \{\top^{\boldsymbol{A}}\}$. ☒

This implies that the notion of an assertional logic can be characterized in a way completely independent of the hierarchy, but which has some consequences for it. A class of matrices is ***unital*** when the filter of each of its matrices is a one-element set. Exercise 6.56 describes a connection of this notion with implicit definability of truth in $\mathsf{Mod}\mathcal{L}$ which is used at the end of the section. It appears that this notion is closely related to that of an assertional logic:

THEOREM 6.111
For any logic $\mathcal{L}$ the following conditions are equivalent:

(i) *$\mathcal{L}$ is an assertional logic.*

(ii) *The class of matrices $\mathsf{Mod}^*\mathcal{L}$ is unital.*

(iii) *$\mathcal{L}$ has theorems and is complete with respect to a unital class of matrices.*

PROOF: (i)⇒(ii) This is actually contained in Corollary 6.110.

(ii)⇒(iii) If $\mathcal{L}$ had no theorems, then for any algebra $\boldsymbol{A}$, the matrix $\langle \boldsymbol{A}, \varnothing\rangle$ would be a model of $\mathcal{L}$. But, for a trivial (one-element) algebra $\boldsymbol{A}$, the matrix $\langle \boldsymbol{A}, \varnothing\rangle$ is always reduced, because then $\boldsymbol{\Omega}^{\boldsymbol{A}}\varnothing = A \times A = \mathrm{Id}_A$. Thus, we would have that $\langle \boldsymbol{A}, \varnothing\rangle \in \mathsf{Mod}^*\mathcal{L}$, against the assumption that this class is unital. This shows that $\mathcal{L}$ has theorems. The second assertion is a consequence of the assumption plus the completeness of $\mathcal{L}$ with respect to $\mathsf{Mod}^*\mathcal{L}$.

(iii)⇒(i) Let M be the unital class of matrices with respect to which $\mathcal{L}$ is complete, and let K be the class of its algebraic reducts. Observe that since $\mathcal{L}$ has theorems, all $\mathcal{L}$-filters are non-empty. Therefore, since the intersection of two $\mathcal{L}$-filters is always an $\mathcal{L}$-filter, and it cannot be empty, there can be at most one one-element $\mathcal{L}$-filter in each (arbitrary) algebra, and in case there is, it is indeed the smallest one (notice that this solves Exercise 2.20). The assumption that M is unital means that algebras in K have indeed an $\mathcal{L}$-filter of this kind, and it is the only one on the algebra producing a reduced matrix. Let $\top$ be a theorem of $\mathcal{L}$ in at most the variable x (which exists by the first assumption and structurality), and let $\boldsymbol{A} \in \mathsf{K}$. Since $\top$ is a theorem, for every every $a \in A$ the point $\top^{\boldsymbol{A}}(a)$ must belong to the mentioned $\mathcal{L}$-filter, therefore this $\mathcal{L}$-filter must be exactly the set $\{\top^{\boldsymbol{A}}(a)\}$, for any $a \in A$. This implies that $\top^{\boldsymbol{A}}(a) = \top^{\boldsymbol{A}}(b)$ for all $a, b \in A$.

Therefore, $\top$ is a constant term of K, and thus the class K is pointed, and $\mathsf{M} = \{\langle \boldsymbol{A}, \{\top^{\boldsymbol{A}}\} : \boldsymbol{A} \in \mathsf{K} \rangle\}$. After this, the completeness of $\mathcal{L}$ with respect to M means that $\mathcal{L}$ is the assertional logic of K. ⊠

It follows, somewhat redundantly, that a truth-equational logic $\mathcal{L}$ is assertional if and only if $\mathsf{Mod}^*\mathcal{L}$ is unital. Proposition 4.58 asserts that this last condition, when demanded of algebraizable logics, produces the class of regularly algebraizable logics; the same happens with the condition that there is a single defining equation and of the form $x \approx \top$ (Theorem 3.52). The class of regularly algebraizable logics is generally considered as belonging to the Leibniz hierarchy. At the end of this section the same conditions, starting from the class of weakly algebraizable logics, produce the class of regularly weakly algebraizable logics, again naturally considered in the Leibniz hierarchy. The conclusion is that assertional logics also deserve the name of *regularly truth-equational*, and that this class can be legitimately considered as belonging to the Leibniz hierarchy. The following result, to be used later on, further reinforces this idea, especially when comparing its point (iii) with conditions 6.124(vii) and 6.142(vi):

THEOREM 6.112
For any logic $\mathcal{L}$, the following conditions are equivalent:

(i) *$\mathcal{L}$ is an assertional logic.*

(ii) *$\mathcal{L}$ has theorems and satisfies the* ***Suszko rules***:

$$x, y, \delta(x, \vec{z}) \vdash_{\mathcal{L}} \delta(y, \vec{z}), \text{ for all } \delta(x, \vec{z}) \in Fm.$$

(iii) *$\mathcal{L}$ has theorems and satisfies that* $x \equiv y \ (\widetilde{\boldsymbol{\Omega}}_{\mathcal{L}} C_{\mathcal{L}}\{x, y\})$.

(iv) *$\mathcal{L}$ has theorems and satisfies that for every algebra $\boldsymbol{A}$ and every $a, b \in A$,* $a \equiv b \ (\widetilde{\boldsymbol{\Omega}}^{\boldsymbol{A}}_{\mathcal{L}} Fg^{\boldsymbol{A}}_{\mathcal{L}}\{a, b\})$.

PROOF: (i)⇒(ii) We know that all assertional logics have theorems; and completeness of $\mathcal{L}$ with respect to any unital class of matrices (Theorem 6.111) directly implies the Suszko rules.

(ii)⇒(i) Let $\langle \boldsymbol{A}, F \rangle \in \mathsf{Mod}^*\mathcal{L}$. Since $\mathcal{L}$ has theorems, $F \neq \varnothing$. Then the Suszko rules imply that F is a one-element set: If $a, b \in F$, then for every $\delta \in Fm$ and every $\vec{c} \in \vec{A}$, $\delta^{\boldsymbol{A}}(a, \vec{c}) \in F$ if and only if $\delta^{\boldsymbol{A}}(b, \vec{c}) \in F$, that is, $a \equiv b \ (\boldsymbol{\Omega}^{\boldsymbol{A}} F)$; since the matrix is reduced, this implies that $a = b$. Thus, all the reduced models of $\mathcal{L}$ are unital, and by Theorem 6.111 this fact implies that $\mathcal{L}$ is an assertional logic.

The equivalence between (ii) and (iii) is a very general fact, and is left to you as Exercise 6.57. That (ii) implies (iv) is shown similarly, using the characterization of the Suszko congruence in Proposition 6.97.3. Finally, it is obvious that (iv) implies (iii). ⊠

The name "Suszko rules" was coined by Rautenberg [211]; a result basically equivalent to Theorem 6.112 (with "assertional" defined as "complete with respect to a unital semantics") is credited by Czelakowski [67] to Suszko, as unpublished.

Observe that any extension of an assertional logic is also assertional; this follows directly, either from Theorem 6.111 (for $\mathcal{L} \leqslant \mathcal{L}'$ implies $\mathsf{Mod}^*\mathcal{L}' \subseteq \mathsf{Mod}^*\mathcal{L}$), or from the syntactic characterization of Theorem 6.112(ii).

Given that truth-equational logics and assertional logics are not necessarily protoalgebraic, it may seem slightly surprising that it is the class $\mathsf{Alg}^*\mathcal{L}$ that plays a role in most of its characterizations. This fact is only apparent: the class $\mathsf{Alg}\,\mathcal{L}$ does indeed play a parallel role, but proving this requires some advanced steps; if you are curious, you can look at Exercise 6.58.

Weakly algebraizable logics

When protoalgebraicity is added to the landscape of definability of truth, the following essential result is obtained.

THEOREM 6.113
If $\mathcal{L}$ is protoalgebraic, then truth is implicitly definable in $\mathsf{Mod}^*\mathcal{L}$ *if and only if it is equationally definable.*

PROOF: If $\mathcal{L}$ is protoalgebraic, then the Leibniz operator and the Suszko operator coincide on its filters (Proposition 6.14), and since the injectivity of each operator on $\mathsf{Mod}^*\mathcal{L}$ characterizes each of the two definability properties (Proposition 6.88 and Theorem 6.106, respectively), they coincide. ⊠

This result, mentioned as "Herrmann's Theorem" in some works, was originally proved in [140], in a different framework, when the Suszko operator did not exist; we realize that using it has made the proof appear as almost trivial.

MODEL-THEORETIC DIGRESSION.
When matrices are viewed as relational structures (as in Section 4.5), *explicit* definability of truth in a class of matrices means definability of (the value of) the symbol D by some set of D-free formulas of the first-order language $\boldsymbol{L}_D$. Theorem 6.113 can be compared with the result of applying the famous Beth's *Definability Theorem* of model theory to the present situation. This theorem (usually presented as a consequence of Craig's *Interpolation Theorem*; see [144, Theorem 6.6.4] or [175, Theorem 2.2.4]), very roughly speaking, states that implicit definability of a relation symbol in an elementary class implies its explicit, first-order definability in the language (without using that symbol). To make a comparison, the class $\mathsf{Mod}^*\mathcal{L}$ should be elementary (in the language $\boldsymbol{L}_D$), but this does not happen in general, not even in all protoalgebraic logics (Theorem 4.77 established some conditions for this to happen). Thus, Beth's Theorem applies only to selected protoalgebraic logics. Moreover, it only proves definability by an unspecified first-order formula, while Herrmann's Theorem proves equational definability, which corresponds to the more restricted case of atomic formulas (which are here equations, as D cannot be used). All these considerations tell us that the result obtained from Herrmann's Theorem 6.113 is stronger than the one that can be obtained from Beth's. Observe, however, that the definability

obtained here uses a *set* of these formulas, while Beth's Theorem obtains a single formula; in this respect, our result is weaker.

As a matter of fact, Example 6.102 describes a logic $\mathcal{L}$ such that $\mathsf{Mod}^*\mathcal{L}$ is elementary, and has its filters first-order definable, but not equationally definable; of course, it is not protoalgebraic. Thus, protoalgebraicity is essential in Theorem 6.113. Herrmann's dissertation and papers [139]–[142] and Raftery's [201] deal with the definability issue from this perspective. ⊠

The following rephrasing of Theorem 6.113, modulo Proposition 6.88, emphasizes that inside protoalgebraic logics, unlike in the general case, the property of the Leibniz operator that characterizes truth-equational logics can be taken to be injectivity (Exercise 6.59 contains yet another equivalent condition):

Corollary 6.114
A protoalgebraic logic $\mathcal{L}$ is truth-equational if and only if for every algebra $\boldsymbol{A}$, the Leibniz operator $\boldsymbol{\Omega}^{\boldsymbol{A}}$ is injective over $\mathcal{F}i_{\mathcal{L}}\boldsymbol{A}$. ⊠

The class of logics that arises in this way turns out to be interesting for other reasons, as the variety of its characterizations from different points of view shows; now it can be introduced in a very simple way:

Definition 6.115
A logic is ***weakly algebraizable*** *when it is protoalgebraic and truth-equational.*

This class of logics was first considered by Czelakowski in unpublished lectures in 1993 under the name "algebraizable in the weak sense", which he also used in [70]. It was first studied in depth by Czelakowski and Jansana in [72] under the name used here, which had been already used in the first edition of [107], in the context of Corollary 6.116. The present perspective immediately locates it in the Leibniz hierarchy:

Corollary 6.116
A logic $\mathcal{L}$ is weakly algebraizable if and only if the Leibniz operator $\boldsymbol{\Omega}^{\boldsymbol{A}}$ is monotone and injective over $\mathcal{F}i_{\mathcal{L}}\boldsymbol{A}$ for every $\boldsymbol{A}$. ⊠

Theorem 6.120 shows that in Corollaries 6.114 and 6.116 it is enough to require the same properties for the formula algebra. One obvious consequence of the injectivity of the Leibniz operator over the filters of weakly algebraizable logics is that all these filters are Leibniz; several equivalent characterizations can then be obtained at once from the properties of Leibniz filters in protoalgebraic logics:

Theorem 6.117
For every logic $\mathcal{L}$ the following conditions are equivalent:

(i) *$\mathcal{L}$ is weakly algebraizable.*

(ii) *For every $\boldsymbol{A}$, the function $F \mapsto (\mathcal{F}i_{\mathcal{L}}\boldsymbol{A})^F$ is a bijection, and hence a dual lattice isomorphism, between the sets $\mathcal{F}i_{\mathcal{L}}\boldsymbol{A}$ and $\mathrm{FCS}_{\mathcal{L}}\boldsymbol{A}$.*

(iii) *For every* $\boldsymbol{A}$, $\boldsymbol{\Omega}^{\boldsymbol{A}}\colon \mathcal{F}i_{\mathcal{L}}\boldsymbol{A} \cong \mathrm{Con}_{\mathsf{Alg}\mathcal{L}}\boldsymbol{A}$ *as complete lattices.*

(iv) *For every* $\boldsymbol{A}$, $\boldsymbol{\Omega}^{\boldsymbol{A}}\colon \mathcal{F}i_{\mathcal{L}}\boldsymbol{A} \cong \mathrm{Con}_{\mathsf{Alg}^*\mathcal{L}}\boldsymbol{A}$ *as complete lattices.*

PROOF: (i)⇒(ii) follows from the previous observation that all filters are Leibniz, applied to Theorem 6.43.

(ii)⇒(iii) The assumption implies that all full g-models of $\mathcal{L}$ have the form $\langle \boldsymbol{A}, (\mathcal{F}i_{\mathcal{L}}\boldsymbol{A})^F \rangle$ for some $F \in \mathcal{F}i_{\mathcal{L}}\boldsymbol{A}$, and by Theorem 6.39 this implies protoalgebraicity. This in turn implies that $\boldsymbol{\Omega}^{\boldsymbol{A}}F = \widetilde{\boldsymbol{\Omega}}^{\boldsymbol{A}}(\mathcal{F}i_{\mathcal{L}}\boldsymbol{A})^F$, and the Leibniz operator $\boldsymbol{\Omega}^{\boldsymbol{A}}$ is the composition of the function of (ii) followed by the Tarski operator $\widetilde{\boldsymbol{\Omega}}^{\boldsymbol{A}}$. By Theorem 5.95, this one is always a dual lattice isomorphism between $\mathrm{FCS}_{\mathcal{L}}\boldsymbol{A}$ and $\mathrm{Con}_{\mathsf{Alg}\mathcal{L}}\boldsymbol{A}$. Therefore, $\boldsymbol{\Omega}^{\boldsymbol{A}}$ is a lattice isomorphism between $\mathcal{F}i_{\mathcal{L}}\boldsymbol{A}$ and $\mathrm{Con}_{\mathsf{Alg}\mathcal{L}}\boldsymbol{A}$, as required.

(iii)⇒(iv) Since $\mathsf{Alg}^*\mathcal{L} \subseteq \mathsf{Alg}\mathcal{L}$, $\mathrm{Con}_{\mathsf{Alg}^*\mathcal{L}}\boldsymbol{A} \subseteq \mathrm{Con}_{\mathsf{Alg}\mathcal{L}}\boldsymbol{A}$. But $\boldsymbol{\Omega}^{\boldsymbol{A}}$ is always onto $\mathrm{Con}_{\mathsf{Alg}^*\mathcal{L}}\boldsymbol{A}$, therefore the assumption implies that $\mathrm{Con}_{\mathsf{Alg}^*\mathcal{L}}\boldsymbol{A} = \mathrm{Con}_{\mathsf{Alg}\mathcal{L}}\boldsymbol{A}$, and the isomorphism in (iii) is actually the same as that in (iv).

(iv)⇒(i) This holds because the assumption in (iv) contains the monotonicity and the injectivity of $\boldsymbol{\Omega}^{\boldsymbol{A}}$ over $\mathcal{F}i_{\mathcal{L}}\boldsymbol{A}$. Hence, by Corollary 6.116, the logic is weakly algebraizable. ⊠

The subtle difference between (iii) and (iv) makes sense because the theorem does not assume that $\mathcal{L}$ is protoalgebraic, and hence one cannot assume that $\mathsf{Alg}\mathcal{L} = \mathsf{Alg}^*\mathcal{L}$. In a particular situation where the logic under study is not known to be protoalgebraic, it may happen that only one of $\mathsf{Alg}^*\mathcal{L}$ and $\mathsf{Alg}\mathcal{L}$ is known, and hence checking one of the conditions may be more convenient than checking the other.

Point (ii) is particularly relevant for the theory of full g-models: It says that weakly algebraizable logics are the ones whose full g-models are perfectly well represented by their ordinary filters. This is observed for some particular logics in Exercise 5.69. However, this does not entail that study of full g-models of weakly algebraizable logics has no interest, for they may have characterizations in terms of Tarski-style conditions or of Gentzen-style rules (as seen in Example 5.91 and Exercises 5.65 and 5.66).

After Theorem 6.120, it is easy to see that the conditions (ii)–(iv) in Theorem 6.117 do characterize the property of being weakly algebraizable even if they are required to hold only for the formula algebra (that is, for theories instead of filters). Thus, there are several *transfer theorems* here. You may find their precise formulations in Exercise 6.61.

Like other classes in the hierarchy, that of weakly algebraizable logics can be characterized in a syntactic way, though not a very nice one due to the presence of parameters everywhere. Recall (Theorem 6.57) that a logic is protoalgebraic if and only if it has a set of congruence formulas with parameters. To handle them, the notation introduced on page 353 is extended:

NOTATION.
If $\Delta(x,y,\vec{z}) \subseteq Fm$, recall that $\Delta\langle\alpha,\beta\rangle := \bigcup\{\Delta(\alpha,\beta,\vec{\delta}) : \vec{\delta} \in Fm\}$. Then, for an arbitrary $\Theta \subseteq Eq$ put $\Delta\langle\Theta\rangle := \bigcup\{\Delta\langle\alpha,\beta\rangle : \alpha \approx \beta \in \Theta\} = \bigcup\{\Delta(\alpha,\beta,\vec{\delta}) : \alpha \approx \beta \in \Theta\ ,\ \vec{\delta} \in Fm\}$. Also, for an algebra $\boldsymbol{A}$ and a set $X \subseteq A \times A$, we set $\Delta^{\boldsymbol{A}}\langle X\rangle := \bigcup\{\Delta^{\boldsymbol{A}}(a,b,\vec{c}) : \langle a,b\rangle \in X\ ,\ \vec{c} \in \vec{A}\}$. The caveats expressed after the introduction of the notation on page 353 apply here as well.

LEMMA 6.118
Let $\mathcal{L}$ be a protoalgebraic logic, with $\Delta(x,y,\vec{z})$ as set of congruence formulas with parameters, and let $E(x) \subseteq Eq$. Truth is equationally definable in $\mathsf{Mod}^*\mathcal{L}$ *by $E(x)$ if and only if $\mathcal{L}$ satisfies*

$$x \dashv\vdash_{\mathcal{L}} \Delta\langle E(x)\rangle. \qquad (\mathrm{ALG3_P})$$

PROOF: It is practical to use Proposition 6.94 and show that it is the property that "for any $\langle \boldsymbol{A},F\rangle \in \mathsf{Mod}\mathcal{L}$ and any $a \in A$, $a \in F$ if and only if $E^{\boldsymbol{A}}(a) \subseteq \boldsymbol{\Omega}^{\boldsymbol{A}}F$" the one that is equivalent to ($\mathrm{ALG3_P}$). But since $\Delta(x,y,\vec{z})$ is a set of congruence formulas with parameters for $\mathcal{L}$, the clause "$E^{\boldsymbol{A}}(a) \subseteq \boldsymbol{\Omega}^{\boldsymbol{A}}F$" can be replaced by "$\Delta^{\boldsymbol{A}}\langle E^{\boldsymbol{A}}(a)\rangle \subseteq F$". The direct implication is straightforward by considering, for each $\Gamma \in \mathcal{T}h\mathcal{L}$, the matrix $\langle \boldsymbol{Fm},\Gamma\rangle \in \mathsf{Mod}\mathcal{L}$: the reformulated assumption says that $x \in \Gamma$ if and only if $\Delta\langle E(x)\rangle \subseteq \Gamma$, and since this holds for all theories, this amounts to the interderivability in ($\mathrm{ALG3_P}$). The converse contains a slightly tricky point in handling parameters. Take any $\langle \boldsymbol{A},F\rangle \in \mathsf{Mod}\mathcal{L}$ and any $a \in A$. We need to prove that $a \in F$ if and only if $\Delta^{\boldsymbol{A}}\langle E^{\boldsymbol{A}}(a)\rangle \subseteq F$. Observe that for any $h \in \mathrm{Hom}(\boldsymbol{Fm},\boldsymbol{A})$ such that $hx = a$, ($\mathrm{ALG3_P}$) implies that $a \in F$ if and only if $\Delta^{\boldsymbol{A}}(h\alpha,h\beta,h\vec{\delta}) \subseteq F$ for all $\alpha \approx \beta \in E(x)$ and all $\vec{\delta} \in \vec{Fm}$. Since only the variable x appears in the equations in $E(x)$, and all (sequences of) formulas $\vec{\delta} \in \vec{Fm}$ appear here, it is not difficult to see (Exercise 6.60) that this implies that $a \in F$ if and only if $\Delta^{\boldsymbol{A}}(h\alpha,h\beta,\vec{c}) \subseteq F$ for all $\alpha \approx \beta \in E(x)$ and all $\vec{c} \in \vec{A}$, that is, if and only if $\Delta^{\boldsymbol{A}}\langle E^{\boldsymbol{A}}(a)\rangle \subseteq F$, as required. ⊠

From this, Theorem 6.57 and Exercise 6.28, the announced characterization follows:

COROLLARY 6.119
A logic $\mathcal{L}$ is weakly algebraizable if and only if there is a set of formulas $\Delta(x,y,\vec{z}) \subseteq Fm$ and a set of equations $E(x) \subseteq Eq$ such that the conditions ($\mathrm{R_P}$), ($\mathrm{MP_P}$), ($\mathrm{Re_P}$) *and* ($\mathrm{ALG3_P}$) *hold.* ⊠

The next application of Lemma 6.118 is an important result: the characterization in Corollary 6.116 holds when the two properties of the Leibniz operator are required only of the formula algebra:

THEOREM 6.120
A logic $\mathcal{L}$ is weakly algebraizable if and only if $\boldsymbol{\Omega}$ is monotone and injective over $\mathcal{T}h\mathcal{L}$.

PROOF: In one direction there is nothing to prove. Now assume that $\boldsymbol{\Omega}$ is monotone and injective over $\mathcal{T}h\mathcal{L}$. By Theorem 6.4, $\mathcal{L}$ is protoalgebraic, and we can use Lemma 6.118 in order to prove that $\mathcal{L}$ is weakly algebraizable. Thus, we want to find some $E(x) \subseteq Fm$ such that (ALG3_P) holds, that is, such that $C_\mathcal{L}\{x\} = C_\mathcal{L}\Delta\langle E(x)\rangle$. Since the Leibniz operator is injective, every theory is Leibniz. Therefore Corollary 6.59 applies, giving

$$\begin{aligned} C_\mathcal{L}\{x\} = (C_\mathcal{L}\{x\})^+ &= C_\mathcal{L} \bigcup\{\Delta\langle\alpha,\beta\rangle : \alpha \equiv \beta \; (\boldsymbol{\Omega} C_\mathcal{L}\{x\})\} \\ &= C_\mathcal{L} \bigcup\{\Delta(\alpha,\beta,\vec{\delta}) : \alpha \equiv \beta \; (\boldsymbol{\Omega} C_\mathcal{L}\{x\}) \,,\; \vec{\delta} \in \vec{Fm}\} \,. \end{aligned} \tag{91}$$

The congruence $\boldsymbol{\Omega} C_\mathcal{L}\{x\}$ is actually a set of pairs of formulas, hence it can be viewed as a set of equations, and (91) actually says that $x \dashv\vdash_\mathcal{L} \Delta\langle\boldsymbol{\Omega} C_\mathcal{L}\{x\}\rangle$; this is almost what we want, save that the equations in $\boldsymbol{\Omega} C_\mathcal{L}\{x\}$ may contain arbitrary variables. So, consider the substitution $\sigma_x z := x$ for all $z \in V$ and take $E(x) := \sigma_x \boldsymbol{\Omega} C_\mathcal{L}\{x\} = \{\sigma_x\alpha \approx \sigma_x\beta : \alpha \equiv \beta \; (\boldsymbol{\Omega} C_\mathcal{L}\{x\})\}$, which is a set of equations in (at most) the variable x. Then, applying structurality to (91), we obtain that

$$\begin{aligned} x = \sigma_x x \in\; & C_\mathcal{L} \bigcup\{\Delta(\sigma_x\alpha,\sigma_x\beta,\sigma_x\vec{\delta}) : \alpha \equiv \beta \; (\boldsymbol{\Omega} C_\mathcal{L}\{x\}) \,,\; \vec{\delta} \in \vec{Fm}\} \\ \subseteq\; & C_\mathcal{L} \bigcup\{\Delta(\sigma_x\alpha,\sigma_x\beta,\vec{\delta}) : \alpha \equiv \beta \; (\boldsymbol{\Omega} C_\mathcal{L}\{x\}) \,,\; \vec{\delta} \in \vec{Fm}\} \\ =\; & C_\mathcal{L} \bigcup\{\Delta(\alpha',\beta',\vec{\delta}) : \alpha' \approx \beta' \in E(x) \,,\; \vec{\delta} \in \vec{Fm}\} \\ =\; & C_\mathcal{L}\Delta\langle E(x)\rangle \,. \end{aligned}$$

From this it follows that $C_\mathcal{L}\{x\} \subseteq C_\mathcal{L}\Delta\langle E(x)\rangle$. To show the reverse inclusion, by the property of the Leibniz operator found in Exercise 6.32 we have that $E(x) = \sigma_x \boldsymbol{\Omega} C_\mathcal{L}\{x\} \subseteq \boldsymbol{\Omega} C_\mathcal{L}\{\sigma_x x\} = \boldsymbol{\Omega} C_\mathcal{L}\{x\}$. Since Δ is a set of congruence formulas (with parameters) for $\mathcal{L}$, this implies that $\Delta\langle E(x)\rangle \subseteq C_\mathcal{L}\{x\}$, and hence also $C_\mathcal{L}\Delta\langle E(x)\rangle \subseteq C_\mathcal{L}\{x\}$, which finishes the proof. ⊠

Putting this together with Corollaries 6.114 or 6.116, we obtain one more *transfer theorem*: when coupled with monotonicity, injectivity of the Leibniz operator on the theories of a logic transfers to all algebras.

If the sets $E(x)$ and $\Delta(x,y,\vec{z})$ are viewed as defining functions between (sets of) formulas and (sets of) equations, then it is possible to prove a characterization of weakly algebraizable logics in the language of *interpretations* (Definition 3.4, but here with parameters), parallel to the characterization of equivalential logics in Corollary 6.83.2, and which has an interesting reading (but which will not be proved here; see [72, Theorem 4.3]).

THEOREM 6.121
A logic $\mathcal{L}$ is weakly algebraizable if and only if there are sets $\Delta(x,y,\vec{z}) \subseteq Fm$ and $E(x) \subseteq Eq$ and there is a class of algebras K such that the following conditions are satisfied:

$$\Theta \vDash_\mathsf{K} \varepsilon \approx \delta \iff \Delta\langle\Theta\rangle \vdash_\mathcal{L} \Delta\langle\varepsilon,\delta\rangle \tag{ALG2$_\text{P}$}$$

for all $\Theta \cup \{\delta \approx \varepsilon\} \subseteq Eq$, that is, Δ interprets $\vDash_\mathsf{K}$ into $\vdash_\mathcal{L}$; and

$$x \dashv\vdash_\mathcal{L} \Delta\langle E(x)\rangle. \tag{ALG3$_\text{P}$}$$

Moreover, the following two conditions are satisfied as well:

$$\Gamma \vdash_{\mathcal{L}} \varphi \iff E(\Gamma) \vDash_{\mathsf{K}} E(\varphi) \tag{ALG1}$$

for all $\Gamma \cup \{\varphi\} \subseteq Fm$*; and*

$$x \approx y \dashv\vDash_{\mathsf{K}} E(\Delta\langle x,y\rangle). \tag{ALG4$_\mathrm{P}$}$$

The class K *can be taken to be* $\mathsf{Alg}^*\mathcal{L}$. ⊠

Recall that by Proposition 3.12 a logic is *algebraizable* if and only if there are $\Delta(x,y) \subseteq Fm$ and $E(x) \subseteq Eq$ and a class K satisfying conditions (ALG2) and (ALG3). Thus, Theorem 6.121 expresses one of the motivations under the name "weakly algebraizable": these logics are exactly those obtained when weakening the conditions for algebraizability by adding parameters to the set Δ, so that the conditions become (ALG2$_\mathrm{P}$) and (ALG3$_\mathrm{P}$) respectively; thus, in some sense, they could also be named *parametrically algebraizable logics*.

EXAMPLES 6.122

1. Any extension of a weakly algebraizable logic is again weakly algebraizable. This follows from either Corollary 6.116 or Theorem 6.120, because if $\mathcal{L} \leqslant \mathcal{L}'$, then $\mathcal{F}i_{\mathcal{L}'}\boldsymbol{A} \subseteq \mathcal{F}i_{\mathcal{L}}\boldsymbol{A}$ and $\mathcal{T}h\mathcal{L}' \subseteq \mathcal{T}h\mathcal{L}$.
2. Trivially, the inconsistent logic in each language is weakly algebraizable.
3. By contrast, the almost inconsistent logic in each language is *not* weakly algebraizable, because it has two theories, $\emptyset$ and Fm, which are different, but $\boldsymbol{\Omega}\emptyset = \boldsymbol{\Omega}Fm = Fm \times Fm$: the Leibniz operator is not injective over the theories of the logic. In fact, this shows that it is not truth-equational either.
4. All algebraizable logics are in particular weakly algebraizable. This can be seen in many different ways, for instance from Theorem 6.117 or from Theorem 6.121, or because we already know that algebraizable logics are both protoalgebraic and truth-equational.
5. There are few examples of logics that are "properly" weakly algebraizable, that is, weakly algebraizable but not algebraizable. One of the most natural group of examples found until now is related to the following class of algebras. An ***ortholattice*** is an algebra $\langle A, \wedge, \vee, \neg, 0, 1\rangle$ of type $\langle 2,2,1,0,0\rangle$[†] whose negation-free reduct $\langle A, \wedge, \vee, 0, 1\rangle$ is a bounded lattice (notice: not necessarily distributive) and where the equations

$$\neg\neg x \approx x \qquad\qquad x \vee \neg x \approx \top$$
$$\neg(x \wedge y) \approx \neg x \vee \neg y \qquad\qquad x \wedge \neg x \approx \bot$$

hold; that is, the operation $\neg$ is a De Morgan negation and a complement. The class of all ortholattices is denoted by OL. A very popular subclass is the class OML of ***orthomodular lattices***, which correspond to *quantum logic*, and

[†] Sometimes the constants are omitted from the language, but then the definition has to state explicitly that the terms $x \vee \neg x$ and $x \wedge \neg x$ are constant in the algebra and the bounds of the lattice.

are characterized by *the orthomodular law*, traditionally expressed as the quasi-equation $(x \preccurlyeq y) \rightarrow (x \vee (y \wedge \neg x) \approx y)$, but which can also be expressed as an equation.

If $\mathsf{K} \subseteq \mathsf{OL}$ is a class of ortholattices that contains at least one non-orthomodular lattice, then its assertional logic $\mathcal{L}^{\top}_{\mathsf{K}}$ is weakly algebraizable but not equivalential, hence it is not algebraizable. The key is the algebraic fact that every ortholattice that is not orthomodular contains the "benzene ring" $\boldsymbol{O}_6$ as a subalgebra; this algebra is described in Example 4.79, where it is also shown that the corresponding matrix $\langle \boldsymbol{O}_6, \{1\} \rangle$ is not reduced. Thus, $\mathsf{Mod}^*\mathcal{L}^{\top}_{\mathsf{K}}$ is not closed under $\mathbb{S}$. This produces a family of examples of weakly algebraizable but non-algebraizable logics, among them $\mathcal{L}^{\top}_{\mathsf{OL}}$, known as *the logic of ortholattices*. The logic $\mathcal{L}^{\top}_{\mathsf{OML}}$ has appeared in Example 3.64. You will find more details in [70, Section 4.7].

6. A simple family of examples of equivalential (in fact, finitely equivalential) logics that fail to be weakly algebraizable are the logics $Ł^{\leqslant}_n$, the n-valued Łukasiewicz logics that preserve degrees of truth, for $n > 2$. The logic $Ł^{\leqslant}_n$ is defined from the MV-algebra $[\mathbf{0,1}]_n$, the n-element subalgebra of $[\mathbf{0,1}]$ described in Example 1.12, by the scheme of preservation of degrees of truth (Example 4.11), that is, for all $\Gamma \cup \{\varphi\} \subseteq Fm$,

$$\Gamma \vdash^{\leqslant}_n \varphi \overset{\text{def}}{\iff} \bigwedge h\Gamma \leqslant h\varphi \text{ for all } h \in \mathrm{Hom}(\boldsymbol{Fm}, [\mathbf{0,1}]_n).$$

Each of these logics is finitely equivalential, with congruence formula $\Delta(x,y) := \{(x \to y)^{n-1} \star (y \to x)^{n-1}\}$. The "fusion" or "intensional conjunction" connective $\star$ is defined by $x \star y := \neg(x \to \neg y)$, and gives the algebra the structure of a commutative monoid. These logics are particular cases of the large family of logics that preserve degrees of truth with respect to a variety of commutative integral residuated lattices mentioned in Example 6.67.5; in general,† the logics in this family are not algebraizable, and by Theorem 6.127 they are not weakly algebraizable either.

7. The logics $\mathcal{E}$ of entailment and of $\mathcal{T}$ ticket entailment and the implication fragment $\mathcal{R}_{\to}$ of relevance logic (shown in Example 6.67.2 to be finitely equivalential) are shown in Example 3.62 to be non-algebraizable by exhibiting two simple algebras, each with several filters of the corresponding logic. The simplicity of the algebra implies that the Leibniz congruence of any non-empty, proper filter is the identity relation, therefore the Leibniz operator is not injective; this shows that these logics are not weakly algebraizable either.

8. The logic E^{MP} associated with the weakest system of classical modal logic E (Example 6.76) is protoalgebraic and non-equivalential. It is not truth-equational, because the class $\mathsf{Mod}^*E^{\mathrm{MP}}$ is not unital: The matrix $\langle \boldsymbol{A}, \{1, \neg a\} \rangle$ is a reduced model of the logic, where $\boldsymbol{A}$ is the algebra considered in Example 6.76.5. Therefore, it is not weakly algebraizable either.

† More precisely: when the variety is not a variety of generalized Heyting algebras.

9. The paraconsistent logics $\mathcal{C}_n$ ($1 \leqslant n \leqslant \omega$) dealt with in Example 6.77, shown there to be protoalgebraic but not equivalential, are not weakly algebraizable either. The proof in [163] that $\mathcal{C}_1$ is not algebraizable actually proves that it is not weakly algebraizable, for it constructs an algebra in which the Leibniz operator is not injective over the $\mathcal{C}_1$-filters. As a consequence, neither of the logics $\mathcal{C}_n$ is weakly algebraizable.

10. The two weak logics of implication $\mathcal{I}$ and $\mathcal{G}_1$ of Example 6.78 are not weakly algebraizable: Exercise 6.35 shows that $\mathcal{I}$ has no algebraic semantics, therefore it is not truth equational, and in particular it is not weakly algebraizable; the case of $\mathcal{G}_1$ is dealt with in [47].

11. The class of truth-equational logics is truly larger than both the class of weakly algebraizable logics and that of assertional logics. Consider the logic in the language $\langle \to, \neg \rangle$ of type $\langle 2, 1 \rangle$ defined by taking as algebraic semantics (with defining equation $x \approx x \to x$) the variety generated by the three-element algebra $\langle \{-1, 0, 1\}, \to, \neg \rangle$ whose operations arise from the structure of this set as an idempotent commutative monoid, with 0 as unit and $-1 \cdot 1 = -1$, negation being the monoid's inverse $\neg x := 1 - x$, and $x \to y := \neg(x \cdot \neg y)$. This logic, which has the same theorems as (but does not coincide with) the implication-negation fragment of $\mathcal{RM}$, is shown in [201, Example 9] to be truth-equational, but neither assertional nor protoalgebraic (hence, it is not weakly algebraizable either). Observe that the term $x \to x$ is a theorem of the logic, but it is not constant over the variety (actually, it is not constant over the generating algebra); this shows that the condition on $\top$ in Theorem 6.109 cannot be weakened to $\top$ being a theorem of the logic.

12. Two natural examples of a similar situation are two logics related to Kleene's three-valued truth-tables, namely the "paraconsistent weak Kleene logic" $\mathcal{PWK}$ and the well-known "logic of paradox" $\mathcal{LP}$ popularized by Priest; these logics are defined from the "weak" and the "strong" versions of Kleene's connectives, and have been studied in [43] and [217], respectively, with an algebraic treatment that is very close to that of Dunn-Belnap's logic $\mathcal{B}$; in fact, the matrix defining $\mathcal{LP}$ is a submatrix of the one that defines $\mathcal{B}$ (Exercise 4.26), therefore $\mathcal{LP}$ is an extension of $\mathcal{B}$, namely the axiomatic extension with the axiom $x \vee \neg x$. The two logics are truth-equational (the defining equation is the unusual $\neg x \preccurlyeq x$) and neither assertional nor protoalgebraic (hence, not weakly algebraizable either). An unusual feature of the first is that $\mathsf{Alg}^*\mathcal{PWK} \subsetneq \mathsf{Alg}\mathcal{PWK} \subsetneq \mathbb{V}\mathcal{PWK}$. The class $\mathsf{Alg}\mathcal{LP}$ is the variety of Kleene lattices, a well-known subvariety of DM.

The logics in 8, 9 and 10 are among the few known protoalgebraic examples that are neither equivalential nor weakly algebraizable; thus, they are located in the leftmost bottom of the Leibniz hierarchy, as depicted in Figure 9 on page 316. The logics in 11 and 12 are located in the rightmost bottom. ⊠

Regularly weakly algebraizable logics

DEFINITION 6.123
A logic $\mathcal{L}$ is ***regularly weakly algebraizable*** *when it is weakly algebraizable and the class* $\mathsf{Mod}^*\mathcal{L}$ *is unital.*

This class of logics was introduced in [70], mostly for the sake of completeness of the graph of the Leibniz hierarchy, and has been little studied. Regularly weakly algebraizable logics can be regarded as a *parametrised version* of regularly algebraizable logics, which are introduced in Definition 3.49; it is not difficult to see that, syntactically, this can be achieved by adding to weakly algebraizable logics a parametrised version of condition (G), which defines regularly algebraizable logics out of the algebraizable ones:

$$x, y \vdash_{\mathcal{L}} \Delta\langle x, y\rangle \tag{$\mathrm{G_P}$}$$

Moreover, and in parallel to what Theorem 6.142 shows for regularly algebraizable logics with respect to equivalential logics, it is possible to see that the regularly weakly algebraizable ones can be defined directly over the class of protoalgebraic logics in a variety of ways. In particular, a syntactic, parametrised characterization can be obtained, as is Corollary 6.119 from Lemma 6.118. All these observations are condensed in the following result:

THEOREM 6.124
Let $\mathcal{L}$ be a protoalgebraic logic with theorems. The following conditions are equivalent:

(i) *$\mathcal{L}$ is regularly weakly algebraizable.*

(ii) *The class* $\mathsf{Mod}^*\mathcal{L}$ *is unital.*

(iii) *The class* $\mathsf{LTMod}^*\mathcal{L}$ *is unital.*

(iv) *$\mathcal{L}$ is complete with respect to a class of unital matrices.*

(v) $x \dashv\vdash_{\mathcal{L}} \Delta\langle x, \top\rangle$, *where $\Delta(x, y, \vec{z})$ is any non-empty set of congruence formulas with parameters and $\top$ is a formula in at most a variable $z \neq x$.*

(vi) *$\mathcal{L}$ satisfies condition* ($\mathrm{G_P}$) *for any Δ as in* (v).

(vii) $x \equiv y \ (\boldsymbol{\Omega} C_{\mathcal{L}}\{x, y\})$.

The formula $\top$ in (v) *is a theorem of $\mathcal{L}$ and an algebraic constant of* $\mathsf{Alg}^*\mathcal{L}$.

SKETCH OF THE PROOF: (i) implies (ii) by definition. That (ii) implies (i) is shown by using the property in Exercise 6.56 and Theorem 6.113. The implications (ii)⇒(iii)⇒(iv) are trivial, and (iv)⇒(ii) follows from Theorem 6.111. The equivalence between (vi) and (vii) is a direct consequence of Δ being a set of congruence formulas with parameters for $\mathcal{L}$. For the rest, see Exercise 6.64. ☒

Notice the slightly unusual form of the formula $\top$ appearing in condition (v), related (but not coinciding) with condition (ALG3) of algebraizability. Condition (ii) suggests the alternative name of *regularly protoalgebraic* for these logics. Their exact location in Figure 9 on page 316 is easily justified:

PROPOSITION 6.125

A logic is regularly weakly algebraizable if and only if it is protoalgebraic and assertional, and if and only if it is weakly algebraizable and assertional.

PROOF: Obviously, every regularly weakly algebraizable logic $\mathcal{L}$ is weakly algebraizable, and hence protoalgebraic, and the condition that $\mathsf{Mod}^*\mathcal{L}$ is unital implies, by Theorem 6.111, that it is an assertional logic. Conversely, if $\mathcal{L}$ is assertional, then by Theorem 6.109 it is truth-equational, with $x \approx \top$ defining truth for some algebraic constant $\top$ of $\mathsf{Alg}^*\mathcal{L}$. If $\mathcal{L}$ is moreover protoalgebraic (a fortiori, if it is weakly algebraizable), then Lemma 6.118 applies, and the condition $x \dashv\vdash_{\mathcal{L}} \Delta\langle x, \top\rangle$ holds; since $\top$ is an algebraic constant, it can be so chosen that the variable x does not occur in it, and then Theorem 6.124.v applies, implying that $\mathcal{L}$ is regularly weakly algebraizable. ⊠

Exercise 6.65 asks you to show a curious fact: the first characterization of being regularly weakly algebraizable in Proposition 6.125 can be formulated in terms of properties of $\mathsf{Alg}^*\mathcal{L}$. Another interesting observation in this direction is that the condition that the assertional logic of a class K of algebras is protoalgebraic can be expressed as a property of K, by transforming conditions (R) and (MP) for the protoimplication set $\Delta(x,y)$ (Theorem 6.7) into conditions on $\vDash_{\mathsf{K}}$:

$$\mathsf{K} \vDash \delta(x,x) \approx \top \quad \text{for all } \delta(x,y) \in \Delta(x,y)$$

$$\{x \approx \top\} \cup \{\delta(x,y) \approx \top : \delta \in \Delta\} \vDash_{\mathsf{K}} y \approx \top$$

The pointed classes of algebras that satisfy these two properties for some set $\Delta(x,y) \subseteq Fm$ are called ***1-protoregular*** in the literature. Clearly, the regularly weakly algebraizable logics are exactly the assertional logics of 1-protoregular classes of algebras. Among these are the ***subtractive varieties***, which are pointed varieties that have a binary term $\alpha(x,y)$ making the equations $\alpha(x,x) \approx \top$ and $\alpha(\top,x) \approx x$ valid; these varieties have an important role in some recent works on the interplay between abstract algebraic logic and universal algebra, as witnessed for instance by [3, 237].

EXAMPLES 6.126

1. Any extension and any expansion of a regularly weakly algebraizable logic is also regularly weakly algebraizable. This follows from the same property for protoalgebraic logics (Proposition 6.11.1) plus condition (v) of Theorem 6.124.
2. All the assertional logics $\mathcal{L}^{\top}_{\mathsf{K}}$ of varieties K of ortholattices that are not all orthomodular, mentioned in Example 6.122.5 as being weakly algebraizable but not algebraizable, are indeed regularly weakly algebraizable, by Proposition 6.125. The "logic of ortholattices" $\mathcal{L}^{\top}_{\mathsf{OL}}$ is one of these.
3. In [72, Theorem 6.10] it is shown that under certain conditions, the assertional logic of a 1-protoregular quasivariety of algebras (which is regularly weakly algebraizable) is not algebraizable. This result, which arises as a generalization of the situation found in the ortholattice case, may provide further examples.

4. The so-called "logic of Andréka and Németi" is an ad hoc logic, defined syntactically in a language $\langle \leftrightarrow, \star \rangle$ of type $\langle 2, 1 \rangle$ by the axioms $\star x$ and $x \leftrightarrow x$, *Modus Ponens* for $\leftrightarrow$ (the rule $x, x \leftrightarrow y \rhd y$), and the following infinite sets of rules: $x \rhd \varphi(x) \leftrightarrow \varphi(\star x)$ and $x \rhd \varphi(\star x) \leftrightarrow \varphi(x)$, for all $\varphi(x) \in Fm$. It is obviously protoalgebraic, and one can show that it is truth-equational using Proposition 6.94, because one can prove that if Γ is a theory of this logic, then $\Gamma = \{\alpha \in Fm : \alpha \equiv \star\alpha \ (\boldsymbol{\Omega}\Gamma)\}$. Thus, it is weakly algebraizable. This logic appears in [32, Appendix B], where it is used to show that monotonicity plus injectivity of the Leibniz operator does not imply its continuity. This shows that it is not finitely equivalential, but in [72] it is shown that it is not even equivalential, because its class of reduced models is not closed under $\mathbb{S}$. Hence it is not algebraizable, and it happens not to be regularly weakly algebraizable either, because when checking the last fact a non-unital reduced model is constructed. Therefore, it provides the needed example to show that the class of regularly weakly algebraizable logics is a real addition to the Leibniz hierarchy.
5. Finally, it is obvious that the class of assertional logics is larger than that of regularly weakly algebraizable ones; any non-protoalgebraic assertional logic witnesses this, for instance the ones mentioned after Theorem 6.109. ⊠

The examples in the two preceding lists show that another good part of the Leibniz hierarchy is really as depicted in Figure 9 on page 316, that is, that inclusions are proper and there are no other implications than those shown there.

Exercises for Section 6.4

EXERCISE 6.39. Prove Lemma 6.84 and use it to show that a class K of algebras is a $\boldsymbol{\tau}$-algebraic semantics for a logic $\mathcal{L}$ if and only if $\mathcal{L}$ is complete with respect to the class of matrices M defined in the lemma, and to obtain Corollary 6.89.

EXERCISE 6.40. Let $\mathcal{L}$ be a logic. Show that truth is implicitly definable in $\mathsf{Mod}\mathcal{L}$ if and only if $\mathcal{L}$ is inconsistent.

EXERCISE 6.41. Let $\mathcal{L}$ be a logic. Following the proof of Proposition 6.88, show that truth is implicitly definable in the class $\mathsf{LTMod}^*\mathcal{L}$ if and only if the Leibniz operator $\boldsymbol{\Omega}$ is injective over $\mathcal{T}h\mathcal{L}$.

EXERCISE 6.42. Let $\mathcal{L}$ be a logic having an algebraic semantics K, with $E(x)$ as set of defining equations. Assume that there is an equation $\varepsilon(x) \approx \delta(x) \in E(x)$ such that $\varepsilon(x) \vdash_{\mathcal{L}} x$ and $\emptyset \vdash_{\mathcal{L}} \delta(x)$. Show that $\mathcal{L}$ is actually truth-equational, with the single equation $\varepsilon(x) \approx \delta(x)$ defining truth in its reduced models.

HINT. Apply Proposition 6.94, with the help of Exercise 4.16.

EXERCISE 6.43. Let K be a class of algebras having a term $\varepsilon(x)$, in at most one variable x, such that $\mathsf{K} \vDash \varepsilon(x) \approx \varepsilon(\varepsilon(x))$, and let $\mathcal{L}$ be the logic having K as algebraic semantics with the single defining equation $x \approx \varepsilon(x)$. Check that the assumptions of Exercise 6.42 are satisfied, and conclude that $\mathcal{L}$ is then truth-equational, and has the same equation as its truth definition.

COMMENT. The situation described here includes a large number of examples of logics having algebraic semantics with a particular defining equation. This covers not only implicative logics (with $\varepsilon(x) = x \to x$) or assertional logics (with $\varepsilon(x) = \top$), but also other substructural logics with less standard $\varepsilon(x)$, such as $x \vee (x \to x)$ or $(x \wedge (x \to x)) \to x$. See [201].

EXERCISE 6.44. Let $\langle \boldsymbol{A}, F \rangle$ be a matrix and $\boldsymbol{\tau}$ a structural transformer. Prove that $\boldsymbol{\tau}$ defines truth in the matrix $\langle \boldsymbol{A}/\boldsymbol{\Omega}^{\boldsymbol{A}}F, F/\boldsymbol{\Omega}^{\boldsymbol{A}}F \rangle$ if and only if $F = \{a \in A : \boldsymbol{\tau}^{\boldsymbol{A}} a \subseteq \boldsymbol{\Omega}^{\boldsymbol{A}} F\}$.

EXERCISE 6.45. The class $\mathsf{K}(\mathcal{L}, \boldsymbol{\tau})$ is shown in Proposition 3.8 to be the largest $\boldsymbol{\tau}$-algebraic semantics of a logic $\mathcal{L}$, provided it has some. Prove that if $\mathcal{L}$ is truth-equational, with truth definition given by $\boldsymbol{\tau}$, then $\mathsf{K}(\mathcal{L}, \boldsymbol{\tau})$ is exactly the class of algebras $\boldsymbol{A}$ such that $Sol_{\boldsymbol{\tau}}^{\boldsymbol{A}} \in \mathcal{F}i_{\mathcal{L}}\boldsymbol{A}$, or, equivalently (see Exercise 6.46), such that $Sol_{\boldsymbol{\tau}}^{\boldsymbol{A}} = \min \mathcal{F}i_{\mathcal{L}}\boldsymbol{A}$.

EXERCISE 6.46. Let $\mathcal{L}$ be a truth-equational logic, with truth definition $\boldsymbol{\tau}$. Check that the proof of Proposition 6.95 actually establishes that for any algebra $\boldsymbol{A}$, $\min \mathcal{F}i_{\mathcal{L}}\boldsymbol{A} = Fg_{\mathcal{L}}^{\boldsymbol{A}} Sol_{\boldsymbol{\tau}}^{\boldsymbol{A}}$. Then extend the result in the proposition by showing that the equality $\min \mathcal{F}i_{\mathcal{L}}\boldsymbol{A} = Sol_{\boldsymbol{\tau}}^{\boldsymbol{A}}$ holds in all $\boldsymbol{A} \in \mathsf{Alg}\mathcal{L}$. Deduce from this that $\mathsf{Alg}\mathcal{L} \subseteq \mathsf{K}(\mathcal{L}, \boldsymbol{\tau})$.

EXERCISE 6.47. Use the last equality in Exercise 6.46 to prove that if $\boldsymbol{\tau}$ and $\boldsymbol{\tau}'$ are two structural transformers that witness the truth-equationality of $\mathcal{L}$, then $\boldsymbol{\tau}x \mathrel{=}\!\mathrel{|}\!\mathrel{\models}_{\mathsf{Alg}\mathcal{L}} \boldsymbol{\tau}'x$.

EXERCISE 6.48. Prove that for every logic $\mathcal{L}$, every algebra $\boldsymbol{A}$ and every $F \in \mathcal{F}i_{\mathcal{L}}\boldsymbol{A}$, the matrix $\langle \boldsymbol{A}/\widetilde{\boldsymbol{\Omega}}_{\mathcal{L}}^{\boldsymbol{A}}F, F/\widetilde{\boldsymbol{\Omega}}_{\mathcal{L}}^{\boldsymbol{A}}F \rangle$ is Suszko-reduced.

EXERCISE 6.49. Prove that for an arbitrary logic $\mathcal{L}$, the Suszko-reduced models of $\mathcal{L}$ are exactly those that are (isomorphic to) subdirect products of the (ordinary) reduced models of $\mathcal{L}$. Deduce from this that the class $\mathsf{Alg}\mathcal{L}$ is the class of algebraic reducts of the Suszko-reduced models of $\mathcal{L}$.

HINT. Look at the proof of Theorem 5.70.

EXERCISE 6.50. Check (recalling that only a finite number of variables occur in each formula) that it is really possible to define the substitution σ' and the formula ψ as required in the proof of Lemma 6.98. Check also that essentially in the same way you can prove that $\sigma \widetilde{\boldsymbol{\Omega}}_{\mathcal{L}} C_{\mathcal{L}} \Gamma \subseteq \widetilde{\boldsymbol{\Omega}}_{\mathcal{L}} C_{\mathcal{L}} \sigma \Gamma$ holds for any finite Γ or for any surjective substitution.

COMMENT. Observe that the result in Exercise 6.32 is actually a consequence of the last property. In Exercise 7.34 you will see that Fregean logics satisfy the same property, but for arbitrary Γ and σ.

EXERCISE 6.51. Prove that for any algebra $\boldsymbol{A}$, the Leibniz operator $\boldsymbol{\Omega}^{\boldsymbol{A}}$ is completely order-reflecting over $\mathcal{F}i_{\mathcal{L}}\boldsymbol{A}$ if and only if for all $F, G \in \mathcal{F}i_{\mathcal{L}}\boldsymbol{A}$, if $\widetilde{\boldsymbol{\Omega}}_{\mathcal{L}}^{\boldsymbol{A}} F \subseteq \boldsymbol{\Omega}^{\boldsymbol{A}} G$, then $F \subseteq G$ (Lemma 6.100).

EXERCISE 6.52. Show that a logic $\mathcal{L}$ is truth-equational if and only if for every $\Gamma \in \mathcal{T}h\mathcal{L}$, the g-matrix $\langle \boldsymbol{Fm}, (\mathcal{T}h\mathcal{L})^{\Gamma} \rangle$ is a full g-model of $\mathcal{L}$. Compare this result with Exercise 6.20.

EXERCISE 6.53. Prove that a logic $\mathcal{L}$ is truth-equational if and only if for every full g-model $\langle \boldsymbol{Fm}, \mathscr{C} \rangle$ of $\mathcal{L}$ on the formula algebra and every $\Gamma \in \mathscr{C}$, the g-matrix $\langle \boldsymbol{Fm}, \mathscr{C}^{\Gamma} \rangle$ is also a full g-model of $\mathcal{L}$.

EXERCISE 6.54.

1. Show, using Lemma 6.100, that if for an algebra $\boldsymbol{A}$ the Leibniz operator is completely order-reflecting over $\mathcal{F}i_{\mathcal{L}}\boldsymbol{A}$, then the Suszko operator $\widetilde{\boldsymbol{\Omega}}_{\mathcal{L}}^{\boldsymbol{A}}$ is order-reflecting, and hence injective, over $\mathcal{F}i_{\mathcal{L}}\boldsymbol{A}$.

2. Prove that the Suszko operator is injective over the $\mathcal{L}$-filters of arbitrary algebras if and only if condition (iii) in Proposition 6.103 holds for all $\boldsymbol{A}$ and all $F \in \mathcal{F}i_{\mathcal{L}}\boldsymbol{A}$.
   HINT. For the direct implication, use Proposition 6.97.5 and the monotonicity of the Suzko operator. The converse follows by combining other results.
3. Add protoalgebraicity to the previous point and compare with Exercise 6.22.
4. Conclude from point 2 that a logic $\mathcal{L}$ is truth-equational if and only if the Suszko operator is injective over the $\mathcal{L}$-filters of arbitrary algebras (thus proving Theorem 6.106).

EXERCISE 6.55. Use Corollary 6.110 and Lemma 4.41 to prove that if $\mathcal{L}$ is an assertional logic, then the class $\mathsf{Alg}^*\mathcal{L}$ is relatively point-regular.

EXERCISE 6.56. Prove that if a logic has theorems, then truth is implicitly definable in any unital class of its models. Use the logic $\mathcal{B}$ to show that this property may fail for logics without theorems.

EXERCISE 6.57. Prove that, for an arbitrary logic $\mathcal{L}$, $x \equiv y \ (\widetilde{\boldsymbol{\Omega}}_{\mathcal{L}} C_{\mathcal{L}}\{x,y\})$ if and only if for every $\delta(x,\vec{z}) \in Fm$, $x,y,\delta(x,\vec{z}) \vdash_{\mathcal{L}} \delta(y,\vec{z})$. Use this property to complete the proof of Theorem 6.112 given in the text.

EXERCISE 6.58. In [201, Proposition 18] it is proved (using Exercise 6.49 and some model theory) that, for an arbitrary logic $\mathcal{L}$, a structural transformer defines truth in $\mathsf{Mod}^*\mathcal{L}$ if and only if it defines truth in the class of the Suszko-reduced models of $\mathcal{L}$. As a consequence, a logic is truth-equational if and only if truth is equationally definable (by the same transformer) in the class of its Suszko-reduced models. Use this fact and Proposition 6.97.6 to show that $\mathsf{Alg}\mathcal{L}$ has the same properties as $\mathsf{Alg}^*\mathcal{L}$ in Lemma 6.91, Corollary 6.92, Proposition 6.95 (this is Exercise 6.46), Theorem 6.109, and Theorem 6.111, replacing $\mathsf{Mod}^*\mathcal{L}$ by the class of the Suszko-reduced models of $\mathcal{L}$, and the Leibniz operator by the Suszko operator, as is appropriate.

EXERCISE 6.59. Let $\boldsymbol{A}$ be any algebra such that the Leibniz operator is monotone on $\mathcal{F}i_{\mathcal{L}}\boldsymbol{A}$. Prove that the Leibniz operator is injective on $\mathcal{F}i_{\mathcal{L}}\boldsymbol{A}$ if and only if it is order-reflecting on $\mathcal{F}i_{\mathcal{L}}\boldsymbol{A}$. Use this to prove that a protoalgebraic logic is truth-equational (hence, weakly algebraizable) if and only if the Leibniz operator is order-reflecting on $\mathcal{F}i_{\mathcal{L}}\boldsymbol{A}$ for all algebras $\boldsymbol{A}$.

COMMENT. It is not true in general that a monotone and injective function between two ordered sets is order-reflecting.

EXERCISE 6.60. Convince yourself that the tricky point spotted in the proof of Lemma 6.118 really works as stated.

HINT. The homomorphism h can be modified by giving all possible values in A to a variable different from x; such a variable is one of the formulas appearing when $\vec{\delta}$ ranges over all (sequences of) formulas.

EXERCISE 6.61. Using Theorem 6.120, prove the following THEOREM, a strengthening of Theorem 6.117: For every logic $\mathcal{L}$, the following conditions are equivalent:

(i) $\mathcal{L}$ is weakly algebraizable.
(ii) The function $\Gamma \mapsto (\mathcal{T}h\mathcal{L})^{\Gamma}$ is a bijection, and hence a dual lattice isomorphism, between the sets $\mathcal{T}h\mathcal{L}$ and $\mathrm{FCS}_{\mathcal{L}}\boldsymbol{Fm}$.
(iii) $\boldsymbol{\Omega}\colon \mathcal{T}h\mathcal{L} \cong \mathrm{Con}_{\mathsf{Alg}\mathcal{L}}\boldsymbol{Fm}$ as complete lattices.
(iv) $\boldsymbol{\Omega}\colon \mathcal{T}h\mathcal{L} \cong \mathrm{Con}_{\mathsf{Alg}^*\mathcal{L}}\boldsymbol{Fm}$ as complete lattices.

Exercise 6.62. Consider the parametrised versions of the conditions (ALG2)–(ALG4) given in Theorem 6.121 (notice that parameters appear only in Δ), and prove that still (ALG1) + (ALG4_p) $\iff$ (ALG2_p) + (ALG3_p).

Exercise 6.63. Discuss the conditions that expansions and fragments of a weakly algebraizable logic have to satisfy in order to be weakly algebraizable as well.

Exercise 6.64. Complete the sketch of the proof of Theorem 6.124 given in the text by proving the implications (ii)$\Rightarrow$(v) and (v)$\Rightarrow$(vi).

Hint. For the first, combine Theorems 6.111 and 6.109 and Lemma 6.118, taking into account that the term $\top$ in these results is assumed to be a theorem of $\mathcal{L}$ in at most the variable x, and is constant over $\mathsf{Alg}^*\mathcal{L}$; from this you can prove that $\vdash_\mathcal{L} \Delta\langle\top(x),\top(z)\rangle$. The second is proved directly, as in Theorem 3.52, but in order to prove $y \dashv\vdash_\mathcal{L} \Delta\langle y,\top\rangle$ from (v) you must handle parameters with care. In both cases you will need the properties of the sets of congruence formulas with parameters mentioned in Exercise 6.28.

Exercise 6.65. Prove that a logic $\mathcal{L}$ is regularly weakly algebraizable if and only if it has $\mathsf{Alg}^*\mathcal{L}$ as an algebraic semantics with $x \approx \top$ as defining equation, where $\top$ is a constant term of $\mathsf{Alg}^*\mathcal{L}$, and $\mathsf{Alg}^*\mathcal{L}$ is closed under subdirect products.

Hint. Use Theorem 6.17.

6.5 Algebraizable logics revisited

One of the strengths of the theory of algebraizability is that there are several approaches to the notion, which turn out to give equivalent characterizations from quite different points of view. While, as seen in the preceding sections, this also happens for other classes in the Leibniz hierarchy, it reaches its peak in the case of the algebraizable ones. To start with, *algebraizable logics* are defined in Section 3.2, in the approach based on the syntactic relations between consequences expressed by structural transformers, as those for which there is a set of formulas $\Delta(x,y) \subseteq Fm$ (without parameters), a set of equations $E(x) \subseteq Eq$ and of a class of algebras K such that the following four conditions are satisfied:

(ALG1) $\Gamma \vdash_\mathcal{L} \varphi \iff E(\Gamma) \vDash_\mathsf{K} E(\varphi)$

i.e., $E(x)$ interprets $\vdash_\mathcal{L}$ into $\vDash_\mathsf{K}$

(ALG2) $\Theta \vDash_\mathsf{K} \varepsilon \approx \delta \iff \Delta(\Theta) \vdash_\mathcal{L} \Delta(\varepsilon,\delta)$

i.e., $\Delta(x,y)$ interprets $\vDash_\mathsf{K}$ into $\vdash_\mathcal{L}$

(ALG3) $x \dashv\vdash_\mathcal{L} \Delta(E(x))$

(ALG4) $x \approx y \mathrel{=\!\mid\!\vDash_\mathsf{K}} E(\Delta(x,y))$

Recall that actually (ALG1) + (ALG4) $\iff$ (ALG2) + (ALG3). When the set Δ is finite, the logic is called *finitely algebraizable*.

Most of the theory of algebraizable logics is developed in Sections 3.2–3.5; the aim of the present section is thus quite limited: to see how they fit into the Leibniz hierarchy, what their relations with other classes in the hierarchy are, how some of the results already obtained there can now be viewed in a larger context, and to prove just a few additional properties. The first result is the central one:

THEOREM 6.127
Let $\mathcal{L}$ be a logic. The following conditions are equivalent:

(i) *$\mathcal{L}$ is algebraizable.*

(ii) *$\mathcal{L}$ is equivalential and weakly algebraizable.*

(iii) *$\mathcal{L}$ is equivalential and truth-equational.*

PROOF: (i)⇒(ii) By Theorem 4.56, the Leibniz operator is in particular monotone and injective over $\mathcal{F}i_{\mathcal{L}}\boldsymbol{A}$ for any $\boldsymbol{A}$, and by Corollary 6.116 this proves weak algebraizability. On the other hand, by Proposition 6.66, every algebraizable logic is equivalential.

(ii)⇒(iii) is trivial.

(iii)⇒(i) Since $\mathcal{L}$ is equivalential, by Corollary 6.83 there is a set $\Delta(x,y)$ of congruence formulas (without parameters) that interprets $\vDash_{\mathsf{Alg}^*\mathcal{L}}$ into $\vdash_{\mathcal{L}}$; thus, condition (ALG2) is satisfied for $\mathsf{K} := \mathsf{Alg}^*\mathcal{L}$. Since $\mathcal{L}$ is truth-equational, it has a set $E(x)$ of equations defining truth in $\mathsf{Mod}^*\mathcal{L}$. By Lemma 6.118, this property is equivalent to saying that condition (ALG3) is also satisfied for any set of congruence formulas for $\mathcal{L}$ (here without parameters). As recalled above, conditions (ALG2) and (ALG3) together establish algebraizability. ⊠

COROLLARY 6.128
A logic is finitely algebraizable if and only if it is both finitely equivalential and weakly algebraizable, and if and only if it is both finitely equivalential and truth-equational. ⊠

These two results settle the relative positions of these classes in the diagram of the Leibniz hierarchy of Figure 9 on page 316. We have already seen examples showing that all the just mentioned classes are different.

It is interesting to realize that these characterizations can be formulated as the conjunction of the two definability conditions that have appeared in the previous sections (recall Corollary 6.56 and Proposition 6.94):

COROLLARY 6.129
For a logic $\mathcal{L}$ the following conditions are equivalent:

(i) *$\mathcal{L}$ is algebraizable.*

(ii) *The class $\mathsf{Mod}^*\mathcal{L}$ has the identity relation definable from the filter (through a set of formulas without parameters) and has its filters definable from the identity relation (through a set of equations).*

(iii) *The class $\mathsf{Mod}\mathcal{L}$ has the Leibniz congruence definable from the filter (through a set of formulas without parameters) and has its filters definable from the Leibniz congruence (through a set of equations).* ⊠

This is an enhanced version of Theorem 4.60, which is proved in a rather ad hoc manner. Exercise 6.66 contains still another, more general form of this result.

After the work in the preceding sections, the purely syntactic characterization of algebraizability by conditions (R), (Sym), (Trans), (Re) and (ALG3) contained in Theorem 3.21 becomes simpler, and more understandable:

PROPOSITION 6.130
A logic $\mathcal{L}$ is algebraizable if and only if it satisfies conditions (R), (Re), (MP) *and* (ALG3) *for some $\Delta(x,y) \subseteq Fm$ and some $E(x) \subseteq Eq$; that is, if and only if it is equivalential with a set $\Delta(x,y)$ of congruence formulas and there is a set of equations $E(x) \subseteq Eq$ such that $x \dashv\vdash_{\mathcal{L}} \Delta(E(x))$.*

PROOF: Left to you as Exercise 6.67. ⊠

This justifies the claim made after Theorem 3.21 that conditions (Sym) and (Trans) could be replaced by (MP) in that result. By combining Definition 6.116 with Theorem 6.127, Corollary 6.69 and Theorem 6.117, which characterize equivalential and weakly algebraizable logics in terms of properties of the Leibniz operator, one obtains a new proof of an already known lattice-theoretic characterization of algebraizability:

THEOREM 6.131
For each logic $\mathcal{L}$ the following conditions are equivalent:

(i) *$\mathcal{L}$ is algebraizable.*
(ii) *The operator $\boldsymbol{\Omega}$ on $\mathcal{T}h\mathcal{L}$ is monotone, injective and commutes with substitutions.*
(iii) *For every $\boldsymbol{A}$, the operator $\boldsymbol{\Omega}^{\boldsymbol{A}}$ on $\mathcal{F}i_{\mathcal{L}}\boldsymbol{A}$ is monotone, injective and commutes with endomorphisms.*
(iv) *For every $\boldsymbol{A}$, $\boldsymbol{\Omega}^{\boldsymbol{A}} \colon \mathcal{F}i_{\mathcal{L}}\boldsymbol{A} \cong \mathrm{Con}_{\mathsf{Alg}\mathcal{L}}\boldsymbol{A}$ is a lattice isomorphism that commutes with endomorphisms.* ⊠

The addition of the result in Theorem 6.71 to the previous one produces a new characterization (recall that continuity implies monotonicity):

THEOREM 6.132
For each logic $\mathcal{L}$ the following conditions are equivalent:

(i) *$\mathcal{L}$ is finitely algebraizable.*
(ii) *The operator $\boldsymbol{\Omega}$ on $\mathcal{T}h\mathcal{L}$ is injective and continuous.*
(iii) *The operator $\boldsymbol{\Omega} \colon \mathcal{T}h\mathcal{L} \cong \mathrm{Con}_{\mathsf{Alg}\mathcal{L}}\boldsymbol{Fm}$ is a continuous lattice isomorphism.*
(iv) *For every algebra $\boldsymbol{A}$, the operator $\boldsymbol{\Omega}^{\boldsymbol{A}}$ on $\mathcal{F}i_{\mathcal{L}}\boldsymbol{A}$ is injective and continuous.*
(v) *For every algebra $\boldsymbol{A}$, the operator $\boldsymbol{\Omega}^{\boldsymbol{A}} \colon \mathcal{F}i_{\mathcal{L}}\boldsymbol{A} \cong \mathrm{Con}_{\mathsf{Alg}\mathcal{L}}\boldsymbol{A}$ is a continuous lattice isomorphism.*

For a finitary $\mathcal{L}$, condition (i) *can be rephrased as "$\mathcal{L}$ is BP-algebraizable".* ⊠

The finitely algebraizable logics whose equivalent algebraic semantics is a quasivariety[†] enjoy a simpler characterization, in which neither commutativity

[†] Recall that this is always the case for finitary logics, but there are some infinitary ones for which this is not the case (Example 3.41).

with endomorphisms nor continuity are needed; since it is formulated for an *arbitrary* quasivariety K, its proof needs an additional, final step.

THEOREM 6.133 (ISOMORPHISM THEOREM, "ORIGINAL" VERSION)
Let $\mathcal{L}$ be a logic (resp., a finitary logic), and let K be a quasivariety. The logic $\mathcal{L}$ is finitely algebraizable (resp., BP-algebraizable) and has the class K as equivalent algebraic semantics if and only if for every algebra $\boldsymbol{A}$, the Leibniz operator is an isomorphism $\boldsymbol{\Omega}^{\boldsymbol{A}} \colon \mathcal{F}i_{\mathcal{L}}\boldsymbol{A} \cong \mathrm{Con}_{\mathsf{K}}\boldsymbol{A}$.

PROOF: The direct implication is a particular case of both Theorems 6.131 and 6.132, since it is assumed that K is the equivalent algebraic semantics of $\mathcal{L}$, hence $\mathsf{K} = \mathsf{Alg}\mathcal{L}$. For the converse, it turns out that, working in the formula algebra, any isomorphism between the lattices $\mathcal{T}h\mathcal{L}$ and $\mathrm{Con}_{\mathsf{K}}\boldsymbol{Fm}$ is automatically continuous; the proof is for $\boldsymbol{\Omega}$, but you will easily realize that the argument is general. In order to prove continuity (Definition 6.70), let $\{\Gamma_i : i \in I\} \subseteq \mathcal{T}h\mathcal{L}$ be an upwards directed family such that $\bigcup \Gamma_i \in \mathcal{T}h\mathcal{L}$. As an order isomorphism between complete lattices, $\boldsymbol{\Omega}$ preserves the supremum of any family, thus $\boldsymbol{\Omega} \bigvee \Gamma_i = \bigvee \boldsymbol{\Omega}\Gamma_i$. Now, the assumption that $\bigcup \Gamma_i \in \mathcal{T}h\mathcal{L}$ implies that $\bigvee \Gamma_i = \bigcup \Gamma_i$ and hence that $\boldsymbol{\Omega} \bigvee \Gamma_i = \boldsymbol{\Omega} \bigcup \Gamma_i$. On the other hand, monotonicity of $\boldsymbol{\Omega}$ implies that the family $\{\boldsymbol{\Omega}\Gamma_i : i \in I\} \subseteq \mathrm{Con}_{\mathsf{K}}\boldsymbol{Fm}$ is also upwards directed, and that K is a quasivariety means that $\vDash_{\mathsf{K}}$ is finitary, and therefore that $\mathrm{Con}_{\mathsf{K}}\boldsymbol{Fm}$, which is its set of theories (Proposition 1.75), is inductive, which implies that $\bigcup \boldsymbol{\Omega}\Gamma_i \in \mathrm{Con}_{\mathsf{K}}\boldsymbol{Fm}$; as a consequence, $\bigvee \boldsymbol{\Omega}\Gamma_i = \bigcup \boldsymbol{\Omega}\Gamma_i$. Putting everything together, we obtain that $\boldsymbol{\Omega} \bigcup \Gamma_i = \bigcup \boldsymbol{\Omega}\Gamma_i$. Thus, $\boldsymbol{\Omega}$ is injective and continuous on $\mathcal{T}h\mathcal{L}$, which by Theorem 6.132 shows that $\mathcal{L}$ is finitely algebraizable (hence, for a finitary $\mathcal{L}$, BP-algebraizable). It remains to prove that the class K is the equivalent algebraic semantics of $\mathcal{L}$: since $\boldsymbol{A} \in \mathsf{K}$ if and only if $\mathrm{Id}_A \in \mathrm{Con}_{\mathsf{K}}\boldsymbol{A}$, and by the assumption this happens if and only if $\mathrm{Id}_A = \boldsymbol{\Omega}^{\boldsymbol{A}}F$ for some $F \in \mathcal{F}i_{\mathcal{L}}\boldsymbol{A}$, which is equivalent to saying that $\boldsymbol{A} \in \mathsf{Alg}^*\mathcal{L}$, we obtain that $\mathsf{K} = \mathsf{Alg}^*\mathcal{L}$, as desired. ⊠

Theorem 6.133 contains one of the exact forms of the Isomorphism Theorem for BP-algebraizable logics originally proved by Blok and Pigozzi in [32]; it is certainly the most famous, and *the most useful* in practice. Most of the logics to be studied are finitary, and most often there are reasons to think that an already determined quasivariety K should be the equivalent algebraic semantics; in this situation, proving this isomorphism shows at the same time that the logic is BP-algebraizable and that the conjectured quasivariety is indeed the equivalent algebraic semantics.

It is interesting to compare the characterizations in Theorems 6.131 and 6.132 with the ones obtained in Theorems 3.58 and 4.56. These two are stronger in the sense that they are formulated for *some* generalized quasivariety K instead of $\mathsf{Alg}\mathcal{L}$, and for *some* isomorphism instead of $\boldsymbol{\Omega}^{\boldsymbol{A}}$; and they actually show that K has to equal $\mathsf{Alg}\mathcal{L}$. By contrast, some of the points in Theorem 6.131 are stronger because they refer only to properties of the Leibniz operator that do not involve its target set. Moreover, the context is different: here the isomorphism appears as a gradual strengthening of the properties of the Leibniz operator through the Leibniz hierarchy, while in previous chapters only the notion of an algebraizable logic is considered, and the isomorphism appears so-to-speak directly as given by the residuals of the evaluated transformers establishing algebraizability. It is possible, even in the present context, to show that the statements of Theorems 6.131 and 6.132 do hold for an arbitrary class of algebras K instead of $\mathsf{Alg}\mathcal{L}$, and afterwards show that they coincide; see Exercise 6.68. Finally, comparing Theorem 6.133 with Theorem 4.56 shows that dealing with a quasivariety (hence, with a finitary $\vDash_{\mathsf{K}}$) dispenses with commutativity with endomorphisms and directly produces finite algebraizability, by proving contintuity.

As a further application of the previous isomorphisms, we can determine how the property that $\mathsf{Alg}\mathcal{L}$ is a quasivariety characterizes in different ways certain modes of algebraizability.

COROLLARY 6.134

1. *If $\mathcal{L}$ is weakly algebraizable and the class $\mathsf{Alg}\mathcal{L}$ is a quasivariety, then $\mathcal{L}$ is finitely algebraizable.*
2. *If $\mathsf{Alg}\mathcal{L}$ is a quasivariety, then $\mathcal{L}$ is weakly algebraizable if and only if $\mathcal{L}$ is finitely algebraizable.*
3. *If $\mathcal{L}$ is finitary and weakly algebraizable, then $\mathcal{L}$ is finitely algebraizable (hence, BP-algebraizable) if and only if $\mathsf{Alg}\mathcal{L}$ is a quasivariety.*

PROOF: 1 is a direct consequence of Theorems 6.117 and 6.133, and 2 and 3 follow from 1, taking into account that $\mathsf{Alg}\mathcal{L}$ is always a quasivariety when $\mathcal{L}$ is BP-algebraizable. ⊠

It is interesting to record here the stronger result obtained by Czelakowski and Jansana in [72, Theorem 5.2] by using more advanced tools:

PROPOSITION 6.135

If $\mathcal{L}$ is a weakly algebraizable logic, then $\mathcal{L}$ is algebraizable if and only if the class $\mathsf{Alg}\mathcal{L}$ is closed under $\mathbb{S}$. ⊠

Full generalized models of algebraizable logics

There is little to say here, since already for weakly algebraizable logics all filters are Leibniz and hence full g-models are determined by arbitrary filters in a canonical way (Theorem 6.117). Hence the same holds for all kinds of algebraizable logics. Perhaps the only novelty is that the characterization of finite algebraizability in terms of the Leibniz operator, contained in Theorem 6.132, has a (rather artificial) counterpart in terms of the Tarski operator. Recall that $\mathcal{FCO}_{\mathcal{L}}\boldsymbol{A}$ denotes the lattice of closure operators of full g-models of a logic $\mathcal{L}$ on the algebra $\boldsymbol{A}$ and that this lattice is dually isomorphic to the lattice $\mathcal{FCS}_{\mathcal{L}}\boldsymbol{A}$ of the corresponding closure systems, which appears in the Isomorphism Theorem 5.95 and related results, and in Theorems 6.43 and 6.117; the order in $\mathcal{FCO}_{\mathcal{L}}\boldsymbol{A}$ is defined by $C \leqslant C'$ if and only if $CX \subseteq C'X$ for all $X \subseteq A$, it is dual to the subset order between the closure systems $\mathscr{C}' \subseteq \mathscr{C}$, and by Proposition 5.96 the infimum in $\mathrm{FCS}_{\mathcal{L}}\boldsymbol{A}$ (which corresponds to the supremum in $\mathrm{FCO}_{\mathcal{L}}\boldsymbol{A}$) is intersection.

THEOREM 6.136
Let $\mathcal{L}$ be a finitary and weakly algebraizable logic. Then $\mathcal{L}$ is finitely algebraizable (hence, BP-algebraizable) if and only if for each algebra $\boldsymbol{A}$ the operator $\widetilde{\boldsymbol{\Omega}}^{\boldsymbol{A}}$ is ***continuous*** *over $\mathrm{FCO}_{\mathcal{L}}\boldsymbol{A}$ in the following sense: If $\{C_i : i \in I\} \subseteq \mathrm{FCO}_{\mathcal{L}}\boldsymbol{A}$ is an upwards directed family, then*

$$\widetilde{\boldsymbol{\Omega}}^{\boldsymbol{A}} \textstyle\bigvee^{\mathrm{f}}_{i\in I} C_i = \bigcup_{i\in I} \widetilde{\boldsymbol{\Omega}}^{\boldsymbol{A}} C_i ,$$

where $\bigvee^{\mathrm{f}}$ denotes the supremum in the lattice $\mathrm{FCO}_{\mathcal{L}}\boldsymbol{A}$.

PROOF: Consider the function $\Phi\colon F \mapsto (\mathcal{F}i_{\mathcal{L}}\boldsymbol{A})^F$. By Theorem 6.117(ii), this is a dual isomorphism between the lattice $\mathcal{F}i_{\mathcal{L}}\boldsymbol{A}$ and the lattice $\mathrm{FCS}_{\mathcal{L}}\boldsymbol{A}$. Protoalgebraicity implies that $\boldsymbol{\Omega}^{\boldsymbol{A}} = \widetilde{\boldsymbol{\Omega}}^{\boldsymbol{A}} \circ \Phi$ and hence that $\widetilde{\boldsymbol{\Omega}}^{\boldsymbol{A}} = \boldsymbol{\Omega}^{\boldsymbol{A}} \circ \Phi^{-1}$. Now let $\{F_i : i \in I\} \subseteq \mathcal{F}i_{\mathcal{L}}\boldsymbol{A}$ be an upwards directed family; then by finitarity $\bigcup F_i \in \mathcal{F}i_{\mathcal{L}}\boldsymbol{A}$ and clearly $(\mathcal{F}i_{\mathcal{L}}\boldsymbol{A})^{\bigcup F_i} = \bigcap (\mathcal{F}i_{\mathcal{L}}\boldsymbol{A})^{F_i}$, that is, $\Phi \bigcup F_i = \bigcap \Phi F_i$. If the function Φ is "dualized", without changing the symbol, so that it produces the closure operator of the full g-model instead of its closure system, then it becomes an isomorphism between the lattices $\mathcal{F}i_{\mathcal{L}}\boldsymbol{A}$ and $\mathrm{FCO}_{\mathcal{L}}\boldsymbol{A}$, and $\Phi \bigcup F_i = \bigvee^{\mathrm{f}} \Phi F_i$. Let us now proceed to the proof:

($\Rightarrow$) If $\mathcal{L}$ is finitely algebraizable, then it is finitely equivalential, and by Theorem 6.71, $\boldsymbol{\Omega}^{\boldsymbol{A}}$ is continuous over $\mathcal{F}i_{\mathcal{L}}\boldsymbol{A}$ for every $\boldsymbol{A}$. Let us show that then $\widetilde{\boldsymbol{\Omega}}^{\boldsymbol{A}}$ is continuous over $\mathrm{FCO}_{\mathcal{L}}\boldsymbol{A}$: If $\{C_i : i \in I\} \subseteq \mathrm{FCO}_{\mathcal{L}}\boldsymbol{A}$ is upwards directed, the family $\{F_i := \Phi^{-1} C_i : i \in I\} \subseteq \mathcal{F}i_{\mathcal{L}}\boldsymbol{A}$ is also upwards directed, and then

$$\begin{aligned} \widetilde{\boldsymbol{\Omega}}^{\boldsymbol{A}} \textstyle\bigvee^{\mathrm{f}} C_i &= \boldsymbol{\Omega}^{\boldsymbol{A}} \Phi^{-1} \textstyle\bigvee^{\mathrm{f}} \Phi F_i = \boldsymbol{\Omega}^{\boldsymbol{A}} \Phi^{-1} \Phi \bigcup F_i \\ &= \boldsymbol{\Omega}^{\boldsymbol{A}} \textstyle\bigcup F_i = \bigcup \boldsymbol{\Omega}^{\boldsymbol{A}} F_i = \bigcup \boldsymbol{\Omega}^{\boldsymbol{A}} \Phi^{-1} C_i = \bigcup \widetilde{\boldsymbol{\Omega}}^{\boldsymbol{A}} C_i . \end{aligned}$$

($\Leftarrow$) If $\{F_i : i \in I\} \subseteq \mathcal{F}i_{\mathcal{L}}\boldsymbol{A}$ is upwards directed, then $\{\Phi F_i : i \in I\} \subseteq \mathrm{FCO}_{\mathcal{L}}\boldsymbol{A}$ is also upwards directed, and the continuity of $\widetilde{\boldsymbol{\Omega}}^{\boldsymbol{A}}$ implies that of $\boldsymbol{\Omega}^{\boldsymbol{A}}$, because $\boldsymbol{\Omega}^{\boldsymbol{A}} \bigcup F_i = \widetilde{\boldsymbol{\Omega}}^{\boldsymbol{A}} \Phi \bigcup F_i = \widetilde{\boldsymbol{\Omega}}^{\boldsymbol{A}} \bigvee^{\mathrm{f}} \Phi F_i = \bigcup \widetilde{\boldsymbol{\Omega}}^{\boldsymbol{A}} \Phi F_i = \bigcup \boldsymbol{\Omega}^{\boldsymbol{A}} F_i$. ⊠

COROLLARY 6.137
A finitary logic $\mathcal{L}$ is finitely algebraizable (hence, BP-algebraizable) if and only if for every $\boldsymbol{A}$, the function $F \mapsto (\mathcal{F}i_{\mathcal{L}}\boldsymbol{A})^F$ is a bijection between $\mathcal{F}i_{\mathcal{L}}\boldsymbol{A}$ and $\mathrm{FCS}_{\mathcal{L}}\boldsymbol{A}$, and the Tarski operator $\widetilde{\boldsymbol{\Omega}}^{\boldsymbol{A}}$ is continuous over $\mathrm{FCO}_{\mathcal{L}}\boldsymbol{A}$ in the sense of Theorem 6.136. ⊠

Concerning the removal of finitarity from the assumptions in the last two results, see Exercise 6.70.

On the bridge theorem concerning the Deduction Theorem

Corollary 3.86 states that, if $\mathcal{L}$ is a BP-algebraizable logic with equivalent quasivariety semantics K, then $\mathcal{L}$ satisfies the DDT if and only if K has EDPRC (equationally definable principal relative congruences). This is partially proved in Section 3.6 as a consequence of the equivalence between $\vdash_{\mathcal{L}}$ and the relative equational consequence $\vDash_{\mathsf{K}}$. The "partially" recalls that the "proof" given there relies on a purely universal algebraic fact whose proof is skipped: That the DDT for the consequence $\vDash_{\mathsf{K}}$ amounts to the quasivariety K having EDPRC. It is interesting to see how the same result can be obtained as a consequence of the bridge Theorem 6.28 for protoalgebraic logics and of the isomorphism between filters and congruences that holds in BP-algebraizable logics. The following result has already an interest by itself, as it can be considered a *bridge theorem* as well, although the algebraic property it involves is certainly not as intuitive as that of having EDPRC.

Theorem 6.138
Let $\mathcal{L}$ be a BP-algebraizable logic, with equivalent quasivariety semantics K. The logic $\mathcal{L}$ satisfies the DDT *if and only if for every $\boldsymbol{A}$, the join-semilattice $\mathrm{Con}_{\mathsf{K}}^{\omega}\boldsymbol{A}$ of the finitely generated K-congruences of $\boldsymbol{A}$ is dually residuated.*

Proof: $\mathcal{L}$ is finitary by assumption, and $\vDash_{\mathsf{K}}$ is finitary because K is a quasivariety. We already know that the closure operator of $\mathcal{L}$-filter generation of any algebra is finitary as well (Theorem 2.23), and it is easy to see that the same holds for the operator of K-congruence generation on any algebra (which by Theorem 1.76 corresponds to the natural notion of a "filter" of $\vDash_{\mathsf{K}}$). Therefore, in their corresponding lattices of closed sets ($\mathcal{F}i_{\mathcal{L}}\boldsymbol{A}$ and $\mathrm{Con}_{\mathsf{K}}\boldsymbol{A}$, respectively), the finitely generated members are the compact ones (Proposition 1.65). Since compactness in a lattice is preserved by lattice isomorphisms (Proposition 1.66), for any algebra $\boldsymbol{A}$, the isomorphism between the lattices $\mathcal{F}i_{\mathcal{L}}\boldsymbol{A}$ and $\mathrm{Con}_{\mathsf{K}}\boldsymbol{A}$ due to the algebraizability of $\mathcal{L}$ restricts to an isomorphism between the join-semilattices of the finitely generated members of each (Proposition 1.64). Since $\mathcal{L}$ is in particular protoalgebraic, by Theorem 6.28 $\mathcal{L}$ satisfies the DDT if and only if for every algebra $\boldsymbol{A}$, the join-semilattice $\mathcal{F}i_{\mathcal{L}}^{\omega}\boldsymbol{A}$ of the finitely generated members of $\mathcal{F}i_{\mathcal{L}}\boldsymbol{A}$ is dually residuated. Since being dually residuated is clearly a property preserved by isomorphisms, it follows that $\mathcal{L}$ satisfies the DDT if and only if for every algebra $\boldsymbol{A}$ the join-semilattice $\mathrm{Con}_{\mathsf{K}}^{\omega}\boldsymbol{A}$ of the finitely generated members of $\mathrm{Con}_{\mathsf{K}}\boldsymbol{A}$ is dually residuated. ⊠

The connection with EDPRC relies again on a purely algebraic property, whose proof is skipped:

Lemma 6.139
A quasivariety K has EDPRC *if and only if for every $\boldsymbol{A} \in \mathsf{K}$, the join-semilattice $\mathrm{Con}_{\mathsf{K}}^{\omega}\boldsymbol{A}$ of the finitely generated K-congruences of $\boldsymbol{A}$ is dually residuated.* ⊠

This lemma is a kind of "algebraic version" of Theorem 6.28; as a matter of fact, its proof is also an algebraic version of the proof of the latter: the role played in the proof of Theorem 6.28 by the Correspondence Theorem of protoalgebraic logics would be played here by the usual Correspondence Theorem of universal algebra; however, this proof hides several subtle, technical details of universal algebraic character that we cannot undertake here (one of these subtleties, comparing with

Theorem 6.28, is that the right-hand side of Lemma 6.139 says "for every $\boldsymbol{A} \in \mathsf{K}$" and not "for every $\boldsymbol{A}$").

THEOREM 6.140
Let $\mathcal{L}$ be a BP-algebraizable logic, with equivalent quasivariety semantics K. The logic $\mathcal{L}$ satisfies the DDT *if and only if* K *has* EDPRC.

PROOF: ($\Rightarrow$) Take $\boldsymbol{A} \in \mathsf{K}$ in Theorem 6.138 and apply Lemma 6.139.

($\Leftarrow$) By Lemma 6.139, if $\boldsymbol{A} \in \mathsf{K}$, then $\mathrm{Con}^{\omega}_{\mathsf{K}}\boldsymbol{A}$ is dually residuated. By algebraizability, this implies that $\mathcal{F}i^{\omega}_{\mathcal{L}}\boldsymbol{A}$ is dually residuated as well. The point is to show that this implies the same property for an arbitrary $\boldsymbol{A}$. Let $F_0 = \bigcap \mathcal{F}i_{\mathcal{L}}\boldsymbol{A}$. Then $\boldsymbol{A}/\boldsymbol{\Omega}^{\boldsymbol{A}}F_0 \in \mathsf{Alg}^*\mathcal{L} = \mathsf{K}$, therefore $\mathcal{F}i^{\omega}_{\mathcal{L}}(\boldsymbol{A}/\boldsymbol{\Omega}^{\boldsymbol{A}}F_0)$ is dually residuated. Since $\boldsymbol{\Omega}^{\boldsymbol{A}}$ is injective over the $\mathcal{L}$-filters, every $\mathcal{L}$-filter is Leibniz, and by Theorem 6.41, $F_0/\boldsymbol{\Omega}^{\boldsymbol{A}}F_0$ is the least $\mathcal{L}$-filter of $\boldsymbol{A}/\boldsymbol{\Omega}^{\boldsymbol{A}}F_0$. Then the Correspondence Theorem for protoalgebraic logics implies that $\mathcal{F}i_{\mathcal{L}}\boldsymbol{A} \cong \mathcal{F}i_{\mathcal{L}}(\boldsymbol{A}/\boldsymbol{\Omega}F_0)$, and this isomorphism restricts to one between the respective join-semilattices of compact members. Therefore, $\mathcal{F}i^{\omega}_{\mathcal{L}}\boldsymbol{A}$ is dually residuated, and by Theorem 6.28, this implies that $\mathcal{L}$ has the DDT. ⊠

Regularly algebraizable logics revisited

Regularly algebraizable logics are defined in Section 3.4 as the algebraizable logics that satisfy the additional property

$$x, y \vdash_{\mathcal{L}} \Delta(x, y) \tag{G}$$

for a nonempty set Δ of congruence formulas. There we saw that they satisfy a particular form of condition (ALG3), namely that $x \dashv\vdash_{\mathcal{L}} \Delta(x, \top)$ for any theorem $\top$ of $\mathcal{L}$ with at most the variable x, and that one can always take $x \approx \top$ as defining equation to witness algebraizability; this is shown in Theorem 3.52, where they are characterized by the conditions (R), (Sym), (Trans), (Re), (MP) and (G), but after Corollary 6.61, this list can be reduced to just (R), (Re), (MP) and (G). Thus, they can be defined directly over the class of equivalential logics:

PROPOSITION 6.141
A logic is regularly algebraizable if and only if it is equivalential and satisfies the set of rules (G) *for some nonempty set of congruence formulas.* ⊠

Although these logics do not enjoy the variety of characterizations of other classes in terms of the Leibniz operator, it is still natural to consider them in the Leibniz hierarchy and within the theory of matrices, as the next result shows.

THEOREM 6.142
Let $\mathcal{L}$ be an equivalential logic with theorems. The following conditions are equivalent:

(i) *$\mathcal{L}$ is regularly algebraizable.*

(ii) *The class $\mathsf{Mod}^*\mathcal{L}$ is unital.*

(iii) *The class $\mathsf{LTMod}^*\mathcal{L}$ is unital.*

(iv) *$\mathcal{L}$ is complete with respect to a unital class of matrices.*

(v) $x \dashv\vdash_{\mathcal{L}} \Delta(x, \top)$, *where* $\Delta(x,y)$ *is any non-empty set of congruence formulas and* $\top$ *is a formula of* $\mathcal{L}$ *in at most a variable* $z \neq x$.

(vi) $x \equiv y \ (\boldsymbol{\Omega} C_{\mathcal{L}}\{x,y\})$.

The formula $\top$ *in* (v) *is a theorem of* $\mathcal{L}$ *and an algebraic constant of* $\mathsf{Alg}^*\mathcal{L}$.

PROOF: Left to you as Exercise 6.72; if you have worked through the book up to this point, I am sure you have enough tools and experience to prove it completely. See also Exercise 6.73. ⊠

This is the parameter-less version of Theorem 6.124, and the absence of parameters makes proofs easier here; as in that case, notice the precise form of the constant $\top$ appearing in condition (v). The presence of (ii) suggests that these logics could be called *regularly equivalential* as well (see also the comment on page 143). By putting Theorem 6.124, Proposition 6.125 and Theorem 6.142 together, we clearly have (see Figure 9 on page 316):

PROPOSITION 6.143

The following classes of logics coincide:

- *The regularly algebraizable.*
- *The equivalential and assertional.*
- *The equivalential and regularly weakly algebraizable.*
- *The algebraizable and assertional.*
- *The algebraizable and regularly weakly algebraizable.*

A parallel result, with "finitely" added at the appropriate places, follows immediately, and other combinations are also possible.

Theorem 6.142 has obvious applications to the determination of whether certain algebraizable logics are regularly algebraizable or fail to be so:

EXAMPLES 6.144

1. Any extension of a regularly algebraizable logic is also regularly algebraizable. And so is any fragment containing the language necessary for the congruence formulas. This is because the notion is characterized syntactically by condition (G) in addition to the conditions for equivalential logics. By the same reason, this also holds for the expansions satisfying (Re) for the new connectives.
2. The algebraizable logics $\mathcal{FL}$, $\mathcal{R}$ and $\mathcal{RM}$ are mentioned in Example 3.50 as not being regularly algebraizable, without proof. By the previous point, it is enough to show it for $\mathcal{RM}$, which is the strongest of the three. In [115] the matrix $\langle \boldsymbol{A}_6, \{e,1\}\rangle$ is presented, where $\boldsymbol{A}_6$ is the algebra described in Figure 10; moreover, it is shown there that this is a reduced model of $\mathcal{RM}$. Thus, the class $\mathsf{Mod}^*\mathcal{RM}$ is not unital, which shows that the logic is not regularly algebraizable.

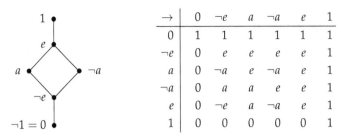

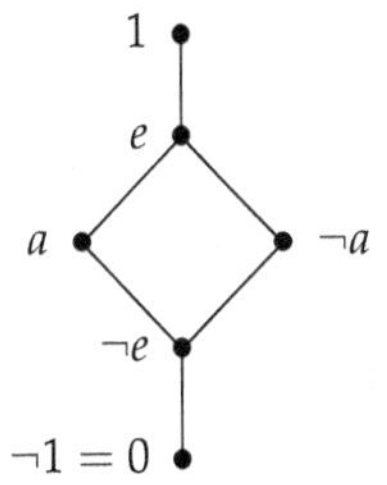

$\to$	0	$\neg e$	a	$\neg a$	e	1
0	1	1	1	1	1	1
$\neg e$	0	e	e	e	e	1
a	0	$\neg a$	e	$\neg a$	e	1
$\neg a$	0	a	a	e	e	1
e	0	$\neg e$	a	$\neg a$	e	1
1	0	0	0	0	0	1

FIGURE 10: The algebra $\boldsymbol{A}_6$ for Example 6.144.2. $\wedge$ and $\vee$ are the lattice operations, the negation is the involution determined by the names of the points, and implication is given explicitly in the table.

3. The paraconsistent logic $\mathcal{J}_3$ described in Example 3.20 and shown to be algebraizable, is not regularly algebraizable: It is not difficult to check that the algebra $\boldsymbol{A}_3$ described there is a simple algebra, that is, one that has no proper congruences. Thus, all the proper matrices it supports are reduced, and in particular $\langle \boldsymbol{A}_3, \{1/2, 1\}\rangle$ is a reduced model of $\mathcal{J}_3$ (actually, it is the matrix that defines the logic). The class $\mathsf{Mod}^*\mathcal{J}_3$ is hence not unital, and the logic is not regularly algebraizable. ⊠

The following result contains a characterization of regularly algebraizable logics among the algebraizable ones.

THEOREM 6.145
Let $\mathcal{L}$ be an algebraizable logic. Then $\mathcal{L}$ is regularly algebraizable if and only if the class $\mathsf{Alg}\mathcal{L}$ is a pointed class of algebras, with pointedness witnessed by some theorem $\top$ of $\mathcal{L}$ with at most the variable x, and such that in each $\boldsymbol{A} \in \mathsf{Alg}\mathcal{L}$, the set $\{\top^{\boldsymbol{A}}\}$ is an $\mathcal{L}$-filter of $\boldsymbol{A}$ (actually, the least one).

PROOF: ($\Rightarrow$) If $\mathcal{L}$ is regularly algebraizable, then it is assertional, and by Lemma 6.107 and Exercise 3.3, $\mathsf{Alg}\mathcal{L}$ is a pointed class, where the constant term is given by some theorem $\top$ of $\mathcal{L}$ with at most the variable x. By Theorem 3.63, it is relatively point-regular. The final condition follows from Proposition 6.95, because $\mathcal{L}$ is truth-equational as well.

($\Leftarrow$) By Theorem 6.142, it is enough to prove that the class of reduced models of $\mathcal{L}$ is unital. Take any $\langle \boldsymbol{A}, F\rangle \in \mathsf{Mod}^*\mathcal{L}$. Since $\top$ is a theorem of $\mathcal{L}$, $\{\top^{\boldsymbol{A}}\} \subseteq F$. But $\boldsymbol{A} \in \mathsf{Alg}^*\mathcal{L} = \mathsf{Alg}\mathcal{L}$, so by assumption $\{\top^{\boldsymbol{A}}\}$ is an $\mathcal{L}$-filter of $\boldsymbol{A}$ (thus, it is the least one). Now, by algebraizability, $\boldsymbol{\Omega}^{\boldsymbol{A}}$ is monotone over the $\mathcal{L}$-filters of $\boldsymbol{A}$, which implies that $\boldsymbol{\Omega}^{\boldsymbol{A}}\{\top^{\boldsymbol{A}}\} \subseteq \boldsymbol{\Omega}^{\boldsymbol{A}}F = \mathrm{Id}_A$, that is, $\boldsymbol{\Omega}^{\boldsymbol{A}}\{\top^{\boldsymbol{A}}\} = \mathrm{Id}_A$, and it is also injective, which implies that $F = \{\top^{\boldsymbol{A}}\}$. The matrix is thus unital, and therefore so is $\mathsf{Mod}^*\mathcal{L}$. ⊠

The additional condition that $\{\top^{\boldsymbol{A}}\}$ be an $\mathcal{L}$-filter, for all algebras in $\mathsf{Alg}\mathcal{L}$, cannot be deleted from the theorem, nor replaced by the condition that $\mathsf{Alg}\mathcal{L}$ is relatively point-regular (something we know is true in regularly algebraizable logics, by Theorem 3.63): in Example 6.144.3 we have just seen that the algebraizable logic $\mathcal{J}_3$ is not regularly algebraizable; however, its equivalent algebraic semantics is a pointed, relatively point-regular variety (see Example 3.20), and indeed pointedness is witnessed by $\top := x \to x$, which is a theorem of the logic; the problem is that $\{\top^{\boldsymbol{A}}\}$ is not a filter of the defining algebra $\boldsymbol{A}_3$. The equivalence, however, can be improved if one assumes that the logic is not just algebraizable, but an assertional one (which $\mathcal{J}_3$ is not); a partial result in this direction appears in Exercise 6.74.

In the literature, relative pont-regularity has been investigated in relation to BP-algebraizability, where $\mathsf{Alg}\mathcal{L}$ is a quasivariety; the following strong and interesting result is obtained:

THEOREM 6.146

1. *A logic is regularly BP-algebraizable if and only if it is the assertional logic of a pointed, relatively point-regular quasivariety.*
2. *For each algebraic language* $\boldsymbol{L}$, *the functions*

$$\begin{array}{ccc} \mathcal{L} & \longmapsto & \mathsf{Alg}\mathcal{L} \\ \mathcal{L}^{\top}_{\mathsf{K}} & \longleftarrow & \mathsf{K} \end{array}$$

establish a bijective correspondence (a dual order isomorphism, in fact) between the families† *of all the regularly BP-algebraizable logics of type* $\boldsymbol{L}$ *and all the pointed, relatively point-regular quasivarieties of* $\boldsymbol{L}$*-algebras.* ⊠

Although we cannot enter into the details of the proof, some reflection on how the process works may be interesting. The implication from left to right is clear from previous results. For the converse, the definition of the assertional logic $\mathcal{L}^{\top}_{\mathsf{K}}$ of a pointed quasivariety K automatically imposes that (ALG1) holds for the transformer given by $E(x) = \{x \approx \top\}$. In order to show that $\mathcal{L}^{\top}_{\mathsf{K}}$ is regularly BP-algebraizable, one needs only show the existence of a set $\Delta(x,y) \subseteq Fm$ satisfying (ALG4), which in this case takes the form

$$x \approx y \mathrel{=\!\mid\!\models_{\mathsf{K}}} \{\delta(x,y) \approx \top : \delta \in \Delta\}. \tag{92}$$

This is sufficient, because this equivalence trivially implies the consequence $x \approx \top, y \approx \top \vDash_{\mathsf{K}} \{\delta(x,y) \approx \top : \delta \in \Delta\}$, which by (ALG1) is condition (G): $x, y \vdash^{\top}_{\mathsf{K}} \Delta(x,y)$. Notice that (92) regards only the consequence $\vDash_{\mathsf{K}}$; thus the purely logical property of regular algebraizability amounts to a purely algebraic property of the class K. The proof that if K is a pointed, relatively point-regular quasivariety, then such a set Δ exists needs more specific universal algebra work, which is beyond the scope of this book; it can be found in [70, Section 5.2] and, more directly, in [75, Theorem 34].

† If you are set-theoretically fussy, you will notice that the first family is clearly a set, as all the logics have the same language, while in principle the second one is a "super-class", as each quasivariety is itself a proper class. This means that the isomorphism itself cannot be a set. However, issues like this one are normally disregarded in abstract algebraic logic.

You may want to compare this with Theorem 3.33, where a similar dual order isomorphism is established between logics and classes of algebras, namely between the extensions of a given algebraizable logic and the sub-generalized quasivarieties of its equivalent algebraic semantics.†

Observe that the above correspondence is independent of the form of the transformer given by the set Δ: if K is a pointed, relatively point-regular quasivariety, then there is one and only one regularly BP-algebraizable logic having K as its equivalent algebraic semantics. This is dual to the fact (known to us since Theorem 3.17) that an algebraizable logic can be algebraized in essentially only one way. One is tempted to say that this kind of quasivarieties can be "logified" in essentially only one way, with the proviso that by "logification" we mean here "regular algebraizability". The situation has become here symmetrical on both sides of the bridge.

Exercises for Section 6.5

EXERCISE 6.66. Prove the following strengthening of Corollary 6.129: A logic $\mathcal{L}$ is algebraizable if and only if it is complete with respect to a class of matrices that has the Leibniz congruence definable from the filter (through a set of formulas without parameters) and has the filters definable from the Leibniz congruence (through a set of equations).

HINT. Of course, only the converse implication has to be proved. You may want to do this by reducing the situation to that of Corollary 6.129, in the context of the present chapter; alternatively, you may prove the syntactic conditions of Theorem 3.21 directly.

EXERCISE 6.67. Prove Proposition 6.130.

EXERCISE 6.68. Reformulate Theorems 6.117, 6.131 and 6.132 so that they refer to the existence of a class K of algebras with the relevant property, instead of referring to $\mathsf{Alg}^*\mathcal{L}$ or to $\mathsf{Alg}\mathcal{L}$. Show that these reformulated statements hold, and that in each case an argument similar to the final step in the proof of Theorem 6.133 also shows that K must forcefully be equal to $\mathsf{Alg}\mathcal{L}$.

EXERCISE 6.69. Revise the proof of Theorem 6.133 and check that it also proves the following modification: Let $\mathcal{L}$ be a logic (resp., a finitary logic). Then $\mathcal{L}$ is finitely algebraizable (resp., BP-algebraizable) if and only if there is a quasivariety K such that the Leibniz operator is an isomorphism $\boldsymbol{\Omega} \colon \mathcal{T}h\mathcal{L} \cong \mathrm{Con}_{\mathsf{K}}\boldsymbol{Fm}$. Explain why the statement has been changed in one essential way.

EXERCISE 6.70. Determine the changes that should be done to the statement and proof of Theorem 6.136 in order to remove the assumption of finitarity of $\mathcal{L}$ in it. Compare with other results on continuity.

HINT. You have to find out how the assumption that the union of an upwards-directed family of filters is a filter can be formulated in terms of a family of closure operators (those corresponding to full g-models).

EXERCISE 6.71. Prove that an equivalential logic $\mathcal{L}$ is regularly algebraizable if and only if $\Gamma/\boldsymbol{\Omega}\Gamma = \{\Gamma\}$ for all $\Gamma \in \mathcal{T}h\mathcal{L}$.

† Theorems 7.28 and 7.54 also establish results of this kind, though in quite a different context (the Frege hierarchy).

EXERCISE 6.72. Prove the equivalence between the six conditions in Theorem 6.142, and the final assertion.

HINT. Some implications or equivalences have already appeared in this book; for instance, the equivalence between (i) and (ii) is hidden in the proof of Proposition 4.58, and that between (i) and (v) in the proof of Theorem 3.52, together with Exercise 3.34.

EXERCISE 6.73. Determine the implications or equivalences between the conditions in Theorem 6.142 that hold for an arbitrary logic.

COMMENT. Of course, points (i) and (v) do not make sense in general, as they involve (implicitly or explicitly) a set of congruence formulas, which in general need not exist.

EXERCISE 6.74. Let $\mathcal{L}$ be an assertional logic such that the class $\mathsf{Alg}^*\mathcal{L}$ is a variety. Prove that $\mathsf{Alg}^*\mathcal{L}$ is point-regular.

HINT. Use Theorem 6.111.

Chapter 7

Introduction to the Frege hierarchy

7.1 Overview

The Frege hierarchy is organized around several *replacement properties* that a logic and its full g-models may have. These properties can be formulated in general, for an arbitrary g-matrix, but their particularizations to logics have clearer interpretations.

You are strongly advised to review the definitions and basic properties of several notions already introduced in Definitions 1.36, 4.21, 5.30 and 5.36 and the ensuing material (see also Lemma 5.37 and Proposition 6.97): the *Frege relation* $\Lambda^A\mathscr{C}$ of a closure system $\mathscr{C}$ on a set A, the *Tarski congruence* $\widetilde{\Omega}^A\mathscr{C}$ of a g-matrix $\langle \boldsymbol{A}, \mathscr{C}\rangle$; and, for each $F \subseteq A$, its *Leibniz congruence* $\boldsymbol{\Omega}^{\boldsymbol{A}}F$, its *Suszko congruence* (relative to $\mathscr{C}$) $\widetilde{\Omega}^{\boldsymbol{A}}_{\mathscr{C}}F := \widetilde{\Omega}^{\boldsymbol{A}}\mathscr{C}^F$, and its *Frege relation* (relative to $\mathscr{C}$) $\Lambda^A_{\mathscr{C}}F := \Lambda^A\mathscr{C}^F$. Recall that $\mathscr{C}^F := \{G \in \mathscr{C} : F \subseteq G\}$. The three last notions define the corresponding *operators* on the set $\mathcal{F}i_{\mathcal{L}}\boldsymbol{A}$, and in the case of the formula algebra, on $\mathcal{T}h\mathcal{L}$. These notations may also be used with the associated closure operator C in the place of the closure system $\mathscr{C}$. In the particular case where $\mathscr{C} = \mathcal{F}i_{\mathcal{L}}\boldsymbol{A}$, since it is the logic $\mathcal{L}$ that actually determines the closure system, the following simpler notations are used: $\widetilde{\Omega}^{\boldsymbol{A}}_{\mathcal{L}}F := \widetilde{\Omega}^{\boldsymbol{A}}_{\mathcal{F}i_{\mathcal{L}}\boldsymbol{A}}F$, and $\Lambda^{\boldsymbol{A}}_{\mathcal{L}}F := \Lambda^{A}_{\mathcal{F}i_{\mathcal{L}}\boldsymbol{A}}F$. As usual, on the formula algebra the superscript is omitted, and the symbol for the logic takes the place of the closure system (or the closure operator), producing $\widetilde{\Omega}\mathcal{L}$ and $\Lambda\mathcal{L}$ (also denoted by $\dashv\vdash_{\mathcal{L}}$) as associated with the logic $\mathcal{L}$, and $\boldsymbol{\Omega}\Gamma$, $\widetilde{\Omega}_{\mathcal{L}}\Gamma$ and $\Lambda_{\mathcal{L}}\Gamma$, as associated with each $\Gamma \subseteq Fm$, the last two being still relative to the logic $\mathcal{L}$.

The next definition is partially anticipated at the end of Section 5.2, but this version goes a little further.

DEFINITION 7.1
A g-matrix $\langle \boldsymbol{A}, \mathscr{C} \rangle$ of type **L** *has* **the property of congruence**, PCONG, *when its Frege relation is a congruence of* $\boldsymbol{A}$ $(\boldsymbol{\Lambda}^{A}\mathscr{C} \in \operatorname{Con}\boldsymbol{A})$. *Equivalently, when* $\boldsymbol{\Lambda}^{A}\mathscr{C} = \widetilde{\boldsymbol{\Omega}}^{\boldsymbol{A}}\mathscr{C}$. *Still equivalently, when $\langle \boldsymbol{A}, \mathscr{C} \rangle$ is a model of the so-called* **rules of congruence**:

$$\frac{\{x_i \rhd y_i \,,\, y_i \rhd x_i \;:\; i < n\}}{\lambda x_0 \ldots x_{n-1} \rhd \lambda y_0 \ldots y_{n-1}} \quad \textit{for each } \lambda \in L,\ n = \operatorname{ar}\lambda \geqslant 1 \qquad \text{(CONG)}$$

A g-matrix $\langle \boldsymbol{A}, \mathscr{C} \rangle$ has **the strong property of congruence**, SPCONG, *when for any $F \in \mathscr{C}$, the g-matrix $\langle \boldsymbol{A}, \mathscr{C}^{F} \rangle$ has the* PCONG. *Equivalently, when* $\boldsymbol{\Lambda}^{A}_{\mathscr{C}}F = \widetilde{\boldsymbol{\Omega}}^{\boldsymbol{A}}_{\mathscr{C}}F$ *for all $F \in \mathscr{C}$, that is, when the Frege and the Suszko operators relative to $\mathscr{C}$ coincide.*

The term "strong property of congruence" is due to Herrmann [140]. There is no Gentzen-style formulation of the SPCONG in general, but Corollary 7.64 provides one for finitary g-matrices.

PROPOSITION 7.2
The PCONG *and the* SPCONG *are preserved by strict surjective homomorphisms between g-matrices, and thus in particular by reductions.*

PROOF: Due to its characterization in terms of Gentzen-style rules, the preservation of the PCONG follows from the general preservation result proved in Theorem 5.63 (Exercise 7.2 asks you to provide an essentially different proof). Now, let h be a strict surjective homomorphism from a g-matrix $\langle \boldsymbol{A}, \mathscr{C} \rangle$ to a g-matrix $\langle \boldsymbol{B}, \mathscr{D} \rangle$. By Theorem 5.48, h induces an order isomorphism between $\mathscr{C}$ and $\mathscr{D}$. By restricting this isomorphism and using Theorem 5.48 again, it follows that for each $F \in \mathscr{C}$, h induces a strict surjective homomorphism from $\langle \boldsymbol{A}, \mathscr{C}^{F} \rangle$ to $\langle \boldsymbol{B}, \mathscr{D}^{hF} \rangle$; and for each $G \in \mathscr{D}$, h induces a strict surjective homomorphism from $\langle \boldsymbol{A}, \mathscr{C}^{h^{-1}G} \rangle$ to $\langle \boldsymbol{B}, \mathscr{D}^{G} \rangle$. Given the definition of the SPCONG in terms of the PCONG, its preservation then follows from that of the PCONG. ⊠

These two congruence properties originate the four classes of logics that constitute the ***Frege hierarchy***:

DEFINITION 7.3
1. *A logic $\mathcal{L}$ is* **selfextensional** *when, viewed as the g-matrix $\langle \boldsymbol{Fm}, \mathcal{T}h\mathcal{L} \rangle$, it has the* PCONG; *that is, when* $\boldsymbol{\Lambda}\mathcal{L} = \widetilde{\boldsymbol{\Omega}}\mathcal{L}$.
2. *$\mathcal{L}$ is* **Fregean** *when as a g-matrix it has the* SPCONG; *that is, when* $\boldsymbol{\Lambda}_{\mathcal{L}}\Gamma = \widetilde{\boldsymbol{\Omega}}_{\mathcal{L}}\Gamma$ *for each $\Gamma \in \mathcal{T}h\mathcal{L}$.*
3. *$\mathcal{L}$ is* ***fully selfextensional*** *when all its full g-models have the* PCONG.
4. *$\mathcal{L}$ is* ***fully Fregean*** *when all its full g-models have the* SPCONG.

We will see that these four classes are different; therefore, there are *no transfer theorems* for the PCONG, nor for the SPCONG, with full generality; however, we will find some transfers under additional restrictions (see Theorems 7.20, 7.53 and 7.71, among others). The obvious relations between the four notions are displayed in Figure 11 (clearly, the SPCONG implies the PCONG, and recall that every logic is a full g-model of itself).

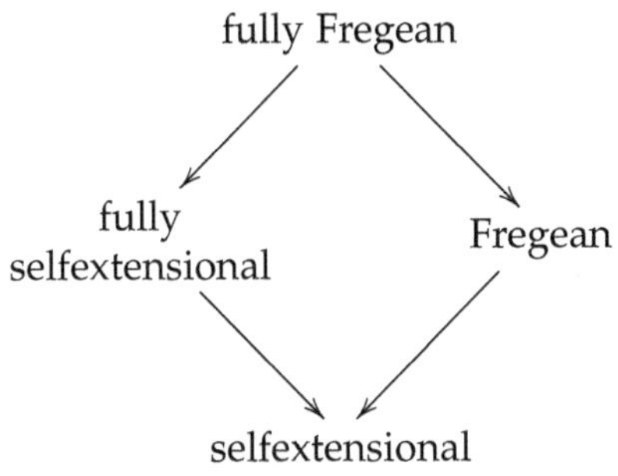

FIGURE 11: The four classes in the Frege hierarchy. Arrows mean inclusion.

Equivalent formulations in terms of the Frege and the Suszko operators, already displayed above for the first two classes, can be given using the specialised notations introduced at the beginning of the section:

PROPOSITION 7.4

1. *A logic* $\mathcal{L}$ *is fully selfextensional if and only if for every algebra* $\boldsymbol{A}$, *the g-matrix* $\langle \boldsymbol{A}, \mathcal{F}i_{\mathcal{L}}\boldsymbol{A} \rangle$ *has the* PCONG; *that is, when* $\boldsymbol{\Lambda}^{A}\mathcal{F}i_{\mathcal{L}}\boldsymbol{A} = \widetilde{\boldsymbol{\Omega}}^{\boldsymbol{A}}\mathcal{F}i_{\mathcal{L}}\boldsymbol{A}$.
2. *A logic* $\mathcal{L}$ *is fully Fregean if and only if for every algebra* $\boldsymbol{A}$, *the g-matrix* $\langle \boldsymbol{A}, \mathcal{F}i_{\mathcal{L}}\boldsymbol{A} \rangle$ *has the* SPCONG; *that is, when* $\boldsymbol{\Lambda}^{\boldsymbol{A}}_{\mathcal{L}} F = \widetilde{\boldsymbol{\Omega}}^{\boldsymbol{A}}_{\mathcal{L}} F$ *for all* $F \in \mathcal{F}i_{\mathcal{L}}\boldsymbol{A}$.

PROOF: By definition a full g-model of $\mathcal{L}$ is a strict inverse image of one of the form $\langle \boldsymbol{A}, \mathcal{F}i_{\mathcal{L}}\boldsymbol{A} \rangle$. By Proposition 7.2 both congruence properties are preserved by strict surjective homomorphisms. Requiring any of them to hold for all full g-models is equivalent to requiring it to hold for basic full g-models. ⊠

Thus, we see that Fregean logics are the ones in which the Frege and the Suszko operators relative to the logic coincide on the formula algebra, while the fully Fregean ones are the ones in which these operators coincide on every algebra. Proposition 7.56 contains further, interesting characterizations of these classes in terms of operators.

The first two classes can be equivalently defined in a more intuitive way:

PROPOSITION 7.5

Let $\mathcal{L}$ *be a logic of type* $\boldsymbol{L}$.

1. $\mathcal{L}$ *is selfextensional if and only if it satisfies the following* (***weak***) ***replacement property***: *For all* $\lambda \in L$, *with* $\mathrm{ar}\lambda = n \geqslant 1$, *and all* $\{\alpha_i, \beta_i : i < n\} \subseteq Fm$,

$$\text{if } \alpha_i \dashv\vdash_{\mathcal{L}} \beta_i \text{ for all } i < n\text{, then } \lambda\alpha_0 \ldots \alpha_{n-1} \dashv\vdash_{\mathcal{L}} \lambda\beta_0 \ldots \beta_{n-1}\,. \qquad \text{(REPL)}$$

2. $\mathcal{L}$ *is Fregean if and only if it satisfies the following* ***strong replacement property***: *For all* $\lambda \in L$, *with* $\mathrm{ar}\lambda = n \geqslant 1$, *and all* $\Gamma \cup \{\alpha_i, \beta_i : i < n\} \subseteq Fm$,

$$\begin{array}{l} \text{if } \Gamma, \alpha_i \dashv\vdash_{\mathcal{L}} \Gamma, \beta_i \text{ for all } i < n\text{,} \\ \text{then } \Gamma, \lambda\alpha_0 \ldots \alpha_{n-1} \dashv\vdash_{\mathcal{L}} \Gamma, \lambda\beta_0 \ldots \beta_{n-1}\,. \end{array} \qquad \text{(SREPL)}$$

PROOF: The equivalence in the first point already appears in Definition 7.1. As to that in the second, it almost does so, but observe that the definition of the SPCONG refers only to theories of the logic, while here an arbitrary $\Gamma \subseteq Fm$ appears; however, by Exercise 7.3, the interderivability relation modulo a set Γ is the same as that modulo the theory $C_{\mathcal{L}}\Gamma$. ⊠

It is even more intuitive to view the two replacement properties in the following simpler ways:

$$\text{If } \alpha \dashv\vdash_{\mathcal{L}} \beta, \text{ then } \delta(\alpha,\vec{z}) \dashv\vdash_{\mathcal{L}} \delta(\beta,\vec{z}) \quad \text{for all } \delta(x,\vec{z}) \in Fm \qquad \text{(REPL)}$$

$$\text{If } \Gamma,\alpha \dashv\vdash_{\mathcal{L}} \Gamma,\beta, \text{ then } \Gamma,\delta(\alpha,\vec{z}) \dashv\vdash_{\mathcal{L}} \Gamma,\delta(\beta,\vec{z}) \quad \text{for all } \Gamma \cup \{\delta(x,\vec{z})\} \subseteq Fm \qquad \text{(SREPL)}$$

These properties may be regarded as formal counterparts of the idea of *substitutivity of equivalents*, and can be considered, to a certain extent, as abstract counterparts of Frege's *principle of compositionality* of denotation (or of truth-value, according to Church) when the notion of logical equivalence is understood as identity of truth-value.† Under this view, selfextensional logics would also deserve to be called "Fregean"; however, starting with [92] this term has been reserved‡ for the more restricted class of logics satisfying the property in its stronger form, i.e., SREPL; the existence of strong relations with classes of algebras also called "Fregean" in the literature (see Theorem 7.78) has reinforced the adequacy of this terminological choice.

The term "selfextensional" was coined by Wójcicki in [247], where he shows that these are the logics that admit a *referential semantics*; see [249, pp. 381–394]. The term "congruential" might be appropriate for any of the four classes of logics in the hierarchy, and it is certainly more descriptive, but then one would have to choose which of the classes it applies to; moreover, using it might prompt some misunderstandings, for it has been previously used in the literature in different senses, namely for Fregean logics in [210], for finitely equivalential logics in [34], and more recently for fully selfextensional logics in [90, 125, 159, 160]. In a few papers, such as [75, 76, 156], the terms "selfextensional" and "Fregean" are also used to denote arbitrary g-matrices having the PCONG and the SPCONG, respectively, but here they are reserved for real logics.

The following examples show that the four classes are distinct and no relations other than those shown hold in general:

EXAMPLES 7.6

1. Being selfextensional and being Fregean are properties inherited by fragments of any kind. The property of being Fregean is also inherited by axiomatic extensions. See Exercise 7.4.

† The abstract algebraic logic approach to the Fregean principles was pioneered by [189], building on the seminal ideas of [231], and is discussed more extensively in [70, pp. 374ff.], [75] and [110, Section 2.1]. The discussion after Corollary 5.33 is also relevant here.

‡ In a few early papers the term "Fregean" is used to mean the same as "protoalgebraic and Fregean" in the present terminology.

2. All two-valued logics, hence all fragments of classical logic, are Fregean. Intuitionistic logic and all its fragments, are Fregean. Some of their fragments (those with at most $\wedge, \vee, \rightarrow$) are fully Fregean. See Example 7.60. Notice that some of these logics have a very high position in the Leibniz hierarchy while others are not even protoalgebraic, or not truth-equational.
3. Logics that are fully selfextensional (hence selfextensional) but not Fregean (hence, not fully Fregean either): Among the protoalgebraic ones, the local logics associated with normal modal systems (see page 461) and two local subintuitionistic logics (Example 7.60.13); among the non-protoalgebraic ones, Belnap-Dunn's logic $\mathcal{B}$ (Example 7.60.9), the "logic of lattices" (Example 7.60.7), the weak relevance logic $\mathcal{WR}$ (Example 7.60.10), positive modal logic $\mathcal{PML}$ (Example 7.60.12) and two other local subintuitionistic logics (Example 7.60.13).
4. The first example of a logic that is Fregean (hence selfextensional) but not fully selfextensional (hence not fully Fregean either) was presented by Babyonyshev in [18]; it is presented by an ad hoc axiomatic system, in a language consisting of four constants and a single unary connective, all with no intended intuitive meaning. This example is non-protoalgebraic.
5. Two examples of logics that are selfextensional but neither fully selfextensional nor Fregean are presented in [7]; they are also constructed solely for this purpose, from three- and four-element matrices, in ad hoc languages. One of these examples has the additional interest of being implicative, hence finitely regularly algebraizable, thus at the top of the Leibniz hierarchy.
6. Logics that are not selfextensional, and hence are outside the Frege hierarchy, but belonging to the Leibniz hierarchy: There are lots of them, even among algebraizable logics, for instance the global logics associated with normal modal systems and the assertional logics of many varieties of commutative integral residuated lattices, among them Łukasiewicz's many-valued logics (see Example 7.47). Another large family of algebraizable and non-selfextensional logics are the three-valued logics of formal inconsistency considered in Example 3.34.10, as shown in [53, Theorem 133]. At the other end of the spectrum, the paraconsistent weak Kleene logic $\mathcal{PWK}$ and Priest's logic of paradox $\mathcal{LP}$ are non-protoalgebraic but truth-equational examples of non-selfextensional logics (see Example 6.122.12).
7. Logics that are neither selfextensional nor protoalgebraic nor truth-equational (that is, logics outside the two hierarchies) are harder to find. One of these is the "logic of distributive bilattices" introduced in [14] and thoroughly studied by Bou and Rivieccio in [48]; it is defined from a four-valued matrix that is an expansion of the matrix $\langle M_4, \{\mathbf{t},\mathbf{b}\}\rangle$, which defines Belnap-Dunn's logic $\mathcal{B}$, with two further lattice-like connectives (written either $\otimes, \oplus$ or $\sqcap, \sqcup$) corresponding to the "knowledge order" in M_4 (see Example 4.12). The algebraic treatment of this logic is strongly parallel to that done for $\mathcal{B}$ in this book. In [48] it is proved that the logic is neither selfextensional (in contrast to

$\mathcal{B}$) nor protoalgebraic; since it has no theorems, it is not truth-equational. It is also the logic that preserves degrees of truth with respect to the variety of distributive bilattices; suprisingly, and in contrast to DL or DM, this variety is the equivalent algebraic semantics of an algebraizable logic. ⊠

Section 7.3 contains some results on the *relations between the two hierarchies*; however, most have complicated formulations and need special assumptions, such as being finitary or having theorems. In general one would say that the two hierarchies are in some sense "orthogonal": as the list of examples above shows, the best behaved classes of each hierarchy contain logics that belong to the best behaved classes of the other, but also logics that are even outside the other. Among others, the following facts are interesting to highlight:

- A Fregean logic is assertional (hence, truth-equational) if and only if it has theorems (Theorem 7.66).
- A truth-equational logic is fully selfextensional if and only if it is fully Fregean (Theorem 7.67).
- All protoalgebraic and Fregean logics with theorems are regularly algebraizable, and the finitary ones are moreover fully Fregean (Theorem 7.75).

An important property proved in Theorems 7.20 and 7.53 is that, for the logics $\mathcal{L}$ each deals with (which together form a very large class of logics), the corresponding class $\mathsf{Alg}\mathcal{L}$ is a variety. This is one step towards a solution of the so-called *variety problem*: to explain why is it so that the algebraic counterpart of so many "real" logics is a variety. The problem is perhaps more striking in the context of BP-algebraizable logics, for which it was originally formulated by Blok and Pigozzi: there, the general theory guarantees only that the equivalent algebraic semantics is a quasivariety, but examples of "real" BP-algebraizable logics whose equivalent algebraic semantics is not a variety are rare.

There are a few *open problems* concerning the basic structure of the Frege hierarchy and its interplay with the Leibniz hierarchy. To mention just two:

- Is the class of fully Fregean logics the intersection of the classes of the Fregean and the fully selfextensional ones? Theorem 7.69 shows that the answer is *yes* for logics with theorems, but the general question remains open.
- Are all protoalgebraic Fregean logics fully Fregean? By Theorem 7.71 the answer is *yes* for all finitary logics, but the non-finitary case remains open.

One of the conclusions of the study of the hierarchy is that being selfextensional is a rather weak property, while being Fregean is a rather strong one; being fully selfextensional seems of an intermediate strength, but it becomes stronger when coupled with other, seemingly weak properties. The Frege hierarchy has not been as thoroughly investigated as the Leibniz hierarchy, and most that is known about it concerns its relations with the Leibniz hierarchy.

7.2 Selfextensional and fully selfextensional logics

By Definition 5.25, a g-matrix has the PCONG if and only if it is a model of certain Gentzen-style rules. By Proposition 5.12, if all g-matrices in a class have the PCONG, then the logic defined by this class has the PCONG, that is, is selfextensional. This fact is very handy in order to construct (counter)examples, especially if one defines a logic from a single, small g-matrix, whose congruences are easily analysed; a particularly simple case of this appears when the Frege relation of the g-matrix is actually the identity relation, which is trivially a congruence.

Since a logic and its finitary companion have the same Frege relation (Proposition 1.49), from the very definition the following is obvious.

PROPOSITION 7.7
A logic is selfextensional if and only if its finitary companion is selfextensional. ⊠

Another elementary fact concerning examples is that selfextensionality is a property preserved by fragments (Exercise 7.4). By contrast, it is not preserved by extensions, not even by axiomatic extensions: Any selfextensional but non-Fregean logic provides a counterexample. As far as I know, the issue of whether full selfextensionality is preserved by fragments and by extensions has not been investigated.

Since $\widetilde{\Omega}\mathcal{L}$ is the largest congruence of the formula algebra below $\boldsymbol{\Lambda}\mathcal{L}$, it follows that $\mathcal{L}$ is selfextensional if and only if $\widetilde{\Omega}\mathcal{L} = \boldsymbol{\Lambda}\mathcal{L}$. Recall that the intrinsic variety of a logic is $\mathbb{V}\mathcal{L} := \mathbb{V}\{\boldsymbol{Fm}/\widetilde{\Omega}\mathcal{L}\} = \mathbb{V}\mathsf{Alg}\mathcal{L} = \mathbb{V}\mathsf{Alg}^*\mathcal{L}$ (Theorem 5.76). Thus, when $\mathcal{L}$ is selfextensional, $\mathbb{V}\mathcal{L} = \mathbb{V}\{\boldsymbol{Fm}/\boldsymbol{\Lambda}\mathcal{L}\}$, and the variety $\mathbb{V}\mathcal{L}$ has a particularly simple characterization:

PROPOSITION 7.8
A logic $\mathcal{L}$ is selfextensional if and only if there is a class K of algebras such that for all $\alpha, \beta \in Fm$, $\alpha \dashv\vdash_{\mathcal{L}} \beta$ if and only if $\mathsf{K} \vDash \alpha \approx \beta$. In this case, $\mathbb{V}\mathcal{L}$ is the largest class K, and the only variety, with this property. Therefore, $\mathbb{V}\mathcal{L} = \mathbb{V}\mathsf{K}$ for any K with these properties; it holds that $\mathbb{V}\mathcal{L} \vDash \alpha \approx \beta$ if and only if $\alpha \dashv\vdash_{\mathcal{L}} \beta$; and the variety $\mathbb{V}\mathcal{L}$ is generated by the algebra $\boldsymbol{Fm}/\boldsymbol{\Lambda}\mathcal{L}$.

PROOF: It is straightforward to see that the relation between formulas α and β given by the property that $\mathsf{K} \vDash \alpha \approx \beta$ is a congruence of the formula algebra; therefore, every logic satisfying the said condition is indeed selfextensional. Conversely, assume that $\mathcal{L}$ is selfextensional. Then, as already pointed out, $\alpha \dashv\vdash_{\mathcal{L}} \beta$ if and only if $\alpha \equiv \beta \; (\widetilde{\Omega}\mathcal{L})$, and by Proposition 5.75.1, this is equivalent to saying that $\mathbb{V}\mathcal{L} \vDash \alpha \approx \beta$. Thus, the condition is satisfied by taking $\mathsf{K} := \mathbb{V}\mathcal{L}$. If there is such a class K, then $\mathbb{V}\mathsf{K}$ also satisfies the same property; that is, $\mathbb{V}\mathsf{K} \vDash \alpha \approx \beta$ if and only if $\alpha \dashv\vdash_{\mathcal{L}} \beta$, therefore $\mathbb{V}\mathsf{K} = \mathbb{V}\mathcal{L}$, which implies that $\mathsf{K} \subseteq \mathbb{V}\mathcal{L}$, and that this one is the only variety with such a property. Finally, since $\boldsymbol{\Lambda}\mathcal{L}$ is a congruence, $\boldsymbol{\Lambda}\mathcal{L} = \widetilde{\Omega}\mathcal{L}$, and therefore $\mathbb{V}\mathcal{L}$ is (by definition) the variety generated by the quotient algebra $\boldsymbol{Fm}/\boldsymbol{\Lambda}\mathcal{L}$. ⊠

Of course, this readily provides a host of examples of selfextensional logics and determines their intrinsic variety; among them are classical logic and all its fragments $\mathcal{C}\ell \restriction \boldsymbol{L}$, which satisfy the above equivalence with $\mathsf{K} = \{\mathbf{2} \restriction \boldsymbol{L}\}$, and hence $\mathbb{V}(\mathcal{C}\ell \restriction \boldsymbol{L}) = \mathbb{V}\{\mathbf{2} \restriction \boldsymbol{L}\}$ (for the corresponding similarity type $\boldsymbol{L}$); this is proved in earlier chapters for some particular fragments.

One of the most usual ways to prove that a logic is *not* selfextensional is to make some ad hoc reasoning, often exhibiting a pair of interderivable formulas that originate, after being operated with a third formula by some connective, a pair of formulas that are not interderivable. You may find simple examples of this procedure in Exercises 7.5 and 7.6, in which you are asked to prove that the "simplest protoalgebraic logic" $\mathcal{I}$ of Example 6.78.1 and Łukasiewicz's many-valued logics are not selfextensional.

Turning to full selfextensionality, more practical characterizations are quickly produced by using some elementary properties. Recall from page 42 that a closure system $\mathscr{C}$ on a set A is said to *separate points* when for all $a, b \in A$, if $a \neq b$, then there is $F \in \mathscr{C}$ such that either $a \in F$ and $b \notin F$ or $a \notin F$ and $b \in F$; this amounts to saying that $\varLambda^A \mathscr{C}$, the Frege relation of $\mathscr{C}$ is the identity relation, but is more intuitive.

Proposition 7.9

For any logic $\mathcal{L}$, the following conditions are equivalent:

(i) *$\mathcal{L}$ is fully selfextensional.*

(ii) *For any $\boldsymbol{A} \in \mathsf{Alg}\mathcal{L}$, the g-matrix $\langle \boldsymbol{A}, \mathcal{F}i_{\mathcal{L}}\boldsymbol{A} \rangle$ has the* PCONG.

(iii) *For any $\boldsymbol{A} \in \mathsf{Alg}\mathcal{L}$ and any $a, b \in A$, if $Fg^{\boldsymbol{A}}_{\mathcal{L}}\{a\} = Fg^{\boldsymbol{A}}_{\mathcal{L}}\{b\}$, then $a = b$. That is, $\varLambda^A \mathcal{F}i_{\mathcal{L}}\boldsymbol{A}$ is the identity relation, or, equivalently, $\mathcal{F}i_{\mathcal{L}}\boldsymbol{A}$ separates points.*

(iv) *For any $\boldsymbol{A} \in \mathsf{Alg}\mathcal{L}$, the relation defined*† *by $a \leqslant^{\boldsymbol{A}}_{\mathcal{L}} b \overset{\text{def}}{\iff} b \in Fg^{\boldsymbol{A}}_{\mathcal{L}}\{a\}$, for any $a, b \in A$, is an order on A.*

Proof: By Proposition 7.4, $\mathcal{L}$ is fully selfextensional if and only if for any $\boldsymbol{A}$, the g-matrix $\langle \boldsymbol{A}, \mathcal{F}i_{\mathcal{L}}\boldsymbol{A} \rangle$ has the PCONG; in particular this shows that (i)⇒(ii). Conversely, (ii)⇒(i) because by Corollary 5.88.4 every full g-model is the inverse image of one of the form $\langle \boldsymbol{A}, \mathcal{F}i_{\mathcal{L}}\boldsymbol{A} \rangle$ with $\boldsymbol{A} \in \mathsf{Alg}\mathcal{L}$, and by Proposition 7.2 the PCONG is preserved by strict surjective homomorphisms. Now, by Corollary 5.88.2, if $\boldsymbol{A} \in \mathsf{Alg}\mathcal{L}$, then the g-matrix $\langle \boldsymbol{A}, \mathcal{F}i_{\mathcal{L}}\boldsymbol{A} \rangle$ is reduced. Then, Proposition 5.55 says that it has the PCONG if and only if its Frege relation is the identity relation; this proves that (ii)⇔(iii). Finally, (iii)⇔(iv) because the relation $\leqslant^{\boldsymbol{A}}_{\mathcal{L}}$ is always reflexive and transitive (this is a general property of closure operators; see Exercise 1.47), therefore it is an order if and only if it is antisymmetric, but this amounts to separating points. ⊠

Property (iv) originates several characterizations of the $\mathcal{L}$-algebras of fully selfextensional logics:

† The relation $\leqslant^{\boldsymbol{A}}_{\mathcal{L}}$ is a particular case of the relation $\leqslant_C$ associated with a closure operator C in (11) on page 41, for $C = Fg^{\boldsymbol{A}}_{\mathcal{L}}$.

COROLLARY 7.10
Let $\mathcal{L}$ be a fully selfextensional logic. For any algebra $\boldsymbol{A}$ the following conditions are equivalent:

(i) $\boldsymbol{A} \in \mathsf{Alg}\mathcal{L}$.

(ii) *$\Lambda^A \mathcal{F}i_{\mathcal{L}} \boldsymbol{A}$ is the identity relation, or, equivalently, $\mathcal{F}i_{\mathcal{L}} \boldsymbol{A}$ separates points.*

(iii) *There is a full g-model $\langle \boldsymbol{A}, \mathscr{C} \rangle$ of $\mathcal{L}$ on $\boldsymbol{A}$ that separates points.*

(iv) *The relation $\leqslant_{\mathcal{L}}^{\boldsymbol{A}}$ defined in point* (iv) *of Proposition 7.9 is an order on A.* ⊠

These conditions give the clue to explain why working with the algebraic models of fully selfextensional logics is particularly convenient: their $\mathcal{L}$-algebras have an order with a clearly logical interpretation, and different points can be separated by subsets with a clear logical significance (different points generate different $\mathcal{L}$-filters). This is particularly fruitful when there is a mathematically workable (usually, purely algebraic or order-theoretic) characterization of the $\mathcal{L}$-filters of the $\mathcal{L}$-algebras, or of the full g-models on these algebras; but it can also be used to obtain general properties, such as those in Exercises 7.7 and 7.8. It should not surprise that some remarkable results have been obtained for fully selfextensional logics under reasonable restrictions, especially concerning topological representations of the involved algebras; see for instance [90, 125].

It seems that to be just selfextensional is a rather weak property; however, it becomes much stronger when coupled with others. The two best known kinds of selfextensional logics are those with conjunction and those with the uniterm DDT (as we will see, all turn out to be fully selfextensional). The first has been extensively studied in [98, 107, 156, 157]; the material that follows is extracted from these works.

Selfextensional logics with conjunction

As becomes clear in this section, having a conjunction is a surprisingly strong property for a logic, and particularly for a selfextensional logic. First a few general facts will be given. Recall from Definition 5.16 and Proposition 5.17 the Property of Conjunction for a closure operator, and some equivalent formulations:

PROPOSITION 7.11
Let C be a closure operator on a set A and let $\wedge$ be a binary term operation on A. The following conditions are equivalent:

(i) *The operation $\wedge$ is a* ***conjunction*** *for C; that is, the following condition holds:*

$$C\{a,b\} = C\{a \wedge b\} \text{ for all } a,b \in A. \qquad \text{PC}$$

(ii) *The g-matrix $\langle \langle A, \wedge \rangle, C \rangle$ is a model of the following sequents or Hilbert-style rules:*

$$x,y \rhd x \wedge y \qquad x \wedge y \rhd x \qquad x \wedge y \rhd y. \qquad \text{(HPC)}$$

(iii) *Every $F \in \mathscr{C}$ is an* ***$\wedge$-filter****; that is, for all $a,b \in A$, $a \wedge b \in F$ if and only if $a \in F$ and $b \in F$.*

When any of these conditions holds, it is said that C satisfies the **Property of Conjunction**, PC, *or that it is* **conjunctive**. *This terminology applies also to g-matrices and to logics in the natural way.* ☒

It is easy to see that either the second or the third rules in (ii) can be replaced by the rule $x \wedge y \rhd y \wedge x$. The next properties are also easy to show (Exercise 7.9) but are nevertheless crucial:

PROPOSITION 7.12
Let C be a closure operator on a set A and let $\wedge$ be a conjunction for C.

1. *For all $a, b \in A$,*

$$a \wedge a \equiv a \ \ (\boldsymbol{\Lambda}^{A} C) \tag{SL1}$$

$$a \wedge b \equiv b \wedge a \ \ (\boldsymbol{\Lambda}^{A} C) \tag{SL2}$$

$$a \wedge (b \wedge c) \equiv (a \wedge b) \wedge c \ \ (\boldsymbol{\Lambda}^{A} C) \tag{SL3}$$

2. *For all $a, b \in A$, $a \in C\{b\}$ if and only if $a \wedge b \equiv a \ (\boldsymbol{\Lambda}^{A} C)$. More generally, for all $n \geqslant 1$ and all $a_0, \dots, a_n \in A$,*

$$a_n \in C\{a_0, \dots, a_{n-1}\} \iff (a_0 \wedge \cdots \wedge a_{n-1}) \wedge a_n \equiv a_0 \wedge \cdots \wedge a_{n-1} \ \ (\boldsymbol{\Lambda}^{A} C).$$

3. *The relation $\boldsymbol{\Lambda}^{A} C$ is a congruence with respect to $\wedge$.* ☒

Property (SL3) makes it possible to use a kind of *notational associativity* as far as closure or equivalence is concerned, and write $a_0 \wedge \cdots \wedge a_{n-1}$, as already done, without caring about any particular binary grouping (which, strictly speaking, should be present, as $\wedge$ is only a binary operation); some parentheses are retained for emphasis. However, perhaps the most interesting characteristic of the closure operators with conjunction is point 2, which says that their finitary part is encoded almost completely in their Frege relation; only the theorems are left undetermined by this property. As a consequence, a finitary closure operator with the PC is characterized by just its Frege relation and its theorems. Exercise 7.10 contains an easy application of this fact.

These observations apply obviously to *logics* with the PC. By structurality, a term $\wedge$ is a conjunction for a logic $\mathcal{L}$ if and only if it satisfies $\{x, y\} \dashv\vdash_{\mathcal{L}} x \wedge y$, and this happens if and only if $\mathcal{L}$ satisfies the three sequents (HPC). Since the operation is given by a binary term, it defines a binary operation on any algebra (which, as usual, is denoted also by $\wedge$ rather than by $\wedge^{\boldsymbol{A}}$), and then all the g-models of $\mathcal{L}$ satisfy these sequents, and hence $\wedge$ is also a conjunction for them. Finally, observe that by Proposition 7.12, if $\mathcal{L}$ has the PC then

$$x \wedge x \dashv\vdash_{\mathcal{L}} x, \tag{SL1}$$

$$x \wedge y \dashv\vdash_{\mathcal{L}} y \wedge x, \tag{SL2}$$

$$x \wedge (y \wedge z) \dashv\vdash_{\mathcal{L}} (x \wedge y) \wedge z, \tag{SL3}$$

and that $\boldsymbol{\Lambda}\mathcal{L}$ is a congruence with respect to $\wedge$. Thus, the quotient $\boldsymbol{Fm}/\boldsymbol{\Lambda}\mathcal{L}$ has a semilattice structure, given by the operation $\wedge$. But what about the other connectives? They will define an algebraic structure in this quotient if and when $\Lambda\mathcal{L}$ is a real congruence of $\boldsymbol{Fm}$, that is, when $\mathcal{L}$ is selfextensional. In this case, having a conjunction is closely related to having a semilattice structure in the algebraic counterpart of the logic; this makes the following notion natural.

Definition 7.13
*Let $\boldsymbol{A}$ be an algebra of type $\boldsymbol{L}$. The algebra $\boldsymbol{A}$ **has a semilattice reduct** when there is a binary term $\wedge$ of $\boldsymbol{L}$ such that the reduct $\langle A, \wedge\rangle$ is a semilattice; that is, when the following equations*

$$x \wedge x \approx x, \tag{SL2}$$

$$x \wedge y \approx y \wedge x, \tag{SL2}$$

$$x \wedge (y \wedge z) \approx (x \wedge y) \wedge z \tag{SL3}$$

*hold in $\boldsymbol{A}$. A class of algebras **has semilattice reducts** when all its algebras have a semilattice reduct for the same binary term. For each type $\boldsymbol{L}$ and term $\wedge$ the class of all $\boldsymbol{L}$-algebras having a semilattice reduct with the term $\wedge$ is denoted by $\mathsf{SL}(\boldsymbol{L}, \wedge)$.*

Thus, $\mathsf{SL}(\boldsymbol{L}, \wedge)$ is the variety of $\boldsymbol{L}$-algebras axiomatized by the equations (SL1)–(SL3), and when $\boldsymbol{L} = \langle \wedge \rangle$ it reduces to the variety SL of semilattices. In these algebras there is an associated order relation, which is denoted simply by $\leqslant$ and is defined by saying that $a \leqslant b$ if and only if $a \wedge b = a$, for all $a, b \in A$; then $a \wedge b = \inf\{a, b\}$, and the symbol $\alpha \preccurlyeq \beta$, when evaluated in one of these algebras, becomes the abbreviation of an equation, namely of $\alpha \wedge \beta \approx \alpha$.

The same labels (SL1)–(SL3) are used for the three parallel groups of properties, either formulated for an arbitrary closure operator, for a logic, or for a class of algebras; clearly, there can be no confusion. Again, since the operation $\wedge$ is associative in algebras where the equation (SL3) holds, no parentheses are needed when writing terms to be evaluated in one of these algebras.

Proposition 7.14
Let $\mathcal{L}$ be a logic, and let $\wedge$ be a binary term of its language. The class $\mathbb{V}\mathcal{L}$ has semilattice reducts given by $\wedge$ if and only if the following conditions hold:

$$x \wedge x \equiv x \ \ (\widetilde{\Omega}\mathcal{L}) \tag{SL1$^+$}$$

$$x \wedge y \equiv y \wedge x \ \ (\widetilde{\Omega}\mathcal{L}) \tag{SL2$^+$}$$

$$x \wedge (y \wedge z) \equiv (x \wedge y) \wedge z \ \ (\widetilde{\Omega}\mathcal{L}) \tag{SL3$^+$}$$

That either of the classes $\mathsf{Alg}^*\mathcal{L}$ or $\mathsf{Alg}\mathcal{L}$ has semilattice reducts is also equivalent to these conditions.

Proof: By Proposition 5.75, these conditions amount to saying that the intrinsic variety $\mathbb{V}\mathcal{L}$ makes the equations (SL1)–(SL3) from Definition 7.13 valid, hence they are equivalent to saying that $\mathbb{V}\mathcal{L}$ has semilattice reducts (given by this $\wedge$). Since by Theorem 5.76 the other two classes generate the same variety, the final statement follows. ⊠

Of course, if $\mathcal{L}$ is selfextensional, then these conditions reduce to the forms of (SL1)–(SL3) in terms of $\dashv\vdash_{\mathcal{L}}$. However, the situation described in the proposition is in general not related to $\wedge$ being a conjunction. On the one hand, not every conjunction satisfies the conditions (SL1$^+$)–(SL3$^+$), as witnessed for instance by the fusion connective of Łukasiewicz's logic and other logics associated with residuated lattices, which is also a conjunction for them but does not determine a semilattice structure in the associated algebras. On the other hand, that these conditions hold does not imply that $\wedge$ is a conjunction for the logic;† the fact that a given connective is or is not a conjunction is rather related to a property of its filters, as Theorem 7.11(iii) shows; in the present framework this has a name:

DEFINITION 7.15
If $\boldsymbol{A}$ has a semilattice reduct and $F \subseteq A$, then F is a **semilattice filter** *when $F \neq \emptyset$, for all $a, b \in A$, if $a, b \in F$, then $a \wedge b \in F$, and if $a \in F$ and $a \leqslant b$, then $b \in F$. The set of semilattice filters of $\boldsymbol{A}$ is denoted by $\mathcal{F}ilt_{\wedge}\boldsymbol{A}$.*

A set is a semilattice filter if and only if it is a non-empty $\wedge$-filter in the sense of Proposition 7.11 (Exercise 7.11). Notice that if $\langle A, \leqslant \rangle$ has a maximum 1 (equivalently, if the operation $\wedge$ has a unit), then $\{1\}$ is a semilattice filter, and indeed the smallest one, and $1 \in F$ for every semilattice filter F, therefore $\mathcal{F}ilt_{\wedge}\boldsymbol{A}$ is a closure system. Due to the condition that a semilattice filter should be non-empty, in algebras without a maximum the family $\mathcal{F}ilt_{\wedge}\boldsymbol{A}$ is not a closure system, because its intersection is empty; the closure system it generates is $\mathcal{F}ilt_{\wedge}\boldsymbol{A} \cup \{\emptyset\}$. This issue, which already appeared in Exercise 1.42 and in Proposition 5.6, makes the empty set require a separate treatment in some results.

DEFINITION 7.16
A logic $\mathcal{L}$ is **semilattice-based** *when there is a class K of algebras having semilattice reducts, with semilattice operation $\wedge$, such that for all $n \geqslant 1$ and all $\varphi_0, \dots, \varphi_n \in Fm$,*

$$\varphi_0, \dots, \varphi_{n-1} \vdash_{\mathcal{L}} \varphi_n \iff \mathsf{K} \vDash \varphi_0 \wedge \dots \wedge \varphi_{n-1} \preccurlyeq \varphi_n. \tag{93}$$

The right-hand side condition means that $h\varphi_0 \wedge \dots \wedge h\varphi_{n-1} \leqslant h\varphi_n$ for all $\boldsymbol{A} \in \mathsf{K}$ and all $h \in \mathrm{Hom}(\boldsymbol{Fm}, \boldsymbol{A})$.

The typical examples of semilattice-based logics are the logics defined by preservation of degrees of truth of Example 4.11, in the cases where the structure of truth degrees has a semilattice reduct; for instance, the logics $Ł_n^{\leqslant}$ defined in Example 6.122.6.

Since $\preccurlyeq$ is shorthand for an equation, the above definition depends only on the equations valid in K; thus, if $\mathcal{L}$ is semilattice-based with respect to K, it is also so with respect to $\mathbb{V}\mathsf{K}$. We will see that a logic can be semilattice-based in essentially only one sense, and with respect to a single variety.

To start with, the following is straightforward:

† A binary term that is a conjunction and moreover satisfies the conditions (SL1$^+$)–(SL3$^+$) is called a *strong conjunction* in [157].

LEMMA 7.17
If $\mathcal{L}$ is semilattice-based with respect to K*, then for any $\alpha, \beta \in Fm$, $\alpha \dashv\vdash_{\mathcal{L}} \beta$ if and only if* $\mathsf{K} \vDash \alpha \approx \beta$. ⊠

Although with a simple proof, the following characterization is fundamental:

THEOREM 7.18
A logic $\mathcal{L}$ is semilattice-based (with respect to some class K*) if and only if $\mathcal{L}$ is selfextensional and has a conjunction. If these conditions hold, then $\mathbb{V}\mathcal{L} = \mathbb{V}\mathsf{K}$, and $\mathcal{L}$ is semilattice-based with respect to $\mathbb{V}\mathcal{L}$, which has semilattice reducts.*

PROOF: ($\Rightarrow$) Assume that $\mathcal{L}$ is a semilattice-based logic with respect to K. By Lemma 7.17 plus Proposition 7.8, $\mathcal{L}$ is selfextensional. The fact that K has semilattice reducts implies that $\mathsf{K} \vDash x \wedge y \preccurlyeq x \wedge y$, $\mathsf{K} \vDash x \wedge y \preccurlyeq x$ and $\mathsf{K} \vDash x \wedge y \preccurlyeq y$. Applying Definition 7.16 to these properties implies that $\mathcal{L}$ satisfies the three sequents (HPC); the formula $x \wedge y$ at the left of $\preccurlyeq$ has to be treated as the conjunction of two formulas in the first case, and as a single formula in the second and third cases. This shows that $\wedge$ is a conjunction for $\mathcal{L}$.
($\Leftarrow$) From the assumption and Propositions 7.8 and 7.12.1 it follows that the equations (SL1)–(SL3) hold in the variety $\mathbb{V}\mathcal{L}$, and therefore it has semilattice reducts. Since $\mathcal{L}$ has a conjunction, Proposition 7.12.2 applies, hence $\varphi_0, \dots, \varphi_{n-1} \vdash_{\mathcal{L}} \varphi_n$ if and only if $(\varphi_0 \wedge \cdots \wedge \varphi_{n-1}) \wedge \varphi_n \dashv\vdash_{\mathcal{L}} \varphi_0 \wedge \cdots \wedge \varphi_{n-1}$, which by selfextensionality means that $\mathbb{V}\mathcal{L} \vDash (\varphi_0 \wedge \cdots \wedge \varphi_{n-1}) \wedge \varphi_n \approx \varphi_0 \wedge \cdots \wedge \varphi_{n-1}$. But $\mathbb{V}\mathcal{L}$ has semilattice reducts, therefore this is equivalent to $\mathbb{V}\mathcal{L} \vDash \varphi_0 \wedge \cdots \wedge \varphi_{n-1} \preccurlyeq \varphi_n$, and this establishes that $\mathcal{L}$ is semilattice-based with respect to $\mathbb{V}\mathcal{L}$.
The remaining assertions also follow from Proposition 7.8. ⊠

After this, it is a simple exercise to show:

COROLLARY 7.19
If $\mathcal{L}$ is semilattice-based with respect to a class K *through a term $\wedge$, and is also semilattice-based with respect to a class* K′ *through a term $\wedge'$, then $\mathbb{V}\mathsf{K} = \mathbb{V}\mathsf{K}' = \mathbb{V}\mathcal{L}$, and for any $\alpha, \beta \in Fm$, $\alpha \wedge \beta \dashv\vdash_{\mathcal{L}} \alpha \wedge' \beta$, and the equation $\alpha \wedge \beta \approx \alpha \wedge' \beta$ holds in the classes* K *and* K′ *and in the variety $\mathbb{V}\mathcal{L}$.* ⊠

This confirms the "absolute" character of being semilattice-based: if a logic is semilattice-based, then it is so with respect to a unique *variety*, which is the intrinsic variety of the logic (and which in this case is determined by the Frege relation of the logic, as witnessed by Lemma 7.17), and through a unique binary operation in this variety (which may be expressed by several but equivalent binary terms).

By Proposition 7.11(iii), if $\mathcal{L}$ is a semilattice-based logic with respect to a class of algebras K having semilattice reducts, then the $\mathcal{L}$-filters of the algebras in K are semilattice filters. In the finitary case the family of $\mathcal{L}$-filters can be characterized, and other very important consequences are also obtained.

Theorem 7.20
Let $\mathcal{L}$ be a finitary selfextensional logic with conjunction.

1. *If $\boldsymbol{A} \in \mathbb{V}\mathcal{L}$, then either $\mathcal{F}i_{\mathcal{L}}\boldsymbol{A} = \mathcal{F}ilt_{\wedge}\boldsymbol{A}$ (in the case that $\mathcal{L}$ has theorems), or $\mathcal{F}i_{\mathcal{L}}\boldsymbol{A} = \mathcal{F}ilt_{\wedge}\boldsymbol{A} \cup \{\emptyset\}$ (in the case that $\mathcal{L}$ has no theorems).*
2. *If $\boldsymbol{A} \in \mathbb{V}\mathcal{L}$, then the g-matrix $\langle \boldsymbol{A}, \mathcal{F}i_{\mathcal{L}}\boldsymbol{A} \rangle$ has the* PCONG *and is reduced.*
3. $\mathsf{Alg}\,\mathcal{L} = \mathbb{V}\mathcal{L}$, *hence* $\mathsf{Alg}\,\mathcal{L}$ *is a variety.*
4. *$\mathcal{L}$ is fully selfextensional.*

Proof: Notice that by Theorem 7.18, $\mathcal{L}$ is semilattice-based with respect to $\mathbb{V}\mathcal{L}$, which means that (93) holds for this class.

1. As said before, if $\boldsymbol{A} \in \mathbb{V}\mathcal{L}$, then every $\mathcal{L}$-filter of $\boldsymbol{A}$ is a semilattice filter, unless it is empty, which happens if and only if $\mathcal{L}$ has no theorems. Now let F be a semilatttice filter of $\boldsymbol{A}$, and assume that $\Gamma \vdash_{\mathcal{L}} \varphi$ and $h \in \operatorname{Hom}(\boldsymbol{Fm}, \boldsymbol{A})$ is such that $h\Gamma \subseteq F$. If $\Gamma = \emptyset$ this means that $\mathcal{L}$ has theorems; then $\emptyset \vdash_{\mathcal{L}} \varphi$ implies $x \vdash_{\mathcal{L}} \varphi$ for any $x \in V \smallsetminus \operatorname{Var}\varphi$, and hence $\boldsymbol{A} \vDash x \preccurlyeq \varphi$, which means that $h\varphi = \max A$; since F is non-empty and an up-set of $\boldsymbol{A}$, this implies that $h\varphi \in F$. If $\Gamma \neq \emptyset$, then by finitarity there is some $n \geqslant 1$ and some $\varphi_0, \dots, \varphi_{n-1} \in \Gamma$ such that $\varphi_0, \dots, \varphi_{n-1} \vdash_{\mathcal{L}} \varphi$, which implies that $h\varphi_0 \wedge \cdots \wedge h\varphi_{n-1} \leqslant h\varphi$. Since $h\varphi_i \in F$ for all $i < n$ and F is a semilattice filter, it follows that $h\varphi_0 \wedge \cdots \wedge h\varphi_{n-1} \in F$, and finally $h\varphi \in F$. In both cases, F is an $\mathcal{L}$-filter.

2. In Exercise 1.42 you showed that the semilattice filter generated by $a \in A$ is the principal up-set $\uparrow a$; this implies that if $a, b \in A$ with $a \neq b$, then $\uparrow a \neq \uparrow b$. Using 1, this implies that $\varLambda^{A}\mathcal{F}i_{\mathcal{L}}\boldsymbol{A} = \operatorname{Id}_A$; notice that the addition of the empty set to a closure system does not alter its interderivability relation (Exercise 7.15). By Proposition 5.55, this means that the g-matrix $\langle \boldsymbol{A}, \mathcal{F}i_{\mathcal{L}}\boldsymbol{A} \rangle$ has the PCONG and is reduced.

3. In general $\mathsf{Alg}\,\mathcal{L} \subseteq \mathbb{V}\mathcal{L}$. By 2, if $\boldsymbol{A} \in \mathbb{V}\mathcal{L}$, then $\boldsymbol{A}$ is the algebraic reduct of a reduced g-model of $\mathcal{L}$ (a full basic one, in fact), and therefore $\boldsymbol{A} \in \mathsf{Alg}\,\mathcal{L}$. By Corollary 5.77, this is equivalent to saying that $\mathsf{Alg}\,\mathcal{L}$ is a variety.

4. By 2 and 3, in any $\boldsymbol{A} \in \mathsf{Alg}\mathcal{L}$ the relation $\varLambda^{A}\mathcal{F}i_{\mathcal{L}}\boldsymbol{A}$ is the identity relation, therefore the condition in Proposition 7.9(iii) is satisfied. This shows that $\mathcal{L}$ is fully selfextensional. ⊠

Actually, point 4 establishes a *transfer theorem*: For finitary logics with conjunction, the PCONG transfers from a logic to all its full g-models.

It is now possible to prove another transfer result, concerning the PIRA, which also shows the power of conjunction. For this, some preliminary properties are needed:

Lemma 7.21
Let $\mathcal{L}$ be a logic satisfying the PIRA *for a unary term $\neg$.*

1. *For every $\varphi \in Fm$, the set $\{\varphi, \neg\varphi\}$ is inconsistent.*
2. *For every $\Gamma \cup \{\varphi, \psi\} \subseteq Fm$, if $\Gamma, \varphi \vdash_{\mathcal{L}} \psi$, then $\Gamma, \neg\psi \vdash_{\mathcal{L}} \neg\varphi$.*

If moreover $\mathcal{L}$ has a conjunction $\wedge$, then:

3. *For every* $\varphi \in Fm$, *the formula* $\varphi \wedge \neg\varphi$ *is inconsistent, and the formula* $\neg(\varphi \wedge \neg\varphi)$ *is a theorem.*
4. *For every* $\varphi, \psi \in Fm$, *it holds that* $\varphi, \neg(\varphi \wedge \psi) \vdash_{\mathcal{L}} \neg\psi$.

PROOF: All these properties are completely straightforward, and are left to you as Exercise 7.13; actually, they hold for an arbitrary closure operator, as formulated in the exercise. ⊠

THEOREM 7.22
Let $\mathcal{L}$ *be a finitary selfextensional logic with conjunction. If* $\mathcal{L}$ *satisfies the* PIRA *for a unary connective* $\neg$, *then all its full g-models satisfy it.*

PROOF: By Theorem 5.63, the PIRA is preserved by strict surjective homomorphisms, therefore it is enough to prove the PIRA for the basic full g-models of the form $\langle \boldsymbol{A}, \mathcal{F}i_{\mathcal{L}}\boldsymbol{A}\rangle$, that is, to prove that for all $X \cup \{a\} \subseteq A$, $\neg a \in Fg^{\boldsymbol{A}}_{\mathcal{L}}X$ if and only if $Fg^{\boldsymbol{A}}_{\mathcal{L}}(X,a) = A$. The implication from left to right is easy: if $\neg a \in Fg^{\boldsymbol{A}}_{\mathcal{L}}X$, then also $\neg a \in Fg^{\boldsymbol{A}}_{\mathcal{L}}(X,a)$, but by Lemma 7.21.1 $x, \neg x \vdash_{\mathcal{L}} y$, therefore $Fg^{\boldsymbol{A}}_{\mathcal{L}}X = A$. Now assume that $Fg^{\boldsymbol{A}}_{\mathcal{L}}(X,a) = A$. In particular, $\neg a \in Fg^{\boldsymbol{A}}_{\mathcal{L}}(X,a)$. Now I claim that there is some $b \in Fg^{\boldsymbol{A}}_{\mathcal{L}}X$ such that $\neg a \in Fg^{\boldsymbol{A}}_{\mathcal{L}}\{a,b\}$: If $X = \emptyset$, then it is enough to take the value of any theorem (existence of theorems has been proved in Lemma 7.21.3), for instance $b := \neg(a \wedge \neg a)$. If $X \neq \emptyset$, then since $\mathcal{L}$ is finitary and has the PC, the operator $Fg^{\boldsymbol{A}}_{\mathcal{L}}$ is finitary and has the PC as well, and then by these properties there are $a_1, \ldots, a_n \in X$, with $n \geqslant 1$, such that $\neg a \in Fg^{\boldsymbol{A}}_{\mathcal{L}}\{a_1, \ldots, a_n, a\} = Fg^{\boldsymbol{A}}_{\mathcal{L}}\{b,a\}$ for $b := a_1 \wedge \cdots \wedge a_n$. This proves the claim. But $\neg a \in Fg^{\boldsymbol{A}}_{\mathcal{L}}\{a,b\}$ implies that $Fg^{\boldsymbol{A}}_{\mathcal{L}}\{a,\neg a\} \subseteq Fg^{\boldsymbol{A}}_{\mathcal{L}}\{a,b\}$, and by Lemma 7.21.3 $x \wedge \neg x \vdash_{\mathcal{L}} y$, therefore it follows that

$$Fg^{\boldsymbol{A}}_{\mathcal{L}}\{a \wedge b\} = Fg^{\boldsymbol{A}}_{\mathcal{L}}\{a,b\} = A = Fg^{\boldsymbol{A}}_{\mathcal{L}}\{a,\neg a\} = Fg^{\boldsymbol{A}}_{\mathcal{L}}\{a \wedge \neg a\}.$$

That is, $a \wedge b \equiv a \wedge \neg a \; (\varLambda^{A}\mathcal{F}i_{\mathcal{L}}\boldsymbol{A})$. But by Theorem 7.20 $\mathcal{L}$ is fully selfextensional, which means that this relation is a congruence of $\boldsymbol{A}$, and as a consequence $\neg(a \wedge b) \equiv \neg(a \wedge \neg a) \; (\varLambda^{A}\mathcal{F}i_{\mathcal{L}}\boldsymbol{A})$. Since by Lemma 7.21.3 $\vdash_{\mathcal{L}} \neg(x \wedge \neg x)$, it follows that $\neg(a \wedge b) \in Fg^{\boldsymbol{A}}_{\mathcal{L}}\{\neg(a \wedge \neg a)\} = Fg^{\boldsymbol{A}}_{\mathcal{L}}\emptyset$. Finally, by Lemma 7.21.4, we conclude that $\neg a \in Fg^{\boldsymbol{A}}_{\mathcal{L}}\{b, \neg(a \wedge b)\} = Fg^{\boldsymbol{A}}_{\mathcal{L}}\{b\} \subseteq Fg^{\boldsymbol{A}}_{\mathcal{L}}X$, as is to be proved. ⊠

In case there are no connectives other than those needed for the relevant properties, things become even easier:

PROPOSITION 7.23
Let $\mathcal{L}$ *be a logic in the language* $\langle \wedge, \neg \rangle$ *of type* $\langle 2,1 \rangle$. *If* $\mathcal{L}$ *satisfies the* PC *with respect to* $\wedge$ *and the* PIRA *with respect to* $\neg$, *then* $\mathcal{L}$ *is selfextensional. If moreover* $\mathcal{L}$ *is finitary, then it is fully selfextensional and all its full g-models satisfy the* PC *and the* PIRA, *and the* PCONG *as well.*

PROOF: By Lemma 5.26, the PC implies that $\varLambda\mathcal{L}$ is a congruence with respect to $\wedge$, and the PIRA implies that $\varLambda\mathcal{L}$ is a congruence with respect to $\neg$. Since these are all the operations of the language, this establishes that $\mathcal{L}$ is selfextensional. The rest of the statement follows from Theorems 7.20 and 7.22. ⊠

Actually, the conclusion of this result can be significantly strengthened by using some results of Section 7.3; see Exercise 7.49.

Now a natural question comes up: Are our logics $\mathcal{L}$ completely determined by their associated variety $\mathbb{V}\mathcal{L}$? As highlighted above, their finite consequences from non-empty assumptions are so: they are encoded in the Frege relation $\Lambda\mathcal{L}$, which corresponds to the equational theory of $\mathbb{V}\mathcal{L}$. Requiring finitarity leaves only the indeterminacy of the theorems. It is easy to see (Exercise 7.15) that if $\mathcal{L}$ has theorems, then another logic, formally different but basically equivalent, can be constructed with the same properties but with no theorems. A uniform criterion to single out one of the two, and which also works when $\mathcal{L}$ has no theorems, arises from the following obscure abstract property:

Definition 7.24
A logic $\mathcal{L}$ is ***pseudo-axiomatic*** *when there exists at least one $\varphi \in Fm$ such that $\emptyset \nvdash_{\mathcal{L}} \varphi$ but $\psi \vdash_{\mathcal{L}} \varphi$ for all $\psi \in Fm$.*

Pseudo-axiomatic logics (first described by Suszko in [166]) are certainly pathological: they have formulas that follow from every formula whatsoever, but still are not theorems themselves (they are "pseudo-theorems" or "pseudo-axioms"). It is not difficult to show (Exercise 7.14):

Proposition 7.25
1. *A logic $\mathcal{L}$ is* ***non-pseudo-axiomatic*** *if and only if $C_{\mathcal{L}}\emptyset = \bigcap\{\Gamma \in Th\mathcal{L} : \Gamma \neq \emptyset\}$.*
2. *A logic is pseudo-axiomatic if and only if it has no theorems and has a smallest non-empty theory.* ⊠

Notice that the intersection in 1 may be empty or not. By 2, all logics with theorems are non-pseudo-axiomatic; however, a non-pseudo-axiomatic logic may have no theorems, in which case that intersection is empty. Adding the empty set to the theories of a logic with theorems produces a pseudo-axiomatic logic, and conversely removing the empty set from the theories of a pseudo-axiomatic logic produces a logic with theorems. Exercise 7.15 shows that this is a bijective correspondence between the sets of logics of the two kinds, and that each corresponding pair of logics are "equivalent" in (almost) all respects, and especially for our algebraic purposes. Thus, restricting the study to non-pseudo-axiomatic logics appears as extremely natural, and produces neat results.

Definition 7.26
Let K be a class of algebras in a similarity type $\boldsymbol{L}$, and fix an order relation $\leqslant$ on each of them.† *The* ***logic of order*** *of K, denoted by $\mathcal{L}^{\leqslant}_{\mathsf{K}} = \langle \boldsymbol{L}, \vdash^{\leqslant}_{\mathsf{K}} \rangle$, is the finitary logic defined in the following way: for all finite $\Gamma \cup \{\varphi\} \subseteq Fm_{\boldsymbol{L}}$,*

$$\Gamma \vdash^{\leqslant}_{\mathsf{K}} \varphi \overset{\text{def}}{\iff} \text{for all } \boldsymbol{A} \in \mathsf{K},\text{ all } a \in A \text{ and all } h \in \operatorname{Hom}(\boldsymbol{Fm}, \boldsymbol{A}),$$
$$\text{if } a \leqslant h\gamma \text{ for all } \gamma \in \Gamma, \text{ then } a \leqslant h\varphi.$$

† A single symbol $\leqslant$ is used for the order relation of each individual algebra in K; this may not lead to any confusion.

This logic can be considered in very general situations, and in this sense it is parallel to the definition of the assertional logic of K, $\mathcal{L}_{\mathsf{K}}^{\top}$, which is defined when all algebras in K are pointed. In principle it is even more general, as Definition 7.26 does not assume that the order relations are uniformly defined in some way. However, a sensible use of this construction requires that the order relation is associated with each algebra in the chosen class in a natural way; for instance, *when* K *has semilattice reducts*, it is tacitly understood that the order is the one associated with the semilattice structure.

Notice that the consequence $\vdash_{\mathsf{K}}^{\leqslant}$ is finitary by definition; in this it differs both from $\vdash_{\mathsf{K}}^{\top}$ and from the consequences $\vdash_{\boldsymbol{A}}^{\leqslant}$ defined in Example 4.11. Actually, the consequence $\vdash_{\mathsf{K}}^{\leqslant}$ agrees with the consequence $\bigcap_{\boldsymbol{A}\in\mathsf{K}} \vdash_{\boldsymbol{A}}^{\leqslant}$ on finitary sets; accordingly, $\mathcal{L}_{\mathsf{K}}^{\leqslant}$ can be called *the finitary logic that preseves degrees of truth* with respect to the class K. Notice also that $\varnothing \vdash_{\mathsf{K}}^{\leqslant} \varphi$ if and only if for all $\boldsymbol{A} \in \mathsf{K}$, all $h \in \operatorname{Hom}(\boldsymbol{Fm}, \boldsymbol{A})$ and all $a \in A$, $a \leqslant h\varphi$; thus, $\mathcal{L}_{\mathsf{K}}^{\leqslant}$ has theorems if and only if all algebras in K have a maximum that is uniformly definable by a term, and has no theorems otherwise.

Proposition 7.27
If K *is a class of algebras with semilattice reducts, then the logic* $\mathcal{L}_{\mathsf{K}}^{\leqslant}$ *is finitary, non-pseudo-axiomatic and semilattice-based with respect to* K*, and is the only logic satisfying these three properties (with respect to* K*).*

Proof: By definition the logic $\mathcal{L}_{\mathsf{K}}^{\leqslant}$ is finitary. It is semilattice-based because a finite set always has an infimum, therefore to say that "for every $a \in A$, if $a \leqslant h\varphi_i$ for all $i < n$, then $a \leqslant h\varphi_n$", is the same as to say that $h\varphi_0 \wedge \cdots \wedge h\varphi_{n-1} \leqslant h\varphi_n$. And it is non-pseudo-axiomatic: If $\psi \vdash_{\mathsf{K}}^{\leqslant} \varphi$ for all ψ, in particular $x \vdash_{\mathsf{K}}^{\leqslant} \varphi$ for any $x \in V \smallsetminus \operatorname{Var}\varphi$, and by the definition of $\mathcal{L}_{\mathsf{K}}^{\leqslant}$, this implies that for every $\boldsymbol{A} \in \mathsf{K}$, every $h \in \operatorname{Hom}(\boldsymbol{Fm}, \boldsymbol{A})$ and every $a \in A$, $a \leqslant h\varphi$, that is, $\varnothing \vdash_{\mathsf{K}}^{\leqslant} \varphi$; this shows that $\mathcal{L}_{\mathsf{K}}^{\leqslant}$ is non-pseudo-axiomatic. Clearly the three properties characterize $\mathcal{L}_{\mathsf{K}}^{\leqslant}$ from K: To be non-pseudo-axiomatic means that the non-empty theories determine the theorems; being semilattice-based with respect to K determines the finitely generated (non-empty) theories, and finitarity means that these determine the other theories. ☒

Therefore:

Theorem 7.28
Let $\boldsymbol{L}$ *be an algebraic type and let* $\wedge$ *represent a binary term in* $\boldsymbol{L}$*. The functions* $\mathcal{L} \mapsto \mathbb{V}\mathcal{L}$ *and* $\mathsf{K} \mapsto \mathcal{L}_{\mathsf{K}}^{\leqslant}$ *establish mutually inverse dual order isomorphisms between the following families:*[†]

- *The family of all the finitary and non-pseudo-axiomatic selfextensional logics with conjunction* $\wedge$ *in the language* $\boldsymbol{L}$*, ordered under the extension relation.*
- *The family of all the subvarieties of the variety* $\mathsf{SL}(\boldsymbol{L}, \wedge)$*, ordered under the subclass relation.*

[†] See footnote [†] on page 410.

PROOF: We have already seen that if $\mathcal{L}$ is selfextensional with conjunction, then the three equations hold in $\mathbb{V}\mathcal{L}$, which is therefore a subvariety of $\mathsf{SL}(\boldsymbol{L},\wedge)$. Conversely, if K is a subvariety of $\mathsf{SL}(\boldsymbol{L},\wedge)$, it is a class with semilattice reducts, and by Proposition 7.27, $\mathcal{L}^{\leqslant}_{\mathsf{K}}$ is finitary, non-pseudo-axiomatic and semilattice-based, hence selfextensional with conjunction. That if $\mathsf{K} \subseteq \mathsf{K}'$, then $\mathcal{L}^{\leqslant}_{\mathsf{K}'} \leqslant \mathcal{L}^{\leqslant}_{\mathsf{K}}$ is trivial. Now assume that $\mathcal{L}$ and $\mathcal{L}'$ are two logics in the family such that $\mathcal{L} \leqslant \mathcal{L}'$. By Lemma 1.38.1, $\boldsymbol{\Lambda}\mathcal{L} \subseteq \boldsymbol{\Lambda}\mathcal{L}'$, and since both logics are selfextensional, by Proposition 5.75.3 this implies that $\mathbb{V}\mathcal{L}' \subseteq \mathbb{V}\mathcal{L}$. So both functions go from the stated family to the other, and reverse order. Since by construction $\mathcal{L}^{\leqslant}_{\mathsf{K}}$ is semilattice-based with respect to K, which is a variety, by Corollary 7.19 $\mathbb{V}\mathcal{L}^{\leqslant}_{\mathsf{K}} = \mathbb{V}\mathsf{K} = \mathsf{K}$. Finally, if $\mathcal{L}$ is a logic in the family, then $\mathbb{V}\mathcal{L}$ is a subvariety of $\mathsf{SL}(\boldsymbol{L},\wedge)$ and by construction $\mathcal{L}^{\leqslant}_{\mathbb{V}\mathcal{L}}$ is finitary, non-pseudo-axiomatic and semilattice-based with respect to $\mathbb{V}\mathcal{L}$; but $\mathcal{L}$ itself has all these properties, therefore by Proposition 7.27 $\mathcal{L}^{\leqslant}_{\mathbb{V}\mathcal{L}} = \mathcal{L}$. Thus, the two functions are mutually inverse bijections, and since both reverse the order, they are dual order isomorphisms. ⊠

Recall that in the case where $\boldsymbol{L} = \langle\wedge\rangle$, $\mathsf{SL}(\boldsymbol{L},\wedge) = \mathsf{SL}$. Then, the smallest logic in the family of logics considered in the theorem is $\mathcal{C\ell}_{\wedge} = \mathcal{L}^{\leqslant}_{\mathsf{SL}}$; see Example 7.34.2.

Theorem 7.20 provides a description of the full g-models of a finitary, selfextensional logic with conjunction: they are the g-matrices that are inverse images by a surjective homomorphism of a g-matrix of the form $\langle \boldsymbol{A}, \mathscr{C}\rangle$, where $\boldsymbol{A} \in \mathbb{V}\mathcal{L}$ and $\mathscr{C}$ is either $\mathcal{F}ilt_{\wedge}\boldsymbol{A}$ or $\mathcal{F}ilt_{\wedge}\boldsymbol{A} \cup \{\varnothing\}$, depending on whether $\mathcal{L}$ has or has not theorems. More workable descriptions depend on those of the variety $\mathbb{V}\mathcal{L}$ and on global descriptions of the closure system generated by the family of its semilattice filters. It turns out that a description by a Gentzen calculus is always possible (although in practice it need not be much better than the one just mentioned).

DEFINITION 7.29
Let $\mathcal{L}$ be a finitary logic. The Gentzen system $\mathfrak{G}^{c}_{\mathcal{L}}$ is the extension of the Gentzen system $\mathfrak{G}_{\mathcal{L}}$ of Definition 5.114 with the "congruence rules" (CONG) *of Definition 7.1, for all connectives of the language.*

Recall that $\mathfrak{G}_{\mathcal{L}}$ is defined either on the set *Seq* of all sequents (when $\mathcal{L}$ has theorems) or on the set Seq° of all sequents with non-empty antecedent (when $\mathcal{L}$ has no theorems), that its initial sequents are the sequents $\varphi_1,\dots,\varphi_n \rhd \varphi$ such that $\varphi_1,\dots,\varphi_n \vdash_{\mathcal{L}} \varphi$, and that its only rules are the structural rules. The superscript "c" in $\mathfrak{G}^{c}_{\mathcal{L}}$ is for "congruence", the property described by the additional rules. You should now revise the notions of a Gentzen system being *adequate* and *fully adequate* for a logic $\mathcal{L}$ (Definitions 5.112 and 5.117). Proposition 5.115 shows that the Gentzen system $\mathfrak{G}_{\mathcal{L}}$ is always adequate for $\mathcal{L}$, and is indeed the weakest adequate Gentzen system for it. It is not difficult to check (Exercise 7.16):

LEMMA 7.30
If $\mathcal{L}$ is a finitary logic, then it is selfextensional if and only if the Gentzen system $\mathfrak{G}^{c}_{\mathcal{L}}$ is adequate for $\mathcal{L}$. ⊠

PROPOSITION 7.31
Let $\mathcal{L}$ be a finitary selfextensional logic with conjunction. A g-matrix is a full g-model of $\mathcal{L}$ if and only if it is a finitary model of $\mathfrak{G}^{c}_{\mathcal{L}}$, without theorems if $\mathcal{L}$ has no theorems.

PROOF: ($\Rightarrow$) The full g-models of a logic without theorems do not have theorems, and they are finitary when the logic is finitary. Besides, full g-models are in particular g-models, so they are models of all the initial sequents of $\mathfrak{G}^{c}_{\mathcal{L}}$, and of course they are models of all structural rules. Finally, by Theorem 7.20, $\mathcal{L}$ is fully selfextensional, that is, its full g-models have the PCONG, which amounts to being a model of the rules (CONG). Hence, all the full g-models of $\mathcal{L}$ are models of the Gentzen system $\mathfrak{G}^{c}_{\mathcal{L}}$.

($\Leftarrow$) Since a g-matrix is a full g-model of a logic if and only if its reduction is, it is enough to prove the statement for reduced g-matrices. Let thus $\langle \boldsymbol{A}, \mathscr{C} \rangle$ be a reduced finitary model of $\mathfrak{G}^{c}_{\mathcal{L}}$, without theorems if $\mathcal{L}$ has no theorems. In order to show that it is a full g-model of $\mathcal{L}$ it is enough to show that $\mathscr{C} = \mathcal{F}i_{\mathcal{L}}\boldsymbol{A}$, because all reduced full g-models are basic. Since $\langle \boldsymbol{A}, \mathscr{C} \rangle$ is a model of all the initial sequents of $\mathfrak{G}^{c}_{\mathcal{L}}$ and $\mathcal{L}$ is finitary, $\langle \boldsymbol{A}, \mathscr{C} \rangle$ is a g-model of $\mathcal{L}$, that is, $\mathscr{C} \subseteq \mathcal{F}i_{\mathcal{L}}\boldsymbol{A}$. Now take any $F \in \mathcal{F}i_{\mathcal{L}}\boldsymbol{A}$. If $F = \emptyset$ this means that $\mathcal{L}$ has no theorems, and then by assumption $\langle \boldsymbol{A}, \mathscr{C} \rangle$ has no theorems either, so $F \in \mathscr{C}$. If $F \neq \emptyset$, we prove that $CF = F$. To this end, take any $a \in CF$. Since C is finitary and F is nonempty, there are $a_1, \dots, a_n \in F$ (for some $n \geqslant 1$) such that $a \in C\{a_1, \dots, a_n\}$. But, being a g-model of $\mathcal{L}$, C satisfies the PC, therefore by Proposition 7.12.2 $(a_1 \wedge \dots \wedge a_n) \wedge a \equiv a_1 \wedge \dots \wedge a_n \; (\boldsymbol{\Lambda}^{A}C)$. Now, the g-matrix is a model of the rules (CONG), which means that it has the PCONG, and is reduced; these two properties imply that $\boldsymbol{\Lambda}^{A}C$ is the identity relation, therefore $(a_1 \wedge \dots \wedge a_n) \wedge a = a_1 \wedge \dots \wedge a_n$. Since C satisfies the PC, F is a $\wedge$-filter, and hence the fact that $a_i \in F$ for all $i = 1, \dots, n$ implies that $a_1 \wedge \dots \wedge a_n \in F$, that is, $a_1 \wedge \dots \wedge a_n \wedge a \in F$, which again by the same reason implies that $a \in F$. This shows that $CF = F$, thus $F \in \mathscr{C}$. This completes the proof that $\mathscr{C} = \mathcal{F}i_{\mathcal{L}}\boldsymbol{A}$. ⊠

THEOREM 7.32
If $\mathcal{L}$ is a finitary logic with conjunction, then it is selfextensional if and only if the Gentzen system $\mathfrak{G}^{c}_{\mathcal{L}}$ is fully adequate for $\mathcal{L}$.

PROOF: If $\mathcal{L}$ is a finitary selfextensional logic with conjunction, then the result of Proposition 7.31, in terms of Definition 5.117, says exactly that $\mathfrak{G}^{c}_{\mathcal{L}}$ is fully adequate for $\mathcal{L}$ (given how $\mathfrak{G}^{c}_{\mathcal{L}}$ and $\mathfrak{G}_{\mathcal{L}}$ are defined). The converse is included in Lemma 7.30, because by Proposition 5.118 a fully adequate Gentzen system is also adequate; notice that conjunction does not play a role in this converse. ⊠

See Exercise 7.17 for an alternative formulation of basically the same result, this time incorporating conjunction to the characterization.

COROLLARY 7.33
Let $\mathcal{L}$ be a finitary and selfextensional logic with conjunction.

1. *$\mathcal{L}$ has a fully adequate Gentzen system, namely $\mathfrak{G}^{c}_{\mathcal{L}}$.*

2. *A g-model of $\mathcal{L}$ is full if and only if it is finitary and has the* PCONG, *and has no theorems if $\mathcal{L}$ has no theorems.* ☒

The investigation of the existence of fully adequate Gentzen systems for a logic, and of necessary or sufficient conditions for this, is one of the more interesting lines of research in abstract algebraic logic, as it merges many topics. Corollary 7.33.1 expresses one of such sufficient conditions: being finitary and selfextensional with a conjunction. It is also interesting to see that, for these logics, the property that tells the full g-models apart, besides finitarity and having or not having theorems like the logic, is just the property of congruence. Similar (but somehow less nice) results are obtained for finitary and selfextensional logics that have the uniterm DDT (Theorem 7.53).

The calculus defining $\mathfrak{G}^{c}_{\mathcal{L}}$ is of course not very interesting: including all finite consequences of $\mathcal{L}$ as initial sequents is cumbersome and definitely not practical. If one has a Hilbert-style presentation of $\mathcal{L}$, then it can replace this set of initial sequents, but still it is not particularly interesting as a Gentzen calculus. The wish is to have one consisting of Gentzen-style rules that reflect truly metalogical properties of the logic, while in $\mathfrak{G}^{c}_{\mathcal{L}}$ this is only so for the congruence rules (still, it is interesting that these are the only non-Hilbert-style rules needed to characterize the full g-models of these logics). Thus it is worthwhile to look for more really-Gentzen-style rules in a calculus defining the fully adequate Gentzen system of a logic (when it has one; recall that if it exists, then it is unique).

Examples 7.34

The preceding results are general statements of facts empirically observed throughout the years in many particular logics that we now recognize as falling under their scope. Some are semilattice-based by their very definition, others are easily shown to be selfextensional and to satisfy the PC. You can check that several of the properties obtained in other places of the book for these particular logics are in fact instances of the general results obtained in this section. Thus, some of the points proved in previous chapters about these logics could be proved in a more direct way using these general results; the given proofs, however, offer a sample of direct work with particular logics, their algebras and models.

1. As a fragment of classical logic, $\mathcal{C\ell}_{\wedge\vee}$ is selfextensional and has the PC, so it is semilattice-based, and hence it is so with respect to its intrinsic variety $\mathbb{V}\mathcal{C\ell}_{\wedge\vee} = \mathsf{Alg}\mathcal{C\ell}_{\wedge\vee} = \mathbb{V}\{\mathbf{2}_{\wedge\vee}\} = \mathsf{DL}$ (this is shown in Example 5.93). Moreover, it is finitary and non-pseudo-axiomatic. By the preceding results we know that this logic is fully selfextensional and has a fully adequate Gentzen system, and that $\mathcal{C\ell}_{\wedge\vee} = \mathcal{L}^{\leqslant}_{\mathsf{DL}}$. By Example 5.93(vi), the full g-models of $\mathcal{C\ell}_{\wedge\vee}$ are the finitary g-matrices without theorems that satisfy the PC and the PDI; since these properties can be expressed by Gentzen-style rules (Proposition 5.17 and Exercise 5.7), the corresponding Gentzen system (described explicitly in Example 3.87) is fully adequate for $\mathcal{C\ell}_{\wedge\vee}$ (see also Example 5.120.3). The

particular case of Corollary 7.33.2 is obtained as the equivalence between points (i) and (vii) of Example 5.93.

2. Parallel results hold for all fragments of classical logic that include conjunction. In particular, $\mathcal{C\ell}_\wedge = \mathcal{L}^{\leqslant}_{\mathsf{SL}}$ and $\mathbb{V}\mathcal{C\ell}_\wedge = \mathsf{Alg}\mathcal{C\ell}_\wedge = \mathbb{V}\{\mathbf{2}_\wedge\} = \mathsf{SL}$.
3. The same can be said of intuitionistic logic and its fragments with conjunction; but in this case the intrinsic variety is not generated by a single two-element algebra.[†] Since $\mathcal{I\ell}$ and all these fragments are characterized by the Tarski-style conditions corresponding to their connectives, as proved in [195], in most cases one would obtain characterizations of their fully adequate Gentzen systems. An early integrated study of all these fragments, still not in the present methodology but clearly prefiguring some of its aspects, is [40].
4. Belnap-Dunn's logic $\mathcal{B}$, the other paradigmatic example of Chapter 5 (Examples 5.82 and 5.92), is also finitary, non-pseudo-axiomatic and selfextensional and has the PC, and is semilattice-based with respect to the variety $\mathsf{Alg}\mathcal{B} = \mathbb{V}\mathcal{B} = \mathbb{V}\{\boldsymbol{M}_4\} = \mathsf{DM}$. As before, it follows that $\mathcal{B} = \mathcal{L}^{\leqslant}_{\mathsf{DM}}$, a fact already present in its first characterization (Example 4.12). A real Gentzen-style calculus for its fully adequate Gentzen system is obtained from the characterizations of its full g-models in points (B19) and (B20) of Example 5.92; see also Example 5.120.4. The particular case of Corollary 7.33.2 is found in (B21).
5. Let L be the variety of lattices, and consider the logic $\mathcal{L}^{\leqslant}_{\mathsf{L}}$ in the language $\langle \wedge, \vee \rangle$ of type $\langle 2,2 \rangle$. In contrast to $\mathcal{C\ell}_{\wedge\vee}$, which is an extension of $\mathcal{L}^{\leqslant}_{\mathsf{L}}$, in this case the variety is not generated by a single algebra. Obviously, the variety L has semilattice reducts, therefore this logic falls under the scope of the constructions on the preceding pages. It is not difficult to find a calculus for its fully adequate Gentzen system; see Exercise 7.19. Further properties of this logic are shown in Example 7.60.7. ⊠

By Proposition 5.121, that $\mathfrak{G}^{c}_{\mathcal{L}}$ is fully adequate for $\mathcal{L}$ implies that its algebraic counterpart coincides with that of $\mathcal{L}$; therefore $\mathsf{Alg}\mathfrak{G}^{c}_{\mathcal{L}} = \mathbb{V}\mathcal{L}$. It turns out that the relation between $\mathfrak{G}^{c}_{\mathcal{L}}$ and the variety $\mathbb{V}\mathcal{L}$ is much stronger. It is expressed in the next result, which uses the notion of *algebraizability of a Gentzen system*, discussed informally in Section 3.7; it is presented here only as a glimpse at a larger area of application of these notions.

THEOREM 7.35
Let $\mathcal{L}$ be a finitary and selfextensional logic with conjunction. The Gentzen system $\mathfrak{G}^{c}_{\mathcal{L}}$ is algebraizable and has the variety $\mathbb{V}\mathcal{L}$ as its equivalent algebraic semantics. The structural transformers between sequents and equations are given by

$$\boldsymbol{\tau}(x_0,\dots,x_{n-1} \rhd x_n) := x_0 \wedge \dots \wedge x_{n-1} \preccurlyeq x_n \quad \text{for } n \geqslant 1,$$

together, when $\mathcal{L}$ has theorems, with $\boldsymbol{\tau}(\varnothing \rhd x) := x \approx \top$ for any theorem $\top$ with at most the variable x, and by $\boldsymbol{\rho}(x \approx y) := \{x \rhd y, y \rhd x\}$.

[†] Except the fragments with only conjunction, and with only conjunction and disjunction, which coincide with those of classical logic.

PROOF: This is left to you as Exercise 7.20; it is a good one to test your understanding of the extension of the idea of algebraizability through structural transformers to more general contexts than just sentential logics. ⊠

This property has been described in the literature by saying that these logics are *G-algebraizable*, in the sense discussed on page 314.

Semilattice-based logics with an algebraizable assertional companion

Now, assume that K is a variety of algebras of some algebraic type $\boldsymbol{L}$ having semilattice reducts, and *moreover* that the associated order in algebras of K has a term-definable maximum. Then K becomes a pointed variety, and it makes sense to consider side-by-side the two logics naturally associated with K: the logic of order $\mathcal{L}_{\mathsf{K}}^{\leqslant}$ of Definition 7.26, and the assertional logic $\mathcal{L}_{\mathsf{K}}^{\top} = \langle \boldsymbol{L}, \vdash_{\mathsf{K}}^{\top} \rangle$, sometimes called informally ***the logic of one*** of K in contexts like the present one. The symbol $\top$ represents any term (with at most one variable) defining the said maximum, hence any theorem of this logic, and as usual this maximum can be denoted by $1 := \top^{\boldsymbol{A}}$ when $\boldsymbol{A} \in \mathsf{K}$. We see that, when the logic of one is algebraizable, the logic of order behaves, in certain respects, like a protoalgebraic logic, without being necessarily so.

Later on several large classes of examples of this situation, already described in the literature, are highlighted.

DEFINITION 7.36
Let K *be a variety of algebras with semilattice reducts and such that the associated order in algebras of* K *has a maximum defined by some term* $\top$ *(which is then constant over* K*). The two associated logics* $\mathcal{L}_{\mathsf{K}}^{\leqslant}$ *and* $\mathcal{L}_{\mathsf{K}}^{\top}$ *are said to be* ***companions*** *of each other:* $\mathcal{L}_{\mathsf{K}}^{\top}$ *is the* ***assertional companion*** *of* $\mathcal{L}_{\mathsf{K}}^{\leqslant}$*, and this one is the* ***semilattice-based companion*** *of* $\mathcal{L}_{\mathsf{K}}^{\top}$*. The logic* $\mathcal{L}_{\mathsf{K}}^{\leqslant}$ ***has an algebraizable assertional companion*** *when the logic* $\mathcal{L}_{\mathsf{K}}^{\top}$ *is algebraizable with defining equation* $x \approx \top$ *and equivalent algebraic semantics the variety* K.

NOTATION.
It is convenient to denote the sets of filters of the logics $\mathcal{L}_{\mathsf{K}}^{\leqslant}$ and $\mathcal{L}_{\mathsf{K}}^{\top}$, on an arbitrary algebra $\boldsymbol{A}$, by $\mathcal{F}i_{\mathsf{K}}^{\leqslant}\boldsymbol{A}$ and $\mathcal{F}i_{\mathsf{K}}^{\top}\boldsymbol{A}$, respectively. ⊠

The assumptions for the next pages, not to be repeated, are those made on K in Definition 7.36, *and* that $\mathcal{L}_{\mathsf{K}}^{\leqslant}$ indeed *has* an algebraizable assertional companion $\mathcal{L}_{\mathsf{K}}^{\top}$; then this logic is automatically regularly algebraizable (Theorem 6.142) and finitely algebraizable, because K is a variety (Theorem 3.37), therefore it is regularly BP-algebraizable, and its equivalent algebraic semantics is the variety K. The logic $\mathcal{L}_{\mathsf{K}}^{\leqslant}$ is a finitary and selfextensional logic with conjunction (hence, fully selfextensional) and theorems (hence, non-pseudo-axiomatic). The basic properties of the situation, following from the preceding pages and from Section 6.5, are easy to check (Exercise 7.21):

PROPOSITION 7.37

1. *The logics* $\mathcal{L}_{\mathsf{K}}^{\leqslant}$ *and* $\mathcal{L}_{\mathsf{K}}^{\top}$ *have the same theorems, namely the formulas* φ *such that* $\mathsf{K} \vDash \varphi \approx \top$.
2. *The logic* $\mathcal{L}_{\mathsf{K}}^{\top}$ *is an extension of* $\mathcal{L}_{\mathsf{K}}^{\leqslant}$*, therefore* $\mathcal{F}i_{\mathsf{K}}^{\top}\boldsymbol{A} \subseteq \mathcal{F}i_{\mathsf{K}}^{\leqslant}\boldsymbol{A}$ *for every* $\boldsymbol{A}$.
3. *In every* $\boldsymbol{A} \in \mathsf{K}$*, the set* $\{1\}$ *is the least filter of the two logics.*
4. $\mathsf{Alg}\mathcal{L}_{\mathsf{K}}^{\leqslant} = \mathbb{V}\mathcal{L}_{\mathsf{K}}^{\leqslant} = \mathsf{Alg}\mathcal{L}_{\mathsf{K}}^{\top} = \mathbb{V}\mathcal{L}_{\mathsf{K}}^{\top} = \mathsf{K}$*, and in each* $\boldsymbol{A} \in \mathsf{K}$, $\mathcal{F}i_{\mathsf{K}}^{\leqslant}\boldsymbol{A} = \mathcal{F}ilt_{\wedge}\boldsymbol{A}$. ⊠

By point 2, the logic $\mathcal{L}_{\mathsf{K}}^{\top}$ satisfies the PC as well, with respect to the same term operation $\wedge$. In the next result, the first outcome of this cross-fertilization, the notion of a Leibniz filter, introduced in Definition 6.40 for a protoalgebraic logic, is used for an arbitrary logic.

THEOREM 7.38

For every $\boldsymbol{A}$ *and every* $F \in \mathcal{F}i_{\mathsf{K}}^{\leqslant}\boldsymbol{A}$ *there exists a unique Leibniz filter* $F^{+} \in \mathcal{F}i_{\mathsf{K}}^{\leqslant}\boldsymbol{A}$ *such that* $\boldsymbol{\Omega}^{\boldsymbol{A}}F^{+} = \boldsymbol{\Omega}^{\boldsymbol{A}}F$*, namely* $F^{+} = \min\{G \in \mathcal{F}i_{\mathsf{K}}^{\leqslant}\boldsymbol{A} : \boldsymbol{\Omega}^{\boldsymbol{A}}G = \boldsymbol{\Omega}^{\boldsymbol{A}}F\} \subseteq F$. *Moreover,* $F^{+} \in \mathcal{F}i_{\mathsf{K}}^{\top}\boldsymbol{A}$ *and it is determined from the congruence* $\boldsymbol{\Omega}^{\boldsymbol{A}}F$ *in the following way:* $F^{+} = \{a \in A : a \equiv \top^{\boldsymbol{A}}(a)\ (\boldsymbol{\Omega}^{\boldsymbol{A}}F)\}$. *In particular, if* $\boldsymbol{A} \in \mathsf{K}$*, then it holds that* $F^{+} = \{a \in A : a \equiv 1\ (\boldsymbol{\Omega}^{\boldsymbol{A}}F)\}$.

PROOF: The proof of the second statement is actually part of the construction performed in order to prove the first one. Consider an arbitrary algebra $\boldsymbol{A}$, and an arbitrary $F \in \mathcal{F}i_{\mathsf{K}}^{\leqslant}\boldsymbol{A}$. The matrix $\langle \boldsymbol{A}/\boldsymbol{\Omega}^{\boldsymbol{A}}F, F/\boldsymbol{\Omega}^{\boldsymbol{A}}F\rangle$ is always reduced, hence $\boldsymbol{A}/\boldsymbol{\Omega}^{\boldsymbol{A}}F \in \mathsf{Alg}^{*}\mathcal{L}_{\mathsf{K}}^{\leqslant} \subseteq \mathsf{Alg}\mathcal{L}_{\mathsf{K}}^{\leqslant} = \mathsf{K}$, therefore actually $\boldsymbol{\Omega}^{\boldsymbol{A}}F \in \mathrm{Con}_{\mathsf{K}}\boldsymbol{A}$. By the algebraizability of $\mathcal{L}_{\mathsf{K}}^{\top}$, there is a unique $F^{+} \in \mathcal{F}i_{\mathsf{K}}^{\top}\boldsymbol{A}$ such that $\boldsymbol{\Omega}^{\boldsymbol{A}}F^{+} = \boldsymbol{\Omega}^{\boldsymbol{A}}F$. In particular, $F^{+} \in \mathcal{F}i_{\mathsf{K}}^{\leqslant}\boldsymbol{A}$. Thus $F^{+} \in \{G \in \mathcal{F}i_{\mathsf{K}}^{\leqslant}\boldsymbol{A} : \boldsymbol{\Omega}^{\boldsymbol{A}}G = \boldsymbol{\Omega}^{\boldsymbol{A}}F\}$. Now let $a \in F^{+}$. Since $x \approx \top$ is the defining equation of $\mathcal{L}_{\mathsf{K}}^{\top}$, but $\boldsymbol{A}$ is not assumed to belong to K, the term $\top$ need not be constant over $\boldsymbol{A}$, so it must be implicitly considered as $\top(x)$ (unless it is a real constant of the language). Then, $a \in F^{+}$ if and only if $\langle a, \top^{\boldsymbol{A}}(a)\rangle \in \boldsymbol{\Omega}^{\boldsymbol{A}}F^{+} = \boldsymbol{\Omega}^{\boldsymbol{A}}F$. Since by assumption $\top$ is a theorem of $\mathcal{L}_{\mathsf{K}}^{\top}$, it is also a theorem of $\mathcal{L}_{\mathsf{K}}^{\leqslant}$, and hence $\top^{\boldsymbol{A}}(a) \in F$; then by compatibility $a \in F$. This shows that $F^{+} \subseteq F$. Now take any $G \in \mathcal{F}i_{\mathsf{K}}^{\leqslant}\boldsymbol{A}$ such that $\boldsymbol{\Omega}^{\boldsymbol{A}}G = \boldsymbol{\Omega}^{\boldsymbol{A}}F$ and consider $G^{+} \in \mathcal{F}i_{\mathsf{K}}^{\top}\boldsymbol{A}$ constructed in the same way as F^{+} but starting from G; then $\boldsymbol{\Omega}^{\boldsymbol{A}}F^{+} = \boldsymbol{\Omega}^{\boldsymbol{A}}F = \boldsymbol{\Omega}^{\boldsymbol{A}}G = \boldsymbol{\Omega}^{\boldsymbol{A}}G^{+}$, and since both are filters of $\mathcal{L}_{\mathsf{K}}^{\top}$ and $\boldsymbol{\Omega}^{\boldsymbol{A}}$ is one-to-one over $\mathcal{F}i_{\mathsf{K}}^{\top}\boldsymbol{A}$, it follows that $F^{+} = G^{+}$. But $G^{+} \subseteq G$, so $F^{+} \subseteq G$. This proves that $F^{+} = \min\{G \in \mathcal{F}i_{\mathsf{K}}^{\leqslant}\boldsymbol{A} : \boldsymbol{\Omega}^{\boldsymbol{A}}G = \boldsymbol{\Omega}^{\boldsymbol{A}}F\}$. Moreover, using the fact that $\boldsymbol{\Omega}^{\boldsymbol{A}}F^{+} = \boldsymbol{\Omega}^{\boldsymbol{A}}F$, the last expression can be equivalently written as $F^{+} = \min\{G \in \mathcal{F}i_{\mathsf{K}}^{\leqslant}\boldsymbol{A} : \boldsymbol{\Omega}^{\boldsymbol{A}}G = \boldsymbol{\Omega}^{\boldsymbol{A}}F^{+}\}$, which says that F^{+} is a Leibniz $\mathcal{L}_{\mathsf{K}}^{\leqslant}$-filter of $\boldsymbol{A}$. ⊠

Thus, the logic $\mathcal{L}_{\mathsf{K}}^{\leqslant}$ has Leibniz filters, like all the protoalgebraic logics, without being necessarily so. Moreover, in this case the property of being a Leibniz filter can be characterized in several ways, some in terms of the associated Leibniz congruence, by much nicer conditions than the ones of Corollary 6.59 (anyway, this one uses a set of congruence formulas with parameters, something not available in general):

Proposition 7.39
For any $\boldsymbol{A}$ and any $F \in \mathcal{F}i^{\leqslant}_{\mathsf{K}}\boldsymbol{A}$, the following conditions are equivalent:

(i) *F is a Leibniz $\mathcal{L}^{\leqslant}_{\mathsf{K}}$-filter of $\boldsymbol{A}$.*

(ii) $F \in \mathcal{F}i^{\top}_{\mathsf{K}}\boldsymbol{A}$.

(iii) $F/\boldsymbol{\Omega}^{\boldsymbol{A}}F = \{1\}$.

(iv) *$F/\boldsymbol{\Omega}^{\boldsymbol{A}}F$ is the smallest $\mathcal{L}^{\leqslant}_{\mathsf{K}}$-filter of $\boldsymbol{A}/\boldsymbol{\Omega}^{\boldsymbol{A}}F$.*

Proof: If F is Leibniz, then $F = F^{+}$, so $F \in \mathcal{F}i^{\top}_{\mathsf{K}}\boldsymbol{A}$. Conversely, if $F \in \mathcal{F}i^{\top}_{\mathsf{K}}\boldsymbol{A}$, then it is also an $\mathcal{L}^{\leqslant}_{\mathsf{K}}$-filter, and by Theorem 7.38, F^{+} is an $\mathcal{L}^{\top}_{\mathsf{K}}$-filter with the same Leibniz congruence as F, and since $\boldsymbol{\Omega}^{\boldsymbol{A}}$ is one-to-one over $\mathcal{F}i^{\top}_{\mathsf{K}}\boldsymbol{A}$, $F = F^{+}$, which implies that F is Leibniz (relatively to $\mathcal{L}^{\leqslant}_{\mathsf{K}}$). This shows that (i) is equivalent to (ii). Due to the way how $\mathcal{L}^{\top}_{\mathsf{K}}$ is algebraized (namely, that the defining equation is $x \approx \top$), the filter of a reduced model of $\mathcal{L}^{\top}_{\mathsf{K}}$ is unique and equal to $\{1\}$. Hence, if F is a $\mathcal{L}^{\top}_{\mathsf{K}}$-filter, then $F/\boldsymbol{\Omega}^{\boldsymbol{A}}F = \{1\}$. Conversely, if $F/\boldsymbol{\Omega}^{\boldsymbol{A}}F = \{1\}$, then $F = \pi^{-1}\{1\}$, where $\pi\colon \boldsymbol{A} \to \boldsymbol{A}/\boldsymbol{\Omega}^{\boldsymbol{A}}F$ is the canonical projection. Since $\{1\}$ is an $\mathcal{L}^{\top}_{\mathsf{K}}$-filter of the quotient algebra, $F \in \mathcal{F}i^{\top}_{\mathsf{K}}\boldsymbol{A}$. This shows that (ii) is equivalent to (iii). Finally, (iii) is equivalent to (iv) because if $F \in \mathcal{F}i^{\leqslant}_{\mathsf{K}}\boldsymbol{A}$, then $\boldsymbol{A}/\boldsymbol{\Omega}^{\boldsymbol{A}}F \in \mathsf{Alg}^{*}\mathcal{L}^{\leqslant}_{\mathsf{K}} \subseteq \mathsf{K}$ and by Proposition 7.37.3 the smallest $\mathcal{L}^{\leqslant}_{\mathsf{K}}$-filter of these algebras coincides with the smallest $\mathcal{L}^{\top}_{\mathsf{K}}$-filter, which is $\{1\}$. ☒

Observe that the equivalence between (i) and (iv) is also present in protoalgebraic logics (Theorem 6.41). One has to be careful here and not to use any reasoning based on the Correspondence Theorem for protoalgebraic logics; for instance, in this proof some steps are required because being Leibniz with respect to one logic is not the same as being so with respect to the other (by algebraizability, all filters of $\mathcal{L}^{\top}_{\mathsf{K}}$ are Leibniz with respect to it). Now, the property in Proposition 7.37.3 can be extended to arbitrary algebras:

Corollary 7.40
For any algebra $\boldsymbol{A}$, $\min \mathcal{F}i^{\leqslant}_{\mathsf{K}}\boldsymbol{A} = \min \mathcal{F}i^{\top}_{\mathsf{K}}\boldsymbol{A}$.

Proof: Put $F^{\circ}_{0} := \min \mathcal{F}i^{\circ}_{\mathsf{K}}$ for $\circ \in \{\leqslant,\top\}$. Since $\mathcal{F}i^{\top}_{\mathsf{K}}\boldsymbol{A} \subseteq \mathcal{F}i^{\leqslant}_{\mathsf{K}}\boldsymbol{A}$, $F^{\top}_{0} \in \mathcal{F}i^{\leqslant}_{\mathsf{K}}\boldsymbol{A}$ and hence $F^{\leqslant}_{0} \subseteq F^{\top}_{0}$. Applying Theorem 7.38 to $F^{\leqslant}_{0}$, we know that there exists $(F^{\leqslant}_{0})^{+} \in \mathcal{F}i^{\top}_{\mathsf{K}}\boldsymbol{A}$ such that $(F^{\leqslant}_{0})^{+} \subseteq F^{\leqslant}_{0}$; but this implies that $F^{\top}_{0} \subseteq (F^{\leqslant}_{0})^{+}$ and hence that $F^{\top}_{0} \subseteq F^{\leqslant}_{0}$, and the equality is proved. ☒

Another property than can be obtained here, similar to the protoalgebraic case and in contrast to the general selfextensional case,[†] is the determination of the reduced models of $\mathcal{L}^{\leqslant}_{\mathsf{K}}$ and of its algebraic reducts:

Proposition 7.41

1. $\mathsf{Alg}^{*}\mathcal{L}^{\leqslant}_{\mathsf{K}} = \mathsf{K}$.
2. *A matrix $\langle \boldsymbol{A}, F\rangle$ is a reduced model of $\mathcal{L}^{\leqslant}_{\mathsf{K}}$ if and only if $\boldsymbol{A} \in \mathsf{K}$ and F is a semilattice filter of $\boldsymbol{A}$ such that $F^{+} = \{1\}$.*

[†] Recall the weird, unpredictable characterizations of the reduced models of some selfextensional logics, such as $\mathcal{C\!\ell}_{\wedge\vee}$ (Proposition 4.49) or $\mathcal{B}$ (point (B13) of Example 5.82).

PROOF: 1. We already know that $\mathsf{Alg}^*\mathcal{L}_{\mathsf{K}}^{\leqslant} \subseteq \mathsf{Alg}\mathcal{L}_{\mathsf{K}}^{\leqslant} = \mathsf{K}$. But $\mathcal{L}_{\mathsf{K}}^{\top}$ is an extension of $\mathcal{L}_{\mathsf{K}}^{\leqslant}$, therefore $\mathsf{K} = \mathsf{Alg}\mathcal{L}_{\mathsf{K}}^{\top} = \mathsf{Alg}^*\mathcal{L}_{\mathsf{K}}^{\top} \subseteq \mathsf{Alg}^*\mathcal{L}_{\mathsf{K}}^{\leqslant}$. This establishes that $\mathsf{Alg}^*\mathcal{L}_{\mathsf{K}}^{\leqslant} = \mathsf{K}$.
2. If $\langle \boldsymbol{A}, F\rangle$ is a reduced model of $\mathcal{L}_{\mathsf{K}}^{\leqslant}$, then by 1, $\boldsymbol{A} \in \mathsf{K}$, which by Theorem 7.20 implies that F is a semilattice filter of $\boldsymbol{A}$. By the construction of F^+ in Theorem 7.38, $a \in F^+$ if and only if $a \equiv 1\ (\boldsymbol{\Omega}^{\boldsymbol{A}}F)$, that is, if and only if $a = 1$, because the matrix is reduced; thus, $F^+ = \{1\}$. The converse follows from the fact that by regular algebraizability of $\mathcal{L}_{\mathsf{K}}^{\top}$, $\langle \boldsymbol{A}, \{1\}\rangle$ is a reduced model of $\mathcal{L}_{\mathsf{K}}^{\top}$ and hence $\boldsymbol{\Omega}^{\boldsymbol{A}}F = \boldsymbol{\Omega}^{\boldsymbol{A}}F^+ = \boldsymbol{\Omega}^{\boldsymbol{A}}\{1\} = \mathrm{Id}_A$, therefore $\langle \boldsymbol{A}, F\rangle$ is reduced as well. ⊠

The result in point 1 is the key to the following interesting result:

THEOREM 7.42
The logic $\mathcal{L}_{\mathsf{K}}^{\top}$ coincides with the logic defined by each of the following classes of matrices:

$$\{\langle \boldsymbol{A}, F\rangle : \boldsymbol{A} \textit{ any algebra, } F \in \mathcal{F}i_{\mathsf{K}}^{\leqslant}\boldsymbol{A} \textit{ and is Leibniz}\} \tag{94}$$

$$\{\langle \boldsymbol{A}, F\rangle : \boldsymbol{A} \textit{ any algebra, } F = \min \mathcal{F}i_{\mathsf{K}}^{\leqslant}\boldsymbol{A}\} \tag{95}$$

PROOF: A matrix and its reduction define the same logic (Theorem 4.34). So, the logic defined by the class (94) is also defined by the class of its reductions, which by Proposition 7.39 is the class $\{\langle \boldsymbol{A}, \{1\}\rangle : \boldsymbol{A} \in \mathsf{Alg}^*\mathcal{L}_{\mathsf{K}}^{\leqslant}\}$. But by Proposition 7.41.1 this is $\{\langle \boldsymbol{A}, \{1\}\rangle : \boldsymbol{A} \in \mathsf{K}\}$, which defines $\mathcal{L}_{\mathsf{K}}^{\top}$ by assumption. The proof that (95) also defines $\mathcal{L}_{\mathsf{K}}^{\top}$ goes along similar lines (Exercise 7.22). ⊠

Thus, one can say that the logic $\mathcal{L}_{\mathsf{K}}^{\top}$ is ***the strong version*** of $\mathcal{L}_{\mathsf{K}}^{\leqslant}$ in three a priori different senses, which here coincide:

- Its assertional companion.
- Its "Leibniz companion", that is, the logic defined by the class (94).
- Its "minimum filter companion", that is, the logic defined by the class (95).

Observe that the two classes of matrices in Theorem 7.42, and the logics they define, can be considered starting from any logic, and they always produce extensions of the original logic, the second being stronger than the first (because every minimum filter is Leibniz).

All these results add to the idea that semilattice-based logics with an algebraizable assertional companion behave particularly well.

In the protoalgebraic case, monotonicity of the Leibniz operator easily implies that in general F^+ is the largest Leibniz filter included in F (Corollary 6.48). Interestingly, this property not only does not hold here in general, but it actually characterizes protoalgebraic logics among the logics of order we are considering:

PROPOSITION 7.43
The logic $\mathcal{L}_{\mathsf{K}}^{\leqslant}$ is protoalgebraic if and only if for each $\boldsymbol{A}$ and each $F \in \mathcal{F}i_{\mathsf{K}}^{\leqslant}\boldsymbol{A}$, F^+ is the largest Leibniz $\mathcal{L}_{\mathsf{K}}^{\leqslant}$-filter included in F.

PROOF: The implication from left to right is Corollary 6.48. To proce the converse implication, let $\boldsymbol{A}$ be any algebra, and let $F, G \in \mathcal{F}i_{\mathsf{K}}^{\leqslant}\boldsymbol{A}$ with $F \subseteq G$. Then $F^+ \subseteq F \subseteq G$ and F^+ is Leibniz. Since by the assumption G^+ is the largest

Leibniz filter contained in G, it follows that $F^+ \subseteq G^+$. But these two are filters of $\mathcal{L}^{\top}_{\mathsf{K}}$, therefore $\boldsymbol{\Omega}^{\boldsymbol{A}}F = \boldsymbol{\Omega}^{\boldsymbol{A}}F^+ \subseteq \boldsymbol{\Omega}^{\boldsymbol{A}}G^+ = \boldsymbol{\Omega}^{\boldsymbol{A}}G$ by the algebraizability of this logic. Thus $\boldsymbol{\Omega}^{\boldsymbol{A}}$ is monotone on $\mathcal{F}i^{\leqslant}_{\mathsf{K}}\boldsymbol{A}$ for any $\boldsymbol{A}$. This is one of the ways of showing that $\mathcal{L}^{\leqslant}_{\mathsf{K}}$ is protoalgebraic. ⊠

It is not difficult to see (Exercise 7.23) that the characterization of reduced models of a protoalgebraic logic contained in Theorem 6.50 can be merged with the one contained in Proposition 7.41, and that the result takes a more pleasant and useful form:

PROPOSITION 7.44
Let $\mathcal{L}^{\leqslant}_{\mathsf{K}}$ be a protoalgebraic logic. A matrix $\langle \boldsymbol{A}, F\rangle$ is a reduced model of $\mathcal{L}^{\leqslant}_{\mathsf{K}}$ if and only if $\boldsymbol{A} \in \mathsf{K}$, F is a semilattice filter of the algebra $\boldsymbol{A}$ and $\{1\}$ is the only $\mathcal{L}^{\top}_{\mathsf{K}}$-filter of $\boldsymbol{A}$ contained in F. ⊠

EXAMPLE 7.45
This proposition clearly generalizes Malinowski's [173, Theorem II.4] characterization of the reduced models of the local logics S^{ℓ} associated with normal modal systems S (which are equivalential) as the matrices $\langle \boldsymbol{A}, F\rangle$ such that $\boldsymbol{A}$ is a modal algebra (Example 7.47.1) in the corresponding variety and F is a Boolean filter of $\boldsymbol{A}$ such that $\{1\}$ is the only open filter contained in it.

A less standard application is the logic E^{MP} associated with the weakest system E of classical modal logic, shown in Example 6.76 to be protoalgebraic but not equivalential. It is mentioned there that $\langle \boldsymbol{A}, F\rangle \in \mathsf{Mod}^{*}E^{\mathrm{MP}}$ if and only if $\boldsymbol{A}$ is a Boolean algebra with an arbitrary unary operation $\Box$ and F is a filter such that $\{1\}$ is the only filter closed under the rule (ER) $x \leftrightarrow y \rhd \Box x \leftrightarrow \Box y$ included in F. To match the present framework, it is enough to realize that E^{MP} is the logic of order of the mentioned variety (which obviously has semilattice reducts); it is not difficult to prove that the corresponding assertional companion is the extension of E^{MP} with (ER), and that this one is algebraizable. Therefore, E^{MP} is fully selfextensional, and so forth. ⊠

The key abstract properties (under our present assumptions) of the logics $\mathcal{L}^{\leqslant}_{\mathsf{K}}$ and $\mathcal{L}^{\top}_{\mathsf{K}}$ are that $\mathcal{L}^{\leqslant}_{\mathsf{K}}$ is always selfextensional while $\mathcal{L}^{\top}_{\mathsf{K}}$ is always algebraizable. One of the more interesting results in this study is that each cannot enjoy the characteristic property of the other unless the two coincide:

THEOREM 7.46
The following conditions are equivalent:

(i) $\mathcal{L}^{\top}_{\mathsf{K}}$ *is selfextensional.*

(ii) $\mathcal{L}^{\leqslant}_{\mathsf{K}}$ *is algebraizable.*

(iii) $\mathcal{L}^{\leqslant}_{\mathsf{K}} = \mathcal{L}^{\top}_{\mathsf{K}}$.

If these equivalent conditions hold, then the (unique) logic is fully selfextensional and regularly BP-algebraizable.

PROOF: (i)⇒(iii) Being an extension of $\mathcal{L}_K^{\leqslant}$, the logic $\mathcal{L}_K^{\top}$ also has a conjunction. Now the assumption that it is in addition selfextensional means that both logics are semilattice-based with respect to the same class of algebras $\mathsf{K} = \mathbb{V}\mathcal{L}_K^{\leqslant} = \mathbb{V}\mathcal{L}_K^{\top}$. Moreover, they are both finitary and have theorems, so they are both forcefully non-pseudo-axiomatic. By Theorem 7.28, in this situation $\mathcal{L}_K^{\top} = \mathcal{L}_K^{\leqslant}$.

(ii)⇒(iii) In order to show that $\mathcal{L}_K^{\leqslant} = \mathcal{L}_K^{\top}$ it is enough to show that in any $\boldsymbol{A}$, $\mathcal{F}i_K^{\leqslant}\boldsymbol{A} = \mathcal{F}i_K^{\top}\boldsymbol{A}$, and for this one needs only prove that $\mathcal{F}i_K^{\leqslant}\boldsymbol{A} \subseteq \mathcal{F}i_K^{\top}\boldsymbol{A}$. The assumption that the logic $\mathcal{L}_K^{\leqslant}$ is algebraizable implies that $\boldsymbol{\Omega}^{\boldsymbol{A}}$ is one-to-one over $\mathcal{F}i_K^{\leqslant}\boldsymbol{A}$, and this implies that for every $F \in \mathcal{F}i_K^{\leqslant}\boldsymbol{A}$, $F = F^{+}$. Therefore, by Proposition 7.39, $F \in \mathcal{F}i_K^{\top}\boldsymbol{A}$.

The converse implications are trivial. The final conclusion just puts the properties of the two logics together. ⊠

Two more equivalent properties can also be added to those in this theorem; that the logic $\mathcal{L}_K^{\leqslant}$ is weakly algebraizable (Exercise 7.24) and that the logic $\mathcal{L}_K^{\top}$ is fully Fregean (Exercise 7.48); the latter requires some of the results in the next section. It is nevertheless interesting to notice already here that in the situation described in this theorem, the (unique) logic is located not only very high in the Leibniz hierarchy, but also at the top of the Frege hierarchy.

EXAMPLES 7.47

Most algebras of logic are ordered and have a maximum described by a constant term; it is thus natural that many examples of the situation studied here have already been identified in the literature. Theorem 7.46 provides, by contraposition, many examples of algebraizable logics that are not selfextensional and an easy way to show this: Often there is some nice characterization of the $\mathcal{L}_K^{\top}$-filters of algebras of K as (semi)lattice filters with some additional property (usually, to be closed under an additional rule) such that clearly not every semilattice filter satisfies it; by Proposition 7.37.4, this implies that the filters of the two logics do not coincide, therefore the two logics do not coincide either, and by Theorem 7.46 this implies that $\mathcal{L}_K^{\top}$ is not selfextensional (and that $\mathcal{L}_K^{\leqslant}$ is not algebraizable). Here are two large families of logics in this situation.

1. The case of the two logics associated with each normal modal system S (Example 1.14) has been analysed in [106]. All these systems are associated with a variety of ***modal algebras***, which are Boolean algebras expanded by a unary operator $\Box$ such that the equations $\Box(x \wedge y) \approx \Box x \wedge \Box y$ and $\Box\top \approx \top$ are valid, and obviously have semilattice reducts. It turns out that the logic of order of one of these varieties is the local logic S^{ℓ} associated with S, while the logic of one is the corresponding global logic S^{g}. By Proposition 7.37.4, $\mathsf{Alg}\,S^{g} = \mathsf{Alg}\,S^{\ell}$ and this is the corresponding variety, which is thus the equivalent algebraic semantics of the algebraizable logic S^{g}; the whole variety of modal algebras corresponds to K, while for $S4$ and $S5$ the varieties of closure algebras and monadic algebras are obtained, respectively. In each case, the S^{g}-filters of the corresponding algebras are the lattice filters (which are the

$\mathcal{S}^{\ell}$-filters) that are *open*, that is, closed under the Necessitation Rule $x \rhd \Box x$. Since (except in trivial cases) not all filters are open, the two logics differ, and therefore $\mathcal{S}^{\ell}$ is selfextensional and equivalential (see Example 6.67.3), but not algebraizable, and $\mathcal{S}^{g}$ is regularly algebraizable (in fact, implicative, see Example 2.17.1) but not selfextensional. Since the logics have conjunction, the logic $\mathcal{S}^{\ell}$ is fully selfextensional; on page 461 it will be shown that it is not Fregean, as an application of Theorem 7.75.

2. The logics associated with varieties of commutative integral residuated lattices have been studied in [46] and already mentioned in Example 6.16.9: with each of these varieties, K, one can associate both a semilattice-based logic $\mathcal{L}_{\mathsf{K}}^{\leqslant}$ (the logic of order, which is the finitary logic that preserves degrees of truth) and a regularly algebraizable one $\mathcal{L}_{\mathsf{K}}^{\top}$ (the logic of one, which is the corresponding assertional logic), so that their mutual relations match the general framework just studied in this subsection. In this case, the filters of $\mathcal{L}_{\mathsf{K}}^{\top}$ are the filters of $\mathcal{L}_{\mathsf{K}}^{\leqslant}$ that are closed under the rule of *Modus Ponens* for $\rightarrow$. As holds in general, $\mathcal{L}_{\mathsf{K}}^{\top}$ is regularly algebraizable but not selfextensional, and $\mathcal{L}_{\mathsf{K}}^{\leqslant}$ is fully selfextensional but not algebraizable; the negative facts hold unless the two logics coincide (again by Theorem 7.46); in [46] it is shown that this happens if and only if K is a variety of generalized Heyting algebras. Thus, this provides a large number of examples of algebraizable but non-selfextensional logics, for instance the logics $Ł_n$ (for $n > 2$) or the logic $Ł_{\infty f}$, and by Proposition 7.7, also $Ł_{\infty}$ (shown to be algebraizable in Example 3.41); incidentally, that these logics are not selfextensional was proved by Wójcicki in [249, p. 390] by the direct method of Exercise 7.6.

   In contrast to the modal case, here the logic of order is not in general protoalgebraic, and some conditions on K are given in the cited paper to ensure this; it appears that the most usual varieties in this family do not satisfy them, therefore they provide further examples of (fairly natural) fully selfextensional but non-protoalgebraic logics (among them, $Ł_{\infty}^{\leqslant}$), namely those detailed in Example 6.16.9. ⊠

Situations like that described in this subsection, in which two companion logics have the same algebraic counterpart and are related by properties involving the Leibniz operator, have been studied from a more abstract point of view in [4, 6, 108].

Selfextensional logics with the uniterm Deduction Theorem

Recall that a logic $\mathcal{L}$ satisfies the *uniterm Deduction-Detachment Theorem* (u-DDT) when there is a binary term $\rightarrow$ such that for all $\Gamma \cup \{\varphi, \psi\} \subseteq Fm$, $\Gamma, \varphi \vdash_{\mathcal{L}} \psi$ if and only if $\Gamma \vdash_{\mathcal{L}} \varphi \rightarrow \psi$. Finitary selfextensional logics with the u-DDT admit a treatment, developed in [107, 155], strongly parallel to that of the finitary selfextensional logics with conjunction; it is exposed here in a summarized way.

Some properties of the logics that satisfy the general form of the Deduction Theorem, found in Section 3.6, are also used (for the particular, uniterm case).

These logics are related to classes of algebras that have a Hilbert algebra reduct. An algebra $\boldsymbol{A} = \langle A, \to \rangle$ of type $\langle 2 \rangle$ is a ***Hilbert algebra*** when the equations

$$x \to x \approx y \to y \tag{H1}$$

$$(x \to x) \to x \approx x \tag{H2}$$

$$x \to (y \to z) \approx (x \to y) \to (x \to z) \tag{H3}$$

$$(x \to y) \to ((y \to x) \to y) \approx (y \to x) \to ((x \to y) \to x) \tag{H4}$$

hold in $\boldsymbol{A}$. From (H1) it follows that we can define $\top := x \to x$ and this is a constant term in the algebra; we put $1 := \top^{\boldsymbol{A}}$. An alternative, quasi-equational presentation is given in Exercise 2.28. If we define $a \leqslant b \overset{\text{def}}{\iff} a \to b = 1$, then this relation is an order and 1 is its maximum; thus, the algebra is in particular an implicative algebra (Proposition 2.14). Hilbert algebras are the algebraic counterpart of the implication fragment of intuitionistic logic (see Table 2 on page 85). The standard references on Hilbert algebras are [83] and [208]; in the latter they are called "positive implication algebras" and are presented in the language $\langle \to, \top \rangle$. We need only very few of their properties, though.

Definition 7.48

Let $\boldsymbol{A}$ be an algebra of type $\boldsymbol{L}$. The algebra $\boldsymbol{A}$ ***has a Hilbert algebra reduct*** *when there is a binary term $\to$ of $\boldsymbol{L}$ such that the reduct $\langle A, \to \rangle$ is a Hilbert algebra, that is, when the equations* (H1)–(H4) *hold in $\boldsymbol{A}$. A class of algebras* ***has Hilbert algebra reducts*** *when all algebras in the class have a Hilbert algebra reduct for the same binary term. For each type $\boldsymbol{L}$ and binary term $\to$ the class of all $\boldsymbol{L}$-algebras having a Hilbert algebra reduct with the term $\to$ is denoted by $\mathsf{HiA}(\boldsymbol{L}, \to)$.*

The class $\mathsf{HiA}(\boldsymbol{L}, \to)$ is obviously a variety, axiomatized by the equations (H1)–(H4), and when $\boldsymbol{L} = \langle \to \rangle$ it reduces to the variety HiA of Hilbert algebras.

Like what is done in implicative algebras, if $\boldsymbol{A}$ has a Hilbert algebra reduct with respect to the term $\to$, one can consider its *implicative filters* (Definition 2.27), the sets $F \subseteq A$ such that $1 \in F$ and F is closed under *Modus Ponens* (if $a \in F$ and $a \to b \in F$, then $b \in F$). The set of implicative filters of $\boldsymbol{A}$ is denoted by $\mathcal{F}i\ell t_{\to} \boldsymbol{A}$.

Definition 7.49

A logic $\mathcal{L}$ is ***Hilbert-algebra-based*** *when there is a class of algebras K of the type of $\mathcal{L}$, with Hilbert algebra reducts with respect to a binary term $\to$, such that $\mathcal{L}$ satisfies the following two conditions, for any $n \geqslant 0$ and any $\varphi_0, \dots, \varphi_n, \psi \in Fm$:*

$$\varphi_0, \dots, \varphi_n \vdash_{\mathcal{L}} \psi \overset{\text{def}}{\iff} \mathsf{K} \vDash \varphi_0 \to (\dots \to (\varphi_n \to \psi)\dots) \approx \top \tag{96}$$

$$\emptyset \vdash_{\mathcal{L}} \psi \overset{\text{def}}{\iff} \mathsf{K} \vDash \psi \approx \top \tag{97}$$

For each language $\boldsymbol{L}$ and each class of algebras K of type $\boldsymbol{L}$ with Hilbert algebra reducts with respect to a binary term $\to$ of $\boldsymbol{L}$, the logic $\mathcal{L}^{\to}_{\mathsf{K}} := \langle \boldsymbol{L}, \vdash^{\to}_{\mathsf{K}} \rangle$ is the finitary logic determined by these conditions.

It is not immediately obvious that the above clauses really determine a logic: for this, clause (96) should be independent of the order in which the formulas to the left of $\vdash_{\mathcal{L}}$ are chosen to be put to the right of $\vDash$, since the expression "$\varphi_0, \dots, \varphi_n$" in that position actually abbreviates the *set* $\{\varphi_0, \dots, \varphi_n\}$ and not a particular sequence of its members. Besides, the four properties of Definition 1.5 should be satisfied. Exercise 7.25 gives the necessary clues to show all this.

Observe that all Hilbert-algebra-based logics have theorems: directly, by (H1) and (97), $\varphi \to \varphi$ for any φ, and in particular $\top$, are theorems of these logics. As with the semilattice-based case, these logics (notice that they are finitary by definition) depend only on the equational theory of the class K, hence they are Hilbert-algebra-based with respect to a single variety, which is the one generated by K (see Corollary 7.52); however, in contrast to what happens in that case, the form of the dependence already determines the logic completely (including the theorems), and the logic can be completely recovered from the interderivability relation.

LEMMA 7.50
If $\mathcal{L}$ is a Hilbert-algebra-based logic, with respect to K and $\to$, then:

1. $\alpha \dashv\vdash_{\mathcal{L}} \beta$ *if and only if* $\mathsf{K} \vDash \alpha \approx \beta$, *for all* $\alpha, \beta \in Fm$.
2. $\mathcal{L}$ *is selfextensional and* $\mathbb{V}\mathcal{L} = \mathbb{V}\mathsf{K}$.
3. $\varphi_0, \dots, \varphi_n \vdash_{\mathcal{L}} \psi$ *if and only if* $\varphi_0 \to (\dots \to (\varphi_n \to \psi)\dots) \equiv \top\ (\boldsymbol{\Lambda}\mathcal{L})$, *for all* $n \geqslant 0$ *and all* $\varphi_0, \dots, \varphi_n, \psi \in Fm$.
4. $\emptyset \vdash_{\mathcal{L}} \psi$ *if and only if* $\psi \equiv \top\ (\boldsymbol{\Lambda}\mathcal{L})$, *for all* $\psi \in Fm$.

PROOF: To prove 1, observe that by (96) $\alpha \dashv\vdash_{\mathcal{L}} \beta$ if and only if $\mathsf{K} \vDash \alpha \to \beta \approx \top$ and $\mathsf{K} \vDash \beta \to \alpha \approx \top$. By the central property (IA4) of implicative algebras, these two conditions imply that $\mathsf{K} \vDash \alpha \approx \beta$, and, trivially, this condition implies the previous ones, modulo (H1). 2 follows from 1 and Proposition 7.8, and 3 and 4 follow from 1 and clauses (96) and (97) of Definition 7.49, respectively. ⊠

The following characterization is parallel to Theorem 7.18.

THEOREM 7.51
A logic $\mathcal{L}$ is Hilbert-algebra-based with respect to some class of algebras K and a binary term $\to$ if and only if it is selfextensional and satisfies the u-DDT *with respect to $\to$. Then $\mathbb{V}\mathcal{L} = \mathbb{V}\mathsf{K}$ and $\mathcal{L}$ is Hilbert-algebra-based with respect to $\mathbb{V}\mathcal{L}$, which has Hilbert algebra reducts, and the term $\to$.*

PROOF: ($\Rightarrow$) Assume that $\mathcal{L}$ is Hilbert-algebra-based with respect to some class K and $\to$. By Lemma 7.50.2, $\mathcal{L}$ is selfextensional and $\mathbb{V}\mathcal{L} = \mathbb{V}\mathsf{K}$. It remains to show that for any $\Gamma \cup \{\varphi, \psi\} \subseteq Fm$, $\Gamma, \varphi \vdash_{\mathcal{L}} \psi$ if and only if $\Gamma \vdash_{\mathcal{L}} \varphi \to \psi$. By finitarity we can assume that either $\Gamma = \emptyset$ or $\Gamma = \{\varphi_0, \dots, \varphi_n\}$ for some $n \geqslant 0$. If $\Gamma = \emptyset$, we should show that $\varphi \vdash_{\mathcal{L}} \psi$ if and only if $\emptyset \vdash_{\mathcal{L}} \varphi \to \psi$, but both statements amount to $\mathsf{K} \vDash \varphi \to \psi \approx \top$ by (96) and (97). In the second case, again by (96), both $\varphi_0, \dots, \varphi_n, \varphi \vdash_{\mathcal{L}} \psi$ and $\varphi_0, \dots, \varphi_n \vdash_{\mathcal{L}} \varphi \to \psi$ amount to $\mathsf{K} \vDash \varphi_0 \to (\dots \to (\varphi_n \to (\varphi \to \psi))\dots) \approx \top$. Thus, the equivalence holds.

($\Leftarrow$) We prove that the logic $\mathcal{L}$ is Hilbert-algebra-based with respect to the variety $\mathbb{V}\mathcal{L}$ and the term $\to$. First, we have to show that this variety has Hilbert algebra reducts. Since $\mathcal{L}$ is assumed to be selfextensional, Proposition 7.8 tells us that this variety is generated by the algebra $\boldsymbol{Fm}/\boldsymbol{\Lambda}\mathcal{L}$, and hence it is enough to prove the four equations (H1)–(H4) in the form of interderivability relations. This is easily done using just the u-DDT; some points already appear in Lemma 3.71, and the rest are similarly straightforward (Exercise 7.26). We realize that some key facts follow from the u-DDT: that the formula $\top := x \to x$ is a theorem of $\mathcal{L}$, that any two theorems are interderivable, and that a formula that is interderivable with a theorem is also a theorem. Using these facts and the u-DDT the necessary number of times, it is easy to see that $\varphi_0, \dots, \varphi_n \vdash_{\mathcal{L}} \psi$ if and only if $\varphi_0 \to (\dots \to (\varphi_n \to \psi)\dots) \dashv\vdash_{\mathcal{L}} \top$, or, equivalently, if and only if $\mathbb{V}\mathcal{L} \vDash \varphi_0 \to (\dots \to (\varphi_n \to \psi)\dots) \approx \top$; this is (96). Similarly one proves that $\emptyset \vdash_{\mathcal{L}} \psi$ if and only if $\psi \dashv\vdash_{\mathcal{L}} \top$ if and only if $\mathbb{V}\mathcal{L} \vDash \psi \approx \top$, which is (97). These two properties show that $\mathcal{L}$ is a Hilbert-algebra-based logic. ⊠

COROLLARY 7.52

If $\mathcal{L}$ is Hilbert-algebra-based with respect to a class K and a term $\to$, and is also Hilbert-algebra-based with respect to a class K' and a term $\to'$, then $\mathbb{V}\mathsf{K} = \mathbb{V}\mathsf{K}' = \mathbb{V}\mathcal{L}$, and for any $\alpha, \beta \in Fm$, $\alpha \to \beta \dashv\vdash_{\mathcal{L}} \alpha \to' \beta$ and the equation $\alpha \to \beta \approx \alpha \to' \beta$ holds in the classes K and K' and in the variety $\mathbb{V}\mathcal{L}$.

PROOF: From Lemma 7.50.1 it follows that $\mathbb{V}\mathsf{K} = \mathbb{V}\mathsf{K}' = \mathbb{V}\mathcal{L}$. Since by Theorem 7.51 $\mathcal{L}$ satisfies the u-DDT both with respect to $\to$ and with respect to $\to'$, by Lemma 3.78 $\alpha \to \beta \dashv\vdash_{\mathcal{L}} \alpha \to' \beta$. This implies the final property. ⊠

As for semilattice-based logics, this confirms the "absolute" character of being Hilbert-algebra-based: if a logic is Hilbert-algebra-based, then it is so with respect to a unique variety, which is the intrinsic variety of the logic, determined in this case by the Frege relation of the logic (Lemma 7.50.1), and through a unique binary operation in this variety (unique up to term-equivalence).

Recall that in Theorem 3.81 it is proved that, in general, the DDT transfers from a logic to all its basic full models, and in Proposition 5.90 that it transfers to all full g-models. Thus, in particular, if $\mathcal{L}$ is Hilbert-algebra-based, then for any algebra $\boldsymbol{A}$ the g-matrix $\langle \boldsymbol{A}, \mathcal{F}i_{\mathcal{L}}\boldsymbol{A} \rangle$ satisfies the u-DDT with respect to $\to$. Adding finitarity, the main results are obtained.

THEOREM 7.53

Let $\mathcal{L}$ be a finitary selfextensional logic satisfying the u-DDT *with respect to a binary connective $\to$.*

1. *If $\boldsymbol{A} \in \mathbb{V}\mathcal{L}$, then $\mathcal{F}i_{\mathcal{L}}\boldsymbol{A} = \mathcal{F}ilt_{\to}\boldsymbol{A}$.*
2. *If $\boldsymbol{A} \in \mathbb{V}\mathcal{L}$, then the g-matrix $\langle \boldsymbol{A}, \mathcal{F}i_{\mathcal{L}}\boldsymbol{A} \rangle$ has the* PCONG *and is reduced.*
3. $\mathsf{Alg}\mathcal{L}$ *is a variety, namely* $\mathsf{Alg}\mathcal{L} = \mathbb{V}\mathcal{L}$.
4. *$\mathcal{L}$ is fully selfextensional.*
5. *A finitary g-model of $\mathcal{L}$ is full if and only if it satisfies the* PCONG *and the* u-DDT.

6. *$\mathcal{L}$ has a fully adequate Gentzen system, namely the extension of the Gentzen system $\mathfrak{G}^c_{\mathcal{L}}$ of Definition 7.29 with the Gentzen-style rule*

$$\frac{\Gamma, \varphi \rhd \psi}{\Gamma \rhd \varphi \to \psi}. \qquad \text{(DT)}$$

PROOF: By Theorem 7.51 we can assume from the start that $\mathcal{L}$ is Hilbert-algebra-based with respect to its intrinsic variety $\mathbb{V}\mathcal{L}$, which has Hilbert algebra reducts.

1. Let $\boldsymbol{A} \in \mathbb{V}\mathcal{L}$ and $F \in \mathcal{F}i_{\mathcal{L}}\boldsymbol{A}$. Since $\top$ is a theorem of $\mathcal{L}$, $1 \in F$. That $\mathcal{L}$ satisfies the u-DDT implies that (MP) for $\to$ is one of its rules, therefore if $a \in F$ and $a \to b \in F$, then $b \in F$. This shows that F is an implicative filter of $\boldsymbol{A}$. Conversely, let F be an implicative filter of $\boldsymbol{A}$ and $h \in \mathrm{Hom}(\boldsymbol{Fm}, \boldsymbol{A})$. From (97) it follows that $h\varphi = 1 \in F$ for all theorems φ of $\mathcal{L}$. Now assume that $\varphi_0, \dots, \varphi_n \vdash_{\mathcal{L}} \psi$ and that $h\varphi_i \in F$ for all $i \leqslant n$. Since by (96) $\mathbb{V}\mathcal{L} \vDash \varphi_0 \to (\dots \to (\varphi_n \to \psi)...) \approx \top$, we know that $h\varphi_0 \to (\dots \to (h\varphi_n \to h\psi)...) = 1 \in F$. Using that F is closed under (MP) as many times as needed we conclude that $h\psi \in F$. Since $\mathcal{L}$ is finitary, this is enough to show that $F \in \mathcal{F}i_{\mathcal{L}}\boldsymbol{A}$.

2. By Proposition 5.55, it is enough to show that $\boldsymbol{\Lambda}^{A}\mathcal{F}i_{\mathcal{L}}\boldsymbol{A}$ is the identity relation. Let $a, b \in A$ and consider the principal up-sets $\uparrow a$ and $\uparrow b$, which by Exercise 7.27 are implicative filters. If $a \neq b$, then either $a \notin \uparrow b$ or $b \notin \uparrow a$, while of course $a \in \uparrow a$ and $b \in \uparrow b$. Since by 1 these up-sets are $\mathcal{L}$-filters, we have found two $\mathcal{L}$-filters separating the points. This means that $a \not\equiv b \ (\boldsymbol{\Lambda}^{A}\mathcal{F}i_{\mathcal{L}}\boldsymbol{A})$, and establishes that $\boldsymbol{\Lambda}^{A}\mathcal{F}i_{\mathcal{L}}\boldsymbol{A}$ is the identity relation.

3. In general $\mathsf{Alg}\mathcal{L} \subseteq \mathbb{V}\mathcal{L}$, but from 2 it follows that $\mathbb{V}\mathcal{L} \subseteq \mathsf{Alg}\mathcal{L}$.

4. From the two preceding points it follows that for any algebra $\boldsymbol{A} \in \mathsf{Alg}\mathcal{L}$, the g-matrix $\langle \boldsymbol{A}, \mathcal{F}i_{\mathcal{L}}\boldsymbol{A} \rangle$ has the PCONG. By Proposition 7.9(ii), this implies that $\mathcal{L}$ is fully selfextensional.

5. By the previous point and Theorem 5.63, any full g-model of $\mathcal{L}$ has the PCONG, and by Proposition 5.90 it has the u-DDT that $\mathcal{L}$ has. To show the converse, let $\langle \boldsymbol{A}, \mathscr{C} \rangle$ be a finitary g-model of $\mathcal{L}$ satisfying the PCONG and the u-DDT. By the PCONG, $\widetilde{\boldsymbol{\Omega}}^{A}\mathscr{C} = \boldsymbol{\Lambda}^{A}\mathscr{C}$; thus, to show that it is full is to show that $\mathscr{C}/\boldsymbol{\Lambda}^{A}\mathscr{C} = \mathcal{F}i_{\mathcal{L}}(\boldsymbol{A}/\boldsymbol{\Lambda}^{A}\mathscr{C})$. Since by assumption $\mathscr{C} \subseteq \mathcal{F}i_{\mathcal{L}}\boldsymbol{A}$ and $\mathscr{C}/\boldsymbol{\Lambda}^{A}\mathscr{C}$ is the reduction of $\mathscr{C}$, we know that $\mathscr{C}/\boldsymbol{\Lambda}^{A}\mathscr{C} \subseteq \mathcal{F}i_{\mathcal{L}}(\boldsymbol{A}/\boldsymbol{\Lambda}^{A}\mathscr{C})$. Now take any $G \in \mathcal{F}i_{\mathcal{L}}(\boldsymbol{A}/\boldsymbol{\Lambda}^{A}\mathscr{C})$ and put $F := \pi^{-1}G$, where $\pi\colon \boldsymbol{A} \to \boldsymbol{A}/\boldsymbol{\Lambda}^{A}\mathscr{C}$. Notice that since π is surjective, $G = \pi F = F/\boldsymbol{\Lambda}^{A}\mathscr{C}$. Thus, in order to show that $G \in \mathscr{C}/\boldsymbol{\Lambda}^{A}\mathscr{C}$ it is enough to show that $F \in \mathscr{C}$, that is, that $CF \subseteq F$. Observe that, since the logic has theorems, G is non-empty, and hence so is F. Take any $a \in CF$; since C is assumed to be finitary, there are $a_0, \dots, a_n \in F$ such that $a \in C\{a_0, \dots, a_n\}$, for some $n \geqslant 0$. Using the u-DDT for C we arrive at $a_0 \to (\dots \to (a_n \to a)...) \in C\varnothing$. Since $\top$ is a theorem of $\mathcal{L}$ and $C\varnothing$ is an $\mathcal{L}$-filter, $1 \in C\varnothing$. But in general all points in $C\varnothing$ are related by the Frege relation (Exercise 1.47), therefore $a_0 \to (\dots \to (a_n \to a)...) \equiv 1 \ (\boldsymbol{\Lambda}^{A}\mathscr{C})$, and this means that $\pi a_0 \to (\dots \to (\pi a_n \to \pi a)...) = 1 \in G$. Since $a_i \in F$, $\pi a_i \in G$, but this set is closed under (MP), therefore we can conclude that $\pi a \in G$, and so $a \in F$. This

proves that $F \in \mathscr{C}$ and hence that $\mathscr{C}/\boldsymbol{\varLambda}^{A}\mathscr{C} = \mathcal{F}i_{\mathcal{L}}(\boldsymbol{A}/\boldsymbol{\varLambda}^{A}\mathscr{C})$, thus completing the proof that $\langle \boldsymbol{A}, \mathscr{C} \rangle$ is full, as desired.

6. Clearly, a finitary g-matrix has the u-DDT with respect to $\rightarrow$ if and only if it is a model of the rule (DT). The previous point implies that the full g-models of $\mathcal{L}$ are exactly the finitary g-matrices that are models of $\mathcal{L}$ and satisfy the PCONG and the u-DDT, or, equivalently, the finitary g-matrices that are models of the Gentzen system $\mathfrak{G}^{c}_{\mathcal{L}}$ supplemented with the rule (DT). This shows that the resulting Gentzen system is fully adequate for $\mathcal{L}$. ☒

Two points of general interest should be highlighted from this result: First, that for finitary logics with the u-DDT, the PCONG *transfers* from the logic to all its full g-models. Second, that all these logics have *a variety* as their algebraic counterpart.

Comparing point 5 with the analogous result for the case of the conjunction (Corollary 7.33.2) we observe that here the PCONG alone is not enough to characterize the full g-models among the g-models of the logic: the u-DDT has to be added, and the corresponding rule (DT) has to be added to the Gentzen system. A result parallel to Theorem 7.35 can also be proved: it is not difficult to show that this Gentzen system is algebraizable (in the extended sense of Section 3.7), with $\mathsf{Alg}\mathcal{L}$ as equivalent algebraic semantics; the transformer $\boldsymbol{\tau}$ is the one suggested by condition (96) above; that is, for $n \geqslant 0$,

$$\boldsymbol{\tau}(x_0, \dots, x_n \rhd x_{n+1}) := x_0 \rightarrow (\dots \rightarrow (x_n \rightarrow x_{n+1})\dots) \approx x_0 \rightarrow x_0$$
$$\boldsymbol{\tau}(\varnothing \rhd x) := x \approx x \rightarrow x,$$

while $\boldsymbol{\rho}$ can be the same as in the case of conjunction, namely $\boldsymbol{\rho}(x \approx y) := \{x \rhd y, y \rhd x\}$. Thus, these logics are *G-algebraizable* in the sense discussed on page 314.

The prototypical and simplest example that falls under the scope of these results is of course the logic $\mathcal{I}\ell_{\rightarrow}$ itself. As the analysis in Example 5.91 directly shows, in this case the u-DDT suffices to characterize the full g-models and the fully adequate Gentzen system; this is because the u-DDT implies congruence for the operation $\rightarrow$ (Lemma 5.26) and there are no other operations.

The parallel of Theorem 7.28 also holds here:

THEOREM 7.54
Let $\boldsymbol{L}$ *be an algebraic type and* $\rightarrow$ *a binary term in* $\boldsymbol{L}$*. The functions* $\mathcal{L} \mapsto \mathbb{V}\mathcal{L}$ *and* $\mathsf{K} \mapsto \mathcal{L}^{\rightarrow}_{\mathsf{K}}$ *establish mutually inverse dual order isomorphisms between the following families:*†

- *The family of all finitary selfextensional logics in the language* $\boldsymbol{L}$ *that satisfy the* u-DDT *for* $\rightarrow$*, ordered under the extension relation.*
- *The family of all the subvarieties of the variety* $\mathsf{HiA}(\boldsymbol{L}, \rightarrow)$*, ordered under the subclass relation.*

† See footnote † on page 410.

Proof: It is very similar to the proof of Theorem 7.28, relying on the basic fact that if $\mathcal{L}$ and $\mathcal{L}'$ are two logics in the family, then $\mathcal{L} \leqslant \mathcal{L}'$ if and only if $\Lambda\mathcal{L} \subseteq \Lambda\mathcal{L}'$; this is established using points 3 and 4 of Lemma 7.50. Exercise 7.29 asks you to check that everything really works. ⊠

A final issue of interest is the comparison between $\mathcal{L}^{\rightarrow}_{\mathsf{K}}$ and other logics that can be naturally associated with any variety K of algebras having Hilbert algebra reducts.[†] On the one hand, since all these algebras are pointed, it makes sense to consider the assertional logic $\mathcal{L}^{\top}_{\mathsf{K}}$. On the other hand, since all these algebras are ordered, it also makes sense to consider the logic $\mathcal{L}^{\leqslant}_{\mathsf{K}}$ of Definition 7.26. The logics $\mathcal{L}^{\leqslant}_{\mathsf{K}}$ and $\mathcal{L}^{\rightarrow}_{\mathsf{K}}$ are finitary by definition, while $\mathcal{L}^{\top}_{\mathsf{K}}$ is finitary because K is a variety. We have:

Proposition 7.55

Let K be a variety of algebras having Hilbert algebra reducts with respect to a binary term $\rightarrow$.

1. *The three logics $\mathcal{L}^{\rightarrow}_{\mathsf{K}}$, $\mathcal{L}^{\leqslant}_{\mathsf{K}}$ and $\mathcal{L}^{\top}_{\mathsf{K}}$ have the same theorems.*
2. $\mathcal{L}^{\rightarrow}_{\mathsf{K}} \leqslant \mathcal{L}^{\leqslant}_{\mathsf{K}} \leqslant \mathcal{L}^{\top}_{\mathsf{K}}$.
3. *Any of the logics $\mathcal{L}^{\leqslant}_{\mathsf{K}}$ or $\mathcal{L}^{\top}_{\mathsf{K}}$ coincides with $\mathcal{L}^{\rightarrow}_{\mathsf{K}}$ if and only if it satisfies the* u-DDT *with respect to $\rightarrow$.*

Proof: That the three logics have the same theorems, which are the formulas φ such that $\mathsf{K} \vDash \varphi \approx \top$, is a trivial consequence of the fact that $1 := \top^{\boldsymbol{A}}$ is the maximum in the algebras in K. That $\mathcal{L}^{\leqslant}_{\mathsf{K}} \leqslant \mathcal{L}^{\top}_{\mathsf{K}}$ is also a consequence of the same fact. To prove that $\mathcal{L}^{\rightarrow}_{\mathsf{K}} \leqslant \mathcal{L}^{\leqslant}_{\mathsf{K}}$ it is enough to use that in Hilbert algebras $\top \rightarrow x \approx x$ holds, and that $a \leqslant b$ if and only if $a \rightarrow b = 1$. Trivially, if $\mathcal{L}^{\leqslant}_{\mathsf{K}}$ or $\mathcal{L}^{\top}_{\mathsf{K}}$ coincides with $\mathcal{L}^{\rightarrow}_{\mathsf{K}}$, then it satisfies the u-DDT with respect to $\rightarrow$, since $\mathcal{L}^{\rightarrow}_{\mathsf{K}}$ satisfies it. Finally, it is straightforward to show (Exercise 7.30) that two finitary logics that have the same theorems and satisfy the u-DDT for the same term coincide; this proves the converse in the last point. ⊠

That in general the three logics can be different is witnessed by the class of Hilbert algebras with infimum, as shown in [90, pp. 133–135]; see also [159, Example 4.1]. Theorem 7.85 shows that the coincidence of the three logics (equivalently, of $\mathcal{L}^{\rightarrow}_{\mathsf{K}}$ and $\mathcal{L}^{\top}_{\mathsf{K}}$) would imply that this single logic has a very special—and prominent—location in the hierarchies.

Further, Exercise 7.31 asks you to see what happens in the case where the class K is a assumed to have at the same time semilattice reducts (of course, relative to the same order relation given by the implicative structure).

Exercises for Sections 7.1 and 7.2

Exercise 7.1. Let $\mathscr{C}$ be a closure system on an algebra $\boldsymbol{A}$. Prove that the Leibniz operator $\boldsymbol{\Omega}^{\boldsymbol{A}}$ is monotone over the family $\mathscr{C}$ if and only if $\boldsymbol{\Omega}^{\boldsymbol{A}}F = \widetilde{\boldsymbol{\Omega}}^{\boldsymbol{A}}_{\mathscr{C}}F$ for all $F \in \mathscr{C}$.

[†] A finer analysis of this situation, in a more general context, can be found in [159].

EXERCISE 7.2. Use Proposition 5.49 to give an alternative proof of the preservation of the PCONG by strict surjective homomorphisms (Proposition 7.2) that does not use the Gentzen-style characterization used in the text.

EXERCISE 7.3. Check that the two descriptions of the notion of a Fregean logic given in Definition 7.3 are indeed equivalent.

EXERCISE 7.4. Prove that the properties of being selfextensional and of being Fregean are preserved by fragments, and that the property of being Fregean is also preserved by axiomatic extensions.

EXERCISE 7.5. Consider the logic $\mathcal{I}$ described in Example 6.78.1.

1. Prove that $\alpha \dashv\vdash_{\mathcal{I}} \beta$ if and only if $\alpha = \beta$ or both α and β are theorems, for any $\alpha, \beta \in Fm$.
2. Show that $x \to x \dashv\vdash_{\mathcal{I}} y \to y$ and that $(x \to x) \to z \not\dashv\vdash_{\mathcal{I}} (y \to y) \to z$, where x, y, z are three distinct variables.
3. Conclude that $\mathcal{I}$ is not selfextensional.

EXERCISE 7.6. Let $\mathcal{L}$ be any of Łukasiewicz's many-valued logics $Ł_\infty$, $Ł_{\infty f}$ or $Ł_n$ with $n > 2$. Using just the definitions of the operations in $[\mathbf{0,1}]$ (Example 1.12), prove that $x \wedge \neg x \dashv\vdash_{\mathcal{L}} \neg(x \to x)$ and that $\nvdash_{\mathcal{L}} (x \wedge \neg x) \to \neg(x \to x)$. Conclude that these logics are not selfextensional.

EXERCISE 7.7. Prove that if $\mathcal{L}$ is a fully selfextensional logic, then the class $\mathsf{Alg}\mathcal{L}$ is closed under subalgebras. Show that if in addition $\mathcal{L}$ has theorems, then the least $\mathcal{L}$-filter of each $\boldsymbol{A} \in \mathsf{Alg}\mathcal{L}$ is a one-element set, and as a consequence the class $\mathsf{Alg}\mathcal{L}$ is pointed and in every $\boldsymbol{A} \in \mathsf{Alg}\mathcal{L}$ the constant element is the maximum of the order $\leqslant_{\mathcal{L}}^{\boldsymbol{A}}$ defined in point (iv) of Proposition 7.9.

HINT. For the first point, use Corollary 7.10.

EXERCISE 7.8. Let $\mathcal{L}$ be a truth-equational and fully selfextensional logic. Prove that $\mathcal{L}$ is assertional.

HINT. Merge the results of Exercises 6.46, 6.47 and 7.7.

EXERCISE 7.9. Prove the three points of Proposition 7.12.

EXERCISE 7.10. Let C, C' be two closure operators on the same set A.

1. Prove that if $C \leqslant C'$, then $C\emptyset \subseteq C'\emptyset$ and $\Lambda^A C \subseteq \Lambda^A C'$.
2. Assume that C and C' are finitary and have the PC with respect to the same binary operation. Prove that the converse of point 1 holds as well.

EXERCISE 7.11. Let $\boldsymbol{A}$ be an algebra with a semilattice reduct, and $F \subseteq A, F \neq \emptyset$. Prove that F is a semilattice filter if and only if it is an $\wedge$-filter in the sense of Proposition 7.11.

EXERCISE 7.12. Prove Corollary 7.19.

EXERCISE 7.13. Let C be a closure operator on a set A satisfying the PIRA with respect to a unary operation $\neg$ on A. Prove, for every $X \cup \{a,b\} \subseteq A$, that the set $\{a, \neg a\}$ is inconsistent and that if $a \in C(X,b)$, then $\neg b \in C(X, \neg a)$. Prove that, if moreover C satisfies the PC with respect to $\wedge$, then $C\{a \wedge \neg a\} = A$ and $\neg(a \wedge \neg a) \in C\emptyset$ for every $a \in A$, and $\neg a \in C(b, \neg(a \wedge b))$ for every $a, b \in A$.

EXERCISE 7.14. Prove the characterizations of pseudo-axiomatic and non-pseudo-axiomatic logics contained in Proposition 7.25.

EXERCISE 7.15.

1. Let $\mathcal{L}$ be a logic with theorems, and let $\mathcal{L}_{\varnothing}$ be the logic defined by the condition that $\mathcal{T}h\mathcal{L}_{\varnothing} := \mathcal{T}h\mathcal{L} \cup \{\varnothing\}$. Show that $\mathcal{L}_{\varnothing}$ is indeed a logic, that it is pseudo-axiomatic, and that its smallest non-empty theory is the set of theorems of $\mathcal{L}$.
2. Let $\mathcal{L}$ be a pseudo-axiomatic logic, and define a logic $\mathcal{L}_{\bullet}$ by the stipulation that $\mathcal{T}h\mathcal{L}_{\bullet} := \mathcal{T}h\mathcal{L} \smallsetminus \{\varnothing\}$. Show that $\mathcal{L}_{\bullet}$ is indeed a logic, and that it has theorems (namely, the smallest non-empty theory of $\mathcal{L}$).
3. Show that if $\mathcal{L}$ has theorems, then $(\mathcal{L}_{\varnothing})_{\bullet} = \mathcal{L}$, and that if $\mathcal{L}$ is pseudo-axiomatic, then $(\mathcal{L}_{\bullet})_{\varnothing} = \mathcal{L}$. Hence, for each type, these constructions establish a bijection between the sets of all pseudo-axiomatic logics and of all logics with theorems.
4. Show that the two logics of each corresponding pair have the same consequences from non-empty sets, the same Frege and Tarski relations, the same class $\mathsf{Alg}\mathcal{L}$ and the same intrinsic variety $\mathbb{V}\mathcal{L}$. Show that for any algebra $\boldsymbol{A}$, the filters of the two logics in $\boldsymbol{A}$ satisfy the same relations as their theories do (points 1 and 2), and hence that the two logics have the same class $\mathsf{Alg}^*\mathcal{L}$.
5. Show that if one of the two logics of each corresponding pair is semilattice-based, then the other is also semilattice-based, and with respect to the same variety.

EXERCISE 7.16. Prove that a finitary logic $\mathcal{L}$ is selfextensional if and only if the Gentzen system $\mathfrak{G}^{c}_{\mathcal{L}}$ of Definition 7.29 is adequate for $\mathcal{L}$.

EXERCISE 7.17. Let $\mathcal{L}$ be a finitary logic, and let $\mathfrak{G}^{cc}_{\mathcal{L}}$ be the Gentzen system defined by extending the Gentzen system $\mathfrak{G}^{c}_{\mathcal{L}}$ of Definition 7.29 with the initial sequents (HPC) of Proposition 7.11(ii). Prove that $\mathcal{L}$ is selfextensional and has the PC if and only if the Gentzen system $\mathfrak{G}^{cc}_{\mathcal{L}}$ is adequate for $\mathcal{L}$, and if and only if $\mathfrak{G}^{cc}_{\mathcal{L}}$ is fully adequate for $\mathcal{L}$. Compare with Theorem 7.32.

EXERCISE 7.18. Check that the logics $\mathcal{C\ell}_{\wedge\vee}$ and $\mathcal{B}$ are non-pseudo-axiomatic.

EXERCISE 7.19. Let L be the variety of lattices, and consider the logic $\mathcal{L}^{\leqslant}_{\mathsf{L}}$ in the language $\langle \wedge, \vee \rangle$ of type $\langle 2, 2 \rangle$ given by Definition 7.26. Characterize its full g-models by Tarski-style conditions, along the lines of Example 5.93(v), and prove that its fully adequate Gentzen system is defined by the calculus that, besides the structural rules, has the sequents (HPC) as initial sequents and the following rules:

$$\frac{\alpha \rhd \varphi \qquad \beta \rhd \varphi}{\alpha \vee \beta \rhd \varphi} \qquad\qquad \frac{\Gamma \rhd \varphi}{\Gamma \rhd \varphi \vee \psi} \qquad \frac{\Gamma \rhd \psi}{\Gamma \rhd \varphi \vee \psi}$$

In particular this implies that $\mathcal{L}^{\leqslant}_{\mathsf{L}}$ is the logic defined by this Gentzen system, as considered in Example 6.16.2.

HINT. Use Exercise 5.22.

EXERCISE 7.20. Prove Theorem 7.35, taking advantage of the fact that, under its assumptions, the Gentzen system $\mathfrak{G}^{c}_{\mathcal{L}}$ is fully adequate for the logic $\mathcal{L}$ (Theorem 7.32).

EXERCISE 7.21. Prove the four points of Proposition 7.37.

EXERCISE 7.22. Prove the second part of Theorem 7.42. Notice that here the logic $\mathcal{L}^{\leqslant}_{\mathsf{K}}$ cannot be assumed to be protoalgebraic.

EXERCISE 7.23. Prove Proposition 7.44, and check that its applications in Example 7.45 work as explained.

Exercise 7.24. Show that condition (ii) of Theorem 7.46 can be relaxed to that of requiring the logic $\mathcal{L}_{\mathsf{K}}^{\leqslant}$ to be just weakly algebraizable.

Exercise 7.25. Prove that Definition 7.49 is unambiguous and really defines a logic.

Hints. For (C), use Exercise 1.15. Gather the necessary properties of Hilbert algebras from equations (H1)–(H4) on page 441 and the presentation in Exercise 2.28; the only non-obvious ones (whose proof you may skip) are $x \to (y \to z) \approx y \to (x \to z)$ and $(x \to y) \to \big((y \to z) \to (x \to z)\big) \approx \top$, and their finite extensions

$$x_0 \to (\ldots \to (x_n \to z)\ldots) \approx x_{p0} \to (\ldots \to (x_{pn} \to z)\ldots), \text{ and}$$

$$\big(x_0 \to (\ldots \to (x_n \to y)\ldots)\big) \to \Big((y \to z) \to \big(x_0 \to (\ldots \to (x_n \to z)\ldots)\big)\Big) \approx \top,$$

for all $n \geqslant 2$ and all permutations p of the set of indices $\{0, \ldots, n\}$.

Exercise 7.26. Let $\mathcal{L}$ be a logic satisfying the u-DDT with respect to a binary term $\to$. Prove that $x \to x$ is a theorem of $\mathcal{L}$, and that the following properties are satisfied:

$$\begin{aligned} x \to x &\dashv\vdash_{\mathcal{L}} y \to y \\ (x \to x) \to x &\dashv\vdash_{\mathcal{L}} x \\ x \to (y \to z) &\dashv\vdash_{\mathcal{L}} (x \to y) \to (x \to z) \\ (x \to y) \to \big((y \to x) \to y\big) &\dashv\vdash_{\mathcal{L}} (y \to x) \to \big((x \to y) \to x\big) \end{aligned}$$

Exercise 7.27. Let $\boldsymbol{A}$ be a Hilbert algebra or, more generally, an algebra with a Hilbert algebra reduct. Prove that for each $a \in A$, the principal up-set $\uparrow a = \{c \in A : a \leqslant c\}$ is an implicative filter.

Exercise 7.28. Give a proof, with a logical flavour, of the following (well-known) algebraic result: If $\boldsymbol{A}$ is a Hilbert algebra and $X \subseteq A$ is non-empty, then the implicative filter generated by X is the set $\{a \in A : \text{there are } c_0, \ldots, c_n \in X, \text{ for some } n \geqslant 0, \text{ such that } c_0 \to (\ldots \to (c_n \to a)\ldots) = 1\}$.

Hint. You may want to use some of the results in Theorem 7.53.

Exercise 7.29. Complete the proof of Theorem 7.54, following the indications in the text.

Exercise 7.30. Let C and C' be two finitary closure operators on an algebra $\boldsymbol{A}$ satisfying the DDT with respect to the same set $I(x,y)$. Prove that $\mathrm{C} = \mathrm{C}'$ if and only if $\mathrm{C}\varnothing = \mathrm{C}'\varnothing$.

Exercise 7.31. Let K be a variety of algebras having Hilbert algebra reducts with respect to $\to$, and at the same time having semilattice reducts with respect to $\wedge$, with the same associated order relation. Prove that the following conditions are equivalent:

1. The logics $\mathcal{L}_{\mathsf{K}}^{\leqslant}$ and $\mathcal{L}_{\mathsf{K}}^{\to}$ coincide.
2. It holds that $(x \wedge y) \to z \dashv\vdash_{\mathcal{L}} x \to (y \to z)$ for $\mathcal{L} = \mathcal{L}_{\mathsf{K}}^{\to}$ (and hence for $\mathcal{L} = \mathcal{L}_{\mathsf{K}}^{\leqslant}$).
3. In algebras of K, the operation $\to$ is the residual of the operation $\wedge$; that is, for every $\boldsymbol{A} \in \mathsf{K}$ and every $a, b, c \in A$, $a \wedge b \leqslant c$ if and only if $a \leqslant b \to c$.

7.3 Fregean and fully Fregean logics

These two kinds of logics are introduced in Definition 7.3. The definition itself presents Fregean logics as the ones in which the Frege and the Suszko operators coincide on the formula algebra, and Proposition 7.4 shows that the fully Fregean logics are the ones for which the two operators coincide on arbitrary algebras.

A further characterization in terms of relations between operators is of some interest here:

PROPOSITION 7.56
1. *A logic* $\mathcal{L}$ *is Fregean if and only if* $\boldsymbol{\Lambda}_{\mathcal{L}}\Gamma \subseteq \boldsymbol{\Omega}\Gamma$ *for every* $\Gamma \in \mathcal{T}h\mathcal{L}$.
2. *A logic* $\mathcal{L}$ *is fully Fregean if and only if* $\Lambda_{\mathcal{L}}^{\boldsymbol{A}}F \subseteq \boldsymbol{\Omega}^{\boldsymbol{A}}F$ *for every* $F \in \mathcal{F}i_{\mathcal{L}}\boldsymbol{A}$ *and every algebra* $\boldsymbol{A}$.

PROOF: By Proposition 7.4, $\mathcal{L}$ is fully Fregean if and only if $\Lambda_{\mathcal{L}}^{\boldsymbol{A}}F = \widetilde{\boldsymbol{\Omega}}_{\mathcal{L}}^{\boldsymbol{A}}F$ for every $F \in \mathcal{F}i_{\mathcal{L}}\boldsymbol{A}$ and every algebra $\boldsymbol{A}$. Since always $\widetilde{\boldsymbol{\Omega}}_{\mathcal{L}}^{\boldsymbol{A}}F \subseteq \boldsymbol{\Omega}^{\boldsymbol{A}}F$, this shows that if $\mathcal{L}$ is fully Fregean, then $\Lambda_{\mathcal{L}}^{\boldsymbol{A}}F \subseteq \boldsymbol{\Omega}^{\boldsymbol{A}}F$ for every $F \in \mathcal{F}i_{\mathcal{L}}\boldsymbol{A}$ and every $\boldsymbol{A}$. For the converse, notice that the Frege operator is monotone (Lemma 1.38.3). Using this fact, the assumption (notice that it is postulated for all the $\mathcal{L}$-filters) and the definition of the Suszko operator, we have that

$$\begin{aligned}\Lambda_{\mathcal{L}}^{\boldsymbol{A}}F &= \bigcap\{\Lambda_{\mathcal{L}}^{\boldsymbol{A}}G : G \in \mathcal{F}i_{\mathcal{L}}\boldsymbol{A}\,,\ F \subseteq G\}\\ &\subseteq \bigcap\{\boldsymbol{\Omega}^{\boldsymbol{A}}G : G \in \mathcal{F}i_{\mathcal{L}}\boldsymbol{A}\,,\ F \subseteq G\} = \widetilde{\boldsymbol{\Omega}}_{\mathcal{L}}^{\boldsymbol{A}}F.\end{aligned}$$

Since in general $\widetilde{\boldsymbol{\Omega}}_{\mathcal{L}}^{\boldsymbol{A}}F \subseteq \Lambda_{\mathcal{L}}^{\boldsymbol{A}}F$, this implies that $\widetilde{\boldsymbol{\Omega}}_{\mathcal{L}}^{\boldsymbol{A}}F = \Lambda_{\mathcal{L}}^{\boldsymbol{A}}F$, showing that $\mathcal{L}$ is fully Fregean. Since the algebra $\boldsymbol{A}$ is arbitrary, this proves 2. The same reasoning, limited to the formula algebra, proves 1. ⊠

It is easy to see (Exercise 7.32) that the characterization in 2 also holds when the condition is restricted to algebras in $\mathsf{Alg}\mathcal{L}$. In a similar vein, the following characterization, parallel to Proposition 7.9 and proved in the same way (Exercise 7.32 again), is also practical:

PROPOSITION 7.57
A logic $\mathcal{L}$ *is fully Fregean if and only if for any* $\boldsymbol{A} \in \mathsf{Alg}\mathcal{L}$, *the g-matrix* $\langle \boldsymbol{A}, \mathcal{F}i_{\mathcal{L}}\boldsymbol{A}\rangle$ *has the* SPCONG. ⊠

To start with a few miscellaneous properties concerning examples, recall that in Exercise 7.4 you are asked to prove:

PROPOSITION 7.58
Being Fregean is a property of a logic that is preserved by fragments and by axiomatic extensions. ⊠

PROPOSITION 7.59
All two-valued logics are Fregean.

PROOF: In this statement any conceivable sense of "being two-valued" is contemplated; thus, $\mathcal{L}$ is assumed to be the logic defined by a two-element algebra **2** and either a matrix or a g-matrix on it. The latter case reduces to the former because the closure systems on a two-element set $2 = \{0,1\}$ are those of the forms $\{2\}, \{\varnothing,2\}, \{\{a\},2\}$ or $\{\varnothing,\{a\},2\}$, for $a = 0$ or $a = 1$. The first two kinds define the inconsistent and the almost inconsistent logics, respectively, and these are Fregean (see below); the other two define the same logics as the matrices $\langle \boldsymbol{2}, \{a\}\rangle$, save for the set of theorems, which has no influence on Fregeanity. So assume that $\mathcal{L}$ is defined by a matrix of the form $\langle \mathbf{2}, \{a\} \rangle$ for either $a = 0$ or $a = 1$. Then, using only the fact that the universe has exactly two points while the designated set has exactly one point, it is easy to see that for any $\Gamma \in \mathcal{T}h\mathcal{L}$ and any $\alpha, \beta \in Fm$, $\alpha \equiv \beta \ (\boldsymbol{\Lambda}_{\mathcal{L}}\Gamma)$ if and only if for every $h \in \mathrm{Hom}(\boldsymbol{Fm}, \mathbf{2})$ such that $h\Gamma \subseteq \{a\}$, $h\alpha = h\beta$. This relation is clearly a congruence, which shows that $\mathcal{L}$ is Fregean. ⊠

EXAMPLES 7.60

1. The inconsistent logics are fully Fregean: Their only g-models have the form $\langle \boldsymbol{A}, \{A\} \rangle$ for an arbitrary $\boldsymbol{A}$, so these are all full (and basic), and their Frege relation is $A \times A$, which is certainly a congruence. But these full g-models have only one closed set, which is their set of theorems, so in fact for them the PCONG and the SPCONG are the same property. This shows that they are fully Fregean.

2. The almost inconsistent logics are fully Fregean: The reason is the same as in the previous case, because the addition of the empty set as a closed set to a g-matrix does not affect its Frege relation.

3. By Propositions 7.58 and 7.59, classical logic and all its fragments are Fregean. Another well-known example of a Fregean logic is intuitionistic logic; one easy way of proving this is to use Proposition 7.74. Proposition 7.58 implies that all its fragments are also Fregean (some of them are not protoalgebraic), as are all their axiomatic extensions; thus, all *intermediate logics* (in the traditional sense) are Fregean.

4. The preceding point can be generalized. Consider the Tarski-style conditions for connectives $\wedge, \vee, \rightarrow, \leftrightarrow, \neg, \bigcirc$ introduced† in Definition 5.16. Proposition 5.26 shows that each of these conditions implies the PCONG for the involved connective; therefore if a logic contains *only* connectives of this set and satisfies the corresponding conditions, then it is selfextensional. But all these conditions, save the PIM for $\bigcirc$, are inherited by axiomatic extensions (Exercise 5.10), therefore in these cases the logic is Fregean. And, finally, some of these conditions transfer to full g-models, such as the PC for $\wedge$, the PDI for $\vee$ and the DDT for $\rightarrow$; therefore in these, more restricted cases the logic is fully Fregean. For instance, this shows that all negation-less fragments of classical logic, and all axiomatic extensions of all the negation-less fragments of intuitionistic logic (not considering the equivalence connective among the primitive ones) are fully Fregean.

5. The logic $\mathcal{C}\ell_{\wedge\vee}$ is not just Fregean (it is a two-valued logic), but fully Fregean, by the previous point (its language contains just $\wedge$ and $\vee$, and it satisfies the PC and the PDI). This is reflected in the characterization of its full g-models in point (v) of Example 5.93: they are the finitary g-matrices without theorems satisfying the PC and the PDI.

† Recall that in the PIM $\bigcirc$ represents a generic unary connective, intended to represent $\Box$, or $\Diamond$, or other modal-like operators.

6. The variety DL of distributive lattices is the only variety K of lattices such that the logic $\mathcal{L}_{\mathsf{K}}^{\leqslant}$, in the language $\langle\wedge,\vee\rangle$, is Fregean. This is shown by proving that if $\mathcal{L}_{\mathsf{K}}^{\leqslant}$ is Fregean, then all lattices in K are distributive; since DL is the smallest variety of lattices, this implies that $\mathsf{K}=\mathsf{DL}$ and shows the statement. By definition, $\mathcal{L}_{\mathsf{K}}^{\leqslant}$ has the PC, which implies both $\{x,y\} \dashv\vdash_{\mathsf{K}}^{\leqslant} \{x, x\wedge y\}$ and $\{x,z\} \dashv\vdash_{\mathsf{K}}^{\leqslant} \{x, x\wedge z\}$. If $\mathcal{L}_{\mathsf{K}}^{\leqslant}$ is Fregean, then $\boldsymbol{\Lambda}_{\mathcal{L}_{\mathsf{K}}^{\leqslant}} C_{\mathcal{L}_{\mathsf{K}}^{\leqslant}}\{x\}$ is a congruence, and this implies that $\{x, y\vee z\} \dashv\vdash_{\mathsf{K}}^{\leqslant} \{x, (x\wedge y)\vee(x\wedge z)\}$. By the PC, this implies that $x\wedge(y\vee z) \vdash_{\mathsf{K}}^{\leqslant} (x\wedge y)\vee(x\wedge z)$, and by the way $\mathcal{L}_{\mathsf{K}}^{\leqslant}$ is defined, this implies that $\mathsf{K}\vDash x\wedge(y\vee z) \preccurlyeq (x\wedge y)\vee(x\wedge z)$, which amounts to saying that the lattices in K are all distributive. Moreover, if $\mathsf{DL}\subseteq\mathsf{K}$, then $\mathcal{L}_{\mathsf{K}}^{\leqslant} \leqslant \mathcal{C\ell}_{\wedge\vee}$, which implies that $\mathcal{L}_{\mathsf{K}}^{\leqslant}$ cannot be protoalgebraic. By Theorem 7.28, different varieties produce different logics. Therefore, this provides a continuum of examples† of logics that are fully selfextensional but neither protoalgebraic nor Fregean.

7. The logic $\mathcal{L}_{\mathsf{L}}^{\leqslant}$ presented in Example 7.34.5 as the finitary semilattice-based logic associated with the variety L of all lattices in the language $\langle\wedge,\vee\rangle$ (without theorems) is a particular case of the last fact mentioned in the previous point. Thus, the logic $\mathcal{L}_{\mathsf{L}}^{\leqslant}$ is also fully selfextensional but neither protoalgebraic nor Fregean.

8. The logic $\mathcal{C\ell}_{\wedge\vee\top\bot}$, considered in Exercise 5.63, is also fully Fregean and in addition has theorems. This can be easily proved from the properties of $\mathcal{C\ell}_{\wedge\vee}$, if you take into account that the addition of constants to a similarity type does not alter the congruences. Since the algebraic counterpart of this logic is the variety DL_{01} of bounded distributive lattices, and the constant $\top$ appears in the type, this variety is pointed, and it is natural to consider the two logics $\mathcal{L}_{\mathsf{DL}_{01}}^{\leqslant}$ and $\mathcal{L}_{\mathsf{DL}_{01}}^{\top}$; actually, $\mathcal{L}_{\mathsf{DL}_{01}}^{\leqslant} = \mathcal{L}_{\mathsf{DL}_{01}}^{\top} = \mathcal{C\ell}_{\wedge\vee\top\bot}$. See Exercise 7.33.

9. Belnap-Dunn's logic $\mathcal{B}$ is selfextensional (actually, fully selfextensional, as we know) but not Fregean. This is shown later on in this section, with the help of Proposition 7.61.

10. The weak relevance logic $\mathcal{WR}$ described in Example 6.16.7 is a conjunctive and selfextensional logic, hence it is fully selfextensional, but it is not Fregean; see [116, Lemma 4.11 and Proposition 5.8].

11. The weak logic of implication $\mathcal{G}_1$ described in Example 6.78.2 is selfextensional but not Fregean, as shown in [47, Theorem 3.17].

12. Positive modal logic $\mathcal{PML}$, described in Example 6.16.5, is selfextensional (hence fully selfextensional, as it is conjunctive) and not Fregean; all this was shown in [153], where its full g-models are characterized in several ways, and its fully adequate Gentzen system is determined.

† It is well known that there is a continuum of varieties of lattices, and that the variety of distributive lattices is the smallest among them; thus, any other variety of lattices contains non-distributive lattices.

13. Among the ten subintuitionistic logics considered in Example 6.16.8, two are fully Fregean ($\mathcal{I}\ell$ and $\mathcal{BPL}$), while the remaining eight are non-Fregean; of these, four are fully selfextensional (the local ones), and the other four are not selfextensional (the global ones). See [4, 6, 45]. ⊠

An interesting and easy property of Fregean logics, which shows their strength, is the following one:

PROPOSITION 7.61
Let $\mathcal{L}$ be a Fregean logic.

1. *$\mathcal{L}$ satisfies the Suszko rules of Theorem 6.112: $x, y, \delta(x, \vec{z}) \vdash_{\mathcal{L}} \delta(y, \vec{z})$, for all $\delta(x, \vec{z}) \in Fm$.*
2. *If $\langle \boldsymbol{A}, F \rangle \in \mathsf{Mod}^*\mathcal{L}$, then either $F = \varnothing$ or F is a one-element set.*
3. *If moreover $\mathcal{L}$ has theorems, then $\mathsf{Mod}^*\mathcal{L}$ is a unital class of matrices.*

PROOF: 1. Trivially, any logic $\mathcal{L}$ satisfies that $x \equiv y \ (\boldsymbol{\Lambda}_{\mathcal{L}} C_{\mathcal{L}}\{x,y\})$. If $\mathcal{L}$ is Fregean, then the relation $\boldsymbol{\Lambda}_{\mathcal{L}} C_{\mathcal{L}}\{x,y\}$ is a congruence, which implies that for all $\delta(x, \vec{z}) \in Fm$, $\delta(x, \vec{z}) \equiv \delta(y, \vec{z}) \ (\boldsymbol{\Lambda}_{\mathcal{L}} C_{\mathcal{L}}\{x,y\})$. It is easy to see that this implies the Suszko rule in the statement.

2. This is an easy application of Theorem 4.23; moreover, it turns out that this property is equivalent to that in point 1, even for an arbitrary logic; it is left to you as Exercise 7.35.

3. Since logics with theorems cannot have empty filters, this follows immediately from point 2. ⊠

If the result in point 3 prompts some obvious consequences in your mind, then go to Theorem 7.66.

Recall that the filter of a reduced matrix may be empty only when the algebra is trivial. Therefore, reduced models of Fregean logics on non-trivial algebras must be unital (irrespectively of whether the logic has theorems or not). This may be a *practical criterion* to disprove that a certain logic, (some of) whose reduced models are known, is Fregean. For instance, this shows that Belnap-Dunn's logic $\mathcal{B}$ is not Fregean, because it has reduced models on a non-trivial algebra with a two-element truth filter, such as those shown at the end of Example 5.82.

The next result provides a simple way to obtain Fregean logics from semantic definitions. It is parallel to the particular case of Proposition 5.12, which provides a way to obtain selfextensional logics from classes of g-matrices that satisfy the PCONG; but here the argument is essentially different, as it does not involve Gentzen-style rules.

PROPOSITION 7.62
If a logic $\mathcal{L}$ is complete with respect to a class of g-matrices M, all of which have the SPCONG, then $\mathcal{L}$ is Fregean.

PROOF: It is not difficult to show (see Exercise 7.36) that if $\mathcal{L}$ is complete with respect to M, then for any $\Gamma \in Th\mathcal{L}$ and any $\alpha, \beta \in Fm$, $\alpha \equiv \beta \ (\boldsymbol{\Lambda}_{\mathcal{L}} \Gamma)$ if and only if $h\alpha \equiv h\beta\ (\Lambda^{A}_{\mathscr{C}} Ch\Gamma)$ for every $\langle \boldsymbol{A}, \mathscr{C} \rangle \in \mathsf{M}$ and every $h \in \mathrm{Hom}(\boldsymbol{Fm}, \boldsymbol{A})$. If all the g-matrices in M have the SPCONG, then all the relations $\Lambda^{A}_{\mathscr{C}} Ch\Gamma$ are congruences, and it is easy to see that this implies that the relation $\boldsymbol{\Lambda}_{\mathcal{L}}\Gamma$ is a congruence. This shows that $\mathcal{L}$ has the SPCONG, and thus that it is Fregean. ⊠

Thus, in order to show that a logic defined by a (hopefully small) set of (hopefully small) g-matrices is Fregean it may be simpler to just check that these g-matrices have the SPCONG.

In the finitary case one can also formulate the Fregean property in terms of Gentzen-style rules, thanks to the following technical property:

Lemma 7.63
A finitary g-matrix $\langle \boldsymbol{A}, \mathscr{C} \rangle$ has the SPCONG *if and only if for every finitely generated $F \in \mathscr{C}$, the g-matrix $\langle \boldsymbol{A}, \mathscr{C}^{F} \rangle$ has the* PCONG.

Proof: Of course only the implication from right to left needs to be proved. Take any $F \in \mathscr{C}$ and any $a, b \in A$ such that $a \equiv b\ (\Lambda^{A}_{\mathscr{C}} F)$, that is, such that $C(F, a) = C(F, b)$. By Lemma 1.40.2 (this is where finitarity of the g-matrix is needed), there is a finite $X \subseteq F$ such that $C(X, a) = C(X, b)$. If we put $G := CX$, then also $C(G, a) = C(G, b)$, which means that $a \equiv b\ (\Lambda^{A}_{\mathscr{C}} G)$. Since $G \in \mathscr{C}$ and is finitely generated, by assumption $\boldsymbol{\Lambda}^{A}\mathscr{C}^{G} = \Lambda^{A}_{\mathscr{C}} G$ is a congruence, and therefore it equals $\widetilde{\Omega}^{\boldsymbol{A}}_{\mathscr{C}} G$. But $G \subseteq F$ and the Suszko operator $\widetilde{\Omega}^{\boldsymbol{A}}_{\mathscr{C}}$ is monotone over $\mathscr{C}$, therefore it follows that $a \equiv b\ (\widetilde{\Omega}^{\boldsymbol{A}}_{\mathscr{C}} F)$. Thus, $\Lambda^{A}_{\mathscr{C}} F = \widetilde{\Omega}^{\boldsymbol{A}}_{\mathscr{C}} F$, which proves that $\Lambda^{A}_{\mathscr{C}} F$ is a congruence. This shows that $\langle \boldsymbol{A}, \mathscr{C} \rangle$ has the SPCONG. ⊠

It is not difficult, using this lemma, to prove the following characterization (Exercise 7.37):

Corollary 7.64
A finitary g-matrix of type $\boldsymbol{L}$ *has the* SPCONG *if and only if it is a model of the rules*

$$\frac{\{\vec{z}, x_i \rhd y_i\ ,\ \vec{z}, y_i \rhd x_i\ :\ i < n\}}{\vec{z}, \lambda x_0 \dots x_{n-1} \rhd \lambda y_0 \dots y_{n-1}} \qquad \text{(SCONG)}$$

for all $\lambda \in L$, with $\mathrm{ar}\lambda = n \geqslant 1$, and for all finite sequences $\vec{z}$ of variables distinct from the x_i and the y_i (including the empty sequence).

In particular, a finitary logic is Fregean if and only if it satisfies the set of Gentzen-style rules (SCONG). ⊠

Notice that, in contrast to the case of the rules (CONG) of Definition 7.1, here for each operation there is a countably infinite number of real rules, because the sequence $\vec{z}$ of "side assumptions" has arbitrary length. The family of rules can be expressed by a single schematic rule, using a metavariable to range over all these sequences. If the logic moreover has a conjunction, then a single rule for each operation is enough (Exercise 7.38).

Investigation of particular characteristics of the full g-models of Fregean logics has not received much attention; the next result is among their few known properties (but it is enhanced in Theorem 7.81 for the protoalgebraic case).

PROPOSITION 7.65
If $\mathcal{L}$ is a Fregean logic with theorems, then for each $\Gamma \in \mathcal{T}h\mathcal{L}$, the g-matrix $\langle \boldsymbol{Fm}, (\mathcal{T}h\mathcal{L})^{\Gamma} \rangle$ is a full g-model of $\mathcal{L}$, and the function $\Phi\colon \Gamma \mapsto (\mathcal{T}h\mathcal{L})^{\Gamma}$ is an order-reversing embedding of $\mathcal{T}h\mathcal{L}$ into $\mathrm{FCS}_{\mathcal{L}}\boldsymbol{Fm}$.

PROOF: Trivially, Φ is always injective and order-reversing. Only the first part needs to be shown: for every $\Gamma \in \mathcal{T}h\mathcal{L}$, the g-matrix $\langle \boldsymbol{Fm}, (\mathcal{T}h\mathcal{L})^{\Gamma} \rangle$ is indeed a full g-model of $\mathcal{L}$, i.e., its reduction is a basic full g-model of $\mathcal{L}$. Writing $\theta := \widetilde{\boldsymbol{\Omega}}(\mathcal{T}h\mathcal{L})^{\Gamma}$ for simplicity, we should show that $(\mathcal{T}h\mathcal{L})^{\Gamma}/\theta = \mathcal{F}i_{\mathcal{L}}(\boldsymbol{Fm}/\theta)$. The $\subseteq$ inclusion always holds, because θ is compatible with all the theories that contain Γ. Let now $F \in \mathcal{F}i_{\mathcal{L}}(\boldsymbol{Fm}/\theta)$. Then $\pi^{-1}F \in \mathcal{T}h\mathcal{L}$, so we only have to show that $\pi^{-1}F \supseteq \Gamma$. Take any $\varphi \in \Gamma$. Since $\mathcal{L}$ has theorems, we can take any $\psi \in C_{\mathcal{L}}\emptyset$, and we find that $\psi \in \Gamma \cap \pi^{-1}F$. Then $C_{\mathcal{L}}(\Gamma, \varphi) = \Gamma = C_{\mathcal{L}}(\Gamma, \psi)$, therefore $\varphi \equiv \psi \ (\boldsymbol{\Lambda}_{\mathcal{L}}\Gamma)$. Now, the assumption that $\mathcal{L}$ is Fregean means that $\theta = \widetilde{\boldsymbol{\Omega}}(\mathcal{T}h\mathcal{L})^{\Gamma} = \widetilde{\boldsymbol{\Omega}}_{\mathcal{L}}\Gamma = \boldsymbol{\Lambda}_{\mathcal{L}}\Gamma$, therefore $\varphi \equiv \psi \ (\theta)$, that is, $\pi\varphi = \pi\psi \in F$. This shows that $\varphi \in \pi^{-1}F$ and hence that $\Gamma \subseteq \pi^{-1}F$, as desired. ⊠

The converse of this result is obviously false, as there are many (weakly) algebraizable logics (hence for them the function Φ is an embedding, and actually a dual lattice isomorphism, by Theorem 6.117) that are not even selfextensional.

It turns out that for fully Fregean logics with theorems a parallel property, this time on arbitrary algebras, can be proved (Exercise 7.39).

The assumption of having theorems seems to be a key point in many results in this section, or at least in their proofs. Concerning Proposition 7.65 and Exercise 7.39, they do not hold for logics without theorems, because in such a case, by Proposition 5.90, their full g-models have no theorems either, while for each non-empty $\Gamma \in \mathcal{T}h\mathcal{L}$, the g-matrix $\langle \boldsymbol{Fm}, (\mathcal{T}h\mathcal{L})^{\Gamma} \rangle$ has as theorems all the formulas in Γ, and hence cannot be a full g-model of $\mathcal{L}$. Modifying Φ by adding $\emptyset$ to $(\mathcal{T}h\mathcal{L})^{\Gamma}$ does not seem to overcome the problem: the proof does not work either, unless the logic is pseudo-axiomatic (Exercise 7.40); however, this trick may work in some particular cases (Exercise 7.41).

Fregean logics and truth-equational logics

It turns out that the class of Fregean logics with theorems is included in one of the classes in the Leibniz hierarchy, namely the class of assertional logics; and that the former class can be characterized in several ways:

THEOREM 7.66
Let $\mathcal{L}$ be a Fregean logic. The following conditions are equivalent:

(i) *$\mathcal{L}$ has theorems.*

(ii) *$\mathcal{L}$ is assertional.*

(iii) *$\mathcal{L}$ is truth-equational.*

(iv) *The Leibniz operator is injective over the $\mathcal{L}$-filters of arbitrary algebras.*

(v) *The Leibniz operator is injective over the theories of $\mathcal{L}$.*

Proof: If $\mathcal{L}$ is a Fregean logic with theorems, then Proposition 7.61.3 shows that $\mathsf{Mod}^*\mathcal{L}$ is unital, and by Theorem 6.111 this means that $\mathcal{L}$ is assertional. This proves that (i) implies (ii). That (ii) implies (iii) is a general fact (Theorem 6.109). That (iii) implies (iv) follows from Theorem 6.101, because being completely order-reflecting implies being injective. Of course, (iv) implies (v), and (v) clearly implies (i) in general, because in a logic without theorems the Leibniz operator is never injective, because then $\emptyset, Fm \in \mathcal{T}h\mathcal{L}$ and $\boldsymbol{\Omega}\emptyset = \boldsymbol{\Omega}Fm = Fm \times Fm$. ⊠

Thus, all Fregean logics with theorems are assertional, and these are the only Fregean and truth-equational logics. There are, however, many assertional logics that are not Fregean, or not even selfextensional. We find many examples of this situation among the regularly algebraizable logics, such as the global logics associated with normal modal systems; or Łukasiewicz's many-valued logics; or, more generally, the logics associated in [123] with any variety of commutative integral residuated lattices that is not a variety of generalized Heyting algebras. All these logics are of the form $\mathcal{L}_{\mathsf{K}}^{\top}$ for the corresponding algebraic counterpart K (a variety of modal algebras, or of MV-algebras, or of residuated lattices, respectively), therefore they are assertional, and as said in Example 7.47, they are not selfextensional.

The class of assertional logics is the smallest class in the Leibniz hierarchy that contains all Fregean logics with theorems. This is because there are Fregean logics with theorems that, while being assertional, are not regularly weakly algebraizable. Examples are $\mathcal{I}\ell^*$, the implication-less fragment of intuitionistic logic, and its axiomatic extensions considered in Example 6.16.6; they are not protoalgebraic, hence in particular not regularly weakly algebraizable. Since it is a fragment of $\mathcal{I}\ell$, the logic $\mathcal{I}\ell^*$ is Fregean, and hence all its axiomatic extensions are Fregean as well; and all these logics have theorems (indeed, they are assertional).

The structure of the Frege hierarchy inside the class of truth-equational logics (in particular, inside the class of assertional logics) can be further clarified.

Theorem 7.67
A truth-equational logic is fully selfextensional if and only if it is fully Fregean. If these conditions hold, then it is assertional.

Proof: Assume that $\mathcal{L}$ is a truth-equational and fully selfextensional logic, let $\langle \boldsymbol{A}, \mathscr{C} \rangle$ be any full g-model of $\mathcal{L}$, and let $F \in \mathscr{C}$. Since $\mathcal{L}$ is assumed to be truth-equational, by Proposition 6.105 the g-matrix $\langle \boldsymbol{A}, \mathscr{C}^F \rangle$ is also a full g-model of $\mathcal{L}$. Since $\mathcal{L}$ is assumed to be fully selfextensional, this g-matrix has the PCONG. This shows that all the full g-models of $\mathcal{L}$ have the SPCONG, and thus that $\mathcal{L}$ is fully Fregean. The converse is trivial. The final observation follows from Theorem 7.66, because truth-equational logics have theorems. ⊠

Thus, for truth-equational logics the Frege hierarchy reduces to exactly three classes: the selfextensional, the Fregean and the fully Fregean. And these are all different: Babyonyshev's logic, already mentioned as a Fregean but not fully Fregean one (Example 7.6.4), has theorems, therefore by Theorem 7.66 it is truth-equational. An example, also highly ad hoc, of a truth-equational and selfextensional logic that is not Fregean is constructed in [7, Example 25]. This example also answers negatively the (more or less natural) question of whether the bottom class in the hierarchy is the union of the two middle classes: it is selfextensional but not Fregean, therefore it is not fully Fregean either, but since it is truth-equational, Theorem 7.67 implies that it cannot be fully selfextensional. Interestingly, this theorem, viewed from the other side, tells us that the structure of the Leibniz hierarchy for fully-selfextensional logics is also somewhat simplified:

COROLLARY 7.68
A fully selfextensional logic is truth-equational if and only if it is assertional. As a consequence, the logic is (weakly/finitely) algebraizable if and only if it is (weakly/finitely) regularly algebraizable. ⊠

Here, unlike in the Fregean case (Theorem 7.66), it is not possible to add the condition of having theorems as an equivalent one; counterexamples are the semilattice-based logics of the form $\mathcal{L}_{\mathsf{K}}^{\leqslant}$, for a variety K of commutative integral residuated lattices not included in that of generalized Heyting algebras, which are fully selfextensional and have theorems, but are not assertional, as shown in Example 7.47.2.

Finally, another natural question on the structure of the Frege hierarchy (whether its top class is the intersection of the two middle ones) can also be clarified, this time affirmatively, for logics with theorems:

THEOREM 7.69
Let $\mathcal{L}$ be a logic with theorems. Then $\mathcal{L}$ is fully selfextensional and Fregean if and only if it is fully Fregean.

PROOF: If a logic is Fregean and has theorems, then by Theorem 7.66 it is truth-equational, therefore we can apply Theorem 7.67 and, using the assumption that the logic is fully selfextensional, conclude that it is fully Fregean. The converse is obvious. ⊠

It is an *open problem* whether the assumption that the logic has theorems can be dispensed with in this result.

Fregean logics and protoalgebraic logics

The algebraic treatment of Fregean logics seems difficult and has not been extensively investigated in full generality. However, when coupled with protoalgebraicity, the property of being Fregean is surprisingly strong, as several results show. The first characterization is easy but significant:

PROPOSITION 7.70
1. *A logic $\mathcal{L}$ is protoalgebraic and Fregean if and only if $\boldsymbol{\Lambda}_{\mathcal{L}}\Gamma = \boldsymbol{\Omega}\Gamma$ (and then both are equal to $\widetilde{\boldsymbol{\Omega}}_{\mathcal{L}}\Gamma$) for every $\Gamma \in \mathcal{T}h\mathcal{L}$; that is, if and only if the Frege, the Leibniz and the Suszko operators coincide on $\mathcal{T}h\mathcal{L}$.*

2. *A logic $\mathcal{L}$ is protoalgebraic and fully Fregean if and only if $\Lambda_{\mathcal{L}}^{\boldsymbol{A}}F = \Omega^{\boldsymbol{A}}F$ (and then both are equal to $\widetilde{\Omega}_{\mathcal{L}}^{\boldsymbol{A}}F$) for every algebra $\boldsymbol{A}$ and every $F \in \mathcal{F}i_{\mathcal{L}}\boldsymbol{A}$; that is, if and only if the Frege, the Leibniz and the Suszko operators coincide on $\mathcal{F}i_{\mathcal{L}}\boldsymbol{A}$ for every algebra $\boldsymbol{A}$.*

PROOF: Recall that a logic $\mathcal{L}$ is protoalgebraic if and only if $\widetilde{\Omega}_{\mathcal{L}}\Gamma = \Omega\Gamma$ for all $\Gamma \in \mathcal{T}h\mathcal{L}$, and also if and only if $\widetilde{\Omega}_{\mathcal{L}}^{\boldsymbol{A}}F = \Omega^{\boldsymbol{A}}F$ for every $F \in \mathcal{F}i_{\mathcal{L}}\boldsymbol{A}$ and every $\boldsymbol{A}$. Coupled with Definition 7.3 and with Proposition 7.56.2, this shows the direct implications and the coincidence of the three operators, in the two cases. As for the converses, the equality $\Lambda_{\mathcal{L}}\Gamma = \Omega\Gamma$ implies the inclusion $\Lambda_{\mathcal{L}}\Gamma \subseteq \Omega\Gamma$, which by Proposition 7.56.1 implies that $\mathcal{L}$ is Fregean, as well as the other inclusion $\Omega\Gamma \subseteq \Lambda_{\mathcal{L}}\Gamma$, which by Lemma 6.2 amounts to the equality $\widetilde{\Omega}_{\mathcal{L}}\Gamma = \Omega\Gamma$, that is, to protoalgebraicity of $\mathcal{L}$. This proves the converse implication in 1; the one in 2 is proved similarly, using Propositions 7.56.1 and 6.14. ⊠

It is interesting to notice that the coincidence of the Leibniz and Suszko operators (either on the formula algebra or on arbitrary algebras) characterizes protoalgebraic logics, and that of the Suszko and Frege operators characterizes Fregean logics (if on the formula algebra) or fully Fregean ones (if on arbitrary algebras), while, by contrast, the coincidence of the Leibniz and Frege operators does not characterize a third class of logics, but just the intersection of the other two; thus, these two operators cannot coincide without implying the coincidence of the three operators.

THEOREM 7.71
A finitary and protoalgebraic logic is Fregean if and only if it is fully Fregean.

PROOF: Observe that the rules (SCONG) form an accumulative set. Thus, Corollary 7.64 says that, for finitary logics, the property of being Fregean is expressible by an accumulative set of Gentzen-style rules. By Theorem 6.37, which holds for finitary protoalgebraic logics, to be a model of an accumulative set of Gentzen-style rules is a property that transfers to all the full g-models of the logic. Since these are all finitary (because $\mathcal{L}$ is finitary), Corollary 7.64 again implies that they have the SPCONG. Thus, the logic is fully Fregean. ⊠

This result shows that all protoalgebraic fragments of $\mathcal{C}\ell$ and of $\mathcal{I}\ell$ are fully Fregean; notice that some of the non-protoalgebraic ones, such as $\mathcal{C}\ell_{\wedge\vee}$, are fully Fregean as well. Some of these examples are also covered by the argument in Example 7.60.4 concerning the Tarski-style conditions that correspond to some particular Gentzen-style rules.

Merging this with Theorem 7.67 we obtain another important result:

THEOREM 7.72
A finitary and weakly algebraizable logic is fully selfextensional if and only if it is Fregean, and if and only if it is fully Fregean. ⊠

Thus, in finitary protoalgebraic logics the Frege hierarchy reduces to at most three classes: The class of selfextensional logics (the largest one), the class of fully selfextensional logics, and the class of Fregean logics (which here turns out to be the smallest). That these classes are indeed different is shown by the corresponding counterxamples. There are many examples of finitary and protoalgebraic logics that are fully selfextensional but not Fregean; Theorem 7.75 provides a practical criterion to identify them. As to the selfextensional but neither fully selfextensional nor Fregean case, the already mentioned logic presented in [7, Example 25], which is finitary and implicative, hence finitely regularly algebraizable and therefore protoalgebraic, sets the issue.

The preceding results and counterexamples can be read as concerning *the transfer problem* of several of the congruence properties. The transfer problem of the SPCONG can be reformulated as the problem of whether every Fregean logic is fully Fregean. Babyonyshev's logic (Example 7.6.4) shows that the answer in general is negative (that is, the property of being Fregean does *not* transfer), while Theorem 7.71 shows that this transfer does hold for finitary protoalgebraic logics. As to the transfer problem of the PCONG, which can be restated as the problem of whether every selfextensional logic is fully selfextensional, the last example mentioned in the previous paragraph shows that it fails even inside a very well behaved class such as the implicative ones. However, recall from Section 7.2 that it does hold for finitary logics satisfying either the PC or the u-DDT.

The next characterization of Fregeanity for protoalgebraic logics brings about one point that is maybe not emphasized enough in Chapter 6: Although in many examples the set of congruence formulas is obtained by "symmetrizing" a protoimplication set, this need not be the case in general for arbitrary equivalential logics and arbitrary protoimplication sets.[†] However, it does hold for protoalgebraic Fregean logics:

THEOREM 7.73
Let $\mathcal{L}$ be a protoalgebraic logic, with $\Delta(x,y)$ as protoimplication set. If $\mathcal{L}$ is Fregean, then the symmetrized set $\Delta^{s}(x,y) := \Delta(x,y) \cup \Delta(y,x)$ is a set of congruence formulas for $\mathcal{L}$, and hence $\mathcal{L}$ is equivalential.

PROOF: By assumption Δ satisfies the properties (R) and (MP), so the set Δ^{s} satisfies them as well. Moreover, by its own construction $\Delta^{s}(x,y) = \Delta^{s}(y,x)$, thus in particular it satisfies (Sym). By Theorem 6.60, in order to show that it is a set of congruence formulas it is enough to show that it satisfies condition (Re). It is convenient and sufficient to show this in the equivalent form

$$\Delta^{s}(x,y) \vdash_{\mathcal{L}} \Delta^{s}\big(\psi(x,\vec{z}),\psi(y,\vec{z})\big) \qquad \text{for all } \psi(x,\vec{z}) \in Fm. \tag{98}$$

Fix some $\psi(x,\vec{z})$ and put $\Psi(x,y,\vec{z}) := \Delta^{s}\big(\psi(x,\vec{z}),\psi(y,\vec{z})\big)$. By (MP) and (Sym),

$$\Delta^{s}(x,y), x \dashv\vdash_{\mathcal{L}} \Delta^{s}(x,y), y$$

[†] Counterexamples are the host of equivalential logics with an implication connective such that $x \to y$ satisfies (R) and (MP) but where $\{x \to y, y \to x\}$ is not the set of congruence formulas; for instance, the local logics associated with normal modal systems dealt with in Examples 6.67.3 and 6.67.4

and now applying that $\mathcal{L}$ is Fregean,

$$\Delta^{s}(x,y),\Psi(x,y,\vec{z}) \dashv\vdash_{\mathcal{L}} \Delta^{s}(x,y),\Psi(y,y,\vec{z}). \tag{99}$$

But $\Psi(y,y,\vec{z}) = \Delta^{s}(\psi(y,\vec{z}),\psi(y,\vec{z}))$, therefore by (R) $\emptyset \vdash_{\mathcal{L}} \Psi(y,y,\vec{z})$. Using this, the entailment from right to left in (99) implies (98). ☒

The converse is in general false: there are many equivalential logics (in fact, algebraizable) that have $\{x \to y\}$ as their protoimplication set, and $\{x \to y, y \to x\}$ as their set of congruence formulas, and are not Fregean (in fact, not even selfextensional); the global logics associated with normal modal systems are but one example.

However, if the assumptions on the set of protoimplication formulas are strengthened, then there is a converse:

PROPOSITION 7.74
Let $\mathcal{L}$ be a logic satisfying the DDT *for a set $I(x,y) \subseteq Fm$. The logic $\mathcal{L}$ is Fregean if and only if the symmetrized set $I^{s}(x,y) := I(x,y) \cup I(y,x)$ is a set of congruence formulas for $\mathcal{L}$.*

PROOF: The assumption implies that $\mathcal{L}$ is protoalgebraic and that I is a protoimplication set, so the implication from left to right is contained in Theorem 7.73. Now assume that I^{s} is a set of congruence formulas. Since $\mathcal{L}$ is protoalgebraic, in order to show that $\mathcal{L}$ is Fregean we can use Proposition 7.70 and show that $\boldsymbol{\Omega}\Gamma = \boldsymbol{\Lambda}_{\mathcal{L}}\Gamma$ for every $\Gamma \in \mathcal{T}h\mathcal{L}$. Using the assumption on I^{s} and the DDT for I it follows that $\varphi \equiv \psi \ (\boldsymbol{\Omega}\Gamma)$ if and only if $I^{s}(\varphi,\psi) \subseteq \Gamma$ if and only if $\Gamma,\varphi \dashv\vdash_{\mathcal{L}} \Gamma,\psi$ if and only if $\varphi \equiv \psi \ (\boldsymbol{\Lambda}_{\mathcal{L}}\Gamma)$. ☒

Thus, we have that the logics that have the DDT and are Fregean are equivalential, and that when an equivalential logic has a set of congruence formulas which, on inspection, happens to be the symmetrization of a set satisfying the DDT, then the logic must be Fregean. Actually, in both cases the conclusion can be much stronger:

THEOREM 7.75
Let $\mathcal{L}$ be a protoalgebraic and Fregean logic with theorems. The logic $\mathcal{L}$ is regularly algebraizable. If moreover $\mathcal{L}$ is finitary, then $\mathcal{L}$ is finitely regularly algebraizable (hence, regularly BP-algebraizable) and fully Fregean.

PROOF: By Theorem 7.73, $\mathcal{L}$ is equivalential. Since $\mathcal{L}$ has theorems, we can use Theorem 6.142, which tells us that in this situation, in order to show that $\mathcal{L}$ is regularly algebraizable it is enough to show that $x \equiv y \ (\boldsymbol{\Omega}C_{\mathcal{L}}\{x,y\})$. But this follows from the trivial fact that $x \equiv y \ (\boldsymbol{\Lambda}_{\mathcal{L}}C_{\mathcal{L}}\{x,y\})$ and Proposition 7.70.

If moreover $\mathcal{L}$ is finitary, then, by Proposition 6.8, the protoimplication set Δ used in Theorem 7.73 can be chosen finite; hence the set Δ^{s} of congruence formulas is finite, which means that $\mathcal{L}$ is finitely equivalential, and hence finitely algebraizable, which in this case amounts to being BP-algebraizable. Moreover, by Theorem 7.71, $\mathcal{L}$ is fully Fregean. ☒

Observe that the addition of the assumption "with theorems" only excludes the almost inconsistent logics, but this is necessary: the almost inconsistent logics are protoalgebraic and Fregean (in fact, they are fully Fregean, as is easy to see) but they are not algebraizable.

Viewed from its negative side, the theorem tells us that all protoalgebraic logics that are not regularly algebraizable, and a fortiori those that are not algebraizable, cannot be Fregean (except the almost inconsistent). This view is helpful because there are many criteria to check whether a logic is or is not (regularly) algebraizable. For instance, the local logics associated with normal modal systems are equivalential but not algebraizable, hence they are not Fregean. Thus, they provide the wanted examples of finitary and protoalgebraic fully selfextensional logics that are not Fregean.

In the finitary case, the equivalent quasivarieties of these logics have been completely characterized. Since they are regularly algebraizable, these quasivarieties are pointed; as usual the constant term is denoted by $\top$ and its value in an algebra $\boldsymbol{A}$ of the quasivariety by $\top^{\boldsymbol{A}}$.

DEFINITION 7.76
A pointed quasivariety K *is* ***congruence-orderable*** *when for all* $\boldsymbol{A} \in \mathsf{K}$ *and all* $a, b \in A$*, if* $\boldsymbol{\Theta}_{\mathsf{K}}^{\boldsymbol{A}}(a, \top^{\boldsymbol{A}}) = \boldsymbol{\Theta}_{\mathsf{K}}^{\boldsymbol{A}}(b, \top^{\boldsymbol{A}})$*, then* $a = b$*, and it is* ***Fregean*** *when it is relatively point-regular and congruence-orderable. The same notions apply to varieties by deleting "relatively" and writing* $\boldsymbol{\Theta}^{\boldsymbol{A}}$ *instead of* $\boldsymbol{\Theta}_{\mathsf{K}}^{\boldsymbol{A}}$.

The reason for the name "congruence-orderable" is that, in any pointed class K, the relation $\leqslant_{\mathsf{K}}^{\boldsymbol{A}}$ defined by putting $a \leqslant_{\mathsf{K}}^{\boldsymbol{A}} b$ if and only if $\boldsymbol{\Theta}_{\mathsf{K}}^{\boldsymbol{A}}(a, \top^{\boldsymbol{A}}) \supseteq \boldsymbol{\Theta}_{\mathsf{K}}^{\boldsymbol{A}}(b, \top^{\boldsymbol{A}})$ is a quasi-order, and K is congruence-orderable if and only if this quasi-order is in fact an order. Initially this property received the name "Fregean" in [189], but Theorem 7.78 justifies reserving this term for the quasivarieties (or varieties) that are both congruence-orderable and relatively point-regular, as is done in more recent literature.

LEMMA 7.77
Let $\mathcal{L}$ *be regularly BP-algebraizable, with* K *as its equivalent algebraic semantics. The quasivariety* K *is congruence-orderable if and only if for all* $\boldsymbol{A} \in \mathsf{K}$, $\boldsymbol{\Lambda}^{A}\mathcal{F}i_{\mathcal{L}}\boldsymbol{A}$ *is the identity relation; that is,* $\mathcal{F}i_{\mathcal{L}}\boldsymbol{A}$ *separates points.*

PROOF: We know that K is a pointed quasivariety. By the main Isomorphism Theorem for algebraizable logics (Theorem 3.58), for every $\boldsymbol{A}$, the function $\boldsymbol{\tau}^{\boldsymbol{A}^{-1}}$ is an isomorphism from $\mathrm{Con}_{\mathsf{K}}\boldsymbol{A}$ to $\mathcal{F}i_{\mathcal{L}}\boldsymbol{A}$. But here $\boldsymbol{\tau}x = \{x \approx \top\}$, where $\top$ is a theorem of $\mathcal{L}$; therefore, if $\boldsymbol{A} \in \mathsf{K}$, then $\boldsymbol{\tau}^{\boldsymbol{A}}a = \{\langle a, \top^{\boldsymbol{A}}\rangle\}$ for any $a \in A$, and hence for any $\theta \in \mathrm{Con}_{\mathsf{K}}\boldsymbol{A}$,

$$\boldsymbol{\tau}^{\boldsymbol{A}^{-1}}\theta = \{a \in A : \boldsymbol{\tau}^{\boldsymbol{A}}a \subseteq \theta\} = \{a \in A : \langle a, \top^{\boldsymbol{A}}\rangle \in \theta\} = \top^{\boldsymbol{A}}/\theta.$$

Therefore, all $\mathcal{L}$-filters of $\boldsymbol{A} \in \mathsf{K}$ have the form $\top^{\boldsymbol{A}}/\theta$ for some $\theta \in \mathrm{Con}_{\mathsf{K}}\boldsymbol{A}$. If $a, b \in A$ are such that $\boldsymbol{\Theta}_{\mathsf{K}}^{\boldsymbol{A}}(a, \top^{\boldsymbol{A}}) = \boldsymbol{\Theta}_{\mathsf{K}}^{\boldsymbol{A}}(b, \top^{\boldsymbol{A}})$, this means that for every

$\theta \in \mathrm{Con}_{\mathsf{K}} \boldsymbol{A}$, $\langle a, \top^{\boldsymbol{A}} \rangle \in \theta$ if and only if $\langle b, \top^{\boldsymbol{A}} \rangle \in \theta$, that is, $a \in \top^{\boldsymbol{A}}/\theta$ if and only if $b \in \top^{\boldsymbol{A}}/\theta$. The previous reformulation of the isomorphism ${\boldsymbol{\tau}^{\boldsymbol{A}}}^{-1}$ tells us that this is the same as saying that $a \in F$ if and only if $b \in F$ for all $F \in \mathcal{F}i_{\mathcal{L}} \boldsymbol{A}$, that is, $Fg^{\boldsymbol{A}}_{\mathcal{L}}\{a\} = Fg^{\boldsymbol{A}}_{\mathcal{L}}\{b\}$. Thus, the equality $\Theta^{\boldsymbol{A}}_{\mathsf{K}}(a, \top^{\boldsymbol{A}}) = \Theta^{\boldsymbol{A}}_{\mathsf{K}}(b, \top^{\boldsymbol{A}})$ is equivalent to the fact that $a \equiv b\ (\boldsymbol{\Lambda}^{A}\mathcal{F}i_{\mathcal{L}}\boldsymbol{A})$. Since to be congruence-orderable is to say that the first equality implies $a = b$, it turns out to be the same as to say that $\boldsymbol{\Lambda}^{A}\mathcal{F}i_{\mathcal{L}}\boldsymbol{A}$ is the identity relation. ⊠

The following result is all the most interesting:

THEOREM 7.78
A logic $\mathcal{L}$ is finitary, protoalgebraic, Fregean and with theorems if and only if it is the assertional logic of a Fregean quasivariety. If these conditions hold, then $\mathcal{L}$ is regularly BP-algebraizable and fully Fregean, and the mentioned quasivariety is $\mathsf{Alg}\,\mathcal{L}$.

PROOF: ($\Rightarrow$) By Theorem 7.75, $\mathcal{L}$ is regularly BP-algebraizable and hence $\mathsf{Alg}\,\mathcal{L}$ is a pointed quasivariety. By Theorem 3.63, $\mathsf{Alg}\,\mathcal{L}$ is relatively point-regular, and it only remains to show that it is congruence-orderable. By Lemma 7.77, it is enough to show that if $\boldsymbol{A} \in \mathsf{Alg}\,\mathcal{L}$, then $\boldsymbol{\Lambda}^{A}\mathcal{F}i_{\mathcal{L}}\boldsymbol{A}$ is the identity relation. But by Theorem 7.71, $\mathcal{L}$ is fully Fregean, and hence fully selfextensional, which means that Corollary 7.10 applies, and shows that $\boldsymbol{\Lambda}^{A}\mathcal{F}i_{\mathcal{L}}\boldsymbol{A}$ is the identity relation in any $\boldsymbol{A} \in \mathsf{Alg}\,\mathcal{L}$.

($\Leftarrow$) Assume that $\mathcal{L}$ is the assertional logic of a Fregean quasivariety K. Since K is a quasivariety, $\mathcal{L}$ is finitary. Since K is relatively point-regular, by Theorem 6.146, $\mathcal{L}$ is regularly BP-algebraizable, with K as its equivalent algebraic semantics, that is, $\mathsf{K} = \mathsf{Alg}\,\mathcal{L}$. In particular, thus, $\mathcal{L}$ is protoalgebraic and has theorems. By Lemma 7.77, we know that $\boldsymbol{\Lambda}^{A}\mathcal{F}i_{\mathcal{L}}\boldsymbol{A}$ is the identity relation in all $\boldsymbol{A} \in \mathsf{K}$. But this is condition (iii) in Proposition 7.9, therefore $\mathcal{L}$ is fully selfextensional. Now, since in particular $\mathcal{L}$ is truth-equational, we can apply Theorem 7.67 and conclude that $\mathcal{L}$ is fully Fregean; thus, it is a fortiori Fregean.

The final statements are proved in each of the two halves of the proof of the main equivalence. ⊠

It is not difficult (Exercise 7.46) to reformulate the previous result in an algebra-centric way:

COROLLARY 7.79
A pointed quasivariety is Fregean if and only if its assertional logic is protoalgebraic, Fregean and has theorems. If these conditions hold, then the logic is regularly BP-algebraizable and fully Fregean. ⊠

The above results are important because Fregean varieties have been deeply studied with universal algebra tools [1, 2, 148, 149, 226]. We know that $\mathcal{C}\ell$ and $\mathcal{I}\ell$ are finitary, protoalgebraic, Fregean, and have theorems, and hence all their axiomatic extensions have the same properties. As a consequence, all varieties of Boolean algebras and of Heyting algebras are Fregean varieties. By contrast, it is possible to see that the variety DL_{01} of bounded distributive lattices is congruence-orderable, but it is not relatively point-regular. As seen in Exercise 5.63 and in Example 7.60.8, this class is $\mathsf{Alg}\mathcal{C\ell}_{\wedge\vee\top\bot}$, and the logic $\mathcal{C\ell}_{\wedge\vee\top\bot}$ is Fregean and has theorems, but is not protoalgebraic.

In the finitary case, putting several results from this and the previous subsection together, one obtains some equivalent characterizations of the same class:

THEOREM 7.80
Let $\mathcal{L}$ be a finitary logic. The following conditions are equivalent:

(i) *$\mathcal{L}$ is protoalgebraic and Fregean with theorems.*

(ii) *$\mathcal{L}$ is algebraizable and fully selfextensional.*

(iii) *$\mathcal{L}$ is weakly algebraizable and fully selfextensional.*

In all these cases $\mathcal{L}$ is regularly BP-algebraizable and fully Fregean.

PROOF: Theorem 7.75 contains (i)⇒(ii) and the final statement; (ii)⇒(iii) is trivial; and (iii)⇒(i) is contained in Theorems 7.67 or 7.72, save for the point of having theorems; but the only protoalgebraic logics without theorems are the almost inconsistent ones, and these are not weakly algebraizable. ⊠

There are several particular readings of this result (and of some previous ones), which help understanding the interaction between the two hierarchies and the strength of these properties a little better:

- For finitary logics, putting two individually weaker properties together, one from each hierarchy, implies stronger properties in each hierarchy: a weakly algebraizable and fully selfextensional logic (as well as a protoalgebraic and Fregean one) must be regularly BP-algebraizable and fully Fregean, thus belonging to very high levels in both hierarchies.
- In the class of finitary weakly algebraizable logics (notice that this includes a very large number of the ordinary logics in the literature), the Frege hierarchy reduces to exactly two classes: the Fregean and the selfextensional. That these are indeed different is shown by some ad hoc examples in [7]; in particular, this paper constructs an implicative finitary logic that is selfextensional but not Fregean. Thus, the two classes are different even inside one of the best behaved classes in the Leibniz hierarchy.
- If $\mathcal{L}$ is fully selfextensional, then $\mathcal{L}$ is weakly algebraizable if and only if $\mathcal{L}$ is algebraizable. And if moreover $\mathcal{L}$ is finitary, then this happens if and only if $\mathcal{L}$ is regularly algebraizable.
- If $\mathcal{L}$ is finitary, Fregean and with theorems, then $\mathcal{L}$ is protoalgebraic if and only if $\mathcal{L}$ is regularly BP-algebraizable. Thus, in an important part of the Frege hierarchy, the Leibniz hierarchy reduces to only two classes: the truth-equational and the regularly BP-algebraizable, which are different: the logic $\mathcal{C\ell}_{\wedge\vee\top\bot}$ is obviously finitary and with theorems, it is Fregean (Example 7.60.8), therefore it is truth-equational (Theorem 7.66), but it is not protoalgebraic (Exercise 5.63).

One of the few known facts about the full g-models of Fregean logics is obtained by adding protoalgebraicity to Proposition 7.65.

THEOREM 7.81
If $\mathcal{L}$ is a protoalgebraic Fregean logic with theorems, then the function $\Phi\colon \Gamma \mapsto (\mathcal{T}h\mathcal{L})^{\Gamma}$ is a dual lattice isomorphism between the lattices $\mathcal{T}h\mathcal{L}$ and $\mathrm{FCS}_{\mathcal{L}}\boldsymbol{Fm}$.

PROOF: After Proposition 7.65, we need only show that Φ is surjective. But since $\mathcal{L}$ is protoalgebraic, Theorem 6.39 says that every full g-model of $\mathcal{L}$ has the form $\langle \boldsymbol{A}, (\mathcal{F}i_{\mathcal{L}}\boldsymbol{A})^{F}\rangle$ for some $F \in \mathcal{F}i_{\mathcal{L}}\boldsymbol{A}$; for g-models on $\boldsymbol{Fm}$ this is $\langle \boldsymbol{Fm}, (\mathcal{T}h\mathcal{L})^{\Gamma}\rangle$ for some $\Gamma \in \mathcal{T}h\mathcal{L}$. ⊠

Notice that, unlike in the case of Proposition 7.65, the assumption of having theorems here leaves out only the almost inconsistent logics.

The second part of Section 7.2 shows that the addition of either the PC or the u-DDT to the property of selfextensionality produces logics with a very good behaviour. It is then no surprise that a similar situation presents itself when these properties are added to that of being Fregean. The first case to address is that of the conjunction.

THEOREM 7.82
Let $\mathcal{L}$ be a finitary and selfextensional logic of type $\boldsymbol{L}$ with conjunction $\wedge$. The logic $\mathcal{L}$ is fully Fregean if and only if for each $\lambda \in L$ with $n = \operatorname{ar}\lambda \geqslant 1$, the quasi-equation

$$\Big(\bigwedge_{i<n} z \wedge x_i \approx z \wedge y_i\Big) \to z \wedge \lambda x_0 \ldots x_{n-1} \approx z \wedge \lambda y_0 \ldots y_{n-1} \qquad \text{(FFR)}$$

holds in the class $\mathsf{Alg}\mathcal{L}$.

SKETCH OF THE PROOF: For the direct implication use that if $\boldsymbol{A} \in \mathsf{Alg}\mathcal{L}$, then $\langle \boldsymbol{A}, \mathcal{F}i_{\mathcal{L}}\boldsymbol{A}\rangle$ is a finitary full g-model of $\mathcal{L}$, and hence has the SPCONG and the PC, and moreover it is reduced. Then, the quasi-equation (FFR) expresses the congruence property for it, for the operation λ. For the converse, recall from Proposition 7.57 that it is enough to show the SPCONG for g-matrices of the form $\langle \boldsymbol{A}, \mathcal{F}i_{\mathcal{L}}\boldsymbol{A}\rangle$ with $\boldsymbol{A} \in \mathsf{Alg}\mathcal{L}$, and that all these are finitary because $\mathcal{L}$ is, have the PCONG by Theorem 7.20, and the PC by Proposition 5.90. Exercise 7.47 asks you to complete the details. ⊠

Then, by restricting the isomorphism found in Theorem 7.28, we get:

COROLLARY 7.83
Let $\boldsymbol{L}$ be an algebraic type and let $\wedge$ represent a binary term in $\boldsymbol{L}$. The functions $\mathcal{L} \mapsto \mathbb{V}\mathcal{L}$ and $\mathsf{K} \mapsto \mathcal{L}^{\leqslant}_{\mathsf{K}}$ establish mutually inverse dual order isomorphisms between the following families:†

- *The family of all finitary and non-pseudo-axiomatic fully Fregean logics in the language $\boldsymbol{L}$ that have $\wedge$ as a conjunction, ordered under the extension relation.*
- *The family of all the subvarieties of the variety $\mathsf{SL}(\boldsymbol{L},\wedge)$ where the quasi-equations (FFR) hold, ordered under the subclass relation.* ⊠

† See footnote † on page 410.

By combining several of the previous results, we obtain one of the most important facts:

THEOREM 7.84
Let $\mathcal{L}$ be a finitary, protoalgebraic and selfextensional logic with conjunction and theorems. The following conditions are equivalent:

(i) *$\mathcal{L}$ is regularly BP-algebraizable.*

(ii) *$\mathcal{L}$ is algebraizable.*

(iii) *$\mathcal{L}$ is weakly algebraizable (or, equivalently, truth-equational).*

(iv) *$\mathcal{L}$ is fully Fregean.*

(v) *$\mathcal{L}$ is Fregean.*

PROOF: The implications (i)⇒(ii)⇒(iii) and (iv)⇒(v) are trivial. (iii)⇒(iv) holds because the general assumptions of the theorem imply that $\mathcal{L}$ is fully selfextensional (Theorem 7.20) and then (iii) plus Theorem 7.67 imply that $\mathcal{L}$ is fully Fregean. Finally, that (v)⇒(i) is contained in Theorem 7.75, again given the general assumptions of the theorem. ⊠

Thus, several of the previous observations on page 463 can be enhanced for logics with conjunction:

- In the class of finitary, protoalgebraic logics with conjunction, the Frege hierarchy reduces to two classes, the selfextensional and the Fregean; there is no need to add "with theorems" because the only logics in this class having no theorems, the almost inconsistent ones, are known to be fully Fregean, hence Fregean.
- A finitary and weakly algebraizable logic with conjunction is selfextensional if and only if it is fully Fregean. Thus, inside this (certainly large) class in the Leibniz hierarchy, the Frege hierarchy is dramatically reduced to just one class! And notice that this includes all BP-algebraizable logics with conjunction, which constitute a large class of the "real" logics encountered in the literature.

Similar results to the three preceding ones hold for finitary selfextensional logics satisfying the u-DDT, but in this case the form of the quasi-equations playing the role of (FFR) is more complicated; see [155] for details. By contrast, the result parallel to Theorem 7.84 can be easily established:

THEOREM 7.85
Let $\mathcal{L}$ be a finitary and selfextensional logic satisfying the u-DDT. *The following conditions are equivalent:*

(i) *$\mathcal{L}$ is regularly BP-algebraizable.*

(ii) *$\mathcal{L}$ is algebraizable.*

(iii) *$\mathcal{L}$ is weakly algebraizable (or, equivalently, truth-equational).*

(iv) *$\mathcal{L}$ is fully Fregean.*

(v) *$\mathcal{L}$ is Fregean.*

PROOF: The implications (i)$\Rightarrow$(ii)$\Rightarrow$(iii) and (iv)$\Rightarrow$(v) are trivial. (iii)$\Rightarrow$(iv) holds because the general assumptions of the theorem imply that $\mathcal{L}$ is fully selfextensional (Theorem 7.53) and then (iii) plus Theorem 7.67 imply that $\mathcal{L}$ is fully Fregean. Finally, since having any form of the DDT implies that the logic is protoalgebraic (Theorem 6.22) and has theorems, Theorem 7.75 establishes that (v)$\Rightarrow$(i). ⊠

Observe that, by Definition 7.49 and Theorem 7.51, the assumption in Theorem 7.85 can rephrased as "$\mathcal{L} = \mathcal{L}^{\rightarrow}_{\mathbb{V}\mathcal{L}}$", and condition (i) as "$\mathcal{L} = \mathcal{L}^{\top}_{\mathbb{V}\mathcal{L}}$"; thus, this settles the issue of comparing the logics $\mathcal{L}^{\rightarrow}_{\mathbb{V}\mathcal{L}}$ and $\mathcal{L}^{\top}_{\mathbb{V}\mathcal{L}}$, discussed at the end of Section 7.2.

From Theorem 7.85 some consequences concerning the relations between the two hierarchies can be obtained:

- In the class of finitary logics with the u-DDT, the Frege hierarchy reduces to two classes, the selfextensional and the Fregean. The two classes are different, as witnessed by the local logics associated with normal modal systems, which have the u-DDT (because they are axiomatic expansions of $\mathcal{C}\ell$), and are selfextensional (Example 7.47.1), but they are not algebraizable, and therefore they are not Fregean.
- Putting Theorems 7.53 and 7.85 together, we see that a finitary, truth-equational logic with the u-DDT is selfextensional if and only if it is fully Fregean. Thus, inside this (fairly large) class in the Leibniz hierarchy, the Frege hierarchy is again dramatically reduced to just one class!

Surprisingly, the class of logics dealt with in Theorem 7.85 actually includes the one dealt with in Theorem 7.84; although these two results need finitarity, even without it we find an interesting result:

THEOREM 7.86
Let $\mathcal{L}$ be a protoalgebraic and Fregean logic with conjunction and theorems. The logic $\mathcal{L}$ satisfies the (multiterm) DDT. *If moreover $\mathcal{L}$ is finitary, then it satisfies the* u-DDT.

PROOF: Let $\Delta(x,y)$ be a protoimplication set for $\mathcal{L}$. Since $\mathcal{L}$ has theorems, Proposition 6.11 tells us that the set Δ can be chosen[†] to be non-empty. We will see that the set $I(x,y) := \Delta(x, x \wedge y)$ satisfies the DDT: for all $\Gamma \cup \{\varphi,\psi\} \subseteq Fm$, $\Gamma,\varphi \vdash_{\mathcal{L}} \psi$ if and only if $\Gamma \vdash_{\mathcal{L}} I(\varphi,\psi)$.

($\Rightarrow$) Assume that $\Gamma,\varphi \vdash_{\mathcal{L}} \psi$. This implies, using the customary trick with the PC, that $\Gamma,\varphi \dashv\vdash_{\mathcal{L}} \Gamma,\varphi \wedge \psi$. Then by Fregeanity $\Gamma,\delta(\varphi,\varphi) \dashv\vdash_{\mathcal{L}} \Gamma,\delta(\varphi,\varphi\wedge\psi)$ for each $\delta \in \Delta$. But by (R), $\emptyset \vdash_{\mathcal{L}} \delta(\varphi,\varphi)$, thus we obtain in particular that $\Gamma \vdash_{\mathcal{L}} \delta(\varphi,\varphi\wedge\psi)$ for all $\delta \in \Delta$, and so $\Gamma \vdash_{\mathcal{L}} I(\varphi,\psi)$.

($\Leftarrow$) Assume that $\Gamma \vdash_{\mathcal{L}} I(\varphi,\psi)$, that is, that $\Gamma \vdash_{\mathcal{L}} \Delta(\varphi,\varphi\wedge\psi)$. Since Δ satisfies (MP), this implies $\Gamma,\varphi \vdash_{\mathcal{L}} \varphi\wedge\psi$, and from this $\Gamma,\varphi \vdash_{\mathcal{L}} \psi$ follows by the PC.

If moreover $\mathcal{L}$ is finitary, we know that the set Δ can be taken finite, which

[†] Notice that the inconsistent logic is not excluded from this result; therefore we cannot ensure that an arbitrary protoimplication set *must* be non-empty.

implies that I can also be taken finite, and using the conjunction one can define $x \to y := \bigwedge I(x,y)$, which is a single term. Then, by the PC, $\Gamma \vdash_{\mathcal{L}} I(\varphi,\psi)$ if and only if $\Gamma \vdash_{\mathcal{L}} \varphi \to \psi$, and then the (multiterm) DDT for $I(x,y)$ implies the uniterm DDT for $x \to y$. ⊠

The set used in the proof need only be a protoimplication set, but in practice the DDT holds for a set obtained in the same way using a set of congruence formulas† (the logic has one, as the assumptions imply it is algebraizable); this generalizes the well-known situation of many logics where $x \to y \dashv\vdash_{\mathcal{L}} x \leftrightarrow (x \wedge y)$.

By Theorem 7.84, this applies in particular to the finitary and algebraizable logics that are selfextensional and have a conjunction (notice that they must have theorems). Thus, these logics belong to the larger, more important class treated in Theorem 7.85 of the finitary Fregean logics with the u-DDT, which in general need not have a conjunction. The logics in this class have been deeply studied in [75, 155], where they have been completely characterized and axiomatized. Observe that all these logics are regularly BP-algebraizable and have a (Fregean) variety as their algebraic counterpart. This last fact is essentially due to the presence of the PC or the u-DDT. A case where this fails is the negation-equivalence fragment $\mathcal{I}\ell_{\neg,\leftrightarrow}$ of intuitionistic logic: it is finitary, algebraizable and Fregean, and its algebraic counterpart $\mathsf{Alg}\mathcal{I}\ell_{\neg,\leftrightarrow}$ is not a variety but a proper (Fregean) quasivariety, as proved in [161]; see also [75, pp. 73ff.] for more information. The fact that the DDT appears in these studies in its uniterm version is also key. The results fail if it is replaced by the multiterm DDT; in particular, in [76, pp. 207ff.] it is shown how to construct an ad hoc finitary logic in an arbitrary language type expanded with two implication connectives that collectively satisfy the multiterm DDT (hence the logic is protoalgebraic), and such that this logic is Fregean (therefore it is regularly BP-algebraizable), but does not satisfy the u-DDT and its algebraic counterpart $\mathsf{Alg}\mathcal{L}$ is not a variety.

Exercises for Section 7.3

Exercise 7.32. Prove that a logic $\mathcal{L}$ is fully Fregean if and only if for any $\boldsymbol{A} \in \mathsf{Alg}\mathcal{L}$, the g-matrix $\langle \boldsymbol{A}, \mathcal{F}i_{\mathcal{L}}\boldsymbol{A} \rangle$ has the SPCONG (Proposition 7.57) and if and only if $\varLambda^{\boldsymbol{A}}_{\mathcal{L}} F \subseteq \varOmega^{\boldsymbol{A}} F$ for every $F \in \mathcal{F}i_{\mathcal{L}}\boldsymbol{A}$ and every $\boldsymbol{A} \in \mathsf{Alg}\mathcal{L}$. Comparing with Proposition 7.9, explain why in principle it does not seem possible to add here the condition that for any $\boldsymbol{A} \in \mathsf{Alg}\mathcal{L}$ and any $X \cup \{a,b\} \subseteq A$, if $Fg^{\boldsymbol{A}}_{\mathcal{L}}(X,a) = Fg^{\boldsymbol{A}}_{\mathcal{L}}(X,b)$, then $a = b$.

Hint. Use Propositions 7.2 and 7.4.

Exercise 7.33. If you did Exercise 5.63, then you can examine the characterizations of the full g-models of $\mathcal{C}\ell_{\wedge\vee\top\bot}$ you presumably found there, and show that all these full g-models have the SPCONG. In this way you will show that this logic is fully Fregean (and, in contrast to $\mathcal{C}\ell_{\wedge\vee}$, it has theorems). Show also that $\mathcal{L}^{\leqslant}_{\mathsf{DL}_{01}} = \mathcal{L}^{\top}_{\mathsf{DL}_{01}} = \mathcal{C}\ell_{\wedge\vee\top\bot}$, where DL_{01} is the variety of bounded distributive lattices.

† Recall that all sets of congruence formulas are, a fortiori, protoimplication sets.

EXERCISE 7.34. Prove that any Fregean logic $\mathcal{L}$ satisfies that $\sigma\widetilde{\Omega}_{\mathcal{L}}C_{\mathcal{L}}\Gamma \subseteq \widetilde{\Omega}_{\mathcal{L}}C_{\mathcal{L}}\sigma\Gamma$ for any $\Gamma \subseteq Fm$ and any substitution σ. Compare with Exercise 6.50. Using the appropriate counterexample(s), prove that this property does not characterize Fregean logics.

HINT. Take Exercise 6.32 into account.

EXERCISE 7.35. Prove that the following conditions are equivalent for any logic $\mathcal{L}$:

(i) $\mathcal{L}$ satisfies the Suszko rules $x, y, \delta(x, \vec{z}) \rhd \delta(y, \vec{z})$, for all $\delta(x, \vec{z}) \in Fm$.

(ii) $\mathcal{L}$ is complete with respect to a class of reduced matrices in which the designated subset has at most one element.

(iii) For all $\langle \boldsymbol{A}, F\rangle \in \mathsf{Mod}^*\mathcal{L}$, F has at most one element.

EXERCISE 7.36. Let $\langle \boldsymbol{A}, \mathscr{C}\rangle$ be a g-matrix, and let $\mathcal{L}$ be the logic defined by it. Prove that for all $\Gamma \in \mathcal{T}h\mathcal{L}$, $\Lambda_{\mathcal{L}}\Gamma = \bigcap\{h^{-1}\Lambda^{A}_{\mathscr{C}}Ch\Gamma : h \in \mathrm{Hom}(\boldsymbol{Fm}, \boldsymbol{A})\}$. Generalize this property to logics defined by a class of g-matrices, and check how this generalization is used in the proof of Proposition 7.62.

EXERCISE 7.37. Prove Corollary 7.64, using Lemma 7.63.

EXERCISE 7.38. Let $\langle \boldsymbol{A}, \mathscr{C}\rangle$ be a finitary g-matrix satisfying the PC. Prove that $\langle \boldsymbol{A}, \mathscr{C}\rangle$ has the SPCONG if and only if for every $a \in A$, the relation $\Lambda^{A}_{\mathscr{C}}C\{a\}$ is a congruence. Reformulate this result in terms of being a model of Gentzen style rules, in the style of Corollary 7.64. See what this result tells us about finitary logics with conjunction.

HINT. Use Theorem 1.46.

EXERCISE 7.39. Let $\mathcal{L}$ be a fully Fregean logic with theorems. Prove directly (without using Theorems 7.66 and 6.104) that for each $\boldsymbol{A}$ and each $F \in \mathcal{F}i_{\mathcal{L}}\boldsymbol{A}$, the g-matrix $\langle \boldsymbol{A}, (\mathcal{F}i_{\mathcal{L}}\boldsymbol{A})^{F}\rangle$ is a full g-model of $\mathcal{L}$, and the function $\Phi^{\boldsymbol{A}}\colon F \mapsto (\mathcal{F}i_{\mathcal{L}}\boldsymbol{A})^{F}$ is an order-reversing embedding of $\mathcal{F}i_{\mathcal{L}}\boldsymbol{A}$ into $\mathrm{FCS}_{\mathcal{L}}\boldsymbol{A}$.

EXERCISE 7.40. Let $\mathcal{L}$ be a Fregean and pseudo-axiomatic logic. Prove that the g-matrix $\langle \boldsymbol{Fm}, (\mathcal{T}h\mathcal{L})^{\Gamma} \cup \{\emptyset\}\rangle$ is a full g-model of $\mathcal{L}$ for each $\Gamma \in \mathcal{T}h\mathcal{L}$, and that the function $\Phi_{\emptyset}\colon \Gamma \mapsto (\mathcal{T}h\mathcal{L})^{\Gamma} \cup \{\emptyset\}$ is an order-reversing embedding of $\mathcal{T}h\mathcal{L}$ into $\mathrm{FCS}_{\mathcal{L}}\boldsymbol{Fm}$.

EXERCISE 7.41. Let $\langle \boldsymbol{A}, \mathscr{C}\rangle$ be a full g-model of the logic $\mathcal{C}\ell_{\wedge\vee}$. Prove that for each $F \in \mathscr{C}$, the g-matrix $\langle \boldsymbol{A}, \mathscr{C}^{F} \cup \{\emptyset\}\rangle$ is also a full g-model of $\mathcal{C}\ell_{\wedge\vee}$, and that in particular the function $\Phi^{\boldsymbol{A}}_{\emptyset}\colon F \mapsto (\mathcal{F}i_{\mathcal{L}}\boldsymbol{A})^{F} \cup \{\emptyset\}$ is an order-reversing embedding of $\mathcal{F}i_{\mathcal{L}}\boldsymbol{A}$ into $\mathrm{FCS}_{\mathcal{L}}\boldsymbol{A}$. Compare this fact with Exercise 7.39 (recall that $\mathcal{C}\ell_{\wedge\vee}$ is fully Fregean but does not have theorems).

EXERCISE 7.42. Let $\mathcal{L}$ be a protoalgebraic logic with theorems. Prove that $\mathcal{L}$ is weakly algebraizable and fully Fregean if and only if $\mathcal{L}$ is fully selfextensional and has the property that for each full g-model $\langle \boldsymbol{A}, \mathscr{C}\rangle$ of $\mathcal{L}$ and each $F \in \mathscr{C}$, the g-matrix $\langle \boldsymbol{A}, \mathscr{C}^{F}\rangle$ is also a full g-model of $\mathcal{L}$.

EXERCISE 7.43. Let $\mathcal{L}$ be a Fregean logic with theorems. Using Lemma 6.100, prove that $\boldsymbol{\Omega}$ completely reflects order on $\mathcal{T}h\mathcal{L}$.

EXERCISE 7.44. Let $\mathcal{L}$ be a fully Fregean logic with theorems. Inspired by Exercises 7.39 and 7.43, prove directly that for each algebra $\boldsymbol{A}$, the Leibniz operator $\boldsymbol{\Omega}^{\boldsymbol{A}}$ completely reflects order over $\mathcal{F}i_{\mathcal{L}}\boldsymbol{A}$.

EXERCISE 7.45. Give an alternative proof of the property that a Fregean logic is truth-equational if and only if it has theorems, included in Theorem 7.66.

HINT. Use Theorem 6.104 and Proposition 7.65.

EXERCISE 7.46. Check that Theorem 7.78 can really be reformulated as Corollary 7.79, and fill in any steps needed to do so.

EXERCISE 7.47. Complete the details in the proof of Theorem 7.82, following the indications given in the text.

HINT. The property found in Exercise 7.38 may be of some help.

EXERCISE 7.48. Show that the condition that the logic $\mathcal{L}_{\mathsf{K}}^{\top}$ is fully Fregean may be added as an equivalent one to the list of conditions in Theorem 7.46.

HINT. You have to combine some of the main results in Section 7.3, not directly related to the situation of two companion logics Theorem 7.46 deals with. Looking at Theorem 7.84 may also be of some help, but notice that protoalgebraicity appears among its assumptions while it does not appear in 7.46.

EXERCISE 7.49. Let $\mathcal{L}$ be a finitary logic in the language $\langle\wedge,\neg\rangle$ of type $\langle 2,1\rangle$ that satisfies the PC and the PIRA for the two operations, respectively. Prove that $\mathcal{L}$ is Fregean, and conclude that it is fully Fregean.

EXERCISE 7.50. Let $\mathcal{L}$ be a finitary and selfextensional logic satisfying the u-DDT. Prove directly that the variety $\mathbb{V}\mathcal{L}$ (which has Hilbert algebra reducts) is point-regular. Use this fact, together with Theorem 6.145, to obtain a direct proof of the implication (iii)$\Rightarrow$(i) of Theorem 7.85.

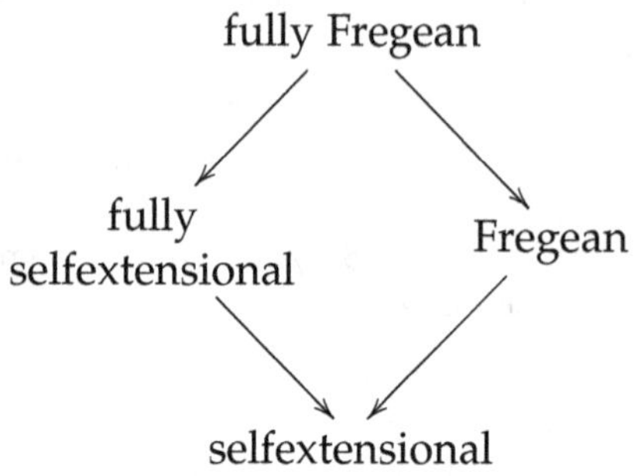

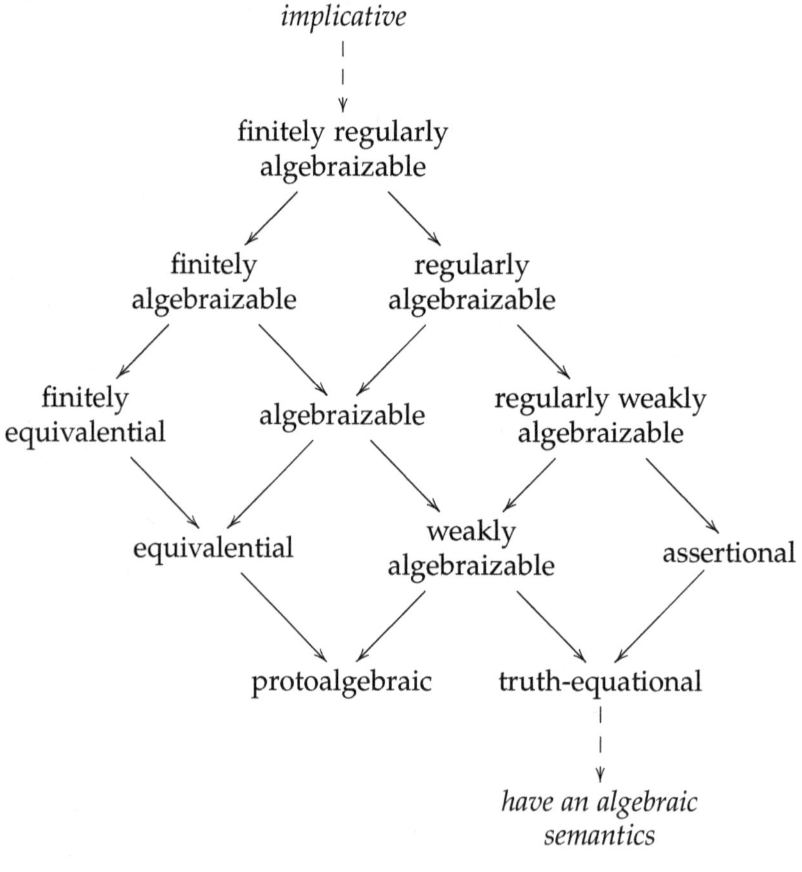

FIGURE 12: The Leibniz and Frege hierarchies (with *other* classes).

Appendix

Summary of properties of particular logics

The properties summarized here are the most relevant to the topic of the book, mainly: classification of a logic $\mathcal{L}$ in the two hierarchies[†] and determination of its reduced models $\mathsf{Mod}^*\mathcal{L}$, of its three (possibly equal) algebraic counterparts $\mathsf{Alg}^*\mathcal{L}, \mathsf{Alg}\,\mathcal{L}$ and $\mathbb{V}\mathcal{L}$, and of its full g-models $\mathsf{FGMod}\,\mathcal{L}$. In the text, the level of detail goes from full proofs to the simple mention, but in all cases either hints or references are given. There is no claim of exhaustivity: besides those mentioned here, further properties may be obtained by applying general theorems to particular logics, and still others can be found in the literature.

It may be interesting to recall the following facts:

- In general $\mathsf{Alg}^*\mathcal{L} \subseteq \mathsf{Alg}\,\mathcal{L} \subseteq \mathbb{V}\mathcal{L} = \mathbb{V}\mathsf{Alg}^*\mathcal{L} = \mathbb{V}\mathsf{Alg}\,\mathcal{L}$.
- If $\mathsf{Alg}^*\mathcal{L}$ is a variety, then $\mathsf{Alg}^*\mathcal{L} = \mathsf{Alg}\,\mathcal{L} = \mathbb{V}\mathcal{L}$.
- If $\mathsf{Alg}\,\mathcal{L}$ is a variety, then $\mathsf{Alg}\,\mathcal{L} = \mathbb{V}\mathcal{L}$.
- If $\mathcal{L}$ is protoalgebraic, then $\mathsf{Alg}^*\mathcal{L} = \mathsf{Alg}\,\mathcal{L}$.
- If $\mathcal{L}$ is algebraizable, then $\mathsf{Alg}^*\mathcal{L} = \mathsf{Alg}\,\mathcal{L}$, and this class is its equivalent algebraic semantics.

These facts are used implicitly to determine some of the classes. Concerning classification, it is useful to review the diagrams of the two hierarchies, reproduced on the facing page.

Notice that whenever *fragments* are mentioned, they should be understood relatively to the language in which the logic is presented; for instance, an expression like "negation-less fragment" should be understood in the context of the given language for intuitionistic logic, here $\langle \wedge, \vee, \to, \neg \rangle$, and should be phrased differently if a falsity constant $\bot$ were included. Occasionally, the connective $\leftrightarrow$ is also considered.

[†] Note that not each and every mentioned logic has been classified in the two hierarchies.

All logics are finitary, save for a few explicitly mentioned ones.

The *classification* adopted in this Appendix is not dogmatic, but practical. For instance, on the ground of certain criteria, many-valued logics should be counted among the substructural logics; some paraconsistent logics are also many-valued; and so forth.

Classical logic and its fragments

Classical logic

Classical logic $\mathcal{C}\ell$ is presented in the language $\langle \wedge, \vee, \rightarrow, \neg \rangle$, both semantically, as a two-valued logic (Example 1.11), and axiomatically (pp. 71ff.). It is implicative (p. 84), hence finitely regularly algebraizable, having $\{x \rightarrow y\,,\, y \rightarrow x\}$ as the set of equivalence formulas and $x \approx x \rightarrow x$ as the defining equation; it is fully Fregean (Example 7.60.4). Its equivalent algebraic semantics is $\mathsf{Alg}^*\mathcal{C}\ell = \mathsf{Alg}\mathcal{C}\ell = \mathbb{V}\mathcal{C}\ell = \mathsf{BA}$, the variety of *Boolean algebras*, where the $\mathcal{C}\ell$-filters coincide with the lattice filters and the Isomorphism Theorem 2.32 restricts to the usual isomorphism between filters and congruences (Exercises 2.31 and 5.50). Its reduced models are $\mathsf{Mod}^*\mathcal{C}\ell = \{\langle \boldsymbol{A}, \{1\} \rangle : \boldsymbol{A} \in \mathsf{BA}\}$ (p. 208). $\mathcal{C}\ell$ has other, nonstandard algebraic semantics, either larger than, or smaller than, or disjoint from BA (pp. 112, 121, and Example 4.79). Its full g-models are characterized algebraically and in terms of Tarksi-style conditions (Exercise 5.66), and these define its fully adequate Gentzen system (Example 5.120.2). $\mathcal{C}\ell$ satisfies the DDT (Exercise 3.54), and using it an explicit form of EDPC for BA is obtained (Exercise 3.63).

The fragments with conjunction and disjunction

The fragment with just conjunction and disjunction $\mathcal{C}\ell_{\wedge\vee}$ is defined by the two-element (distributive) lattice $\mathbf{2}_{\wedge\vee}$ and has no theorems (Exercise 1.29), hence it is not truth-equational (Corollary 6.92). It is not protoalgebraic (Example 6.16.1) and is non-pseudo-axiomatic (Exercise 7.18). The Leibniz congruence in its models is characterized, and $\mathsf{Mod}^*\mathcal{C}\ell_{\wedge\vee}$ and $\mathsf{Alg}^*\mathcal{C}\ell_{\wedge\vee}$ are determined and found to be weird classes (Example 4.47), while $\mathsf{Alg}\mathcal{C}\ell_{\wedge\vee} = \mathbb{V}\mathcal{C}\ell_{\wedge\vee} = \mathsf{DL} = \mathbb{V}\{\mathbf{2}_{\wedge\vee}\}$, the variety of *distributive lattices;* and its full g-models are characterized in a variety of ways, algebraic or proof-theoretic (Example 5.93). The last conditions define the Gentzen system fully adequate for $\mathcal{C}\ell_{\wedge\vee}$ (Example 5.120.3), which is algebraizable and has DL as equivalent algebraic semantics (Example 3.87), hence $\mathcal{C}\ell_{\wedge\vee}$ is a G-algebraizable logic that is not algebraizable (p. 314). This Gentzen system satisfies a form of the DDT and DL has EDPC (p. 172), but DL is not the (equivalent or not) algebraic semantics of any logic whatsoever (Example 4.61 and Exercise 4.38). An analysis of the Isomorphism Theorem 5.95 for $\mathcal{C}\ell_{\wedge\vee}$ shows that congruences of distributive lattices can be represented as certain families of prime filters, and for an arbitrary algebra $\boldsymbol{A}$ a representation of the relative congruences $\mathrm{Con}_{\mathsf{DL}}\boldsymbol{A}$ as the closed sets of certain topology over A is obtained

(Example 5.99). $\mathcal{C}\ell_{\wedge\vee}$ equals $\mathcal{L}^{\leqslant}_{\mathsf{DL}}$, the logic of order of DL (Example 7.34.1), and is the only Fregean logic among all the logics of order of subvarieties of L, the variety of lattices (Example 7.60.6).

The fragment with conjunction, disjunction and truth and falsity constants, $\mathcal{C}\ell_{\wedge\vee\top\bot}$, has similar properties, except that it is assertional, hence truth-equational; $\mathsf{Alg}\mathcal{C}\ell_{\wedge\vee\top\bot} = \mathbb{V}\mathcal{C}\ell_{\wedge\vee\top\bot} = \mathsf{DL}_{01}$, the variety of *bounded distributive lattices* (Exercise 5.63, Example 6.16.4 and p. 384). The logic $\mathcal{C}\ell_{\wedge\vee\top\bot}$ coincides with $\mathcal{L}^{\leqslant}_{\mathsf{DL}_{01}}$, the logic of order of DL_{01}, and also with $\mathcal{L}^{\top}_{\mathsf{DL}_{01}}$, the assertional logic of the same class.

The fragment with only conjunction

The fragment of classical logic with conjunction $\mathcal{C}\ell_{\wedge}$ is defined from the two-element semilattice $\mathbf{2}_{\wedge}$ and has no theorems (Exercise 1.29), hence it is not truth-equational (Corollary 6.92). It is not protoalgebraic (Example 6.16.1). The Leibniz congruence of its models is characterized; $\mathsf{Alg}^*\mathcal{C}\ell_{\wedge} = \mathbb{I}\{\mathbf{1}_{\wedge}, \mathbf{2}_{\wedge}\}$; $\mathsf{Mod}^*\mathcal{C}\ell_{\wedge} = \mathbb{I}\{\langle \mathbf{1}_{\wedge}, \emptyset\rangle, \langle \mathbf{1}_{\wedge}, \{1\}\rangle, \langle \mathbf{2}_{\wedge}, \{1\}\rangle\}$ (Example 4.46). Its algebraic counterpart is $\mathsf{Alg}\mathcal{C}\ell_{\wedge} = \mathbb{V}\mathcal{C}\ell_{\wedge} = \mathsf{SL} = \mathbb{V}\{\mathbf{2}_{\wedge}\}$, the variety of *semilattices* (Exercises 5.50 and 5.53). $\mathcal{C}\ell_{\wedge} = \mathcal{L}^{\leqslant}_{\mathsf{SL}}$, the logic of order of SL (Example 7.34.2).

Miscellaneous fragments of classical logic

- All the fragments of $\mathcal{C}\ell$ are Fregean (Example 7.60.3), and all the negation-less fragments are fully Fregean (Example 7.60.4). All protoalgebraic fragments of $\mathcal{C}\ell$ are fully Fregean (p. 458).
- All the fragments where $\top$ is definable have an algebraic semantics, and it is a quasivariety (p. 114).
- All the fragments with implication are implicative (Theorem 2.16). An axiomatization of the implication fragment $\mathcal{C}\ell_{\to}$ is given (p. 84).
- The negation fragment $\mathcal{C}\ell_{\neg}$ has no algebraic semantics (p. 112), while all fragments with $\wedge$ or with $\vee$ have one (p. 113). $\mathcal{C}\ell_{\neg}$ satisfies the PIRA but its basic full g-models need not satisfy it; thus it is a counterexample to the transfer of the PIRA (Example 5.24).
- The equivalence fragment $\mathcal{C}\ell_{\leftrightarrow}$ is finitely regularly algebraizable but not implicative (Example 3.50) and has the variety of *Boolean groups* as equivalent algebraic semantics (Exercise 3.26).

Intuitionistic logic, its fragments and extensions, and related logics

Intuitionistic logic

Intuitionistic logic $\mathcal{I}\ell$ is presented axiomatically in the language $\langle\wedge, \vee, \to, \neg\rangle$; it is implicative (p. 84) and fully Fregean (Example 7.60.4). Its equivalent algebraic semantics is $\mathsf{Alg}^*\mathcal{I}\ell = \mathsf{Alg}\mathcal{I}\ell = \mathbb{V}\mathcal{I}\ell = \mathsf{HA}$, the variety of *Heyting algebras*, in which the $\mathcal{I}\ell$-filters coincide with the lattice filters and the Isomorphism

Theorem 2.32 restricts to the usual isomorphism between filters and congruences (Exercises 2.31 and 5.50). $\mathsf{Mod}^*\mathcal{I}\ell = \{\langle \boldsymbol{A}, \{1\}\rangle : \boldsymbol{A} \in \mathsf{HA}\}$ (p. 208). HA is a nonstandard algebraic semantics for $\mathcal{C}\ell$ (p. 112). The full g-models of $\mathcal{I}\ell$ are characterized algebraically and in terms of Tarksi-style conditions (Exercise 5.65), and these define its fully adequate Gentzen system (Example 5.120.2). $\mathcal{I}\ell$ satisfies the DDT (p. 164), and using it an explicit form of EDPC for HA is obtained (Exercise 3.63).

The implication fragment of intuitionistic logic

The fragment of intuitionistic logic with only implication $\mathcal{I}\ell_{\to}$ is an implicative logic, and an axiomatization is given (p. 84). $\mathsf{Alg}\mathcal{I}\ell_{\to} = \mathbb{V}\mathcal{I}\ell_{\to} = \mathsf{HiA}$, the class of *Hilbert algebras* (Exercises 2.28 and 5.50), presented first quasi-equationally (p. 97) and then equationally (p. 441). $\mathsf{Mod}^*\mathcal{I}\ell_{\to} = \{\langle \boldsymbol{A}, \{1\}\rangle : \boldsymbol{A} \in \mathsf{HiA}\}$ (p. 208). $\mathcal{I}\ell_{\to}$ is the weakest logic in the language $\langle \to \rangle$ that satisfies the classical DDT (Corollary 3.73). The g-models and the full g-models of $\mathcal{I}\ell_{\to}$ are studied in relation to the DDT, and several characterizations are given (Example 5.91). On every algebra there is an isomorphism between the full g-models of $\mathcal{I}\ell_{\to}$ and its filters on that algebra (Example 5.98). The Gentzen system fully adequate for $\mathcal{I}\ell_{\to}$ is the one defined by the rules corresponding to the DDT (Example 5.120.1); this Gentzen system is algebraizable, hence $\mathcal{I}\ell_{\to}$ is a logic that is both algebraizable and G-algebraizable (p. 445).

Miscellaneous fragments and extensions of intuitionistic logic

- All the fragments of $\mathcal{I}\ell$ where $\top$ is definable have an algebraic semantics, and it is a quasivariety (p. 114). The negation fragment has no algebraic semantics (p.112), while all fragments with $\wedge$ or with $\vee$ have one (p. 113).
- All *superintuitionistic* or *intermediate* logics, that is, all the extensions of intuitionistic logic, are algebraizable, with same transformers as $\mathcal{I}\ell$ (Example 3.34.1).
- All the fragments with implication are implicative (Theorem 2.16).
- All the fragments of $\mathcal{I}\ell$, and all their axiomatic extensions (the *superintuitionistic* or *intermediate logics* in the traditional sense) are Fregean (Example 7.60.3). All the protoalgebraic fragments are fully Fregean (p. 458).
- All the negation-less fragments of $\mathcal{I}\ell$, and all their axiomatic extensions, are fully Fregean (Example 7.60.4). An axiomatization is given for the full negation-less fragment $\mathcal{I}\ell^+$ (*positive logic*); $\mathsf{Alg}\mathcal{I}\ell^+$ is the variety of *generalized Heyting algebras* (p. 84).
- The implication-less fragment $\mathcal{I}\ell^*$ is not protoalgebraic (Example 6.16.6), and is assertional, hence truth-equational, and belongs to the Leibniz hierarchy (p. 384). $\mathsf{Alg}^*\mathcal{I}\ell^*$ is a weird class, while $\mathsf{Alg}\mathcal{I}\ell^*$ is the variety PCDL of *pseudo-complemented distributive lattices* (Example 6.16.6). $\mathcal{I}\ell^*$ is Fregean but

not regularly weakly algebraizable (p. 456). The same results hold for a certain denumerable chain of its axiomatic extensions, corresponding to all the subvarieties of PCDL (same places).

- The equivalence fragment $\mathcal{I}\ell_{\leftrightarrow}$ is algebraizable, with equivalent algebraic semantics the variety of *intuitionisic equivalence algebras*; an axiomatization is given (Example 3.34.9). An explicit form of the PLDDT for this logic is found in the literature (Example 6.25.4).
- The fragment $\mathcal{I}\ell_{\neg,\leftrightarrow}$ is algebraizable and Fregean, and its algebraic counterpart is a Fregean quasivariety, but not a variety (p. 467).
- Both $\mathcal{I}\ell_{\leftrightarrow}$ and $\mathcal{I}\ell_{\neg,\leftrightarrow}$ are examples of finitely regularly algebraizable but not implicative logics (Example 3.50).

Some related logics

- *Minimal logic* is an implicative logic, given axiomatizatically; its algebraic counterpart is the class of *contrapositionally complemented lattices* (p. 84).
- The variety K of *Hilbert algebras with infimum* provides an example of a class of algebras that defines, in natural ways, three different logics $\mathcal{L}_{\mathsf{K}}^{\rightarrow}$, $\mathcal{L}_{\mathsf{K}}^{\leqslant}$ and $\mathcal{L}_{\mathsf{K}}^{\top}$; each has different properties, but they have the same theorems (p. 446).
- There are ten different *subintuitionistic logics* ($\mathcal{I}\ell$ is included as a limit case); five of them, including Visser's $\mathcal{BPL}$, are non-protoalgebraic; two of these are among the few known logics $\mathcal{L}$ such that $\mathsf{Alg}^*\mathcal{L} \subsetneq \mathsf{Alg}\,\mathcal{L} \subsetneq \mathbb{V}\mathcal{L}$; among the protoalgebraic ones, three are finitely regularly algebraizable (including $\mathcal{I}\ell$) and two are finitely equivalential but not weakly algebraizable (Example 6.16.8). $\mathcal{BPL}$ is truth-equational (p. 384), while others are not (Example 6.93.1). Two are fully Fregean ($\mathcal{I}\ell$ and $\mathcal{BPL}$) while the remaining eight are non-Fregean; of these, four are fully selfextensional but non-Fregean (the local ones), and the other four are not selfextensional (the global ones) (Example 7.60.13).

Other logics of implication

- The logic $\mathcal{I}mp$, the weakest implicative logic in the language $\langle\rightarrow\rangle$, is axiomatized, and its theorems are characterized; $\mathsf{Alg}^*\mathcal{I}mp = \mathsf{IA}$, the class of *implicative algebras* (Exercise 2.13). This class is a quasivariety that is not a variety (Exercise 2.14), so that $\mathsf{Alg}^*\mathcal{I}mp = \mathsf{Alg}\mathcal{I}mp = \mathsf{IA} \subsetneq \mathbb{V}\mathcal{I}mp$ (p. 279). The $\mathcal{I}mp$-filters need not coincide with the implicative filters, therefore $\mathcal{I}mp$ cannot be axiomatized with (MP) as the only proper rule (Exercise 2.29).
- The logic $\mathcal{BCK}$ is algebraizable in the same way as all implicative logics (though it is not one of them), with equivalent algebraic semantics the (proper) quasivariety of *BCK algebras* (Example 3.34.3). It does not satisfy any DDT (p. 172), but satisfies a Local DDT (p. 173, and Example 6.25.1).
- The logic $\mathcal{BCI}$ is not algebraizable (Example 3.34.3), and is deductively equivalent to its Gentzen-style presentation, which is not algebraizable either (p. 179).

- The logic $\mathcal{I}$, defined axiomatically in the language $\langle \to \rangle$, is characterized as the *simplest protoalgebraic logic* in a natural sense (p. 327). Its theorems are determined; it is not equivalential and has no algebraic semantics (hence it is not truth-equational), so that it is one of the logics located in the leftmost bottom of the Leibniz hierarchy; $\mathbb{V}\mathcal{I}$ is the class of *all algebras of the type*, while $\mathsf{Alg}^*\mathcal{I}$ is a smaller class, hence it is not a variety (Example 6.78.1). This logic is one of the two known examples of a protoalgebraic logic that does not satisfy the PDDT (Example 6.25.6). The logic $\mathcal{I}$ is not selfextensional (Exercise 7.5).
- The logic $\mathcal{G}_1$, an extension of $\mathcal{I}$ associated with a weak version of the DDT, is defined from a Gentzen-style calculus; it is protoalgebraic but neither equivalential (Example 6.78.2) nor weakly algebraizable (Example 6.122.10); thus, it is another logic in the leftmost bottom of the Leibniz hierarchy. It is selfextensional and not Fregean (Example 7.60.11).

Modal logics

In this book (Example 1.14), a *normal modal system* is the set of formulas that are valid in a class of (Kripke-style) frames; examples are the well-known $K, T, S4, S5$, etc., K being the smallest one. For each such a system S, two logics are defined semantically from the corresponding class of frames: the *global* one S^g and the *local* one S^ℓ. In general $S^\ell \leqslant S^g$ but they have the same theorems, which are the formulas of S; the main difference is in the *Necessitation Rule* $x \rhd \Box x$, which is only admissible in S^ℓ while is derivable in S^g. All these logics are expansions of the classical, non modal $\mathcal{C\ell}$.

The two logics of each pair have the same algebraic counterpart $\mathsf{Alg}\, S^\ell = \mathsf{Alg}\, S^g$, which is a variety of modal algebras; in the cases of K, $S4$ and $S5$ the varieties of all *modal algebras*, *closure algebras* and *monadic algebras* are obtained, respectively. In each case, S^ℓ and S^g are respectively the logic of order and the logic of one of the corresponding variety, and on one of these algebras the S^ℓ-filters are the lattice filters while the S^g-filters are the open filters (Example 7.47.1). Moreover, S^g is the strong version (in the sense of p. 349) of S^ℓ; the filters of S^g on any algebra are exactly the Leibniz filters of S^ℓ, which are the filters of S^ℓ that are closed under the Necessitation Rule (Example 6.53.1).

The local normal modal logics

- For each normal modal system S, the logic S^ℓ is not algebraizable (Example 3.61) but is equivalential; K^ℓ and T^ℓ are not finitely equivalential (Example 6.67.3) while $K4^\ell$ and all the S^ℓ extending it are finitely equivalential (Example 6.67.4). All the S^ℓ satisfy the classical DDT (p. 164) and are fully selfextensional (Example 7.47.1), but not Fregean (p. 461).
- The class Mod^*S^ℓ is determined (Example 7.45). The class Mod^*K^ℓ is not closed under ultraproducts, hence is not elementary (p. 229).

- The $\Box$ fragment of K^ℓ has no fully adequate Gentzen system (Example 5.123).

The global normal modal logics

- For each normal modal system S, the logic S^g is implicative (Example 2.17.1); it is not selfextensional (Example 7.47.1); and it does not satisfy the classical DDT (p. 164).
- $S4^g$ and $S5^g$ satisfy the DDT for the term $\Box x \to y$ (Example 6.25.1), and using it (and algebraizability) an explicit form of EDPC for closure algebras and monadic algebras is obtained (Exercise 3.63). These logics have a fully adequate Gentzen system (p. 313).
- K^g and T^g do not satisfy any DDT (p. 172), but they satisfy the Local DDT for an infinite family of terms (Example 6.25.1). These logics do not have a fully adequate Gentzen system, and are examples of algebraizable logics that are not G-algebraizable (p. 314).

Other modal logics

- *Positive modal logic* $\mathcal{PML}$ is the fragment of K^ℓ without negation and implication, but including $\top$ and $\bot$. It is not protoalgebraic; $\mathsf{Alg}^*\mathcal{PML} \subsetneq \mathsf{Alg}\mathcal{PML}$; the latter is the variety of *positive modal algebras* (Example 6.16.5). It is not truth-equational (Example 6.27.1). It is fully selfextensional but not Fregean; its full g-models and its fully adequate Gentzen system are characterized (Example 7.60.12).
- The logic E^{MP} associated with *the weakest classical system* E of modal logic is protoalgebraic and not equivalential; the classes $\mathsf{Mod}^*E^{\mathrm{MP}}$ and $\mathsf{Alg}E^{\mathrm{MP}} = \mathsf{Alg}^*E^{\mathrm{MP}}$ are described; the latter is the variety of *Boolean algebras expanded with an arbitrary unary operator* (Example 6.76). It is not truth-equational, hence not weakly algebraizable; thus, it is another of the logics in the leftmost bottom of the Leibniz hierarchy (Example 6.122.8). This logic is the logic of order of the mentioned variety, and is fully selfextensional; its assertional companion is its extension with the Extensionality Rule $x \leftrightarrow y \rhd \Box x \leftrightarrow \Box y$, which is algebraizable (Example 7.45).
- Herrmann's *Last Judgement* logic $\mathcal{LJ}$ is defined axiomatically in the standard modal language; it is algebraizable, with defining equation $\neg x \approx \neg(x \to x)$ and with $\Delta(x,y) = \{\Box^n(x \to y), \Box^n(y \to x) : n \geqslant 0\}$ as set of equivalence formulas, but not finitely algebraizable; its equivalent equational consequence is not finitary, and its equivalent algebraic semantics is not a quasivariety; its largest τ-algebraic semantics is a quasivariety, thus larger than the equivalent one (Example 3.42). The logic $\mathcal{LJ}$ is an example of a logic $\mathcal{L}$ for which $\mathsf{Alg}\mathcal{L} = \mathsf{Alg}^*\mathcal{L}$ without this class being a quasivariety (p. 276). $\mathcal{LJ}$ is not regularly algebraizable (p. 143).

- *Dellunde's logic* $\mathcal{L}_D$ is defined axiomatically in the language $\langle \leftrightarrow, \square \rangle$ of type $\langle 2, 1 \rangle$; it is finitary and regularly algebraizable (Example 3.53), but not finitely algebraizable (Exercise 3.35).
- The family of *propositional dynamic logics* $\mathcal{PDL}$ is described in general terms. These logics are equivalential; they are finitely equivalential if and only if the algebra of programs of their similarity type is finitely generated; they are not algebraizable (Example 6.67.6).

Many-valued logics

Łukasiewicz's infinitely-valued logics

In the literature, the above name is applied to the logic $Ł_\infty$ and to its finitary companion $Ł_{\infty f}$; their semilattice-based companion $Ł_\infty^{\leqslant}$, a much less popular logic, would also deserve to bear that name.

- The logic $Ł_\infty$ is defined semantically, in the language $\langle \wedge, \vee, \rightarrow, \neg \rangle$ of type $\langle 2, 2, 2, 1 \rangle$, from the matrix $\langle [\mathbf{0}, \mathbf{1}], \{1\} \rangle$, where $[\mathbf{0}, \mathbf{1}]$ is the algebra that generates the variety MV of *MV-algebras*, also known as *Wajsberg algebras* (Example 1.12). It is not finitary (Example 1.15), and it is not the infimum of the family $\{Ł_n : n \geqslant 2\}$, neither in the lattice of all logics, nor in the lattice of all finitary logics (Example 4.10). It is implicative, and it is finitely regularly algebraizable with respect to the algebra $[\mathbf{0}, \mathbf{1}]$; hence its equivalent algebraic semantics is $\mathbb{GQ}\{[\mathbf{0}, \mathbf{1}]\}$, a subclass of MV; the consequence $\vDash_{[\mathbf{0}, \mathbf{1}]}$ is not finitary, which implies that $\mathbb{GQ}\{[\mathbf{0}, \mathbf{1}]\}$ is not a quasivariety; thus, this is an example of an algebraizable logic with both transformers finite, but with both consequences non-finitary (Example 3.41). It is also an example of a logic $\mathcal{L}$ for which $\mathsf{Alg}\mathcal{L} = \mathsf{Alg}^*\mathcal{L}$ but this class is not a quasivariety (p. 276). Nevertheless, $\mathbb{V}Ł_\infty = \mathsf{MV}$ (Example 5.83). $Ł_\infty$ is an example of an algebraizable logic that is not G-algebraizable (p. 314). $Ł_\infty$ is not selfextensional (Example 7.47.2).
- The finitary companion $Ł_{\infty f}$ of the logic $Ł_\infty$ has a finite Hilbert-style axiomatization (p. 48), and is the infimum of the family $\{Ł_n : n \geqslant 2\}$ in the lattice of all finitary logics (Example 4.10). It is implicative, hence finitely regularly algebraizable (Example 3.34.6) and its equivalent algebraic semantics is MV (Example 3.41). The original proof of algebraizability with respect to MV used the Isomorphism Theorem, in the formulation of Corollary 3.59 (p. 153). $\mathbb{V}Ł_{\infty f} = \mathsf{MV}$ (p. 277 and Example 5.83). $Ł_{\infty f}$ satisfies the LDDT (Example 6.25.1) but does not satisfy the DDT for any set of binary terms (Example 6.29). $Ł_{\infty f}$ is not selfextensional (Example 7.47.2).
- The logic $Ł_\infty^{\leqslant}$ that preserves degrees of truth with respect to the algebra $[\mathbf{0}, \mathbf{1}]$, or equivalently with respect to the variety MV (hence, the semilattice-based companion of the two preceding logics) is finitary and not protoalgebraic (Example 6.16.9) and is fully selfextensional (Example 7.47.2).

Łukasiewicz's finitely-valued logics

- These logics are defined semantically from $[\mathbf{0,1}]_n$, the n-element subalgebra of $[\mathbf{0,1}]$, and are finitary (Example 1.12). $Ł_2 = \mathcal{C}\ell$. They are extensions of $Ł_\infty$ (p. 186). They are implicative (Example 3.34.6) and their equivalent algebraic semantics is $\mathsf{Alg}^*Ł_n = \mathsf{Alg}Ł_n = \mathbb{V}Ł_n = \mathbb{V}\{[\mathbf{0,1}]_n\}$ (Example 5.83). These logics satisfy the DDT for the term $x \to^n y$ (p. 165); $\mathbb{V}\{[\mathbf{0,1}]_n\}$ is a discriminator variety, and thus has EDPC, related to the DDT of $Ł_n$ (p. 172). For $n > 2$ the logics $Ł_n$ are not selfextensional (Example 7.47.2).
- The infimum $Ł^\infty$ of the family $\{Ł_n : n \geqslant 2\}$ in the lattice of all logics is a finitely approximable logic that is not finitary; moreover, it is different from $Ł_\infty$; actually, it is stronger than it (Example 4.10). This shows that the lattice of finitary logics (in a fixed language) is not a complete sublattice of the lattice of all logics (p. 54).
- For $n > 2$, the logics $Ł_n^\leqslant$ that preserve degrees of truth with respect to $[\mathbf{0,1}]_n$, hence with respect to $\mathbb{V}\{[\mathbf{0,1}]_n\}$, are finitely equivalential but not truth-equational, hence not weakly algebraizable (Example 6.122.6); these logics are fully selfextensional (Example 7.47.2). $Ł_n$ is the strong version (in the sense of p. 349) of $Ł_n^\leqslant$, and the filters of $Ł_n$ are those of $Ł_n^\leqslant$ that are closed under (MP) for $\to$ (Example 6.53.2).

Łukasiewicz's three-valued logic, $Ł_3$

An alternative definition of this logic is given in the language $\langle \to, \neg \rangle$ of type $\langle 2,1 \rangle$, from a three-element matrix $\langle \boldsymbol{A}_3, \{1\} \rangle$, or using $\boldsymbol{A}_3$ an algebraic semantics through suitable transformers; this logic is deductively equivalent to $\mathcal{J}_3$, the paraconsistent logic of Da Costa and D'Ottaviano, thus providing an example of two algebraizable logics with the same equivalent algebraic semantics (Example 3.20 and p. 179). $Ł_3$ is regularly algebraizable (Example 3.50), and its equivalent algebraic semantics is point-regular (p. 156).

Post's m-valued logics

The logics in this family (one for each $m \geqslant 2$) are among the implicative logics considered by Rasiowa in [208]; their algebraic counterpart is the class of *Post algebras* of order m (p. 84).

Belnap-Dunn's four-valued logic

The logic $\mathcal{B}$ is presented semantically in the language $\langle \wedge, \vee, \neg \rangle$ of type $\langle 2,2,1 \rangle$, first from epistemic considerations (Example 1.13) and then algebraically from a g-matrix associated with the four-element De Morgan lattice M_4 (Example 4.12), and equivalently from its order relation (p. 240). The Leibniz congruence on its models is characterized (Exercise 4.36). The classes $\mathsf{Mod}^*\mathcal{B}$ and $\mathsf{Alg}^*\mathcal{B}$ are determined and found to be rather weird; $\mathsf{Alg}\mathcal{B} = \mathbb{V}\mathcal{B} = \mathsf{DM} = \mathbb{V}\{M_4\}$, the variety of *De Morgan lattices*; its non-empty filters on these algebras are the lattice filters (Example 5.82 and Exercises 5.50 and 5.53). An analysis of the Isomorphism Theorem 5.95 in this case shows that congruences of De Morgan lattices can be represented by certain families of prime filters (Example 5.100). The full g-models of $\mathcal{B}$ are characterized in several ways, either of a semantic or of a syntactic flavour (Example 5.92), and its fully adequate Gentzen system is obtained (Example 5.120.4). $\mathcal{B}$ is an example of a G-algebraizable logic that is not algebraizable (p. 314); actually, it is not even protoalgebraic (Example 6.16.3). The logic $\mathcal{B}$ is fully selfextensional (Example 7.6.3) but not Fregean (Example 7.60.9), and is non-pseudo-axiomatic (Exercise 7.18).

The expansion of $\mathcal{B}$ with a truth constant $\top$ is also considered. It is fully selfextensional and not protoalgebraic; its algebraic counterpart is the variety of *De Morgan algebras*; and its full g-models are characterized (Exercise 5.64 and Example 6.16.4).

Two logics related to Kleene's three-valued truth-tables

The *paraconsistent weak Kleene logic* $\mathcal{PWK}$ is a three-valued logic that is truth-equational (with $\neg x \preccurlyeq x$ as defining equation) but neither assertional nor protoalgebraic; its algebraic counterparts satisfy $\mathsf{Alg}^*\mathcal{PWK} \subsetneq \mathsf{Alg}\mathcal{PWK} \subsetneq \mathbb{V}\mathcal{PWK}$ (Example 6.122.12). $\mathcal{PWK}$ is not selfextensional (Example 7.6.6). Priest's *logic of paradox* $\mathcal{LP}$ has very similar properties, except that $\mathsf{Alg}\mathcal{LP} = \mathbb{V}\mathcal{LP}$, the variety of *Kleene lattices* (same places). These two logics are among the few known logics in the rightmost bottom of the Leibniz hierarchy.

The logic of distributive bilattices

The logic of order of the variety of distributive bilattices is originally defined as a four-valued logic in the expanded language $\langle \wedge, \vee, \otimes, \oplus, \neg \rangle$ of type $\langle 2, 2, 2, 2, 1 \rangle$, and its algebraic study is also close to that of $\mathcal{B}$; it is one of the few known logics that are neither protoalgebraic, nor truth-equational, nor selfextensional, that is, that are outside the two hierarchies of abstract algebraic logic (Example 7.6.7).

Substructural logics

The term *substructural logics* may be used in a technical, restricted sense to refer, as in [123], to the logics associated with varieties of *residuated lattices* in a certain, defined way. The weakest is $\mathcal{FL}$, corresponding to the variety of all residuated lattices, and related to a Gentzen system called *Full Lambek calculus*; therefore, here all its axiomatic extensions, corresponding to the subvarieties, are considered, along with fragments of these extensions. However, the landscape of substructural logics is certainly much larger. For instance, it comprises the logics preserving degrees of truth (i.e., the logics of order) associated with each of these varieties. Finally, several *relevance logics* may be considered as belonging to the group of substructural logics under its original conception, related to the presence or absence of structural rules in Gentzen calculi.

- $\mathcal{FL}$ and all its axiomatic extensions are finitely algebraizable, with a variety of residuated lattices as equivalent algebraic semantics, defining equation $x \wedge \top \approx \top$ and equivalence formulas $\{x \backslash y, y \backslash x\}$; the defining equation can be read as $\top \preccurlyeq x$; the fragments containing at least $\wedge, \backslash, \top$ are also algebraizable, with the corresponding class of subreducts as equivalent algebraic semantics (Example 3.34.2). The logics of a variety of *integral residuated lattices* are assertional (i.e., of the form $\mathcal{L}_{\mathsf{K}}^{\top}$) and indeed implicative (Example 2.17.2), hence finitely regularly algebraizable, while those whose variety contains a non-integral residuated lattice are not regularly algebraizable (Example 3.50); in particular, $\mathcal{FL}$ is not regularly algebraizable (Example 6.144.2). All these logics satisfy a (very complicated) form of the PLDDT; parameters can be eliminated for extensions of $\mathcal{FL}_e$, an extension of $\mathcal{FL}$ with the exchange rule; these logics satisfy the LDDT and the DD sets have a simpler form (Example 6.25.3).
- For each variety K of commutative integral residuated lattices, the logic of order $\mathcal{L}_{\mathsf{K}}^{\leqslant}$ and the logic of one $\mathcal{L}_{\mathsf{K}}^{\top}$ are considered (the latter is the one in the previous point); the filters of $\mathcal{L}_{\mathsf{K}}^{\top}$ are the filters of $\mathcal{L}_{\mathsf{K}}^{\leqslant}$ that are closed under (MP) for $\rightarrow$; as a semilattice-based logic, $\mathcal{L}_{\mathsf{K}}^{\leqslant}$ is fully selfextensional; $\mathcal{L}_{\mathsf{K}}^{\top}$ is not selfextensional and $\mathcal{L}_{\mathsf{K}}^{\leqslant}$ is not algebraizable, unless the two logics coincide, and this happens if and only if K is a variety of generalized Heyting algebras (Example 7.47.2).
- Many logics $\mathcal{L}_{\mathsf{K}}^{\leqslant}$ are non-protoalgebraic (Example 6.16.9); those that are protoalgebraic, are automatically finitely equivalential as well (Example 6.67.5). Among the non-protoalgebraic are those corresponding to the classes of all commutative integral residuated lattices, MV-algebras ($Ł_{\infty}^{\leqslant}$), product algebras, BL-algebras, MTL-algebras, $\mathrm{FL}_{\mathrm{ew}}$-algebras, and any variety generated by a family of continuous t-norms over the real unit interval that is not the variety of Gödel algebras. The non-algebraizable are not truth-equational, hence they are not weakly algebraizable (Example 6.122.6).

Some logics in this large class are mentioned explicitly: *Fuzzy logics*, understood in [132] as the extensions of *Hájek's basic logic* $\mathcal{BL}$, are implicative (Example 2.17.2). The expansions with the "Baaz-Monteiro Delta" connective $\triangle$ satisfy the DDT for the term $\triangle x \rightarrow y$ (p. 165). Several *linear logics* without exponentials are algebraizable (Example 3.34.5). *Product logic* Π and *Gödel-Dummet's logic* (this one is also an intermediate logic) are algebraizable (Example 3.34.6).

Relevance logics

- *Relevance logic* $\mathcal{R}$ is finitely algebraizable, with $x \wedge (x \rightarrow x) \approx x \rightarrow x$ as defining equation; its equivalent algebraic semantics $\mathsf{Alg}\mathcal{R} = \mathsf{Alg}^*\mathcal{R} = \mathbb{V}\mathcal{R}$ is the variety of *R-algebras* (Example 3.34.7). This logic is not regularly algebraizable

(Example 6.144.2). It satisfies an explicitly given PLDDT but does not satisfy the LDDT (Example 6.25.2). $\mathcal{R}$ is an example of an algebraizable logic that is not G-algebraizable (p. 314).

- The implication fragment $\mathcal{R}_{\to}$ is finitely equivalential (Example 6.67.2) but not algebraizable (Example 3.62). Therefore, it cannot be truth-equational; this fact is also shown explicitly (Example 6.122.7).
- The *weak relevance logic* $\mathcal{WR}$, defined from $\mathcal{R}$ by Wójcicki's criteria, has a natural implication connective satisfying (MP) but not (R), because it has no theorems; consequently, it is not protoalgebraic; its algebraic counterpart is $\mathsf{Alg}\mathcal{WR} = \mathsf{Alg}^*\mathcal{WR} = \mathsf{Alg}\mathcal{R}$ (Example 6.16.7). $\mathcal{WR}$ is fully selfextensional and not Fregean (Example 7.60.10).
- The logic called *R-mingle* $\mathcal{RM}$, an axiomatic extension of $\mathcal{R}$, is finitely algebraizable with the simpler defining equation $x \approx x \to x$; its equivalent algebraic semantics is the variety of *Sugihara algebras* (Example 3.34.7). It is not regularly algebraizable (Example 6.144.2). $\mathcal{RM}$ satisfies the DDT for a non-standard binary term (p. 165).
- A logic defined algebraically from a three-element algebra, having the same theorems as (but not coinciding with) the implication-negation fragment of $\mathcal{RM}$ is truth-equational; the defining equation is $x \approx x \to x$, but the term $x \to x$ is not an algebraic constant in its defining algebra, therefore the logic is not assertional; it is not protoalgebraic either (Example 6.122.11). Thus, this logic is one of the few known logics located in the rightmost bottom of the Leibniz hierarchy.
- The *logic of entailment* $\mathcal{E}$ is equivalential, because its implication fragment is so (Example 6.67.2), but is not algebraizable (Example 3.62).
- The *logic of ticket entailment* $\mathcal{T}$ is not algebraizable (Example 3.62).
- The relevance logic denoted by $\mathcal{PW}$ (and in other ways) in the literature has no algebraic semantics (p. 112).

Other logics

- Da Costa and D'Ottaviano's *paraconsistent logic* $\mathcal{J}_3$ is introduced semantically from a three-element matrix $\langle \boldsymbol{A}_3, \{1/2, 1\}\rangle$, in parallel to Łukasiewicz's three-valued logic $Ł_3$; the deductive equivalence of the two logics is established; they are both algebraizable with the class $\mathbb{Q}\{\boldsymbol{A}_3\} = \mathbb{ISP}\{\boldsymbol{A}_3\}$, which is actually a variety, as equivalent algebraic semantics (Example 3.20). $\mathcal{J}_3$ is not regularly algebraizable (Example 6.144.3); however, its equivalent algebraic semantics is a pointed, relatively point-regular variety (pp. 156 and 410).
- Da Costa's *paraconsistent systems* $\mathcal{C}_n$ are protoalgebraic but not equivalential (Example 6.77). They are not truth-equational, hence not weakly algebraizable (Example 6.122.9). Hence, these logics are located in the leftmost bottom of the Leibniz hierarchy.

- A very large group of three-valued *logics of formal inconsistency*, including Sette's logic P^1, are algebraizable, with defining equation $(x \to x) \to x \approx x \to x$ and either $(x \leftrightarrow y) \wedge (\circ x \leftrightarrow \circ y)$ or $\sim\sim((x \leftrightarrow y) \wedge (\circ x \leftrightarrow \circ y))$ as equivalence formula (Example 3.34.10), and are not selfextensional (Example 7.6.6).
- The most prominent *quantum logic*, the assertional logic of the variety OML of *orthomodular lattices*, is finitely regularly algebraizable, and its set of equivalence formulas can be given using several defined implication connectives (Example 3.64 and Exercise 3.51). This logic satisfies no DDT (p. 173) and no Local DDT (p. 174). It is the strong version (in the sense of p. 349) of a weaker system also considered in the literature as associated with OML (Example 6.53.3). The assertional logic of any class of *ortholattices* containing some non-orthomodular one (in particular, the class OL of all ortholattices) is not equivalential (Example 6.122.5) but is regularly weakly algebraizable (Example 6.126.2).
- The *logic of lattices* $\mathcal{L}^{\leqslant}_{\mathsf{L}}$ is the logic of order (i.e., preserving degrees of truth) of the *variety of lattices* L; it is weaker than $\mathcal{C}\ell_{\wedge\vee}$ and is not protoalgebraic (Example 6.16.2). It is fully selfextensional and can be defined from its fully adequate Gentzen calculus, which is determined, and its full g-models are characterized (Example 7.34.5). This logic is not Fregean (Example 7.60.7).

Ad hoc examples

- Example 3.9 shows that the logic $\mathcal{L}$ defined in the language $\langle \Box \rangle$ of type $\langle 1 \rangle$ by just the rule $x \rhd \Box x$ has an algebraic semantics (namely, the class of all algebras of the type). It has also the classes $\mathsf{Alg}^*\mathcal{L}$ and $\mathsf{Alg}\mathcal{L}$ as algebraic semantics; the logic has no theorems, therefore it is neither truth-equational nor protoalgebraic (Example 6.93.2).
- Example 3.43 presents a logic, defined algebraically from a suitable variety in a language of type $\langle 2,1,1,1 \rangle$, which is non-finitary, is finitely algebraizable, but is not BP-algebraizable: its algebraizability needs an infinite number of defining equations. This logic shows that the finitarity of the equational relative consequence and the finiteness of the set of equivalence formulas do not imply the finitarity of the logic, and do not imply the finiteness of the set of defining equations either.
- Example 6.25.5 considers the logic defined by the class of matrices consisting of a ring with its additive unit as designated element. This logic is finitary and protoalgebraic, hence it satisfies the PLDDT, and also satisfies a PDDT where parameters cannot be eliminated.
- Example 6.102 presents a logic defined by an axiomatic system formulated in a language of type $\langle 1,1,0 \rangle$ and which has an algebraic semantics but is not truth-equational; truth turns out to be first-order definable in its reduced models, but not equationally definable.

- In Example 6.126.4, the so-called *logic of Andréka and Németi* is defined by an axiomatic system, in a language of type $\langle 2,1 \rangle$. It is protoalgebraic and truth-equational, hence weakly algebraizable, but not regularly; it is not equivalential, hence not algebraizable.
- Example 7.6.4 considers a logic defined by an axiomatic system in a language of type $\langle 1,0,0,0,0 \rangle$. It is Fregean but not fully selfextensional, and not protoalgebraic. It is truth-equational, thus establishing that inside truth-equational logics, to be Fregean does not imply to be fully Fregean (p. 456); that is, that the strong property of congruence does not transfer to full g-models (p. 459). It is among the few known examples of logics where $\mathsf{Alg}^*\mathcal{L} \subsetneq \mathsf{Alg}\mathcal{L} \subsetneq \mathbb{V}\mathcal{L}$.
- Example 7.6.5 mentions a logic defined from a four-element matrix in a language of type $\langle 2,1,0,0,0,0 \rangle$, which is selfextensional but neither fully selfextensional nor Fregean, and is implicative. Hence, it provides a counterexample showing that the property of congruence does not transfer to full g-models, even inside implicative logics (pp. 456 and 459).

Bibliography

[1] Agliano, P. Congruence quasi-orderability in subtractive varieties. *Journal of the Australian Mathematical Society* 71 (2001), 421–446.

[2] ——— Fregean subtractive varieties with definable congruences. *Journal of the Australian Mathematical Society* 71 (2001), 353–366.

[3] Agliano, P., and Ursini, A. On subtractive varieties iii: from ideals to congruences. *Algebra Universalis* 37 (1997), 296–333.

[4] Albuquerque, H. *Operators and strong versions in abstract algebraic logic.* Ph. D. Dissertation, University of Barcelona, March 2016.

[5] Albuquerque, H., Font, J. M., and Jansana, R. Compatibility operators in abstract algebraic logic. *The Journal of Symbolic Logic.* To appear.

[6] ——— The strong version of a sentential logic. Manuscript in preparation (2016).

[7] Albuquerque, H., Font, J. M., Jansana, R., and Moraschini, T. Assertional logics, truth-equational logics, and the hierarchies of abstract algebraic logic. In *Don Pigozzi on Abstract Algebraic Logic and Universal Algebra*, J. Czelakowski, Ed., series *Outstanding Contributions to Logic*. Springer-Verlag. 25 pp. To appear.

[8] Anderson, A. R., and Belnap, N. D. *Entailment. The logic of relevance and necessity*, vol. i. Princeton University Press, 1975.

[9] Anderson, A. R., Belnap, N. D., and Dunn, J. M. *Entailment. The logic of relevance and necessity*, vol. ii. Princeton University Press, 1992.

[10] Andréka, H., and Németi, I. General algebraic logic: A perspective on 'What is logic'. In *What is a logical system?*, D. M. Gabbay, Ed., vol. 4 of *Studies in Logic and Computation*. Oxford University Press, Oxford, 1994, pp. 393–443.

[11] Andréka, H., Németi, I., and Sain, I. Abstract model theoretic approach to algebraic logic. CCSOM Working paper 92-92, University of Amsterdam, 1992. 45 pp. (Version December 14, 1992; first version 1984).

[12] ——— Algebraic logic. In *Handbook of Philosophical Logic, Second Edition*, D. M. Gabbay and F. Guenthner, Eds., vol. 2. Kluwer Academic Publishers, 2001, pp. 133–248.

[13] Anellis, I. H., and Houser, N. Nineteenth century roots of algebraic logic and universal algebra. In *Algebraic logic*, H. Andréka, J. D. Monk, and I. Németi, Eds., vol. 54 of *Colloquia Mathematica Societatis János Bolyai*. North-Holland, Amsterdam, 1991, pp. 1–36.

[14] Arieli, O., and Avron, A. Reasoning with logical bilattices. *Journal of Logic, Language and Information* 5 (1996), 25–63.

[15] Avron, A. The semantics and proof theory of linear logic. *Theoretical Computer Science* 57 (1988), 161–184.

[16] ——— Natural 3-valued logics – characterization and proof theory. *The Journal of Symbolic Logic* 56 (1991), 276–294.

[17] ——— The method of hypersequents in the proof theory of propositional non-classical logics. In *Logic: from foundations to applications*, W. Hodges, M. Hyland, C. Steinhorn, and J. Truss, Eds. Oxford University Press, Oxford, 1996, pp. 1–32.

[18] Babyonyshev, S. V. Fully Fregean logics. *Reports on Mathematical Logic* 37 (2003), 59–78.

[19] Balbes, R., and Dwinger, P. *Distributive lattices*. University of Missouri Press, Columbia (Missouri), 1974.

[20] Běhounek, L., and Cintula, P. Fuzzy logics as the logics of chains. *Fuzzy Sets and Systems* 157 (2006), 604–610.

[21] Běhounek, L., Cintula, P., and Hájek, P. Introduction to Mathematical Fuzzy Logic. In *Handbook of Mathematical Fuzzy Logic, vol. 1*, P. Cintula, P. Hájek, and C. Noguera, Eds., vol. 37 of *Studies in Logic*. College Publications, London, 2011, pp. 1–101.

[22] Bell, J. L., and Slomson, A. B. *Models and ultraproducts: An introduction*. North-Holland, Amsterdam, 1971. Second revised printing.

[23] Belnap, N. D. A useful four-valued logic. In *Modern Uses of Multiple-Valued Logic*, J. M. Dunn and G. Epstein, Eds. Reidel, Dordrecht-Boston, 1977, pp. 8–37. Also published as a chapter of [9].

[24] Bergman, C. *Universal Algebra. Fundamentals and Selected Topics*. CRC Press, 2011.

[25] Birkhoff, G. *Lattice Theory*, 3rd ed., vol. xxv of *Colloquium Publications*. American Mathematical Society, Providence, 1973. (1st ed. 1940).

[26] Blackburn, P., de Rijke, M., and Venema, Y. *Modal logic*, vol. 53 of *Cambridge Tracts in Theoretical Computer Science*. Cambridge University Press, Cambridge, 2001.

[27] Blackburn, P., van Benthem, J., and Wolter, F., Eds. *Handbook of Modal Logic*, vol. 3 of *Studies in Logic and Practical Reasoning*. Elsevier, Amsterdam, 2007.

[28] Blok, W. J., and Hoogland, E. The Beth property in algebraic logic. *Studia Logica* 83 (2006), 49–90.

[29] Blok, W. J., and Jónsson, B. Equivalence of consequence operations. *Studia Logica (Special issue in memory of Willem Blok)* 83 (2006), 91–110.

[30] Blok, W. J., and Pigozzi, D. Protoalgebraic logics. *Studia Logica* 45 (1986), 337–369.

[31] ——— Alfred Tarski's work on general metamathematics. *The Journal of Symbolic Logic* 53 (1988), 36–50.

[32] ——— *Algebraizable logics*, vol. 396 of *Memoirs of the American Mathematical Society* A.M.S., Providence, January 1989. Out of print. Scanned copy available from `http://orion.math.iastate.edu:80/dpigozzi/`.

[33] ——— Local deduction theorems in algebraic logic. In *Algebraic Logic*, H. Andréka, J. D. Monk, and I. Németi, Eds., vol. 54 of *Colloquia Mathematica Societatis János Bolyai*. North-Holland, Amsterdam, 1991, pp. 75–109.

[34] ——— Algebraic semantics for universal Horn logic without equality. In *Universal Algebra and Quasigroup Theory*, A. Romanowska and J. D. H. Smith, Eds. Heldermann, Berlin, 1992, pp. 1–56.

[35] ——— Abstract algebraic logic and the deduction theorem. Manuscript, available from `http://orion.math.iastate.edu:80/dpigozzi/`, 2001.

[36] Blok, W. J., and Raftery, J. G. Ideals in quasivarieties of algebras. In *Models, algebras and proofs*, X. Caicedo and C. H. Montenegro, Eds., vol. 203 of *Lecture Notes in Pure and Applied Mathematics*. Marcel Dekker, New York, 1999, pp. 167–186.

[37] ——— Assertionally equivalent quasivarieties. *International Journal of Algebra and Computation* 18 (2008), 589–681.

[38] Blok, W. J., and Rebagliato, J. Algebraic semantics for deductive systems. *Studia Logica (Special issue on Abstract Algebraic Logic, Part II)* 74 (2003), 153–180.

[39] Bloom, S. L. Some theorems on structural consequence operations. *Studia Logica* 34 (1975), 1–9.

[40] ——— A note on Ψ-consequences. *Reports on Mathematical Logic* 8 (1977), 3–9.

[41] Bloom, S. L., and Brown, D. J. Classical abstract logics. *Dissertationes Mathematicae (Rozprawy Matematyczne)* 102 (1973), 43–51.

[42] Blyth, T. S. *Lattices and ordered algebraic structures*, series *Universitext*. Springer-Verlag, 2005.

[43] Bonzio, S., Gil-Férez, J., Paoli, F., and Peruzzi, L. On paraconsistent weak Kleene logic. Submitted manuscript (2016).

[44] Boole, G. *The mathematical analysis of logic. Being an essay towards a calculus of deductive reasoning*. Macmillan, Cambridge, 1847.

[45] Bou, F. *Implicación estricta y lógicas subintuicionistas*. Master Thesis, University of Barcelona, 2001.

[46] Bou, F., Esteva, F., Font, J. M., Gil, A. J., Godo, L., Torrens, A., and Verdú, V. Logics preserving degrees of truth from varieties of residuated lattices. *Journal of Logic and Computation* 19 (2009), 1031–1069.

[47] Bou, F., Font, J. M., and García Lapresta, J. L. On weakening the deduction theorem and strengthening Modus Ponens. *Mathematical Logic Quarterly* 50 (2004), 303–324.

[48] Bou, F., and Rivieccio, U. The logic of distributive bilattices. *Logic Journal of the IGPL* 19 (2011), 183–216.

[49] Brown, D. J., and Suszko, R. Abstract logics. *Dissertationes Mathematicae (Rozprawy Matematyczne)* 102 (1973), 9–42.

[50] Burris, S., and Legris, J. The algebra of logic tradition. In *The Stanford Encyclopedia of Philosophy*, E. N. Zalta, Ed., Spring 2015 ed. Archived as: `http://plato.stanford.edu/archives/spr2015/entries/algebra-logic-tradition/`.

[51] Burris, S., and Sankappanavar, H. P. *A Course in Universal Algebra. The Millenium Edition*. 2012 update. Electronic version freely available at `http://www.math.uwaterloo.ca/~snburris/htdocs/ualg.html`. 1st ed. published as vol. 78 of *Graduate Texts in Mathematics*, Springer-Verlag, 1981.

[52] Carnielli, W., and Coniglio, M. E. *Paraconsistent Logic: Consistency, Contradiction and Negation*, vol. 40 of *Logic, Epistemology, and the Unity of Science*. Springer-Verlag, 2016.

[53] Carnielli, W., Coniglio, M. E., and Marcos, J. Logics of formal inconsistency. In *Handbook of Philosophical Logic, Second Edition*, D. M. Gabbay and F. Guenthner, Eds., vol. 14. Springer-Verlag, 2007, pp. 1–94.

[54] Celani, S., and Jansana, R. A closer look at some subintuitionistic logics. *Notre Dame Journal of Formal Logic* 42 (2001), 225–255.

[55] Chagrov, A., and Zakharyashev, M. *Modal Logic*, vol. 35 of *Oxford Logic Guides*. Oxford University Press, Oxford, 1997.

[56] CHANG, C. C., AND KEISLER, H. J. *Model Theory*, 3rd ed., vol. 73 of *Studies in Logic and the Foundations of Mathematics*. North-Holland, Amsterdam, 1990. Reprinted by Dover Publications, New York, 2012.

[57] CHELLAS, B. *Modal Logic: an introduction*. Cambridge University Press, Cambridge, 1980.

[58] CIGNOLI, R. Quantifiers on distributive lattices. *Discrete Mathematics* 96 (1991), 183–197.

[59] CIGNOLI, R., MUNDICI, D., AND D'OTTAVIANO, I. M. L. *Algebraic foundations of many-valued reasoning*, vol. 7 of *Trends in Logic - Studia Logica Library*. Kluwer Academic Publishers, Dordrecht, 2000.

[60] CINTULA, P., HÁJEK, P., AND NOGUERA, C., Eds. *Handbook of Mathematical Fuzzy Logic*, vols. 37–38 of *Studies in Logic*. College Publications, London, 2011.

[61] CINTULA, P., AND NOGUERA, C. Implicational (semilinear) logics I: a new hierarchy. *Archive for Mathematical Logic* 49 (2010), 417–446.

[62] ——— A general framework for Mathematical Fuzzy Logic. In *Handbook of Mathematical Fuzzy Logic, vol. 1*, P. Cintula, P. Hájek, and C. Noguera, Eds., vol. 37 of *Studies in Logic*. College Publications, London, 2011, pp. 103–207.

[63] ——— The proof by cases property and its variants in structural consequence relations. *Studia Logica* (*Special issue on Abstract Algebraic Logic*) 101 (2013), 713–747.

[64] ——— A Henkin-style proof of completeness for first-order algebraizable logics. *The Journal of Symbolic Logic* 80 (2015), 341–358.

[65] CLEAVE, J. P. *A study of logics*, vol. 18 of *Oxford Logic Guides*. Oxford University Press, Oxford, 1991.

[66] CZELAKOWSKI, J. Reduced products of logical matrices. *Studia Logica* 39 (1980), 19–43.

[67] ——— Equivalential logics, I,II. *Studia Logica* 40 (1981), 227–236 and 355–372.

[68] ——— Algebraic aspects of deduction theorems. *Studia Logica* 44 (1985), 369–387.

[69] ——— Local deductions theorems. *Studia Logica* 45 (1986), 377–391.

[70] ——— *Protoalgebraic logics*, vol. 10 of *Trends in Logic - Studia Logica Library*. Kluwer Academic Publishers, Dordrecht, 2001.

[71] ——— The Suszko operator. Part I. *Studia Logica* (*Special issue on Abstract Algebraic Logic, Part II*) 74 (2003), 181–231.

[72] CZELAKOWSKI, J., AND JANSANA, R. Weakly algebraizable logics. *The Journal of Symbolic Logic* 65 (2000), 641–668.

[73] CZELAKOWSKI, J., AND MALINOWSKI, G. Key notions of Tarski's methodology of deductive systems. *Studia Logica* 44 (1985), 321–351.

[74] CZELAKOWSKI, J., AND PIGOZZI, D. Amalgamation and interpolation in abstract algebraic logic. In *Models, algebras and proofs*, X. Caicedo and C. H. Montenegro, Eds., vol. 203 of *Lecture Notes in Pure and Applied Mathematics*. Marcel Dekker, New York and Basel, 1999, pp. 187–265.

[75] ——— Fregean logics. *Annals of Pure and Applied Logic* 127 (2004), 17–76.

[76] ——— Fregean logics with the multiterm deduction theorem and their algebraization. *Studia Logica* 78 (2004), 171–212.

[77] DA COSTA, N. C. A. On the theory of inconsistent formal systems. *Notre Dame Journal of Formal Logic* 15 (1976), 497–510.

[78] Davey, B. A., and Priestley, H. A. *Introduction to lattices and order*, 2nd ed. Cambridge University Press, Cambridge, 2002.

[79] Dellunde, P. A finitary 1-equivalential logic not finitely equivalential. *Bulletin of the Section of Logic* 24 (1995), 120–122.

[80] ——— *Contributions to the model theory of equality-free logic*. Ph. D. Dissertation, University of Barcelona, 1996.

[81] Dellunde, P., and Jansana, R. Some characterization theorems for infinitary universal Horn logic without equality. *The Journal of Symbolic Logic* 61 (1996), 1242–1260.

[82] Di Nola, A., and Leuştean, I. Łukasiewicz logic and MV-algebras. In *Handbook of Mathematical Fuzzy Logic, vol.* II, P. Cintula, P. Hájek, and C. Noguera, Eds., vol. 38 of *Studies in Logic*. College Publications, London, 2011, pp. 469–583.

[83] Diego, A. *Sur les algèbres de Hilbert*. Gauthier-Villars, Paris, 1966.

[84] Dummett, M. *Elements of intuitionism*, 2nd ed., vol 39 of *Oxford Logic Guides*. Oxford University Press, Oxford, 2000.

[85] Dunn, J. M. Contradictory information: Too much of a good thing. *Journal of Philosophical Logic* 39 (2010), 425–452.

[86] Dunn, J. M., and Hardegree, G. M. *Algebraic methods in philosophical logic*, vol. 41 of *Oxford Logic Guides*. Oxford University Press, Oxford, 2001.

[87] Elgueta, R. *Algebraic model theory for languages without equality*. Ph. D. Dissertation, University of Barcelona, 1994.

[88] ——— Characterizing classes defined without equality. *Studia Logica* 58 (1997), 357–394.

[89] Enderton, H. B. *A mathematical introduction to logic*, 2nd ed. Academic Press, New York, 2001. 1st ed., 1972.

[90] Esteban, M. *Duality theory and abstract algebraic logic*. Ph. D. Dissertation, University of Barcelona, 2013.

[91] Fitting, M. Bilattices and the semantics of logic programming. *Journal of Logic Programming* 11 (1991), 91–116.

[92] Font, J. M. On the Leibniz congruences. In *Algebraic Methods in Logic and in Computer Science*, C. Rauszer, Ed., vol. 28 of *Banach Center Publications*. Polish Academy of Sciences, Warszawa, 1993, pp. 17–36.

[93] ——— Belnap's four-valued logic and De Morgan lattices. *Logic Journal of the IGPL* 5 (1997), 413–440.

[94] ——— On the contributions of Helena Rasiowa to mathematical logic. *Multiple-Valued Logic* 4 (1999), 159–179.

[95] ——— An abstract algebraic logic view of some multiple-valued logics. In *Beyond two: Theory and applications of multiple-valued logic*, M. Fitting and E. Orlowska, Eds., vol. 114 of *Studies in Fuzziness and Soft Computing*. Physica-Verlag, Heidelberg, 2003, pp. 25–58.

[96] ——— Generalized matrices in abstract algebraic logic. In *Trends in Logic. 50 years of Studia Logica*, V. F. Hendriks and J. Malinowski, Eds., vol. 21 of *Trends in Logic - Studia Logica Library*. Kluwer Academic Publishers, Dordrecht, 2003, pp. 57–86.

[97] ——— Beyond Rasiowa's algebraic approach to non-classical logics. *Studia Logica* 82 (2006), 172–209.

[98] ——— On semilattice-based logics with an algebraizable assertional companion. *Reports on Mathematical Logic* 46 (2011), 109–132.

[99] ——— Atoms in a lattice of theories. *Bulletin of the Section of Logic* 42 (2013), 21–31.

[100] ——— The simplest protoalgebraic logic. *Mathematical Logic Quarterly* 59 (2013), 435–451. With an *Erratum* in vol. 60 (2014), p. 91.

[101] ——— Ordering protoalgebraic logics. *Journal of Logic and Computation*. To appear. Published online on March 2014. DOI: `10.1093/logcom/exu015`.

[102] ——— Abstract algebraic logic – an introductory chapter. In *Hiroakira Ono on Residuated Lattices and Substructural Logics*, N. Galatos and K. Terui, Eds., series *Outstanding Contributions to Logic*. Springer-Verlag. 68 pp. To appear.

[103] Font, J. M., Gil, A. J., Torrens, A., and Verdú, V. On the infinite-valued Łukasiewicz logic that preserves degrees of truth. *Archive for Mathematical Logic* 45 (2006), 839–868.

[104] Font, J. M., Guzmán, F., and Verdú, V. Characterization of the reduced matrices for the conjunction-disjunction fragment of classical logic. *Bulletin of the Section of Logic* 20 (1991), 124–128.

[105] Font, J. M., and Jansana, R. On the sentential logics associated with strongly nice and semi-nice general logics. *Bulletin of the IGPL* 2 (1994), 55–76.

[106] ——— Leibniz filters and the strong version of a protoalgebraic logic. *Archive for Mathematical Logic* 40 (2001), 437–465.

[107] ——— *A general algebraic semantics for sentential logics. Second revised edition*, vol. 7 of *Lecture Notes in Logic*. Association for Symbolic Logic, 2009. Electronic version freely available through Project Euclid at `http://projecteuclid.org/euclid.lnl/1235416965`. 1st ed., Springer-Verlag, 1996.

[108] ——— Leibniz-linked pairs of deductive systems. *Studia Logica* (*Special issue in honor of Ryszard Wójcicki on the occasion of his 80th birthday*) 99 (2011), 171–202.

[109] Font, J. M., Jansana, R., and Pigozzi, D. Fully adequate Gentzen systems and the deduction theorem. *Reports on Mathematical Logic* 35 (2001), 115–165.

[110] ——— A survey of abstract algebraic logic. *Studia Logica* (*Special issue on Abstract Algebraic Logic, Part* II) 74 (2003), 13–97. With an update in vol. 91 (2009), 125–130.

[111] ——— On the closure properties of the class of full g-models of a deductive system. *Studia Logica* (*Special issue in memory of Willem Blok*) 83 (2006), 215–278.

[112] Font, J. M., and Moraschini, T. Logics of varieties, logics of semilattices, and conjunction. *Logic Journal of the IGPL* 22 (2014), 818–843.

[113] ——— M-sets and the representation problem. *Studia Logica* 103 (2015), 21–51.

[114] Font, J. M., and Rodríguez, G. Note on algebraic models for relevance logic. *Zeitschrift für Mathematische Logik und Grundlagen der Mathematik* 36 (1990), 535–540.

[115] ——— A note on Sugihara algebras. *Publicacions Matemàtiques* 36 (1992), 591–599.

[116] ——— Algebraic study of two deductive systems of relevance logic. *Notre Dame Journal of Formal Logic* 35 (1994), 369–397.

[117] Font, J. M., and Verdú, V. Algebraic logic for classical conjunction and disjunction. *Studia Logica* (*Special issue on Algebraic Logic*) 50 (1991), 391–419.

[118] ——— The lattice of distributive closure operators over an algebra. *Studia Logica* 52 (1993), 1–13.

[119] Gabbay, D. M., and Guenthner, F., Eds. *Handbook of Philosophical Logic, Second Edition*. Kluwer Academic Publishers, 2001ff. 18 vols.

[120] Gabbay, D. M., and Maksimova, L. L. *Interpolation and definability: Modal and Intuitionistic Logics*, vol. 46 of *Oxford Logic Guides*. Oxford University Press, Oxford, 2005.

[121] ——— Interpolation and definability. In *Handbook of Philosophical Logic, Second Edition*, D. M. Gabbay and F. Guenthner, Eds., vol. 15. Springer-Verlag, 2011, pp. 67–123.

[122] GALATOS, N., AND GIL-FÉREZ, J. Modules over quantaloids: Applications to the isomorphism problem in algebraic logic and π-institutions. *Journal of Pure and Applied Algebra*. To appear.

[123] GALATOS, N., JIPSEN, P., KOWALSKI, T., AND ONO, H. *Residuated lattices: an algebraic glimpse at substructural logics*, vol. 151 of *Studies in Logic and the Foundations of Mathematics*. Elsevier, Amsterdam, 2007.

[124] GALATOS, N., AND TSINAKIS, C. Equivalence of consequence relations: an order-theoretic and categorical perspective. *The Journal of Symbolic Logic* 74 (2009), 780–810.

[125] GEHRKE, M., JANSANA, R., AND PALMIGIANO, A. Canonical extensions for congruential logics with the deduction theorem. *Annals of Pure and Applied Logic* 161 (2010), 1502–1519.

[126] GINSBERG, M. L. Multivalued logics: A uniform approach to inference in artificial intelligence. *Computational Intelligence* 4 (1988), 265–316.

[127] GISPERT, J., AND MUNDICI, D. MV-algebras: a variety for magnitudes with archimedean units. *Algebra Universalis* 53 (2005), 7–43.

[128] GOTTWALD, S. *A treatise on many-valued logics*, vol. 9 of *Studies in Logic and Computation*. Research Studies Press, Baldock, 2001.

[129] GRÄTZER, G. *Lattice Theory. Foundation*. Birkhäuser, Basel, 2011.

[130] GRZEGORCZYK, A. An approach to logical calculus. *Studia Logica* 30 (1972), 33–43.

[131] GYURIS, V. *Variations of algebraizability*. Ph. D. Dissertation, University of Illinois at Chicago, 1999.

[132] HÁJEK, P. *Metamathematics of fuzzy logic*, vol. 4 of *Trends in Logic - Studia Logica Library*. Kluwer Academic Publishers, Dordrecht, 1998.

[133] HALMOS, P. R. *Algebraic logic*. Chelsea Publ. Co., New York, 1962.

[134] ——— How to write mathematics. *L'enseignement mathématique* 16 (1970), 123–152. Reprinted in *How to write mathematics*, American Mathematical Society (6th printing, 2000), pp. 19–48.

[135] ——— *I want to be a mathematician. An automathography*. Springer-Verlag, Heidelberg and New York, 1985.

[136] HAREL, D., KOZEN, D., AND TIURYN, J. Dynamic logic. In *Handbook of Philosophical Logic, Second Edition*, D. M. Gabbay and F. Guenthner, Eds., vol. 4. Kluwer Academic Publishers, Dordrecht, 2002, pp. 99–218.

[137] HENKIN, L., MONK, J. D., AND TARSKI, A. *Cylindric algebras. Part I*, vol. 64 of *Studies in Logic and the Foundations of Mathematics*. North-Holland, Amsterdam, 1971.

[138] ——— *Cylindric algebras. Part II*, vol. 115 of *Studies in Logic and the Foundations of Mathematics*. North-Holland, Amsterdam, 1985.

[139] HERRMANN, B. Algebraizability and Beth's theorem for equivalential logics. *Bulletin of the Section of Logic* 22 (1993), 85–88.

[140] ——— *Equivalential logics and definability of truth*. Ph. D. Dissertation, Freie Universität Berlin, 1993. 61 pp.

[141] ——— Equivalential and algebraizable logics. *Studia Logica* 57 (1996), 419–436.

[142] ——— Characterizing equivalential and algebraizable logics by the Leibniz operator. *Studia Logica* 58 (1997), 305–323.

[143] Hilbert, D., and Bernays, P. *Grundlagen der Mathematik* ii. Springer-Verlag, 1939. 2nd ed., 1970.

[144] Hodges, W. *Model theory*, vol. 42 of *Encyclopedia of Mathematics and its Applications*. Cambridge University Press, Cambridge, 1993.

[145] Hoogland, E. Algebraic characterizations of various Beth definability properties. *Studia Logica (Special issue on Abstract Algebraic Logic, Part* i*)* 65 (2000), 91–112.

[146] Hsieh, A., and Raftery, J. G. Conserving involution in residuated structures. *Mathematical Logic Quarterly* 53 (2007), 583–609.

[147] Humberstone, L. *The Connectives*. The MIT Press, Cambridge, 2011.

[148] Idziak, P., Słomczyńska, K., and Wroński, A. Fregean varieties. *International Journal of Algebra and Computation* 19 (2009), 595–645.

[149] ——— The commutator in equivalential algebras and Fregean varieties. *Algebra Universalis* 65 (2011), 331–340.

[150] Jansana, R. Propositional consequence relations and algebraic logic. In *The Stanford Encyclopedia of Philosophy*, E. N. Zalta, Ed., Spring 2011 ed. Archived as: `http://plato.stanford.edu/archives/spr2011/entries/consequence-algebraic/`.

[151] ——— *Una introducción a la lógica modal*. Tecnos, Madrid, 1990.

[152] ——— Los fragmentos □ de la lógica modal K. In *Actas del* vii *Congreso de Lenguajes Naturales y Lenguajes Formales* (Vic, Barcelona, 1991), C. Martín-Vide, Ed., pp. 409–413.

[153] ——— Full models for positive modal logic. *Mathematical Logic Quarterly* 48 (2002), 427–445.

[154] ——— Leibniz filters revisited. *Studia Logica* 75 (2003), 305–317.

[155] ——— Selfextensional logics with implication. In *Logica Universalis*, J.-Y. Béziau, Ed. Birkhäuser, Basel, 2005, pp. 65–88.

[156] ——— Selfextensional logics with a conjunction. *Studia Logica* 84 (2006), 63–104.

[157] ——— Algebraizable logics with a strong conjunction and their semi-lattice based companions. *Archive for Mathematical Logic* 51 (2012), 831–861.

[158] ——— Bloom, Brown and Suszko's work on abstract logics. In *Universal Logic: An Anthology. From Paul Hertz to Dov Gabbay*, J.-Y. Béziau, Ed., series *Studies in Universal Logic*. Birkhäuser, Basel, 2012, pp. 251–260.

[159] ——— On the deductive system of the order of an equationally orderable quasivariety. *Studia Logica*. To appear. Published online on 8 february 2016. doi: `10.1007/s11225-016-9650-7`.

[160] Jansana, R., and Palmigiano, A. Referential semantics: duality and applications. *Reports on Mathematical Logic* (*Special issue in memory of Willem Blok*) 41 (2006), 63–93.

[161] Kabziński, J., Porębska, M., and Wroński, A. On the $\{\leftrightarrow,\sim\}$-reduct of the intuitionistic consequence operation. *Studia Logica* 40 (1981), 55–66.

[162] Lewin, R. A., Mikenberg, I. F., and Schwarze, M. G. Algebraization of paraconsistent logic P^1. *Journal of Non-Classical Logics* 7 (1990), 79–88.

[163] ——— C_1 is not algebraizable. *Notre Dame Journal of Formal Logic* 32 (1991), 609–611.

[164] ——— $P1$ algebras. *Studia Logica* 53 (1994), 21–28.

[165] Łoś, J. *O matrycach logicznych*, vol. 19 of *Prace Wrocławskiego Towarzystwa Naukowege, Ser. B*. University of Wrocław, 1949.

[166] Łoś, J., and Suszko, R. Remarks on sentential logics. *Indagationes Mathematicae* 20 (1958), 177–183. See J. Zygmunt's corrections in [251].

[167] Makinson, D. *Bridges from classical to nonmonotonic logic*, vol. 5 of *Texts in Computing*. King's College Publications, London, 2005.

[168] ——— Logical friendliness and sympathy. In *Logica Universalis*, J.-Y. Béziau, Ed. Birkhäuser, Basel, 2005, pp. 195–205.

[169] ——— Completeness theorems, representation theorems: what's the difference? In *Hommage à Wlodek: philosophical papers dedicated to Wlodek Rabinowicz*, T. Rønnow-Rasmussen, B. Petersson, J. Josefsson, and D. Egonsson, Eds. Lunds Universitet, Lund, 2007.

[170] Maksimova, L. L. Definability and interpolation in non-classical logics. *Studia Logica* 82 (2006), 271–291.

[171] Mal'cev, A. I. *Algebraic systems*, vol. 152 of *Grundlehren der Mathematischen Wissenschaften*. Springer-Verlag, Berlin, 1971.

[172] Malinowksi, J. *Equivalence in intensional logics*. Ph. D. Dissertation, Polish Academy of Sciences, Institute of Philosophy and Sociology, Warsaw, 1989.

[173] ——— Modal equivalential logics. *Journal of Non-Classical Logics* 3 (1986), 13–35.

[174] Moisil, G. C. Recherches sur l'algèbre de la logique. *Annales Scientifiques de l'Université de Jassy* 22 (1935), 1–117.

[175] Monk, J. D. *Mathematical logic*, vol. 37 of *Graduate Texts in Mathematics*. Springer-Verlag, New York, 1976.

[176] ——— The contributions of Alfred Tarski to algebraic logic. *The Journal of Symbolic Logic* 51 (1986), 899–906.

[177] Monteiro, A. Matrices de Morgan caractéristiques pour le calcul propositionnel classique. *Anais da Academia Brasileira de Ciências* 33 (1960), 1–7.

[178] ——— La sémisimplicité des algèbres de Boole topologiques et les systèmes déductifs. *Revista de la Unión Matemática Argentina* 25 (1971), 417–448.

[179] Moraschini, T. The semantic isomorphism theorem in abstract algebraic logic. Submitted manuscript, March 2015.

[180] ——— *Investigations into the role of translations in abstract algebraic logic*. Ph. D. Dissertation, University of Barcelona, June 2016.

[181] Mostowski, A. Proofs of non-deducibility in intuitionistic functional calculus. *The Journal of Symbolic Logic* 13 (1948), 204–207.

[182] Németi, I. Algebraization of quantifier logics, an introductory overview. *Studia Logica* 50 (1991), 485–569.

[183] Nowak, M. A characterization of consequence operations preserving degrees of truth. *Bulletin of the Section of Logic* 16 (1987), 159–166.

[184] ——— Logics preserving degrees of truth. *Studia Logica* 49 (1990), 483–499.

[185] Nurakunov, A., and Stronkowski, M. M. Relation formulas for protoalgebraic equality free quasivarieties: Pałasińska's theorem revisited. *Studia Logica (Special issue on Abstract Algebraic Logic)* 101 (2013), 827–847.

[186] Pałasińska, K. Finite basis theorem for filter-distributive protoalgebraic deductive systems and strict universal Horn classes. *Studia Logica (Special issue on Abstract Algebraic Logic, Part II)* 74 (2003), 233–273.

[187] Paoli, F. *Substructural logics: a primer*, vol. 13 of *Trends in Logic - Studia Logica Library*. Kluwer Academic Publishers, Dordrecht, 2002.

[188] Peckhaus, V. The mathematical origins of nineteenth-century algebra of logic. In *The Development of Modern Logic*, L. Haaparanta, Ed. Oxford University Press, Oxford, 2009, pp. 159–195.

[189] Pigozzi, D. Fregean algebraic logic. In *Algebraic Logic*, H. Andréka, J. D. Monk, and I. Németi, Eds., vol. 54 of *Colloquia Mathematica Societatis János Bolyai*. North-Holland, Amsterdam, 1991, pp. 473–502.

[190] ——— Abstract algebraic logic: past, present and future. A personal view. In *Workshop on Abstract Algebraic Logic*, J. M. Font, R. Jansana, and D. Pigozzi, Eds., vol. 10 of *Quaderns*. Centre de Recerca Matemàtica, Bellaterra, 1998, pp. 122–138.

[191] ——— Abstract algebraic logic. In *Encyclopaedia of Mathematics, Supplement* iii, M. Hazewinkel, Ed. Kluwer Academic Publishers, Dordrecht, 2001, pp. 2–13.

[192] Pogorzelski, W. On Hilbert's operation on logical rules i. *Reports on Mathematical Logic* 12 (1981), 35–50.

[193] Pogorzelski, W., and Słupecki, J. Basic properties of deductive systems based on nonclassical logics (i,ii). *Studia Logica* 9,10 (1960), 77–95 and 163–176, resp. In Polish with English summaries.

[194] Pogorzelski, W., and Wojtylak, P. *Completeness Theory for Propositional Logics*, series *Studies in Universal Logic*. Birkhäuser, Basel, 2008.

[195] Porębska, M., and Wroński, A. A characterization of fragments of the intuitionistic propositional logic. *Reports on Mathematical Logic* 4 (1975), 39–42.

[196] Porte, J. *Recherches sur la théorie générale des systèmes formels et sur les systèmes connectifs*. Gauthier-Villars, Paris, 1965.

[197] ——— Fifty years of deduction theorems. In *Proceedings of the Herbrand Symposium*, J. Stern, Ed. North-Holland, Amsterdam, 1982, pp. 243–250.

[198] Priest, G. Paraconsistent logic. In *Handbook of Philosophical Logic, Second Edition*, D. M. Gabbay and F. Guenthner, Eds., vol. 6. Kluwer Academic Publishers, Dordrecht, 2002, pp. 287–393.

[199] Prucnal, T., and Wroński, A. An algebraic characterization of the notion of structural completeness. *Bulletin of the Section of Logic* 3 (1974), 30–33.

[200] Raftery, J. G. Correspondences between Gentzen and Hilbert systems. *The Journal of Symbolic Logic* 71 (2006), 903–957.

[201] ——— The equational definability of truth predicates. *Reports on Mathematical Logic (Special issue in memory of Willem Blok)* 41 (2006), 95–149.

[202] ——— A non-finitary sentential logic that is elementarily algebraizable. *Journal of Logic and Computation* 20 (2010), 969–975.

[203] ——— Contextual deduction theorems. *Studia Logica (Special issue in honor of Ryszard Wójcicki on the occasion of his 80th birthday)* 99 (2011), 279–319.

[204] ——— A perspective on the algebra of logic. *Quaestiones Mathematicae* 34 (2011), 275–325.

[205] ——— Inconsistency lemmas in algebraic logic. *Mathematical Logic Quarterly* 59 (2013), 393–406.

[206] ——— Relative congruence schemes and decompositions in quasivarieties. Submitted manuscript, 2015.

[207] ——— Admissible rules and the Leibniz hierarchy. *Notre Dame Journal of Formal Logic*. To appear.

[208] Rasiowa, H. *An algebraic approach to non-classical logics*, vol. 78 of *Studies in Logic and the Foundations of Mathematics*. North-Holland, Amsterdam, 1974.

[209] Rasiowa, H., and Sikorski, R. *The mathematics of metamathematics*. Państwowe Wydawnictwo Naukowe, Warszawa, 1963. (3rd ed., 1970).

[210] Rautenberg, W. 2-element matrices. *Studia Logica* 40 (1981), 315–353.

[211] ——— On reduced matrices. *Studia Logica* 52 (1993), 63–72.

[212] Rebagliato, J., and Verdú, V. On the algebraization of some Gentzen systems. *Fundamenta Informaticae (Special issue on Algebraic Logic and its Applications)* 18 (1993), 319–338.

[213] ——— A finite Hilbert-style axiomatization of the implication-less fragment of the intuitionistic propositional calculus. *Mathematical Logic Quarterly* 40 (1994), 61–68.

[214] ——— Algebraizable Gentzen systems and the deduction theorem for Gentzen systems. Mathematics Preprint Series 175, University of Barcelona, June 1995.

[215] Restall, G. Subintuitionistic logics. *Notre Dame Journal of Formal Logic* 35 (1994), 116–129.

[216] ——— *An introduction to substructural logics*. Routledge, London and New York, 2000.

[217] Rivieccio, U. Some extensions of the Belnap-Dunn logic. Manuscript in preparation, 2016.

[218] Rodríguez, A. J., Torrens, A., and Verdú, V. Łukasiewicz logic and Wajsberg algebras. *Bulletin of the Section of Logic* 19 (1990), 51–55.

[219] Russo, C. An order-theoretic analysis of interpretations among propositional deductive systems. *Annals of Pure and Applied Logic* 164 (2013), 112–130.

[220] Rybakov, V. V. *Admissibility of logical inference rules*, vol. 136 of *Studies in Logic and the Foundations of Mathematics*. Elsevier, Amsterdam, 1997.

[221] Sain, I. Beth's and Craig's properties via epimorphisms and amalgamation in algebraic logic. In *Algebraic Logic and Universal Algebra in Computer Science*, C. Bergman, R. Maddux, and D. Pigozzi, Eds., vol. 425 of *Lecture Notes in Computer Science*. Springer-Verlag, 1990, pp. 209–226.

[222] Sayed Ahmed, T. Algebraic logic, where does it stand today? *The Bulletin of Symbolic Logic* 11 (2005), 465–516.

[223] Schlechta, K. *Nonmonotonic logics. Basic concepts, results, and techniques*, vol. 1187 of *Lecture Notes in Artificial Intelligence*. Springer-Verlag, Berlin, 1997.

[224] Scott, D. Completeness and axiomatizability in many-valued logic. In *Proceedings of the Tarski Symposium*, L. Henkin et al., Eds., vol. 25 of *Proceedings of Symposia in Pure Mathematics*. American Mathematical Society, Providence, 1974, pp. 411–436.

[225] Shoesmith, D. J., and Smiley, T. J. *Multiple-conclusion logic*. Cambridge University Press, Cambridge, 1978.

[226] Słomczyńska, K. Unification and projectivity in Fregean varieties. *Logic Journal of the IGPL* 20 (2012), 73–93.

[227] Strasser, C., and Antonelli, G. A. Non-monotonic logic. In *The Stanford Encyclopedia of Philosophy*, E. N. Zalta, Ed., Fall 2015 ed. Archived as: `http://plato.stanford.edu/archives/fall2015/entries/logic-nonmonotonic/`.

[228] Stronkowski, M. M. Defining subdirect product closed classes in infinitary logic. *Algebra Universalis* 69 (2013), 231–235.

[229] Surma, S. On the origin and subsequent applications of the concept of the Lindenbaum algebra. In *Logic, Methodology and Philosophy of Science* vi, *Hannover* 1979. North-Holland, Amsterdam, 1982, pp. 719–734.

[230] Suszko, R. Identity connective and modality. *Studia Logica* 27 (1971), 7–39.

[231] ——— Abolition of the Fregean axiom. In *Logic Colloquium* 1972–73, R. Parikh, Ed., vol. 453 of *Lecture Notes in Mathematics*. Springer-Verlag, Berlin, 1975, pp. 169–239.

[232] Suzuki, Y., Wolter, F., and Zacharyaschev, M. Speaking about transitive frames in propositional languages. *Journal of Logic, Language and Information* 7 (1998), 317–339.

[233] Tarski, A. Fundamentale Begriffe der Methodologie der deduktiven Wissenschaften I. *Monatshefte für Mathematik und Physik* 37 (1930), 360–404. English translation, revised and annotated by the author, in [236], pp. 60–109.

[234] ——— Über einige fundamentale Begriffe der Metamathematik. *Comptes Rendus des Séances de la Société des Sciences et des Lettres de Varsovie, Classe* III 23 (1930), 22–29. English translation, revised and annotated by the author, in [236], pp. 30–37.

[235] ——— Grundzüge der Systemenkalkül. *Fundamenta Mathematicae* 25,26 (1935,1936), 503–526 and 283–301, resp. English translation, revised and annotated by the author, in [236], pp. 342–383.

[236] ——— *Logic, Semantics, Metamathematics. Papers from* 1923 *to* 1938. Clarendon Press, Oxford, 1956. 2nd ed., Hackett Pub. Co., Indianapolis, 1983.

[237] Ursini, A. On subtractive varieties I. *Algebra Universalis* 31 (1994), 204–222.

[238] van Dalen, D. Intuitionistic logic. In *Handbook of philosophical logic, Second Edition*, D. M. Gabbay and F. Guenthner, Eds., vol. 5. Kluwer Academic Publishers, Dordrecht, 2001, pp. 1–114.

[239] Verdú, V. Some algebraic structures determined by closure operators. *Zeitschrift für Mathematische Logik und Grundlagen der Mathematik* 31 (1985), 275–278.

[240] Visser, A. A propositional logic with explicit fixed points. *Studia Logica* 40 (1981), 155–175.

[241] Voutsadakis, G. Categorical abstract algebraic logic: equivalent institutions. *Studia Logica (Special issue on Abstract Algebraic Logic, Part* II*)* 74 (2003), 275–311.

[242] Wallman, H. Lattices and topological spaces. *Annals of Mathematics* 39 (1938), 112–126.

[243] Wansing, H. Displaying as temporalizing. Sequent systems for subintuitionistic logic. In *Logic, Language and Computation*, S. Akama, Ed. Kluwer Academic Publishers, 1997, pp. 159–178.

[244] Wójcicki, R. Logical matrices strongly adequate for structural sentential calculi. *Bulletin de l'Académie Polonaise des Sciences, Classe* III, 17 (1969), 333–335.

[245] ——— Some remarks on the consequence operation in sentential logics. *Fundamenta Mathematicae* 68 (1970), 269–279.

[246] ——— Matrix approach in the methodology of sentential calculi. *Studia Logica* 32 (1973), 7–37.

[247] ——— Referential matrix semantics for propositional calculi. *Bulletin of the Section of Logic* 8 (1979), 170–176.

[248] ——— *Lectures on propositional calculi.* Ossolineum, Wrocław, 1984.

[249] ——— *Theory of logical calculi. Basic theory of consequence operations*, vol. 199 of *Synthèse Library*. Reidel, Dordrecht, 1988.

[250] ——— A three element matrix whose consequence operation is not finitely based. *Bulletin of the Section of Logic* 8 (1979), 68–71.

[251] Zygmunt, J. Structural consequence operations and logical matrices adequate for them. In *Universal Logic: An Anthology. From Paul Hertz to Dov Gabbay*, J.-Y. Béziau, Ed., series *Studies in Universal Logic*. Birkhäuser, Basel, 2012, pp. 163–184.

Author index

Occurrences in the Appendix and the Bibliography and in names of mathematical notions are not listed.

Index of logics

Only particular logics are listed; for general *kinds* or *classes* of logics, see the General Index and the Table of Contents. Occurrences in the Appendix are not listed. Numbers in boldface point to explicit definitions of the logic.

Index of classes of algebras

Only occurrences of the algebras as a class are listed; occurrences in the Appendix are excluded. Explicit definitions of some classes appear in boldface. An "n" after the page number refers to a footnote.

General index

Only definitions and very relevant or non-standard occurrences are listed. The detailed treatment of the main notions studied in the book is best found through the Table of Contents. Particular logics are listed in the Index of Logics. An "n" after the page number refers to a footnote.

Index of acronyms and labels

Symbol index

Common mathematical symbols having their expected meaning are not listed. Symbols denoting fragments of logics are not listed.

Many of the listed symbols contain variable components, which appear in the given explanation. These complex symbols are used throughout the book with other "values" of these components; for instance, to find the meaning of $\boldsymbol{\Omega}^{\boldsymbol{B}} G$ you should look for $\boldsymbol{\Omega}^{\boldsymbol{A}} F$ and make the appropriate replacements.

Almost all listed symbols denote an object, but some symbol complexes denoting a fact have also been listed.

Alphabetical order has been slightly altered.

This book was typeset by the author using LaTeX
with Hermann Zapf's *Palatino* typeface,
and was completed in Barcelona
on the day of Sant Jordi
of the year 2016

www.ingramcontent.com/pod-product-compliance
Lightning Source LLC
LaVergne TN
LVHW012326060326
832902LV00011B/1735